STROMATOLITES: INTERACTION OF MICROBES WITH SEDIMENTS

Cellular Origin, Life in Extreme Habitats and Astrobiology

Volume 18

Series Editor:

Joseph Seckbach
The Hebrew University of Jerusalem, Israel

For other titles published in this series, go to
www.springer.com/series/5775

Stromatolites: Interaction of Microbes with Sediments

Edited by

Vinod C. Tewari
Wadia Institute of Himalayan Geology, Dehradun, Uttarakhand, India

and

Joseph Seckbach
Hebrew University of Jerusalem, Israel

Springer

Editors
Vinod C. Tewari
Wadia Institute of Himalayan Geology
Dehradun, Uttarakhand
India
vinodt1954@yahoo.co.in

Joseph Seckbach
Hebrew University of Jerusalem
Israel
seckbach@huji.ac.il

ISBN 978-94-024-0541-5 ISBN 978-94-007-0397-1 (eBook)
DOI 10.1007/978-94-007-0397-1
Springer Dordrecht Heidelberg London New York

Printed on acid-free paper

Cover illustration: Stromatolites at Shark Bay (Hamelin Pool), 700 km. north-northwest of Perth, Western Australia. *Photo by J. William Schopf, UCLA, USA*

Springer is part of Springer Science+Business Media (www.springer.com)

TABLE OF CONTENTS

PART 1:
ARCHAEAN: PROTEROZOIC STROMATOLITES AND MICROBIOTA

PART 2: PHANEROZOIC STROMATOLITES

PART 3: MODERN STROMATOLITES (MARINE, LACUSTRINE, HOTSPRINGS)

PART 4: MODERN INSTRUMENTAL TECHNIQUES FOR THE STUDY OF STROMATOLITES AND MICROBIOTA

PART 5: GEOCHEMISTRY AND GEOMICROBIOLOGY OF STROMATOLITES AND MICROBIOTA

PART 6: ASTROBIOLOGY

PART 7: SUMMARY, CONCLUSIONS AND FUTURE PROSPECTS

INTRODUCTION TO STROMATOLITES[1]

Stromatolites, layered sedimentary structures produced by mat-building phototrophic organisms (usually cyanobacteria), are considered the most ancient biological record and the earliest evidence of the emergence of life on Earth. The early stromatolites are the major constituent of the fossil record for billions of years, beginning around 3.5 bya. In the Precambrian era, the "primitive" environments of Earth were too hostile to support life as we know it, and the stromatolites could thrive without competition. They declined in the Phanerozoic as victims of grazing animals.

Today, stromatolites continue to form, although they are nearly extinct in normal marine environments and live a precarious existence in only a few localities worldwide. Modern stromatolites exist in areas that most other life forms consider less desirable, in extreme environments containing hypersaline water, high alkalinity, and high or low temperatures zones. Such places exclude grazing snails and other animals which consume the cyanobacteria. Recent formations of stromatolites were discovered in Shark Bay (Australia) as well as throughout Western Australia, the Bahamas (such as Exuma Cays), the Indian Ocean, various places in the USA (such as in Yellowstone National Park), Laguna Salgada (Brazil), the Mexican Desert, Glacier National Park (Montana and Canada), and the Solar Lake in Sinai, which is heliothermally heated and contains hypersaline water. The study of modern stromatolites assists in the interpretation of ecology and environment of ancient stromatolites as well as possible life on extraterrestrial planets or moons.

Eighty four authors contributed the 34 chapters to this volume. They are from 27 countries (Argentina, Australia, Belgium, Brazil, Canada, China, Croatia, Germany, France, Israel, India, Italy, Japan, Mexico, New Zealand, Norway, Poland, Portugal, Russia, Slovenia, South Korea, Spain, Sweden, Switzerland, Taiwan, UK, USA). We divided this book into seven sections, covering Achaean and modern stromatolites and Astrobiology. We hope that this book will enrich our readers by making a wide range of data on stromatolites accessible.

Joseph Seckbach Hebrew University of Jerusalem

December 16, 2010

[1] I thank Fern Seckbach and Professors David Chapman (UCSB) and Maud Walsh (LSU) for proofreading the introduction.

Professor Joseph Seckbach is the Founder and Chief Editor of Book Series ***Cellular Origins, Life in Extreme Habitats and Astrobiology (COLE; see www.springer.com/series/5775)***. He is the author of several chapters in this series. Dr. Seckbach earned his Ph.D. from the University of Chicago, IL, USA (1965) and spent his postdoctoral years in the Division of Biology at Caltech (Pasadena, CA, USA). Then he headed a team for searching for extraterrestrial life at the University of California at Los Angeles (UCLA). Later, he has been appointed to the faculty of the Hebrew University (Jerusalem, Israel) where he performed algal research and taught biological courses until his retirement. Among his publications are books, scientific articles in the lines of phytoferritin, cellular evolution, acidothermophyllic algae and life in extreme environments. Dr. Seckbach is the co-editor of ***Proceeding of Endocytobiology VII Conference*** (Frieburg, Germany, 1998) and the ***Proceedings of Algae and Extreme Environments Meeting*** (Trebon, Czech Republic, 2000). His edited volume (with Richard Gordon) entitled ***Divine Action and Natural Selection: Science, Faith, and Evolution*** has been published by World Scientific Publishing Company (2008). His recent interest is in the field of enigmatic microorganisms and life in extreme environments.

E-mail: **seckbach@huji.ac.il**

FOREWORD

Stromatolites: The Present as a Key to the Past

"Everybody will agree that these are very basic points that 'stromatologists' should know at first: they will provide us with the necessary and fundamental botanical background to neatly investigate the algal mats that we find on our tidal flats. Moreover if, by understanding and deciphering the dynamics of recent freshwater structures, we can thoroughly account for the mode of growth, of formation, for the leading ecological requirements and environmental responses of fossil freshwater stromatolites…, then the field is open to **sound**, **natural** and **realistic** understanding of Paleozoic and Precambrian stromatolites…; even if [those] stromatolites originated in different chemical environments, they appear to me not only as analogous but furthermore as **homologous** structures, the growth form and building up of which was in equilibrium with homologous general environmental factors" (Monty, 1972).

These words can be found in an account of the recent algal stromatolitic deposits, Andros Island, Bahamas, by the late Claude Monty (1937–1999). The modern stromatolites in the Bahamas were extensively studied ever since and became a popular model for studies aiming at an understanding of the microbial processes that led to the formation of those Paleozoic and Precambrian stromatolites mentioned by Monty, many of which feature in the chapters in this volume.

Stromatolites (from Greek στρῶμα, *strōma*, mattress, bed, stratum, and λιθος, *lithos*, rock) can be defined as layered accretionary structures formed in shallow water by the trapping, binding and cementation of sedimentary grains by biofilms of microorganisms, especially cyanobacteria (http://en.wikipedia.org/wiki/Stromatolites). Extensive information on the nature of stromatolites can be found in the chapters in the current volume and in an earlier volume in this book series (Seckbach and Walsh, 2009).

I first learned about the connection between ancient stromatolites and modern microbial mat structures when I first visited Solar Lake (Sinai Peninsula, now in Egypt) in the beginning of 1975. The layered structure of the microbial mat dominated by cyanobacteria in this small hypersaline heliothermally heated lake and the calcification processes of the mat show how a stromatolite-like geological record is generated before our eyes, a record that extends for more than two thousand years (Krumbein and Cohen, 1974; Cohen et al., 1977).

Other modern "living" stromatolites are found in Western Australia while the most famous site is of course Hamelin Pool. A visit to Western Australia brought me in September 2005 to the no less interesting site of Lake Clifton (Yalgorup National Park) with its thrombolites 'living rocks', microbialites that grow in brackish water. In comparison with the Hamelin Pool stromatolites, the Lake Clifton thrombolites have been very little explored (Friedman, 1995).

Sites such as Solar Lake in Sinai, Andros Island in the Bahamas and Hamelin Pool and Lake Clifton in Western Australia can be studied as model systems towards understanding of the processes that in the past led to the formation of the extensive stromatolites found in the fossil record. This book contains many excellent examples of such studies. However, it should be stressed that extrapolation of the phenomena occurring at such sites in the present to obtain an understanding of events that happened in the Precambrian is not always simple and straightforward. The problems were formulated very clearly by Claude Monty, and therefore the final paragraph from his 1972 paper deserves to be quoted here in full:

"All the small scale processes and factors responsible for the given features and behaviour of a stromatolitic flat, as well as the complex flow of biological, social, chemical…phenomena within a single algal dome or mat have always frightened me when I consider the simple (too simple!) resulting laminated structures that the palaeontologist has to study; I wonder then how much of the natural history of a stromatolite is left in the thin compacted residual laminae found in the base of a Recent deposit; how much then in a fossil stromatolite? Well, very little most probably. But we shall miss or misunderstand this "very little message" if we do not know the rules of the game, if we do not know where the message might come from, if we do not analyse it in full detail, if we do not sharpen our concepts. To this purpose, the Present may be a key to the Past provided:

(1) We have a good critical knowledge of the present processes at every single level of organization
(2) We dig out from the **right** present the **right** key to the **right**, **analogous** or **homologous** Past."

These words are as true today as they were when they were written, nearly 40 years ago. I am convinced that the chapters in this book fulfil these two provisions set by Monty, so that a proper understanding of the true nature of ancient stromatolites can be achieved.

References

Cohen, Y., Krumbein, W.E. and Shilo, M. (1977) Solar Lake (Sinai). 4. Stromatolitic cyanobacterial mats. Limnol. Oceanogr. **22**: 635–656.

Friedman, G.M. (1995) Microatoll microbialites of Lake Clifton, Western Australia: the morphological analogues of *Crytozoön proliferum* Hall, the first formally-named stromatolite. Facies **32**: 255–258.

Krumbein, W.E. and Cohen, Y. (1974) Biogene, klastische und evaporitische Sedimentation in einem mesothermen monomiktischen ufernahen See (Golf von Aqaba). Geol. Rundsch. **63**: 1035–1065.
Monty, C.L.V. (1972) Recent algal stromatolitic deposits, Andros Island, Bahamas. Preliminary Report. Geol. Rundsch. **61**: 742–783.
Seckbach, J. and Walsh, M. (2009) *From Fossils to Astrobiology. Records of Life on Earth and Search for Extraterrestrial Biosignatures. Cellular Origin, Life in Extreme Habitats and Astrobiology 12*. Springer, Berlin.

Aharon Oren

The Hebrew University of Jerusalem,
91904 Jerusalem,
Israel

Biodata of **Aharon Oren**, author of foreword: ***"Stromatolites: The Present as a Key to the Past"***

Aharon Oren is a professor of microbial ecology at the Institute of Life Sciences, The Hebrew University of Jerusalem. He obtained his Ph.D. at the Hebrew University in 1978 and spent a post-doctoral period at the University of Illinois at Urbana-Champaign. He has authored over 360 scientific publications and he has written or edited 11 books. His current functions include: president, the International Society for Salt Lake Research; general secretary/treasurer, the International Committee on Systematics of Prokaryotes, member of the board of the Israel Society for Microbiology and others. He is editor of *FEMS Microbiology Letters*, *International Journal of Systematic and Evolutionary Microbiology, Extremophiles,* and *Saline Systems*. His research centres on the ecology, physiology, biochemistry and taxonomy of halophilic microorganisms and the microbiology of hypersaline environments. In 2010 he received an honorary doctorate from the University of Osnabrück, Germany.

E-mail: **orena@cc.huji.ac.il**

Professor Vinod Chandra Tewari is currently the Head of the Sedimentology Group at Wadia Institute of Himalayan Geology, Dehradun, Uttarakhand, India. He has been Regular and Senior Associate and also TRIL Fellow of International Centre for Theoretical Physics, Trieste, Italy, between 1998 and 2009. He was born in Almora, India, in 1954 and obtained his higher education including Ph.D. from the University of Lucknow in **Geology** in 1986. He joined Wadia Institute in 1978 and continued his research in the Lesser Himalayn stromatolites, microbiota and isotope chemostratigraphy. Dr. Tewari taught geology at Kumaon Univerisity, Nainital, Uttarakhand (U.K.), India, as Professor of Geology. Dr. Tewari worked on stable isotopes in Biogeochemistry Department of the Max-Planck Institute, Mainz, Germany, as Visiting Professor in 1995. Professor Tewari's scientific interests are in the areas of Precambrian and Phanerozoic stromatolites, sedimentation, carbon isotope chemostratigraphy, genesis, early evolution and diversification of life and its astrobiological significance. He has been associated with several International Geological Correlation Programme (IGCP) Projects such as Stromatolites, Biosedimentology of the Microbial Build-ups, Phosphorites, Precambrian–Cambrian Boundary, Global Bioevents and The Rise and Fall of Vendian Biota. He has 80 research papers published in reputed journals to his credit, and edited several volumes of *Himalayan Geology Journal*, India, and *Journal of Nepal Geological Society*, Kathmandu, Nepal. His current interests are Cretaceous–Paleogene boundary, global extreme and complex climatic events from Neoproterozoic Snowball Earth to Holocene period, India–Asia collision and evolution of Himalaya. Professor Tewari has organized first ***Indo-Soviet Symposium on Stromatolites and Stromatolitic Deposits*** and other IGCP meetings in India and abroad related to stromatolites and phosphorites, etc. He has been one of the organizers of the ***World Summit on Ancient Microscopic Fossils*** held in University of California, Los Angeles, USA, in 2008. Professor Tewari is the member of the Editorial Board of the *International Journal of Astrobiology* published from New York, USA.

E-mail: **vtewari@wihg.res.in**

LIST OF AUTHORS FOR *"STROMATOLITES: INTERACTION OF MICROBES WITH SEDIMENTS"*

All Senior Authors Are Underlined

ALBARRACIN V. HELENA
LIMILA, PLANTA PILOTO DE PROCESOS INDUSTRIALES Y MICROBIOLÓGICOS (PROIMI) CCT, CONICET,
AV. BELGRANO Y PASAJE CASEROS, 4000 TUCUMÁN,
ARGENTINA

ANDRES MIRIAM S.
CHEVERON ENERGY TECHNOLOGY COMPANY,
SAN RAMON, CA, USA

ARROUY M. JULIA
CENTRO DE INVESTIGACIONES GEOLÓGICAS (CIG),
UNIVERSIDAD NACIONAL DE LA PLATA-CONICET,
CALLE 1N 644, 1900 LA PLATA, ARGENTINA

ASPER A. PATRICIA
DIVISION OF HYDROGRAPHY, PORT OF LISBON AUTHORITY,
LISBON, PORTUGAL

AWRAMIK M. STANLEY
DEPARTMENT OF EARTH SCIENCE, UNIVERSITY
OF CALIFORNIA, SANTA BARBARA, CA 93106, USA

BASKAR SUSMITHA
CENTRE FOR GEOBIOLOGY, UNIVERSITY OF BERGEN,
P.O. BOX 7803, 5020 BERGEN, NORWAY
AND
DEPARTMENT OF ENVIRONMENTAL SCIENCE AND ENGINEERING, GURU JAMBHESHWAR UNIVERSITY OF SCIENCE AND TECHNOLOGY,
HISAR 125001, HARYANA, INDIA

BASKAR RAMANATHAN
CENTRE FOR GEOBIOLOGY, UNIVERSITY OF BERGEN,
P.O. BOX 7803, 5020 BERGEN, NORWAY
AND
DEPARTMENT OF ENVIRONMENTAL SCIENCE AND
ENGINEERING, GURU JAMBHESHWAR UNIVERSITY
OF SCIENCE AND TECHNOLOGY,
HISAR 125001, HARYANA, INDIA

BONACCORSI ROSALBA
NASA AMES RESEARCH CENTRE / SETI INSTITUTE, SPACE
SCIENCE AND ASTROBIOLOGY DIVISION MS. 245-3,
MOFFETT FIELD, CA 94035, USA

BOWLIN EMILY
UNIVERSITY OF MIAMI/RSMAS- MGG, 4600 RICKENBACKER
CAUSEWAY, MIAMI, FL 33149, USA

BROOKEFIELD MICHEL E.
INSTITUTE OF EARTH SCIENCES, ACADEMIA SINICA,
P.O. BOX 1-55 NANKANG, TAIPEI 11529, TAIWAN

BROWNE M. KATHLEEN
RIDER UNIVERSITY, 2083 LAWRENCEVILLE ROAD,
LAWRENCEVILLE, NJ 08619, USA

CAMPBELL A. KATHLEEN
GEOLOGY, SCHOOL OF ENVIRONMENT, UNIVERSITY OF
AUCKLAND, PRIVATE BAG 92019, AUCKLAND 1142, NEW ZEALAND

CHACÓN ELIZABETH B.
FACULTAD DE CIENCIAS DE LA TIERRA, UNIVERSIDAD
AUTÓNOMA DE NUEVO LEÓN, CARRETERA CERRO
PRIETO KM 8 LINARES, NUEVO LEÓN 67700, MEXICO

CHELA FLORES JULIAN
THE ABDUS SALAM INTERNATIONAL CENTRE FOR THEORETICAL
PHYSICS, STRADA COSTIERA 11, 34014 TRIESTE, ITALY

CONIGLIO M.
DEPARTMENT OF EARTH AND ENVIRONMENTAL SCIENCES,
UNIVERSITY OF WATERLOO, WATERLOO, ON, CANADA

CUSTAL LILLIAN
UNIVERSITY OF MIAMI/RSMAS- MGG, 4600 RICKENBACKER CAUSEWAY, MIAMI, FL 33149, USA

DELAMETTE MICHEL
PARC NATUREL RÉGIONAL DE CHARTREUSE,
F- 38380 SAINT–PIERRE–DE CHARTREUSE, FRANCE

DELPOMDOR FRANCK
DEPARTMENT OF EARTH AND ENVIRONMENTAL SCIENCES, UNIVERSITY OF BRUSSELES, 50 AV. F.D. ROOSEVELT,
B- 1050 BRUSSELES, BELGIUM

DILLON G. JESSE
DEPARTMENT OF BIOLOGICAL SCIENCES, CALIFORNIA STATE UNIVERSITY, LONG BEACH, CA 90840, USA

DROBNE KATICA
IVAN RAKOVEC INSTITUTE OF PALAEONTOLOGY, ZRC SAZU, NOVI TRG 2, P.O.BOX 306, 1001 LJUBLJANA, SLOVENIA

FAIRCHILD RICH THOMAS
DEPARTMENT OF SEDIMENTARY AND ENVIRONMENTAL GEOLOGY, INSTITUTE OF GEOSCIENCES,
UNIVERSITY OF SÃO PAULO, RUA DO LAGO,
562, 05508-080 SÃO PAULO, BRAZIL

FARÍAS MARIA EUGENIA
LIMILA, PLANTA PILOTO DE PROCESOS INDUSTRIALES Y MICROBIOLÓGICOS (PROIMI) CCT, CONICET,
AV. BELGRANO Y PASAJE CASEROS, 4000 TUCUMÁN, ARGENTINA

FILHO SALLUN WILLIAM
DIVISION OF GEOLOGY, GEOLOGICAL INSTITUTE, ENVIRONMENTAL SECRETARIAT OF SÃO PAULO STATE,
AV. MIGUEL STEFANO 3900, 04301-903 SÃO PAULO, BRAZIL

FÖLLMI KARL B.
UNIVERSITÉ DE LAUSANNE, ANTHROPOLE, CH-1015 LAUSANNE, SWITZERLAND

FOSTER S. JAMIE
DEPARTMENT OF MICROBIOLOGY AND CELL SCIENCE, SPACE LIFE SCIENCES LAB, UNIVERSITY OF FLORIDA, KENNEDY SPACE CENTER, FL 32899, USA

GILLAN C. DAVID
DEPARTMENT OF BIOLOGY, MONS UNIVERSITY, 6 AV. DU CHAMP DE MARS, 7000 MONS, BELGIUM

GLASAUER S.
DEPARTMENT OF LAND RESOURCE SCIENCE, UNIVERSITY OF GUELPH, GUELPH, ON N1G 2W1, CANADA

GOH FALICIA
GENOME INSTITUTE OF SINGAPORE, 60 BIOPOLIS STREET, SINGAPORE 138672, SINGAPORE
AND
SCHOOL OF BIOTECHNOLOGY AND BIOMOLECULAR SCIENCES, UNIVERSITY OF NEW SOUTH WALES, SYDNEY, NSW 2052, AUSTRALIA

GÓMEZ BERRENDERO ESTHER
LABORATORIO DE FICOLOGÍA, FACULTAD DE CIENCIAS, UNAM, APDO. POSTAL 70-620, CIUDAD UNIVERSITARIA COPILCO, COYOACÁN 04510, MEXICO

GREEN J. STEFAN
RESEARCH RESOURCES CENTER, UNIVERSITY OF ILLINOIS AT CHICAGO, CHICAGO, IL 60612, USA

GUHEY RAJEEVA
GOVERNMENT NAGARJUNA P.G. SCIENCE COLLEGE, RAIPUR, CHHATTISGARH, INDIA

HANDLEY KIM. M.
DEPARTMENT OF EARTH AND PLANETARY SCIENCE, UNIVERSITY OF CALIFORNIA AT BERKELEY, BERKELEY, CA, USA

HIKIDA YOSHINORI
NAKAGAWA MUSEUM OF NATURAL HISTORY, 28-9 YASUKAWA, NAKAGAWA TOWN, HOKKAIDO 098-2626, JAPAN

IGISU MOTOKO
DEPARTMENT OF EARTH AND PLANETARY SCIENCES,
TOKYO INSTITUTE OF TECHNOLOGY, O-OKAYAMA 2-12-1,
MEGURO-KU, TOKYO 152-8551, JAPAN
AND
DEPARTMENT OF EARTH AND SPACE SCIENCE, OSAKA
UNIVERSITY, 1-1 MACHIKANEYAMA-CHO, TOYONAKA-SHI,
OSAKA 560-0043, JAPAN
AND
DEPARTMENT OF EARTH SCIENCE AND ASTRONOMY,
THE UNIVERSITY OF TOKYO, KOMABA, MEGURO-KU,
TOKYO 153-8902, JAPAN

ISOZAKI YUKIO
DEPARTMENT OF EARTH SCIENCE AND ASTRONOMY,
THE UNIVERSITY OF TOKYO, KOMABA, MEGURO,
TOKYO 153-8902, JAPAN

JENKINS G. ROBERT
FACULTY OF EDUCATION AND HUMAN SCIENCES
YOKOHAMA NATIONAL UNIVERSITY, 79-2 TOKIWADAI,
HODOGAYA-KU, YOKOHAMA, 240-8501, JAPAN
AND
BAVARIAN STATE COLLECTION FOR PALAEONTOLOGY
AND GEOLOGY, RIENARD-WAGNER STR., 10, 80333 MUNICH,
GERMANY

JONG DE T.M. JEROEN
DEPARTMENT OF EARTH AND ENVIRONMENTAL
SCIENCES, UNIVERSITÉ LIBRE DE BRUXELLES,
50 AV. F.D. ROOSEVELT, 1050 BRUSSELES, BELGIUM

JUAN M. MALDA BARRERA
FACULTAD DE CIENCIAS NATURALES, UNIVERSIDAD AUTÓNOMA
DE QUERÉTARO, AVENIDA DE LAS CIENCIAS S/N COL,
JURIQUILLA, QUERÉTARO 76230, MEXICO

JURKOVŠEK B.
GEOLOGICAL SURVEY OF SLOVENIA, DIMIČEVA 14,
1000 LJUBLJANA, SLOVENIA

KILBURN R. MATT
CENTRE FOR MICROSCOPY, CHARACTERISATION
AND ANALYSIS, THE UNIVERSITY OF WESTERN AUSTRALIA,
35 STIRLING HIGHWAY, CRAWLEY, WA 6009, AUSTRALIA

KOLO KAMAL
DEPARTMENT OF GEOLOGY, VRIJE UNIVERSITEIT BRUSSEL,
PLEINLAAN 2, B – 1050 BRUSSELS, BELGIUM

KRAJEWSKI KRZYSZTOF P.
INSTITUTE OF GEOLOGICAL SCIENCES, POLISH ACADEMY
OF SCIENCES, TWARDA 51/ 55, 00-818, WARSZAWA, POLAND
AND
DEPARTMENT OF ANTARCTIC BIOLOGY, POLISH ACADEMY
OF SCIENCES, USTRZYCKA 10/12, 02-141, WARSZAWA, POLAND

LEE CHOON KWANG
DEPARTMENT OF MINERAL AND MINING ENGINEERING, SANGJI
UNIVERSITY, WONJU-SI, GANGWON-DO 220-702, SOUTH KOREA

LEE, M. NATUSCHKA
DEPARTMENT OF MICROBIOLOGY, TECHNISCHE UNIVERSITÄT,
MÜNCHEN 85354, FREISING, GERMANY

LEIMING YIN
NANJING INSTITUTE OF GEOLOGY AND PALAEONTOLOGY,
CHINESE ACADEMY OF SCIENCES, NANJING 210008, CHINA

LOKHO KAPESA
WADIA INSTITUTE OF HIMALAYAN GEOLOGY,
DEHRADUN 248001, UTTARAKHAND, INDIA

MARUYAMA SHIGENORI
DEPARTMENT OF EARTH AND PLANETARY SCIENCES, TOKYO
INSTITUTE OF TECHNOLOGY, O-OKAYAMA 2- 12-1, MEGURO –KU,
TOKYO 152-8551, JAPAN

MASSARI FRANCESCO
GEOSCIENCE DEPARTMENT, PADOVA UNIVERSITY, VIA GIOTTO 1,
PADOVA, ITALY

MONTEJANO GUSTAVO
LABORATORIO DE FICOLOGÍA, FACULTAD DE CIENCIAS,
UNAM, APDO. POSTAL 70-620, CIUDAD UNIVERSITARIA COPILCO,
COYOACÁN 04510, MEXICO

MIMURA K.
GRADUATE SCHOOL OF ENVIRONMENTAL STUDIES,
NAGOYA UNIVERSITY, NAGOYA 464-8601, JAPAN

NAKASHIMA SATURO
DEPARTMENT OF EARTH SCIENCE, UNIVERSITY OF CALIFORNIA, SANTA BARBARA, CA 93106, USA

OGORELEC B
GEOLOGICAL SURVEY OF SLOVENIA, DIMIČEVA 14, 1000 LJUBLJANA, SLOVENIA

OREN AHARON
DEPARTMENT OF PLANT AND ENVIRONMENTAL SCIENCES, THE INSTITUTE OF LIFE SCIENCES, AND THE MOSHE SHILO MINERVA CENTRE FOR MARINE BIOGEOCHEMISTRY.
THE HEBREW UNIVERSITY OF JERUSALEM, 91904, JERUSALEM, ISRAEL

OUWEHAND J. PIETER
WARNER AG, CH - 4501, SOLOTHURN, SWITZERLAND

ØVREÅS LISE
CENTRE FOR GEOBIOLOGY, UNIVERSITY OF BERGEN,
P.O. BOX 7803, 5020 BERGEN, NORWAY
AND
DEPARTMENT OF BIOLOGY, UNIVERSITY OF BERGEN,
P.O. BOX 7803, 5020 BERGEN, NORWAY

PERRI EDOARDO
DIPARTIMENTO DI SCIENZE DELLA TERRA, UNIVERSITÀ DELLA CALABRIA, 87036 RENDE, ITALY

POIRE DANIEL GUSTAVO
CENTRO DE INVESTIGACIONES GEOLÓGICAS (CIG),
UNIVERSIDAD NACIONAL DE LA PLATA–CONICET,
CALLE 1N 644, 1900 LA PLATA, ARGENTINA

PRÉAT R. ALAIN
DEPARTMENT OF EARTH AND ENVIRONMENTAL SCIENCES,
UNIVERSITÉ LIBRE DE BRUXELLES, 50 AV. F.D. ROOSEVELT,
1050, BRUSSELS, BELGIUM

PRIAN PIERRE JEAN
BUREAU RECHERCHES GÉOLOGIQUES ET MINIÈRES,
3 AV, CLAUDE GUILLEMIN, BP 360009, F-45060 ORLÉANS CÉDEX 2, FRANCE

PUGLIESE NEVIO
DEPARTMENT OF GEOSCIENCES, UNIVERSITY OF TRIESTE, VIA WEISS 2, 34127 TRIESTE, ITALY

RAMOS SANCHEZ MARCO
FACULTAD DE CIENCIAS NATURALES, UNIVERSIDAD AUTÓNOMA DE QUERÉTARO, AVENIDA DE LAS CIENCIAS S/N COL, JURIQUILLA, QUERÉTARO 76230, MEXICO

REID R. PAMELA
UNIVERSITY OF MIAMI/RSMAS- MGG, 4600 RICKENBACKER CAUSEWAY, MIAMI, FL 33149, USA

RICCAMBONI RODOLFO
DEPARTMENT OF GEOSCIENCES, UNIVERSITY OF TRIESTE, VIA WEISS 2, 34127 TRIESTE, ITALY

RIDDER DE CHANTAL
DEPARTMENT OF MARINE BIOLOGY, UNIVERSITÉ LIBRE DE BRUXELLES, 50 AV. F.D. ROOSEVELT, 1050 BRUSSELS, BELGIUM

RIEU R.
TOTAL E&P NEDERLAND, BORDEWIJKLAAN 18, DEN HAAG, THE NETHERLANDS

ROUTH JOYANTO
DEPARTMENT OF GEOLOGY AND GEOCHEMISTRY, STOCKHOLM UNIVERSITY, S10691, STOCKHOLM, SWEDEN

RUIJI CAO
NANJING INSTITUTE OF GEOLOGY AND PALAEONTOLOGY, CHINESE ACADEMY OF SCIENCES, NANJING 210008, CHINA

SECKBACH JOSEPH
THE HEBREW UNIVERSITY OF JERUSALEM, ISRAEL. HOME ADDRESS: MEVO HADAS 20, P.O BOX 1132, EFRAT, 90435, ISRAEL

SINHA DEEPIMA
GOVERNMENT NAGARJUNA P.G. SCIENCE COLLEGE, RAIPUR, CHHATTISGARH, INDIA

SPADAFORA ALESSANDRA
DIPARTIMENTO PROVINCIALE DI COSENZA, ARPACAL.
AGENZIA REGIONALE PROTEZIONE AMBIENTE
CALABRIA, 87100 COSENZA, ITALY

SRIVASTAVA PURNIMA
CENTRE OF ADVANCED STUDY IN GEOLOGY, UNIVERSITY
OF LUCKNOW, LUCKNOW 226007, UTTAR PRADESH, INDIA

STRAUSS HARALD
INSTITUT FÜR GEOLOGIE UND PALÄONTOLOGIE,
WESTFÄLISCHE WILHELMS-UNIVERSITÄT MÜNSTER,
CORRENSSTRAßE 24, 48149, MÜNSTER, GERMANY

SUGITANI KENICHIRO
GRADUATE SCHOOL OF ENVIRONMENTAL STUDIES,
NAGOYA UNIVERSITY, NAGOYA 464-8601, JAPAN

SUMINA L. EVGENIA
BORISSYAK PALAEONTOLOGICAL INSTITUTE RAS,
PROSOUZNAYA STR., 123, MOSCOW, RUSSIA

SUMIN D.L.
INDEPENDENT RESEARCHER, ZHEBRUNOVA STR., 5. 44,
MOSCOW, RUSSIA

TEWARI VINOD C.
WADIA INSTITUTE OF HIMALAYAN GEOLOGY,
DEHRADUN 248001, UTTARAKHAND, INDIA

THORSETH INGUNN H.
CENTRE FOR GEOBIOLOGY, UNIVERSITY OF BERGEN,
P.O. BOX 7803, 5020 BERGEN, NORWAY
AND
DEPARTMENT OF EARTH SCIENCE, UNIVERSITY OF BERGEN,
P.O. BOX 7803, 5020 BERGEN, NORWAY

TUCKER MAURICE E.
DEPARTMENT OF EARTH SCIENCES, DURHAM UNIVERSITY,
DURHAM DH1 3LE, UK

TUNIS G.
DEPARTMENT OF GEOSCIENCES, UNIVERSITY
OF TRIESTE, VIA WEISS 2, 34127, TRIESTE, ITALY

UENO YUICHIRO
GLOBAL EDGE INSTITUTE, TOKYO INSTITUTE
OF TECHNOLOGY, POST NO. 12- 21, MEGRO - KU,
TOKYO, 152 - 8551, JAPAN

WACEY DAVID
SCHOOL OF EARTH AND ENVIRONMENT,
THE UNIVERSITY OF WESTERN AUSTRALIA, 35
STIRLING HIGHWAY, CRAWLEY, WA 6009, AUSTRALIA

WALTER MALCOLM R.
AUSTRALIAN CENTRE FOR ASTROBIOLOGY,
UNIVERSITY OF NEW SOUTH WALES, SYDNEY,
NSW, 2052, AUSTRALIA

WESTPHAL HILDEGARD
LEIBNIZ CENTER FOR MARINE TROPICAL ECOLOGY, FAHREN-
HEITSTRASSE 6, 28357 BREMEN, GERMANY

YAMAMOTO ATSUSHI
DEPARTMENT OF EARTH SCIENCE AND ASTRONOMY,
THE UNIVERSITY OF TOKYO, KOMABA, MEGURO,
TOKYO - 153- 8902, JAPAN
AND
DEPARTMENT OF LIFE SCIENCE, KINKI UNIVERSITY, KOWAKAE,
OSAKA 577-8502, JAPAN

ACKNOWLEDGEMENTS

The chapters of the book *Stromatolites: Interaction of Microbes with Sediments* has been peer reviewed by experts in different disciplines covered in the book. We are grateful to them for sparing their valuable time in peer review of the various chapters and acknowledging their names in alphabetical order: Agustin Martin-Algarra, Spain; Giovanni Aloisi, Germany; Wlady Altermann, Germany; Brendan Burns, Australia; Kathleen Browne, USA; Nick Butterfield, UK; Sherry Cady, USA; David Chapman, USA; J.C. Dillo, USA; Julian Chela Flores, Italy; Falicia Goh, Singapore/Australia; Kirsten Habicht, Denmark; Stepfan Green, USA; Karl Foellmi, Switzerland; Peter Hoff, Germany; Robert Jenkins, Japan; David Johnsten, USA; Anne Jungblut, Canada; Jozef Kazmierczak, Poland; Noffke Nora, USA; Krzysztof Krajewski, Poland; Luca Martire, Italy; Edoardo Perri, Italy; Alain Preat, Belgium; Nevio Pugliese, Italy; D.S.N. Raju, India; V.P. Rao, India; Pamela Reid, USA; Robin Renaut, Canada; J.W. Schopf, USA (for cover photo of stromatolites in Australia); Graham Shields, UK; David Saudry, Israel; Fern Seckbach, Israel; Joseph Seckbach, Israel; Harald Strauss, Germany; Ken Sugitani, Japan; Aharon Oren, Israel; Jennifer Tait, UK; Andreas Teske, Italy; V.C. Tewari, India; Maurice Tucker, UK; Malgorzata Moczydlowska Vidal, Sweden; Suhai Xiao, USA; Xunlai Yuan, PR China; Atsushi Yamamoto, Japan; David Wacey, Australia; Maud Walsh, USA.

Vinod C. Tewari and **Joseph Seckbach** (Eds.)

16th August 2010

PART 1:
ARCHAEAN: PROTEROZOIC STROMATOLITES AND MICROBIOTA

Filho
Fairchild
Guhey
Sinha
Tewari
Préat
Delpomdor
Kolo
Gillan
Prian
Ruiji
Leiming
Srivastava
Sugitani
Mimura
Walter
Tucker

Biodata of **Dr. William Sallun Filho** and **Professor Thomas Rich Fairchild**, authors of ***"Proterozoic Stromatolites of the Itaiacoca Group, Southeast Brazil"***

Dr. William Sallun Filho, geologist, is currently a researcher at the Geological Institute, Environmental Secretariat of the State of São Paulo, Brazil. He obtained his master degree and Ph.D. from the University of São Paulo, Brazil. His scientific interests are in the areas of Precambrian paleobiology, stromatolites, carbonate rocks, and karst terrains.

E-mail: **wsallun@gmail.com**

Professor Thomas Rich Fairchild is a member of the Faculty of the Instituto de Geociências of the University of São Paulo, Brazil, where he teaches courses on paleontology and historical geology. He obtained his Ph.D. from the University of California at Los Angeles in 1975 and returned to Brazil in 1976, where he had been a Peace Corps Volunteer from 1966 to 1969. His research interests center upon Proterozoic life, including the transition to microfossils and stromatolites.

E-mail: **trfairch@hotmail.com**

William Sallun Filho

Thomas Rich Fairchild

V.C. Tewari and J. Seckbach (eds.), *STROMATOLITES: Interaction of Microbes with Sediments*, Cellular Origin, Life in Extreme Habitats and Astrobiology 18, 3–19
DOI 10.1007/978-94-007-0397-1_1, © Springer Science+Business Media B.V. 2011

PROTEROZOIC STROMATOLITES OF THE ITAIACOCA GROUP, SOUTHEAST BRAZIL

WILLIAM SALLUN FILHO[1]
AND THOMAS RICH FAIRCHILD[2]
[1]*Division of Geology, Geological Institute, Environmental Secretariat of São Paulo State, Av. Miguel Stefano 3900, 04301-903 São Paulo, Brazil*
[2]*Department of Sedimentary and Environmental Geology, Institute of Geosciences, University of São Paulo, Rua do Lago, 562, 05508-080 São Paulo, Brazil*

Keywords Proterozoic stromatolites • Itaiacoca Group • Southeast Brazil • Paleoenvironmental • Correlations • Morphotypes • *Conophyton*

1. Introduction

The first stromatolites to be described in Latin America were identified in slightly metamorphosed dolostones of the Itaiacoca Group south of Itapeva, state of São Paulo (SP), Southeast Brazil (Fig. 1), by Almeida (1944), who named them *Collenia itapevensis*. Although he did not use the word "stromatolite" anywhere in his text, Almeida correctly interpreted these structures as microbial in origin and recognized both their paleoenvironmental significance and Precambrian age. Few other stromatolites were described in Brazil (see Fairchild and Sallun Filho, 2004) until the decade of 1970, when Precambrian geology and life became an important subject of study worldwide.

This chapter synthesizes present knowledge of the stromatolites of the Itaiacoca Group, especially in the region of Itapeva, previously published in Portuguese, bringing together detailed descriptions, paleoenvironmental interpretations, and suggested correlations. The classical criteria of Hofmann (1969) served as the basis for objective description of stromatolitic morphotypes (Fairchild, 1993). We chose not to attempt taxonomic classification of the stromatolites, given the lack of a biologically significant hierarchy of criteria for stromatolite classification, exception for the conically laminated stromatolites in the area (*Conophyton*), whose geometrical laminar form provides more objective operational criteria for differentiation.

Most of the fossils illustrated and used in this study are deposited in the collections of Systematic Paleontology Laboratory from Geosciences Institute, University of São Paulo.

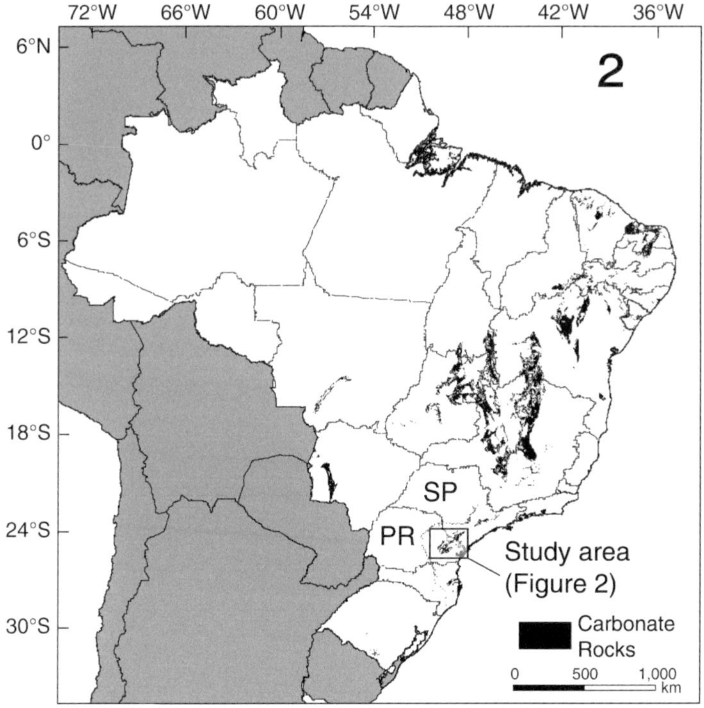

Figure 1. Study area with main occurrences of carbonate rocks in Brazil [modified from Bizzi et al. (2001)].

1.1. GEOLOGICAL SETTING

The Itaiacoca Group comprises a metavolcanosedimentary association that crops out for about 170 km NE–SW in a narrow band in the states of São Paulo and Paraná (Southeast Brazil) (Figs. 1 and 2), within the Ribeira fold belt. A detailed regional stratigraphic framework relating these two groups is lacking because of structural and faciologic complexities and the lack of geochronologic data and regional stratigraphic markers common to both areas. The majority of the publications recognize a metavolcanosedimentary overlain by a fine terrigenous-chemical succession, including stromatolitic limestones (Trein et al., 1985; Theodorovicz et al., 1986; Souza, 1990; Reis Neto, 1994; Prazeres Filho et al., 1998). Otherwise, other authors (Siga Jr. et al., 2003, 2009) suggest the opposite sequence.

The group is limited to the northwest and southeast by late Neoproterozoic to Early Paleozoic granitic complexes and covered, in angular unconformity, by Paleozoic sedimentary rocks of the Paraná basin at its northeastern and southwestern extremities. It is distinct from the Açungui Group to the southeast and also known for its stromatolites (Bigarella and Salamuni, 1958; Marini et al., 1967; Marini and Bósio, 1971; Fairchild, 1977, 1982).

The age of the Itaiacoca Group was recently obtained using SHRIMP in zircons from metabasic bodies concordant with stromatolitic dolomites indicating

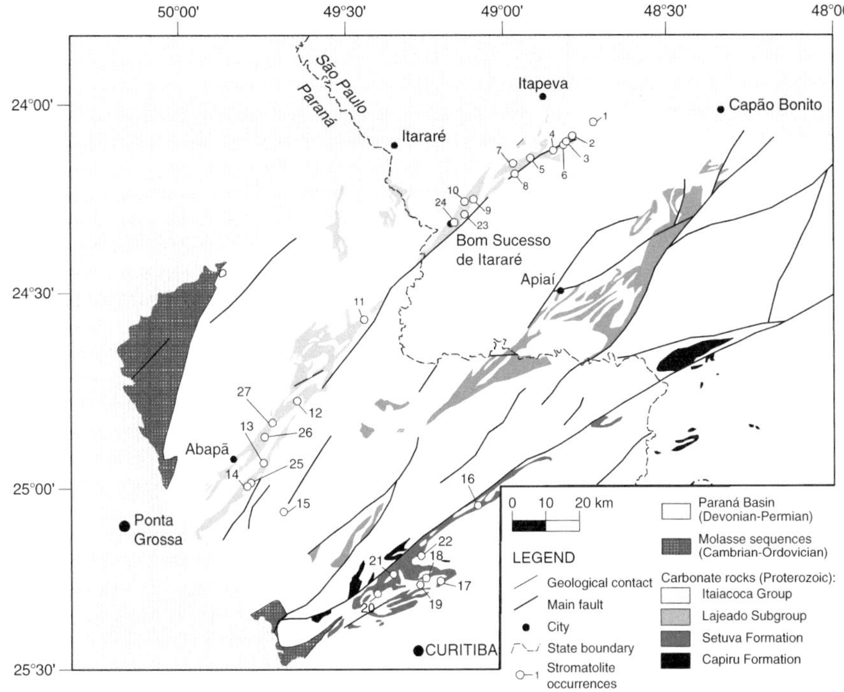

Figure 2. Stromatolite occurrences in carbonate rocks of the Ribeira Belt, São Paulo, and Paraná, Southeast Brazil. Occurrences based on Almeida (1944, 1957), Bigarella and Salamuni (1956), Marini and Bósio (1971), Fairchild (1977, 1982), Theodorovicz et al. (1986), Fairchild and Theodorovicz (1989), Souza (1990), Sallun Filho (1999), Guimarães et al. (2002), Santos (2008), and geology based on Bistrichi et al. (1981), Mineropar (1989), Campanha et al. (1987).

depositional ages related to Mesoproterozoic (1,030–908 Ma) (Siga Jr. et al., 2009). On the basis of the similarity of coniform stromatolites in the Itaiacoca Group to *Conophyton garganicum*, a form known from lower Neoproterozoic and Mesoproterozoic successions in other parts of the world, Fairchild (1977) suggested an age between 1,700 and 850 Ma for the stromatolitic carbonates. The volcanic rocks over the dolomites provide the ages related to Neoproterozoic (645–628 Ma) (Siga Jr. et al., 2003, 2009). Deformation and metamorphism (greenschist facies) affected the Itaiacoca Group in Neoproterozoic, with emplacement of granitic intrusions (630-590 Ma) representative of magmatic arcs (Prazeres Filho et al., 2001).

2. Stromatolites of the Itaiacoca Group

At the time Almeida (1944) described the stromatolites in the Itaiacoca Group south of Itapeva, very little was known about this category of fossils. In fact, nowhere in Almeida's paper is the word "stromatolite" used, which is undoubtedly why Riding (1999) did not count it among the 11 papers published in the 1940s containing this word. In 1957, Almeida described what he judged to be forms distinct from his *C. itapevensis* at new stromatolite localities in the Itaiacoca Group further to the southwest in the state of Paraná (PR). From this, he concluded that these structures could prove useful as stratigraphic markers and in correlation.

Fairchild (1977) reexamined most of these localities and recognized two distinctly different forms among the stromatolites Almeida had named from south of Itapeva as *C. itapevensis* more than 30 years earlier, much prior to the development of stromatolite studies in the former Soviet Union, Australia, Africa, and Canada. One of these he attributed to *C. garganicum*, the other, a branched columnar form characterized by platy columns. He also interpreted them as having formed in a quiet, subtidal depositional environment and identified the presence of *Conophyton* near Abapã (PR), about 150 km from Itapeva.

At present, at least 20 stromatolite occurrences are known from three regions in the nearly 200-km-long band of outcrops of the Itaiacoca Group in São Paulo and Paraná (Fig. 2): (1) in the northwest, near Itapeva and Nova Campina (SP); (2) southeast of region 1, near Bom Sucesso de Itararé (SP); (3) in the southeast, near Abapã (PR). The northwest sector has been the most studied sector.

2.1. REGION OF ITAPEVA

At least nine localities with stromatolites in dolostones and limestones occur south of Itapeva (Fig. 2, points 1–8). Of these, the most important localities are the quarries at Lavrinhas (Fig. 2, point 2) and Indumine/Chiquinho de Barros (Fig. 2, point 4) and the outcrop at Nova Campina (Fig. 2, point 5).

Unbranched columnar coniform stromatolites of the group *Conophyton* predominate. At Indumine, they form a domical or lenticular body of impure (silty) limestone.

The Lavrinhas quarry is mostly made up of stratified dolostones, with about 50 m of stromatolitic dolostones at the base of the section. Due to partial effacement by incipient metamorphism, stromatolitic laminar structure is not immediately evident to the observer but may be discerned locally by the alternation of light to medium gray tones; the carbonate between stromatolites is darker. Because of the tight packing and large size (meter scale) of stromatolite aggregates, bedding planes are absent from this part of the section. The majority of the stromatolites are decimeter-sized, unbranched cylindrical forms with conical lamination (Fig. 3a). The axes of the stromatolites are not perpendicular to the inferred bedding plane. During growth, columns tended to expand laterally until they abutted against others to form aggregates. Merger of the mat-forming communities of contiguous columns then led to the development of partial envelopment of the aggregate by a continuous laminated mantle with the same laminar characteristics as the original columns.

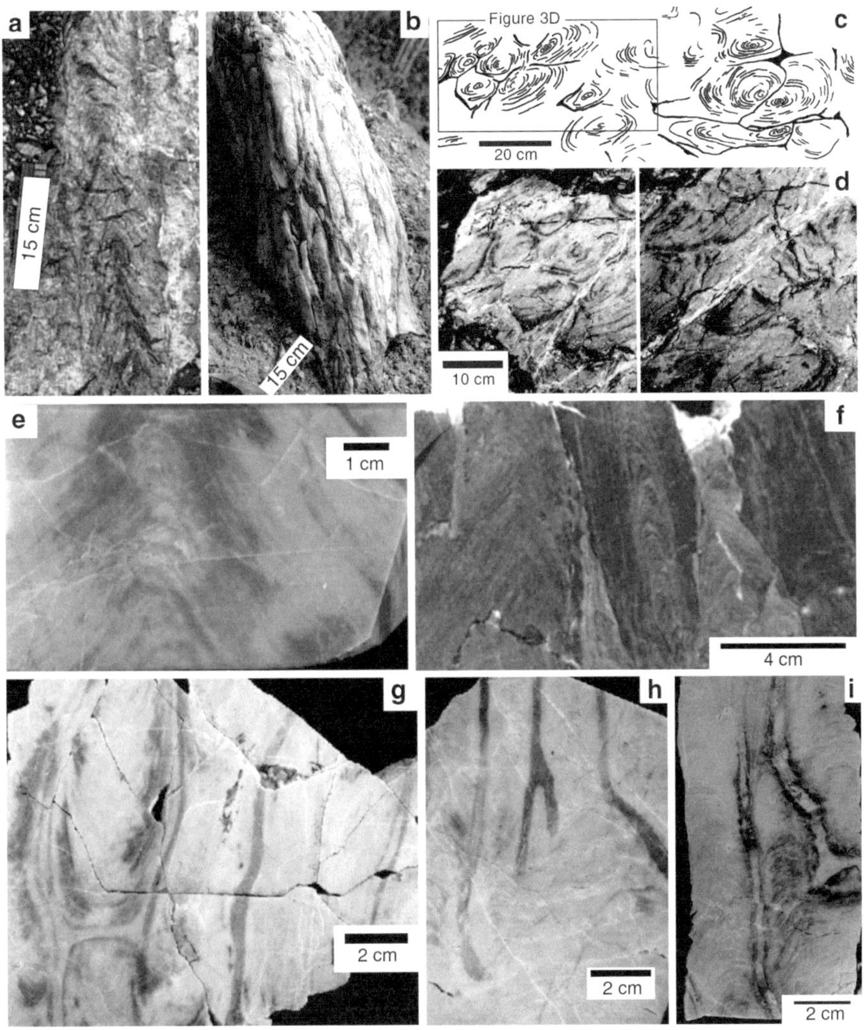

Figure 3. Region of Itapeva (SP). *Conophyton*: (**a**) long column with conical lamination, Lavrinhas; (**b**) columnar deformed stromatolites; (**c**, **d**) Photo (**d**) and interpretative sketch (**c**) of a group of columns of *Conophyton*, in oblique section, showing asymmetric lateral growth, Lavrinhas; (**e**) longitudinal cross-section of column showing conical lamination and axial zone, Lavrinhas; (**f**) longitudinal cross-section through three closely spaced deformed columns, with conical lamination and axial zone evident in the column on the *right*, Indumine. *Morphotype 1*: (**g**) columns with parallel, elongate, elliptical to polygonal transverse contours, Lavrinhas; (**h**) branched columnar stromatolites in longitudinal oblique view with columns with fine, convex to parabolic lamination, Lavrinhas; (**i**) paratype of *Collenia itapevensis*, showing a small projection.

The active quarry at Indumine and the inactive Chiquinho de Barros quarry are parts of the same dark gray to black argillaceous, stromatolitic limestones. This limestone lens has undergone tectonic shearing so that, locally, stromatolites may exhibit considerable deformation with foliation commonly developed in the more argillaceous carbonate between columns (Fig. 3b). The limestone body represents a large bioherm (lithoherm), at least 20 m thick, intercalated between fine-grained terrigenous rocks, now phyllites. Weathering clearly highlights the compositional differences between stromatolites and the more argillaceous interarea sediment. The stromatolites are predominantly erect, unbranched columns with coniform lamination. The axes of the stromatolites are not perpendicular to the inferred bedding plane, although intrastratal rotation related to shearing may have occurred.

At Indumine, stromatolites are found in impure limestones, whereas at Lavrinhas, they occur in relatively pure dolostones. For example, at Indumine insoluble impurities (quartz, clay minerals) may comprise 10% of the columns of *Conophyton* and between 13% and 54% of intercolumnar sediment, in contrast to the 2% of impurities in the stromatolites and 5% of intercolumnar sediment at Lavrinhas. Regionally, stromatolites occur predominantly in the purer dolostones, like those at Lavrinhas. The Indumine–Chiquinho de Barros limestones are the only impure carbonates.

Conophyton in this region occurs as smooth cylinders with generally elliptical to oblong, but also circular to equidimensional polygonal, transverse outlines. Columns are parallel, unbranched, erect, and contiguous to closely spaced, commonly forming aggregates of columns partially enveloped by a laminated mantle. The outermost columns in these aggregates often exhibit asymmetrical, centrifugal development once their proximity to more internal columns impedes their centripetal growth (Fig. 3c, d). Spaces between columns are very narrow, with values less than 0.1 for the ratio interspace/column diameter. At Lavrinhas, stromatolites are large, commonly 30 cm across (longer axis) and varying from 10 to 70 cm. At Indumine and Chiquinho de Barros, columns are generally smaller, between 4 and, exceptionally, 60 cm. Laminar profiles of Conophyton from Lavrinhas, Indumine, and Chiquinho de Barros, are parabolic to pointed, with a narrow axial zone exhibiting medium to high degree of laminar inheritance of successive, according to criteria proposed by Walter (1972) (Fig. 3e, f). Laminae do not present secondary undulations. Lamination is defined by an alternating succession of light and dark laminae of submillimetric to millimetric thickness.

Despite the difficulty in discerning lamination in hand samples and outcrops of the generally pale dolostones at Lavrinhas, lamination may be well preserved (class 1), at least locally, in petrographic thin sections of the best samples, again according to criteria developed by Walter (1972). Measurements made in thin section of samples from Lavrinhas furnished values for the thickness of the light (L1) and dark laminae (L2) between 0.01 and 0.46 mm, with most dark laminae within the range of 0.01 and 0.18 mm and light laminae between 0.01 and

0.26 mm. In thin section, laminae in the stromatolites from Indumine may be well defined and relatively well preserved (preservational class 2 of Walter, 1972), but laminar thickness could not be reliably measured because of recrystallization. Petrographic study also reveals irregular, commonly tectonized (sheared) column margins. Fairchild (1977) classified these unbranched columnar stromatolites with coniform lamination as *Conophyton* cf. *C. garganicum*, on the basis of the ratio of light to dark laminae, the striated aspect of darker laminae, and characteristics of the axial zone (Fig. 6).

A second form called morphotype 1 occurs in this region as transversely elongated, branched columnar stromatolites may be observed in the dolostones at Lavrinhas, point 3 abandoned quarry (Fig. 3g–i), and other smaller localities but appear to be especially abundant at Nova Campina outcrop. Field relationships between *Conophyton* and morphotype 1 are not preserved in available outcrops. They are not found in the limestones at Indumine/Chiquinho de Barros. Much of Almeida's (1944) description of *C. itapevensis* appears to refer to this stromatolitic morphotype, which consists of relatively tall (up to 35 cm in height in an incomplete specimen), transversely elongate, smooth-walled, very closely spaced (separated by no more than a few millimeters), parallel columns oriented normal to inferred bedding. In transverse section, they are generally oblong to elliptical, or less frequently equidimensional, and commonly exhibit polygonal outlines. Branching is infrequent and α-parallel (Walter, 1972). Projections in niches and coalescence may be observed. Laminae are symmetrical to asymmetrical with a relatively high degree of laminar inheritance and slightly convex profiles parallel to the longer transverse axis and very convex, or even parabolic profiles with relatively high synoptic relief, parallel to the shorter transverse axis. At the column margins, laminae may or may not inflect downward to form a wall over previous laminae. Despite their very close spacing, columns are not connected by bridges. Lamination is visible as thin bands of alternating tones of light and medium gray, with no evident rhythmicity. Laminae may be darker at column margins or in patches within the columns. As also observed in the coniform stromatolites in the dolostones, laminae are better preserved in these parts of the columns. Light laminae (L1) vary from 0.1 to 0.55 mm in thickness and dark laminae (L2) from 0.05 to 0.25 mm.

A third columnar morphotype (morphotype 2), similar to morphotype 1, occurs in this region, principally at point 6 (Fig. 2) but also present at Lavrinhas and Indumine. Different from morphotype 1, the stromatolites of morphotype 2 are unbranched and larger with more elliptical transverse cross-sections and more convex (parabolic) lamination. Columns are straight, smooth walled, closely spaced, parallel to one another, and oriented normal to inferred bedding. They may reach a meter in height. In transverse cross-section, they are rounded to elliptical with greater axes between 10 and 20 cm in length and lesser axes between 5 and 15 cm. Laminar profiles are symmetrical to asymmetrical and moderately to very convex, and commonly parabolic (but not conical), with a high degree of laminar inheritance. At column margins, laminae are inflected downward without overlapping. Synoptic relief varies from 2 to 10 cm.

2.2. REGION OF BOM SUCESSO DE ITARARÉ

There are at least four occurrences of stromatolites in dolostones in the region of Bom Sucesso de Itararé (SP) (Fig. 2, points 9, 10, 23, 24), of which the most important are Cal Sinhá (Fig. 2, point 10) and Geovidro (Fig. 2, point 24) quarries. The stromatolites and depositional environment here are very different from those near Itapeva and Abapã. Smaller branched and unbranched columnar stromatolites predominate and stratiform microbialites and oolitic dolostones are present, but none of the morphotypes (including *Conophyton*) evident near Itapeva have been observed.

According to Santos (unpublished research, 2008), the most distinctive stromatolites in this area form an extensive bioherm in the Cal Sinhá quarry at Taquaruçu da Serra. These stromatolites, here designated morphotype 3, are simple and branched, straight to sinuous, cylindrical to subcylindrical columnar forms with very convex laminar profiles and a high degree of laminar inheritance (Fig. 4). The columns occur in clusters parallel to commonly divergent columns, and these clusters, closely spaced, make up the bioherm. Stratiform microbialites are finely laminated and mildly undulose with or without bird's eye structures.

2.3. REGION OF ABAPÃ

At least five localities with stromatolites are known in the region of Abapã (PR) in dolostones and limestones (Fig. 2, points 11–14 and 25–27), most notably at Rio do Morcego, Ribeirão das Areias (Fig. 2, points 13 and 14), and the Calpar and Itatinga (Fig. 2, points 26 and 27) quarries.

Unbranched columnar stromatolites having parabolic to coniform lamination and an axial zone predominate and are so similar to the *Conophyton*

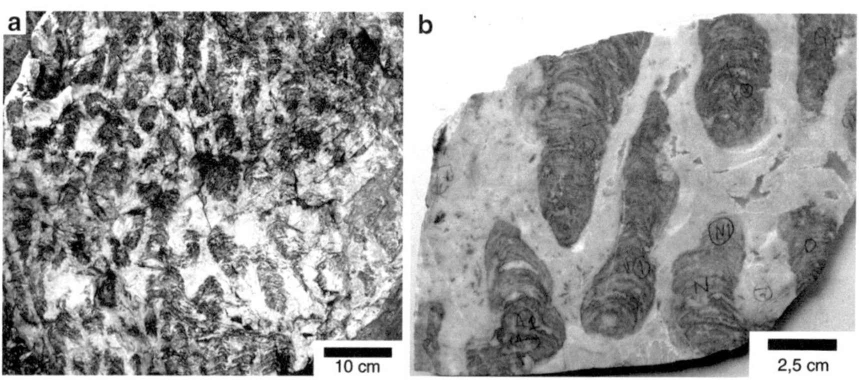

Figure 4. Region of Bom Sucesso de Itararé (SP). *Morphotype 3*: (**a, b**) simple and branched columnar forms forming clusters of divergent columns.

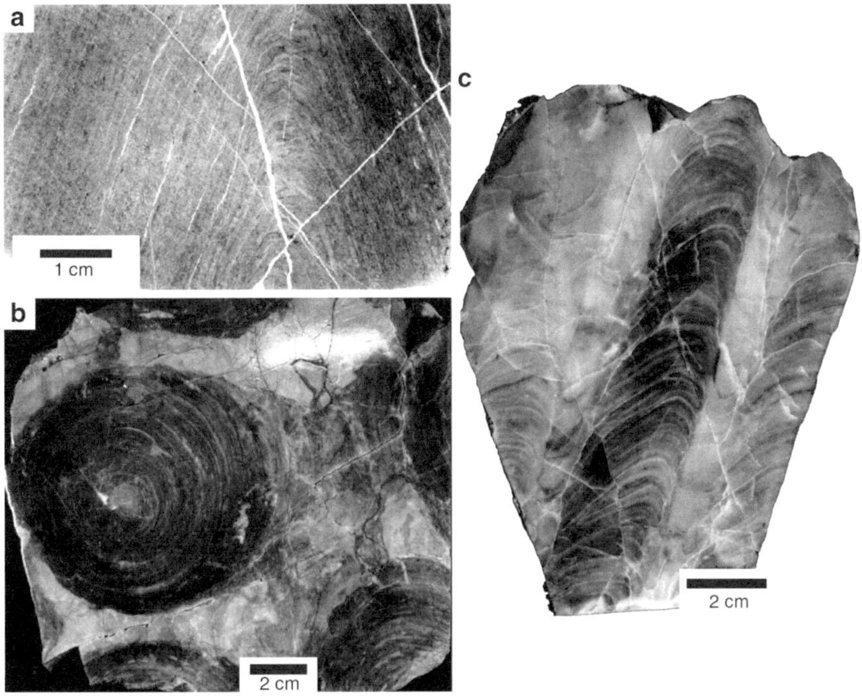

Figure 5. Region of Abapã (PR). *Conophyton*: (**a**) longitudinal cross-section of column showing conical lamination and axial zone, Rio Morcego; (**b**) circular transverse cross-section, Ribeirão das Areias; (**c**) unbranched, nonconiform columnar form with convex to parabolic lamination.

near Itapeva, although less deformed tectonically and with better preserved lamination, as to be considered the same form – *Conophyton* cf. *C. garganicum* (Fig. 5a, b). Columns are cylindrical, with circular to elliptical transverse cross-sections, up to 13 and 18 cm along their minor and major axes. Columns are closely spaced, separated by no more than 1–4 cm, and with high synoptic relief and degree of laminar inheritance of the coniform lamination clearly evident in longitudinal section. The axial zone is narrow with a medium to highly regular axial trace.

In thin-section, laminae are very thin and poorly preserved, with the dark laminae particularly degraded. Dark laminae (L2) vary between 0.04 and 0.5 mm, and light laminae (L1) between 0.04 and 0.76 mm, with a greater concentration of values between 0.04 and 0.3 mm and peaks at 0.14 and 0.26 mm for L2, and values between 0.04 and 0.4 mm with a peak at 0.15 mm for L1.

Unbranched, nonconiform columnar stromatolites with convex to parabolic lamination also occur in this region but are not associated with the coniform stromatolites (Fig. 5c).

3. Discussion

Stromatolites have been used for decades as indicators of ancient coastlines and water depth, particular depositional environments, stratigraphic facing in deformed rock successions, and even as stratigraphic markers for intrabasinal and rough biostratigraphic correlation in Proterozoic rocks. Although their utility as guide fossils remains controversial, stromatolites have become important tools in basin analysis, especially for intrabasinal correlation, characterization of facies, paleocurrents, basin paleogeography, and other paleoenvironmental parameters, as well for eustatic sea level changes (see Hoffman, 1967; Preiss, 1973; Hoffman 1974; Bertrand-Sarfati and Trompette, 1976; Trompette and Boudzoumou, 1981; Bertrand-Sarfati and Moussine-Pouchkine, 1985; Fairchild and Herrington, 1989; Altermann and Herbig, 1991; Bertrand-Sarfati et al., 1991). In the opinion of Bertrand-Sarfati and Trompette (1976), the precision of intrabasinal stromatolite correlation should be tested against other methods of correlation, whereas its reliability will depend on the care taken in the description and identification of the stromatolites. Soviet workers established an empirical biostratigraphic scheme based on stromatolites of the Mesoproterozoic and Neoproterozoic (Riphean and Vendian) that allowed correlation among various Precambrian regions of the former Soviet Union (Semikhatov, 1976), Africa (Bertrand-Sarfati, 1972), and Australia (Walter, 1972), but it proved to be difficult to apply and not very precise, which has led to skepticism as to its validity. It is well known, for instance, that gross stromatolite morphology commonly reflects differential facies controls so that changes in stromatolite forms within a stratigraphic section do not necessarily indicate temporal transitions between different stratigraphic units (Haslett, 1976).

Despite the controversy regarding the validity of stromatolite biostratigraphy, the simple, quasi-geometric characteristics of the coniform stromatolites make the members of the group *Conophyton* possibly the easiest to identify of the putatively biostratigraphically useful stromatolites and hence one of the most reliable for broad correlation. In fact, study of *Conophyton* worldwide has demonstrated that certain forms, including the three forms tentatively identified in the Brazilian Precambrian, have restricted temporal distribution (Preiss, 1976). Thus, *C. cylindricum* and *C. metulum* (Moeri, 1972; Cloud and Moeri, 1973; Fairchildm, 1977) in the Vazante Formation (Minas Gerais) would likely indicate a Mesoproterozoic age (Preiss, 1976). In the Itaiacoca Group, the coniform stromatolites at Itapeva (SP) and Abapã (PR) are sufficiently similar in their lamination and axial zones to be placed within the same stromatolite form and classified as *C. garganicum* (Fig. 6).

Preiss (1976) identified the stratigraphic range of this form as Lower to lower Upper Riphean, which would correspond roughly to the upper Mesoproterozoic to lower Neoproterozoic, or approximately to the period from 1,700 to 850 Ma. If so, argued Fairchild (1977), then the age of the *Conophyton*-bearing carbonates in the Itaiacoca Group should fall within this interval. Geochronologic studies since then have tended to corroborate Fairchild's suggestion, as evidenced by ages of Siga Jr. et al. (2009).

Conophyton is also considered generally as an indicator of relatively deep, calm marine environments, probably below normal wave base at depths up to

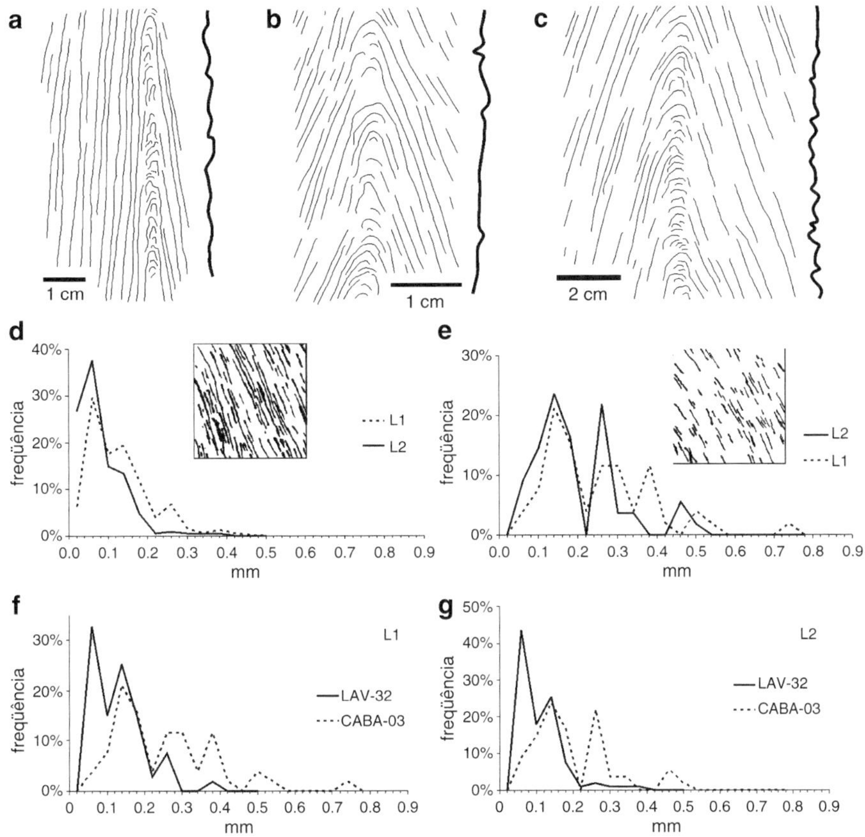

Figure 6. Axial zone and lamination in *Conophyton* from Indumine (**a**), Lavrinhas (Itapeva) (**b**), and Rio Morcego (Abapã) (**c**); graphs of relative frequency of thickness of dark laminae (L2) and light laminae (L1) in *Conophyton* from Lavrinhas (Itapeva) (**d**) and Rio Morcego (Abapã) (**e**), and comparative graphs of both localities (**f**, **g**) [modified from Sallun Filho and Fairchild (2005)].

several tens of meters, as based on the delicate, fine conical lamination, high synoptic profile, abundance and regularity of form in *Conophyton*, and generally a complete absence of sediments and sedimentary structures suggestive of shallow water deposition associated with *Conophyton* (Donaldson and Taylor, 1972). When *Conophyton* passes upward into shallower environments, abundant small, divergent columns begin to branch off from its uppermost surface in Christmas-tree fashion, which is called *Jacutophyton* in stromatolite taxonomy (Hoffman, 1976). Moreover, the apical warp at the tip of the conical lamination (which is responsible for the axial zone in the lithified stromatolite) is interpreted as resulting from concentration of the photosynthetic microbial mat-forming community at the highest possible point of the stromatolite in response to minimal levels of luminosity (Walter, 1977). Thus, *Conophyton* is envisaged as representing an ecological adaptation to the deepest possible conditions for abundant stromatolite development within the photic zone. The massive accumulations of *Conophyton* at Itapeva and Abapã are thus interpreted as evidence of such deeper water environments within the Itaiacoca Group. Conditions may have been slightly shallower and/or more brightly illuminated in the area of Lavrinhas than at Indumine, south of Itapeva. This is suggested by the tendency for lateral stromatolite growth of *Conophyton*, association with smaller, branched stromatolites (morphotype 1), dominance of light-colored dolostone, and upward passage into a very thick succession of thin-bedded dolostone evident at Lavrinhas and the narrower, more regular columns, dark-colored calcareous composition of *Conophyton*, significant amount of associated terrigenous sediment between columns (murky waters), and development of the bioherm between underlying and overlying pelitic sediments at Indumine (Fig. 7).

The sedimentological and morphological analysis of the stromatolites near Bom Sucesso reveals a much different paleoenvironmental setting. The abundance of oolitic carbonates, the presence of stratiform microbialites with bird's eye structures, and the abundance, small size, and divergent branching of the stromatolites here, together with the apparent absence of *Conophyton*, all point to a much shallower peritidal depositional environment. By the same token, some of

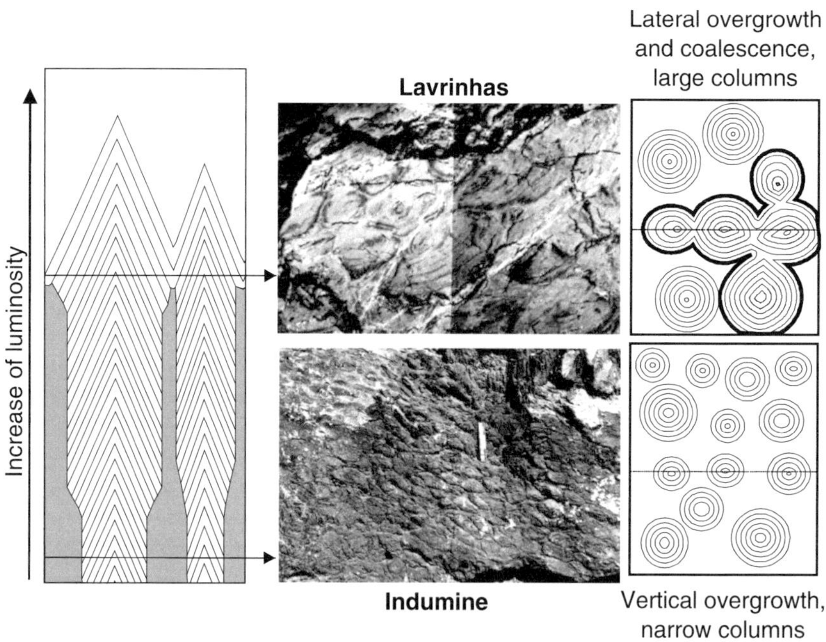

Figure 7. Model for development of *Conophyton* in well-lit waters (Lavrinhas) and in poorly illuminated waters (Indumine) [adapted from Donaldson (1976), by Sallun Filho and Fairchild (2004)].

the nonconiform stromatolites locally evident near Itapeva and Abapã may represent shallower environments at the top of, marginal to, or not contemporaneous with the *Conophyton* bioherms.

4. Acknowledgments

The authors are thankful to Professor Vinod C. Tewari and Professor Joseph Seckbach for invitation to contribute this chapter for the book. The anonymous reviewer and Professor Vinod C. Tewari are thanked for the suggestions.

5. References

Almeida, F.F.M. (1944) *Collenia itapevensis* sp. n. – um fóssil pré-cambriano do Estado de São Paulo. Bol. FFLCH USP Geologia **45**(1): 89–106.

Almeida, F.F.M. (1957) Novas ocorrências de fósseis no pré-cambriano brasileiro. An. Acad. Bras. Ciênc. **29**: 63–72.

Altermann, W. and Herbig, H.G. (1991) Tidal flats deposits of the Lower Proterozoic Campell Group along the southwestern margin of the Kaapvaal Craton, Northern Cape Province, South Africa. J. Afr. Earth Sci. **13**(3/4): 415–435.

Bertrand-Sarfati, J. (1972) Stromatolites columnaires du Precambrien superieur du Sahara nord-occidental Inventaire, morphologie et microstructure des laminations Correlations stratigraphiques. Centre de Recherches sur les zones arides – Series Geologie **14**, 245 p.

Bertrand-Sarfati, J. and Moussine-Pouchkine, A. (1985) Evolution and enviromental conditions of Conophyton-Jacutophyton associaton in the Atar dolomite (upper Proterozoic, Mauritania). Prec. Res. **29**: 207–234.

Bertrand-Sarfati, J. and Trompette, R. (1976) Use of stromatolites for intrabasinal correlation: example from the Late Proterozoic of the northwestern margin of the Taudinni Basin, In: M.R. Walter (ed.) *Stromatolites*. Elsevier, Amsterdam, pp. 517–522.

Bertrand-Sarfati, J., Moussine-Pouchkime, A. and Semikhatov, M.A. (1991) Conical stromatolites Conophyton-Jacutophyton as indicators of eustatic movements in Proterozoic cratonic sedimentary areas, In: Symp. on Fossil Algae, V, Capri, pp. 51.

Bigarella, J.J. and Salamuni, R. (1956) Estudo preliminares na Série Açunguí. V – Estruturas organógenas nos dolomitos da Formação Capirú (Estado do Paraná). Dusenia, **7**(6): 317–323.

Bigarella, J.J. and Salamuni, R. (1958) Estudos preliminares na série Açunguí VIII – A Formação Votuverava. Bol. Inst. Hist. Nat. Geol. **2**: 6.

Bistrichi, C.A., Carneiro, C.D.R., Dantas, A.S.L., Ponçano, W.L., Campanha, G.A.C., Nagata, N., Almeida, M.A., Stein, D.P., Melo, M.S. and Creminini, O.A. (1981) *Mapa geológico do Estado de São Paulo, escala 1:500.000*. IPT/PRÓ-MINÉRIO 2, geological map.

Bizzi, L.A., Schobbenhaus, C., Gonçalves, J.H., Baars, F.J., Delgado, I.M., Abram, M.B., Leão Neto, R., Matos, G.M.M. and Santos, J.O.S. (2001) *Geologia, tectônica e recursos minerais do Brasil: Sistema de Informações Geográficas – SIG e mapas na escala 1:2.500000*. CPRM, Brasília, CD-ROM.

Campanha, G.A.C., Bistrichi, C.A. and Almeida, M.A. (1987) Considerações sobre a organização litoestratigráfica e evolução tectônica da Faixa de dobramentos Apiaí, In: Simp. Sul-bras. Geol., III, Curitiba, Atas **2**: 725–742.

Cloud, P. and Moeri, E. (1973) Conophyton in the Bambui Group: what form and age? Geology **1**: 127.

Donaldson, J.A. and Taylor, A.H. (1972) Conical-columnar stromatolites and subtidal enviroment. Am. Assoc. Pet. Geol. Bull. **56**: 614.

Donaldson, J.A. (1976) Paleoecology of Conophyton and associated stromatolites in the Precambrian Dismal Lakes and Rae Groups, Canada. In: M.R. Walter (ed.) *Stromatolites*, Elsevier , pp. 523 534.

Fairchild, T.R. (1977) *Conophyton* and other columnar stromatolites from the Upper Precambrian Açungui Group near Itapeva, SP, Brazil, In: Simp. Reg. Geol., 1, São Paulo, Atas, pp. 179–198.
Fairchild, T.R. (1982) New stromatolites from the Upper Precambrian Açungui Group, eastern Paraná, Brazil, and their potential stratigraphic use. Bol. IGc USP **13**: 43–50.
Fairchild, T.R. (1993) Um método prático de descrição de estromatólitos para o geólogo de campo, In: Congr. Bras. Paleo. 13. São Leopoldo, Resumos, pp. 207.
Fairchild, I.J. and Herrington, P.M. (1989) A tempestite-stromatolite-evaporite association (Late Vendian, east Greenland): a shoreface-lagoon model. Prec. Res. **43**: 101–127.
Fairchild, T.R. and Sallun Filho, W. (2004) *Collenia itapevensis*, o primeiro fóssil pré-cambriano brasileiro e sua importância no estudo de estromatólitos no Brasil, In: V. Mantesso Neto, A. Bartorelli, C.D.R. Carneiro and B.B. Brito Neves (eds.) *O desvendar de um continente: a moderna geologia da América do Sul e o legado da obra de Fernando Flávio Marques de Almeida*. Beca, São Paulo, pp. 177–186.
Fairchild, T.R. and Theodorovicz, A. (1989). Novas ocorrências de estromatólitos no Grupo Itaiacoca (Proterozóico médio a superior), sul do Estado de São Paulo, In: Congr. Brás. de Paleo., XI, Curitiba, Resumos, pp. 4.
Guimarães, S.B., Reis Neto, J.M. and Siqueira, R.B.L. (2002) Caracterização dos estromatólitos da Formação Capiru (Proterozóico) nas regiões de Morro Azul e Morro Grande: leste do Paraná. Bol. Paran. Geoc. **51**: 77–88.
Haslett, P.G. (1976) Lower Cambrian stromatolites from open and sheltered intertidal environments, Wirrealpa, South Australia, In: M.R. Walter (ed.) *Stromatolites*. Elsevier, Amsterdam, pp. 565–584.
Hoffman, P. (1967) Algal stromatolites: use in stratigraphic correlation and paleocurrent determination. Science **157**: 1043–1045.
Hoffman, P. (1974) Shallow and deepwater stromatolites in lower Proterozoic platform-to-basin facies change, Great Slave Lake, Canada. Am. Assoc. Pet. Geol. Bull. **58**(4): 856–867.
Hoffman, P. (1976) Environmental diversity of middle Precambrian stromatolites, In: M.R. Walter (ed.) *Stromatolites*. Elsevier, Amsterdam, pp. 599–611.
Hofmann, H.J. (1969) Atributes of stromatolites. Geol. Surv. Canada Paper **69-39**: 58.
Marini, O.J. and Bósio, N.J. (1971) Estromatólitos em dolomitos do Grupo Açungui. An. Acad. Bras. Ciênc. **43**(1): 161–175.
Marini, O.J., Trein, E. and Fuck, R. (1967) O Grupo Açungui no estado do Paraná. Bol. Par. Geoc. **23-25**: 43–104.
Mineropar (1989) *Mapa geológico do Estado do Paraná, escala 1:650.000*. Mineropar, Curitiba, CD-ROM.
Moeri, E. (1972) On a columnar stromatolite in the Precambrian Bambui Group of central Brazil. Eclogae Geol. Helv. **65**(1): 185–195.
Prazeres Filho, H.J., Basei, M.A.S., Harara, O.M., Passarelli, C.R., Siga Jr., O., Reis Neto, J.M. and Sato, K. (2001) The Alto Ribeira magmatic arc (Paraná State – Southern Brazil): geochemical and isotopic evidences of Magmatic Focus Migration and its tectonic implications, In: Simp. Sul-Am. Geol. Isot., 3, Pucon, Chile, Extend Abstract, pp. 213–216.
Prazeres Filho, H.J., Guimarães, G., Basei, M., Siga Jr., Reis Neto, J.M., Campanha, G. and Sallun Filho, W. (1998) Mapa geológico 1:50.000 da porção centro-sul da Faixa Itaiacoca – PR, In: Cong. Bras. Paleo., 40, Belo Horizonte, Anais, pp. 36.
Preiss, W.V. (1973) Palaeocological interpretations of South Australian Precambrian stromatolites. J. Geol. Soc. Aust. **19**: 501–532.
Preiss, W.V. (1976) Intercontinental correlations, In: M.R. Walter (ed.) *Stromatolites*. Elsevier, Amsterdam, pp. 359–370.
Reis Neto, J.M. (1994) *Faixa Itaiacoca: registro de uma colisão entre dois blocos continentais no Neoproterozóico*. Ph.D. Thesis, IGC-USP, São Paulo, Brazil, 253 p.
Riding, R. (1999) The term stromatolite: towards an essential definition. Lethaia **32**: 321–330.
Sallun Filho, W. (1999) Análise dos estromatólitos do Grupo Itaiacoca (Proterozóico), ao sul de Itapeva, SP. Monograph (unpublished research), IGc-USP, São Paulo, Brasil.
Sallun Filho, W. and Fairchild, T.R. (2004) Estromatólitos do Grupo Itaiacoca ao sul de Itapeva, SP. Rev. Bras. Paleo. **7**(3): 359–370.

Sallun Filho, W. and Fairchild, T.R. (2005) Estudo comparativo entre estromatólitos do tipo *Conophyton* das faixas Ribeira (Grupo Itaiacoca, SP-PR) e Brasília (grupos Vazante, MG e Paranoá, GO). Rev. Inst. Geol. **26**(1/2): 1–18.

Santos, A.A. (2008) *Reinterpretação dos estromatólitos do Grupo Itaiacoca entre Abapã-PR e Bom Sucesso-SP e seu significado paleoambiental e estratigráfico.* Monograph (unpublished research), IGc-USP, São Paulo, Brasil.

Semikhatov, M.A. (1976) Experience in stromatolite studies in the U.S.S.R., In: M.R. Walter (ed.) *Stromatolites*. Elsevier, Amsterdam, pp. 337–357.

Siga Jr., O., Basei, M.A.S., Sato, K., Prazeres Filho, H.J., Cury, L.F., Weber, W., Passarelli, C.R., Harara, O.M. and Reis Neto, J.M. (2003) U-Pb (Zircon) ages of metavolcanic rocks from the Itaiacoca Group: tectonic implications. Rev. IGc USP **3**: 39–49.

Siga Jr., O., Basei, M., Passarelli, C., Sato, K , Cury, L. and Mcreath, I. (2009) Lower and Upper Neoproterozoic magmatic records in Itaiacoca Belt (Paraná-Brazil): Zircon ages and lithostratigraphy studies. Gond. Res. **15**: 197–208.

Souza, A.P. (1990) *Mapa geológico na escala 1:50.000 e esboço da evolução tectônica e sedimentar do Grupo Itaiacoca, nas folhas Barra do Chapéu e Ouro Verde – SP/PR.* M.Sc. Thesis, IGc USP, São Paulo, Brazil, 200 p.

Theodorovicz, A., Câmara, M.M., Morais, S.M., Godoy, H.K. and Takahashi, A.T. (1986) Projeto Engenheiro Maia-Ribeirão Branco – Relatório Final. CPRM/PROMINÉRIO, unpublished report.

Trein, E., Reis Neto, J.M., Biondi, J.C. and Monastier, M.S. (1985) Revisão da Formação Itaiacoca: identificação de uma sequência metavulcano-sedimentar em Abapã (PR), In: Simp. Reg. Geol., 5, São Paulo, Atas 1, pp. 169–185.

Trompette, R. and Boudzoumou, F. (1981) Palaeogeographic significance of stromatolitic buildups on late Proterozoic plataforms: the example of the west Congo Basin. Palaeog. Palaeoclim. Palaeoe. **66**: 101–112.

Walter, M.R. (1972) Stromatolites and the biostratigraphy of the Australian Precambrian and Cambrian. Palaeont. Assoc. London Spec. Pap. Palaeont. **11**: 190.

Walter, M.R. (1977) Interpreting stromatolites. Am. Sci. **65**: 563–571.

Biodata of Professor **Rajeev Guhey** and Professor **Vinod Chandra Tewari,** authors of ***"Meso-Neoproterozoic Stromatolites from the Indravati and Chhattisgarh Basins, Central India."***

Professor Rajeeva Guhey, is currently in the Department of Geology, at Government Nagarjuna Post Graduate College, Raipur, Chhattisgarh affiliated under Pt. Ravishankar University, Raipur, India. He obtained his Ph.D. in ***Geology*** from Pt. Ravishankar Shukla University, Raipur in 1993. He has completed several research projects on Proterozoic sedimentary carbonates from Chhattisgarh and Indravati Basins, India. He was a recipient of ***Young Scientists Award*** for best paper presentation in 1992 from Devi Ahilya University, Indore, M.P. India. He organized a National Seminar on "***Proterozoic Basins of India with special reference to Chhattisgarh Basin***". He has participated in ***32nd IGC, Florence, Italy, 2004***. Professor Guhey is teaching sedimentology, Remote sensing, hydrogeology, in Graduate classes.

E-mail: **rajeevgeol@gmail.com**

Professor Vinod C. Tewari is currently the Head of the Sedimentology Group at Wadia Institute of Himalayan Geology, Dehradun, and a Senior Associate of International Centre for Theoretical Physics, Trieste, Italy. He obtained his Ph.D. from the University of Lucknow in *Geology* in 1986 and continued his research in Wadia Institute. Dr. Tewari taught Geology at Kumaon University, Nainital, Uttarakhand (U.K.), India, as Professor of Geology. Professor Tewari's scientific interests are in the areas of Precambrian stromatolites, sedimentation, carbon isotope chemostratigraphy, genesis, early evolution and diversification of life and its astrobiological significance. He is associated with the International Geological Correlation Programme (IGCP) Project 493 on *The Rise and Fall of Vendian Biota*. He has 80 research papers published to his credit and edited several volumes of Himalayan Geology, India, and *Journal of Nepal Geological Society*, Kathmandu,

Rajeeva Guhey

V. C. Tewari

V.C. Tewari and J. Seckbach (eds.), *STROMATOLITES: Interaction of Microbes with Sediments*, Cellular Origin, Life in Extreme Habitats and Astrobiology 18, 21–42
DOI 10.1007/978-94-007-0397-1_2,

Nepal. Professor Tewari has organized first *Indo-Soviet Symposium on Stromatolites and Stromatolitic Deposits* and other IGCP meetings in India. He has been one of the organizers of the World Summit on Ancient Microscopic Fossils held in University of California, Los Angeles, USA, in 2008. He is a member of the editorial board of the *International Journal Astrobiology*, New York, USA.

E-mail: **vtewari@wihg.res.in**

Deepima Sinha is a research scholar, department of Geology, at Government Nagarjuna Post Graduate College, Raipur, Chhattisgarh, affiliated to Pt. Ravishankar Shukla University, Raipur, C.G. India. She obtained her Ph.D. from Pt. Ravishankar Shukla University, Raipur, India (2009), on ***Proterozoic Columnar stromatolites from Jagdalpur Formation, Indravati Basin, Chhattisgarh: Palaeoenvironment and Stratigraphic implications*** under supervision of Dr. Rajeeva Guhey, Professor, Geology, Government NPG College, Raipur and Dr. Mukund Sharma, Scientist-E, Birbal Sahni Institute of Palaeobotany, Lucknow, India. She was a recipient of the ***Young Scientist's Award*** for best paper presentation in 2006 from Chhattisgarh Council of Science and Technology, at Indira Gandhi Agricultural University, Raipur, CG India. This book chapter is a part of her Ph.D. Thesis.

E-mail: **deepimasinha21@gmail.com**

MESO-NEOPROTEROZOIC STROMATOLITES FROM THE INDRAVATI AND CHHATTISGARH BASINS, CENTRAL INDIA

RAJEEVA GUHEY[1], DEEPIMA SINHA[1], AND VINOD CHANDRA TEWARI[2]
[1]*Government Nagarjuna P.G. Science College, Raipur, Chhattisgarh, India*
[2]*Wadia Institute of Himalayan Geology, Dehradun 248001, Uttarakhand, India*

Keywords Stromatolites • Meso-Neoproterozoic • Indravati • Chhattisgarh basins • India • Biostratigraphy • Correlation • Southern Urals • Lesser Himalaya

1. Introduction

The Meso-Neoproterozoic period is very significant for the diversification of the stromatolite taxa on Earth. Stromatolites have been used for the regional and global biostratigraphic correlations of the Proterozoic successions. In India, Proterozoic stromatolites are well preserved in the sedimentary basins of the Peninsular and Himalayan sedimentary basins (Valdiya, 1989; Tewari, 1989, 2004; Raha and Das, 1989). This chapter mainly deals with the Meso-Neoproterozoic stromatolites from the Indravati and Chhattisgarh basins of Central India (Figs. 1 and 2). We have briefly described the stromatolite taxa from these two very significant basins with special reference to their occurrence, stratigraphic distribution, age, and correlation with other equivalent basins in Peninsular and Lesser Himalayan basins. The Meso-Neoproterozoic stromatolites are mainly confined to the younger Raipur Formation of Chhattisgarh Basin and Jagdalpur Formation of Indravati Basin. These two sedimentary basins situated between two other major Paleoproterozoic to Mesoproterozoic basins, namely the Vindhyans to its north and the Cuddapah to its south. Chhattisgarh Basin is spread over an area of about 35,000 km^2 while Indravati Basin covers approximately 9,000 km^2 area. Both the basins are 300 km apart. The identification and description of stromatolites of these basins are based on their morphology and branching pattern of columns following mainly Raaben (1969) and Cloud and Semikhatov (1969).

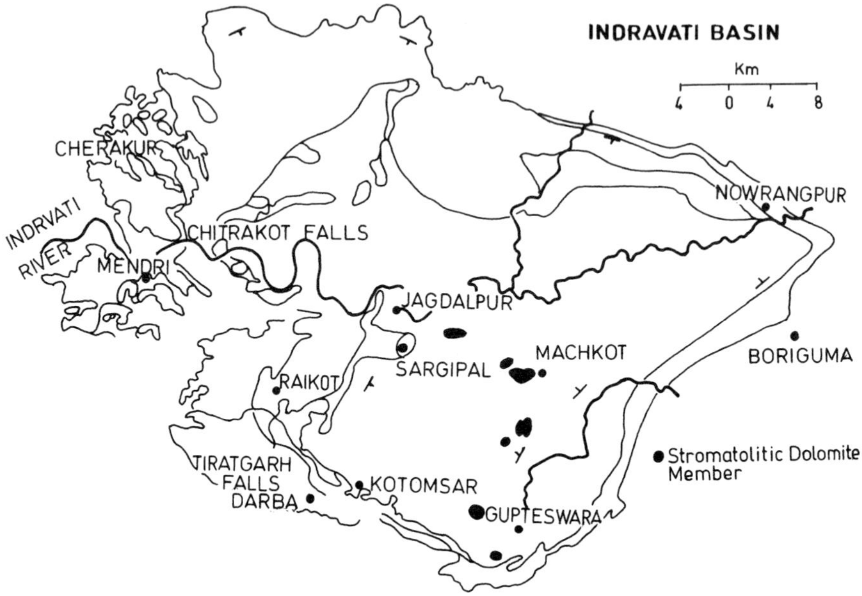

Figure 1. Locality map of the stromatolite occurrences in Indravati Basin [after Ramakrishnan (1987)].

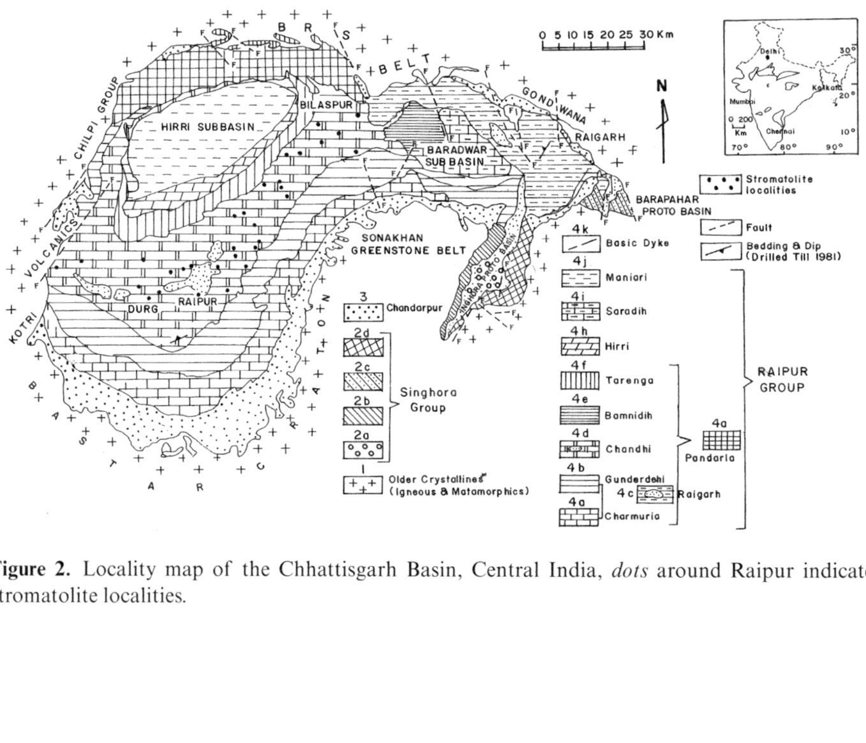

Figure 2. Locality map of the Chhattisgarh Basin, Central India, *dots* around Raipur indicate stromatolite localities.

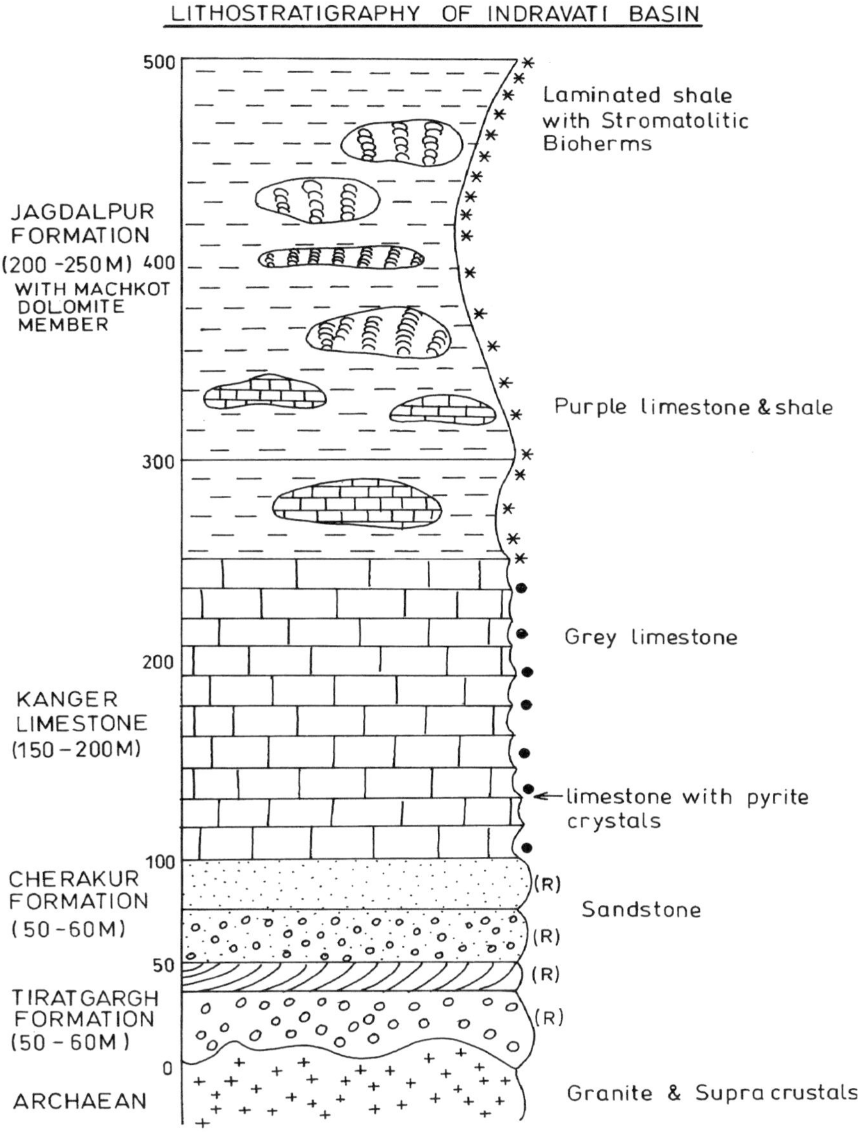

Figure 3. Lithostratigraphy of the Indravati Basin [modified after Ramakrishnan (1987)]. *Asterisk* Stromatolite sample, *Bullet* Sample, *(R)* Reference sample.

2. Stromatolites from the Indravati and Chhattisgarh Basins

2.1. PREVIOUS WORK

The stromatolites were reported earlier as concretions by King (1881). Sen (1966) briefly described their morphology. Vishwanathaiya and Sastri (1973) have reported stromatolites while studying the occurrence of phosphorites. Chandra and Bhattacharya (1973) studied the stromatolites from Raipur Limestone at Mandir Hasaud. Ghosh and Shah (1965), Schnitzer (1971), Shrivastava (1977), Murti (1978), Jairaman and Banerjee (1980), Moitra and Pal (1985), Moitra (1984, 1996, 1999), Chatterjee et al. (1990), and Guhey and Wadhwa (1993) described their occurrence in the Raipur Group. Kreuzer et al. (1977) carried out geochronological studies of Chandarpur Sandstone. Das et al. (1990) described gypsum band in Tarenga Shale and suggested a Sabkha type environment during that time. A tentative correlation of the Chhattisgarh stromatolites has been attempted by previous workers (Raaben and Tewari, 1987; Valdiya, 1989; Tewari, 1989; Raha and Das, 1989).

2.2. PRESENT STUDY AND SYSTEMATIC DESCRIPTION OF STROMATOLITES FROM THE INDRAVATI BASIN

Columnar and branching stromatolites have been recognized on the basis of vertical and horizontal biostromes and buildups in the outcrop; systematic samples were collected for serial sectioning. Morphological features such as shape of columns, type of branching, lamina shape, nature of lateral surface wall, and so on, are described following Raaben (1969), Semikhatove and Raaben (1969), and Grey (1989). Microstructures of the stromatolites have been studied to confirm their identifications.

2.2.1. Group Colonnella Komar, 1964
Type form: *Colonnella cormosa* Komar

Diagnosis:

Columns are straight, narrow, cylindrical, mostly linked with less or narrow intercolumnar space commonly uniform in width. Laminae are gently convex. The microstructure is smooth.

Colonnella laminata Komar:
(Fig. 4).

Materials:

Seven specimens. This variety is known from three outcrops. The material described has been collected from NW of Koliapara, Kerlakonta, Machkot, and Mundapal localities.

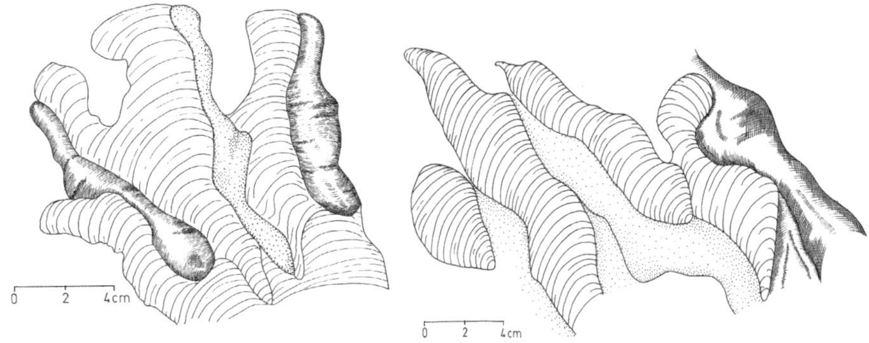

Figure 4. Three-dimensional reconstruction of *Colonnella laminata*.

Description:

Regular subcylindrical parallel columns with rounded ovate cross profile. The diameter ranges from 5 to 15 cm and height up to a meter. The stromatolites outcrop from deeply convex bell-shaped and lensoid morphology. Columns are semispherical and when approaching the margins they are sometime sharply conical.

Outcrop:

This stromatolite form is well developed in the middle part of the Jagdalpur Shale Formation. The stromatolitic bioherms are well exposed near the Bastar Village NW of Koliapara, Kerlakonta, in the Ganeshbahar Greek, of Machkot Area and in the East of Mundapal Village. All the stromatolitic outcrops are red-colored dolomitic limestone. The carbonate units are surrounded by shales. In the Mundapal exposure, several calcite veins and stylolites are present.

Mode of Occurrence:

Stromatolitic bioherms occur as isolated outcrop in between the shales. In some of the outcrops, stromatolites are large, complete, and compact, while in others the stromatolites are broken on the margin.

Fascicle Morphology:

Fascicles are made up of columns without any branching. Fine laminae alternate with mud and limestone. Laminae profile is hemispherical and convex overlapping of laminae forms a clear wall structure.

Branching Habit:

Branching is absent.

Column Shape and Margin Structure:

Columns are cylindrical in shape and ovate to semicircular in transverse section. Mostly formed by flexuous, dome-shaped, or hemispherical laminae which are convex stacked one above the other, margin is compact and thin.

Microstructure and Texture:

Laminae of stromatolites occur in pair of light and dark colors. The light and dark laminae are thicker toward the margins of the column and in central portion the light and dark laminae are equally thick. Dark laminae produce streaky to uniform texture. Light laminae range in thickness from 400 to 250 μm and dark laminae range in size from 80 to 120 μm and are made up of equigranular dolomites. Both the dark and light laminae have well-defined boundary. This boundary can be marked on the stromatolitic columns owing to very thin admixture of dark argillaceous material accumulated or trapped on the microcrystalline surface.

Interspace Filling:

Interspaces between the columns are very narrow and contain a mixture of clay and dolomite. The grain size is almost identical to those clay particles within the laminae of the stromatolite column.

Secondary Alteration:

Stromatolites are present in the red-colored dolomites which are part of red-colored Jagdalpur Shale Formation. Presence of ghost rhombs and equigranular dolomite in the laminae suggests the primary trapping and binding of the carbonate grains. Stylolites are noted in few places.

Comparison:

The present stromatolite form differs from other forms of *Colonnella* group by a very simple microstructure of the microbial strata and their deeply convex laminae.

Distribution:

These stromatolites have been identified in Jagdalpur Shale Formation. They are noticed at the middle level of this formation.

Age:

Mesoproterozoic to Neoproterozoic.

2.2.2. Group Gymnosolen Steinmann, 1911
Type form: *Gymnosolen ramsayi* Steinmann, 1911

Diagnosis:

Smooth, straight, subcylindrical, actively branching columns which expended or swollen before branching subparallel to slightly divergent. Sometimes bridging and coalescing of columns at few places. Laminae are convex. Transverse section of the column is subcircular oval to polygonal.

Gymnosolen furcatus Komar:

(Figs. 5 and 6).

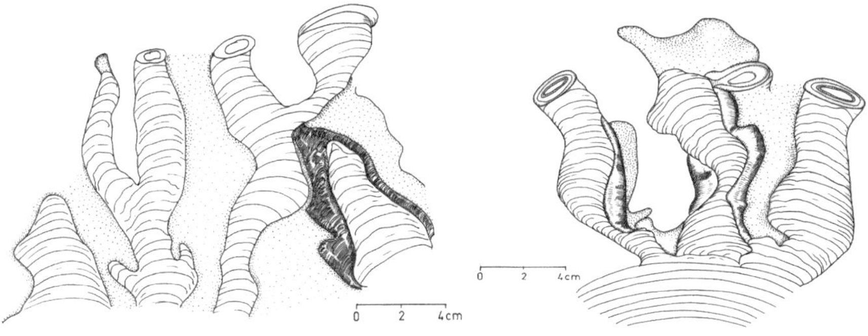

Figure 5. Three-dimensional reconstruction of *Gymnosolen furcatus*.

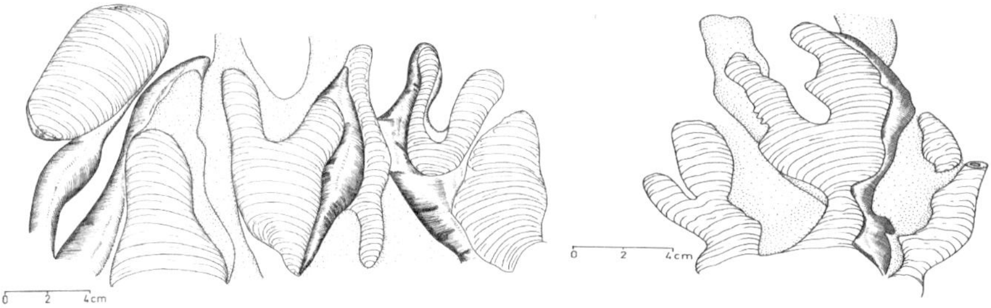

Figure 6. Three-dimensional reconstruction of *Gymnosolen furcatus*.

Description:

The structure of stromatolite consists of vertical subcylindrical columns of 10–15 cm long cross profile rounded or ovate of 2–5 cm diameter. Repeated branching observed. Columns swollen before branching. Intervals between the section of branching at 6–9 cm and up.

Materials:

Five specimens were collected from Machkot, Koliapara, Mundapal, Bharni, Ghatlohanga, and Raykera Nala areas.

Outcrop:

The dolomitic hummocks containing this stromatolite are present in Machkot Reserve Forest. The specimens are collected from Ganeshbahar Creek and from the roadside cuttings of Machkot-Tiria forest. Nearly all the hummocks have similar type of stromatolites. In between two hummocks are red colors shales, which terminate by the side of the hummocks.

Mode of Occurrence:

Columnar branching stromatolites make the fascicle. Columns are made up of hemispherical convex lamina stocked one over the other. Fascicle broadens before branching.

Fascicle Morphology:

Fascicles are made up of columns with frequent branching. Fine lamina alternates with shale and limestone. Lamina profile is hemispherical and where branching is present, concavity and convexity of the laminae can be seen to continue. Wall structure is clear and rarely some part of the stromatolites does not form clear wall structure.

Branching Habit:

Branching frequent and divergent, columns are bulbous before the branching.

Column Shape and Margin Structure:

Columns are subcylindrical in shape and ovoidal in transverse section. They are formed by flexuous, dome-shaped/hemispherical laminae which are convex, stocked one above the other. Overlapping of the laminae forms a wall structure.

Lamina:

Lamina profile varies from gentle to flexuous, convex, dome-shaped/hemispherical set of dark and light laminae. Laminae at the junction of branching are continuous and convex and concave at thin at the margin. Sometimes the laminae are found disrupted because of the secondary alteration.

Microstructure and Texture:

Laminae of the stromatolites occur in light and dark pairs. The lighter laminae are thicker in middle and dark laminae are equally thick and produce streaky texture. Light laminae usually range in thickness from 200 to 300 μm and dark and light laminae range in size from 50 to 100 μm. Both the dark and light laminae have well-defined boundary. Dark laminae are composed of brownish-gray non-oriented rounded lumps of kerogen. Lighter laminae are made up of finely granular dolomite. Sometimes streak of shale is noticed in the lighter laminae.

Interspace Fillings:

Interspaces between columns vary and are filled with shale and carbonate that are very fine grained almost equal to those adjust shales.

Secondary Alteration:

Stromatolites are present in light red-colored dolomite, suggesting their secondary alterations.

Comparison:

A markedly subcylindrical column with divergent branching and typical laminae profile permits its assignment to *Gymnosolen furcatus* Komar which is different from other known *Gymnosolen* varieties.

Distribution:

These stromatolites have been identified in Jagdalpur Shale Formation in Machkot, Koliapara, Mundapal, Bharni, Ghatlohanga, and Raykera Nala areas.

Age:

Neoproterozoic.

2.2.3. Group Kussiella Krylov, 1963
Type form: *Kussiella kussiensis* Maslov

Diagnosis:

Straight cross-ribbed columnar stromatolites, passive branching of columns with rounded cross-section, laminae in the columns drop freely on the margin; rarely do the successive laminae drop over the underlying laminae.

Kussiella enigmatica Raaben:

(Fig. 7).

Materials:

Four specimens from Mongrapal, Mundapal, Junaguda, Machkot, and Gupteshwar areas.

Description:

Straight cross-ribbed subcylindrical columnar stromatolite parallel branches, width of the column ranges from 20 cm in the lower portion of the structure to 1–3 cm in the upper portion. Columns form edges and fringes. Connecting bridges occur rarely.

Outcrop:

Kussiella enigmatica occurs in Machkot Area and NW of Mundapal locality where Jagdalpur Shale Formation is exposed. The stromatolites are present in the red-colored dolomite. The outcrop is lensoid 10 × 2.5 m bioherms. The strata containing stromatolites are 2–3 m thick and may be traced over 2 km.

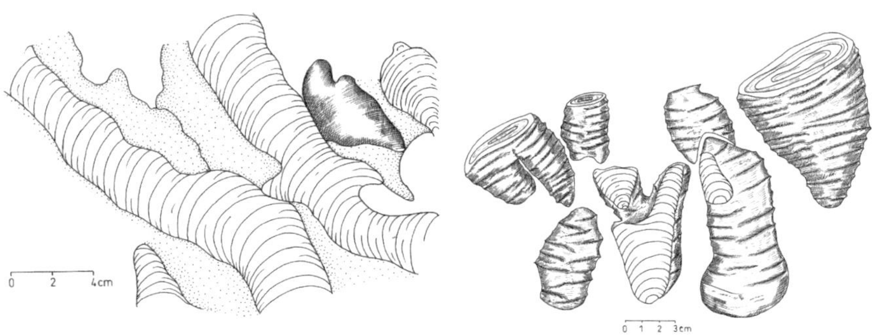

Figure 7. Three-dimensional reconstruction of *Kussiella enigmatica*.

Mode of Occurrence:

Kussiella enigmatica occurs in the mode of bioherm in the Machkot and NW of Mundapal areas.

Fascicle Morphology:

Fascicles are made up of columns with frequent branching. Fine laminae alternate with shale and limestone. Laminae freely hang and do not form clear wall structure.

Branching Habit:

Branching is frequently passive and parallel.

Column Shape and Margin Structure:

Columns are rounded in cross-section and are cylindrical in shape at margin. Formed by laminae which have peaks and cornices falling freely with almost straight, subparallel columns; constrictions at the base are infrequent. Bridging is noticed between the columns.

Lamina Profile:

Lamina profile varies from gentle to flexuous, convex, dome-shaped/hemispherical set of dark and white carbonate laminae. Toward the top, laminae are sometimes straight and thickened at the center and thin at the margin.

Microstructure and Texture:

Laminae of the stromatolite occur in dark and light pairs. Lighter laminae are 100–1,000 μm thick and the dark carbonate laminae are 40–200 μm. The carbonate is of 3–5 μm grain size, xenotopic to hypidiotopic equidimensional, and polygonal. Lighter pale laminae vary from 90% carbonate to 90% clay. The carbonate has the same texture as in the dark laminae except the grain size is large.

Interspace Filling:

Interspaces between the columns are narrow and contain a mixture of clay and carbonate with same microstructure and texture. Nature of most of the filling is dust form.

Secondary Alteration:

Carbonate at NW of Mundapal is cleaved and stromatolites are deformed; those reconstructed are from Machkot Area. Mundapal stromatolites are well exposed due to Joints making possible the observations in the field. This Joint pattern may be due to crystallization of silicic component present in the interspaces.

Comparison:

Markedly cylindrical columns with rounded to ovate cross-section and convex hemispherical laminae profile with cornices and peaks along with bridging phenomenon permits that stromatolite is properly classified as *Kussiella enigmatica.* It differs from

Kussiella kussiensis described from the Shali, Deoban, and Buxa dolomites of the Lesser Himalaya in overhanging laminae and bridging pattern (Tewari, 1989, 1994).

Distribution:

Jagdalpur Formation, in the Machkot Area and NW of Mundapal, Mongrapal, Junaguda, Gupteshwar areas of Indravati Group in Jagdalpur District, Chhattisgarh.

Age:

Neoproterozoic.

2.2.4. Group Boxonia Korolyuk, 1960
Type form: *Boxonia gracilis* Korolyuk

Diagnosis:

Stromatolites with subparallel, subcylindrical columns which are walled with few bumps and bridges. Branching α-β parallel and frequent.

Age:

Neoproterozoic–Terminal Proterozoic.

2.2.5. Boxonia pertaknurra Walter
(Fig. 8).

Material:

One specimen from Mundapal, Koliapara.

Description:

Columns are erect, branching predominantly and parallel continuous wall except in the lower region. Laminae are rhombic or rectangular with streaky and clotty microstructure.

Outcrop:

The *B. pertaknurra* are exposed in Mundapal immediately below the *K. enigmatica* bioherms, mound size range between 2.3×10 m. The two adjacent mounds are filled with shale.

Mode of Occurrence:

B. pertaknurra occurs in the mode of bioherms extending for 1 km in the Mundapal locality.

Fascicle Morphology:

Fascicle is made up of elongated columns and infrequent branches. Fine laminae show admixture of clay and limestone. Branches are mostly parallel and divide at angle which is less than 5° seldom 1 cm apart.

Branching Habit:

Branching frequent, dichotomous, predominantly α-parallel with moderate β-parallel. Bridging between narrow column occurs infrequently.

Column Shape and Margin Structure:

The columns are 2–4 cm wide and 14–30 cm long with irregular transverse section, bridging and coalescing noticed. Small columns are straight and subcylindrical. In transverse section, these are rounded or oblong. The columns are smooth with short thin transverse ribs. Walls are formed (Fig. 8).

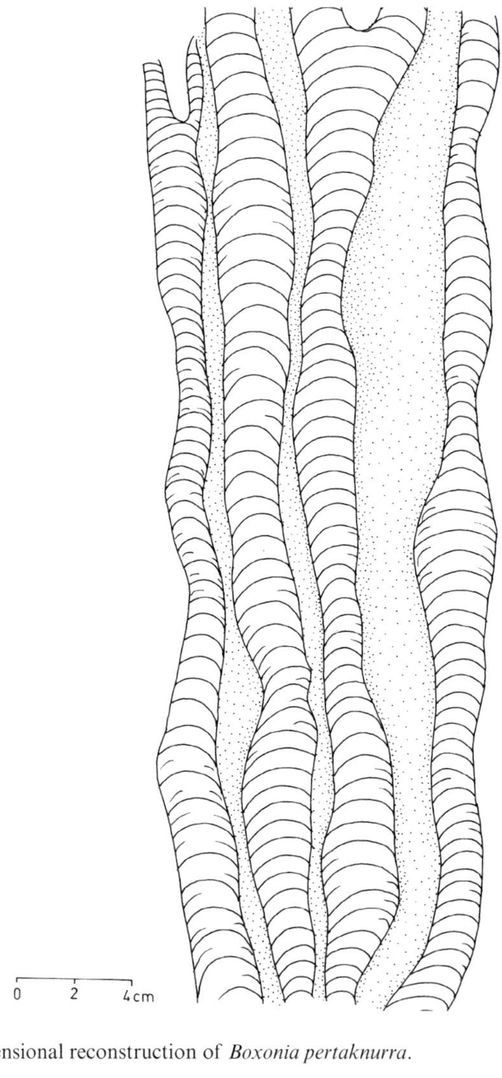

Figure 8. Three-dimensional reconstruction of *Boxonia pertaknurra*.

Lamina Shape:

Laminae are hemispherical to rectangular, margins forming a multilaminate wall on the columns margin. Toward the top of the column, there are some laminae which are flattened or concave. A wall is always present.

Microstructure and Texture:

Dark and light laminae alternate in the column. Dark laminae are made up of xenotopic equigranular carbonate grains of 10–25 μm size. Colorless light laminae are made up of xenotopic to idiotopic carbonate grains and the thickness of the laminae varies between 50 and 350 μm. The grain size is similar to dark laminae.

Interspace Fillings:

Interspaces between the columns are filled with homogeneous dolomite and clay. Sometimes the remnants of clast have been noticed floating in the clay material.

Secondary Alteration:

Stylolite Formation has demarcated and differentiated the columns from interspace filling.

Comparison:

In gross morphology, the *B. pertaknurra* resembles *Acaciella*, *Kussiella*, *Boxonia* but the presence of clear wall structure (Fig. 8) differentiates it from other forms and justifies its consideration as *Boxonia*. Microstructure and texture make it a clear identification in *pertaknurra*. It is different from *Boxonia gracilis* described from the Lower Tal Formation of the Lesser Himalaya by Tewari (1989) in having narrow large columns and thick wall.

Distribution:

Jagdalpur Limestone Formation at Mundapal bears the bioherms.

Age:

Terminal Proterozoic.

3. Paleobathymetry and Paleoenvironment

The present identified stromatolitic assemblages of Indravati Basin, Chhattisgarh, mainly belong to *Colonnella laminate* Komar, *Gymnosolen furcatus* Komar, *Kussiella enigmatica* Raaben, and *Boxonia Pertaknurra* Walter. These stromatolitic assemblages indicate marine subtidal to intertidal flat environment. The shallow intertidal environment is also supported by the occurrence of intraformational conglomerate in Jagdalpur shale and limestone. Based on these stromatolites, a schematic depositional model showing different stromatolite varieties from intertidal to supratidal condition is shown in Fig. 9.

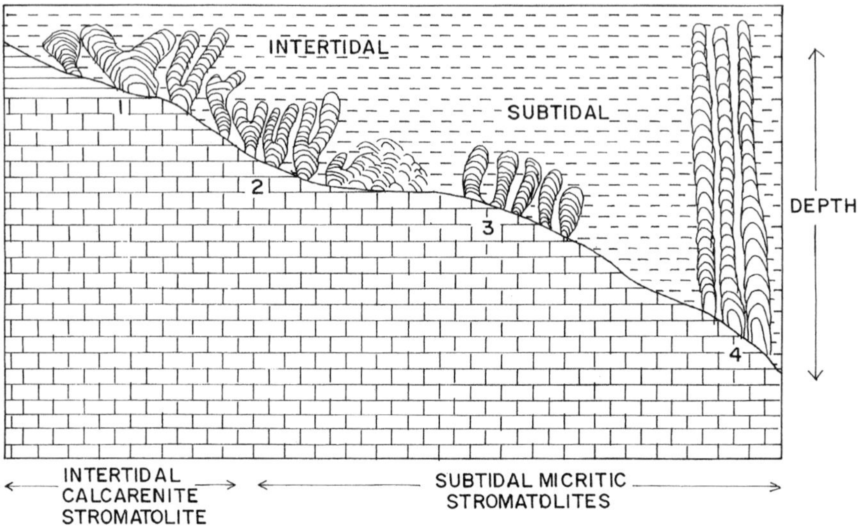

Figure 9. Relationship between stromatolite morphology and the depositional environment.

4. Stromatolites from the Chhattisgarh Basin

In Meso to Neoproterozoic Chhattisgarh Basin (Fig. 2), stromatolites are mainly confined to Chandi Formation of Raipur Group. The columnar stromatolites of Chandi Limestone (Raipur Limestone) identified belonged to *Colonnella* and *Baicalia* Groups. The domal nonbranching stromatolites are identified as *Nucleella* (Komar, 1966; Guhey and Wadhava, 1993). Moitra (1999) studied the stromatolites in detail and identified 24 forms classified under 12 groups. The stromatolite groups mainly identified belonging to *Baicalia*, *Gymnosolen*, *Jacutophyton*, *Colonnella*, *Tungussia*, *Conophyton*, *Kussiella*, *Inzeria*, *Linella*, *Anabaria*. *Omechtenia, and Acaciella* are rare forms.

5. Correlation of Meso-Neoproterozoic Stromatolitic Formations

The correlation of the Chhattisgarh Basin with other Meso-Neoproterozoic basins of the Peninsular India (Vindhyan Basin, Cuddapah Basin, Pakhal Basin), Lesser Himalaya (Jammu, Shali, Simla), and that of the type area in Southern Urals (former Soviet Union) has been shown in the Fig. 10 (Tewari, 1989).

6. Discussions and Conclusions

Stromatolites have been studied from the Indravati and Chhattisgarh basins by various workers (Schnitzer, 1971; Jairaman and Banerjee, 1980; Jha et al., 1990; Guhey and Wadhawa, 1993; Moitra, 1999). We have reviewed the available data and

Figure 10. Field occurrence of *Kussiella enigmatica* showing nature of columns, margins, variation in size, and pinkish color (**a**–**e**).

found that there is tremendous scope for the systematic study of these microbial structures. We have given systematic description of some characteristic stromatolite taxa and their global comparison and correlation. The stromatolite assemblages of the Chhattisgarh Basin (Raipur Group) have an Upper Riphean (Neoproterozoic) affinity. The Raipur Group has been isotopically dated at 750–700 Ma (Schnitzer, 1971; Kreuzer et al., 1977) and can be correlated with the Upper Vindhyans

(Fig. 10). The stromatolite assemblage of the Raipur Group is comparable with *Tungussia nodosa*, *Inzeria tjomusi*, and *Gymnosolen cf. furcatus* of the assemblage of Sinian System, China (Tewari, 1989 and the references therein).

Sikdar (1989) recorded the stromatolite taxa *Baicalia* sp., *Colonnella* sp., *Gymnosolen* sp., and *Anabaria* sp. between the areas Nandini and Jamul. A tidal

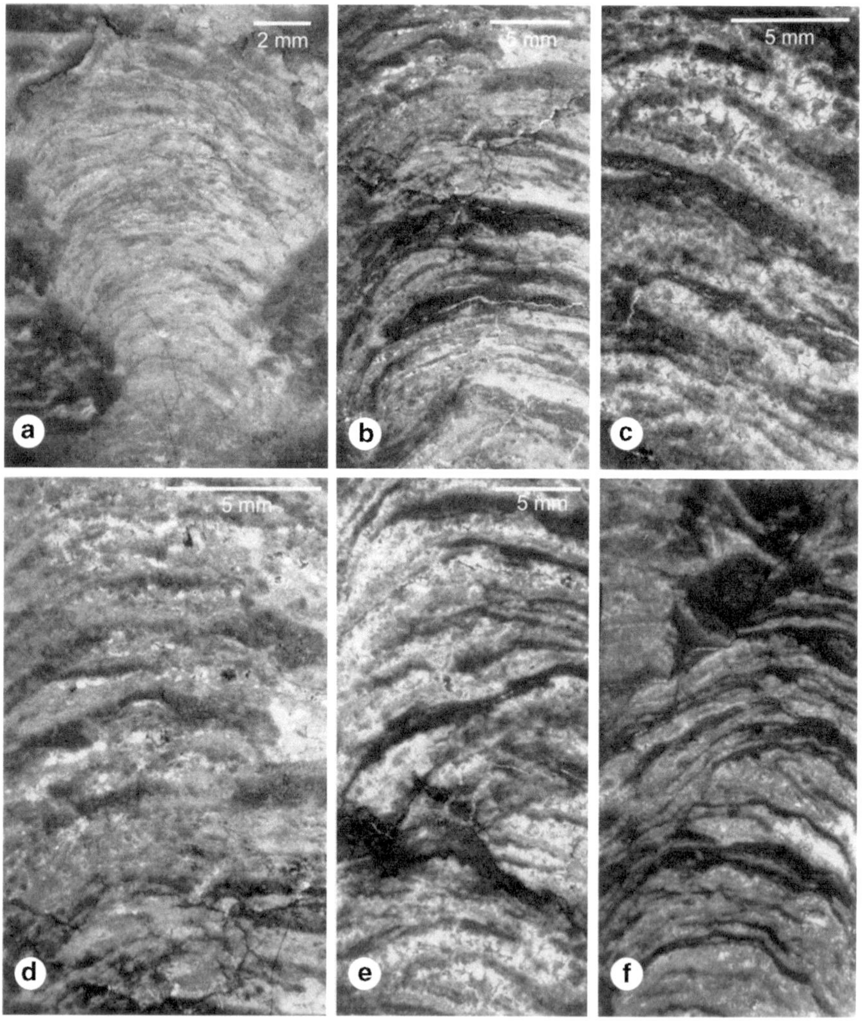

Figure 11. Microstructures of the stromatolite forms. (**a**) *Gymnosolen furcatus* showing bulbous nature, (**b**) *Gymnosolen furcatus* with light and dark laminae, (**c**) *Gymnosolen furcatus* showing margin of column, (**d–f**) showing the enlarged columns, dolomite grains, and intercolumnar spaces.

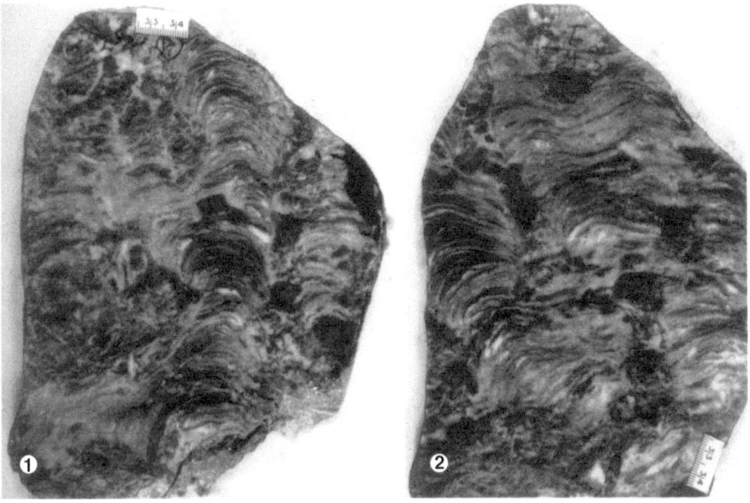

Figure 12. Polished slabs of *Gymnosolen furcatus*, locality Koliapara.

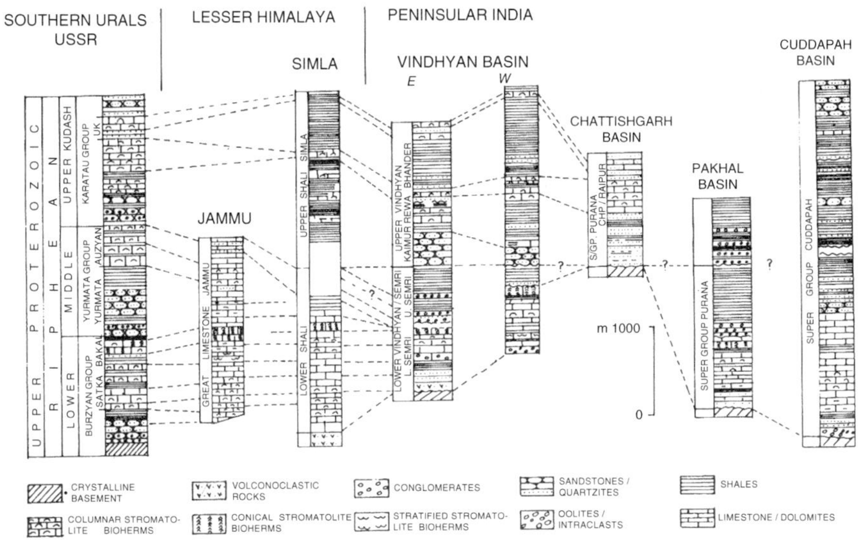

Figure 13. Correlation of Chhattisgarh Basin with other equivalent basins in the Peninsular shield and Lesser Himalaya, India, and the type area of Riphean (Upper Proterozoic) in Russia (former USSR; Tewari, 1989).

flat, Raipur (subtidal to intertidal) depositional environment was suggested for the Raipur Formation. This is in agreement with the present authors' observation from both the basins, Indravati and Chhattisgarh. High-energy conditions during the formation of stromatolite are indicated by the action of waves and crenulation of laminae.

It has been observed in the recent Shark Bay stromatolites from Australia that the height of the stromatolite columns is controlled by the tidal range and the position of the structure in the intertidal zone.

Moitra (1999) has reported filamentous cyanobacteria *Siphonophycus*, *Oscillatoriopsis*, *Taeniatum*, and *Caudiculophycus* from the Raipur Formation. These prokaryotic microfossils are abundantly found in the Meso-Neoproterozoic Deoban and Buxa dolomite formations of the Lesser Himalaya (Tewari, 1989, 2004, 2009; Schopf et al., 2008 and the references therein). The carbon isotope chemostratigraphy and the laser Raman spectroscopic studies of the stromatolitic carbonates of Indravati and Chhattisgarh basins will generate additional data for its global correlation, biotic diversification, and paleoclimatic events during Meso-Neoproterozoic period.

7. Acknowledgments

The authors are thankful to the Principal, Government Nagarjuna P.G. Science College Raipur, Chhattisgarh and Director Birbal Sahni Institute of Palaeobotany, Lucknow, India for lab facilities. We are grateful to Dr. Mukund Sharma, Scientist E of BSIP, Lucknow, for critically reviewing the chapter and suggestions. Thanks are also due to Director, Wadia Institute of Himalayan Geology, Dehradun, Uttarakhand, India for permission to publish the book Chapter.

8. References

Ball, V. (1877) On the geology of Mahanadi Basin and its vicinity. Rec. Geol. Surv. India **10**(4): 167–186.

Chanda, S.K. and Bhattacharya, A. (1973) Geochronologic and environmental significance of stromatolites from Raipur limestone, Chhattisgarh Basin, Central India. Proceedings 60th Ind. Sc. Cong. Abstract 201–202.

Chatterjee N, Das N. Ganguly M. and Chhatterjee B (1990) Stromatolite based biostratigraphic zonation of Chandi Formation, Raipur Group, Chhattisgarh Supergroup, in and around Dhamdha-Nandini area, Durg Distt. M.P.

Cloud, P.E. and Semikhatov M.A. (1969) Proterozoic stromatolite zonation, Ame. Jour. Sci. 267, 1017–1061.

Crookshank, H. (1963) Geology of Southern Bastar and Jaypore from Bailadila range to Eastern Ghat, Mem. Geol. Surv. India. **87**: 149.

Das, N, Ganguli, M. and Arora (1990) Microfacies assemblages of Gypsum from Chhattisgarh Basin: A sabkha model of Evaporite Formation. Precambrian of Central India, Geol. Surv. Ind. Spl, Publ. 28, pp 639–647.

Dutt, NVBS (1963) Stratigraphy and correlation of Indravati Series (Purana Group) Bastar District M.P. Jour. Geol. Soc. Ind. **4**: 35–49.

Dutt, N. K., Dutta, S.M., Mathur Setti, D. N., Sarkar, S.C., Thampi, C. J. and Joshi, V. B. (1983) An appraisal of the natural resources of the Indravati Basin, Orissa, Madhya Pradesh Maharasstra, India. National Natural Resources Management System, Hyderabad Session III **35**: 1–20.

Ghosh, P.K. and Shah S.C. (1965) On the occurrence of stromatolites from the Raipur limestone (M.P.) Proc. 51st and 52nd Ind. Sci. Cong. Abstract 196.

Grey, K. (1989) Handbook for study of stromatolite and associated structures (second draft): In Kennard J.M. and Burne, R. V. (eds. Stromatolite Newsletter, **14**: 82–140.
Guhey R. and Wadhawa, N.P. (1993) Stromatolites from Raipur limestone around Nandini District Durg, M.P. Ind. Jour. Earth Sci.Vol. **20**(1): 42–49. **54**
Jairaman, R.G. and Banerjee, D.M. (1980) Preliminary studies of stromatolites from Raipur area, Chhattisgarh Basin, Geol. Surv. Ind. Misc. Publ. No. 44, 57–67.
King, W. (1881) Geology of the Pranhita-Godavari Valley. Mem. Geol. Surv. India, **18**(3): 73–77.
Komar, V. A. (1964) Columnar stromatolites from the north of the Siberian Platform. Uch.Zap. Nauchno Issled. Inst. Geol. Ar Palaeontol. Biostratigra. **6**: 84–105 (In Russian)
Korolyuk I.K. (1960) Stromatolites of the lower Cambrian and Proterozoic of the Irkustsk Amphitheater Tr. Inst. Geol. Razrab, Goryuck Iskop Akad Nauk SSSR **1**
Kreuzer, H., Harre, W., Schnitzer W.A., Murti, K.S. and Srivastava, N.K. (1977) K/Ar Dates of two glauconites from the Chandarapur Series (Chhattisgarh / India): On the stratigraphic status of the late Precambrian basin in Central India. Geol. Jb. **28**: 23–36.
Moitra, A.K. (1984) Microbiota from Raipur Formation, Chhattisgarh Group, M.P., Rec. Geol. Surv. Ind., **116**: 163–171.
Moitra, A.K. (1996) Preliminary study of stromatolites of Raipur Formation, M.P., Proc. xi Colloq. On Micropaleont. And Stratigraphy Bull., **54**: 124–130.
Moitra, A.K. (1999) Biostratigraphic study of stromatolites and microbiota of Chhattisgarh basin, M.P., India, Palaeontologia Indica, Geol. Surv. India. **51**: 95 pp
Murti, K.S. (1978) A study of stromatolites of Chhattisgarh Basin, Proc. Symp. On Purana Formations of Peninsular India. Univ. Saugar, 268–275.
Raaben, M.E. (1969) Columnar stromatolites and late Precambrian stratigraphy. Amer. Jour. Sci. **267**:1–18.
Raaben, M.E and Tewari, V.C. (1987) Riphean stromatolites in India. Izv. Akad. Nauk. SSSR. Ser. Geol. **3**, (in Russian).
Raha, P.K. and Das, D.P. (1989) Correlation of stromatolite bearing Upper Proterozoic basins of India and paleocgeographic significances. Him. Geol. **13**: 119–142.
Ramakrishnan, M. (1987) Stratigraphy, sedimentary environment and evolution of the late Proterozoic Indravati Basin, Central India, Geol. Surv. Ind. 139–160.
Schnitzer, W.A. (1971) Das jung prakambring Indians, Negliederung stromatolithen furthering and litho fazie 1, Verghi eche. *Erlanges Geol. Abh.*, **85**: 1–44.
Schnitzer, W. A. (1969) Die Jung-algonkischen sedimentation sraume peninsula Indians N. J. b. Geol Palaeont. Abn. **33**:191–198.
Schopf, J.W., Tewari, V.C. and Kudryavtsev, A.B. (2008) Discovery of a new chert – permineralised microbiota in the Proterozoic Buxa Formation of the Ranjit Window, Sikkim, NE Lesser Himalaya, India and its astrobiological implications. Astrobiology, **8**(4): 231–242.
Sen, S (1966) Stromatolites in Raipur limestone M.P. Ind. Min. **20**: 57–58
Sharma M. (2001) In Book Review "Biostratigraphic study of stromatolites and microbiota of the Chhattisgarh Basin MP. India, by A. K. Moitra, Geol. Soc. Ind. Mem Geol Surv Ind *Palaeontologia Indica* New Series Vol. 51, 1999" Jour Geol. Soc. Ind. **57**: 287–289 pp.
Sharma, V. P. (1975) Note on stratigraphic classification of Purana rocks of Bastar distt. M.P. Geol. Surv. Ind. Misc. Publi. **25**(1): 171–175
Shrivastava, N.K. (1977) Sediment-petrographical and geochemical studies of the Late Precambrian stromatolites of India, "Fossil Alga" Recent results and developments. (Ed. E. Fluegel). 107–112.
Sikdar, P.K. (1989) Stromatolites and depositional environment of Raipur Formation, Chhattisgarh Group, Nandini – Jamul, area, Durg district, M.P. Him. Geol. **13**: 87–92.
Steinmann F. (1911) Uber Gymnosolen ramsayieine coelenterate von der Halbinsel Kanin Bull. Soc. Georg. Finl., **31**: 18–23.
Tewari, V.C. (1984) Discovery of Lower Cambrian Stromatolite from Mussoorie, Tal Phosphorite, India. Curr. Sci. **53**(6)

Tewari, V.C. (1989) Upper Proterozoic - Lower Cambrian stromatolites and the Indian stratigraphy. Him. Geo. **13**: 143–180.

Tewari, V.C. (1994) Sedimentology of the rocks of Deoban Basin, Dhuraphat area, Sarya Valley, Estern Kumaun, Lesser Himalaya, geosciences Journal **15** (2); 117–162

Tewari, V.C. (2004) Microbial diversity in Meso-Neoproterozoic formations with particular refernce to the Himalaya., India. In : J. Seckbach (ed.) *Origins* Kluwer Dordrecht, pp. 515–528.

Tewari, V.C. (2009) Proterozoic unicellular and multicellular fossils from India and their implicatins. J. Seckbach and M. Walsh (eds.) *From Fossils to Astrobiology*, Springer – Science + Business Media B.V., pp. 119–139.

Valdiya, K.S. (1989) Precambrian stromatolite biostratigraphy of India- A review. Him. Geo. **13**: 181–213.

Vishwanathaiya, M.N. and Sastri J.C.V. (1973) High phosphorous stromatolitic limestone from Nandini Durg District M.P. India. Curr. Sci. **42**: 738–741

Walter M. R. (1976). Stromatolites: Development in Sedimentology, Elsevier Amsterdan, **20**: 790.

Biodata of **Professor Alain R. Préat, Mr. Franck Delpomdor, Dr. Kamal Kolo, Dr. David C. Gillan,** and **Dr. Jean-Pierre Prian**, authors of ***"Stromatolites and Cyanobacterial Mats in Peritidal Evaporative Environments in the Neoproterozoic of Bas-Congo (Democratic Republic of Congo) and South Gabon"***

Professor Alain R. Préat is currently a co-director of the Biogeochemistry-Earth System in the Université Libre de Bruxelles (ULB), Belgium. He obtained his Ph.D. from the ULB in Geology on sedimentology of Paleozoic carbonates in 1985 and became Professor in 2000. His scientific interests are in the areas of carbonate sedimentology and geochemistry, particularly in the study of the origin of the red pigmentation of matrices of the Phanerozoic limestones. He carries out this research with marine biologists and microbiologists of the ULB. Recently, he started new research on African Precambrian carbonates.

E-mail: **apreat@ulb.ac.be**

Franck Delpomdor is a Ph.D. assistant at the Department of Earth Sciences and Environmental Sciences, University of Brussels, Belgium. His research concerns the Neoproterozoic rocks surrounding the Congo craton in DRC. His MSc thesis (in association with the Royal Museum of the Central Africa, Belgium) focused on the Upper Diamictite Formation and the Schisto-Calcaire Subgroup in the West Congolian Group (DRC). Present Ph.D. research activities are on Early Neoproterozoic microbial carbonates of Mbuji-Mayi Supergroup from the southern DRC.

E-mail: **fdelpomd@ulb.ac.be**

Alain R. Préat

Franck Delpomdor

V.C. Tewari and J. Seckbach (eds.), *STROMATOLITES: Interaction of Microbes with Sediments*, Cellular Origin, Life in Extreme Habitats and Astrobiology 18, 43–63
DOI 10.1007/978-94-007-0397-1_3, © Springer Science+Business Media B.V. 2011

Dr. Kamal Kolo is currently a researcher at the Vrije Universiteit Brussel, Department of Geology, Belgium. He obtained his Ph.D. from the University of Bucharest in 1985 and from Vrije Universiteit Brussel in 2009. Dr. Kolo's scientific interests are in the areas of experimental studies on microbial–mineral surface interactions, especially fungal, leading to biomineralization, mineral substrate alteration, and bioweathering in relation with ancient microbial biosignatures. He currently investigates fungal interaction with natural minerals surface as a thigmotropic behavior.

E-mail: **kakolo@vub.ac.be**

Dr. David C. Gillan is currently Professor in microbiology at the Mons University (UMons) in Belgium. Since 1994, his research focuses on marine microbial ecology, including metazoan–microbial symbioses, with a particular interest for biogeochemistry and microbial communities living in sediments (past and present-day environments). He is presently studying the relationship between benthic microbial communities and fluxes of metallic contaminants.

E-mail: **david.gillan@umons.ac.be**

Dr. Jean-Pierre Prian is a senior geologist at the BRGM (Bureau Recherches Geologiques Minières, Orléans, France). He worked in and around the cartography, on regional basin studies, and for mining around the world. He published numerous geological maps particularly in Gabon.

E-mail: **p.prian@brgm.fr**

Kamal Kolo

David C. Gillan

Jean-Pierre Prian

STROMATOLITES AND CYANOBACTERIAL MATS IN PERITIDAL EVAPORATIVE ENVIRONMENTS IN THE NEOPROTEROZOIC OF BAS-CONGO (DEMOCRATIC REPUBLIC OF CONGO) AND SOUTH GABON

ALAIN R. PRÉAT[1], FRANCK DELPOMDOR[1], KAMAL KOLO[2], DAVID C. GILLAN[3], AND JEAN-PIERRE PRIAN[4]

[1]*Department of Earth Science and Environmental Sciences, University of Brussels, 50 av. FD Roosevelt, 1050 Brussels, Belgium*
[2]*Department of Geology, Vrije Universiteit Brussel, Pleinlaan 2, 1050 Brussels, Belgium*
[3]*Department of Biology, University of Mons, 6 av. du Champ de Mars, 7000 Mons, Belgium*
[4]*Bureau Recherches Géologiques et Minières, 3 av. Claude Guillemin, BP 360009, 45060 Orléans Cédex 2, France*

Abstract The "Schisto-Calcaire Subgroup" is a muddy predominantly subtidal shelf succession that crops in the West Congolian Belt. The approximately 300-m-thick CI (Bas-Congo) and approximately 20-m-thick SCIII (Gabon) series were deposited in a very shallow marine evaporitic environment. The evidence for this interpretation includes sedimentology of dolomite and sulfate minerals and oxygen isotopes. Cyanobacteria (probable Oscillatorians) formed mats on the inland marshes fringing ponds of channeled belts. In Gabon, they are associated with abundant domal stromatolites deposited in shallow to lower intertidal settings. While diagenetic overprints (dolomicrosparitization, sulfate growth, silica replacement) may be significant, several microbial laminar mudstones retained their original fabric. SEM analysis revealed well-preserved three-dimensional (3D) cyanobacterial communities associated with the stromatolites. During progressive lithification in the upper part of shallowing-upward evaporitic sequences, the stromatolites constituted a favorable substrate which has been invaded and colonized by probable fungal hyphae. These produced characteristic features that have been reproduced in vitro in experiments.

Keywords Stable isotopes (C, O) • Bacterial mats • Precambrian • Paleoenvironments • Microfacies • Diagenesis

1. Introduction

Detailed sedimentary facies analyses of African Precambrian carbonate series have been problematic because of regional deformation and high metamorphic grade. Emphasis has generally been placed on identifying sedimentary successions and their geodynamic settings, rather than on depositional processes and facies organization. The stratigraphy of the Schisto-Calcaire Subgroup (part of the West Congolian Group belonging to the West Congolian Belt) of Democratic Republic of Congo and Gabon has been relatively well documented (Cahen, 1978; Alvarez, 1995; Chevallier et al., 2002; Frimmel et al., 2006), but no detailed interpretation of the carbonate depositional environments of the series has been established. This lack of study is amazing since the carbonate shelves (platforms and ramps) recognized in Precambrian basins closely match those defined for Phanerozoic and modern successions. The fundamental controls on platform evolution appear to have been established since the Paleoproterozoic (Grotzinger, 1989). Studies on modern and fossil bioconstructions have shown that their growth is affected by a multitude of environmental factors, such as temperature, illumination, turbidity, substrate, and nutrient levels, most of these factors varying systematically with depths (Narbonne et al., 2000).

Well-preserved laminated limestones in the upper part of the Schisto-Calcaire Subgroup are observed in drill cores from the Bas-Congo and in outcrop in South Gabon. The stromatolites are easily recognized in Gabon on the basis of their macromorphology and microtextural properties despite alteration by diagenetic recrystallization (mainly dolomitization). Microbial mats are present in the laminated limestones of the drill cores of Bas-Congo and have been affected by diagenetic processes. Nevertheless, perfectly preserved remains of microorganisms are recognizable in Bas-Congo and help to analyze the stromatolitic limestones in Gabon. In this case, a fungal invasion has followed the mat development and is related to stressful conditions during early lithification. In the current contribution, we present two cases where Neoproterozoic microbial growth has escaped partly diagenesis and induced well-preserved laminated sedimentary structures. This chapter focuses on the depositional setting of the stromatolitic levels of the upper part of the Schisto-Calcaire Subgroup.

2. Regional Setting and Stratigraphic Framework

2.1. THE WEST CONGO SUPERGROUP IN BAS-CONGO (MAYUMBIAN BASIN)

The Bas-Congo is located in the western part of the Democratic Republic of Congo (DRC). It belongs to the West Congolian Belt (WCB) between 1°S and 12°S and extends for more than 1,400 km in length and 150–300 km in width. The belt extends from southwestern of Gabon and crosses the western part of the Republic of Congo and of the DRC to northwestern of Angola (Fig. 1).

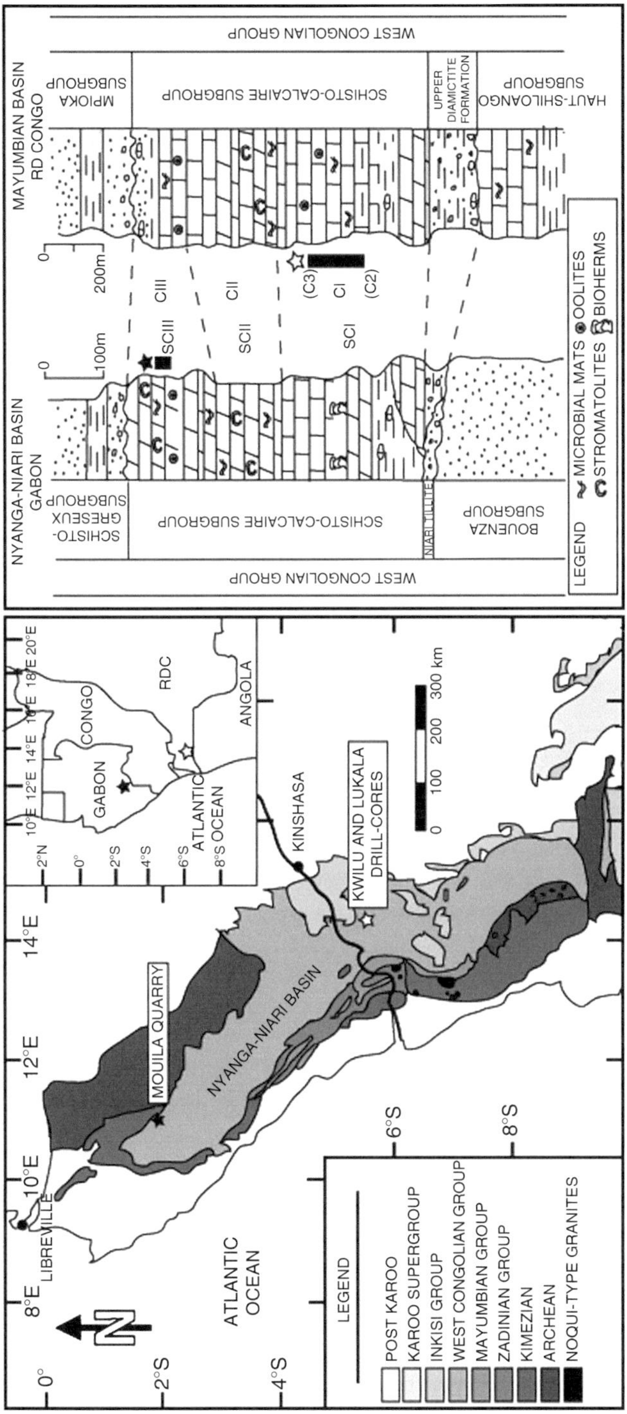

Figure 1. Locations and stratigraphic correlations of the studied carbonate series (*black star* for Gabon and *white star* for Bas-Congo) in the West Congolian Belt (modified after Frimmel et al., 2006; Prian, 2008). The studied intervals are indicated by *black vertical bars*.

The Kimezian basement (2.5–1.6 Ga) is overlain by folded and unmetamorphosed Neoproterozoic rocks of the West Congo Supergroup, this latter being subdivided into the Zadinian (around 999? Ma, Mayumbian (around 910? Ma), and West Congolian groups (from 910 to 566 Ma; Tack et al., 2001). The West Congolian Group is constituted of more than 6,000 m of siliciclastics and carbonates, consisting from base to top, of the Sansiskwa Subgroup, the Lower Diamictite Formation, the Haut-Shiloango Subgroup, the Upper Diamictite Formation, the Schisto-Calcaire Subgroup, and the Mpioka Subgroup. The Upper Diamictite Formation (150–200 m thick) consisting of poorly sorted conglomerates with striated clasts corresponds probably to the Marinoan glaciation (635 ± 1.2 Ma in Namibia; Hoffmann et al., 2004) and is overlain by a cap carbonate belonging to the Schisto-Calcaire Subgroup. This subgroup (1,100 m thick) is divided into five subunits (C1–C5; Delhaye and Sluys, 1923) and regrouped into four units (CI, CII, CIII, CIV; Lepersonne, 1951; Cahen and Lepersonne, 1967; Fig. 1). The CIV unit does not appear in the studied area. The subunits are, from base to top, C1 zoned pink dolomites (cap carbonate), C2 laminated argillaceous limestones, C3 laminated and oolitic limestones, C4 stromatolitic limestones and dolostones, and C5 dolomitic limestones with coarse oolites ("Oolite of Kisantu"). In our study, we collected samples from drillcores (preserved since 1949 in Belgium at the Royal Museum of the Central Africa, Tervuren) crossing the C2 and C3 subunits (or CI unit) in the Kwilu and Lukala areas (Fig. 1).

2.2. THE NEOPROTEROZOIC OF GABON (NYANGA BASIN)

The Nyanga synclinal is an important geotectonic unit of the southern Gabon consisting of a vast deformed basinal structure over 250 km long belonging to the WCB. The Gabonese part of the belt, averaging some 100 km in width, runs parallel to the Atlantic coast and the main part terminates in the north of Gabon (Fig. 1). Among the five major lithostratigraphical units of the Gabonese part of the WCB, the West Congolian Group outcrops in the Nyanga synclinal (Gérard, 1958) and consists of the succession of several informal units. The Schisto-Calcaire Subgroup overlies directly the "Niari Tillite" (Fig. 1), which has been correlated with the Cryogenian glacial event (about 635 Ma, Hoffmann et al., 2004; Condon et al., 2005). Saylor et al. (1998) on the basis of chronostratigraphic and sedimentological studies in Namibia have proposed a maximum age of 564 Ma for the Cryogenian glaciation. Consequently the precise age of the formations of the West Congo Supergroup is still debatable.

The "Schisto-Calcaire Subgroup" (or "Carbonate Subgroup," Prian, 2008) is predominantly a carbonate sequence composed of four formations ("Nsc1 to Nsc4" or "SCI to SCIII," see Prian, 2008), consisting of calcareous to dolomitic shales with an uppermost sandy shale-siltstone unit with interbedded limestone. The formations are well mapped and the series of the studied Mouila quarry, belonging to the "Nsc3" or "SCIII" Formation of the eastern Nyanga flank, consist of gray or cream-colored cherty stromatolitic limestones or dolostones characterizing the lower part or "Lower Member" of the Formation. Its upper part or "Upper Member" consisting of oolitic dolostones does not outcrop in the studied area.

The Nsc3 "Lower Member" exposed in the old Mouila quarry represents an exceptional outcrop which escaped the heavy lateritization of the area and has not suffered important burial or metamorphism. It consists of a conformable section (~20 m) of Neoproterozoic rocks providing a continuous record of carbonate (dolomite) sedimentation with several metric stromatolitic layers affording a limited view of a probable extensive carbonate depositional system. The strata are unmetamorphosed, flat lying, and well exposed along extensive cliff faces.

3. Results

3.1. THE CYANOBACTERIAL MATS IN BAS-CONGO

3.1.1. Depositional Setting (C2 and C3 Subunits)

One-hundred-ninety-nine samples have been collected for petrography (thin sections) and SEM observations (JEOL JSM-6400) in the C2 and C3 subunits of the Schisto-Calcaire Subgroup. The C2 subunit (200 m thick) consists of calcite-cemented, homogeneous gray dolomites alternating with millimeter- to centimeter-thick green bindstones, including detrital materials, microbial mats, evaporitic nodules, and lenses. The overlying C3 subunit (thickness of 100 m), subdivided into C3a and C3b, contains flat- to wavy-laminated dolomites with millimeter- and centimeter-thick green bindstones and gray dolomudstones. The C3b consists mainly of massive gray dolomites with intraformational conglomerates, oolites, pisolites, and stratiform black chert nodules.

Our petrographical analysis revealed nine microfacies (MF1-9, predominantly flat-laminated carbonates, Delpomdor, 2007) forming a standard sequence from semi-restricted to restricted peritidal environments associated with shallow subtidal zones submitted to detrital fluxes and containing benthic cyanobacterial mats. This sedimentation evolved to saline supratidal environments with inter- and supratidal channels in a sabkha. Flat-laminated carbonates examined on the SEM revealed abundant microbial filaments (Fig. 2a) forming colonies still in their growth position as suggested by their common orientation in various bundles (Fig. 2a). The filaments are thin (0.7–2 μm in diameter, length 1–50 μm) in the microbial mats and thicker (diameter 5–10 μm, length 50–100 μm) in the microsparitized matrix. They are well preserved both in the laminae and in the altered matrix. They are segmented, unbranched, and slightly lobed suggesting spheroidal to ellipsoidal vesicles of 1–2 μm in diameter (Fig. 2a–c). The colonies rest on a carbonate sole representing probably relics of "EPS" (extracellular polymeric substances; Fig. 2b). Free small solitary spheroidal unicells or vesicles are randomly distributed in the mats. Their external mucilaginous wall is not visible and coated by calcite cement.

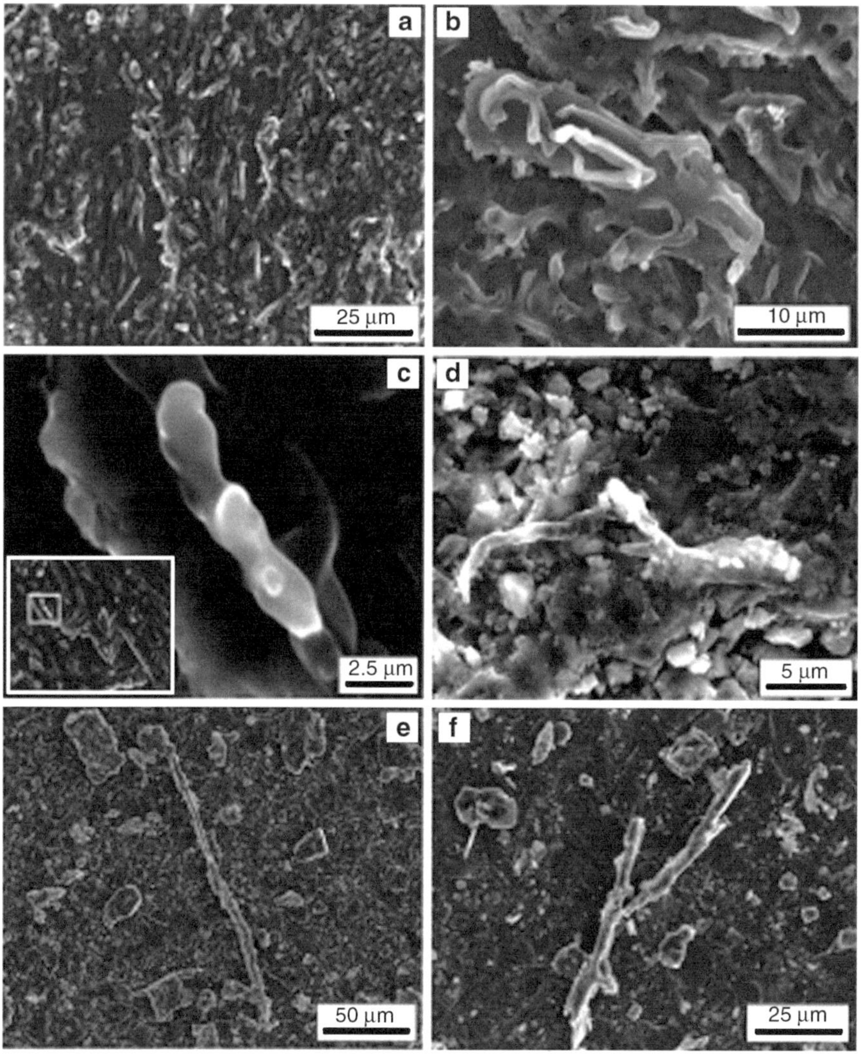

Figure 2. Bas-Congo. Microfossils from the subtidal–intertidal carbonate facies (SEM images). (**a**) View of probable *Siphonophycus septatum* mat; (**b**) preserved *S. septatum* filaments on EPS in a jelly calcite cement; (**c**) detail of segmented and unbranched filament; (**d**) altered *S. septatum* filament in a dolomicrosparitized matrix; (**e**, **f**) *Siphonophycus* (larger diameter) in a dolomicrosparitized mudstone.

The carbonates are predominantly constituted by pervasive hypidiotopic and idiotopic dolomites associated with anhydrite in nodules and lenses in the mats. Other evaporitic minerals are idiomorphic polyhalite or pseudomorphs of sulfates (laths, rosette-like aggregates, castellated crystals, see Clark, 1980) and "swallow tail" crystals (see Amieux, 1980). Microquartz and megaquartz constitute the last diagenetic phase.

The sedimentological evolution of C2 and C3 subunits records plurimetric shallowing-upward sequences in an evaporitic lagoon (Delpomdor, 2007). The lower part of the sequences consists of flat-laminated cyanobacterial mats (MF1 to MF4) and evolves into intertidal–supratidal or subaerial evaporitic facies in the upper part (MF5 and MF6). Channels with oolites and pisolites (MF7) are present in the middle part of the sequences and are typical of a back-margin setting (Hardie, 1977). Micrite is strongly replaced by early dolospar and late silica in the sabkha environment (MF8 and MF9). The cyanobacterial mats are embedded within extensively dolomitized sediments. The mats are formed primarily by calcification of the cyanobacteria in upper subtidal and lower intertidal environments on a semi-arid to arid tidal flat. The calcitization would have resulted from the precipitation of CO_2 uptake during photosynthesis (Pentecost and Riding, 1986; Freytet and Verrecchia, 1999; Riding, 2000) or from the increasing alkalinity during sulfate reduction and Ca^{2+} release from the EPS by heterotrophic bacteria (Visscher et al., 1998; Reid et al., 2000). Pervasive early dolomites replaced calcite in the cyanobacterial mats (Fig. 2d–f) near sediment–water interfaces affected by periodic emergence and/or hypersaline conditions. A few dolomudstones escaped this strong diagenetic alteration and retained original microbial mats with abundant well-preserved tubules or filaments and spheroidal unicells enveloped in gelatinous-like calcite cements (Fig. 2a–c).

3.1.2. Microbial Description and Discussion

The filamentous microfossils observed in Bas-Congo (Fig. 2) are probably ancient photosynthetic bacteria resembling the present-day cyanobacteria. Although morphology alone is of little use to identify bacteria, the present-day cyanobacteria are usually classified into broad groups using morphological characteristics (van den Hoek et al., 1995). Five morphological subgroups are usually recognized (1) unicellular cyanobacteria dividing by binary fission, (2) unicellular cyanobacteria dividing by multiple fission, (3) filamentous cyanobacteria containing differentiated cells called heterocysts that are used for nitrogen fixation (Nostocalean cyanobacteria), (4) filamentous cyanobacteria with no heterocysts (Oscillatorian cyanobacteria), and finally (5) branching filamentous cyanobacteria.

As the filaments observed in the present study are unbranched and that no heterocysts were observed, the microfossils might represent ancient Oscillatorians. An important point to consider is the observation that the filamentous microfossils are enveloped by jelly calcite cements that may constitute ancient "EPS" or mucilaginous sheaths. The Oscillatorians feature genera such as *Oscillatoria*, *Arthrospira (Spirulina)*, *Lyngbya*, *Phormidium*, and *Microcoleus* (van den Hoek et al., 1995). In *Oscillatoria*, individual cells are disk shaped and filaments do not produce mucilaginous sheaths. *Arthrospira* filaments are usually large, screw-like coils and do not produce mucous sheaths either. Consequently, the microfossils are probably not related to *Oscillatoria* and *Arthrospira*. In *Microcoleus*, individual filaments resemble *Oscillatoria* but bundles of filaments are held together by a stiff mucilaginous sheath, a situation that is not observed in the microfossils.

The trichomes of *Lyngbya* are enclosed in a robust mucilaginous sheath. *Lyngbya* is thus a good modern analog of the filamentous microfossils. In *Lyngbya*, the filaments are cylindrical, usually wider than 6 μm, and cells are shorter than they are wide.

These cyanobacteria usually form large layered mats of various thicknesses. Another possibility is the genus *Phormidium*. This organism usually forms flat, slimy mats of filaments and is similar in morphology to *Lyngbya* and *Oscillatoria*. *Phormidium* is also producing EPS (Hoiczyk, 1998). However, the sheaths on *Phormidium* are looser than the rigid sheaths of *Lyngbya*. The mats are usually attached to benthic substrates and can detach and float to the surface. In *Phormidium*, trichomes range from 2 to 12 μm in diameter, a width which is comparable with that of the microfossils. In addition, cells are not always distinctly separated, a situation also observed in the microfossils. Although microfossils resemble *Phormidium*, it should be remembered that morphological criteria are not always reliable because they may change, for example, due to environmental factors (Teneva et al., 2005).

Finally, the filaments found in Bas-Congo seem comparable (in terms of morphology and size) with the mat builders of the fossil species *Siphonophycus septatum* (Schopf, 1968) and therefore could indicate a quiet-water lower subtidal to intertidal environment in subtropical to tropical zones (Noffke et al., 2003). Recent marine cyanobacteria (*Plectonema*) associated with near-surface environments in the carbonates of Cabo Rojo, Puerto Rico (Chacon et al., 2006), display a similar morphology of that of our microbes.

3.1.3. Geochemistry

Carbon and oxygen isotope compositions were analyzed in the most homogeneous facies (Fig. 6). Between 1 and 2 mg of powdered sample (whole rock) were reacted with 100% H_3PO_4 at 75°C to extract the CO_2 from the dolomites. The amount of extracted CO_2 at −196°C and its $\partial^{13}C_{V\text{-}PDB}$ and $\partial^{18}O_{V\text{-}PDB}$ were measured using a Thermo-Finnigan 252 mass spectrometer at Erlangen University (Germany). The reproducibility of the $\partial^{13}C_{V\text{-}PDB}$ and the $\partial^{18}O_{V\text{-}PDB}$ measurements is 0.06‰.

The $\partial^{13}C$ values vary between −2‰ and −5‰ in the microbial dolomudstones (MF1 to MF3) and between −2‰ and −3‰ in the microsparitized dolomudstones (MF8). Weak negative values (−2‰ to 0‰) are measured in the oolitic/pisolitic carbonates. The $\partial^{18}O$ values lie between −8‰ and −12‰ in the same samples. The C2 and C3 cyanobacterial carbonates present lower $\partial^{13}C$ values with regard to oolitic and dolosparitized rocks. There is no agreement on the original carbon isotope seawater composition during Neoproterozoic times. Here, we retained the two most widely used models for isotope seawater composition of Veizer and Hoefs (1976) and Jacobsen and Kaufman (1999). In this context, the Bas-Congo carbonates have a normal marine $\partial^{13}C$ value (Fig. 6). The microbial carbonates having more negative $\partial^{13}C$ value (up to 2‰) than the associated oolites (representing a normal marine environment) imply a strong biological activity in the shallow subtidal environment. The $\partial^{18}O$ values are homogeneous and suggest a temperature increase during the burial (Frimmel et al., 2006; Delpomdor, 2007). The burial temperature has been estimated at 70°C by Dewaele (personal communication, RMCA) from the highest negative values ($-11.5‰\ \partial^{18}O$).

3.2. THE STROMATOLITIC LEVELS IN THE NYANGA BASIN

3.2.1. Mouila Section

In the Mouila old quarry, 20 m of thin- to medium-bedded homogeneous dolomudstones interstratified with massive laminated peloidal and stromatolitic grayish dolomites are exposed along two extensive cliff faces (Fig. 3a). The rocks are partly silicified with stratiform black chert nodules, a few centimeters thick, and thin siliceous impregnations molding the laminae of the stromatolites (Fig. 3b–d). Small-sized (less than 5 cm in length) elliptic whitish nodules are observed in undulous and discontinuous beds which pinch out laterally on a plurimetric scale. These beds are slightly yellowish. Rare slumped beds are observed and associated with tepee-like structures in the upper part of the section. Clays are not abundant and consist of subcentimetric levels capping thin decimetric beds. The stromatolites

Figure 3. Gabon. (**a**) Mouila old quarry. Lower level of the SCIII series (10 m). (**b**) Close lateral linkage stromatolitic hemispheroids with undulating blackish chert laminae. (**c**) and (**d**) Domal stromatolite and thin-bedded-laminated cyanobacterial grayish dolomudstones with undulating blackish chert laminae.

are hemispherical (height of 10–20 cm, length up to 50 cm). They are typically laterally linked hemispheroids, smooth domal laminated hemispheroids, and wavy stromatolites (LLH-C type of Logan et al., 1964, Fig. 3b–d). The facies are fine grained, mud supported, grayish to slightly whitish, and particularly homogeneous (Fig. 3a).

3.2.2. Facies Assemblages and Stromatolitic Paleoenvironment

Sixty samples have been systematically collected in the profile (Fig. 3a) for petrography to constrain the paleoenvironment. They enable to recognition of five major "diagenetic" dolomitic microfacies (MF1-5) whose succession from 1 to 5 constitutes a standard sequence recording evolution from shallow subtidal quiet environments colonized by cyanobacteria to supratidal settings with a pronounced evaporitic (sulfate and dolomite) diagenesis (study in progress).

Microfacies 1 is a fine-grained laminated clayey detrital dolomudstone with a large amount (up to 30%) of framboidal pyrite (5–30 μm). It consists of flat-laminated microbial mats or "cryptomicrobialites" (Fig. 3d) and low domal stromatolites without intercolumn space (Fig. 3b). The stromatolite laminae are curved and form overlapping domes, with younger laminae truncating against older ones (Fig. 3c, d). Parallel laminite consists of millimeter-scale planar, wavy, and very low-angle cross-stratified laminae composed of well-sorted carbonate mud to very fine clay and silt. Laminae are primarily produced by alternation of organic-rich and organic-poor horizons. The brighter layers are micropeloidal micrite or clotted mudstone draping over dense mat layer (Fig. 4a). Mat constructors in thin section are poorly preserved (filamentous ghosts are occasionally present) but are still visible on the SEM despite the dolomitization process (Fig. 4c) and are similar to those observed in Bas-Congo (Fig. 2). Smooth flat laminated dolostone associated with disrupted fenestral and crinkled fabrics are common. These latter exhibit typical near horizontal sheet cracks associated with vertical and step-like thin mud cracks isolating micritic lumpy patches.

Other microfacies (microfacies 2–5, study in progress) are not described here and exhibit different dolomicrosparites and dolosparites with saddle dolomite and anhydrite replacing the stromatolitic levels (Fig. 4b).

3.2.3. Sedimentary and Diagenetic History

The Mouila series are composed of dolomitized cyanobacterial mudstones and stromatolitic bindstones. The major diagenetic alteration consists of a thin pervasive hypidiotopic dolomitization probably related to episodes of anhydritization. Consequently, the former grayish microlaminated micritic sediment of microfacies 1 (Figs. 3a and 4a) is progressively replaced by a relatively fine-grained homogeneous whitish dolomicrosparite, which may still contain thin discontinuous microbial mat relicts (Fig. 4b, c). At the beginning of the replacement process, the dolomite is a mimetic fabric-preserving dolomite with crystal size varying between micrite and microsparite (<50 μm). Evaporite minerals are common in the matrix and consist of lath-shaped, rosette-like aggregates, enterolithic small nodules, and castellated crystals (*sensu* Clark, 1980), which often grow inside the mat levels

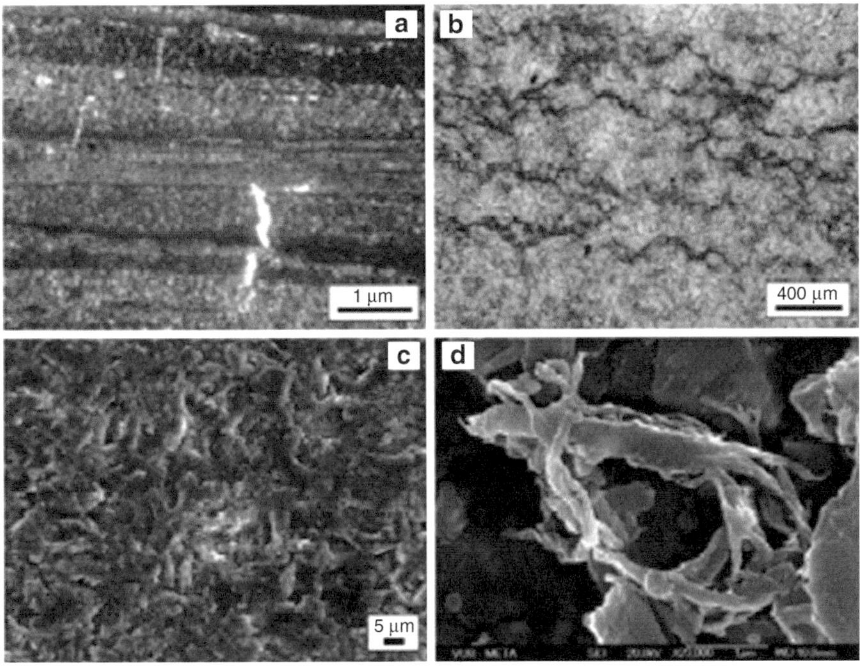

Figure 4. Gabon. (**a**) Smooth flat laminated dolostone showing the alternation of well-sorted peloidal dolopackstone and thinner homogenous dolomudstone (thin section). (**b**) Irregularly dolomicrosparitized mudstone with remnants of thin layers of blackish cyanobacterial mats (thin section). (**c**) Filamentous dichotomous bacteria 2–5 μm in diameter (probable cyanobacteria, partly destroyed by a fine-grained dolomicrospar) (SEM). Compare with Fig. 2 of Bas-Congo. Sample from a domal stromatolite (Fig. 3). (**d**) Probable hyphae (diameter of 0.1–0.2 μm) associated with EPS and filling a pit in a stromatolitic hemispheroid (SEM) (see Fig. 5).

(intrasedimentary growth) in the phreatic or capillary zones. Small-sized fenestrae ("microloferite") are similar to those illustrated by Nédélec et al. (2007) and suggest the temporal development of microbial mats limiting gas outflow from the sediment. Larger fenestrae related to crinkled microbial mats indicate that the substrate was temporarily exposed to quiet water conditions. Hypersaline conditions developed and microenterolithes replaced progressively the muddy matrix. "Elephant skin texture" with micropinnacles and net-like structures (Gerdes et al., 2000) are common in the facies associated with the stromatolitic layers. Coarse saddle dolomite replaced partly or totally the evaporite minerals and silica is the last diagenetic phase observed. The ultimate phase consisted of irregular pressure solution seams or stylolites which have concentrated the insolubles (clays, pyrite).

Micritic, millimeter-scale laminae interlaminated with organic-rich thin levels (benthic microbial mats, Fig. 4a) associated with small autobrecciated layers and numerous mud cracks point to a tidal-flat environment (Purser, 1973; Hardie, 1977; Sellwood, 1986; Clough and Goldhammer, 2000). LLH stromatolites point to a shallow to lower intertidal setting. Their strong diagenetic imprint could be related to shallowing of water depth due to relative sea-level fall and/or upward stromatolite construction (Hoffman, 1974). The soft peloidal mud is layered into wavy and discontinuous lenticular laminae. Crinkled fenestral laminae being either flat or domal, even or pinching are probably related to the form of cyanobacterial mats that could have developed in inland marshes similar to those present on the Andros Island in the low "algal" marshes fringing the ponds of channeled belts (Hardie and Ginsburg, 1977). The depositional paleoenvironment is interpreted as peritidal possibly representing small intertidal and supratidal ponds. Fenestrae and mud cracks indicate episodic drying conditions in upper intertidal and supratidal settings. Laterally linked domal stromatolites were deposited in shallow to lower intertidal settings in quiet environments as they do not exhibit particular elongation (Logan et al., 1964). Evaporitic conditions were also present as indicated by various tepee structures or nodules (on the field) and dolomitic pseudomorphs after sulfate crystals with anhydrite relicts. The "elephant skin texture" characterizes salina playas, "anchialine" pools, or hypersaline pertitidal lagons and is typical of calm environments below sheets of shallow surface brine (Gerdes et al., 2000).

Dolomitization of the former laminar microbial sediments is the main diagenetic alteration in the series erasing progressively the mats (Fig. 4b) and forming a homogeneous dolomicrosparite. The origin of dolomite is still a controversial subject in sedimentary geology since its discovery over 200 years ago, because it involves a large number of interacting factors, such as thermodynamics, chemical kinetics, hydrology, host-rock mineralogy, and texture (Hardie, 1987). In addition, other parameters such as microbial activity could also be considered. For example, Vasconcelos and McKenzie (1997) and Sanz-Montero et al. (2006) have shown that the formation of dolomite is associated with microbial mats under anoxic hypersaline conditions in the Lagoa Vermelha, Brazil (for the first) or has taken place in relatively diluted lake water in the Miocene of the Madrid Basin in Spain (for the second). In this last case, further evaporative concentration resulted in precipitation of sulfates sealing the mats and creating endoevaporitic microenvironments.

An early diagenesis, possibly mediated by microbial activity, is suggested for the Mouila Neoproterozoic series; dolomicrite follows at an inframillimeter scale the microbial laminae and developed progressively from the cyanobacteria which are partly or totally engulfed and still recognizable; they form a 3D-network, some are dichotomic, their average diameter is comprised between 1 and 2 μm, and their minimal length is 20 μm (Fig. 4c). The association of fine-grained dolomite with mud cracks and sheet cracks (disrupted flat lamination) together with their very fine grain size and the presence of former sulfates suggests that dolomite is most likely precipitated from hypersaline waters during the dry season (Préat et al., 2010).

3.2.4. Fungal Diagenesis

Numerous hyphae and spores in the upper part of metric shallowing-upward evaporitic peritidal sequences have been observed in the stromatolitic layers under SEM analysis (Kolo, 2009; Préat et al., 2010). The SEM examination revealed abundant micrometer-sized oval-circular pits, relatively deep and densely colonized with 0.1–3.0 μm-thick microbial filamentous structures, which are frequently embedded in or engulfed by EPS-like material (Fig. 4d), on which a number of small (0.5 μm) spherical grains of dolomite were observed. Quadratic-shaped crystals and rhombohedra were also observed in the pits either attaching to the filaments or forming part of the pits' walls. These minerals (now dolomitic) may originally have been oxalates. Fig. 5a–h shows several structural features which were produced in vitro by fungi (Fig. 5b, d, f, h) (Kolo et al., 2007; Préat et al., 2010) and their analogs observed in the stromatolitic level of Mouila (Fig. 5a, c, e, g). The first Neoproterozoic feature consists of tiny dolomite crystals, mixed with permineralized filaments, filling small pits with the formation of a collar (Fig. 5a). A similar structure has been obtained using a fungal culture on a dolomitic substrate; the recent pits show a mixture of neominerals (here Ca- and Mg-oxalates) and fungal hyphae (Fig. 5b, d). The surface of the filaments (Neoproterozoic) is rough and shows many blebs and attachments (Fig. 5c) representing crystal aggregates (<1 μm) adhering to their surface. This is a typical fungal phenomenon (Fig. 5d). In Fig. 5d, the colonizing fungi have already produced a large quantity of crystals (<1 μm) inside and outside the pit. These micron-sized dolomitic crystals and the filaments litter the bottom, as well as the walls of the Neoproterozoic pits (Fig. 5e), while in Fig. 5f a typical recent pit is filled by fungally biomineralized prismatic and bi-pyramidal crystals of weddellite. The last common feature is the development of an honeycomb-alveolar structure produced by fungi on the Neoproterozoic stromatolitic substrate (Fig. 5g) or on the experimental carbonate substrate (Fig. 5h). Quadratic-rhombic pore space, after the dissolution of dolomite crystals, and formation of a filamentous-EPS mat, cover on old crystal boundaries associated with new crystal formation, suggest a similar colonization pattern and diagenetic pattern. These results allow for a definition of an eogenetic sequence driven by fungal invasion of the Neoproterozoic stromatolitic substrate. It also provides a detailed evaluation of the fungal role in an attempt to understand the mechanisms involved in the paleoweathering of a Neoproterozoic carbonate substrate.

3.2.5. Geochemistry

The stable isotope analyses (Fig. 6) support the hypothesis of a dolomitization most likely precipitated from hypersaline waters during the dry season, the oxygen isotopic values varying by a few permil (on average 2‰) to values as much as 3‰ more positive than those of the co-occurring non (or less) evaporitic sediment (dolomudstone) of the lower part of the shallowing-upward cycles (study in progress). This lower part displays carbon and oxygen isotope compositions typical to those of normal Neoproterozoic seawater (Fig. 6) (Jacobsen and Kaufman, 1999). Given the oxygen isotope fractionation factor for co-occurring dolomite and calcite (Land, 1980), this suggests that the former microbial carbonates and the dolomite in these dolomudstones, dolomicrosparites, and stromatolitic dolostones

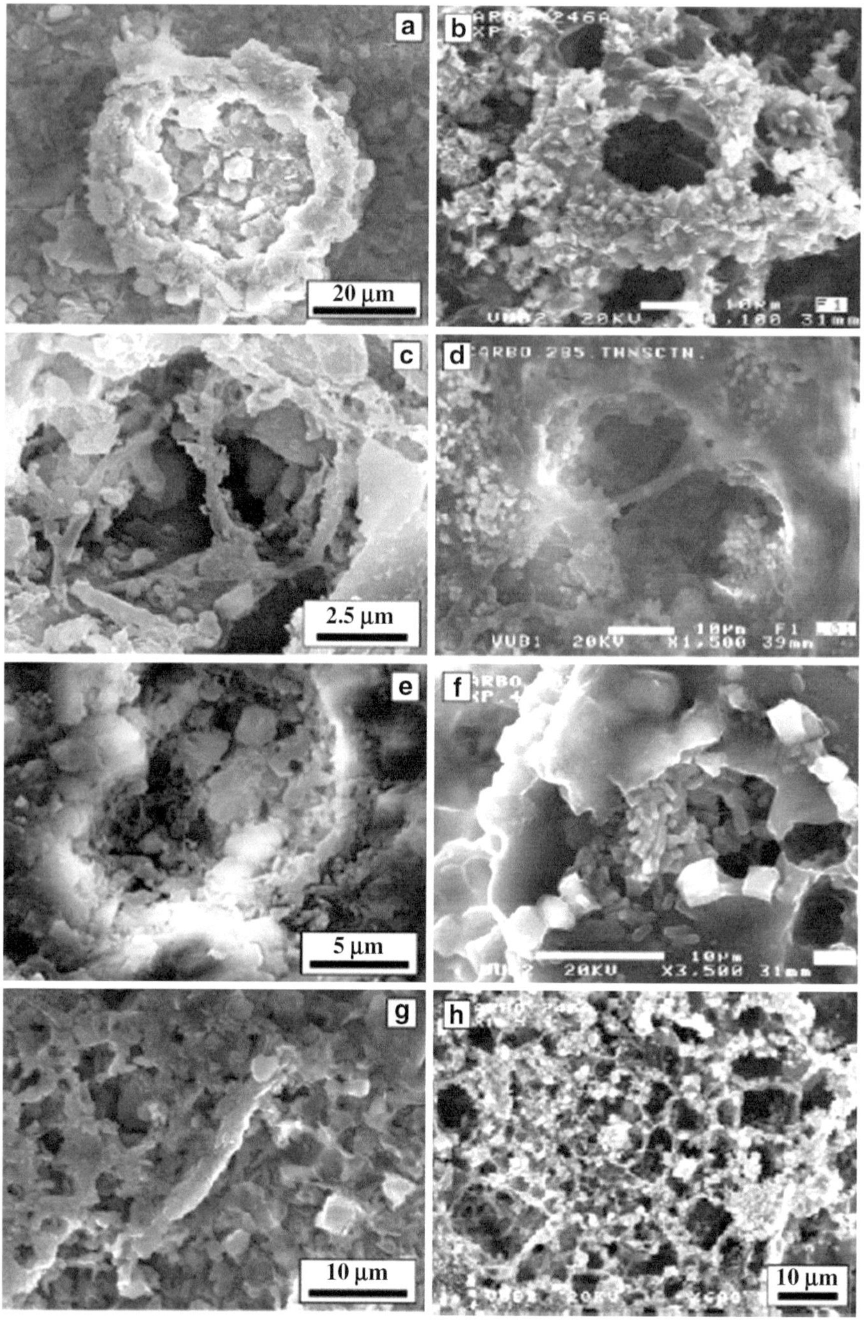

Figure 5. Gabon **a**, **c**, **e**, **g** and in vitro experiments **b**, **d**, **f**, **g**. SEM images. Comparison of "equivalent" fungal features associated with the filling of pits in the Mouila stromatolite and pits formed through in vitro fungal culture on a dolomitic substrate (sample from the Lower Carboniferous of northern France, Kolo et al., 2007). (**a–d**) Elevated mineral "collar" surrounding a pit. (**e**, **f**) Tiny crystals and filaments littering the bottom and the walls of the pits. (**g**, **h**) Honeycomb alveolar structure produced by fungal activity.

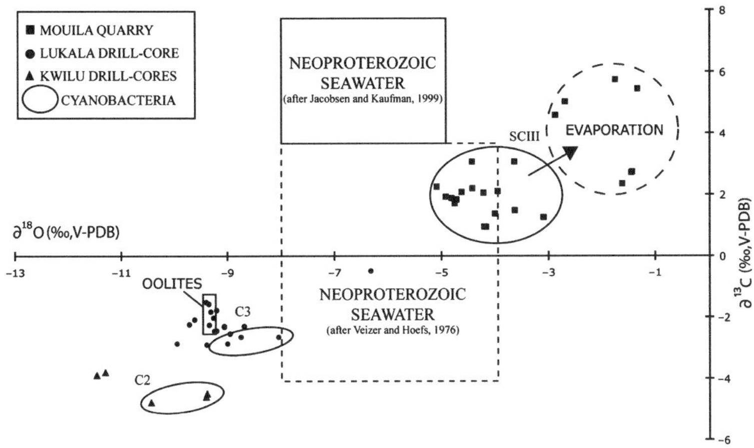

Figure 6. Binary plots of stable isotope compositions ($\partial\ ^{18}O/\partial^{13}C$) of the Schisto-Calcaire Subgroup C2–C3 subunits of Bas-Congo and SCIII unit of Gabon.

formed in a fluid of similar oxygen isotope composition and that dolomitization was early diagenetic. It formed soon after deposition as suggested by the delicate interlayering with microbial mats and its very fine grain size.

The model emphasizes relatively shallow water formation under hypersaline conditions with $CaCO_3$ (limestone and/or primary dolomite) precipitation accompanied by evaporite formation and early dolomitization. The observed sedimentary features point to a general peritidal environment (temporarily emerged subtidal, intertidal, and also subaerially exposed supratidal zones), analogous to present littoral lagoons or sabkhas formed under warm semi-arid conditions, such as can be seen along the Persian Gulf.

The variations in $\partial^{13}C$ isotope signature are not significant at any one level of the Mouila series and indicate that the bulk of the values could be representative of original marine waters (Fig. 6). Most of the values are around +2‰ and +3‰ and do not appear to have been altered by postdepositional diagenetic processes. As for the Bas-Congo facies, the microbial mats have more negative $\partial^{13}C$ value than the oolites which are associated with the series (study in progress, data not presented here).

4. Conclusion

While diagenetic overprints may be significant, petrography, SEM analysis, and isotope relationships show that the general Neoproterozoic depositional environment can still be discerned. This is particularly the case with the microbial content which has been preserved in the shallow subtidal and supratidal zones of the carbonate shelf. Preservation of benthic microbial communities in growth position, in a context rich in autochtonous carbonate precipitates, is highly favorable to study the relations between biological and mineral components of stratiform, laminated organo-sedimentary structures. This will help in the understanding of the diagenetic processes.

The Bas-Congo and Mouila series point to a shallow marine water subjected to hypersaline conditions with $CaCO_3$ (limestone and/or primary dolomite) precipitation accompanied by evaporite formation and early dolomitization. The environments were constituted by subtidal and intertidal lagoons colonized by stromatolitic domes (Gabon) and cyanobacterial mats (Gabon and Bas-Congo). During exposure periods, fungi invaded and colonized the semi-lithified sediment which underwent dolomitization in the supratidal zones of the carbonate platform in the Mouila series. Although the exact mechanisms of dolomitization are difficult to determine, evaporitive pumping of seawater is consistent with the sedimentology, fabric preservation, and isotopic signatures within the upper parts of the shallowing sequences. The important development of the upper part of the shallowing-upward cycles highlighted in our study reveals the incompleteness of the sedimentation. Numerous hiatuses and condensations will occur as a result of sedimentation constantly near emersion. In this context, maximum microbial development can take place concurrently with the lowest sedimentation rates and during brief sedimentary pauses as suggested by Seong-Joo and Golubic (2000) in the Mesoproterozoic of China.

An equivalent shelf domain has already been reported for Neoproterozoic carbonates of the Central African craton. Alvarez and Maurin (1991) and Alvarez (1995) described microbial mats with stromatolites on an extensive shelf from the "Schisto-Calcaire Supergroup" in the Comba Basin (Bas-Congo Province) and suggested they developed on beach and tidal flat environments.

5. Acknowledgments

In Gabon, the field work was done under the terms of the "Programme Sysmin, 8ème Fonds Européen du Développement au groupement NRGM-CGS-SANDER-MRAC." This study was partly financially supported by the Belgium Fonds National de la Recherche Scientifique (FNRS) (FRFC grant no 2.4.578.08 F). We thank Dr. Max Fernandez Alonso and Dr. Luc Tack for the access to the samples in the Royal Museum for Central Africa (Belgium).

6. References

Alvarez, Ph. (1995) Evidence for a Neoproterozoic carbonate ramp on the northern edge of the Central African craton: relations with Late Proterozoic intracratonic troughs. Geol. Rundschau **84**: 636–648.

Alvarez, Ph. and Maurin, J.C. (1991) Evolution sédimentaire et tectonique du bassin protérozoïque supérieur de Comba (Congo): stratigraphie séquentielle du Supergroupe Ouest-Congolien et modèle d'amortissement sur décrochement dans le contexte de la tectogenèse panafricaine. Precambrian Res. **50**: 137–171.

Amieux, P. (1980) Exemple d'un passage des 'black shales' aux évaporites dans le Ludien (Oligocène inférieur) du bassin de Mormoiron (Vaucluse, Sud-Est de la France). Bull. Centres Rech. Explor.-Prod. Elf-Aquitaine **4/1**: 281–307.

Cahen, L. (1978) La stratigraphie et la tectonique du Supergroupe Ouest-Congolien dans les zones médiane et externe de l'orogénèse Ouest-Congolien (pan-african) au Bas-Zaïre et dans les regions voisines. Annals of Royal Museum of the Central Africa, in 8, Sci. Geol. **83**: 150.

Cahen, L. and Lepersonne, J. (1967) The precambrian of the Congo, Rwanda and Burundi, In: K. Rankama (ed.) *The Precambrian*, Vol. 3. Interscience Publishers, New York, pp. 143–290.

Chacon, E., Berrendero, E. and Pichel, F.G. (2006) Biogeological signatures of microboring cyanobacterial communities in marine carbonates from Cabo Rojo, Puerto Rico. Sed. Geol. **185**: 215–228.

Chevallier, L., Makanga, J.F. and Thomas, R.J. (2002) Carte géologique de la république gabonaise, 1: 1 000 000. Notice explicative. Council for Geoscience, South Africa, pp. 195.

Clark, D.N. (1980) The diagenesis of Zechstein carbonate sediments, In: H. Füchtbauer and T. Peryt (eds.) *The Zechstein Basin with Emphasis on Carbonate Sequences. Contribution to Sedimentology*, Vol. 9. E. Schweizerbart'sche Verlagsbuchandlung, Stuttgart, pp. 167–203.

Clough, J.G. and Goldhammer, R.K. (2000) Evolution of the Neoproterozoic Katakturuk Dolomite Ramp Complex, Northeastern Brooks Range, Alaska, In: J.P. Grotzinger and N.P. James (eds.) *Carbonate Sedimentation and Diagenesis in the Evolving Precambrian World*, Vol. 67. SEPM Special Publication, Tulsa, OK, pp. 209–241.

Condon, D., Zhu, M., Bowring, S., Wang, W., Yang, A. and Jin, Y. (2005) U-Pb ages from the Neoproterozoic Doushanto Formation, China. Science **308**: 95–98.

Delhaye, F. and Sluys, M. (1923) Esquisse géologique du Congo Occidental. Etude du Système du Schisto-Calcaire: missions géologiques de 1914 et 1918-19. Etablissement Cartographique E. Patesson, Bruxelles-Uccle, pp. 1923–1924.

Delpomdor, F. (2007) *Lithostratigraphie et sédimentologie de la chaîne Ouest Congolienne du Néoprotérozoïque supérieur (Formation de la Diamictite supérieure et Sous-groupe du Schisto-Calcaire) Bas-Congo, République Démocratique du Congo*. Unpubl. M.Sc. thesis, Free University of Brussels, Brussels, Belgium, 138 pp.

Freytet, P. and Verrecchia, E.P. (1999) Calcitic radial palisadic fabric in freshwater stromatolites: diagenetic and recrystallized feature or physicochemical sinter crust? Sediment. Geol. **126/1-4**: 97–102.

Frimmel, H.E., Tack, L., Basei, M.S., Nutman, A.P. and Boven, A. (2006) Provenance and chemostratigraphy of the Neoproterozoic West Congolian Group in the Democratic Republic of Congo. J. Afr. Earth Sc. **46**: 221–239.

Gérard, G. (1958) *Carte géologique de l'Afrique Equatoriale Française au 2 000 000 avec notice explicative*. Brazza. Dir. Mines et Géol, Afrique Equatoriale Française, 198 pp., 4 feuilles.

Gerdes, G., Krumbein, W. and Noffke, N. (2000) Evaporite microbial sediments, In: R. Riding and S. Awramik (eds.) *Microbial Sediments*. Springer, Berlin, pp. 196–208.

Grotzinger, J.P. (1989) Facies and evolution of Precambrian carbonate depositional systems: emergence of the modern platform archetype, In: P.D. Crevello, J.L. Wilson, J.F. Sarg and J.F. Read (eds.) *Controls on Carbonate Platform and Basin Development*, Vol. 44. Special Publication. Society of Economic Paleontologists and Mineralogists, Tulsa, OK, pp. 79–106.

Hardie, L.A. (ed.) (1977) Sedimentation on the modern carbonate tidal flats of Northwest Andros Island, Bahamas. John Hopkins Univ. Stud. Geol. **22**: 202.

Hardie, L.A. (1987) Dolomitization, a critical view of some current views. J. Sed. Res. **57**: 166–183.

Hardie, L.A. and Ginsburg, R.N. (1977) Layering: the origin and environmental significance of lamination and thin bedding, In: Hardie, L.A. (ed.) *Sedimentation on the Modern Carbonate Tidal Flats of Northwest Andros Island, Bahamas*, Vol. 22. John Hopkins University Studies in Geology, Baltimore, pp. 50–123.

Hoffman, P.F. (1974) Shallow and deepwater stromatolites in Lower Proterozoic facies change, Great Slave Lake, Canada. Am. Assoc. Petrol. Geol. **58**: 856–867.

Hoffmann, P.F., Condon, D.J., Bowring, S.A. and Crowley, J.L. (2004) U-Pb zircon date from the Neoproterozoic Ghaub Fomration, Namibia: constraints on Marinoan glaciation. Geology **32**: 817–820.

Hoiczyk, E. (1998) Structural and biochemical analysis of the sheath of *Phormidium uncinatum*. J. Bacteriol. **180**: 3923–03932.

Jacobsen, S.B. and Kaufman, A.J. (1999) The Sr, C and O isotopic evolution of Neoproterozoic seawater. Chem. Geol. **161**: 37–57.

Kolo, K. (2009) Microbial alteration of mineral substrates. Experimental and fossil microbe-mineral interactions. Department of Geology, Vrije Universiteit Brussels, Unpublished Ph.D. thesis, 221p.

Kolo, K., Keppens, E., Préat, A. and Claeys, P. (2007) Experimental observations on fungal diagenesis of carbonate substrates. J. Geophys. Res **112**: G01007, doi: 10.1029/2006JG000203.

Land, L.S. (1980) The isotopic and trace element geochemistry of dolomite: the state of the art, In: D.H. Zenger, J.B. Dunham and R.L. Ethington (eds.) *Concepts and Models of Dolomitization*, Vol. 28. Society of Economic Paleontologists and Mineralogists, Special Publication, Tulsa, OK, pp. 87–110.

Lepersonne, J. (1951) Données Nouvelles sur la stratigraphie des terrains anciens du Bas-Congo. Bull. Soc. Belg. Paléont. Hydr. **LX(2)**: 169–189.

Logan, B.W., Rezak, R. and Ginsburg, R.N. (1964) Classification and environmental significance of algal stromatolites. J. Geol. **72**: 68–83.

Narbonne, G.M., James, N.P., Rainbird, R.H. and Morin, J. (2000) Early neoproterozoic (Tonian) patch reef complexes, Victoria Island, Arctic, Canada, In: J.P. Grotzinger and N.P. James (eds.) *Carbonate Sedimentation and Diagenesis in the Evolving Precambrian World*, Vol. 67. SEPM Special Publication, Tulsa, OK, pp 163–177.

Nédélec, A., Affaton, P., France-Lanord, C., Charriere, A. and Alavaro, J. (2007) Sedimentology and chemostratigraphy of the Bwipe Neoproterozoic cap dolostones (Ghana, Volta Basin): a record of microbial activity in a peritidal environment. C. R. Geosci. **339**: 223–239.

Noffke, N., Gerdes, G. and Klenke, T. (2003) Benthic cyanobacteria and their influence on the sedimentary dynamics of peritidal depositional systems (siliciclastic, evaporitic salty, and evaporitic carbonatic). Earth-Sci. Rev. **62**: 163–176.

Pentecost, A. and Riding, R. (1986) Calcification in cyanobacteria, In: B. Leadbeater and R. Riding (eds.) *Biomineralization in Lower Plants and Animals*, Clarendon Press, Amsterdam, pp. 73–90.

Préat, A., Kolo, K., Prian, J.P. and Delpomdor, F. (2010) A peritidal evaporite environment in the Neoproterozoic of South Gabon (Schisto-Calcaire Subgroup, Nyanga basin). Precambrian Res. **177**: 253–265, doi: 10.1016/j.precamres.2009.12.003.

Prian, J.P. (2008) Notice géologique et ressources minerales de la carte de N'Dendé à 1:200000. Document provisoire, BRGM, 79p.

Purser, B.H. (ed.) (1973) *The Persian Gulf. Holocene Carbonate Sedimentation and Diagenesis in a Shallow Epicontinental Sea*. Springer, Berlin, pp. 471

Reid, R.P., Visscher, P.T., Decho, A., Stolz, J.K., Bebout, B.M., Dupraz, C., MacIntyre, I.G., Paerl, H.W., Pinchney, J.L., Prufert-Bebout, L., Steppe, T.F. and DesMarais, D.J. (2000) The role of microbes in accretion, lamination and early lithification of modern marine stromatolites. Nature **406**: 989–992.

Riding, R. (2000) Microbial carbonates: the geological record of calcified bacterial-algal mats and biofilms. Sedimentology **47**: 179–214.

Sanz-Montero, M.E., Rodriguez-Aranda, J.P. and Calvo, J.P. (2006) Mediation of endoevaporitic microbial communities in early replacement of gypsum by dolomite: a case study from Miocene lake deposits of the Madrid Basin, Spain. J. Sed. Res. **76**: 1257–1266.

Saylor, B.Z., Kaufman, A.J., Grotzinger, J.P. and Urban, F. (1998) A composite reference section for terminal Proterozoic strata of Southern Namibia. J. Sed. Res. **68**: 1223–1235.

Schopf, J.W. (1968) Microflora of the bitter springs formation, Late Precambrian, Central Australia. J. Paleontol. **42**: 651–688.

Sellwood, B.W. (1986) Shallow-marine carbonate environments, In: H.G. Reading (ed.) *Sedimentary Environments and Facies*. Blackwell Scientific Publications, Oxford, pp. 283–342.

Seong-Joo, L. and Golubic, S. (2000) Biological and mineral component of an ancient stromatolite: Gaoyuzhuang Formation, Mesoproterozoic of China, In: J.P. Grotzinger and N.P. James (eds.)

Carbonate Sedimentation and Diagenesis in the Evolving Precambrian World, Vol. 67. SEPM Special Publication, Tulsa, OK, pp. 91–102.
Tack, L., Wingate, M.T.D., Liégeois, J.-P., Fernandez-Alonzo, M. and Deblond, A. (2001) Early Neoproterozoic magmatism (1000-910 Ma) of the Zadinian and Mayumbian Groups (Bas-Congo), onset of Rodinia rifting at the western edge of the Congo craton. Precambrian Res. **110**: 277–306.
Teneva, I., Dzhambazov, B., Mladenov, R. and Schirmer, K. (2005) Molecular and phylogenetic characterization of *Phormidium* species (cyanoprokaryota) using the CPCB-IGS-CPCA locus. J. Phycol. **41**: 188–194.
van den Hoek, C., Mann, D.G. and Jahns, H.M. (1995) *Algae, An Introduction to Phycology*. Cambridge University Press, Cambridge, pp. 623.
Vasconcelos, C. and McKenzie, J. (1997) Microbial mediation of modern dolomite precipitation and diagenesis under anoxic conditions (Lagoa Vermelha, Rio de Janeiro, Brazil). J. Sed. Res. **67**: 378–390.
Veizer, J. and Hoefs, J. (1976) The nature of $^{18}O/^{16}O$ and $^{13}C/^{12}C$ secular trends in sedimentary carbonate rocks. Geochim. Cosmochim. Acta **40**: 1387–1395.
Visscher, P.T., Reid, R.P., Bebout, B.M., Hoeft, S.E., Macintyre, I.G. and Thompson, J.A. (1998) Formation of lithified micritic laminae in modern marine stromatolites (Bahamas): the role of sulfur cycling. Am. Mineral. **83**: 1482–1493.

Biodata of **Professor Cao Ruiji** and **Professor Yin Leiming**, authors of ***"Microbiota and Microbial Mats Within Ancient Stromatolites in South China"***

Professor Cao Ruiji graduated in Geology from Geology Department, Nanjing University of China in 1958. He worked at the Nanjing Institute of Geology and Palaeontology, Chinese Academy of Sciences from 1991 to 1995, and also served as Director of the Nanjing Institute of Geology and Paleontology, Chinese Academy of Sciences. He has visited Australia and University of California, Los Angeles, USA, as a visiting scholar. He has carried out field work in Australia, North Korea, Japan, and the Bahamas. His research focuses on late Precambrian biostratigraphy and stromatolites.

Professor Yin Leiming graduated in Geology, especially in Stratigraphy and Paleontology, from Nanjing University in 1963. He worked at the Nanjing Institute of Geology and Palaeontology, Chinese Academy of Sciences. He has visited Sweden (Lund and Uppsala Universities), Belgium (Paleontological Institute of the Royal Museum), Italy (Pisa University), and the USA (Harvard University) as a Visiting Researcher. He has been a member of paleontological delegations to both Australia and North Korea. His research focuses on Precambrian microfossils and early Paleozoic acritarchs. Professor Yin Leiming has received two First-Class Awards and one Third-Class Award of the Chinese Academy of Sciences.

E-mail: **leimingyin@yahoo.com.cn**

Cao Ruiji

Yin Leiming

V.C. Tewari and J. Seckbach (eds.), *STROMATOLITES: Interaction of Microbes with Sediments*, Cellular Origin, Life in Extreme Habitats and Astrobiology 18, 65–86
DOI 10.1007/978-94-007-0397-1_4,

MICROBIOTA AND MICROBIAL MATS WITHIN ANCIENT STROMATOLITES IN SOUTH CHINA

CAO RUIJI AND YIN LEIMING
Nanjing Institute of Geology and Palaeontology, Chinese Academy of Sciences, Nanjing 210008, China

Abstract Abundant stromatolites have been found from the Neoproterozoic Jiudingshan and Niyuan formations in northern Jiangsu Province and northern Anhui Province, South China. The stromatolites are mostly stratiform, rarely small domed and conical (*Conophyton*-like), with more or less clearly laminated structures. Well-preserved, silicified microbial mats containing mat-building microfossils have been found in small domed, conical, and stratiform stromatolites of the Neoproterozoic Jiudingshan Formation and Niyuan Formation. Main mat-producing species are *Gloeodiniopsis suxianensis* among the coccoids, and *Siphonophycus inornatum* and *Siphonophycus* sp. among the filamentous cyanophytes, although *Myxococcoides* sp., *Leiosphaeridia* sp., and *Eoentophysalis robusta* are often included in the community. Present study indicates that the morphology of a microbial mat, particularly first microbial mat, plays an important role in the morphogenesis of stromatolites, and the community dominated by *Gloeodiniopsis suxianensis*, which is largely confined to stratiform stromatolites, may represent an intertide setting.

Keywords Microbiota • Microbial mats • Neoproterozoic • Stromatolites • South China • Metabolic activity • Blue-green algae • Benthic microbial deposits

1. Introduction

Stromatolites are usually interpreted as organosedimentary structures produced by sediment trapping and binding and/or precipitation resulting from metabolic activity and the growth of organisms such as blue-green algae (Awramik et al., 1976). Recently, stromatolites have been defined as laminated benthic microbial deposits (Riding, 2000). Stromatolites are laminated sedimentary rocks with predominantly upward convex laminations. Zhang and Hofmann (1992) considered stromatolites to be lithified or fossilized microbial mats and mat layers that are equivalent to stromatolitic laminae. Microbial mats are accretionary cohesive microbial communities that are often laminated and found growing at the sediment–water interface and occasionally at the sediment–air interface (Pierson et al., 1992).

Fossil stromatolites were widespread in the Precambrian and reached a high abundance and diversity by the Proterozoic. Stromatolites have been used for environmental and paleoecologic analysis and for biostratigraphic correlation of the Precambrian. Many microfossils have been discovered within Proterozoic stromatolites, although there have also been a few Archean and Phanerozoic occurrences.

Since 1965, many studies concerning microfossils from silicified stromatolites found worldwide have been conducted; however, little is known about the structure and taxonomic composition of microbial mats in ancient stromatolites due to their scarcity and poor preservation. Abundant stromatolites have been found in the Neoproterozoic Jiudingshan and Niyuan formations in northern Jiangsu Province and northern Anhui Province, South China. These stromatolites are mostly stratiform, rarely small domed and conical (*Conophyton*-like), and have more or less clearly laminated structures. Some of these stromatolites were partially silicified under the conditions of extremely early lithification; thus, the primary composition and molds of microbial mats in these stromatolites are well preserved. This study was conducted to evaluate the microbiotas and benthic microbial mats that occur in these ancient stromatolitic black cherts. The data presented here provide significant information regarding the microbiotas, mat builders, mat dwellers, and allochthonous microorganisms in Precambrian stromatolites.

2. Geological Setting

The Neoproterozoic strata exposed in northern Jiangsu Province and northern Anhui Province are composed of unmetamorphosed or slightly metamorphosed carbonates that contain abundant stromatolites and detrital rocks and have an average thickness of 3,000 m. The strata are unconformably overlain by the Lower Cambrian Houjianshan Formation and lie unconformably above the metamorphic rocks of the Fengyang Group. The Neoproterozoic strata may be divided, in descending order, into the Huinan, Xuhuai, and Suxian groups. The Xuhuai Group includes the Zhaowei, Niyuan, Jiudingshan, Zhangqu, Shijia, and Wangshan formations (Fig. 1).

Although there is a lack of reliable isotopic dating for the Neoproterozoic strata, carbonates of the Xuhuai and Suxian groups yield numerous stromatolite forms, including *Jurusania cylindrica*, *Inzeria intia*, *Gymnosolen* cf. *furcatus*, *Boxonia jinshanzhaiensis*, and *Acaciella* sp., which are correlated with the upper part of the Cryoginian (Cao et al., 1985). In the Huainan area of Anhui Province, abundant macroscopic fossils have been found in the Huinan Group. The principal taxa are *Chuaria circularis* Walcott, *Chuaria annularis* Zheng, *Chuaria multirugosa* Du et Tian, *Shouhsienia shouhsienensis* Xing, *S. longa* Xing, *Ovidiscina bagongshania* Zheng, *O. longa* Du et Tian, *Bipatinella cervicalis* Zheng, *Lingulites folirormis* Duan et Du, *Pumilibaxa huaiheiana* Zheng, *P. symmetrica* Fu, *Nephroformia liulaobeiensis* Zheng, *Tawuia dalensis* Hofman, *T. sinensis* Duan, *T. striatia* Zheng, *T. afflata* Fu, *Pumilibaxa longa* Yan, *Mesonactus imparilis* Fu,

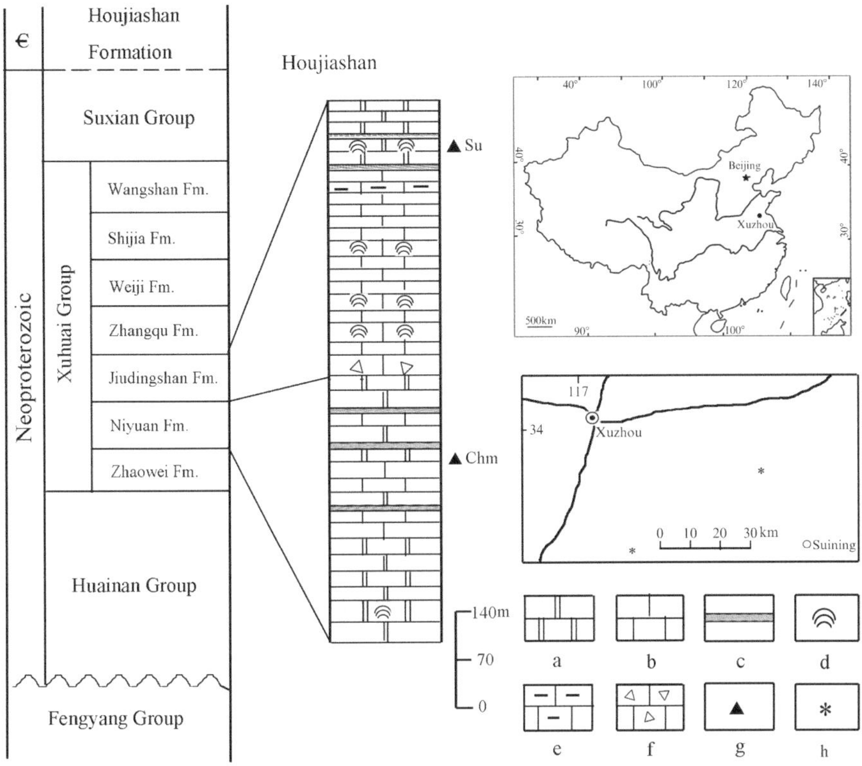

Figure 1. Map of the schematic stratigraphic column of the Neoproterozoic Jiudingshan and Niyuan formations in N. Jiangsu Province and N. Anhui Province, China. (**a**) Dolostone, (**b**) limestone, (**c**) chert, (**d**) stromatolite, (**e**) marlite, (**f**) breccia limestone, (**g**) fossiliferous horizon, (**h**) locality of section.

M. luoyingensis Fu, *Eurycyphus lycotropus* Fu, *Eurycyphus altilis* Fu, *Stenocyphus subtilis* Fu, *Huainanella cylinddrica* Wang, *Sabllidites* sp., *Tyrasotaenia* sp., and *Morania* sp. Some of them were previously interpreted as worm-like animals, but recent work indicated that they are remains of coenocytic algae (Dong et al., 2008). The Huainan macroscopic fossil assemblage, dominated by aquicolous algae, occurred prior to the well-known Ediacara fauna and should have been flourishing before the latest Precambrian glacial epoch, with an age ranging from about 850 to 700 Ma (Fu, 1989).

The fossiliferous stromatolitic cherts examined in this study were collected from the Jiudingshan Formation at Liulou Village, which is approximately 25 km northwest of Suining County in N. Jiangsu, and from the Niyuan Formation at Qingtongshan of Suxian, Anhui Province. The Jiudingshan and Niyuan formations are represented by a series of dark gray limestones or dolostones with cherty bands and stromatolitic bioherms. The sedimentary rocks are relatively unmetamorphic and have a maximum thickness of greater than 700 m. Two formations are characterized by the development of ripple marks, birds eye pores, wave laminae, and scouring structures. The breccia limestones, which developed in the base of the Jiudingshan Formation, have a thickness of approximately 30 m and a very wide geographical distribution. These sedimentary features are indicative of high-energy conditions in a shallow water environment.

3. Materials and Preservation

The strata evaluated in this study generally consist of thin- to medium-bedded gray dolostones and thin-bedded gray siliceous limestones with partially silicified stratiform, as well as small domed, conical, and undulating forms of stromatolites. Abundant organic matter makes the chert dark or black in stromatolitic hand-size specimens. These stromatolites were silicified in the early diagenetic stage. Well-preserved microbial mats, which are described in this paper, were found in black cherts that occur as stromatolitic layers, bands, and lenses in limestones. These original limestones have often been replaced by dolomite and obscured by secondary recrystallization. It is likely that an early diagenetic replacement by chert occurred prior to recrystallization of the carbonates. In the stromatolites under consideration here, chertification seems to always start in the laminae with a high organic content. In general, chertification follows the dark, organism-rich layer of the stromatolite. Primary stromatolitic fabrics have been replaced by chert prior to neomorphic recrystallization. The chert is aphanitic and composed of chalcedony and microcrystalline quartz. In the thin section, the black chert contains brown or amorphous organic matter disseminated throughout the silica matrix. The microbial mats were primarily embedded in the silica, and crystallization of the chert did not alter the form of the microbial mats. The fossil mat forms were usually more or less cohesive fabrics of intertwined filaments or gelatinous matter produced by both filamentous and coccoid microorganisms. Cellular and filamentous structures within the mat are well preserved and fossilized. Most microfossils in microbial mats are preserved in their original growth position.

4. Microbial Mats of Ancient Stromatolites

4.1. DOMED STROMATOLITE-BUILDING MICROBIAL MATS

Sample No. Su-01 is a small domed stromatolite. The laminae simply arch upward, without any overlapping margins. The column is approximately 5–8 mm wide and 9 mm high. It is generally gray-black in color and contains reduced siliceous compounds and organic matter. This specimen was studied in vertical thin sections. The stromatolite is composed of laminated domed microbial mats that consist of only one species, *Siphonophycus inornatum*. The microbial mats were constructed by a series of domed laminae. Each lamina consists of alternating light, thick, algae-poor layers that are generally 0.4–1 mm thick, and dark, thinner, algal-rich layers that are generally 0.1–0.4 mm thick. The light layers are composed of anastomosing bundles of filaments that are erect or inclined at varies angles and form a three-dimensional reticulated framework. The dark layers consist of filamentous bundles that are prostrate or inclined at various angles. The bundles are closely crowded into a generally thin opaque partition and form a planar reticulated framework (Fig. 2a, c). The growth pattern of the filaments and the superposition

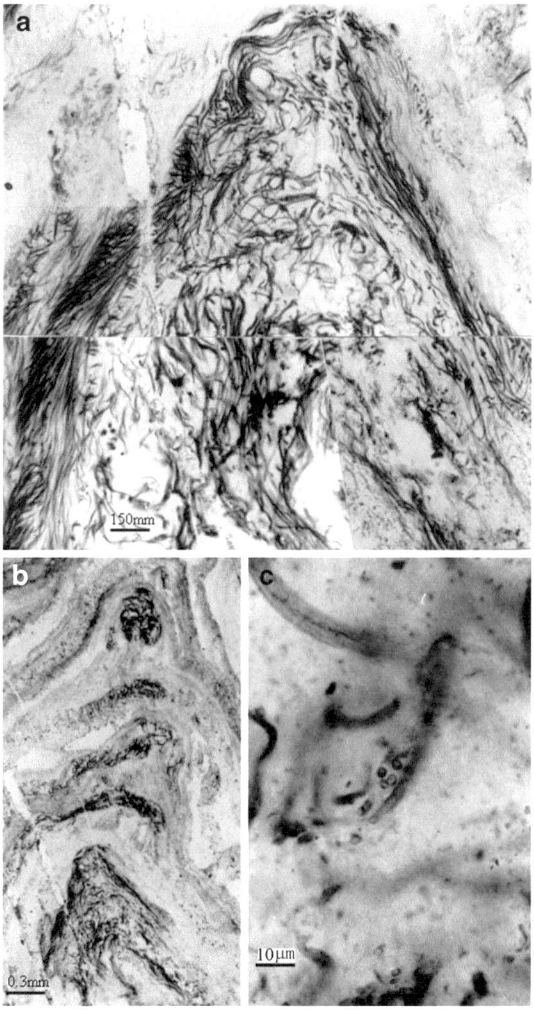

Figure 2. Thin section photos showing the domed microbial mats composed of the unique filamentous microfossils *Siphonophycus inornatum* in a domed stromatolite. (**a**) Filamentous layers in a domed stromatolite; (**b**) bridge-like lamella; (**c**) enlargement of **a**.

of the convex laminae in the stromatolite display characteristics of organisms in search of light. The filaments are especially abundant and densely packed in the dark layers; however, they are much rarer in the light layers. Openings of irregular fenestra caused by internal solution occur commonly in the light microcrystalline layer. These microbial mats are almost entirely composed of filamentous microorganisms, although a few microbial trappings and binding of gravel-sized micrograins have been observed around the filaments (Fig. 2d). However, the trap sediment mechanism was not the primary mode of formation of the stromatolite, which likely formed through in situ precipitation of carbonate. Some bridge-like laminae occur occasionally at the wall rock on the right side (Fig. 2b). The biological imprint on the lamina texture is created through the orientation of filaments and unicells, the motility of major filamentous forms, and the adhesion and abundance of microbial sheath material (Semikhatov et al., 1979). The laminated domed *S. inornatum* microbial mats are characterized by a single composition, cyclic growth pattern, and rhythm change of filamentous abundance.

To date, there are two different interpretations of the cycle of filamentic growth orientation: (1) a modern non-lithified stromatolite, such as one from Andros Island, Bahamas (Monty, 1967, 1976), also exhibits this type of growth pattern. A laminated pattern can result from the alternation of phases showing horizontally growing filaments with phases characterized by vertical bundles of filaments. Golubic and Focke (1978) concluded that these filaments belonged to the same microbial species, now classified as *Phormidium hendersonii* Howe. This species is motile and characterized by daily movements in accordance with diurnal light variation. Such phototactic movement and subsequent production of a common hard gel can induce an alternating arrangement of vertical and horizontal filaments, which together with entrapped sedimentary particles, produces a primary noctidiurnal lamination. (2) Recently, Lee et al. (1999, 2000) evaluated microfossil populations from the Mesoproterozoic Gaoyuzhuang Formation, China. They suggested that the sedimentary processes and microbial growth followed reciprocal rhythms. Specifically, they believed that low sedimentation rates permitted copious microbial growth with biological diversification, whereas high sedimentation rates suppressed microbial growth and challenged survivorship strategies. Accordingly, high sedimentation rates correlate with scattered colonies of coccoids and loose webs of predominantly upright filaments, whereas low sedimentation rates correlate with dense, laterally connected colonies of coccoids and changes in the filament orientation from vertical to horizontal.

The mats discussed in the present study retain a filamentous growth pattern, as well as changes in abundance. Furthermore, biolamination doublets in the mats are usually well developed. Since the biolamination originated from the combined effects of filamentous abundance and growth patterns, its development depended partially on the alternation of relatively favorable and unfavorable climates. In the dark layers, the filaments often have prodigious numbers, which may indicate that microorganisms flourished under a warm climate and low sedimentation rates. In the light layers, the filaments are comparatively rare. The growth of filamentous microorganisms probably occurred in cold or cool climates under high sedimentation rates. Each biolamination in the stromatolite is approximately 0.5–1.4 mm in thickness. The maximum growth rate observed from the living stromatolites in Shark Bay, Western Australia, amounted to approximately 1 mm per year (Playford and Cockbain, 1976). Thus, it appears that the biolamination in the present stromatolite is due to seasonal influence and not the daily circadian rhythm.

4.2. CONICAL STROMATOLITE-BUILDING MICROBIAL MATS

Specimen No. su-10 is a small *Conophyton*-like stromatolite. The stromatolite is commonly composed of dark cherts (silica) with occasional dolomite. The cone in this specimen is approximately 3–10 mm wide and 12 mm high, with prominent macrolaminae and frequent lenticular thickening in the laminae. The laminae are mostly conical and have a considerably higher synoptic relief than those associated with modern mats. The axial zone recognized in Precambrian *Conophyton* stromatolites (Komar et al., 1965) is present in the cone. Two morphological types of microbial mats have been identified in the longitudinal thin section of cone (Fig. 3b).

1. A pinnacle-like *Siphonophycus* sp. microbial mat occurs in the base of stromatolite. Based on the description by Cao et al. (2001), this pinnacle-like mat is the first microbial mat. This mat was constructed by a distinctive net-like complex of growing filaments dominated by *Siphonophycus* sp. and also contained some trapped detrital grain. A few *Myxococcoides* sp. were also occasionally observed within the mat (Fig. 3a, c). These coccoids are probably planktonic algae that descended into the microbial mat community and were preserved. Based on the observations of this stromatolite, the authors can speculate on how the pinnacle-like mat (first microbial mat) of the developing stromatolite was constructed. Initially, filaments gliding on the substrate surface encountered bedrock irregularities and were then deflected upward above them, toward the light source (the sun). The erect tufts of filaments are the initial accretions of these mats, and the processes and mechanisms involved in the erect growth of the filamentous tufts were likely similar to those described by Walter et al. (1976). Next, a regular conical reticulated framework consisting of tangled filaments with entrapped sediments was formed on the tufts of the erect filaments. However, in the flanks of the conical framework, the filaments are oblique at high angles to the horizontal line and usually arranged parallel to the rim. This may be due to the upward movement of the filaments along the inclined slope of the mat flanks and toward light. Studies of modern and fossil stromatolites have shown that the orientation of cyanophyte filaments within stromatolites can be controlled by light (Monty, 1967; Gebelein, 1969; Zhang, 1986). Indeed, the modern, non-lithified, conical stromatolite data from Yellowstone National Park suggest that many species of *Phormidium* are present in Yellowstone waters, but apparently only *P. tenue* var. *granuliferum* constructs

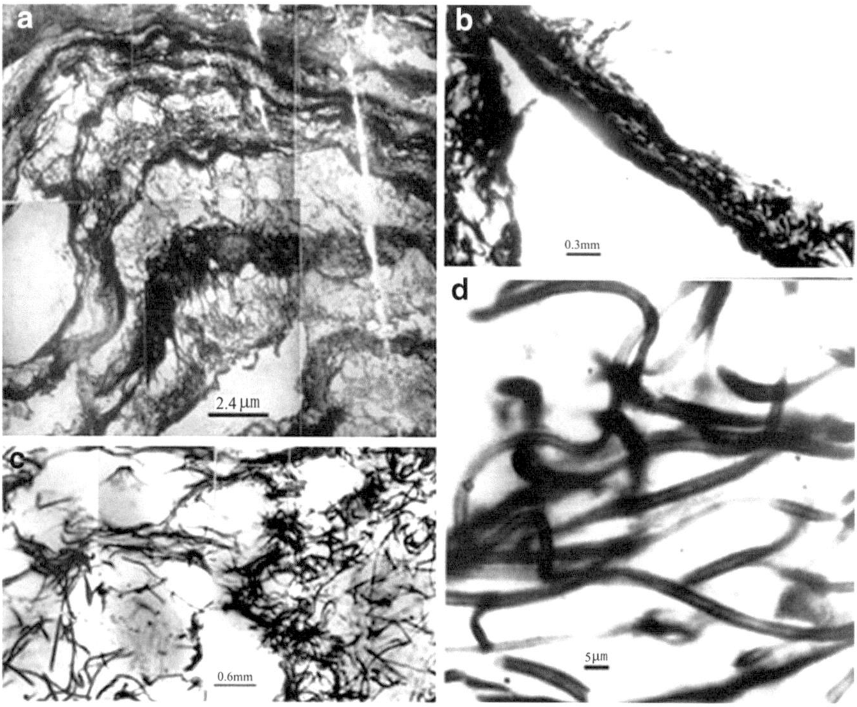

Figure 3. Pinnacle-like microbial mat in the longitudinal section of a small conophyton-like stromatolite. (**a**) Enlargement of **b**, showing lamella composited by *Siphonophycus*-like filaments; (**b**) longitudinal section of a small conophyton-like stromatolite; (**c**) tubular filaments mixed with some coccoid microfossils; (**d**) *Siphonophycus inornatum* Zhang, 1981.

conical stromatolites (*Conophyton weedii*). This indicates that the morphology of the present mat results from the phototaxis and entangled behavior of the principal constructing microorganisms, *Siphonophycus* sp.

2. The bamboo shoot coat-like *Siphonophycus* sp. microbial mat grew around the pinnacle-like microbial mat described above. These mats are similar to succedent mats described by Cao et al. (2001). These mats and sediment layers contain a series of laminae, each of which consists of a light layer (sediment layer) and a dark layer (mat) and a few detrital particles that were trapped and bound by the microbial community. The mechanically deposited particles in the light layer of each lamina were likely washed over the algal mats by tides and storms. The dark layer likely represents favorable conditions for algal growth and a low, more or less constant rate of sedimentation. The central part of each dark layer (mat) is primarily composed of vertical filaments, but the filaments are usually arranged prostrate along both sides of the dark layer (mat). Most dark layers (mats) are thickened more than 1–3 times at the central part. The convex dark layer (mat) was composed of organisms that were in search of light.

In addition, a well axial zone is present due to superposition of the convex layers (mats) and an oval knot of coiled filaments occurs between the two laminae. This zone was originally a mat that broke and slid upward in response to strong currents before the covering laminae were formed. The broken fragments and slidden traces can be clearly seen in the thin section.

Conophyton are non-ramified, relatively regular conical stromatolites. *Conophyton* are formed by a series of conic microstrata and a well-pronounced axial zone where the thickness increases by 2–10 times. However, there is currently little information available describing the morphogenesis of *Conophyton*, with the exception of a study conducted by Walter (1976). The results of the present study indicate that the first microbial mat appears to be the "embryo" of conical stromatolites. The succedent microbial mats were then controlled by the morphology of the first microbial mat. Two types of microbial mats may have played different roles in the formation of the conical stromatolite. The first microbial mat may be regarded as the growth bud of the stromatolite and morphological mold for construction of the conical stromatolitic laminae. The succedent microbial mats merely increased the height and volume of the stromatolites.

4.3. STRATIFORM STROMATOLITE: BUILDING MICROBIAL MATS

Samples No. Su-6, Su-11, and Su-13 are three stratiform stromatolites collected at different horizons in the upper part of the Jiudingshan Formation. The following types of fossilized microbial mats occur in some laminae of stratiform stromatolites.

Ripple or flat microbial mats mixed with the spheroid microorganisms, *Gloeodiniopsis suxianensi* and *Eoentophysalis cumulus*, and the filamentous microorganism, *Siphonophycus inornatum*,extend parallel to the lamination of the stromatolite and can be traced laterally over a considerable distance (tens of centimeters).

1. *Gloeodiniopsis suxianensis* is a dominant coccoid microfossil that is often distributed in the lower portion of a single mat (Fig. 4a) or grouped as lump within a given space (Fig. 4b), whereas *Siphonophycus inornatum* is primarily distributed in the middle and upper portions of a single mat. Despite a single mat being less than 0.5 mm thick, the growth of filaments and spheroid microorganisms within a mat appears to follow a regular pattern. This may be related to the microenvironmental differences within the mat, such as the factors of illumination and oxygen supplementation. These two microorganisms may have persisted because conditions were favorable for their growth.
2. Single flat *Gloeodiniopsis suxianensis* Yin microbial mats usually occur only in the base of stratiform stromatolites. These organisms lie on the underlying surface of mechanically deposited and scoured sediments and reach a thickness

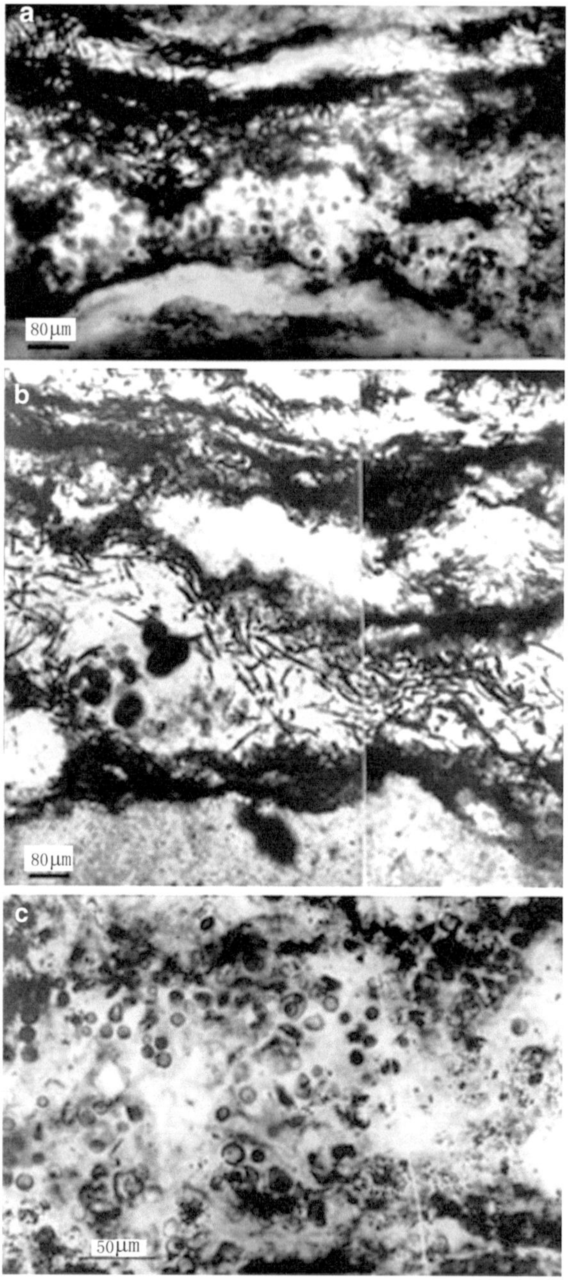

Figure 4. Microbial mats in the stratifera stromatolite. (**a**) Spheroidal microorganisms often distributed in the lower portion of a single mat; (**b**) grouped as lump of coccoid microfossils; (**c**) a single flat mat of *Gloeodiniopsis suxianensis* in the stratifera stromatolite.

of almost 0.16–0.25 mm (Fig. 4c). Occasionally, some weak sediment laminae are visible on the top of the mat. The mats are dominated by abundant *Gloeodiniopsis suxianensis* Yin among the coccoid Cyanophyte. The *G. suxianensis* community in the mat evaluated here was embedded in a fossilized gelatinous matrix that was likely excreted by the organisms themselves. The sediments are trapped around the colonies and on the surface of the mats. Monospecific microfossils of *G. suxianensis* account for an overwhelming majority of the total community preserved in this mat, which may indicate that they are mat builders. These microbial mats have often been covered by laminated tabular *Siphonophycus inornatum* microbial mats.

3. Laminated wavy *Siphonophycus inornatum* Zhang Y. microbial mats primarily built the present stratiform stromatolites. Different types of structures that have been contracted by the arranged pattern of filaments were observed in the mats evaluated here (Fig. 5). The mats are dominated by *S. inornatum*, but occasionally contain a few thin filaments belonging to *Eomycetopsis robusta* within (Fig. 6a, b). The fossil record of stromatolitic microbiotas consists of mat-building cyanobacteria and mat-dwelling taxa that occupied selected microenvironments in the surface or near-surface mat layers but did not contribute structurally to the mat construction (Knoll, 1982). If anoxic microenvironments form within the photic zone, a distinct layer of photosynthetic bacteria may develop beneath the cyanobacterial surface community (Grotzinger and Knoll, 1999). As a result, there may be dwelling organisms that occur in anoxic microenvironments within the mats. These mats retain the particle-trapping behavior of the filaments within their microfabric. The filaments of *S. inornatum* were densely packed in bundles and primarily arranged in prostate orientation or inclined at various angles to the laminae of the stromatolite, which may indicate that they are mat builders. The mats also contain some coccoids, including *Eoentophysalis cumulus* and *Gloeodiniopsis suxianensis*. These organisms occur as local populations on a bedding plane (Fig. 6c), which suggests that they may have been mat dwellers.
4. Unlaminated loosely lumped *Siphonophycus inornatum* Zhang Y. microbial mats occur as thick layers or lenses (approximately 10 mm) in stratiform stromatolites. These mats are characterized by high porosity, no clearly laminated structure, and prostrate or inclined arrangement of filaments with entrapped sediments. The mats of *S. inornatum* also contain some other microfossils, such as *Leiosphaeridia* sp. and *Myxococcoides* sp. These organisms were preserved in the mats by chance and may have been planktic organisms or dwellers. Occasionally, a transitive state from an unlaminated microbial mat to a laminated microbial mat is present and important in the same stromatolite. Generation of a laminated structure in the rock required iterative changes in environmental variables, including physical and chemical conditions during the sedimentation process. Periods of low sedimentation rates and sedimentary pauses (sedimentary stasis) enable uninhibited growth and differentiation of microbial mats, which is recognized in the fossil as organic-rich layers of stromatolites. In contrast, only highly motile

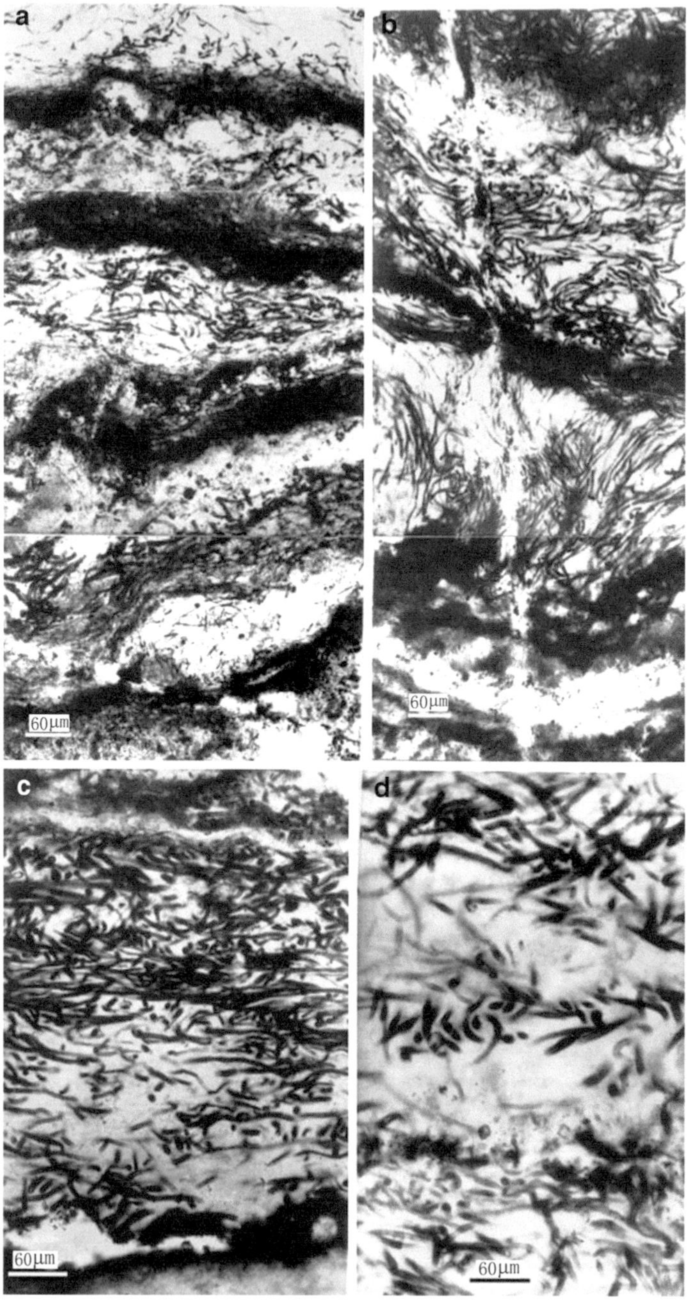

Figure 5. Photos showing four structure types in laminated wavy *Siphonophycus inornatum* microbial mats.

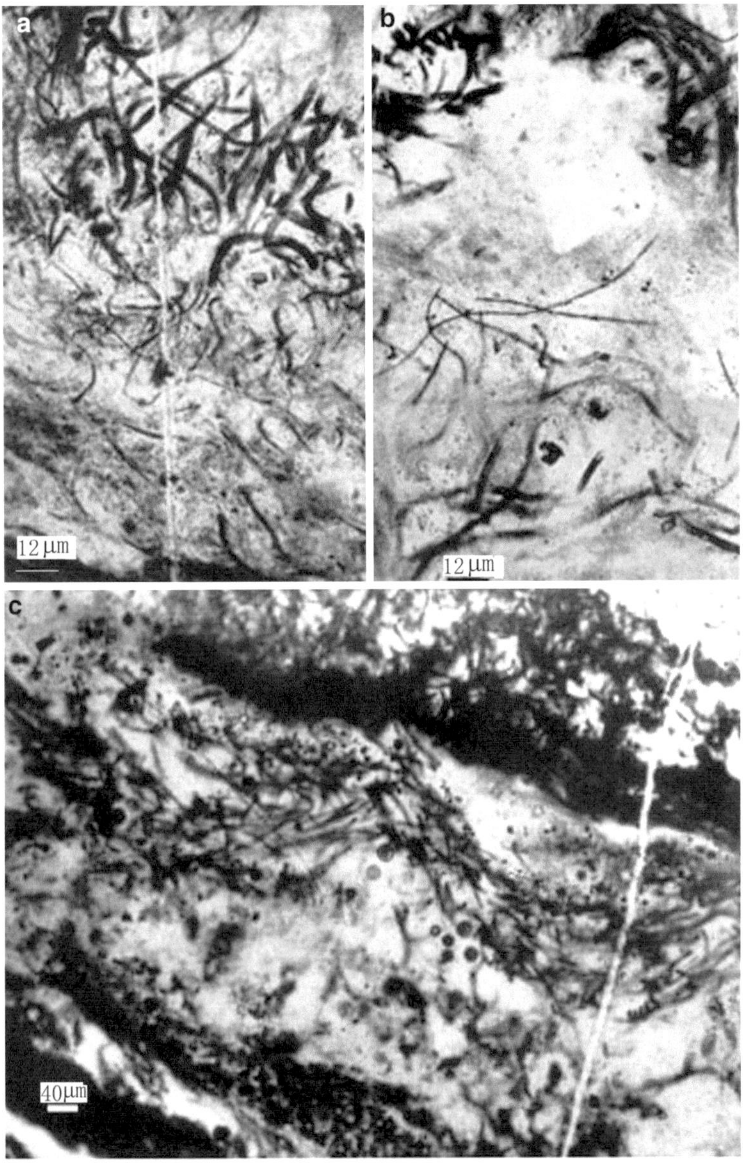

Figure 6. (**a**, **b**) Laminated wavy *Siphonophycus inornatum* microbial mats contained a few thin filaments belonging to *Eomycetopsis robusta*; (**c**) the mats were dominated by *Siphonophycus inornatum*, but occasionally contained coccoids.

pioneer species can overcome periods of high sedimentation rates (sedimentary kinetics) and respond behaviorally to changes in the sedimentary conditions (Lee and Golubic, 1999). The aforementioned transition from an unlaminated to a laminated mat likely reflects a change in the balance between algal growth and the sedimentation rate. The development of unlaminated loose mats relied on a harmonious balance between algal growth and sedimentation rate.

5. Conclusions

1. Silicified microbial mats containing mat-building microfossils have been found in small domed, conical, and stratiform stromatolites of the Neoproterozoic Jiudingshan Formation and stratiform stromatolites of the Niyuan Formation. Evaluation of the abundance and distribution of microfossils within microbial mats clearly demonstrated that the primary mat-producing species were *Gloeodiniopsis suxianensis* among the coccoids and *Siphonophycus inornatum* and *Siphonophycus* sp. among the filamentous cyanophytes, although *Myxococcoides* sp., *Leiosphaeridia* sp., and *E. robusta* were also often observed in the community.

   The microfossils that could be assigned to the tubular Oscillatoriaceae *Siphonophycus* genus account for an overwhelming majority of the total community preserved in the microbial mats. The filaments of the *Siphonophycus* genus were densely packed in the laminae of the stromatolites, particularly small domed and conical stromatolites. These organisms may represent chemically altered remnants of the external polysaccharide sheath of *Lynybya*- or *Phromidium*-type blue-green algae. The diameter and thickness of the tubes and the variety of growth habits exhibited by different species suggest that *Siphonophycus* represents a number of different filamentous algae that may or may not be closely related. Nevertheless, in the absence of identifiable trichomes, these forms can only be treated as morphological species.
2. The morphology of a microbial mat, particularly the first microbial mat, plays an important role in the morphogenesis of stromatolites.
3. The community dominated by *Gloeodiniopsis suxianensis*, which is largely confined to stratiform stromatolites, may represent an intertide setting. This is because the stratiform (planar laminated) stromatolites from the pre-Phanerozoic and Phanerozoic are commonly formed in periodically exposed settings (Awramik, 1984).

6. Systematics of Microbiota of Ancient Stromatolites

Stromatolites from both the Jiudingshan Formation and the Niyuan Formation are characterized by the occurrence of cyanobacterial mat-forming filaments and several types of colonial unicells.

Systematic description of microfossils

Division: Cyanophyta
Class: Cocogoneae
Order: Chroococcales
Family: Chroococcaceae

Genus *Gloeodiniopsis* (Schopf)em. Knoll and Golubic, 1979
Type species: *Gloeodiniopsis lamellosa* (Schopf)em. Knoll and Golubic, 1979

Gloeodiniopsis suxianensis Yin, 1990
(Fig. 7a–e, h, k)

1990 *Gloeodiniopsis suxianensis* Yin, p. 282, pl. 2, figs. 1–6, 8–13; pl. 3, figs. 2–6;
pl. 4, figs. 1–3, 5.

Description: Spheroids and ellipsoids with single, double, or multiple outlines, solitary or in groups of 2, 3, 4 up to 8 individuals within a common envelope. Surface of the envelope punctate. Groups within particular microenvironments showing less variability in size, shape, and state of preservation. Average diameters of spheroids and ellipsoids, expressed as means ± standard deviation, 13 ± 2 μm for the internal outlines (n = 125), 17 ± 2.5 μm for the external outlines (n = 125), and 18 ± 3.5 μm for the single outline (n = 40). Single inclusion spherical, elongated, or star shaped to reticulate, 6.8–17.4 μm (average = 10.5 μm, n = 15).

Remarks: This species is similar to *G. lamellosa* (Schopf)em. Knoll and Golubic, but the latter is generally smaller and has fewer tetrads, and is distinguished from *G. magna* Nyberg and Schopf (1984) in that it contains smaller individuals. The species differs from both *G. pangjiapuensis* Zhang, 1981 and *G. hebeiensis* Zhang, 1981 in that it exhibits a larger size and thicker envelopes.

Occurrence: The Niyuan Formation, at Qingtongshan of Suxian, Anhui Province, and Suining County, N. Jiangsu Province; the Jiudingshan Formation, at Suining County, N. Jiangsu Province, China.

Family: Entophysalidaceae
Genus: *Eoentophysalis* Hofmann, 1976
Type species: *Eoentophysalis belcherensis* Hofmann, 1976

Eoentophysalis cumulus Knoll and Golubic, 1979
(Fig. 7g, i, l, m)

1979 *Eoentophysalis cumulus* Knoll and Golubic, pp. 148–149, figs. 2e.

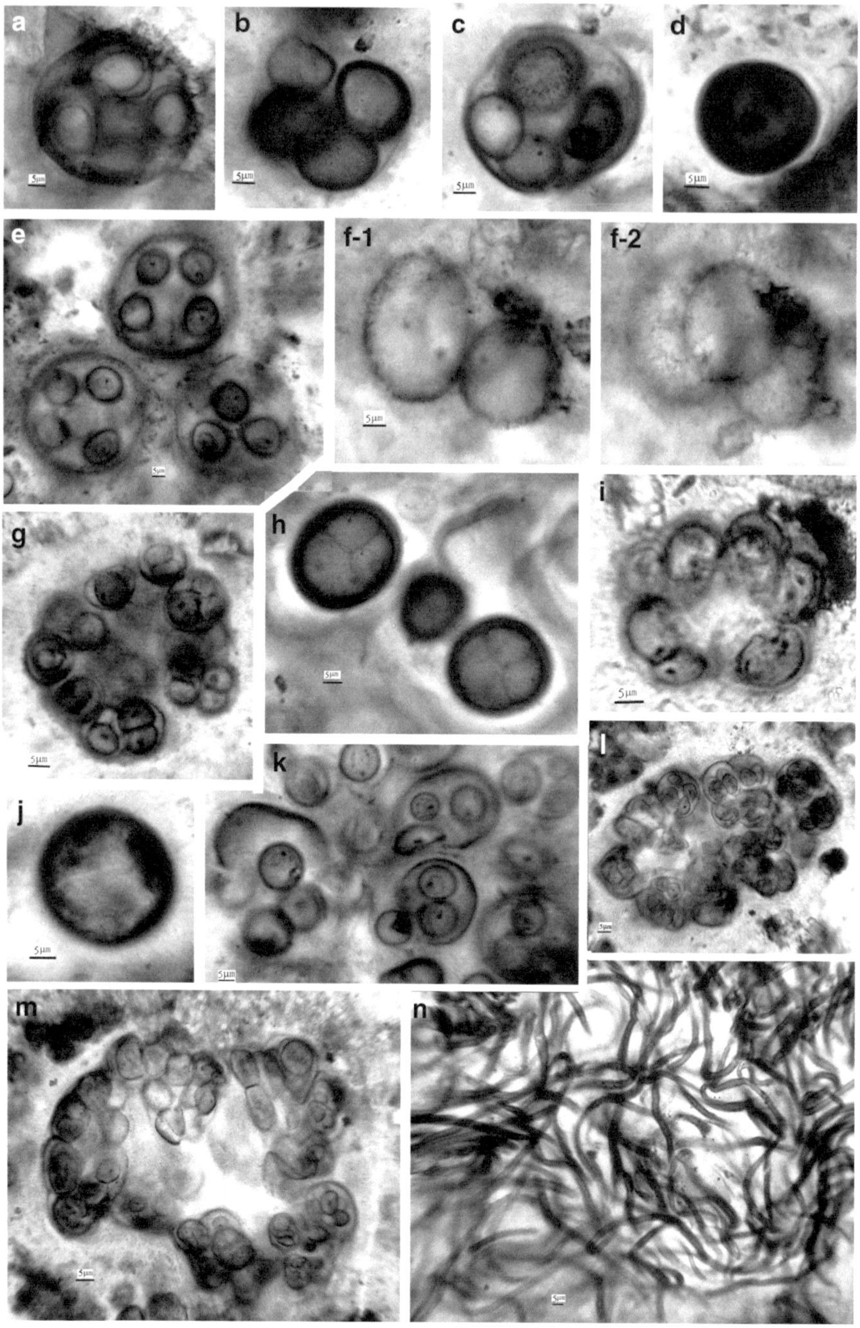

Figure 7. Microfossils found in stromatolites from both the Jiudingshan Formation and the Niyuan Formation. (**a–e**, **h**, **k**) *Gloeodiniopsis suxianensis* Yin, 1990, (**a–c**, **h**) from the Niyuan Formation at Qingtongshan of Suxian, Anhui Province, (**d**, **e**, **k**) from the Jiudingshan Formation at Suining County, N. Jiangsu Province; (**f**) *Myxococcoides* sp., from the Niyuan Formation at Qingtongshan of Suxian, Anhui Province; (**g**, **i**, **l**, **m**) *Eoentophysalis cumulus* Knoll and Golubic, 1979, from the Jiudingshan Formation at Suining County, N. Jiangsu Province; (**j**) *Leiosphaeridia* sp., from the Jiudingshan Formation at Suining County, N. Jiangsu Province; (**n**) *Siphonophycus inornatum* Zhang, 1981, from the Jiudingshan Formation at Suining County, N. Jiangsu Province, China.

Description: Irregular colonial form; the clusters consists of three-dimensional cellular packages with varying arrangements, often surrounded or held together by a common pigmented substance. Individual cell-like units are marked with single or double outlines and often contain single or double granular inclusions. Dimensions of individual cell-like units are 3.5–5 μm. 8–16 cell-like units are packaged to form larger irregular clusters, which are often enveloped by a pigmented substance.

Remarks: Present specimens are well preserved, and individual cell-like units are little larger than those described by Knoll and Golubic (1979).

Occurrence: The Niyuan Formation and the Jiudingshan Formation, at Suining County, N. Jiangsu Province, China.

Family: Chroococcaceae
Genus: *Myxococcoides* Schopf, 1968
Type species: *Myxococcoides minor* Schopf, 1968

Myxococcoides sp.
(Figs. 3c and 7f)

Description: Cells spheroidal or ellipsoidal; surface texture varies from psilate to granular-like degradation. A few cells are composed of clusters that have not been enveloped by the visible pigmented substance. Their size ranges from 25 to 35 μm.

Remarks: Only a few specimens have been found in stratiform stromatolites at both the Niyuan Formation and the Jiudingshan Formation. These specimens differ from known species in that they are much larger in size and form simple clusters.

Class: Hormogoneae
Order: Oscillatoriales
Family: Oscillatoriaceae

Genus: *Siphonophycus* Schopf, 1968
Type species: *Siphonophycus kestron* Schopf, 1968

Siphonophycus inornatum Zhang, 1981
(Figs. 2d, 5d, and 7n)

1981 *Siphonophycus inornatum* Zhang, p. 492, pl. 1, figs.1, 3, 4, 5.

Description: Unbranched tubular structures, non-septate, commonly circular in cross-section; straight, bent, or folded, sometimes twisted; surface texture psilate or shown granular by degradation, up to 190 μm long (incomplete specimen). Filaments commonly gathered in bundles or twisted around each other; 3.5–7 μm wide, wall 0.5 μm thick.

Remarks: The specimens described here are quite similar to those described from the Gaoyuzhuang Formation of North China (Zhang, 1981). This species differs from *S. kestron* (Schopf, 1968) in that it is smaller and has a psilate surface. This species differs from *S. capitaneum* Nyberg and Schopf, 1984 in that it has a smaller size and thinner wall.

Occurrence: The Niyuan Formation, at Qingtongshan of Suxian, Anhui Province, and Suining County, N. Jiangsu Province; the Jiudingshan Formation, at Suining County, N. Jiangsu Province, China.

Siphonophycus sp.
(Fig. 3a, c)

Description: Unbranched tubular structures, non-septate, commonly circular in cross-section; straight, bent, or folded; irregularly degraded organic materials are present within the tubular structure. Filaments, with 4–4.5 μm wide, are commonly twisted around each other; single tubular filaments up to 250 μm long (incomplete specimen).

Remarks: Since the present specimens commonly have organic residues and are quite long, it is difficult to determine if they are a known species or a novel species.

Occurrence: The Jiudingshan Formation, at Suining County, N. Jiangsu Province, China.

Genus: *Eomycetopsis* (Schopf)em. Knoll and Golubic, 1979
Type species: *Eomycetopsis robusta* (Schopf) em. Knoll and Golubic, 1979

Eomycetopsis robusta (Schopf) em. Knoll and Golubic, 1979
(Fig. 6a, b)

1968 *Eomycetopsis robusta* Schopf, p. 685, pl. 82, figs. 2, 3; pl. 83, figs. 1–4.
1979 *Eomycetopsis robusta* Schopf, emend. Knoll and Golubic, p. 149, fig. 4a, b.

Description: Unbranched tubular structures, occasionally solitary, commonly sinuously intertwined, circular to elliptical in cross-section, partially flattened. Surface, coarsely to irregularly granular in texture; filaments tubular with diameter of 1.5–2.5 μm.

Remarks: Specimens described here are consistent with *Eomycetopsis robusta* (Schopf)em. Knoll and Golubic, 1979. Most species show granular texture, which could be caused by degradation.

Occurrence: The Niyuan Formation and the Jiudingshan Formation at Suining County, N. Jiangsu Province, China.

Incertae sedis
Group: Acritarcha Evitt, 1963
Subgroup: Sphaeromorphitae Downie et al., 1963
Genus: *Leiosphaeridia* (Eisenack) em. Turner, 1984
Type species: *Leiosphaeridia baltica* Eisenack, 1958

Leiosphaeridia sp.
(Fig. 7j)

Description: Individual cell-like vesicle, surface texture psilate or granular-like degradation; thicker wall and no folds caused by compression. Its diameter is 45 μm.

Occurrence: The Niyuan Formation and the Jiudingshan Formation at Suining County, N. Jiangsu Province, China.

7. Acknowledgments

We thank editors Professor Vinod C. Tewari and Professor J. Seckbach for kindly inviting us to contribute this chapter and their valuable suggestions for modification of the chapter. We thank Drs. M. Moczydłowska Vidal and Xunlai Yuan for reviewing the manuscript and their comments. This study was supported by the National Science Foundation of China (no. 40839910, 40625006) and State Key Laboratory of Palaeobiology and Stratigraphy (Nanjing Institute of Geology and Palaeontology, CAS) (no. 083102).

8. References

Awramik, S.M. (1984) *Ancient Stromatolites and Microbial Mats Microbial Mats: Stromatolites*. Alan R. Liss, New York, NY, pp. 1–22.

Awramik, S.M., Margulis, L. and Barghoorn, E.S. (1976) Evolutionary processes in the formation of stromatolites, In: M.R. Walter (ed.) *Sromatolites. Developments in Sedimentology 20*. Elsevier, Amsterdam, pp. 149–162.

Cao, R.J., Yuan, X.L. and Xiao, S.H. (2001) On morphogenesis of *Conophyton* stromatolites – analysis of a *Conophyton*-like specimen from Neoproterozoic Jiudingshan Formation in N. Jiangsu Province, China. Acta Palaeontol. Sin. **40**(3): 318–329.

Cao, R.J., Zhao, W.J. and Xia, G.S. (1985) Lat Precambrian stromatolites from North Anhui Province. Mem. Nanjing Inst. Geol. Paleont. Acad. Sin. **21**: 1–84.

Dong, L., Xiao, S.H., Shen, B., Yuan, X.L., Yan, X.Q. and Peng, Y.B. (2008) Restudy of the worm-like carbonaceous compression fossils *Protoarenicola*, *Pararenicola* and *Sinosabellidites* from early Neoproterozoic successions in North China. Palaeogeogr. Palaeoclimatol. Palaeoecol. **258**: 138–161.

Downie, C., Evill, W.R. and Sarjeant, W.A.S. (1963) Dinoflagellates, hystrichospheres, and the classification of the acritarchs. Stanford Univ. Publ. Geol. Sci. **7**: 1–16.

Evitt, W.R. (1963) A discussion and proposals concerning fossil dinoflagelletes, hystrichospheres, and acritarchs. Proc. Natl. Acad. Sci. U.S.A. **49**: 158–164.

Fu, J.H. (1989) Assemblage of huainan biota and its characteristics. Acta Palaeontol. Sin. **28**(5): 642–652.

Gebelein, C.D. (1969) Distribution morphology, and accretion rate of recent sub-tidal algal stromatolites, Bermuda. J. Sediment. Petrol. **39**: 46–69.
Golubic, S. and Focke, J.W. (1978) Phormidium hendersonii Howe: identity and significance of a modern stromatolite building microorganism. J. Sediment. Petrol. **48**: 751–764.
Grotzinger, J.P. and Knoll, A.H. (1999) Stromatolite in Precambrian carbonates: evolutionary mileposts or environmental dipsticks? Annu. Rev. Earth Planet. Sci. **27**: 313–358.
Hofmann, H.J. (1976) Precambrian microflora, Belcher Islands, Canada: significance and systematics. J. Paleontol. **50**: 1040–1073.
Knoll, A.H. (1982) Microfossils from the late Precambrian Draken Conglomerate, Ny Friesland, Spitsbergen. J. Paleontol. **56**: 755–790.
Knoll, A.H. and Golubic, S. (1979) Anatomy and taphonomy of a Precambrian algal stromatolite. Precambrian Res. **10**: 115–151.
Komar, V.A., Rasben, M.E. and Semikhatov, M.A. (1965) Conophyton in the Riphean of the USSR and their stratigraphy importance. Acad. Sci. USSR Geol. Inst. Trudy **131**: 1–72.
Lee, S.J. and Golubic, S. (1999) Microfossil populatons in the context of synsedimentary micrite deposition and acicular carbonate precipitation: Mesoproterozoic Gaoyuzhuang Formation, China. Precambrian Res. **96**: 183–208.
Lee, S.J., Kathleen, M.B. and Golubic, S. (2000) On stromatolite lamination, In: R.E. Riding and S.M. Awramik (eds.) *Microbial Sediments*. Springer-Verlag, Heidelberg, pp. 16–24.
Monty, C.L.V. (1967) Distribution and structure of recent stromatolitic algal mats, eastern andros Island, Bahamas. Soc. Géol. Belg. Ann. **90**: 55–100.
Monty, C.L.V. (1976) The origin and development of cryptalgal fabrics, In: M.R. Walter (ed.) *Stromatolites. Development in Sedimentology 20*. Elsevier, Amsterdam, pp. 193–249.
Nyberg, A.V. and Schopf, W. (1984) Microfossil in stromatolitic cherts from the Upper Proterozoic Minyar Formation, southern Ural mountains. Journal of Paleotology **58**: 738–772.
Pierson, B.K., Bauld, J., Castenholz, R.W., Damelio, E., Marais, D.J., Farmer, J.D. and Grotzinger, J.P. (1992) Modern mat-building microbial communities: a key to the interpretation of Protozoic stromatolitic communities, In: J.W. Schopf and C. Kllein (eds.) *The Proterozoic Biosphere, A Multidisciplinary Study*. Cambridge University Press, New York, NY, pp. 247–342.
Playford, P.E. and Cockbain, A.E. (1976) Modern algal stromatolites at Hamelin Pool, a hypersaline barred basin in Shark Bay, Western Australia, In: M.R. Walter (ed.) *Stromatolites. Developments in Sedimentology 20*. Elsevier, Amsterdam, pp. 389–411.
Riding, R. (2000) Microbial carbonates: the geologycal record of calcified bacterial – algal mats and biofilms. Sedimentology **47**(Suppl. 1): 179–214.
Schopf, J.W. (1968) Microflora of the Bitter Springs Formation, Late Precambrian, central Australia. J. Paleontol. **42**: 651–688.
Semikhatov, M.A., Gebelein, C.D., Cloud, P., Awramik, S.M. and Benmore, W.C. (1979) Stromatolite morphogenesis – progress and problems. Can. J. Earth Sci. **16**: 992–1015.
Turner, R.E. (1984) Acritarchs from the type area of the Ordovician Caradoc Series, Shropshire, England. Palaeontographica Abt. B **190**: 87–187.
Walter, M.R., Bauld, J. and Brock, T.D. (1976) Microbiology and morphogenesis of columnar stromatolites (Conophyton, Vacerrilla) from the hot springs in Yellowstone National Park, In: M.R. Walter (ed.) *Stromatolites. Developments in Sedimentology 20*. Elsevier, Amsterdam, pp. 273–310.
Walter, M.R. (1976) Stromatolites, Developments in sedimentology, 20(Monograph), Amsterdam; Elsevier, pp. 1–790.
Yin, L.M. (1990) Coccoid microfossils from the Niyan Formation (Upper Proterozoic) at Qingtongshan of Suxian, Anhui and their significance. Palaeontol. Cathayana **5**: 277–294.
Zhang, Y. (1981) Proterozoic stromatolitic microfloras of the Gaoyuzhuang Formation (Early Sinian; Riphean), Hebei, China. J. Paleontol. **55**(3): 485–506.
Zhang, Z.Y. (1986) Solar cyclicity in the Precambrian microfossil record. Palaeontology **29**: 101–111.
Zhang, Y. and Hofmann, L. (1992) Blue-green algal mats of the salinas in San-Ya, hai-nan Island (China): structure, taxonomic composition, and implications for the interpretation of Precambrian stromatolites. Precambrian Res. **56**: 275–290.

Biodata of **Professor Vinod Chandra Tewari** and **Dr. (Mrs) Purnima Srivastava**, authors of ***"Morphological Changes in Microscopic–Megascopic Life and Stromatolites Recorded During Late Palaeoproterozoic–Neoproterozoic Transiton: The Vindhyan Supergroup, India"***

Professor Vinod C. Tewari is currently the Head of the Sedimentology Group at Wadia Institute of Himalayan Geology, Dehradun, and a Senior Associate of International Centre for Theoretical Physics, Trieste, Italy. He obtained his Ph.D. from the University of Lucknow in *Geology* in 1986 and continued his research in Wadia Institute. Dr. Tewari taught Geology at Kumaon Univerisity, Nainital, Uttarakhand, India, as Professor of Geology. Professor Tewari's scientific interests are in the areas of Precambrian stromatolites, sedimentation, carbon isotope chemostratigraphy, genesis, early evolution and diversification of life and its astrobiological significance. He is associated with the International Geological Correlation Programme (IGCP) Project 493 on *The Rise and Fall of Vendian Biota*. He has 80 research papers published to his credit and edited several volumes of Himalayan Geology, India, and *Journal of Nepal Geological Society*, Kathmandu, Nepal. Professor Tewari has organized first *Indo-Soviet Symposium on Stromatolites and Stromatolitic Deposits* and other IGCP meetings in India related to stromatolites, phosphorites, and so on. He participated in the Indian and International field conferences on Vindhyans in the year 1999 and 2002. He has been one of the organizers of the *World Summit on Ancient Microscopic Fossils* held in University of California, Los Angeles, USA, in 2008.

E-mail: **vtewari@wihg.res.in**

V.C. Tewari and J. Seckbach (eds.), *STROMATOLITES: Interaction of Microbes with Sediments*, Cellular Origin, Life in Extreme Habitats and Astrobiology 18, 87–114
DOI 10.1007/978-94-007-0397-1_5,

Dr. Purnima Srivastava obtained her Ph.D. in Geology from University of Lucknow, India, in 1991. Presently she is working as Guest Faculty in Centre of advanced Study in Geology. She has 22 years of research experience in various aspects of Proterozoic palaeobiology especially taphonomy, evolutionary palaeobiology biodiversity and palaeoecology of microfossils and megafossils in Indian sections. She discovered number of fossil horizons in Vindhyans and Deoban Formation of the Lesser Himalaya. She has published more than 25 research papers in national and international journals. She has participated in many international conferences including the *World Summit on Ancient Microscopic Fossils* organized by Prof. J.W. Schopf in University of California, Los Angeles, USA in 2008.

E-mail: **purnimasrivastava_51@rediffmail.com**

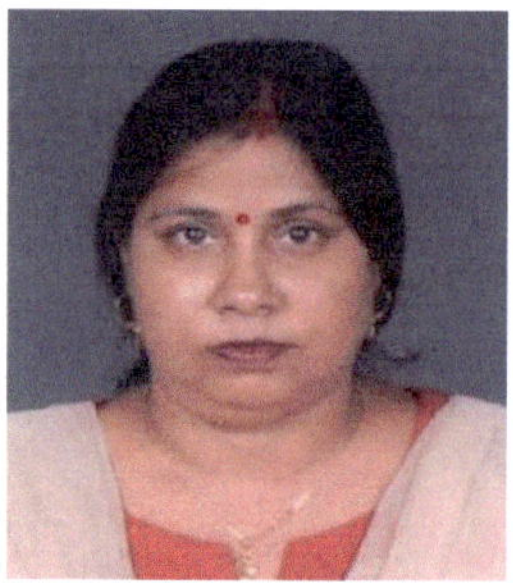

MORPHOLOGICAL CHANGES IN MICROSCOPIC–MEGASCOPIC LIFE AND STROMATOLITES RECORDED DURING LATE PALAEOPROTEROZOIC–NEOPROTEROZOIC TRANSITION: THE VINDHYAN SUPERGROUP, INDIA

PURNIMA SRIVASTAVA[1] AND VINOD CHANDRA TEWARI[2]
[1]*Centre of Advanced Study in Geology, University of Lucknow, Lucknow 226007, Uttar Pradesh, India*
[2]*Wadia Institute of Himalayan Geology, Dehradun 248001, Uttarakhand, India*

Abstract The Palaeoproterozoic to Neoproterozoic era (1,700–542 Ma) is a significant time of transition, reflecting major biotic events in the evolution of life at global level. This time has been marked by the domination of cyanobacterial prokaryotic community, emergence of eukaryotes and their subsequent radiation, transition from microscopic to megascopic life, emergence of metazoan and metaphytes (evolution from plant to animal clades) and evolution from unicellular to multicellular organization. Recent palaeobiological studies in the Vindhyan Supergroup, India, provide substantial data to establish evolutionary history of the Proterozoic life. The available fossil record throws light on biological/morphological changes at microscopic to megascopic level in different time frames in the Vindhyan Supergroup. Palaeoproterozoic to Mesoproterozoic fossil record of the Lower Vindhyans is marked by the diversity and domination of benthic cyanobacterial communities such as *Eoentophysalis* and *Siphonophycus* (this is also valid to some extent for rarely well-preserved Neoarchean microfossil assemblages; Kazmierczak and Altermann, Science 298:2351, 2002). In Semri Group, Lower Vindhyans, *Glenobotrydion*, *Glaeodiniopsis*, *Eosynechococcus and Sphaerophycus* are the other commonly occurring benthic coccoid forms. On the contrary, the Neoproterozoic life in the Bhander Group, Upper Vindhyans, is dominated by the planktic communities, like *Myxococcoides* and acritarchs of variable morphologies and dimensions. Presence of akinites/*Archaeoellipsoides* is very common in Palaeoproterozoic–Mesoproterozoic Vindhyan microfossil assemblages, which are totally absent in Neoproterozoic assemblages. Diverse cellular filaments of cyanobacterial affinity and domination of planktic coccoidal form genera are the common features of Lower Vindhyans. Rapid precipitation instantly entombed fragile trichomes and preserving them as organic-walled fossils, which were silicified prior to the neomorphic alteration of host carbonates. Intermediate carbonaceous fossil forms, exhibiting super-imposed size range of microscopic and megascopic fossils and inferred as the missing link between the evolution from microscopic to megascopic life, have already been recorded (Srivastava, Curr. Sci. 86:644–646, 2004) from the Rewa Group, Upper Vindhyans (Mesoproterozoic age).

Neoproterozoic (Upper Vindhyan) microfossil assemblages are marked by the presence of highly diversified eukaryotes, presence of very small-sized, helically coiled filamentous cyanobacteria *Obruchevella*, a possible *Volvox* colony, emergence of large-sized acanthomorphs and exceptionally large-sized dividing cell-like unit and many other unidentified complex morphologies. The most important is the recently reported, process-bearing age-marker acanthomorph (of Cryogenian 850–630 Ma), *Trachyhystrichosphaera* from the Sirbu Shale Formation, Bhander Group, of the Vindhyan Supergroup, Rajasthan (Srivastava, J. Earth Syst. Sci. 118:575–582, 2009).

Among carbonaceous megafossils, branched filaments at meso- to megascopic level, ensheathed *Chuaria*, *Chuaria* with a cluster of small spheroids in centre, with excystment structures, association of *Chuaria* and *Tawuia* (Kumar, Precambarian Res. 106:187–211, 2001), dichotomously branched megascopic alga and possibly a pre-metazoan organism are the morphologies of Neoproterozoic Vindhyan assemblages. Presence of Ediacaran fossils, algal mat textures and *Arumberia* (an organosedimentary structure) are the forms already reported from the Bhander Group, suggesting an Ediacaran (630–542 Ma) age for it. Stromatolites are also extensively developed and well preserved in the Vindhyan Supergroup (Valdiya, J. Geol. Soc. India 10:1–25, 1969; Valdiya, Himalayan Geol. 13:181–214, 1989; Kumar, J. Palaeontol. Soc. India 19:24–27, 1976a; Kumar, J. Palaeontol. Soc. India 18:13–21, 1976b; Kumar, J. Palaeontol. Soc. India 23–24:166–184, 1980; Kumar, 1984; Tewari, Himalayan Geol. 13:143–180, 1989; Tewari, J. Palaeontol. Soc. India 48:155–165, 2003a; Tewari, Gondwana Geol. Mag. 7:383–394, 2003b; Raaben and Tewari, Izvestia Acad. Nauk. CCCP Ser. Geol. 7:17–25, 1987). They have been used successfully in the intra- and interbasinal correlation. The Semri Group is characterized by the dominance of *Cyathotes* coniform stromatolites and *Conophyton*, while the Upper Vindhyan shows actively branched stromatolites *Baicalia* sp. and complete absence of coniform stromatolite (Misra and Kumar, 2005; Misra, Stromatolite biostratigraphy of the Vindhyan Basin, Unpublished Ph.D. Thesis, 2004; Tewari, Gondwana Geol. Mag. 7:383–394, 2003b). Microscopic as well as megascopic fossils, inferred as eukaryotes (acanthomorphs, multicellular algae and Ediacaran fossils), exhibit maximum diversity in the Bhander Group, the Uppermost Vindhyans. Thus, Vindhyan fossil record indicates gradual evolution and morphological complexity from Palaeoproterozoic to Neoproterozoic time. There is no doubt that palaeobiological, sedimentological and palaeoenvironmental changes collectively prepared the ground for the next major shift in the biosphere near the end of the Proterozoic Eon, i.e. "the Cambrian Explosion" (Knoll and Sergeev, Neues Jahrb. Geol. Palaontol. 195:289–302, 1995).

Keywords Stromatolites • Palaeoproterozoic • Neoproterozoic • Vindhyan Supergroup • Cyanobacteria • Prokaryotic • Metazoan • Metaphyte • Multicellular • Organic-walled microfossils • Acritarchs • Riphean • Ediacaran • Palaeobiology • Chuaria • Tawuia • Grypania • *Obruchevella* • Semri Group • Bhander Group

1. Introduction

The Palaeoproterozoic–Neoproterozoic time (1,700–542 Ma) is a significant era from the evolutionary palaeobiological point of view. It reflects the major biological changes in microscopic as well as megascopic life. On the basis of palaeobiological evidences, it is now well established that the evolution acts continuously and the ecosystems are also changing in time and space. In the investigation of life's history, analysis of changing patterns of diversity has provided insight into macroevolutionary dynamics of evolving clades and ecosystems and into the physical processes on earth that govern the evolutionary change (Schopf and Klein, 1992). Detailed study in the field of evolutionary palaeobiology, mainly during the last two decades, has led to better understanding of the Palaeo-Neoproterozoic biosphere. Globally, morphological/biological diversity among prokaryotic microfossils increases, which may be a result of cumulative increase in the number of sampled formations and a preservational bias. In India, due to scarcity of authentic fossil findings, these biological changes are still inadequately defined. According to Schopf and Klein (1992), true diversity may have been essentially stable over this interval. It has been assured that along with evolution of life, there has been a simultaneous evolution of the atmosphere and the earth's crust also. Detailed studies and comparatively much reliable fossil record have now made us capable to better understand the Proterozoic biotic events and bio-geochemical changes in time and space. The Archaean and the Proterozoic time is often called the age of the cyanobacteria/cyanophyta (Altermann and Kazmierczak, 2003; Altermann et al., 2006). These organisms are held responsible for changing the composition of the atmosphere as well as the color of the earth's sediments. They were the dominating microbial community on earth, until the eukaryotes replaced them tentatively (?) during Mesoproterozoic time. Thus, cyanobacteria can be credited of making the earth prepared for further diversification and stabilization of life. On the basis of data available in India and in other parts of the world, it has been interpreted that near the Mesoproterozoic–Neoproterozoic transition, planktonic eukaryotes exhibited patterns of increasing diversity, which appears to be independent of the number of sampled fossiliferous formations and a preservational bias (Schopf and Klein, 1992). The morphological diversity and turnover rates of acritarhs-producing metaphytes and metazoans increased significantly (Knoll, 1994). There is a major long-term decline in the diversity of planktonic eukaryotes through the late Proterozoic at about 900–800 Ma and evidently reaching minimum in the Vendian/Ediacaran (630–542 Ma). The explosive diversification of metazoans during Proterozoic–Phanerozoic transition appears to have been a heterogeneous event, occurring in many discrete phases (Schopf and Klein, 1992).

The Vindhyan Supergroup, India (1,700–550 Ma in age), has been globally acknowledged as one of the best repositories to mark the evidences of biological changes at microscopic to megascopic level. It is a huge sedimentary basin, well studied for biosedimentological aspects. However, a number of problems, especially those related with Vindhyan's age, sequential representation of evolution of

Proterozoic life or transition from microscopic to megascopic life, acritarchs diversification, time of emergence of metazoans, taxonomical, taphonomical and environmental consequences and palaeoecological details during Palaeoproterozoic–Neoproterozoic times, are still in a state of infancy and require more efforts and data for better understanding of these aspects. This chapter deals with the morphological changes noticed among fossils and stromatolite assemblages recorded by other workers and author's own observations.

2. The Vindhyan Supergroup

The Vindhyan Supergroup is about 4,000-m-thick, least metamorphosed sedimentary succession distributed in Central India in a sickle-shaped basin around Bundelkhand Massif, spreading from Agra in northwest through southeastern Rajasthan, to eastward in Son Valley up to Sasaram in Bihar to Hoshangabad in south (Fig. 1). It occupies an area of about 104,000 km^2. The exposures occur in patches, forming elevated hillocks and extended ridges on flat terrains in parts of Madhya Pradesh, Uttar Pradesh, Bihar and Rajasthan States. The supergroup is subdivided into four groups, viz. the Semri Group, the Kaimur Group,

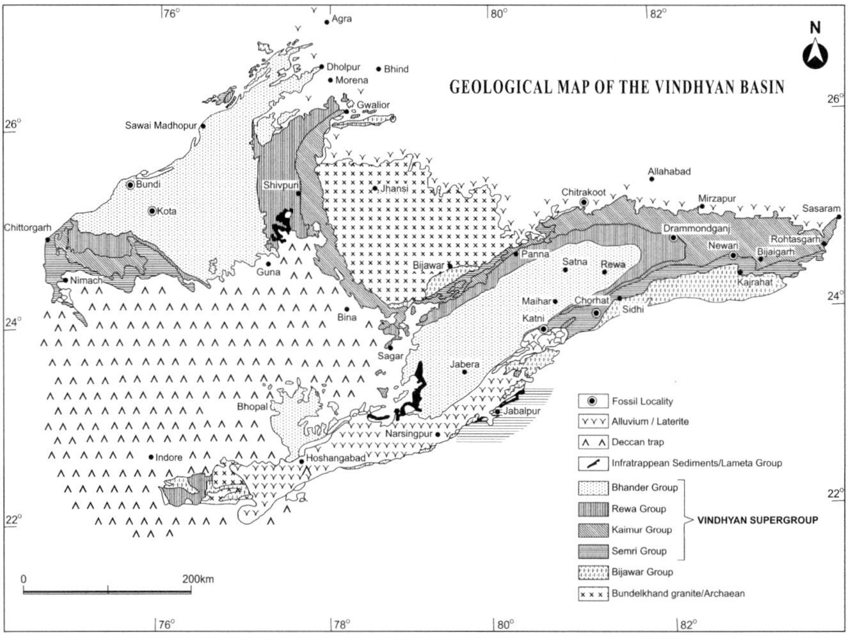

Figure 1. Geological map of the Vindhyan Basin, simplified after Soni et al. (1987).

Table 1. Generalized lithostratigraphy of the Vindhyan Supergroup, India.

Group	Formations
Bhander Group	Dholpura Shale
	Balwan Limestone
	Maihar Sandstone
	Sirbu Shale
	Bundi Hill Sandstone
	Samaria Shale
	Lakheri Limestone/Bhander Limestone
	Ganurgarh Shale
Rewa Group	Upper Rewa Sandstone
	Jhiri Shale
	Lower Rewa Sandstone
	Panna Shale
Kaimur Group	Dhandraul Quartzite
	Scarp Sandstone and Conglomerate
	Bijaigarh Shale
	Susnai Breccia
	Upper Quartzite
	Lower Quartzite
Unconformity	
Semri Group	Rohtas Formation
	Kheinjua Formation/Chorhat Formation
	Porcellanite Formation
	Kajrahat Limestone
	Basal Formation
Bijawar Group	Phyllites

the Rewa Group, and the Bhander Group (Auden, 1938; Soni et al., 1987; Sastry and Moitra, 1984; Prasad, 1984). Each group is further divided into Formations and Members (Table 1). Traditionally the Semri Group is considered as Lower Vindhyans, whereas the other three groups are categorized under the Upper Vindhyans. The overall lithology is represented by the calcareous, argillaceous and arenaceous sedimentary rocks (Fig. 2). The supergroup unconformably overlies the Bundelkhand Granite and metasediments of the Bijawar Group ~2,500 Ma in age (Crawford and Compston, 1970; Mandal et al., 2002).

2.1. AGE OF THE VINDHYAN SUPERGROUP

Age of the Vindhyan Supergroup is still a matter of debate. Conventionally, it is considered to be Palaeo-Neoproterozoic. Record of 1.1-billion-year-old triploblastic animal traces (Seilacher et al., 1998) from the Churhat Sandstone, Lower Vindhyans, and Small Shelly Fauna (earliest Cambrian in age) from the Rohtas

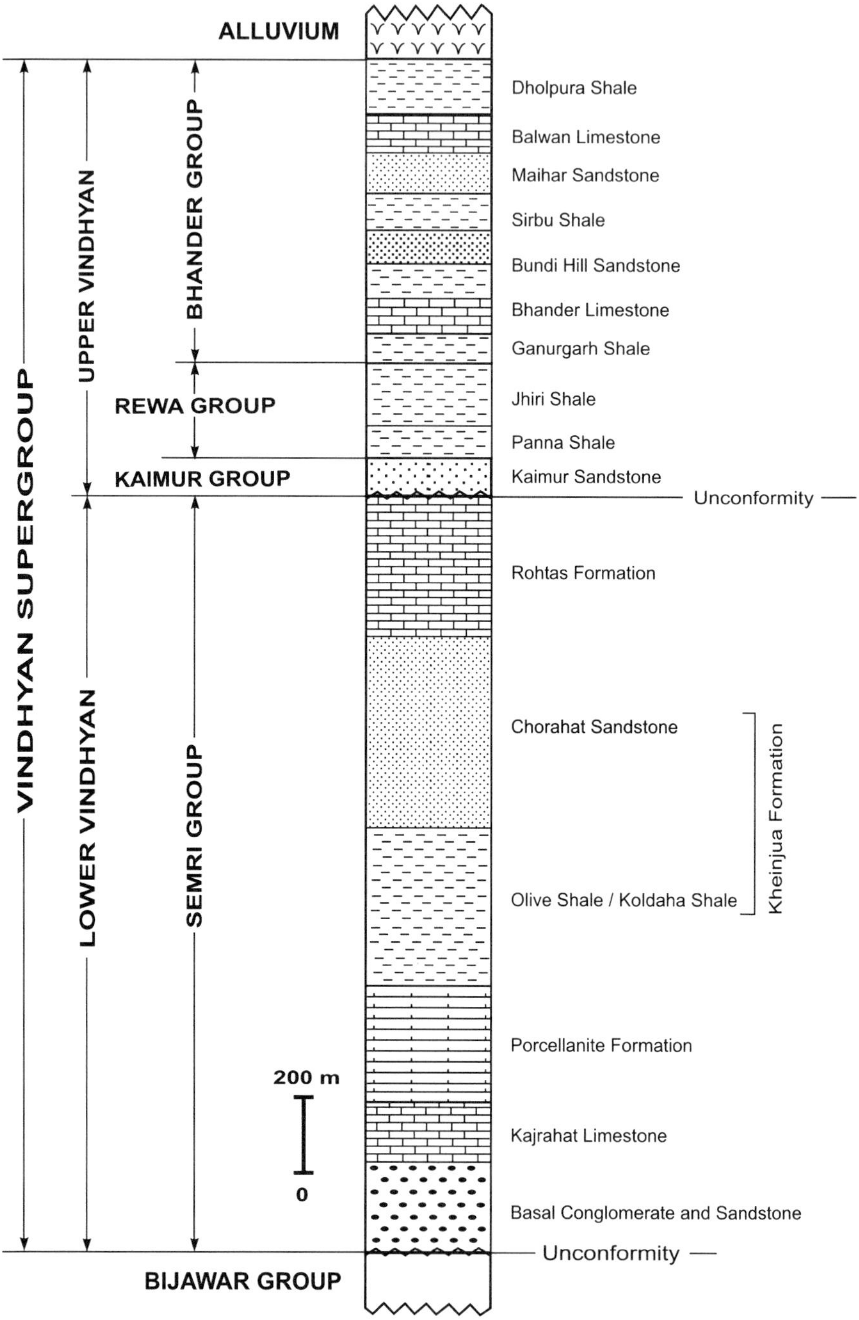

Figure 2. Generalized lithostratigraphy of the Vindhyan Basin.

Formation, Semri Group (Azmi, 1998), suggested a far younger age than the traditional age to the Vindhyan Supergroup (Azmi et al., 2007). These findings created a hot debatable issue of abiotic origin (Brasier, 1999; Kerr, 2002; Morris et al., 1998; Hofmann, 2005). The age, biogenicity and reproducibility of these fossils remained inconclusive to some extent, thus generated interest among geochronologists to resolve the age especially the upper age and time for deep metazoan origin in the Vindhyan Supergroup. Recently, Bengtson et al. (2009) confirmed the reproducibility of small shelly fauna and preferred to bring the evolution far behind rather than changing the conventional age of the Vindhyan Supergroup. However, in view of strong geochronological evidence of Palaeoproterozoic age of the Lower Vindhyan, we need to consider the mounting indication that the Palaeoproterozoic biota was more diversified than is generally assumed. However, the Indian scientific community is still reluctant to accept the authenticity of Azmi's fossils. A number of reports for the age of the Vindhyan Supergroup, based on different methodologies and parameters, are now available (Ray, 2006), which can be summarized as follows:

Maihar Sandstone	Ediacara fauna	630–542 Ma	De (2003, 2006)
Bundi Hill Sandstone	Ediacara fauna	630–542 Ma	Srivastava (2005, 2008a)
Bhander Group	Detrital zircon dates	~1,028 Ma	Malone et al. (2008)
Sirbu Shale, Bhander Group	*Trachyhystrichosphaera*	Cryogenian (850–630 Ma)	Srivastava (2009)
Dholpura Shale, Bhander Group	*Chuaria–Tawuia*	Upper Riphean	Srivastava (2002)
Lakheri Limestone (Rajasthan)	Sr/Sr	~650 Ma	Ray et al. (2003)
Bhander Limestone	*Chuaria–Tawuia*	1,100–700 Ma	Kumar and Srivastava (1997)
Jhiri Shale, Rewa Group	*Chuaria–Tawuia*	1,100–700 Ma	Rai et al. (1997)
Kaimur Group, Maihar	Rb/Sr	>1,067 Ma	Kumar et al. (2001)
Rohtasgarh Limestone	Pb/Pb isochron	~1,601 ± 130 Ma	Ray et al. (2003)
Rampur Shale	U/Pb, Zr (TIMS) (SHRIMP) dates	~1,599 ± 8 Ma ~1,628 ± 8 Ma	Rasmussen et al. (2002) and Ray et al. (2002)
Glauconite (Chitrakut, Semri Group)	Rb/Sr	~1,504–1,409 Ma	Kumar et al. (2001)
Kajrahat Limestone	Pb/Pb isochron	~1,721 ± 90 Ma	Sarangi et al. (2004)
Basement Rocks	Pb/Pb zircon (SIMS)	~2,492 ± 10 Ma	Mandal et al. (2002)

3. Stromatolite Occurrences in the Vindhyan Supergroup

The Vindhyan Supergroup shows excellent development of stromatolites (Figs. 3–6), which have been used quite successfully in the intrabasinal and interbasinal correlations (Valdiya, 1969, 1989; Kumar, 1976a, b, 1978, 1980; Raaben and Tewari, 1987;

Figure 3. Longitudinal section of the *Conophyton cylindricus*, Bhagwanpura Limestone, Chittorgarh, Rajasthan (Tewari, 1989).

Figure 4. Microphotograph of the axial zone of *Conophyton cylindricus* from the Bhagwanpura Limestone, Rajasthan (Tewari, 1989).

Figure 5. Field outcrop of *Ephyaltes myriocranus*, transverse section, Fawn Limestone, Lower Vindyan Salkhn, Uttar Pradesh (Tewari, 2003b).

Tewari, 1989, 2003a, b; Raha and Das, 1989). There are number of stromatolite-bearing horizons both in Lower and Upper Vindhyans, viz. upper portion of Kajrahat Limestone Formation, where alternate cycles of *Jcutophyton–Thyssagates* and *Calypso* have been recorded. Mathur (1981) reported *Cryptozoon*-like stromatolites from the Fawn Limestone of the Kheinjua Formation. Development of stromatolites at Salkhan, Pataudh and Bargawan in Son Valley, Central India, has also been reported by Misra and Kumar (2005) and Misra (2004), which can be represented by *Siren* at Newari, *Misstissania* at Pataudh and *Ephyaltes* at Salkhan Hill. Kumar (1976a, b, 1980) reported development of collophane in association with stromatolites from the Tirohan Limestone in Chitrakut area, Uttar Pradesh. In Chambal Valley, the Bhagwanpura Limestone of Semri Group also exhibits the development of stromatolites (Rao et al., 1977). The Semri Group is characterized by the dominance of coniform stromatolites. In all, there are 11 types of stromatolites reported from the Semri Group, which are *Kussiella kussiensis*, *Colonnella columnaris*, *C. kajrahatensis*, *Patella* sp., *Khutesaria misreae*, *Ephyaltes myriocranus*, *Siren pylodes*, *Calypso moneres*, *Thyssagates odontophytes*, *Cyathotes phorbadicia* and *Misstassania wabassinon* (Misra, 2004). The Bhander Group shows complete absence of coniform stromatolites. Columnar stromatolites are profusely developed in Bhander Limestone, exposed near Sajjanpur Village on Satna-Rewa Road. In Chambal Valley, stromatolites are reported from the Samria Shale, the Sirbu Shale and the Balwan Limestone Formations. There are only five types of stromatolites reported from the Bhander Group, the Uppermost Vindhyans, which show active branching. These are *Baicalia baicalica*, *B. burra*, *Patomia ossica*, *Cryptozoon* sp. and *Maiharia maiharensis* (Kumar, 2009, personal communication).

Figure 6. Field outcrop (*above*) and polished slab (*below*) of the new form *Maiharia maiharensis* from Maihar area Kumar (1976a, b).

4. The Fossil Assemblage from the Vindhyan Supergroup

The Vindhyan Supergroup (1,700–550 Ma in age) has been globally acknowledged worldwide as one of the best repositories to mark the evidences of biological changes at microscopic to megascopic level (Fig. 7). A number of reports comprising direct evidences of life (in form of microfossils, megafossils, stromatolites) and indirect evidences such as trace fossils and some geochemical evidences are also available.

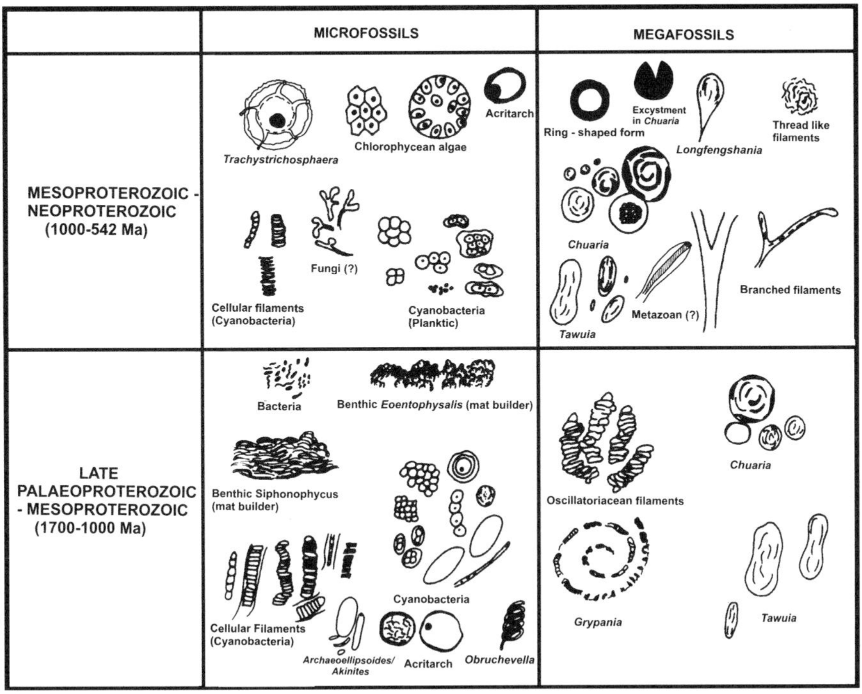

Figure 7. Morphological changes in Palaeoproterozoic–Neoproterozoic microfossils and megafossils of the Vindhyan Supergroup.

As far as affinity of these fossils is concerned, bacterial, cyanobacterial, algal, fungal, acritarchean and metazoan forms have been identified and published simultaneously. All these evidences have served significantly in biostratigraphic correlation. The most significant biological changes through time (Fig. 7) observed in the Lower–Upper Vindhyan Supergroup are as follows:

(a) Domination of benthic mat-building cyanobacterial prokaryotic community (Fig. 8b, e, i), followed by domination of planktic, clast-bound communities (organisms occurring within clasts) in preceding era.
(b) Presence of akinites (Fig. 9a, c–e) in Lower Vindhyan and absolute absence in Upper Vindhyans. These fossils set a minimum date for the evolution of derived cyanobacteria, capable of marked cell differentiation, and they corroborate geochemical evidence, indicating that atmospheric oxygen level was above 1% of present-day level during Mesoproterozoic time. In the presence of oxygen, a protected environment for nitrogenase is produced by these akinites, which were abundant in coastal communities (Srivastava, 2005). Their presence also indicates atmospheric evolution (see Srivastava, 2005).

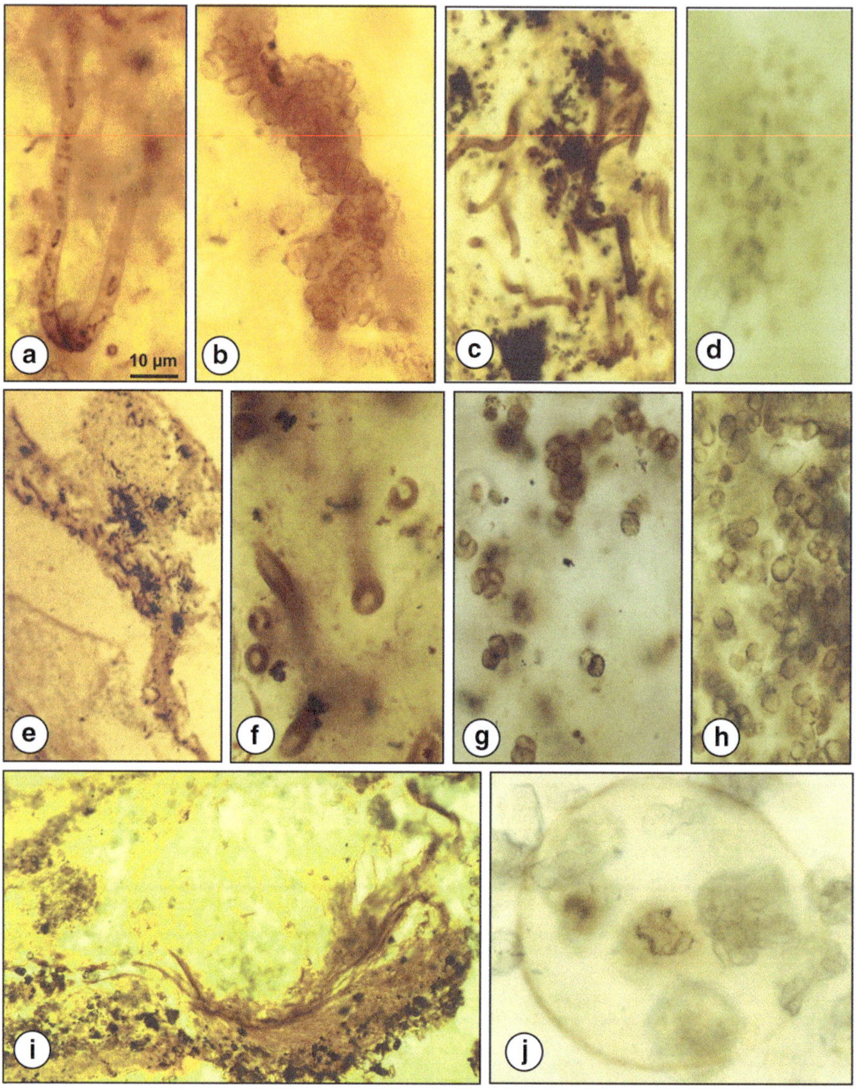

Figure 8. Photomicrographs of microfossils in petrographic thin sections of chert under microscope. Co-ordinates (in mm) and slide numbers are given. Scale bar given is 10 µm for **a**, **b**, **g**, 20 µm for **c**, **f**, **h**, **j** and 50 µm for **e** and **i**. (**a**) Cellular filament, identified as *Talakania uluksanella*, slide no. N/93/3, co-ordinates 35.6/69.4. (**b**) *Eoentophysalis belcherensis*, slide no. N/93/3, co-ordinates 16.2/7.5. (**c**) Tubular filaments of *Siphonophycus robustum,* slide no. Bn-a, co-ordinates 35.7/70.2. (**d**) *Eosynechocuus isolatus,* slide no. N/93/6, 16.1/ 66.3. (**e**) Filamentous mat builders, slide no. Bn-a, co-ordinates 34.2/70.5. (**f**) Cross sections of filaments, slide no. Bn-a, co-ordinates 36.3/70.5. (**g**) *Sphaerophycus* and *Tetraphycus*, slide no. N4-C, co-ordinates 14.4/76.2. (**h**) *Myxococcoides minor*, slide no. N4-C, co-ordinates 15.5/ 80.1. (**i**) Mat-building filamentous community, slide no. N4-C, co-ordinates 9.9/65.3. (**j**) Large Sphaeromorph/Leiosphaeroid, slide no. N4-C, co-ordinates10.2/73.2.

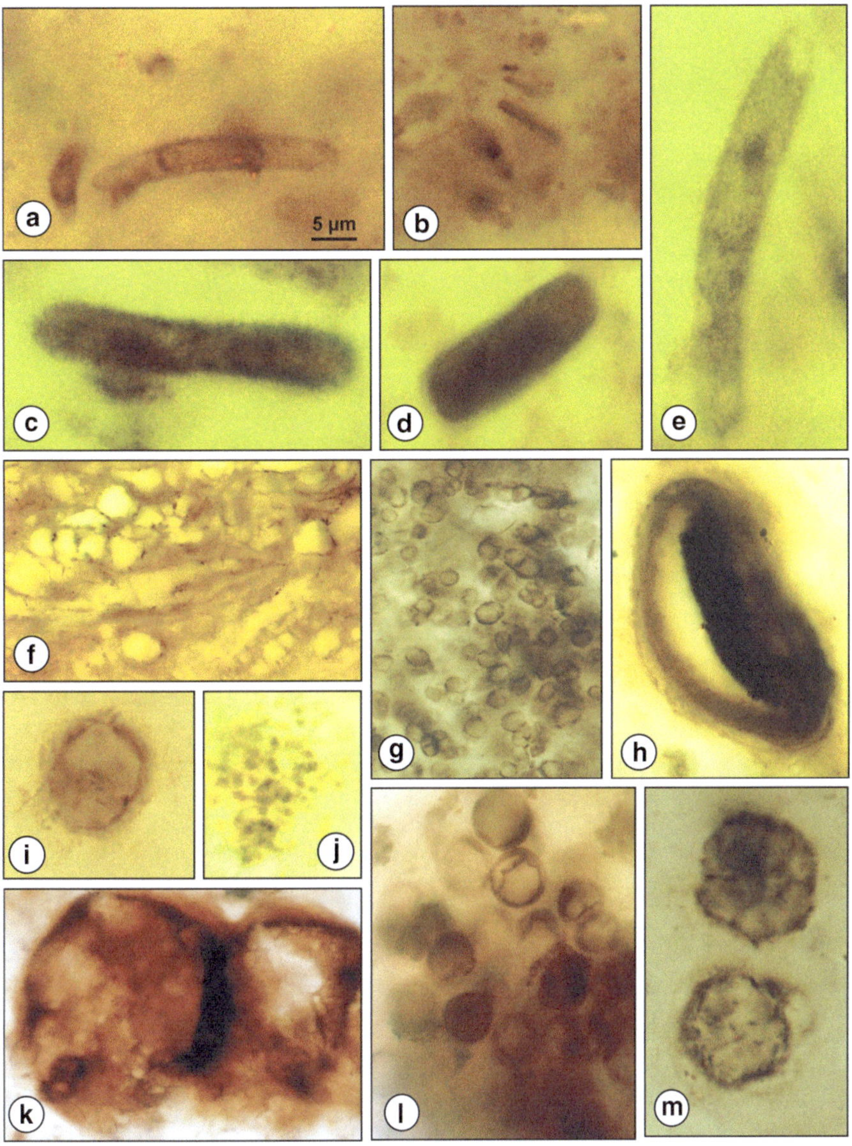

Figure 9. Photomicrographs of microfossils in petrographic thin sections of chert under microscope. Co-ordinates (in mm) and slide numbers are given. Scale bar given in **a** represents 5 μm for **a**–**e**, **i**, **j**, **m**, 20 μm for **f**–**h**, **l** and 50 μm for **h**, **k**. (**a**) Small-sized akinites, slide no. N-IV, co-ordinates 9.0/74. (**b**) Possibly bacterial forms <1 μm, slide no. N/93/8, co-ordinates 35.6/69.4. (**c**) *Archaeoellipsoides*, slide no. N-IV, co-ordinates 13.6/73.4. (**d**) Akinites, slide no. N-IV, co-ordinates 13.2/72.8. (**e**) Akinite with terminal opening-like structure, slide no. N/93/3, co-ordinates 7.7/60.9. (**f**) Mixed population of filaments and coccoids, slide no. N-4-C, co-ordinates 8.7/68.3. (**g**) Isolated and attached coccoids, slide no. N-4-C, co-ordinates 15.5/80.1. (**h**) Double-walled, large acritarch with inner body, slide no. N4-C, co-ordinates 13.1/78.0. (**i**, **m**) Small acanthomorphs, slide no. N4-C, co-ordinates 10.8/84.2. (**j**) Bacterial forms, slide no. N/93/6, co-ordinate 12.2/81.4. (**k**) Large dividing cell-like unit, attributable to eukaryote, slide no. Bn-1, co-ordinates 27.0/73.9. (**l**) Cluster of *Gloeodiniopsis, Caryosphaeroides, Melasmatosphaera,* slide no. Bn-3, co-ordinates 20.0/74.8.

(c) Presence of transitional/intermediate life forms between the evolution from microscopic to megascopic life from the Panna Shale, Upper Vindhyans (see Srivastava, 2004). These carbonaceous fossils exhibit overlapped size range of microscopic and megascopic organisms.
(d) Radiation of eukaryotes, represented by ornamented acritarchs, acanthomorphs (Fig. 10a, b), chlorophycean forms and metazoans in the Bhander Group (Srivastava, 2009).
(e) Increase in size of acritarchs, large-sized dividing cell-like units in Uppermost Vindhyans (Fig. 9k).
(f) Emergence of complex and advanced morphologies (Fig. 10c, f–h).
(g) Emergence of metazoans/Ediacara Fauna or evolution of animal clade from the Bundi Hill Sandstone (Fig. 10j, k by Srivastava, 2006) and from the Maihar Sandstone (De, 2003, 2006).

Microfossils reported from the petrographic thin sections of chert and megafossils preserved in shales and sandstones have been considered for the present study (macerated material has not been considered, as possibility of contamination is always there, if due precautions have not been taken into account). Microfossil and megafossil reports from the Semri Group (Tandon and Kumar, 1977; McMenamin et al., 1983; Maithy and Shukla, 1984; Maithy and Babu, 1988a, b; Kumar, 1995; Kumar and Srivastava, 1989, 1991, 1992, 1995; Srivastava, 2006; Srivastava and Bali, 2006; Sharma, 2003, 2006a, b, c; Rai and Singh, 2004; Sergeev et al., 2008) have been considered for the present study. In Upper Vindhyans, reports from the Jhiri and Panna Shale of the Rewa Group (Rai et al., 1997; Srivastava, 2004), from the Sirbu Shale (Kumar and Srivastava, 2003; Kumar and Pandey, 2008a; Srivastava, 2009), from the Samaria Shale and Bhander Limestone (Srivastava, 2008), from the Bundi Hill Sandstone (Srivastava, 2006, 2008), from the Balwan Limestone, from the Maihar Sandstone (Kumar and Pandey, 2008b) and from the Dholpura Shale, the Uppermost Vindhyans (Srivastava, 2002) have been considered.

4.1. LIFE DURING PALAEOPROTEROZOIC–MESOPROTEROZOIC TIME (1,700–1,000 MA)

The fossil record of this period is represented by the domination of cyanobacterial communities at microscopic level. Evolution of life from microscopic to megascopic level and development of nucleus in cells are the landmark evolutionary events of Palaeoproterozoic–Mesoproterozoic time. There is also possibility that these events might have happened much earlier but a good fossil record is lacking. On the basis of available records of micro-, meso- and megafossils of the Vindhyan Supergroup, these biotic events seem to have occurred by late Palaeoproterozoic to Mesoproterozoic and later on radiate extensively during Neoproterozoic. Palaeobiological evidences and inferences of morphological changes among microfossils and megafossils in the Vindhyan Supergroup are

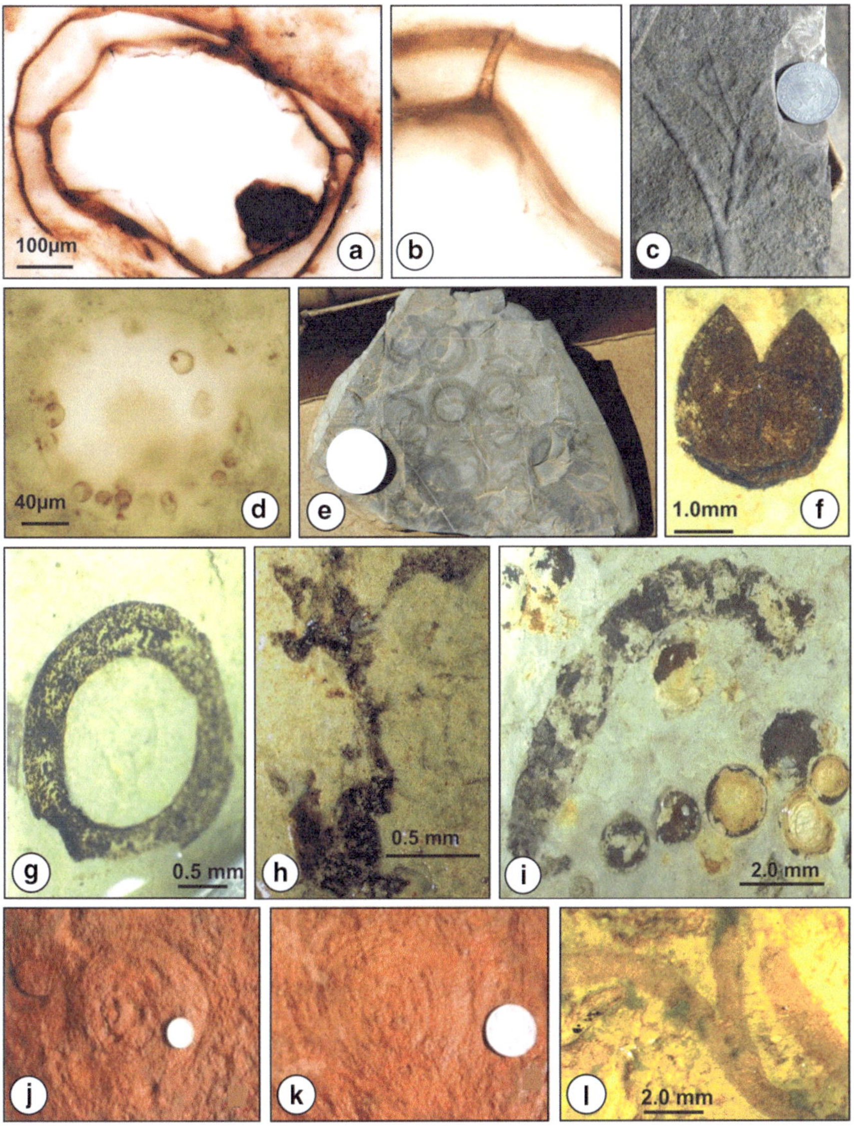

Figure 10. Photomicrographs of microfossils in petrographic thin sections of chert under microscope. Co-ordinates (in mm) and slide numbers are given. Scale bar is given in every photograph. Scale bar given in (**a**) is same for (**b**) also. (**a**, **b**) Process-bearing acritarch (acanthomorph) *Trachyhystrichosphaera,* slide no. Bn-3, co-ordinates 164/71.4. (**c**) Megascopic branching and associated *Grypania*-like structure in Rohtas Formation, Lower Vindhyan. (**d**) *Volvox* colony-like structure, slide no. Bun-14, co-ordinates 9.5/84.0. (**e**) *Grypania* in a cluster, Rohtas Formation, Lower Vindhyans. (**f**) Excystment in *Chuaria*, Sirbu Shale, Bhander Group, Uppermost Vindhyans. (**g**) Ring-shaped carbonaceous megafossil, Sirbu Shale, Upper Vindhyan, may be a hold fast of *Chuaria–Tawuia.* (**h**) A multicellular plant-like carbonaceous megafossil, Uppermost Vindhyans. (**i**) A unique association of *Chuaria–Tawuia,* Bhander, Sirbu Shale, Bhander Group, Upper Vindhyans. (**j**, **k**) Ediacaran fossils, *Aspidella*-like medusoids, Bundi Hill Sandstone, Bhander Group, Upper Vindhyan (for scale, coins are given). (**k**) Branching in carbonaceous megafossil in Sirbu Shale, Bhander Group, Upper Vindhyan.

based on the study of microfossils and megafossils from the Semri and the Rewa Group of the Vindhyan Supergroup. The changes observed are:

- Mat-building prokaryotic cyanobacterial benthic community dominated the Palaeoproterozoic–Mesoproterozoic microfossil assemblages. Mats of densely interwoven cyanobacterial filaments, represented by *Siphonophycus* and *Gunflintia,* are the dominating mat builders (Fig. 8c, e, i).
- In addition to monospecific population, mixed population of filaments and coccoids both belonging to cyanobacterial community also occurs as mat builders (Fig. 9f).
- Billowy monospecific colonies of *Eoentophysalis* (*Entophysalidacean*) occur as mat-building benthic community in most of the microfossil assemblages during the Palaeoproterozoic–Mesoproterozoic times (Fig. 8b).
- *Gloeodiniopsis*, *Globophycus* (Fig. 9l), *Palaeoanacystis*, *Glenobotrydion*, *Eosynechococcus* (Fig. 8d), *Sphaerophycus*, *Tetraphycus* (Fig. 8g) and *Myxococcoides* (Figs. 8h and 9g) are some commonly occurring cyanobacterial genera of the Lower Vindhyan microfossil assemblages.
- Cellular filaments of cyanobacterial affinity (Fig. 8a) exhibit marked diversity in Palaeoproterozoic–Mesoproterozoic, which are represented by *Oscillatoriopsis*, *Palaeolyngbya*, *Cephalophytarion*, *Obconicophycus*, *Caudiculophycus*, *Cyanonema* and *Rhicnonema*, which are the commonly occurring components of Lower Vindhyan microfossil assemblages.
- *Archaeoellipsoides*, is a large sausage-shaped planktic form (Fig. 9a–e) interpreted as an akinite of nostocalean or Rivularian affinity, which has also been described as; detached cyanophycean spores or akinites. The most common akinite forming modern cyanobacteria belong to the family Nostocaceae. *Anabaena*, *Cylindrospermum* and *Aulosira* are the genera very common among them. Akinites are resting non-motile spores, differentiable from the vegetative cells by the absence of cell division, storage of metabolic products and absence of gas vesicles. Once they detach from the vegetative cells, they germinate and form hormogones and new vegetative trichome in favorable conditions (Stulp and Stam, 1984). As a fossil, they are reported globally in Mesoproterozoic assemblages only. It is totally absent in Neoproterozoic microfossil assemblages (Kumar and Srivastava, 1995; Knoll and Sergeev, 1995; Srivastava, 2005).
- Among carbonaceous megafossils, *Chuaria* and *Tawuia* exhibit very simple morphologies.
- *Grypania* is considered as the megascopic cyanobacteria and according to some workers, it is considered to be a eukaryotic alga (Knoll et al., 2006). It is a spirally coiled form, reported from the Rohtas Formation (Katni area, Tandon and Kumar, 1977; Srivastava, 2009), Semri Group, Lower Vindhyans (Fig. 10e), which is considered to be 1,700 Ma old. In other parts of the world, it is recorded from the 1,450-million-year-old rocks (Walter et al., 1990).
- A very unusual megascopic form exhibiting prominent dichotomous branching with a small *Grypania*-like spiral form (Fig. 10c) is also preserved as a mold.

The specimen belongs to the Rohtas Formation, Lower Vindhyans. As far as affinity of this form is concerned, it is inferred that it can be a eukaryotic algal form or a thallophytic megascopic alga (Srivastava, 2008).

- A number of plausible initial and intermediate fossil forms have been observed while studying the metaphytes from the Panna Shale, Rewa Group, Upper Vindhyans. They exhibit the overlapping size range of microfossils and megafossils. These forms have been interpreted as the intermediate forms/mesoforms or a missing link between the evolutions from microscopic to megascopic life (Srivastava, 2004).

4.2. LIFE DURING NEOPROTEROZOIC TIMES (1,000–542 MA)

Life gradually evolved during Neoproterozoic, which is well evidenced by microscopic and megascopic fossil record of the Bhander Group, the Vindhyan Supergroup. Morphologies became more complex and advance in comparison with Palaeo-Mesoproterozoic fossil record, reflecting a picture of more diverse and dynamic ecosystem than that of proceeding era. The following morphological changes are observed during Neoproterozoic Bhander Group, the Vindhyan Supergroup.

- Mat-building benthic colonies of cyanobacterial population occur as minor component (Fig. 8f) in Neoproterozoic microfossil assemblages.
- *Eoentophysalis* occurs mostly as mat dwellers or an allochthonous population even if it occurs in abundance (Fig. 8b).
- Eukaryotes diversified globally during Neoproterozoic, viz. green alga represented by *Volvox* colony-like morphologies (Fig. 10d), fungi, acritarhs (Figs. 8j and 9h), especially the acanthomorphs (Fig. 10a, b). An exceptionally well-preserved acanthomorph of quite large size (~600 μm) identified as *Trachyhystrichosphaera* has recently been reported from the Bhander Group, Rajasthan (Srivastava, 2009). It is considered to be an age-marker microfossil of the Upper Riphean (Fig. 10a, b). It is being reported for the first time from any Indian microfossil assemblage of Proterozoic age.
- Among carbonaceous megafossils, complex morphologies represented by branched filaments (Fig. 10l), internal structures and presence of enveloping sheath around *Chuaria* indicate morphological advancement in life forms during Neoproterozoic times.
- A unique association of *Chuaria* and *Tawuia* in a single specimen (Fig. 10i) indicates their common affinity and both appear to be part of a single plant as inferred by other workers (Kumar, 2001, Kumar & Srivastava, 2003).
- Excystment among *Chuaria* (Fig. 10f) is the feature reported only from the Neoproterozoic Vindhyan fossil assemblages. This particular feature is considered to be a characteristic of eukaryotic affinity (Srivastava, 2008).
- Presence of microbial mats and *Arumberia* (Fig. 11a–d) has also been reported from the Maihar Sandstone Formation of the Bhander Group,

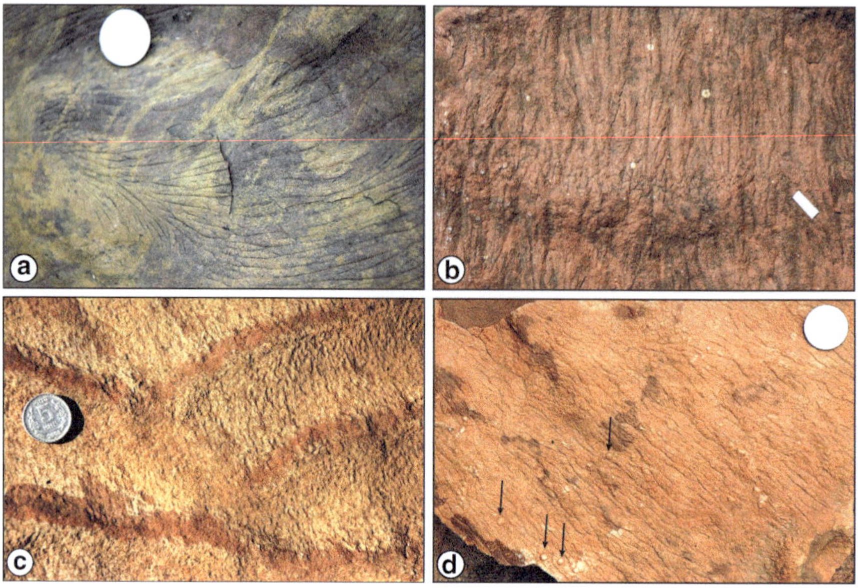

Figure 11. Algal mat textures in Upper Vindhyans (samples of Kumar and Pandey, 2008) exhibiting *Arumberia* and *Beltanelliformis*.

Son Valley (Kumar and Pandey, 2008a). It is suggested that Ediacara fauna is associated with the microbial mats; hence, the presence of microbial mats can be considered as an indication of presence of Ediacara fauna. The association of *Beltanelliformis* (Fig. 11d, marked with arrows) with *Arumberia* (considered as an organosedimentary structure of Ediacaran age) is an additional supporting parameter for assigning the Ediacaran age for fossil-bearing horizon of the Maihar Sandstone, the Bhander Group, Uppermost Vindhyans.

- Ediacaran fossils have been reported from the Maihar Sandstone and Bundi Hill Sandstone formations of the Bhander Group (Fig. 10j–k), exposed in Son Valley and Chambal Valley, respectively (De, 2003, 2006; Srivastava, 2006, 2008). Preservation of these fossils is not of very good quality and the number of specimens for identified genera and species are also very limited. More efforts and search for new Ediacaran fossil-bearing horizons are required to draw some meaningful conclusions.

5. Role of Taphonomy in Morphological Diversity

Taphonomy is the post-mortem changes in an organism after its death, till its recovery as a fossil. This includes decomposition, burial, compaction and other chemical, biological or physical changes or post-mortem information loss among the organisms. Being able to recognize taphonomic processes that have taken place can lead to

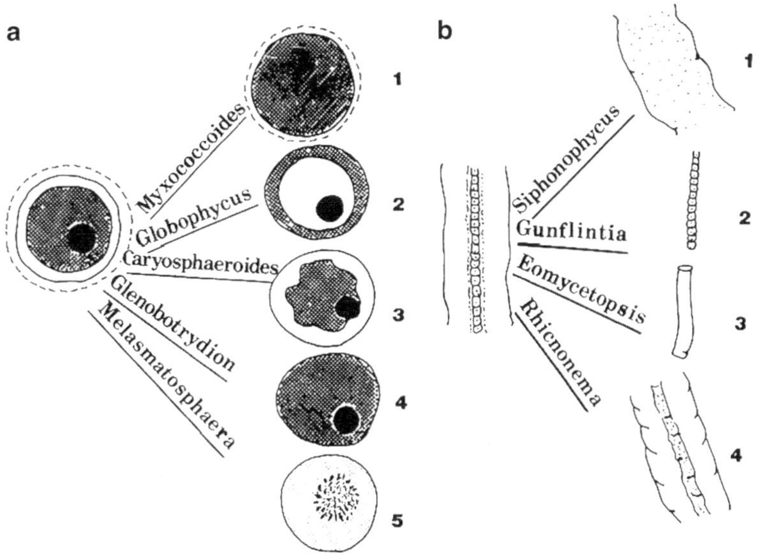

Figure 12. Role of taphonomy in morphological diversity among microfossils, after Kumar and Srivastava (1992).

a better understanding of palaeoenvironments and even life history of the once living organism. Many genera assigned to different taxonomic positions may possibly represent the degradational variants of a common biologic entity (Figs. 12 and 13). This may be true for both coccoids and filamentous forms (Kumar and Srivastava, 1992). *Glenobotrydion*, *Myxococcoides*, *Globophycus*, *Gloeodiniopsis*, *Caryosphaeroides* and *Melasmatosphaera* can be placed among the degradational variants of a common biologic entity (Hofmann, 1976). *Glaeodiniopsis* with intracellular mass can give rise to all above-mentioned form genera. Presence or absence of intracellular mass and sheath may also be governed by the taphonomy. Similarly, a single biologic entity *Eoentophysalis* may give rise to morphologically variant forms assignable to *Sphaerophycus*, *Myxococcoides*, *Palaeoanacystis*, *Eosynechococcus* and *Myxococcoides*. For filamentous forms, *Gunflintia*, *Siphonophycus*, *Oscilatoriopsis* and *Rhicnonema* may represent degradational variants of *Paleoscytonema*-, *Palaeolyngbya*- or *Rhicnonema*-like fossil form (Kumar and Srivastava, 1992).

6. Discussion and Conclusion

- A period between Palaeo-Neoproterozoic transitions reflects some very significant morphological changes which can be justified by micro-, meso- and megascopic life of the Lower and Upper Vindhyan sequences.
- During Palaeoproterozoic–Mesoproterozoic period, domination of mat-building cyanobacterial population can be represented by the exceptionally well-preserved

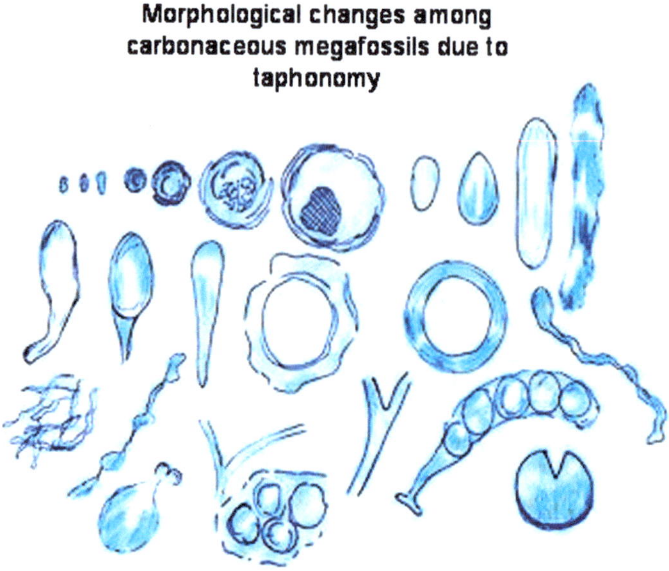

Figure 13. Morphological changes among carbonaceous megafossils due to taphonomy.

microfossil assemblages of the Semri Group. This paleomicrobial community formed the stromatolites in the photic zone due to photosynthesis. At the same time, if we compare the complexity and diversity among eukaryotes, many division stages (Fig. 9k), representing green algal affinity, are reported only from the Bhander Group, Upper Vindhyans during Neoproterozoic. The occurrence of typical Riphean stromatolite taxa from the Lower and the Upper Vindhyans such as *Kussiella kussiensis*, *Colonnella columnaris*, *Conophyton garganicus*, *Conophyton cylindricus*, *Baicalia* sp., *Baicalia satnaensis*, *Tungussia* sp. and *Gymnosolen* sp. is identical to the Southern Urals type area in Russia and shows a distinct morphological variation from columnar to conical and branching forms (Valdiya, 1969, 1989; Kumar, 1976, 1980; Tewari, 1989, 2003a, b; Raaben and Tewari, 1987; Raha and Das, 1989). It is very significant that no stromatolite taxa of younger age (Lower Cambrain) have been recorded from the Vindhyan so far and the existing stromatolite assemblage suggests Palaeoproterozoic to Neoproterozoic age.

- Diversity among cellular filaments of cyanobacterial affinity (Fig. 8a) in Palaeoproterozoic–Mesoproterozoic microfossil assemblages is probably due to rapid silicification prior to neomorphic alteration of host carbonate, which resulted in retention of fragile trichomes in Mesoproterozoic cherts as well as in retention of marked sphericity of *Myxococcoides*, a chroococcalean form genera (Figs. 8h and 9g). The organisms were encased in cement, at or near the sediment–water interface and as a result escaped compression (Knoll and Sergeev, 1995).

- Presence of akinite-forming cyanobacteria or *Archaeoellipsoides* (Fig. 9a, c–e) in Mesoproterozoic and its absence in the Neoproterozoic assemblages can be explained logically that akinite forming cyanobacteria do not thrive in sea water of normal salinity; it is likely that these pools hosted the organisms, which produced the abundant akinites, observed in Mesoproterozoic peritidal assemblages. Such pools support growth of *Cladophora* and other alga, suggesting absence of *Archaeoellipsoides* in Neoproterozoic may be a consequence of eukaryotic radiation (Knoll and Sergeev, 1995). Presence of these fossils sets a minimum date for the evolution of derived cyanobacteria, capable of marked cell differentiation, and corroborates geochemical evidence indicating that atmospheric oxygen level was above 1% of present-day level during Mesoproterozoic time. In the presence of oxygen, a protected environment for nitrogenase is produced by these akinites, which were abundant in coastal communities of Mesoproterozoic shallow marine carbonates. It was interpreted that presence of these forms indicates biospheric evolution (Srivastava, 2005). These fossils also suggest that sexual mode of reproduction by means of sporangia or akinites may have developed by the close of Mesoproterozoic (~1,200 Ma) time (Sharma, 2006).
- Presence of microbial mats and *Arumberia* in association with *Beltanelliformis* (indicating Ediacaran age) reported from the Maihar Sandstone Formation of the Bhander Group, Son Valley (Fig. 11a–d), serves as an additional support for Ediacaran age for fossil-bearing horizon (Kumar and Pandey, 2008a).
- Presence of large-sized leiosphaerids and an age-marker acanthomorph *Trachyhystrichosphaera* in Sirbu Shale Formation, Bhander Group, suggests Cryogenian (850–630 Ma) age for the fossil-bearing Sirbu Shale assemblage (Srivastava, 2009).
- A ring-shaped carbonaceous megafossil (Fig. 10g) reported from the Sirbu Shale, Central India (Kumar and Srivastava, 2003), may represent a basal part or hold fast of *Chuaria–Tawuia* multicellular plant or an unidentified complex morphology of this assemblage.
- Presence of Ediacaran fossils (Fig. 10j, k) in the Bhander Group (Central India and Rajasthan; De, 2003, 2006; Srivastava, 2006) is the most significant evidence in support of Ediacaran age as upper age limit for the Vindhyan Supergroup.
- Abundantly occurring *Eoentophysalis* (Fig. 8b) in Mesoproterozoic times seems to be related with rapid carbonate precipitation and cementation in peritidal Palaeoproterozoic and Mesoproterozoic cherts.
- Presence of branched carbonaceous megafossils (Fig. 10l) exclusively in Neoproterozoic Bhander Group supports the radiation of eukaryotes during Neoproterozoic times. However, presence of dichotomously branched megascopic alga (Fig. 10c) in Rohtas Formation puts a question mark on this interpretation and its affinity.

On the basis of data available till date, Late Palaeoproterozoic (Statherian, 1,700–1,600 Ma) to Stenian (1,200–1,000 Ma) age for the Semri, Kaimur and

Rewa Groups and Tonian–Ediacaran (850–542 Ma) age for the Bhander Group, Upper Vindhyan, are suggested. Knoll and Sergeev (1995) explained in detail and very logically the possible reasons for these gradual biological changes among microfossil and megafossil assemblages globally. Interpretations based on the study of Palaeo-Mesoproterozoic biological changes in the Vindhyan fossil assemblages are well comparable with the reports from the Billyakh Group, northern Siberia by Golovenok and Belova (1984), Yakschin (1991), and Sergeev et al. (1995); from the Jixian Group, China, by Yakschin (1990), from the Dismal Lakes Group and the Arctic Canada by Horodyski and Donaldson (1980). Recent molecular phylogenies imply the rapid diversification of higher eukaryotes relatively late in Meso-Neoproterozoic boundary (Knoll, 1992). Expanding eukaryotic participation in microbial mat communities may contribute to the differences between Mesoproterozoic and Neoproterozoic stromatolites. Rapid precipitation instantly entombed fragile microorganisms, preserving them as casts and molds and in favorable conditions as organic-walled remains that were silicified prior to the neomorphic alteration of host carbonates. This explains the retention of fragile trichomes in Mesoproterozoic cherts, as well as the other characters of coccoid forms (Knoll and Sergeev, 1995).

For the Neoproterozoic biological changes, data from Siberia, German (1990), from Svalbard (Knoll et al., 1991; Butterfield et al., 1994), from the Bitter Springs Formation, Australia (Schopf, 1968; Schopf and Blacic, 1971; Zang and Walter, 1992), from China (Yin, 1985; Zhang, 1989; Zang, 1992) are biostratigraphically comparable with the coeval Vindhyan assemblages. The controversial records of the Cambrian fossils from the Lower Vindhyan (Azmi, 1998; Azmi et al., 2007) are doubtful and not acceptable (for a detail discussion refer Tewari, 1998; Bengtson et al., 2007, 2009; Kumar, 2009 and the references therein). Besides many evidences and data available in India, we are still in state of infancy and require more authentic data and a more systematic approach (in sample collection) for the better understanding of rate of transition, whether it was fast or slow, as well the role of taphonomy in the process of changes. At the same time, there is no doubt that palaeobiological, sedimentological and palaeoenvironmental changes collectively prepared the ground for the next major shift in the biosphere near the end of the Proterozoic Eon (Knoll and Sergeev, 1995).

7. Acknowledgements

The authors are grateful to Prof. N.L. Chhabra, Professor and Head Department of Geology, Lucknow University, Lucknow, Uttar Pradesh, India and Prof. B.R. Arora, Director, W.I.H.G., Dehradun, Uttarakhand, India, for providing the facilities and the permission to publish the book chapter. Professors Wladislaw Altermann, Ludwig Maximilians University, Munich, Germany, and Nichlas J. Butterfield, Department of Earth Sciences, University of Cambridge, Cambridge, UK, critically reviewed the earlier draft of the chapter and we are indebted to them for constructive criticism and valuable suggestions which improved the present chapter.

8. References

Altermann, W. and Kazmierczak, J. (2003) Archean microfossils: a reappraisal of early life on Earth. Res. Microbiol. **154**: 611–617.

Altermann, W., Kazmierczak, J., Oren, A. and Wright, D. (2006) Microbial calcification and its impact on the sedimentary rock record during 3.5 billion years of Earth history. Geobiology **4**: 147–166.

Auden, J.B. (1938) Vindhyan sedimentation in Son valley, Mirzapur District. Mem. Geol. Surv. India **62**(2): 141–250.

Azmi, R.J. (1998) Discovery of Lower Cambrian small shelly fossils and Brachiopods from the Lr. Vindhyan of Son Valley, Central India. J. Geol. Soc. India **52**: 381–389.

Azmi, R.J., Joshi, D., Tewari, B.N., Joshi, M.N., Mohan, K. and Srivastava, S.S. (2007) Age of the Vindhyan Supergroup of Central India: an exposition of biochronology vs radiochronology, micropaleontology: application in stratigraphy and palaeooceanography. D.K. Sinha (ed.) Narosa Publishing House, New Delhi, India, pp. 29–62.

Bengtson, S., Belivanova, V. Rasmussen, B. and Whitehouse, M. (2007) GSA. Denver Annual Meeting, Colorado Convention Center.

Bengtson, S., Belivanova, V., Rasmusin, B. and Whitehouse, M. (2009) The controversial Cambrian fossils of the Vindhyan are real but more than a billion years older. www.pnas.org/cgi/doi/10.1073./pnas.0812460106.

Brasier, M. (1999) Discovery of Lower Cambrian small shelly fossils and Brachiopods from the Lr. Vindhyan of Son Valley, Central India. Discussion. J. Geol. Soc. India **53**: 727–730.

Butterfield, N.J., Knoll, A.H. and Swett, K. (1994) Palaeobiology of the Neoproterozoic Svanbergfjellet Formation, Spitsbergen. Fossil Strata **34**: 1–84.

Crawford, A.R. and Compston, W. (1970) The age of the Vindhyan System, Peninsular India. Q. J. Geol. Soc. London **125**: 351–371.

De, C. (2003) Possible organisms similar to Ediacaran forms from the Bhander Group, Vindhyan Supergroup, Late Neoproterozoic of India. J. Asian Earth Sci. **21**: 387–395.

De, C. (2006) Ediacara fossil assemblage in the upper Vindhyans of Central India and its significance. J. Asian Earth Sci. **27**: 660–683.

Golovenok, V.K. and Belova, M.Y. (1984) *Obruchevella* from the Patom Highland and the Vendian of southern Kazakhstan. Dokl. Akad. Nauk SSSR **272**: 1462–1468.

Hofmann, H.J. (1976) Precambrian microflora, Belcher Islands, Canada: significance and systematics. J. Paleontol. **50**: 1040–1073.

Hofmann, H.J. (2005) Palaeoproterozoic dubiofossils from India. Revisited-Vindhyan triploblastic animal burrows or pseudofossils? J. Palaeontol. Soc. India **50**(2): 113–120.

Horodyski, R.J. and Donaldson, J.A. (1980) Microfossils from the Middle Proterozoic Dismal Lakes Group, Arctic Canada. Precambrian Res. **11**: 125–159.

Kazmierczak, J. and Altermann, W. (2002) Neoarchean biomineralisation by benthic cyanobacteria. Science **298**: 2351.

Kerr, R.A. (2002) Earliest animals tracks or just mud cracks? Science **295**: 1209–1210.

Knoll, A.H. (1992) The early evolution of eukaryotes: a geological perspective. Science **256**: 622–627.

Knoll, A.H. (1994) Neoproterozoic evolution and environmental change, In: S. Bengtson (ed.) *Early Life on Earth*. Nobel Symposium No 84. Columbia, New York.

Knoll, A.H. and Sergeev, V.N. (1995) Taphonomic and evolutionary changes across the Mesoproterozoic–Neoproterozoic transition. Neues Jahrb. Geol. Palaontol. **195**: 289–302.

Knoll, A.H., Swett, K. and Mark, J. (1991) Palaeobiology of a Neoproterozoic tidal flat/lagoonal complex: the Draken Conglomerate Formation, Spitsbergen. J. Paleontol. **65**: 531–570.

Knoll, A.H., Javaux, E.J., Hewitt, D. and Cohen, P. (2006) Eukaryotic organisms in Proterozoic oceans. Phill. Trans. R. Soc. B. **361**: 1023–1038.

Kumar, S. (1976a) Significance of stromatolites in the correlation of Semri Series (Lower Vindhyans) of Son Valley and Chitrakut area, U.P. J. Palaeontol. Soc. India **19**: 24–27.

Kumar, S. (1976b) Stromatolites from the Vindhyan rocks of Son Valley-Maihar area, district Mirzapur (U.P.) and Satna (M.P.). J. Palaeontol. Soc. India **18**: 13–21.

Kumar, S. (1978) Stromatolites and environment of deposition of the Vindhyan Supergroup of Central India. Jour. Palaeont. Soc. India, 21–22: 33–43

Kumar, S. (1980) Stromatolites and Indian biostratigraphy: a review. J. Palaeontol. Soc. India **23–24**: 166–184.

Kumar, S. (1984) Present study of stromatolite biostratigraphy in India, Geophytology **14**(1): 96–110

Kumar, S. (1995) Megafossils from the Mesoproterozoic Rohtas Formation (The Vindhyan Supergroup) Katni area, Central India. Precambrian Res. **72**: 171–184.

Kumar, S. (2001) Mesoproterozoic megafossil Chuaria–Tawuia association may represent parts of a multicellular plant, Vindhyan Supergroup, Central India. Precambrian Res. **106**: 187–211.

Kumar, S. (2009) Controversy concerning Cambrian fossils from the Vindhyan sediments: a re-assessment. J. Palaeontol. Soc. India **54**(1): 115–117.

Kumar, S. and Pandey, S.K. (2008a) *Arumberia* and associated fossils from the Neoproterozoic Maihar Sandstone, Vindhyan Supergroup, Central India. J. Palaeontol. Soc. India **53**(1): 83–97.

Kumar, S. and Pandey, S.K. (2008b) Discovery of organic-walled microbiota from the black- bedded chert, Balwan Limestone, the Bhander Group, Lakheri area, Rajasthan. Curr. Sci. **94**(6): 797–800.

Kumar, S. and Srivastava, P. (1989) Indian Precambrian Microbiotas from the bedded chert: a review. J. Palaeontol. Soc. India **34**: 68–77.

Kumar, S. and Srivastava, P. (1991) Record of microfossils from the nonstromatolitic middle Proterozoic Vindhyan Chert, Chitrakut area, Uttar Pradesh. J. Geol. Soc. India **38**: 511–515.

Kumar, S. and Srivastava, P. (1992) Microfossils from the black chert of the Bhagwanpura Limestone (middle Proterozoic) Vindhyan Supergroup, Chittorgarh Area, Rajasthan. Curr. Sci. **62**(4): 371–372.

Kumar, S. and Srivastava, P. (1995) Microfossils from the Kheinjua Formation, Middle Proterozoic Semri Group (Lower Vindhyan) Newari Area, Central India. Precambrian Res. **74**: 91–117.

Kumar, S. and Srivastava, P. (1997) A note on the carbonaceous megafossils from the Neoproterozoic Bhander Group, Maihar Area, M. P. J. Paleontol. Soc. India **42**: 141–146.

Kumar, S. and Srivastava, P. (2003) Carbonaceous megafossils from the Neoproterozoic Bhander Group, Central India, J. Palaeontol. Soc. India **48**: 139–154.

Kumar, A., Gopalan, K. and Rajagopalan, G. (2001) Age of the Lower Vindhyan sediments, Central India. Curr. Sci. **81**(7): 806–809.

Maithy, P.K. and Babu, R. (1988a) The mid-Proterozoic Vindhyan microbiota from the Chopan, Southeast Uttar Pradesh. J. Geol. Soc. India **33**: 584–590.

Maithy, P.K. and Babu, R. (1988b) Chitinozoa like remains from the Vindhyan Supergroup of Son Valley. Palaeobotanist **37**(1): 77–80.

Maithy, P.K. and Shukla, M. (1984) Biological remains from the Suket Shale Formation. Geophytology **14**: 212–215.

Malone, S.J., Meert, J.G., Banerjee, D.M., Pandit, M.K., Tamrat, E., Kamenov, G.D., Pradhan, V.R. and Sohl, L.E. (2008) Palaeomagnetism and detrital Zircon geochronology of the Upper Vindhyan sequence, Son Valley and Rajasthan, India: a ca. 1000 Ma closure age for the Purana Basins? Precambrian Res. **164**: 137–159, doi: 10.1016/jprecamres.2008.04.004.

Mandal, M.E.A., Goswami, J.N., Deomurari, M.P. and Sharma, K.K. (2002) Ion microprobe Pb/Pb ages of zircons from the Bundelkhand Massif, northern India: implications for crustal evolution of the Bundelkhand-Aravalli Supercontinent. Precambrian Res. **117**: 85–100.

Mathur, S.M. (1981) Stromatolites in the Precambrian Vindhyan Supergroup of Central India. A review. Ann. Soc. Geol. Belg. **104**: 87–91.

McMenamin, D.S., Kumar, S. and Awramik, S.M. (1983) Microbial fossils from the Kheinjua Formation, Middle Proterozoic Semri Group (Lower Vindhyan), Son Valley Area, Central India. Precambrian Res. **21**: 247–271.

Misra, Y. (2004) *Stromatolite biostratigraphy of the Vindhyan Basin*. Unpublished Ph.D. Thesis.

Misra, Y. and Kumar, S. (2005) Coniform stromatolites and the Vindhyan Supergroup, Central India: implication for Basinal correlation and age. J. Palaeontol. Soc. India **50**(2): 153–167.

Morris, S.C., Jensen, S. and Butterfield, N.J. (1998) Fossil discoveries in India: continued. Science **282**: 1265.

Prasad, B. (1984) Geology, sedimentation and palaeogeography of the Vindhyan Supergroup, southeastern Rajasthan. Part II: Vindhyan stratigraphy. Mem. Geol. Surv. India **116**: 148.

Raaben, M.E. and Tewari, V.C. (1987) Riphean stromatolites in India. Izvestia Acad. Nauk. CCCP Ser. Geol. **7**: 17–25 (in Russian).

Raha, P.K. and Das, D.P. (1989) Correlation of stromatolite bearing Upper Proterozoic basins of India and paleogeographic significance. Himalayan Geol. **13**: 119–142. K.S. Valdiya and V.C. Tewari (eds.) Wadia Institute of Himalayan Geology, Dehradun.

Rai, V. and Singh, V.K. (2004) Discovery of *Obruchevella* Reitlinger 1948 from the Late Palaeoproterozoic Lower Vindhyan succession and its significance. J. Palaeontol. Soc. India **49**: 189–196.

Rai, V., Shukla, M. and Gautam, R. (1997) Discovery of carbonaceous megafossils (*Chuaria–Tawuia* assemblage) from the Neoproterozoic Vindhyan succession (Rewa Group), Allahabad-Rewa Area, India. Curr. Sci. **73**: 783–788.

Rao, K.S., Lal, C. and Ghosh, D.B. (1977) Algal stromatolites in the Bhander Group of Formations, Vindhyan Supergroup, Satna district, Madhya Pradesh. Rec. Geol. Surv. India **109**(2): 38–48.

Rasmussen, B., Bose, P.K., Banerjee, S., Fletcher, I.R. and Mc Naughton, N.J. (2002) 1.6 Ga U-Pb zircon age for the Chorhat Sandstone, Lower Vindhyan, India: possible implications for early evolution of animals. Geology **30**: 103–106.

Ray, J. (2006) Age of the Vindhyan Supergroup: a review of recent findings. J. Earth Syst. Sci. **115**(1): 149–160.

Ray, J.S., Martin, M.W., Veizer, J. and Bowring, S.A. (2002) U-Pb zircon dating and Sr isotope systematics of the Vindhyan Supergroup, India. Geology **30**: 131–134.

Ray, J.S., Veizer, J and Davis, W.J. (2003) C, O, Sr and Pb isotope systematics of carbonate sequences of the Vindhyan Supergroup, India: age, diagenesis, correlation and implication for global events. Precambrian Res. **121**: 103–140.

Sarangi, S. Gopalan, K. Kumar, S. (2004) Pb–Pb age of the earliest megascopic eukaryotic alga bearing Rohtas Formation, Vindhyan Supergroup, India: implication for Precambrian atmospheric oxygen evolution. Precambrian Res. **132**: 107–121.

Sastry, M.V.A. and Moitra, A.K. (1984) Vindhyan stratigraphy: a review. Mem. Geol. Surv. India **116**(Part II): 109–148.

Schopf, J.W. (1968) Microflora of the Bitter Springs Formation Late Precambrian Central Australia, J. Paleontol. **42**: 651–688.

Schopf, J.W. and Blacic, J.M. (1971) New microorganisms from the Bitter Springs Formation (Late Precambrian) of the North Central Amadeus Basin, Australia. J. Paleontol. **45**: 925–960.

Schopf, J.W. and Klein, C. (1992) *The Proterozoic Biosphere: A Multidisciplinary Study*. Cambridge University Press, Cambridge, p. 1348.

Seilacher, A. Bose, P.K. and Pfluger, F. (1998) Triploblastic animals more than 1 billion years ago: trace fossil evidence from India. Science 282: 80–83.

Sergeev, V.N., Knoll, A.H. and Grotzinger, J.P. (1995) Palaeobiology of the Mesoproterozoic Billyakh Group, Anabar uplift, Northern Siberia. Palaeontol. Soc. Mem. **39**: 1–37.

Sergeev, V.N. Sharma, M. and Shukla, Y. (2008) Mesoproterozoic silicified microbiotas of Russia and India – characteristics and contrasts. Palaeobotanist **57**: 323–358.

Sharma, M. (2003) Age of Vindhyans-Palaeobiological evidence: a paradigm shift (?). J. Palaeontol. Soc. India **48**: 191–214.

Sharma, M. (2006a) Late Palaeoproterozoic (Statherian) carbonaceous films from the Olive Shale (Koldaha Shale), Semri Group, Vindhyan Supergroup, India. J. Palaeontol. Soc. India **51**(2): 27–35.

Sharma, M. (2006b) Palaeobiology of Mesoproterozoic Salkhan Limestone, Semri Group, Rohtas, Bihar, India: systematics and significance. J. Earth Syst. Sci. **115**(1): 67–98.

Sharma, M. (2006c) Small-sized akinites from the Mesoproterozoic Salkhan Limestone, Semri Group, Rohtas, Bihar, India: systematic and significance. J. Earth Syst. Sci. **115**: 67–99.

Soni, M.K., Chakraborty, S. and Jain, V.K. (1987) Vindhyan Supergroup – a review. In Purana Basins of India. Mem. Geol. Soc. India **6**: 87–138.

Srivastava, P. (2002) Carbonaceous megafossils from the Dholpura Shale, Uppermost Vindhyan Supergroup, Rajasthan: an age implication. J. Palaeontol. Soc. India **47**: 97–105.

Srivastava, P. (2004) Carbonaceous fossils from the Panna Shale, Rewa Group (Upper Vindhyans), Central India: a possible link between evolution from micro-megascopic life. Curr. Sci. **86**(5): 644–646.

Srivastava, P. (2005) Vindhyan akinites: an indicator of mesoproterozoic biospheric evolution. Orig. Life Evol. Biosph. **35**: 175–185.

Srivastava, P. (2006) Possible Ediacaran fossils from the Bundi Hill Sandstone, Upper Vindhyans, Rajasthan: their, global significance. Diamond Jubilee International Conference on Changing Scenario in Palaeobotany and allied subjects held at BSIP, Lucknow in Nov. 2006, pp 188.

Srivastava, P. (2008) Some bizarre forms of the Meso-Neoproterozoic Vindhyan Supergroup, India (Abstract). In: World Summit on Ancient Microscopic Fossils, Los Angeles, California, USA. 27 July–2 Aug, 2008.

Srivastava, P (2009) *Trachyhystrichosphaera*: an age marker acanthomorph from the Bhander Group, Upper Vindhyans, Rajasthan. J. Earth Syst. Sci. 118(5): 575–582.

Srivastava, P. and Bali, R. (2006) Proterozoic carbonaceous remains from the Chorhat Sandstone: oldest fossils of the Vindhyan Supergroup. Geobios **39**: 873–878.

Stulp, B.K. and Stam, W.T. (1984) Genotypic relationship between strains of Anabaena (cyanophyceae) and their correlation with morphological affinities. Br. Phycological J. **19**: 287–301.

Tandon, K.K. and Kumar, S. (1977) Discovery of annelid and arthropod remains from the Lower Vindhyan rocks (Precambrian) of Central India. Geophytology 7: 126–128.

Tewari, V.C. (1989) Upper Proterozoic-Lower Cambrian stromatlites and Indian stratigraphy. Himalayan Geol. **13**: 143–180. K.S. Valdiya and V.C. Tewari (eds.) Wadia Institute of Himalayan Geology, Dehradun.

Tewari, V.C. (1998) Discovery of Lower Cambrian small shelly fossils and Brachiopods from the Lower Vindyan of Son Valley, Central India (Discussion). J. Geol. Soc. India **52**: 381–389.

Tewari, V.C. (2003a) Correlation and paleobiology of the Vindhyan and Lesser Himalayan stratigraphic successions. J. Palaeontol. Soc. India **48**: 155–165.

Tewari, V.C. (2003b) Correlation of Central Indian Vindhyans and Lesser Himalayan sequences, stratigraphy and paleobiology. Gondwana Geol. Mag. **7**: 383–394.

Valdiya, K.S. (1969) Stromatolites of the Himalayan carbonate formations and the Vindhyans. J. Geol. Soc. India **10**: 1–25.

Valdiya, K.S. (1989) Precambrian stromatolite biostratigraphy of India – a review. Himalayan Geol. **13**: 181–214.

Walter, M.R., Du, R. and Horodyski, R.J. (1990) Coiled carbonaceous megafossils from the middle Proterozoic of Jixian (Tianjin) and Montana. Am. Jour. Sci. 290-A: 133–148.

Yakschin, M.S. (1990) Origin of microstructures in the Early Riphean flat-laminated stromatolites, In: *Iskopaemye problematiki SSSR (The Bizarre Fossils of the USSR) Moscow, Nauka*, pp. 5–13.

Yakschin, M.S. (1991) Vodoroslevayab mikrobiota nizhnego rifeya anabarskogo podnyatia (Algal microbiota from the Lower Riphean deposits of Anabar Uplift): Novosibirsk, Nauka, Sibirskoe Otdelelenie, 61 pp.

Yin, L. (1985) Microfossils of the Doushantuo Formation in the Yangtze George District, Western Hebei, China. Palaeontol. Cathayana **2**: 229–249.

Zang, W.L. (1992) Sinian and early Cambrian floras and biostratigraphy on the South China Platform. Palaeontographica B **224**: 75–119.

Zang, W.L. and Walter, M.R. (1992) Late Proterozoic and early Cambrian microfossils and biostratigraphy, Amadeus Basin, central Australia. Mem. Assoc. Aust. Palaeontol. **12**: 132.

Zhang, Y. (1989) Multicellular thallophytes with differentiated tissues from Late Proterozoic phosphate rocks of south China. Lethaia **22**: 113–132.

Biodata of **Kenichiro Sugitani, Koichi Mimura**, and **Malcolm R. Walter**, authors of ***"Farrel Quartzite Microfossils in the Goldsworthy Greenstone Belt, Pilbara Craton, Western Australia: Additional Evidence for a Diverse and Evolved Biota on the Archean Earth"***

Dr. Kenichiro Sugitani is currently a professor at Graduate School of Environmental Studies of the Nagoya University, Japan. He obtained his DSc degree from the Nagoya University in 1992. He had field works in Pilbara Craton ten times since 1989. His scientific interests are taxonomy and phylogeny of Precambrian microfossils, early evolution of Earth's surface environment and its interaction with life, and impact of human activities on biogeochemical cycles of nutrients and biotic diversity.

E-mail: **sugi@info.human.nagoya-u.ac.jp**

Dr. Koichi Mimura is currently an associate professor at Graduate School of Environmental Studies of the Nagoya University, Japan. He obtained his DSc degree from the Nagoya University in 1995. His scientific interests are evolution of organic materials in meteorites, shock reaction of organic materials, and origins of life.

E-mail: **mimura@eps.nagoya-u.ac.jp**

Dr. Kenichiro Sugitani

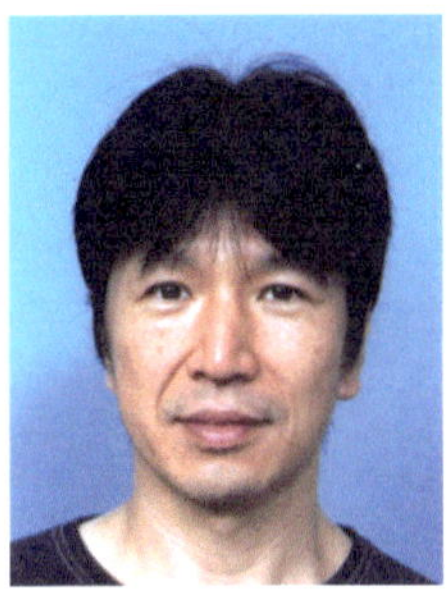

Dr. Koichi Mimura

V.C. Tewari and J. Seckbach (eds.), *STROMATOLITES: Interaction of Microbes with Sediments*, Cellular Origin, Life in Extreme Habitats and Astrobiology 18, 115–132
DOI 10.1007/978-94-007-0397-1_6, © Springer Science+Business Media B.V. 2011

Dr. Malcolm Walter is Professor of Astrobiology at the University of New South Wales, Australia. He is Director of the Australian Centre for Astrobiology. He has worked for 40 years on the geological evidence of early life on Earth, including the earliest convincing evidence of life. During 1999 his book "The Search for Life on Mars" was published by Allen & Unwin. He has published more than 120 articles and several other books. In 2004 he was elected a Fellow of the Australian Academy of Science, and in 2005 he was made a Fellow on the Geological Society of Australia.

E-mail: **Malcolm.walter@unsw.edu.au**

FARREL QUARTZITE MICROFOSSILS IN THE GOLDSWORTHY GREENSTONE BELT, PILBARA CRATON, WESTERN AUSTRALIA
Additional Evidence for a Diverse and Evolved Biota on the Archean Earth

KENICHIRO SUGITANI[1], KOICHI MIMURA[1], AND MALCOLM R. WALTER[2]
[1]*Graduate School of Environmental Studies, Nagoya University, Nagoya 464-8601, Japan*
[2]*Australian Centre for Astrobiology, University of New South Wales, Sydney, NSW 2052, Australia*

Abstract A microfossil assemblage recently discovered from the Farrel Quartzite (ca. 3.0 Ga) in the Goldsworthy greenstone belt in the Pilbara Craton comprises five main morphological types (thread like, film-like, spheroidal [>15 and <15 mm], and lenticular to spindle-like) and some elaborate morphologies that might represent reproducing or resting stages of the spheroidal and lenticular to spindle-like structures. This assemblage is abundant within a stratiform black chert associated with evaporite, suggesting an early evolution of a diverse and complex ecosystem in an Archean shallow water environment, although taxonomic and phylogenetic studies have only just begun. So far, from more than ten sites of occurrence of black cherts in the Goldsworthy greenstone belt, similar assemblages of microstructures have been found.

Keywords Farrel Quartzite • Microfossils • Goldsworthy greenstone belt • Pilbara Craton • Spheroidal • Spindle-like • Archean • Microstructures • Western Australia • Diverse and complex ecosystem

1. Introduction

A morphologically diverse assemblage of carbonaceous microstructures was recently reported from chert in the Farrel Quartzite of the ca. 3.0 Ga Gorge Creek Group in the Pilbara Craton, Western Australia (Sugitani et al., 2007). Their biogenicity was claimed based on a rigorous examination of occurrence, composition, morphological complexity, size distributions, and taphonomy, which has been supported by following studies including palynological extraction of microfossils (Grey and Sugitani, 2009), reconstructed in-focus and 3-D images (Sugitani et al., 2009a), Nano-SIMS analyses (Oehler et al., 2009), description of elaborate morphologies suggestive of reproducing cells and spores (Sugitani et al., 2009b), and discoveries of similar assemblage of microstructures from more than ten sites at the same stratigraphic horizon (Sugitani et al., 2008). Thus, the biogenicity of the Farrel Quartzite microstructures seems to be established.

In this study, we review main morphological types and several types of elaborate morphologies of the Farrel Quartzite assemblage, referring to their modes of occurrences and positions in thin sections. Most of specimens described here are from two thin sections made from hand specimens collected from Mount Grant, demonstrating prolific occurrence of the assemblage and the possibility of future taxonomic and phylogenetic studies.

2. Geology and Stratigraphy

The Goldsworthy greenstone belt (at Mount Goldsworthy and Mount Grant) is located approximately 100 km east of Port Hedland in the northeastern Pilbara Craton (Fig. 1). The volcanic sedimentary succession of this belt is composed of an older, dominantly volcanic sequence and a younger, dominantly sedimentary succession. The older volcanic sequence is part of the 3.53–3.17 Ga Pilbara Supergroup, whereas a younger sedimentary succession is part of the 3.02–2.93 Ga De Grey Supergroup (Hickman and Van Kranendonk, 2008). The lower part of the De Grey Supergroup is composed of the 3.02–3.01 Ga Gorge Creek Group, which comprises the Farrel Quartzite and the conformably overlying Cleaverville Formation (Fig. 1b). On the basis of regional stratigraphy and geochronology (Van Kranendonk et al., 2006; Hickman and Van Kranendonk, 2008; Sugitani et al., 2003), the age of the putative microfossils is suggested to be older than 3.01 Ga.

The Farrel Quartzite at the Goldsworthy greenstone belt is composed of fine to very coarse-grained sandstone with minor conglomerate, mafic to ultramafic volcaniclastic layers, evaporite beds, barite beds, and black chert layers (Fig. 1b, c). The thickness and lithology of the Farrel Quartzite are laterally variable, and the rocks have been metamorphosed to greenschist facies and pervasively silicified. The stratiform black cherts containing putative microfossils and associated evaporite beds in the uppermost portion are laterally continuous (CE2 in Fig. 1b). They can be traced approximately 7 km along the strike in the central and western Mount Grant area. The Farrel Quartzite is considered to have deposited at a continental or continental margin setting (Sugitani et al., 2003, 2006). Sugitani et al. (2007) suggests that the Farrel Quartzite assemblage probably represents microbial communities that inhabited in a shallow, closed to semi-closed basin on the early Earth.

3. Sedimentary Petrology of Black Chert

Microfossil-bearing black cherts in the Farrel Quartzite are mostly less than 20 cm thick and are generally massive in appearance, although parallel lamination can be seen locally. The black chert displays a wide range of microscopic fabrics, which

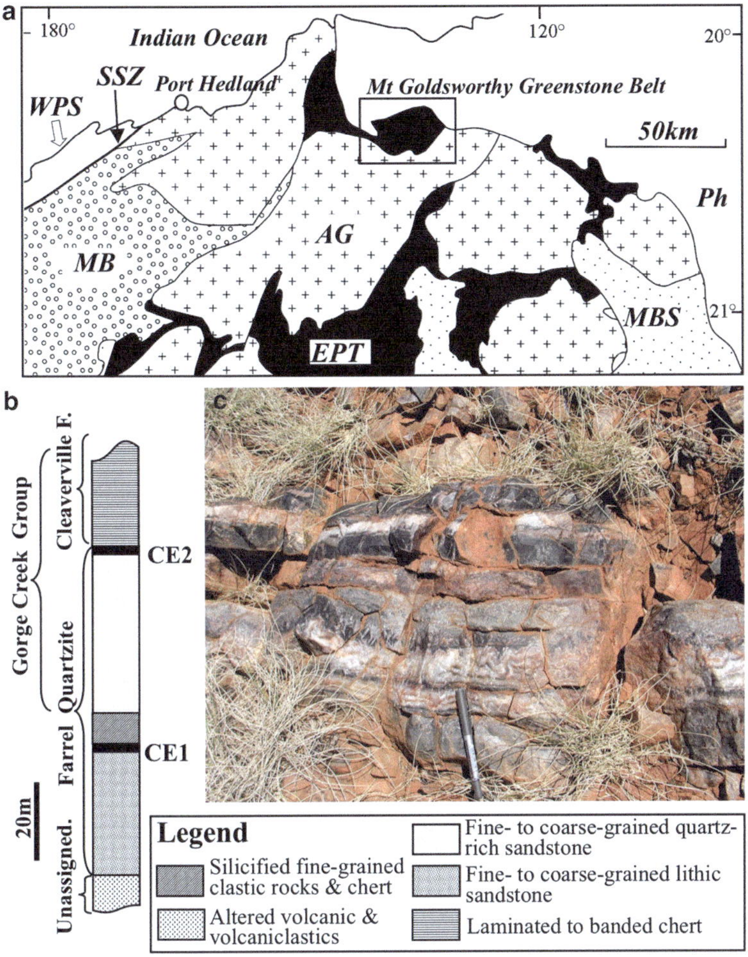

Figure 1. (**a**) Geology of the northeastern part of the Pilbara Craton (modified from Sugitani et al., 2007). *WPS* West Pilbara Superterrane, *SSZ* Sholl Shear Zone, *MB* Mallina Basin, *AG* Archean granitoid rocks, *EPT* East Pilbara Terrane, *MBS* Mount Bruce Supergroup, *Ph* Phanerozoic cover. (**b**) Simplified stratigraphic column for the type locality at Mount Grant. CE1 and CE2 show an association of black chert and an evaporite layer. (**c**) Photograph of the upper portion of CE2 at the type locality.

tend to be distributed stratigraphically (Fig. 2a). The massive fabric is characterized by a mostly homogeneous distribution of carbonaceous particles with dispersed irregularly shaped pure chert masses (Mf in Fig. 2a, d). In some horizons, carbonaceous particles are heterogeneously distributed, resulting in well-developed irregularly shaped pure chert masses (fenestrae-like fabric: Fs in Fig. 2a, c). The chert masses are composed of a microcrystalline quartz mosaic that often displays domain to chalcedonic extinction figure. In some cases, carbonaceous

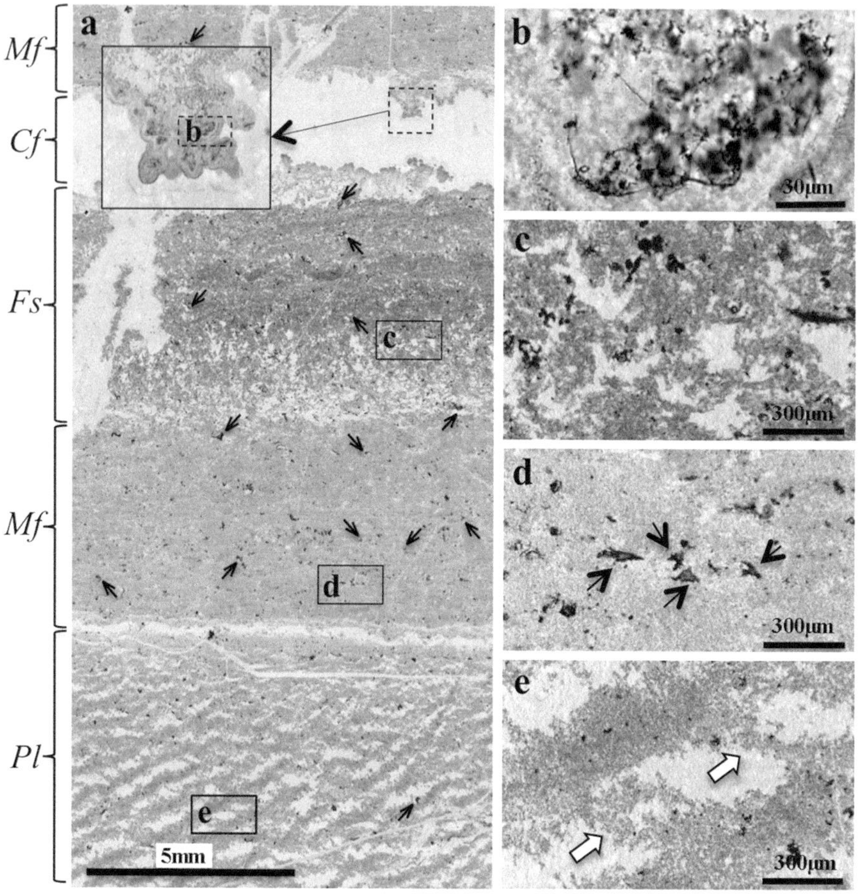

Figure 2. (**a**) *Left*: photomicrograph of GFSV6. *Mf*, *Cf*, *Fs*, and *Pl* show massive, cavity-fill, fenestrae-like, and pseudo-laminated fabrics, respectively. The *solid arrows* show film-like structures. Photomicrographs on the *right* (**b–e**) show enlargements of areas in *small rectangles* (**b–e**) on this image. (**b**) Thread-like structures in colloform-textured portion. (**c**) Fenestrae-like irregularly shaped pure chert masses. (**d**) Magnification of massive portion. The *solid arrows* indicate film-like structures. (**e**) Magnification of pseudo-laminated portion. The *white arrows* indicate that laminae are connected with each other.

dusts lie on the bottom of a chert mass. These masses possibly originated from fenestral fillings. The term pseudo-laminated fabric (Pl in Fig. 2a, e) is used for areas where carbonaceous laminae of massive fabric 100–300 μm thick alternate with pure cherts of nearly the same thickness. In places carbonaceous laminae are connected or branched, producing lenticular pure chert masses (the white arrows in Fig. 2e). Namely, this texture is not a result of particle settling. The lamination is more likely a secondary product, although the formation process is not known.

In addition to these three carbonaceous chert fabrics, the black chert often contains stratiform white chert layers of various thicknesses and widths (cavity-fill fabric: Cf in Fig. 2a), in which colloform textures are developed. These textures often contain thread-like carbonaceous structures (Fig. 2b). This fabric probably represents post-depositional cavity-fill silica precipitates that formed after dissolution of soluble mineral layers.

In the Goldsworthy greenstone belt, black chert veins and dikes are rare and their occurrence is spatially independent of the stratiform black chert layer that is the main host for the carbonaceous microstructures. An association of the stratiform black chert and silicified evaporite bed (chert evaporite band) can be traced along the strike ca. 7 km at Mount Grant and ca. 100 m at Mount Goldsworthy. Additionally, (1) stratiform black chert layers and interbedded clastic layers are cut by a fault; (2) the chert itself contains laminae composed of terrigeneous detrital materials or volcaniclastics; (3) locally, the chert is brecciated and brecciated chert clasts are set in a fine- to medium-grained sandstone bed; and (4) lamination that can be identified as an alignment of very fine-grained carbonaceous particles and carbonaceous microstructures is present. These convergent lines of evidence indicate that the stratiform black chert layer formed primarily as a sedimentary deposit.

4. Samples and Methods

Samples for this study were collected from two sites at Mount Grant. Two hand specimens (GFSV3 and GFSV6) collected from CE2 in the vicinity of the type locality at Mount Grant (S20°22′43.6″, E119°26′54.3″) were used for this study. These samples were cut into slabs, perpendicular to the bedding plane. Standard thin sections (2.5×3.4 cm wide and 30–35 μm thick) were examined with a petrographic microscope (Leitz-DMRP). Several specimens described here are identical to those in Sugitani et al. (2007, 2009a, b).

The block samples, slabs, and thin sections are currently in the collection at Nagoya University, Japan, while studies are underway. They will be reposited in an Australian collection as required under the Australian Protection of Movable Cultural Heritage Act, if the microstructures in question are proved to be fossils.

5. Occurrences and Features of the Five Main Morphologies and Related Structures

The five main morphological types of the Farrel Quartzite assemblage are thread-like, film-like, spheroidal (>15 and <15 μm), and lenticular to spindle-like structures. The biogenicity of these structures has been argued based on geological context, syngenicity, chemical and isotopic composition, physical properties, internal complexity, size range and distribution, taphonomic features, morphological diversity and elaboration, and morphological similarity to modern microorganisms (Sugitani et al., 2007, 2009a, b; Oehler et al., 2009; Grey and Sugitani, 2009). The analyzed new thin sections contain all representative morphotypes previously defined. In the following, we describe representative specimens and related elaborate morphologies and occurrences.

5.1. THREAD-LIKE STRUCTURES

Thread-like structures are generally less than 1 μm in diameter and highly entangled. It is unclear whether they are branched or not. Subtypes such as a relatively thick granular type and a narrower smooth one may be present. They were first discovered from colloform-textured white to translucent chert layer (Fig. 2a, b) and quartz vein, representative of post-depositional silica phase. Based on these occurrences, they were interpreted as dubiofossils (Sugitani et al., 2007), whereas other Archean carbonaceous threads in a chert dike elsewhere are interpreted as possible fossilized subsurface chemoautotrophic microbes (Ueno et al., 2001, 2004).

On the other hand, thread-like structures present in the Farrel Quartzite black chert with a massive fabric may more likely be biogenic. The carbonaceous threads are distributed densely in a layer less than 5 mm thick (indicated by the white arrow in Fig. 3). They occur within irregularly shaped white chert masses (Fig. 4a). The masses are composed of a mosaic of microcrystalline quartz like the surrounding carbonaceous portions, suggesting that they are syndepositional or very early diagenetic structures such as fenestrae. The threads are less than 0.5 μm in diameter. They tend to be extensively entangled and occasionally penetrate into the carbonaceous portions (Fig. 4b). Aggregates of threads are often set in the bottoms of chert masses. Such occurrences are inconsistent with diagenetic segregation of carbonaceous matter and suggest the biogenicity of the threads. While many examples of Archean thread-like putative microfossils are found in literature (for instance, Schopf and Walter, 1983; Schopf, 2006), occurrences similar to the present case have never been reported.

5.2. FILM-LIKE STRUCTURES

The film-like structures are composed of sheets of carbonaceous particles and are generally less than 1 μm in thickness. Some are still preserved as membranous objects as indicated by the results of palynological extraction (Grey and Sugitani, 2009). Their sizes are various and can be up to >500 μm along the major axis. As shown in Fig. 2a, they are ubiquitous and abundant in carbonaceous chert of the Farrel Quartzite. The structures are mostly single sheets, although some appear to be branched (Fig. 3a) and/or have appendages such as clusters of opaque particles

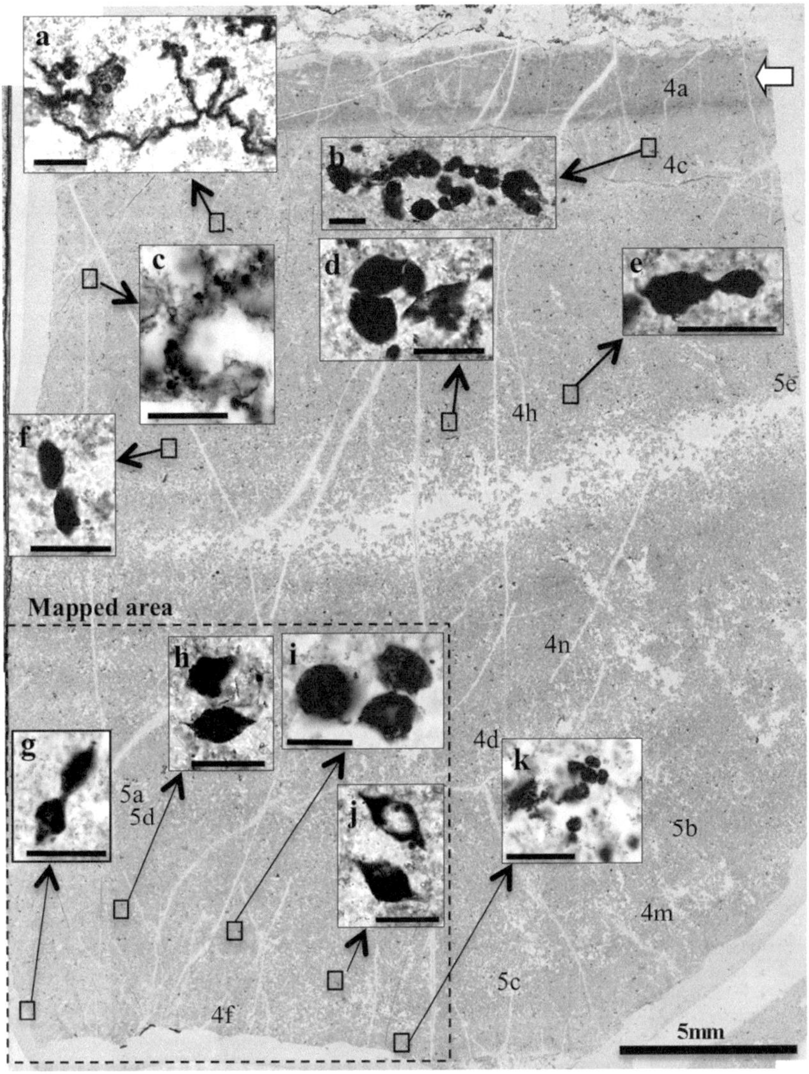

Figure 3. Photomicrograph of GFSV3 showing massive fabric. The *numbers with alphabets* (e.g., 4a) correspond to microfossils shown in Figs. 4 and 5. Microfossils in *small boxes* are referred to in the text. Scale bars in the boxes are 50 μm.

(probably sulfides) or small carbonaceous spheroids less than 5 μm in diameter. They are wrinkled or folded to various degrees (Fig. 4c). Edges of the structures are often notched and rolled up. These features suggest that the structures were flexible but breakable. Many of the structures, if not all, are likely fragments of larger structures. The structures can be interpreted as reworked and transported

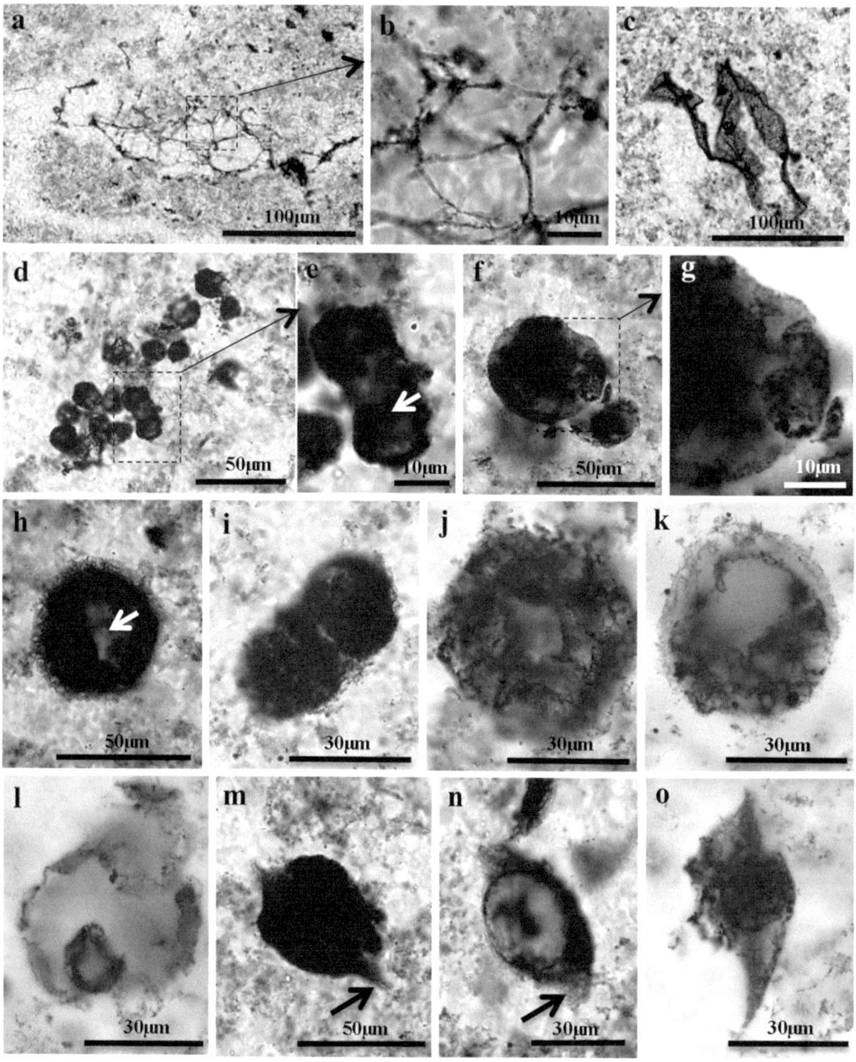

Figure 4. Photomicrographs of main five morphological types. The locations of specimens shown in **a**, **c**, **d**, **f**, **h**, **m**, and **n** in the thin section are indicated in Fig. 3. (**a**) Entangled thread-like structures within irregularly shaped pure chert mass in a thin layer (indicated by the *white arrow* in Fig. 3). Some threads penetrate into surrounding carbonaceous portions. (**b**) Magnification of (**a**). (**c**) Folded film-like structure. (**d**) Cluster of small spheroids. (**e**) Magnification of (**d**) showing fused pair. The portion indicated by the *white arrow* is broken and hollow. (**f**) Simple single-walled spheroid. (**g**) Magnification of (**f**) showing folds of wall. (**h**) Thin-walled spheroid having a diffuse envelope. The *arrow* shows a broken portion of the wall. (**i**) Paired thin-walled spheroids having a diffuse envelope. Slide GFWE3-F6, position R-Z48/2 (England Finder). (**j**) Spheroid having an extensively folded wall. Slide GFWE3-S6, position R-N54/1. (**k**) Thick-walled spheroid. GFWE2-G6, position R-B61/3. (**l**) Broken spheroid with an inner hollow body. Slide NGWM1X, position R-N40. (**m**) Spindle-like structure with a symmetric flange (*the arrow*). (**n**) Spindle-like structure with asymmetric flange (*the arrow*). (**o**) Spindle-like structure with an inner spherical body. Slide GFWE2-S5, position R-W62.

biofilms, that is, microbial communities encapsulated within a polymeric matrix (e.g., O'Toole et al., 2000). Similar film- or mat-like structures have been reported from Archean carbonaceous cherts in the Barberton Greenstone Belt (e.g., Tice and Lowe, 2006; Walsh, 1992; Westall et al., 2001).

5.3. SPHEROIDAL STRUCTURES

Spheroidal structures, including nearly complete spheres, range from 2.5 to 60 μm in maximum diameter. Small spheroids with diameters ranging from 5 to 15 μm predominate. As a whole, the size distribution is approximately unimodal with a skewness toward the larger size of 60 μm. Sugitani et al. (2007) classified the spheroidal structures into small (<15 μm) and large (>15 μm).

5.3.1. Small Spheroids (<15 μm)

Small spheroids tend to occur as colony-like clusters, occasionally associated with fluffy envelopes or film-like structures (Fig. 3c). Clusters are generally irregularly shaped and are composed of less than 20 spheroids. Some clusters are spherical or rod like. Some are composed of more than 200 spheroids. The walls of the spheroids vary from hyaline to granular. These variations are consistent with taphonomic alteration. Although rare, fused pairs of two small spheroids are present (Fig. 4e). Small spheroids are quite abundant. In the examined area of ca. 150 mm^2 (GFSV3), 13 clusters are identified (Figs. 3 and 6). Each cluster is composed of 5–15 spheroids (Figs. 3k and 4d), giving rise to more than 100 spheroids that comprise clusters in this small area.

Small carbonaceous spheroids have already been reported from some Archean sedimentary successions (e.g., Dunlop et al., 1978; Knoll and Barghoorn, 1977; Walsh, 1992). However, their biogenicity was not always fully established or has been doubted (Oehler, 1976; Fox et al., 1983). This is attributed partly to their poorly identifiable wall features and/or lack of obvious colony-like clustering. In contrast, small spheroids of the Farrel Quartzite assemblage have clearly identifiable walls with taphonomic features and occur in colony-like clusters, some of which have been extracted by palynological procedures (Grey and Sugitani, 2009). These features are common for both modern microbes and younger microfossils (e.g., Awramik and Barghoorn, 1977; Hofmann and Jackson, 1991; Knoll, 1982; Oehler, 1978; Schopf, 1968). Thus, cluster-forming small spheroids of the Farrel Quartzite assemblage are likely biogenic. It is also reasonable to suggest that solitary small spheroids are liberated cells from colonies, despite the absence of independent evidence for their biogenicity.

5.3.2. Large Spheroids (>15 μm)

Large spheroidal structures range from 15 to 60 μm in maximum diameter and include several subtypes. Some of them have recently been classified taxonomically by Sugitani et al. (2009a, b), proposing four types (1) simple single-walled spheroids (Fig. 4f), (2) thin-walled spheroids having a diffuse envelope (Fig. 4h, i), (3) thick-walled spheroids (Fig. 4k), and (4) spheroids having an extensively folded wall (Fig. 4j). Simple single-walled spheroids, the most common and representative type, are described in detail here.

The walls of simple single-walled spheroids are generally less than 1 μm in thickness. Some walls are granular in appearance, whereas others are hyaline (Grey and Sugitani, 2009). Heterogeneously distributed carbonaceous clots or a reticulate pattern can be seen on the hyaline walls (Sugitani et al., 2009a, b). The wall is wrinkled to various degrees, dimpled, blistered, or partially double layered (Fig. 4g). Specimens that appear to be broken or torn are common. The wall features suggest that they were originally composed of flexible but breakable material like organic membranes, providing evidence for biogenicity. Similar wall morphologies are common in Proterozoic organic-walled microfossils (e.g., Knoll et al., 1991; Hofmann and Jackson, 1991; Samuelsson et al., 1999).

Simple single-walled spheroids sometimes contain a single inner hollow spheroid or multiple spheroids or are associated with multiple small spheroids inside and/or outside (Figs. 4l and 5a). The inner objects are hollow and in a few specimens appear to be expelled from the outer spheroid. Thus, some large spheroids possibly represent reproductive mother cells (multiple fissions) or resistant outer wall containing endospores inside, providing compelling evidence for the biogenicity of the large spheroids in the Farrel Quartzite assemblage (Sugitani et al., 2009a, b). Relatively large carbonaceous spheroids have been reported from older Archean successions (Nagy and Nagy, 1969; Schopf and Barghoorn, 1967; Schopf and Packer, 1987; Walsh, 1992). The present study may provide new insights into these spheroids whose biogenicity has not yet been widely accepted.

5.4. LENTICULAR TO SPINDLE-LIKE STRUCTURES

Lenticular to spindle-like structures (hereafter abbreviated as spindle-like structures) involve a wide range of morphologies that are composed of a generally hollow lens, oval or spheroid with spear- to flange-like appendage (Fig. 4m–o). Most of them are 20–60 μm in maximum length. Broken specimens are common.

Flange-like appendages characterize this morphological type. Two styles of flange are recognized. The flange surrounds the lenticular body at its equatorial plane (Fig. 4m) or is attached peripherally at one hemisphere (Fig. 4n). Sugitani et al. (2009a, b) called these two types as symmetric and asymmetric flanges. In both types, the width, thickness, outline, and texture are variable between specimens. The width of the flanges varies from less than 5 to 20 μm, while it is relatively constant in a given specimen and does not exceed the diameter of body. The flange is sometimes very thin, less than 1 μm, and appears to be constant in thickness, whereas in other cases it thickens toward the body and its thickness reaches

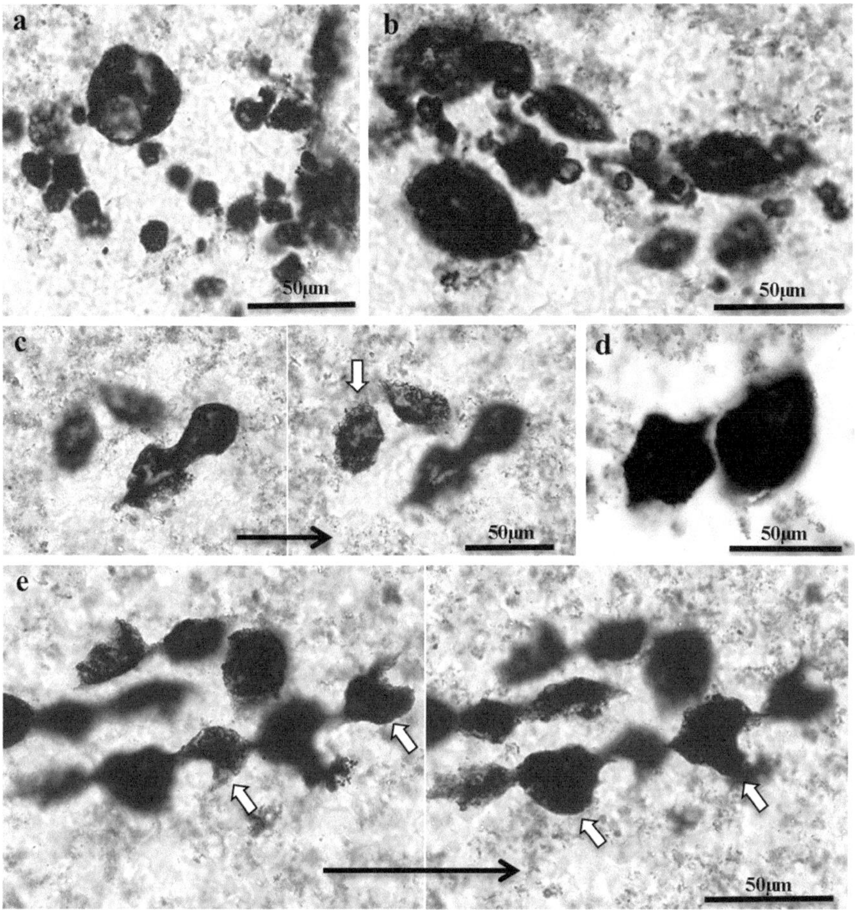

Figure 5. Photomicrographs of elaborate morphologies and occurrence in GFSV3. Localities of all specimens in the thin section are indicated in Fig. 3. (**a**) Cluster composed of a single large spheroid and multiple small ones. (**b**) Cluster composed of large lenticular structures and small spheroids. (**c**) Dumbbell-like structure composed of connected pair of spheroids and obliquely arranged pair of lenticular structures (the *white arrow*). The *solid arrow* shows a deepening focal depth. (**d**) Nearly parallel arranged pair of spindle-like structures. (**e**) Chain-like arrangement of spindle-like structures. The *solid arrow* shows a deepening focal depth. The spindle units are not linearly arranged, but alternately. This is inferred from the focusing pattern of the units (*white arrows*) depending on the focal depth.

more than 5 μm. The flange has diffuse or smooth margins and has textural variation such as being hyaline, reticulated, or striped. In some specimens, the texture of the flange and that of the body are continuous, whereas in others they are discontinuous. Lateral extension of the spheroid or lenticular body by sediment compaction cannot explain this substructure, because in colony-like clusters, spindles are often oriented randomly and spindle-like structures often occur as obliquely aligned, detached, or attached pairs (Fig. 3b, d, h, i, j). Consequently, these flange-like appendages are primary. Although any direct extant homeomorph for this structure is difficult to identify, it is intriguing that the mature phycoma of Pterosperma, a prasionphyte alga, has a flange-like ala (Tappan, 1980). Several Proterozoic microfossils have structures equivalent to flanges (Samuelsson et al., 1999).

Some of spindle-like structures are probably related to spheroidal structures, because spindle-like structures containing spheroidal objects (Fig. 4o) and colony-like clusters composed of spindles and relatively small spheres (Fig. 5b) are present. On the basis of these occurrences, Sugitani et al. (2009a, b) suggest that some of spindle-like structures may represent robust outer wall of cell at resting stage with spheroidal endospore inside, like similar-shaped microstructures in Archean carbonaceous chert in the Barberton Greenstone Belt (Walsh, 1992). Spindle-like structures, on the other hand, also comprise other elaborate morphologies including linearly or obliquely arranged pairs and chain-like structures (Figs. 3e–h, j and 5c–e). These morphologies are common and their variations are unlikely taphonomic. Rather they represent diversity in spindle-like structures and/or morphological variation within the life cycle of a single species, although the taxonomy of spindle-like structures is poorly understood.

6. Significance and Unresolved Problems of the Farrel Quartzite Assemblage

When life emerged and how life evolved during the early history of the Earth have long been topics of interdisciplinary investigation. However, most fossil records reported from Paleoarchean and Mesoarchean rocks have been treated as being controversial or even artifacts (e.g., Brasier et al., 2002; Buick, 1984; García-Ruiz et al., 2003; Hofmann, 2004; Moorbath, 2005; Schopf et al., 2002; Schopf and Walter, 1983). This is partly due to the small size, morphological simplicity, and poor preservation of many of described putative microfossils. Their wall and internal textures are generally poorly preserved or invisible under the standard petrographic microscope, which also gives rise to skepticism about these putative microfossils.

In this context, the Farrel Quartzite assemblage is inconsistent with the known Archean record of putative microfossils. In addition to the thread-like and film-like structures and small spheroidal structures morphologically analogous to prokaryotic organisms and their products, the assemblage contains three-dimensionally preserved spheroidal and spindle-like structures of large size (>15 μm) that have clearly identifiable, membranous walls. These spheroidal and spindle-like structures often comprise colony-like clusters or occur as pairs. As shown in Fig. 6, such elaborate morphologies and occurrences are occasionally abundant. In the relatively prolific portion of ca. 150 mm^2 in area and 30 μm in thickness, ten colony-like clusters of spindles, ten large spheroids, eight clusters composed of spheroids and spindles, 16 spindles, and five paired structures are

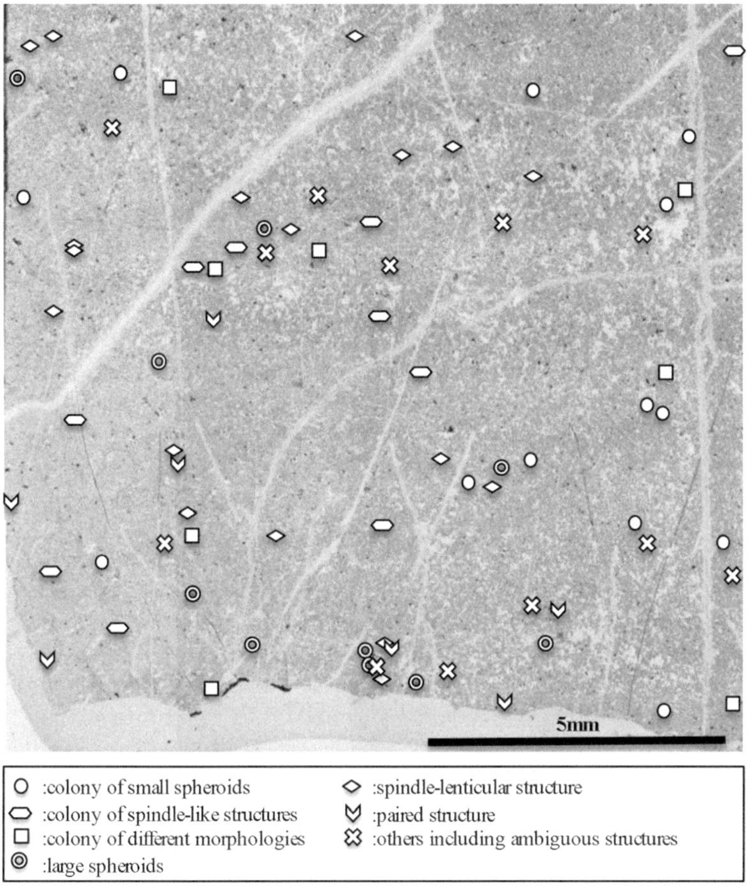

Figure 6. Composite photomicrograph of selected area (see Fig. 3) of GFSV3, showing distribution of microfossils roughly classified into seven categories.

contained. Such occurrences are comparable with well-preserved organic-walled microfossils in the Proterozoic (e.g., Knoll et al., 1991; Hofmann and Jackson, 1991; Samuelsson et al., 1999).

While these accumulating lines of evidence and data demonstrate biogenicity of the Farrel Quartzite assemblage and its potential as evidence of the early evolution of life on Earth, several problems need to be resolved:

1. In order to advance investigations of possible phylogeny and affinity, taxonomy needs to be established. However, only provisional taxonomy has been proposed for part of large spheroidal structures (Sugitani et al., 2009a, b). Taphonomic alteration and possible life cycle variants make it difficult to interpret morphological diversity straightforward in the taxonomic context.

2. Large spheroidal and spindle-like structures are >15 μm and occasionally up to 80 μm along their major axis, and comprise elaborate morphologies that are difficult to compare directly with other fossils and modern organisms. Their sizes are more consistent with eukaryotic than prokaryotic microorganisms, which however should be a critical issue, considering the age of the host rocks and current debate about the timing of the rise of the eukaryotes (e.g., Knoll et al., 2006).

In order to resolve these problems, continuing search of well-preserved specimens and data accumulation are indispensable. In addition to standard paleontological approaches, introducing new technologies such as Nano-SIMS (Oehler et al., 2009) and 3D Raman analyses (Schopf and Kudryavtsev, 2005) to them could be a help to make a breakthrough in taxonomy of this enigmatic microfossil assemblage in the Archean.

7. Summary

The microfossil assemblage recently discovered from the Farrel Quartzite (ca. 3.0 Ga) in the Goldsworthy greenstone belt (the Mount Goldsworthy and Mount Grant) in the Pilbara Craton is composed of diverse morphological types, suggesting an early evolution of a complex ecosystem in the Archean shallow water environment. In addition to preliminarily defined main five morphological types (thread like, film like, spheroidal [>15 and <15 μm], and lenticular to spindle-like), highly elaborate morphologies and occurrences that might represent reproducing or resting stages have recently been recognized and some taxonomic interpretations are also presented. Although the assemblage involves some unresolved problems regarding taxonomy, phylogeny, and affinities, accumulation of high-quality data of well-preserved specimens and introducing new technologies to them could be a help to make a breakthrough in revealing a whole picture of the assemblage. This is not unrealistic, considering that the Goldsworthy greenstone belt is quite prolific for this microfossil assemblage.

8. Acknowledgments

We especially wish to express our gratitude to Dr. Martin J. Van Kranendonk and Dr. Kathleen Grey, Geological Survey of Western Australia, for their help in our fieldwork and constructive discussions. Financial support to KS from the Japan Society for the Promotion of Science (the Joint Research Program, Japan–Australia and a grant-in-aid no. 19340150) is gratefully acknowledged. MRW acknowledges funding from the University of NSW and the Australian Research Council. An invitation to contribute and editorial support by Dr. V. Tewari and helpful comments by Dr. M.M. Walsh are greatly appreciated.

9. References

Awramik, S.M. and Barghoorn, E.S. (1977) The Gunflint microbiota. Precambrian Res. **5**: 121–142.

Brasier, M.D., Green, O.R., Jephcoat, A.P., Kleppe, A.K., Van Kranendonk, M.J., Lindsay, J.F., Steele, A. and Grassineau, N.V. (2002) Questioning the evidence for Earth's oldest fossils. Nature **416**: 76–81.

Buick, R. (1984) Carbonaceous filaments from North Pole, Western Australia: are they fossil bacteria in Archaean stromatolites? Precambrian Res. **24**: 157–172.

Dunlop, J.S.R., Milne, V.A., Groves, D.I. and Muir, M.D. (1978) A new microfossil assemblage from the Archaean of Western Australia. Nature **274**: 676–678.

Fox, S.W., Syren, R.M., Ingram, M., Price, B.J. and Costello, J. (1983) Ancient microspheres: abiogenic, protobiogenic, or biogenic? Precambrian Res. **23**: 1–8.

García-Ruiz, J.M., Hyde, S.T., Carnerup, A.M., Christy, A.G., Van Kranendonk, M.J. and Welham, N.J. (2003) Self-assembled silica-carbonate structures and detection of ancient microfossils. Science **302**: 1194–1197.

Grey, K. and Sugitani, K. (2009) Palynology of Archean microfossils (c.3.0 Ga) from the Mount Grant area, Pilbara Craton, Western Australia: further evidence of biogenicity. Precambrian Res. **173**: 60–69.

Hickman, A.H. and Van Kranendonk, M.J. (2008) Archean crustal evolution and mineralization of the northern Pilbara Craton – a field guide. Geological Survey of Western Australia, Record 2008/13.

Hofmann, H.J. (2004) Archean microfossils and abiomorphs. Astrobiology **4**: 135–136.

Hofmann, H.J. and Jackson, G.D. (1991) Shelf-facies microfossils from the Uluksan Group (Proterozoic Bylot Supergroup), Baffin Island, Canada. J. Paleontol. **65**: 361–382.

Knoll, A.H. (1982) Microfossils from the late Precambrian Draken Conglomerate, Ny Friesland, Svalbard. J. Paleontol. **56**: 755–790.

Knoll, A.H. and Barghoorn, E.S. (1977) Archean microfossils showing cell division from the Swaziland System of South Africa. Science **198**: 396–398.

Knoll, A.H., Swett, K. and Mark, J. (1991) Paleobiology of a Neoproterozoic tidal flat/lagoonal complex: the Draken Conglomerate Formation, Spitsbergen. J. Paleontol. **65**: 531–570.

Knoll, A.H., Javaux, E.J., Hewitt, D. and Cohen, P. (2006) Eukaryotic organisms in Proterozoic oceans. Phil. Trans. Roy. Soc. **361B**: 1023–1038.

Moorbath, S. (2005) Dating earliest life. Nature **434**: 155.

Nagy, B. and Nagy, L.A. (1969) Early Pre-Cambrian Onverwacht microstructures: possibly the oldest fossils on Earth? Nature **223**: 1226–1229.

Oehler, J.H. (1976) Hydrothermal crystallization of silica gel. Geol. Soc. Am. Bull. **87**: 1143–1152.

Oehler, D.Z. (1978) Microflora of the middle Proterozoic Balbirini Dolomite (McArthur Group) of Australia. Alcheringa **2**: 269–309.

Oehler, D.Z., Robert, F., Walter, M.R., Sugitani, K., Allwood, A., Meibom, A., Mostefaoui, S., Selo, M., Thomen, A. and Gibson, E.K. (2009) NanoSIMS: insights to biogenicity and syngeneity of Archaean carbonaceous structures. Precambrian Res. **173**: 70–78.

O'Toole, G., Kaplan, H.B. and Kolter, R. (2000) Biofilm formation as microbial development. Ann. Rev. Microbiol. **54**: 49–79.

Samuelsson, J., Dawes, P.R., Vidal, G. (1999) Organic-walled microfossils from the Proterozoic Thule Supergroup, Northwest Greenland. Precambrian Res. **96**: 1–23.

Schopf, J.W. (1968) Microflora of the Bitter Springs Formation, late Precambrian, central Australia. J. Paleontol. **42**: 651–688.

Schopf, J.W. (2006) Fossil evidence of Archaean life. Phil. Trans. Roy. Soc. **361B**: 869–885.

Schopf, J.W. and Barghoorn, E.S. (1967) Alga-like fossils from the early Precambrian of South Africa. Science **156**: 508–512.

Schopf, J.W. and Kudryavtsev, A.B. (2005) Three-dimensional Raman imagery of Precambrian microscopic organisms. Geobiology **3**: 1–12.

Schopf, J.W. and Packer, B.M. (1987) Early Archean (3.3-billion to 3.5-billion-year-old) microfossils from Warrawoona Group, Australia. Science **237**: 70–73.

Schopf, J.W. and Walter, M.R. (1983) Archean microfossils: new evidence of ancient microbes, In: J.W. Schopf (ed.) *Earth's Earliest Biosphere, Its Origin and Evolution*. Princeton University Press, Princeton, pp. 214–239.

Schopf, J.W., Kudryavtsev, A.B., Agresti, D.G., Wdowiak, T.J. and Czaja, A.D. (2002) Laser-Raman imagery of Earth's earliest fossis. Nature **416**: 73–76.

Sugitani, K., Mimura, K., Suzuki, K., Nagamine, K. and Sugisaki, R. (2003) Stratigraphy and sedimentary petrology of an Archean volcanic-sedimentary succession at Mt. Goldsworthy in the Pilbara Block, Western Australia: implications of evaporite (nahcolite) and barite deposition. Precambrian Res. **120**: 55–79.

Sugitani, K., Yamashita, F., Nagaoka, T., Yamamoto, K., Minami, M., Mimura, K. and Suzuki, K. (2006) Geochemistry and sedimentary petrology of Archean clastic sedimentary rocks at Mt. Goldsworthy, Pilbara Craton, Western Australia: evidence for the early evolution of continental crust and hydrothermal alteration. Precambrian Res. **147**: 124–147.

Sugitani, K., Grey, K., Allwood, A., Nagaoka, T., Mimura, K., Minami, M., Marshall, C.P., Van Kranendonk, M.J. and Walter, M.R. (2007) Diverse microstructures from Archaean chert from the Mount Goldsworthy–Mount Grant area, Pilabra Craton, Western Australia: microfossils, dubiofossil, or pseudofossils. Precambrian Res. **158**: 228–262.

Sugitani, K., Grey, K., Nagaoka, T., Mimura, K. and Walter, M.R. (2008) Highly elaborate putative microfossils from >2.97 Ga chert, Pilbara Craton: indications of cell division? *Australian Earth Science Convention 2008, Abstract*, pp. 235.

Sugitani, K., Grey, K., Nagaoka, T. and Mimura, K. (2009a) Three-dimensional morphological and textural complexity of Archean putative microfossils from the northeastern Pilbara Craton: indications of biogenicity of large (>15 μm) spheroidal and spindle-like structures. Astrobiology **9**: 603–615.

Sugitani, K., Grey, K., Nagaoka, T., Mimura, K. and Walter, M.R. (2009b) Taxonomy and biogenicity of Archaean spheroidal microfossils (c. 3.0 Ga) from the Mount Goldsworthy-Mount Grant area in the northeastern Pilbara Craton, Western Australia. Precambrian Res. **173**: 50–59.

Tappan, H. (1980) *The Paleobiology of Plant Protists*. W.H. Freeman, San Francisco, CA, pp. 1028.

Tice, M.M. and Lowe, D.R. (2006) The origin of carbonaceous matter in pre-3.0 Ga greenstone terrains: a review and new evidence from the 3.42 Ga Buck Reef Chert. Earth Sci. Rev. **76**: 259–300.

Ueno, Y., Isozaki, Y., Yurimoto, H. and Maruyama, S. (2001) Carbon isotopic signatures of individual Archean microfossils (?) from Western Australia. Int. Geol. Rev. **43**: 196–212.

Ueno, Y., Yoshioka, H., Maruyama, S. and Isozaki, Y. (2004) Carbon isotopes and petrography of kerogens in ~3.5-Ga hydrothermal silica dikes in the North Pole area, Western Australia. Geochim. Cosmochim. Acta **68**: 573–589.

Van Kranendonk, M.J., Hickman, A.H., Smithies, R.H., Williams, I.R., Bagas, L. and Farrell, T.R. (2006) Revised lithostratigraphy of Archean supracrustal and intrusive rocks in the northern Pilbara Craton, Western Australia. Geological Survey of Western Australia, Record **2006/15**.

Walsh, M.M. (1992) Microfossils and possible microfossils from the Early Archean Onverwacht Group, Barberton Mountain Land, South Africa. Precambrian Res. **54**: 271–293.

Westall, F., de Wit, M.J., Dann, J., van der Gaast, S., de Ronde, C.E.J. and Gerneke, D. (2001). Early Archean fossil bacteria and biofilms in hydrothermally-influenced sediments from the Barberton greenstone belt, South Africa. Precambrian Res. **106**: 93–116.

Biodata of **Professor Vinod Chandra Tewari** and **Professor Maurice E. Tucker**, authors of ***"Ediacaran Krol Carbonates of the Lesser Himalaya, India: Stromatolitic Facies, Depositional Environment and Diagenesis"***

Professor Vinod C. Tewari is currently the Head of the Sedimentology Group at Wadia Institute of Himalayan Geology, Dehradun and a Senior Associate of the International Centre for Theoretical Physics, Trieste, Italy. He obtained his Ph.D. from the University of Lucknow in *Geology* in 1986 and continued his research in the Wadia Institute. Dr. Tewari taught Geology at Kumaon University, Nainital, Uttarakhand (U.K.), India as Professor of Geology. Professor Tewari's scientific interests are in the areas of Precambrian stromatolites, sedimentation, carbon isotope chemostratigraphy, genesis, early evolution and diversification of life and its astrobiological significance. His current interests are carbonate sedimentology of the Lesser and Tethyan carbonates, speleothems, a paleoclimate and monsoon study of Himalayan cave systems. He is associated with the International Geological Correlation Programme (I.G.C.P.) Project 493 on The Rise and Fall of Vendian Biota. He has 80 research papers published to his credit and has edited several volumes of Himalayan Geology, India and the Journal of Nepal Geological Society, Kathmandu, Nepal. Professor Tewari organised the first *Indo-Soviet Symposium on Stromatolites and Stromatolitic Deposits* and other IGCP meetings in India. He was one of the organizers of the World Summit on Ancient Microscopic Fossils held at the University of California, Los Angeles, USA in 2008.

E-mail: **vtewari@wihg.res.in**

V.C. Tewari and J. Seckbach (eds.), *STROMATOLITES: Interaction of Microbes with Sediments*, Cellular Origin, Life in Extreme Habitats and Astrobiology 18, 133–156
DOI 10.1007/978-94-007-0397-1_7,

Professor Maurice E. Tucker is Professor of Geological Sciences at Durham University, following earlier appointments at the Universities of Newcastle Upon Tyne, Cardiff and Sierra Leone, and UC Berkeley. His interests are in all rocks that fizz and he has worked on limestones from all geological periods, from many parts of the world. His current interests are in the origin of bedding in shelf limestones, Quaternary highstand carbonates of western Australia, and the role of microbes in the formation of tufa, stromatolites and dolomite. His work on Precambrian carbonates in the Lesser Himalaya and eastern Ghats, India continues through collaboration with Professor Vinod C. Tewari of the Wadia Institute of Himalayan Geology, Dehradun and Sarbani Patranabis Deb of the Indian Statistical Institute, Kolkata.

E-mail: **m.e.tucker@durham.ac.uk**

EDIACARAN KROL CARBONATES OF THE LESSER HIMALAYA, INDIA: STROMATOLITIC FACIES, DEPOSITIONAL ENVIRONMENT AND DIAGENESIS

VINOD CHANDRA TEWARI[1] AND MAURICE E. TUCKER[2]
[1]*Wadia Institute of Himalayan Geology, Dehradun 248001, Uttarakhand, India*
[2]*Department of Earth Sciences, Durham University, Durham DH1 3LE, UK*

Abstract Upper Krol Formation (Krol D Member) is a typical Ediacaran (Terminal Neoproterozoic) microbial carbonate facies. It is well developed in the Lesser Himalayan Krol belt extending from the Solan in the Himachal to the Nainital in Uttarakhand Lesser Himalaya, India. Krol D contains Ediacaran metazoan fossils and abundant microbial structures in the form of columnar, domal and stratified stromatolites. Peloids, oncoids and microphytolites are also microbially formed structures. Carbonate grainstones and packstones are common in the Krol D, and the allochemical constituents are intraclasts, ooids, coated grains, peloids, microbial grains, and catagraphs. Sedimentary structures present in the non-stromatolitic Krol carbonates include bird's eyes, cross bedding, symmetrical ripples and shallow channel structures. Mud cracks are also present. An intertidal to supratidal (shallow water) depositional environment with some moderate currents and intermittent periods of exposure has been suggested for the Krol D Member of the Lesser Himalaya. Petrography, cathodoluminescence (CL) and electron probe micro analysis (EPMA) has been done to study the diagenetic and chemical changes in the microfacies of the Krol D Member. Intraclasts are rectangular in section and composed of coarse dolomite crystals. CL shows that these have growth zones. Peloidal structures are composed entirely of micrite. Ooids in the Krol D Member are mostly spheroidal to ellipsoidal grains and range in diameter from 0.1 to 0.4 mm. Composite ooids are larger grains and consist of smaller ooids. The CL of the ooids shows non-luminescence. The diagenetic history of the Krol D ooids and intraclasts suggest that many of these grains may have precipitated as micrite and possibly microbial micritization has taken place. The Krol D dolomite shows abundant evidence of marine cementation as the first diagenetic stage. A second type of cement is fibrous crust. Under CL, the fibrous crusts are brightly luminescent. These fibrous and micritic cements are of shallow marine origin. The diagenetic textures suggest that originally fine-grained carbonate was replaced by fine dolomite and that coarse fibrous high Mg calcite cements were replaced by dolomite too, preserving the original fabrics of the crystals. Terminal Neoproterozoic sea water chemistry must have contributed to the dolomitization in the Krol D.

Keywords Ediacaran • Lesser Himalaya • India • Stromatolitic facies • Isotope Stratigraphy • Diagenesis • Microbial carbonates • Uttarakhand • Krol D • Grainstones • Packstones • Intertidal • Cathodoluminescence • Microfacies • Intraclasts • Ooids • Cementation • Dolomitisation

1. Introduction

The basement of the Lesser Himalaya is not exposed; however, sedimentological evolution of the Lesser Himalaya initiated with late Palaeoproterozoic rifting event (1,800 Ma) followed by a shelf cycle of tidal flat sedimentation during Mesoproterozoic. The North-Western Lesser Himalaya is characterized by two major sedimentary belts in sequence stratigraphic order. The older sequence is late Palaeoproterozoic to early Neoproterozoic (750 Myr) in age and major lithofacies of this belt are (I) Volcano-siliciclastic (Rampur–Manikaran–Chamoli–Berinag Quartzite), (II) Clastic-argillaceous (Sundernagar–Hurla–Rautgara–Rudraprayag–Bhawali Quartizite and volcanics), (III) Microbial-Stromatolitic carbonate–phosphorite (Shali–Larji–Deoban–Lameri–Pipalkoti–Gangolihat–Dharchula) and (IV) Argillo-calcareous (Sataun, Mandhali–Sor–Thalkedar) in stratigraphic order. The glacial boulder beds have not been reported from the inner belt.

The Deoban–Jaunsar Group in the Lesser Himalaya represents preglacial, preEdiacaran older carbonate-siliciclastic shallow marine sedimentation. The complete absence of Ediacaran biota, non-glacial sedimentation and characteristic Mesoproterozoic stromatolite taxa and microfossils in black cherts confirm a preEdacaran age. The unconformably overlying Balini (diamictite) Formation is a Neoproterozoic (Marinoan/Blainian) glacial deposit corresponding to Snowball Earth palaeo latitude (Cryogenian Period, Tewari, 2001a). The overlying pink cap microbial carbonate of the Blaini Formation represents the base of the Terminal Neoproterozoic. The younger carbonate sedimentary succession of the Krol belt is Terminal Proterozoic (Ediacaran, 650–540) Myr in age and stretching over a distance of 350 km showing major facies variations at Solan, Nigalidhar, Korgai, Mussoorie (Fig. 1), Garhwal and Nainital synclines (Tewari, 2002a, b). The Krol carbonates represent passive continental marginal (carbonate ramp) facies variations as cyclic para sequences such as shaly limestone and calcareous shale facies and purple green shales with lenticular bands of limestone and gypsum. Brecciated cherty, oolitic dolomite facies characterized by various types of ooid, bird's eye structure, microbial laminated and stromatolitic build ups, and oncolites indicate that the depositional environment was tidal falt (high energy peritidal, Tewari and Qureshi, 1985; Tewari and Joshi, 1993; Tewari, 2002a, b; Jiang et al., 2003).

The Ediacaran microbial mat and metazoan facies of the Krol is composed of finely laminated siltstone, shale and sandy layers with wave ripples. The shaly

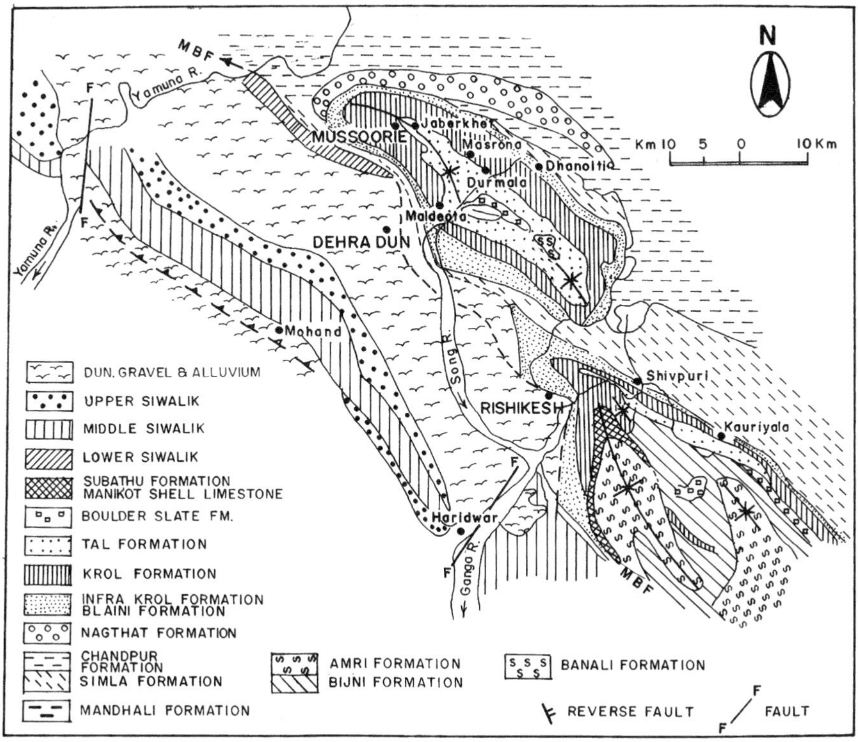

Figure 1. Geological map of the Ediacaran sedimentary basins of the Lesser Himalaya, India.

limestone facies of the Upper Krol grades into the Chert–Phosphorite Facies of the Lower Tal characterized by black chert, shale and phosphorite associated with pyrite, oncolites and stromatolites (Tewari, 1984a, b, 1989, 1990, 1992, 1993a, b, 1999, 2001a, b, 2002a, b, 2004, 2007). The Precambrian/Cambrian boundary transitional facies shows evidence of upwelling and stratification of ocean as revealed by carbon isotope excursions in Krol-Tal basin strata. The other Lower Cambrian facies of the Tal Formation include bioturbated purple grey siltstone (trace fossil) and channel sandstone, orthoquartzite of fluvio-deltaic and marine shelf facies at the top of the sequence. The Krol-Tal basin was possibly obliterated during Lower Cambrian period due to Pan African epeirogenic movements around 550 Myr (by well-documented granites of this age, Almora and Mandi granites) in the Lesser Himalayan region. The post Ediacaran sedimentation of the Lesser Himalaya is represented by chert–phosphorite, siltstone and sandstone facies (lagoonal tidal flat environment). The stratigraphic position of the Tal Formation is restricted to only the central part (Nigalidhar–Korgai–Mussoorie and Garhwal synclines). The Upper Tal Formation (Quartzite) marks the end of sedimentation and regression of the sea from the Lesser Himalaya.

The Lesser Himalayan sediments are covered by three main Phanerozoic marine transgressions of (1) Permian, (2) Late Cretaceous and (3) Late Palaeocene to Eocene age in different areas of the Lesser Himalaya from Jammu in the NW to Arunachal Pradesh in the NE. Recent sedimentological studies and palaeobiological discoveries from the Shergaon–Buxa-Miri Group sediments, Arunachal Pradesh suggest that they are equivalents of the Blaini–Krol-Tal sediments and, therefore, quite significant for Proterozoic-Cambrian palaeoclimate and basin evolution in the Eastern Himalaya (Tewari, 2003). The northern margin of the Indian Gondwanaland must have included parts of Tibet and China, during Neoproterozoic–Cambrian times forming a megacrustal block. It is supported by the close facies relationships and identical biotic and isotopic events from the Sino-Indian region (Fig. 2).

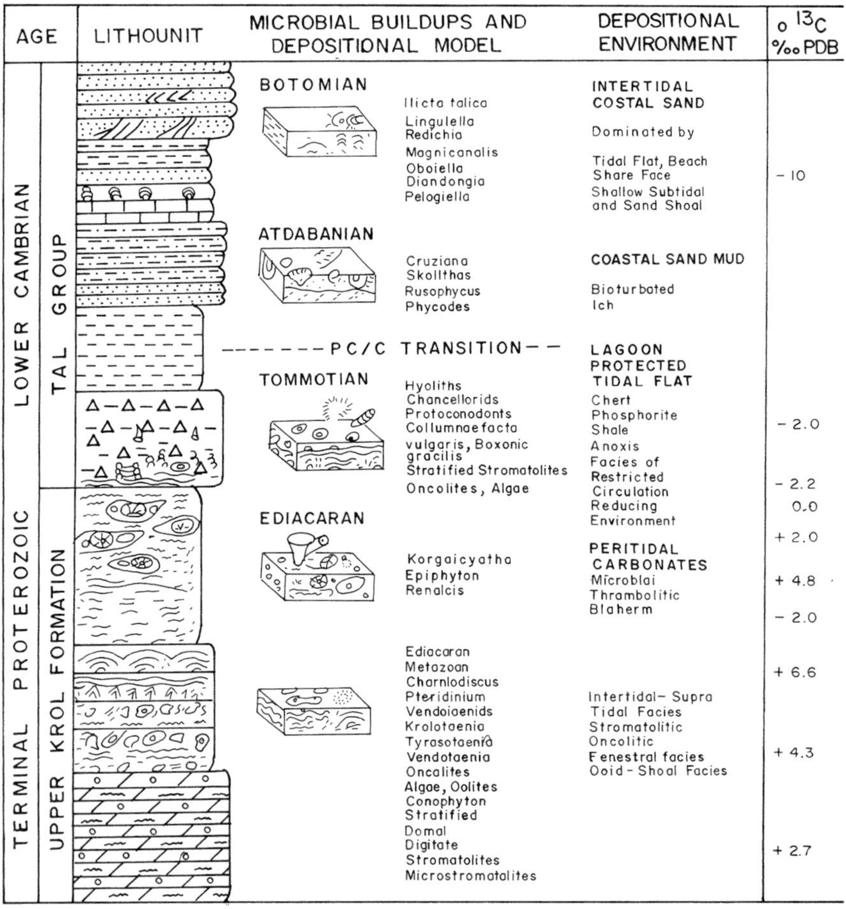

Figure 2. Depositional environment of the Ediacaran Krol carbonates, Lesser Himalaya.

2. Krol D Member Facies

Stromatolites are a feature in the Krol Formation and vary from planar microbial laminites, with desiccation features, laminoid fenestrae, polygonal cracks and intraclasts, to columnar and domal stromatolites on a 10-cm to metre-scale (see Fig. 9a). In addition, there are lime mudstones, locally with laminae from mechanical deposition of mud, but without microbial lamiae. These lime muds were largely deposited from suspension with some traction current activity. The origin of the mud itself may well have been microbial of course.

Carbonate grainstones and packstones are common in the Krol and the grains here are intraclasts, ooids and coated grains, peloids, microbial grains and aggregates (catagraphs/microphytolites, Figs. 3–9).

Microbial grains and aggregates are composed of clots of micrite, irregular in shape, and range from 10s to 200 μm in size (Figs. 3, 4). They have lighter and

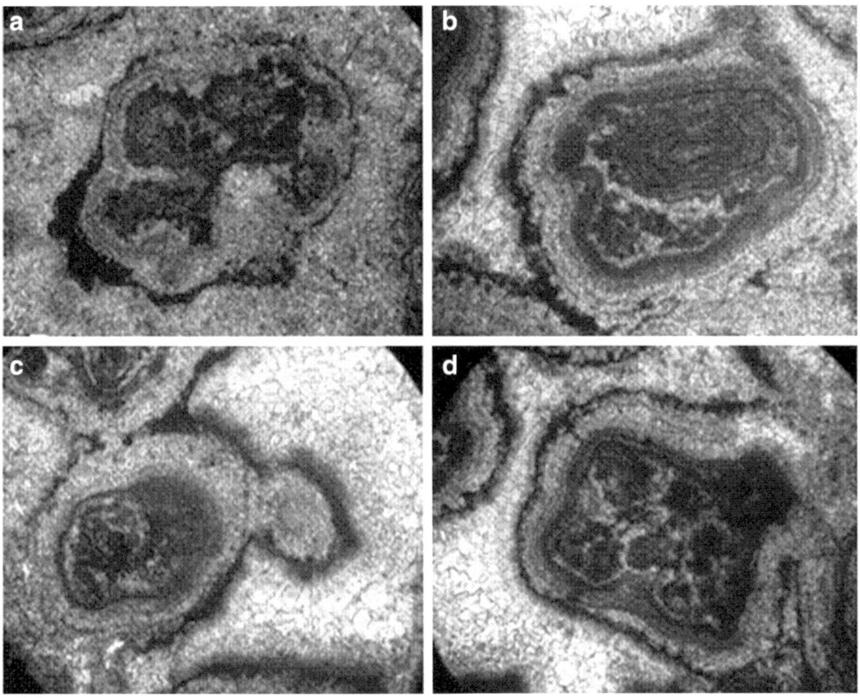

Figure 3. (**a**–**d**) Microbial grains composed of clotted micrite forming large compound grains which are also coated with marine cement fringes. These aggregate grains are showing the effects of etching or minor disolution at their margins, before being overgrown by an irregular micrite layer and then cemented by marine fibrous carbonate. The cavities are finally cemented by clear sparry dolomite cement. For all photos Fields of view 7 x 3 mm.

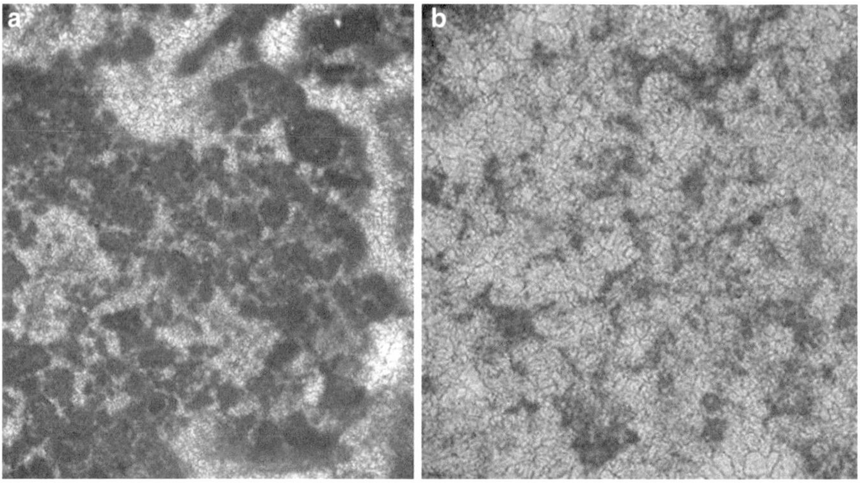

Figure 4. Microbial structures. (**a**) Microbial clots and peloids; (**b**) Lacy strings of micrite with knotted or beaded structures along their length. Fields of view 3 x 2 mm.

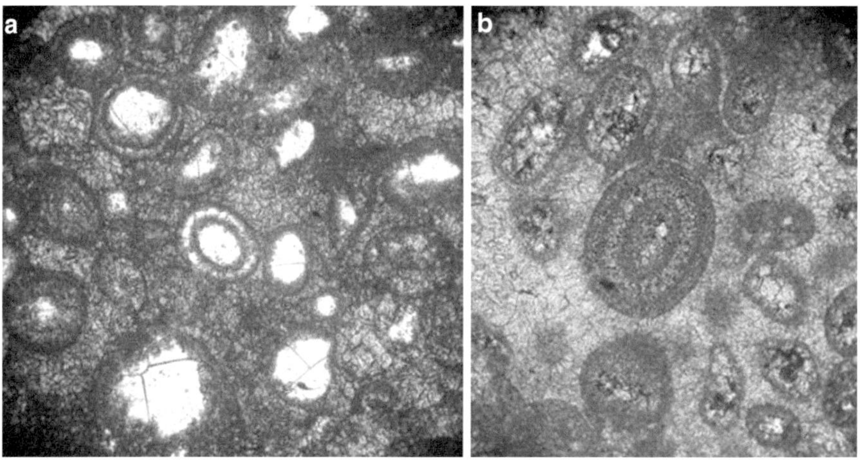

Figure 5. (**a**,**b**) Ooids with a variety of structure: thin coat around an intraclast; thicker cortical layer around a sparry dolomite centre (formerly the nucleus) or empty pore (dissolved out nucleus); micritic-microsparitic ooid with no nucleus; ooids with several thick cortical layers; ooid with dark micritic and clear sparry cortical layers. Both ooid examples are totally dolomite. Fields of view 5 x 3 mm.

darker (dense/less dense) patches, and so appear "lumpy". Sometimes there is a hint of short filamentous, worm-like micritic structures in the clots; in less dense areas, the micrite locally forms a lacy texture of "strings", in some cases with a knotted or beaded texture along the strings (Fig. 4b). The clots are commonly aggregated together to form larger grains several to many mm, even cm, in diameter. In some cases, there are cement crusts and coatings around the grains too. Micritic clots also occur within some of the oolitic or coated-grain wacke-

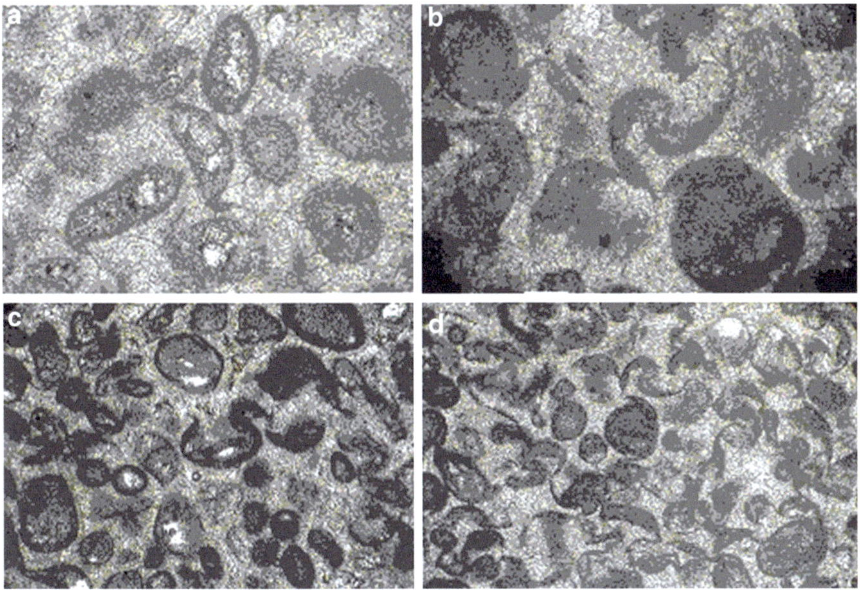

Figure 6. **(a–d)** Ooids showing effects of compaction from dissolution of metastable carbonate nucleus or cortical layers (presumably aragonite), generating duck-like structures or elephant trunk-to-tail texture. a and b Fields of view 7 x 5 mm. c and d Field view 5 x 3 mm.

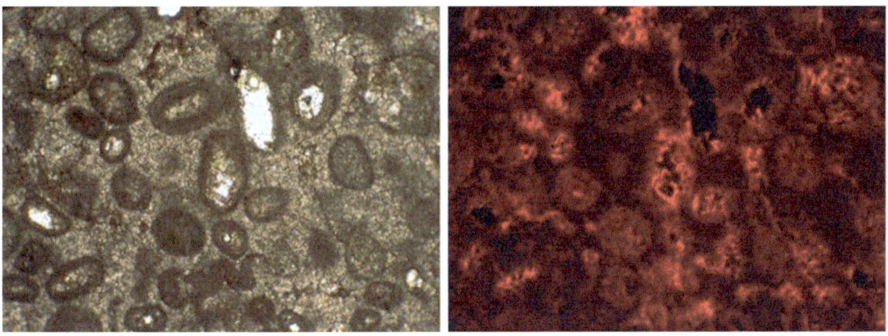

Figure 7. Oolite under PPL (*left*) and CL (*right*). Note the variety of CL response; some ooids are dull, others are non-luminescent, or just the outer cortical layer is dull/black. Some ooids have bright and zoned central parts, indicating they are occupied by cement and the ooid centres have suffered dissolution. Fields of view 5 x 3 mm.

packstones; they may also form laminae in some of the stromatolites and planar microbial laminites. It is quite likely that some of the lime mudstones were originally composed of this clotted micritic material too. These clots, which in other formations form the well-known thrombolitic texture (also texture grumuleuse), are probably of microbial origin, being mineralised microbes and organic matter, such as EPS (mucus). The microbes would appear to have been mostly of coccoid-type bacteria, but some filamentous bacteria may also have been present.

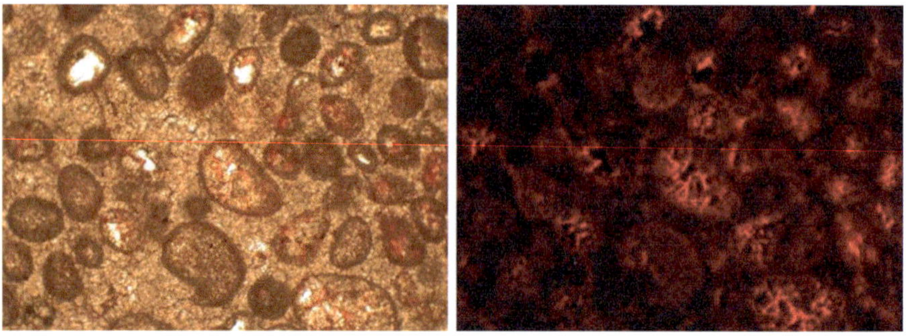

Figure 8. Coated grains with thick dark organic coating in the outer margins. PPL (*left*) and CL (*right*).

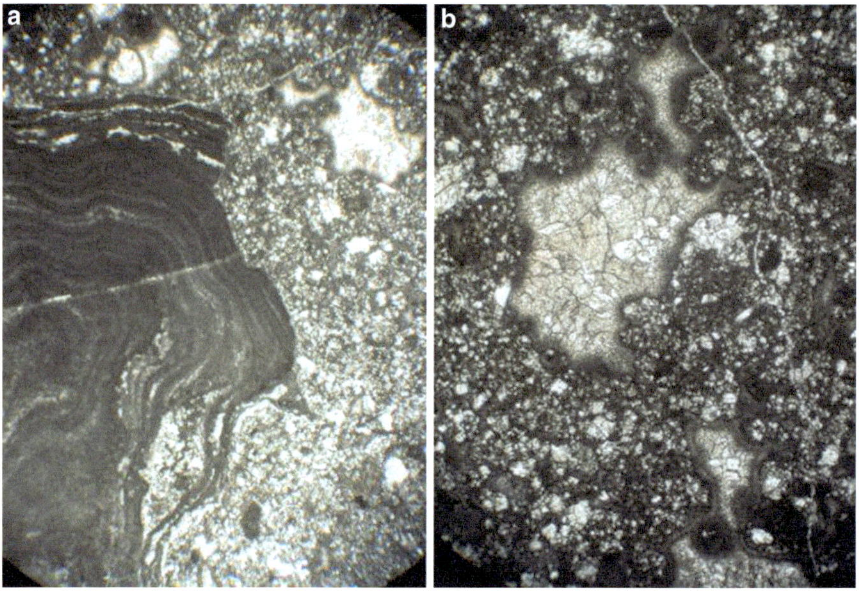

Figure 9. (**a**) Stromatolitic head with laminae extending into adjacent sediment, also evidence of erosion. (**b**) Fenestrae lined with micrite and fibrous carbonate cement fringe, then sparry dolomite cement, within a microbial-clot pack-grainstone. Field of view for 9a: 7 x 5 mm; Field of view for 9b: 5 x 5 mm.

Some of these microbial grains appear well structured and organised; hence, the terms catagraph or microphytolite have been applied by other workers.

Intraclasts are rectangular in section, ranging from several 100 μm to mm in length. They are generally composed of coarse dolomite crystals; cathodoluminescence (CL) shows that these have growth zones and so they are cement. It is likely that these intraclasts were composed of a metastable carbonate precursor, which dissolved out during early diagenesis. This is likely to have been aragonite (see later). Peloidal structures are composed entirely of micrite and have a uniform

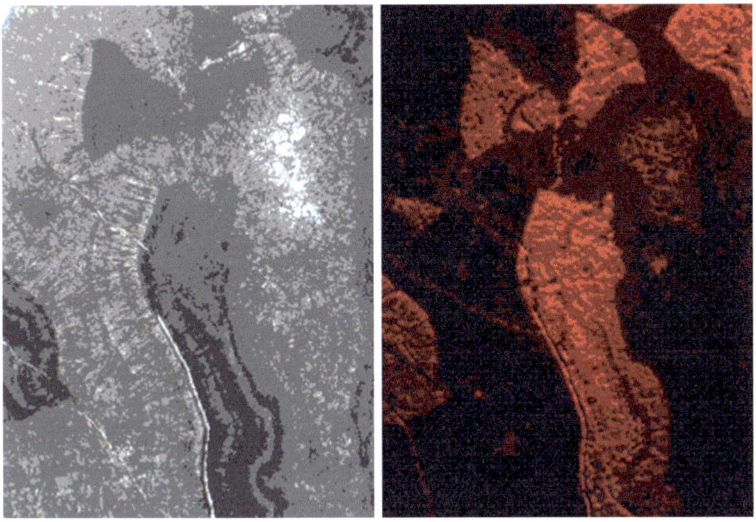

Figure 10. Thick fibrous carbonate cement fringe under PPL (*left*) and CL (*right*). The first fibrous cement crust is darker under PPL but bright under CL, and has been fractured and brecciated. The next generation fibrous cement is much more transparent under PPL but is non-luminescent. The final cavity filling dolomite cement is white under PPL and dull under CL. Fields of view 7 x 5 mm.

shape and size, being spheroidal to ellipsoidal, and up to 0.5 mm in diameter. Peloids may have been faecal matter, since it is likely that soft-bodied organisms were around at the time, or reworked microbial grains. The ooids and coated grains are described later.

Sedimentary structures present in non-stromatolitic Krol carbonates include bird's, eyes, mostly on the mm-scale (Fig. 9b), occurring within peloidal pack/grain-stones and mudstones (Fig. 9b). Cross laminae and cross bedding are present too, as well as symmetrical ripples and shallow channel structures. Desiccation cracks locally occur. Shallow-water deposition, with some moderate current activity, accounts for much of the succession, with intermittent periods of exposure (Fig. 2).

3. Original Mineralogy and Diagenesis of Krol D Member Oolites

3.1. OOIDS AND COATED GRAINS

Ooids in the Krol D Member are mostly spheroidal to ellipsoidal grains of diameter 0.1–0.4 mm (see Figs. 6, 7, 8). The majority are simple (an extremely thin or single cortical layer) to normal (a thicker cortex with several growth zones) ooids with a nucleus. However, some ooids do not seem to have a nucleus, at least in the plane of section. Compound (or composite) ooids, larger grains consisting of several smaller ooids enclosed by cortex, are rare.

3.2. ORIGINAL MINERALOGY OF GRAINS

The preservation of original detail in the Krol ooids is generally good, although the whole rock in all cases is composed of dolomite. The ooids are of several types: (a) wholly composed of micritic–microsparitic dolomite with a vague concentric lamination; (b) a cortex of micritic–microsparitic laminated dolomite around a coarse sparry dolomite nucleus; (c) as (b) but with an empty pore where the nucleus would have been; (d) a cortex of micrite–microsparite with a cortical layer of clear sparry dolomite or porosity. The nuclei consist of intraclasts, rectangular to rounded in section. Some ooids show the effects of compaction (see later) generating duck-like features or elephantine structures, as a result of dissolution of the nucleus or cortical layers (Fig. 6).

The CL of the ooids invariably shows non-luminescence for the ooid cortex, which is composed of dark micritic–microsparitic dolomite. Sparry dolomite crystals in the nucleus or in sparry cortical layers mostly have a dull (red) luminescence. However, within the central part of the nucleus area, there may be bright red growth zones.

3.3. INTERPRETATION

The structure of the ooids with micritic–microsparitic dolomite cortices and clear, coarse dolomite crystals forming the nuclei suggests different original mineralogies for the cortex and nucleus. Most likely would be that the cortex was originally composed of high-Mg calcite, since this is commonly replaced by dolomite with good textural preservation (Tucker, 2001). The cortex may well have had a micritic texture originally too, possibly from microbial micritisation (see below), and this then would have aided the dolomitisation by providing numerous nucleation points for dolomite crystal growth (Tucker, 2001). The coarse grey to white sparry dolomite crystals occurring in the ooids' centres are a replacement of the original nucleus. The CL evidence for inward growth suggests nucleation upon already-formed dolomite crystals within the cortex, with these dolomite crystals first replacing the nucleus carbonate (grey crystals – the colour due to retained impurities) and then dolomite being precipitated directly in the centre of the ooids, effectively as a cement, rather than a replacement, as the nucleus carbonate had dissolved away by this stage. It is difficult to say what the nuclei were originally, since their structure has gone, but some form of aragonitic intraclast is most likely. The presence of collapsed and compacted ooids (see below) supports this interpretation, and one is reminded of compacted bivalve shells preserved though the presence of micrite envelopes, which are so common in Phanerozoic limestones. Ooids with a sparry carbonate layer as part of the cortex are interpreted as originally bi-mineralic ooids. It is likely that they consisted of alternating high-Mg calcite and aragonite layers originally. Such ooids have been described from Late Precambrian carbonate strata before (Tucker, 1984, 1992). The presence of collapsed and compacted ooids (see below) supports this interpretation.

4. Diagenesis

4.1. MICRITE ENVELOPES AND MICRITISATION

Many of the grains in the Krol D are composed of micritic carbonate (ooids, peloids, microbial clots), and although some of this may have been precipitated as micrite, there is also the possibility that microbial micritisation has taken place. This could be the case with the ooids, which may have had coarser crystals, even radial ones, originally, although evidence for this was not seen. However, some of the larger coated grains do have thin dark zones on the outer margins of the grains (Fig. 8). In a Phanerozic limestone, these would be attributed to microbial micritisation. Sea-floor recrystallisation/neomorphism to give micrite is also a possibility (Reid and Macintyre, 1998).

4.2. MARINE CEMENTS

The Krol D member dolomites contain abundant evidence for marine cementation as the first diagenetic stage. Fibrous carbonate fringes and crusts up to several mm even a cm thick occur upon and within stromatolites, and also around ooids. Fenestrae within the granular-micritic sediment of stromatolitic facies are lined by isopachous fibrous carbonate (Fig. 9b). Fragments of the fibrous carbonate crusts occur within the sediment too, as intraclasts (Fig. 10), broken off through erosion or possibly desiccation.

Two types of fibrous crust are evident. One consists of elongate columnar crystals with a well-defined growth lamination (Fig. 10). Some of the fibrous crust layers are darker than others, as if they contain disseminated micritic carbonate or even clay. Under CL, the darker, probably more clayey fibrous crusts are brightly luminescent, often with the lamination showing clearly as varying CL colours. Along some laminae, there are variations too, dark and light zones; the brighter CL areas are probably areas of recrystallisation/replacement. Some clearer fibrous dolomite crusts are often non-luminescent. Also with well-developed fibrous to acicular crystals are brown isopachous cement fringes around grains, which are overlain syntaxially by cavity filling sparry dolomite (Figs. 11, 12).

A second type of cement crust is less clearly fibrous, but does show a vague growth lamination. This occurs around coated grains and under CL has a mottled to dull luminescence.

Isopachous cement fringes of micrite, a few tens of microns thick, also occur around fenestrae and microbial grains (Figs. 3, 9b). There is also a suggestion in some oolites that micritic cements precipitated at grain contacts in meniscus-type geometry, indicating a vadose diagenetic environment. Micrite may coat several grains indicating a post-depositional origin; these fringes may be asymmetric.

These fibrous and micritic cements are likely of marine origin. They are the first cement generation, so clearly precipitated early. Although now dolomite, the well-developed fibrous cements were probably originally high-Mg calcite. There is no evidence of major dissolution as would be expected if the cement was originally composed of aragonite. Similar cement crusts are well described from French Polynesia by Aissaoui (1988), also in high-Mg calcite. The second type of marine crust, less clearly fibrous, may well have been aragonite originally. The dolomite it now consists of is showing a more replacement, sparry texture.

Dissolution of carbonate did take place as indicated by dolomite cement crystals filling cavities once occupied by the ooid nuclei and the compacted ooids with the curious "elephantine" and "duck" structures formed as a result of dissolution during compaction (Fig. 6). These deformed ooids occur along thin zones and are related to the development of pressure dissolution seams.

There is some evidence of minor dissolution or etching around grains while in the marine environment. The outer margin of some cement layers is often very irregular and pitted before the succeeding micrite cement (Fig. 11).

During shallow to deep burial diagenesis, dolomite cement was precipitated in remaining cavities. It occurs within fenestrae, and in the intergranular porosity of oolitic and microbial-clot grainstones. This dolomite displays growth zones under CL, and in some cases a non-luminescent to bright to dull luminescence is recorded. This pattern is similar to that known for calcite precipitated during burial and linked to changing redox conditions of porewater as organic matter decomposed and Mn and Fe were made available (Tucker and Wright, 1990).

Some dolomites are composed of a coarser, microsparry dolomite mosaic with crystals ranging from several 10s to 100 μm in diameter. These crystals have a greyish appearance and a drusy texture is not apparent. They have a dull luminescence, with local bright patches. These dolomites could be the result of

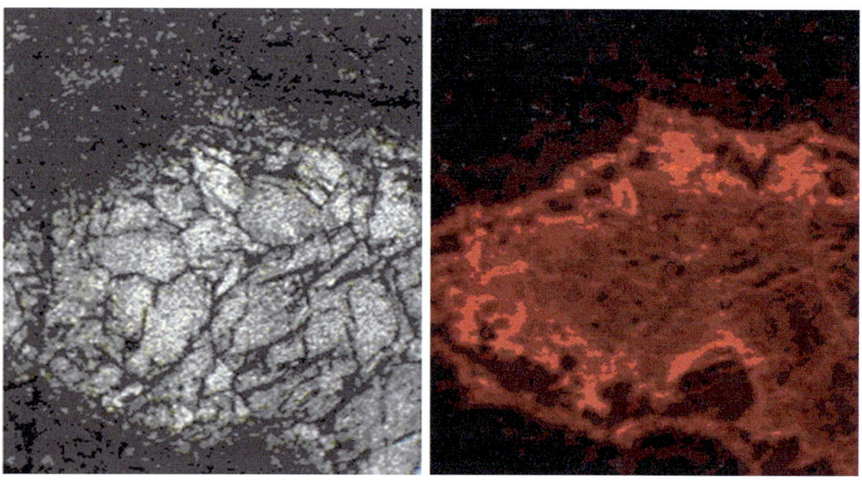

Figure 11. Cavity filled with fibrous carbonate and sparry dolomite cement under PPL (*left*) and CL (*right*). Fields of view 7 x 5 mm.

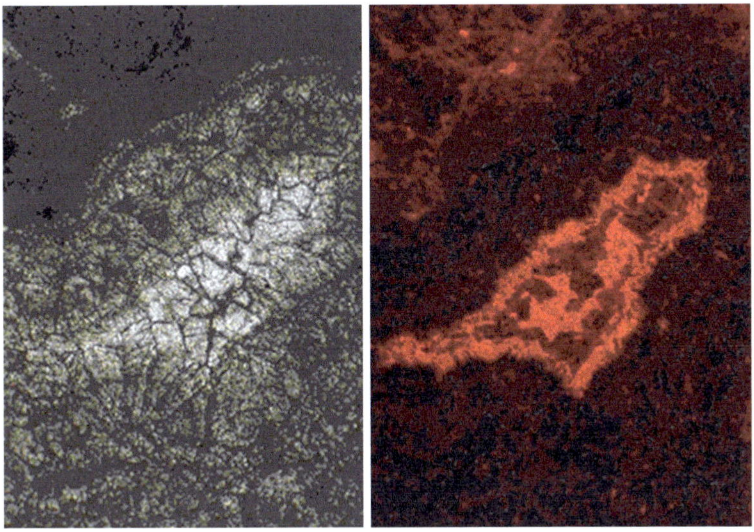

Figure 12. Cavity filled with fibrous carbonate and sparry dolomite cement under PPL (*left*) and CL (*right*). Note that most of the fibrous cement is non-luminescent, then the outer part is bright, followed by dull sparry dolomite cement, and finally the cavity is filled by bright dolomite spar. Fields of view 7 x 5 mm.

replacement of $CaCO_3$ but an origin through recrystallisation of earlier dolomite is also possible.

Dolomite rhombs on a scale of a few 10s to several 100 μm also occur within the Krol carbonates. They are scattered throughout the sediment of some facies, with larger rhombs occurring within the centres of cavities. CL shows an early dull luminescence and later bright zones. These dolomites are growing within the earlier-formed micritic dolomite and so are replacements.

There are fractures and veins of dolomite cutting through the rocks. These are usually filled with brightly luminescing dolomite, and in some cases the vein dolomite connects with the later generations of dolomite cements occurring within cavities. This arrangement suggests that the later phases of burial dolomite cement are related to fracturing (Fig. 13).

5. Krol Dolomitisation and Dolomite Precipitation

The Krol D is a typical Precambrian carbonate: the rock is totally dolomite, the preservation of the fabrics is generally good and the formation contains abundant microbial structures in the form of stromatolites and grains, which could be of microbial origin (peloids, microphytolites). The contrast of Precambrian dolomites with many Phanerozoic dolomites in terms of original textural preservation has led to speculations that the Precambrian was different in some way (see review in Tucker and Wright, 1990). The features of the dolomite described here

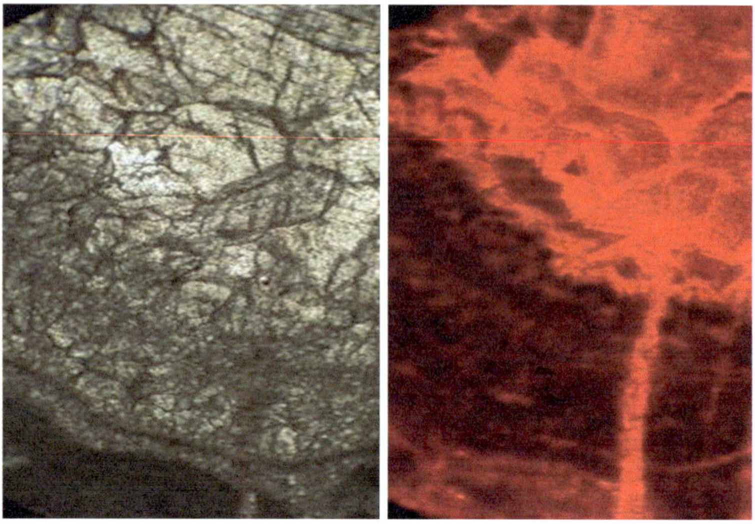

Figure 13. Cavity filled with fibrous carbonate and sparry dolomite cement under PPL (*left*) and CL (*right*). Note the fracture filled with same colour luminescing cement as sparry dolomite in centre of cavity. Fields of view 7 x 5 mm.

are interpreted in terms of very early dolomitisation of the sediment, whether it be lime mud or micritic grains, and of the marine cements. The textures would suggest that originally fine-grained carbonate was replaced by fine dolomite, and that coarse, fibrous, high-Mg calcite cements were replaced by dolomite too, faithfully preserving the original fabrics of the crystals. This is known from other Precambrian dolomites of course (e.g., the Beck Spring, Tucker, 1983). There is some evidence too of aragonite being present, in intraclasts, ooid nuclei, ooid cortical layers and as a cement; this was replaced by dolomite with coarse crystals or dissolved out and the voids filled by dolomite cement.

It would appear that there was some aragonite in the original Krol D sediments, in intraclasts, forming the nuclei to some ooids, and rarely occurring in cortical layers of ooids. This aragonite suffered dissolution, and then dolomite cement was precipitated in the moulds. There is some evidence of aragonite cement too.

During later diagenesis, dolomite continued to be precipitated but now as the main cement, filling pore spaces with drusy equant dolomite crystals with well-defined growth zones, in the same way that calcite does in a typical Phanerozoic limestone.

So there are two issues: how was the dolomite precipitated in the first place, and why did it continue to be precipitated into burial diagenesis. Two factors are immediately implicated: microbial processes and seawater chemistry.

It is generally agreed that dolomite will not precipitate all by itself, directly from seawater; some help is required. In recent years, a lot of evidence has been assembled implicating microbes as having a hand in dolomite precipitation. It has been shown by several authors (e.g., Van Lith et al., 2003; Vasconcelos et al., 2006) that during the degradation of microbes, Mg-rich calcite may be precipitated and this may lead to dolomite precipitation too, notably within the extracellular polymeric substance (EPS, mucus) generated by the bacteria. Indeed a model has been presented by Mastandrea et al. (2006) for microbial dolomitisation and dolomite precipitation in the Dolomia Principale (Triassic, Italy), a 100% dolomitised platform carbonate with excellent preservation of original textures, just like a Precambrian dolomite. Taking this further, Perri and Tucker (2007) reported the discovery of mineralised bacteria and EPS in stromatolites of the Dolomia Principale, providing evidence for a microbial origin of the dolomite. There is abundant evidence for microbes in the Krol D member; indeed, they were ubiquitous. Thus, it is extremely likely that these organisms provided the micro-geochemical environment for the early precipitation of dolomite and dolomite replacement of the pre-existing dominantly high-Mg calcite sediment of lime mud, stromatolites, microbial clots, peloids and ooids.

Another factor that may have contributed to the very early dolomitisation in the Krol D, and also the majority of other Precambrian dolomites, is seawater chemistry. It is well known that two factors therein promote dolomite precipitation: a higher Mg/Ca ratio and a lower SO_4 content (Tucker and Wright, 1990), and indeed, in the Precambrian the Mg/Ca ratio was higher and the SO_4 content was lower, than they were during much of the Phanerozoic (Tucker, 1992; Hardie, 1996, 2003; Schlager, 2005). Therefore, seawater chemistry in the Precambrian would also have promoted dolomite precipitation. It is worth noting here that during the Permian and Triassic, when dolomites are also common and many are fabric retentive like the Precambrian ones (e.g., Tucker and Marshall, 2004; Perri and Tucker, 2007), the Mg/Ca ratio of seawater was also high (Hardie, 1996, 2003). But the micritic early dolomite in the Krol D Member is often a small proportion of the dolomite rock as a whole. There are the replaced marine, probably high-Mg calcite cements, and all the later cavity-filling dolomite cements too. Although the identification of dolomitised bacteria in microbialites (Perri and Tucker, 2007) may implicate microbes in the syn-sedimentary precipitation of the dolomite, an explanation is still required for all the other dolomite. The precipitation of later diagenetic dolomite spar cement in wholly dolomitic carbonate successions, in an analogous manner to calcite spar cement in non-dolomitic carbonate formations, may indicate some chemical differences of the formation fluids. As noted earlier, seawater was certainly different in the Precambrian, and thus one would expect burial fluids to have also had a higher Mg/Ca ratio and lower SO_4, since they are commonly derived from seawater, so that burial dolomite cementation would be favoured over calcite cementation. The abundance of microbes within the sediment may have helped too with the burial precipitation. One further likely factor promoting dolomite cementation during burial is the presence of the early dolomite, which would have provided nuclei and substrates for continued dolomite precipitation.

6. Carbon Isotope Chemostratigraphy of the Ediacaran Krol Carbonates

Carbon isotope stratigraphies of Proterozoic and Early Cambrian sequences have been established in recent years from different parts of the world. For example, Siberian platform (Magaritz et al., 1986; Knoll et al., 1995), Morocco (Tucker, 1986), Greenland-Spitsbergan (Fairchild and Spiro, 1987), China (Hsu et al., 1985; Shen and Schidlowski 2000; Lambert et al., 1987; Brasier et al., 1992), Krol-Tal sequence of the Lesser Himalaya (Aharon et al., 1987; Tewari, 1991, 1998a, b, 2007; Kumar and Tewari, 1995; Brasier et al., 1992; Kaufman and Knoll (1995); Kaufman et al., 2006; Tewari and Sial, 2007), Iran (Brasier et al., 1990), Namibia (Kaufman et al., 1991) and from Brazil (Santos et al., 2000). Analysis of global Proterozoic and Cambrian boundary sequences by Ripperdan (1994) suggest that all these localities preserve the secular variations in the carbon isotope composition of the global ocean system. The carbon isotope excursions across the Precambrian and Cambrian boundary have been interpreted to explain evolutionary diversification and extinction events, palaeoenvironmental and palaeooceanic conditions, changes in sea water chemistry, extraterrestrial impacts and to establish chemostratigraphy (Aharon et al., 1986; Brasier, 1992 and the references therein). Tucker (1996) also attempted to explain the role of the formation of rifted basins and passive continental margin in the late Precambrian-Cambrian transition on ocean circulation and the carbon affecting the isotopic composition of seawater.

$\delta^{13}C$ chemostratigraphic data for the Tehri and Pauri Garhwal areas of the Uttarakhand Lesser Himalaya, India have been recorded. A composite $\delta^{13}C$ profile for the Krol Formation (Ediacaran System) was prepared based on new data from the Kauriyala and Rikhnikhal sections of the Garhwal syncline (Fig. 14) and previously published $\delta^{13}C$ data from Mussoorie syncline (Tewari, 2007) have been compared. The Lower Krol Formation is composed of shaly limestone and characterized by Vendotaenids. $\delta^{13}C$ values rise to +4.93‰ PDB in the overlying Krol C dolomite in Rikhnikhal area, which rise to +6‰ PDB in the Mussoorie syncline. Two carbon isotope trends can be recognized in the Upper Krol dolomite of the Kauriyala section of the Garhwal syncline. A shift from negative to positive values within the Krol E carbonate just below the Krol–Tal contact indicates a period of oxygen deficient bottom waters and an enrichment of ^{13}C in the shallow water carbonate basin. A marked positive excursion in the upper part of the Krol (+2.5‰) and sharp decline to (−4‰) in the phosphatic dolomite is very characteristic. The $\delta^{13}C$ carb and $\delta^{18}O$ curves show excellent co-variance suggesting extensive alteration and modification of the primary isotopic signals (Fig. 14). The Upper Krol carbonates are cherty, and, so oolitic, stromatolitic-oncolitic and zebra fabric is well developed. Krol D/E Member has yielded well-preserved Ediacaran medusoids and frondose forms in the upper silty layers (Tewari, 1989, 1990, 1996, 2004; Mathur and Shanker, 1989; Maithy and Kumar, 2007; Shanker et al., 1997; Kumar et al., 1997). In the northeastern Lesser Himalaya, the positive shift in the carbon isotope values of the Buxa Dolomite is also correlated with the Ediacaran carbonate sediments (Tewari, 2003, 2007; Tewari and Sial, 2007). A negative excursion has been recorded ($\delta^{13}C$ −10.52‰ PDB) just

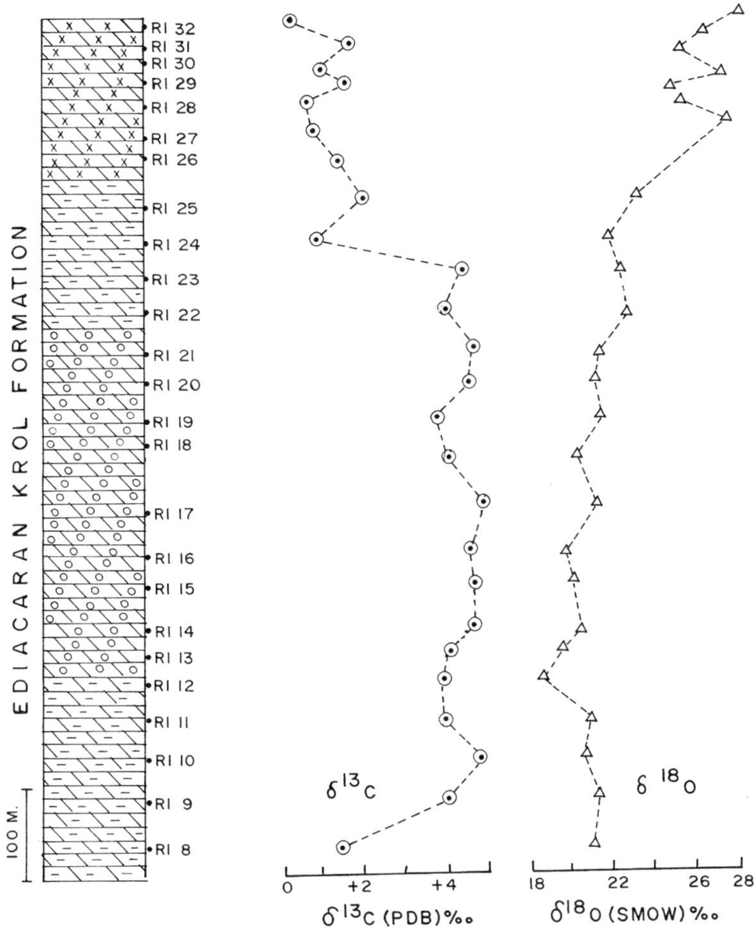

Figure 14. Carbon and oxygen isotope chemostratigraphy of the Ediacaran Krol carbonates, Lesser Himalaya, India.

below the Ediacaran-Lower Cambrian boundary. The Ediacaran C-isotope stratigraphy of the Krol Formation (*Krolian*) of the Lesser Himalaya is identical to the Doushantuo, Lower Dengying and Upper Dengying Formations of South China. The palaeogeographic reconstruction of the palaeo-continents suggests that around 650–540 Ma, Lesser Himalaya and South China were very close to each other.

7. Discussions and Conclusion

The Krol Formation is Ediacaran (Terminal Neoproterozoic) with dominantly microbial carbonate facies. It is well developed in the Lesser Himalayan Krol belt extending from the Solan in the Himachal to the Nainital in Uttarakhand Lesser

Himalaya. Upper Krol contains Ediacaran metazoan fossils and abundant microbial structures in the form of columnar, domal and stratified stromatolites. Peloids, oncoids and microphytolites are also microbially formed structures. Carbonate grainstones and packstones are common in the Upper Krol, and the allochemical constituents are intraclasts, ooids, coated grains, peloids, microbial grains and catagraphs. Sedimentary structures present in the non-stromatolitic Krol carbonates include bird's eyes, cross bedding, symmetrical ripples and shallow channel structures. Mud cracks are also present. An intertidal to supratidal (shallow water) depositional environment with some moderate currents and intermittent periods of exposure has been suggested for the Krol D Member of the Lesser Himalaya.

Petrography and cathodoluminescence of Krol carbonates show wide variations in texture. Intraclasts are rectangular in section and composed of coarse dolomite crystals. CL shows that these have growth zones. Peloidal structures are composed entirely of micrite. Ooids in the Krol D Member are mostly spheroidal to ellipsoidal grains and range in diameter from 0.1 to 0.4 mm. Composite ooids are larger grains and consist of smaller ooids. The CL of the ooids shows non-luminescence. The diagenetic history of the Krol D ooids and intraclasts suggests that many of these grains may have precipitated as micrite and possibly microbial micritization has taken place. The Krol D dolomite shows abundant evidence of marine cementation as the first diagenetic stage. A common type of cement is fibrous crust. Under CL the fibrous crusts are brightly luminescent. These fibrous and micritic cements are of shallow marine origin. The bright luminescent stromatolitic laminae are observed (Fig. 9).

The diagenetic textures suggest that originally fine-grained carbonate was replaced by fine dolomite and that coarse fibrous high Mg calcite cements were replaced by dolomite too, preserving the original fabrics of the crystals. Terminal Neoproterozoic sea water chemistry must have contributed to the dolomitization in the Krol D. $\delta^{13}C$ chemostratigraphic data for the Mussoorie, Tehri and Pauri Garhwal areas of the Uttarakhand Lesser Himalaya, India has been compared. A composite $\delta^{13}C$ profile for the Krol Formation (Ediacaran System) from the Kauriyala Rikhnikhal and Luxmanjhula sections of the Garhwal syncline (Fig. 14) substantiates previously from the Mussoorie syncline (Tewari, 1991, 2007; Tewari and Kumar, 1995; Tewari and Sial, 2007). The Blaini diamictites are well developed in both sections representing the Neoproterozoic glacial *Marinoan/Blainian* event. The Blaini pink cap microbial carbonate shows a negative $\delta^{13}C$ excursion characteristic of cap carbonates globally. At the Luxmanjhula section, near Rishikesh, the $\delta^{13}C$ values for a stromatolitic pink cap limestone of the Cryogenian Blaini Formation varies from −1.73 to −1.86 ‰ PDB. Oxygen isotopes range from −9.13 to −11.63‰ PDB in carbonates. The Lower Krol Formation is composed of shaly limestone and marl. $\delta^{13}C$ values rise to +4.93‰ PDB in the overlying Krol C dolomite in Rikhnikhal area, which rise to +6‰ PDB in the Mussoorie syncline. The Upper Krol carbonates are cherty, oolitic and stromatolitic-oncolitic, and zebra fabric is well developed. Krol D/E Member has yielded well-preserved Ediacaran medusoids and frondose forms in the upper silty layers.

A negative excursion has been recorded ($\delta^{13}C$ −10.52‰ PDB) just below the Ediacaran-Lower Cambrian boundary. The Ediacaran C-isotope stratigraphy of the Krol Formation (*Krolian*) of the Lesser Himalaya is identical to the Doushantuo, Lower Dengying and Upper Dengying Formations of South China and also comparable with the global data (Tewari, 2008, 2009). The palaeogeographic reconstruction of the palaeo-continents suggests that around 650–540 Ma, Lesser Himalaya and South China were very close to each other. The published a paleogeographic and palaeomagnetic data during this period show strong evidence for the existence and breakup of the Rodinia Supercontinent followed by the Pan African/Pan Indian orogenic event, which assembled the Gondwana Supercontinent.

8. Acknowledgements

Dr. Edoardo Perri, Italy is thanked for critically reviewing the chapter and suggestions. V.C. Tewari is thankful to Professor B.R. Arora, Director, Wadia Institute of Himalayan Geology, Dehradun, India for facilities and permission to publish the chapter.

9. References

Aharon, P., Schidlowski, M. and Singh, I.B. (1987) Chronostratigraphic carbon isotopic record of the Lesser Himalaya. Nature **327**: 699–702.

Aissaoui, D.M. (1988) Magnesian calcite cements and their diagenesis: dissolution and dolomitisation. Sedimentology **35**: 821–841.

Brasier, M.D. (1992) Palaeooceanography and changes in the biological cycling of phosphorus across the Precambrian-Cambrian boundary, In: J.L. Lipps and P.W. Signor (eds.) *Origin and Early Evolution of the Metazoa*. Plenium Press, New York, pp. 483–515.

Braiser, M.D., Magaritz, M., Corfield, R., Hullin, Luo, Xiche, Wu, Lin, Ougang, Zhiwen, Jhang, Hamadi, B., Tinggin, He, and Frazier, A.G. (1990). The carbon and oxygen isotope record in the Precambrian-Cambrian boundary interval in China and Iran and their correlation. Geol. Mag. **127**: 319–332.

Fairchild, I.J. and Spiro, B. (1987) Petrological and isotopic implications of some contrasting late Precambrian Carbonates. NE Spitsbergan. Sedimentology **34**: 973–989.

Hardie, L.A. (1996) Secular variations in seawater chemistry: an explanation for the coupled secular variations in the mineralogy of marine limestones and potash evaporites over the past 600 m.y. Geology **24**: 279–283.

Hardie, L.A. (2003) Secular variations in Precambrian seawater chemistry and the timing of Precambrian aragonite seas and calcite seas. Geology **31:** 785–788.

Hsu, K.J., Orberhansli, H., Gao, J.Y., Shu, S., Haihong, C. and Krahenbuhl, U. (1985) Strange love ocean before the Cambrian explosion, Nature **316**: 809–811.

Jiang, G., Christir Blick, N., Kaufman, A.J., Banerjee, D.M. and Rai, V. (2003) Carbonate platform growth and cyclicity at a terminal Proterozoic passive margin, Infra Krol Formation and Krol Group, Lesser Himalaya, India. Sedimentology **50**: 921–952.

Kaufman, A.J. and Knoll, A.H. (1995) Neoproterozoic variations in the carbon isotopic composition of sea water: stratigraphic and biogeochemical implications. Precambrian Res. **73**: 27–49.

Kaufman, A.J., Hayes, J.M., Knoll, A.H. and Germs, G.J.B. (1991) Isotopic composition of carbonates and organic carbon from Upper Proterozoic succession in Namibia: stratigraphic variation and the effects of diagenesis and metamorphism. Precambrian Res. **49**: 301–327.

Kumar, B. and Tewari, V.C. (1995) Carbon and oxygen isotope trends in late Precambrian–Cambrian carbonates from the Lesser Himalaya, India. Curr. Sci. **69**: 929–931.

Knoll, A.H., Kaufman, A.J., Semikhatov, M.A., 1995. The carbon isotopic composition of Proterozoic carbonates; Riphean successions from Northeastern Siberia (Anabar massif, Turukhansk uplift). Am. J. Sci. **295**: 823–850.

Kumar, B. and Tewari, V.C. (1995) Carbon and Oxygen isotope trends in late Precambrian-Cambrian carbonates from the Lesser Himalaya, India. Curr. Sci. **69**: 929–931.

Kumar, G., Shanker, R., Maithy, P.K., Mathur, V.K. and Bhattacharyya, S.K. (1997) Terminal Proterozoic-Cambrian sequences in India. A review with special reference to Precambrian-Cambrian boundary, Palaeobotanist **46** (1.2): 19–31.

Lambert, I.B., Walter, M.R., Welong, Z., Songian, L. and Guogan, M. (1987) Palaeoenvironment and carbon isotope stratigraphy of Upper Proterozoic carbonates of the Yangtze Platform. Nature **325**: 140–142.

Magaritz, M. Holser, W.T. and Kirishvink, J.L. (1986) Carbon isotope events across the Precambrian–Cambrian boundary on the Siberian Platform. Nature **320**: 258–259.

Maithy, P.K. and Kumar, G. (2007) Biota in the terminal Proterozoic successions on the Indian subcontinent: a review, In: P. Vickers-Rich and P. Komarower (eds.) *The Rise and Fall of the Ediacaran Biota*. Geological Society of London, Special Publication **286**: pp. 315–330.

Mastandrea, A., Perri, E., Russo, F., Spadafora, A. and Tucker, M.E. (2006) Microbial primary dolomite from a Norian carbonate platform, northern Calabria, southern Italy. Sedimentology **53**: 465–480.

Mathur, V.K. and Shanker, R. (1989) First record of ediacaran fossils from the Krol Formation, Nainital syncline. J. Geol. Soc. India **34**: 245–254.

Perri, E. and Tucker, M.E. (2007) Bacterial fossils\and microbial dolomite in Triassic stromatolites. Geology **35**: 207–210.

Reid, R.P. and Macintyre, I.G. (1998) Carbonate recrystallization in shallow-marine environments: a widespread diagenetic process forming micritized grains. J. Sediment. Res. **68**: 928–946.

Ripperdan, R.L. (1994) Global variations in carbon isotope composition during the latest Neoproterozoic and earliest Cambrian. Annu. Rev. Earth Planet, Sci. **22**: 385–417.

Santosh, R.V., Alvarenga, C.J.S., de Dardenne, M.A. and Sial, A.N. (2000) Carbon and Oxygen isotope profiles across Meso-Neoproterozoic limestone from Central Brazil: Bambui and Paranoa groups. Precambrian Res. **104**: 107–122.

Schlager, W. (2005) Secular variations in the stratigraphic record. Facies **51**: 13–17.

Shanker, R., Mathur, V.K., Kumar, G. and Srivastava, M.C. (1997) Additional Ediacaran biota from the Krol Group, Lesser Himalaya, India and their significance. Geosci. J. **18**: 79–94.

Shen, Y. and Schidlowski, M. (2000) New C isotope Stratigraphy from southwest China: implications for the placement of the Precambrian-Cambrian boundary on the Yangtze platform and global correlations, Geology **28**(7): 623–626.

Tewari, V.C. (1984a) Discovery of Lower Cambrian stromatolites from Mussoorie Tal Phosphorite, India. Curr. Sci. **53**(6): 19–21.

Tewari, V.C. (1984b) Stromatolites and Precambrian – Lower Cambrian biostratigraphy of the Lesser Himalaya, In: *Proceedings Vth Indian Geophytological Conference Special Publication*, pp. 71–97.

Tewari, V.C. (1989) Upper Proterozoic-Lower Cambrian stromatolites and Indian Stratigraphy. Himalayan Geol. **13**: 143–180.

Tewari, V.C. (1990) Ediacaran biota, stromatolites, and isotopic studies from Lesser Himalaya, with special reference to Precambrian–Cambrian Boundary, In: *National Seminar on Precambrian Geology, 14–16th February*, 1990, Department of Geology, University of Madras, India, p. 79.

Tewari, V.C. (1991) The Carbon and Oxygen isotope trend of the Deoban-Blaini-Krol-Tal microbial carbonate from the Lesser Himalaya, India. Geosci. J. **12**(1): 13–16.

Tewari, V.C. (1992) Global decline of Pre-Ediacaran (Riphean stromatolites and the emergence of Ediacaran biota: paleobiological and stable isotope evidences from the Lesser Himalaya. Geol. Soc. India **39**: 260–261.

Tewari, V.C. (1993a) Precambrian and Lower Cambrian stromatolites of the Lesser Himalaya, India. Geophytology **23**(1): 19–39.

Tewari, V.C. (1993b) Ediacaran metaphytes from the Lower Krol Formation, Lesser Himalaya, India. Geosci. J. **14**(12): 143–148.

Tewari, V.C. (1996) Discovery of pre Ediacaran acritarch Chuaria circularis (Walcott, 1899, Vidal and Ford, 1985) from the Deoban Mountains, Lesser Himalaya, India, Geosci. J. **17**(1): 25–39.

Tewari, V.C. (1998a) Regional correlation of the Lesser Himalayan and Tethyan basin sediments of Kali Valley, Indo-Nepal area. J. Nepal Geol. Soc. **18**: 37–57.

Tewari, V.C. (1998b) Lower Cambrian carbon isotope evidence of photosynthesis from Krol-Tal carbonates of the Lesser Himalaya and global stratification in early Cambrian oceans. Mineralogical Magazine (Goldschmidt Conference, 1998, Toulouse, France) **62A**: 1506.

Tewari, V.C. (1999) Vendotaenids: earliest megascopic multicellular alge on Earth. Geosci. J. **20**(1): 77–85.

Tewari, V.C. (2001a) Neoproterozoic glaciation in the Uttaranchal Lesser Himalaya and the global Palaeoclimatic change. Geol. Surv. India. Spl. Publ. **65**(3): 49–56.

Tewari, V.C. (2001b) Discovery and sedimentology of microstromatolites from Menga Limestone (Neoproterozoic/Vendian). Upper Subansiri district, Arunachal pradesh, Northeastern Himalaya, India. Curr. Sci. **80**(11): 1440–1444.

Tewari, V.C. (2002a) Lesser Himalayan stratigraphy. Sedimentation and correlation from Uttaranchal to Arunachal. Gyanodaya Prakashan, Nainital, India, 63–88.

Tewari, V.C. (2002b) Proterozoic-Cambrian sedimentation and associated natural resources of Uttaranchal: Proceedings of the National Workshop on Natural Wealth of Uttaranchal Technology Publishers, Dehradun, 189–222.

Tewari, V.C. (2003) Sedimentology, palaeobiology and stable isotope chemostratigraphy of the Terminal Neoproterozoic Buxa Dolomite, Arunachal Pradesh NE Lesser Himalaya. J. Himalayan Geol. **24**(2): 1–18

Tewari, V.C. (2004) Microbial diversity in Meso-Neoproterozoic formations, with particular reference to the Himalaya, In: J. Seckbach (ed.) *Origins*. Kluwer, Dordrecht, pp. 515–528.

Tewari, V.C. (2007) The rise and the decline of the Ediacaran biota: paleobiological and stable isotopic evidence from the NW and NE Lesser Himalaya, India, In: P.R. Vickers and P. Komarower (eds.) *The Rise and Fall of Ediacaran Biota*. Geological Society, London, Special Publications, pp. 77–102.

Tewari, V.C. (2008) Neoproterozoic glaciation: climate–tectonic relationship. Himalayan Geol. **29**(3): 126–128.

Tewari, V.C. (2009) Neoproterozoic snowball Earth and the sedimentological evolution of the Lesser Himalaya, India, In: *2nd International Conference on Precambrian Continental Growth and Tectonism, Jhansi, India* (abstract), 19.

Tewari, V.C. and Joshi, M. (1993) Stromatolite microstructures: a new tool for biostratigraphic correlation of the Lesser Himalayan carbonates. Himalayan Geol. **4**(2): 19–29.

Tewari, V.C. and Qureshi, M.F. (1985) Algal structures from the Upper Krol-Lower Tal Formation of Garhwal and Mussoorie synclines and their paleoenvironmental significance. J. Geol. Soc. India **26**: 111–117.

Tewari, V.C. and Sial, A.N. (2007) Neoproterozoic – Early Cambrian isotopic variation and chemostrarigraphy of the Lesser Himalaya, India, Eastern Gondwana. Chem. Geol. **237**: 64–88.

Tucker, M.E. (1983) Diagenesis, geochemistry and origin of a Precambrian dolomite, the Beck Spring Dolomite of eastern California. J. Sediment. Petrol. **53**: 1097–1119.

Tucker, M.E. (1984) Calcite, aragonite and mixed calcitic-aragonitic ooids from the mid-Proterozoic Belt Supergroup, Montana. Sedimentology **31**: 627–644.

Tucker, M.E. (1986) Carbon isotope excursions in Precambrian/Cambrian boundary beds, Morocco. Nature **319**: 48–50.

Tucker, M.E. (1992) The Precambrian–Cambrian boundary: seawater chemistry, ocean circulation and nutrient supply in metazoan evolution and extinction. J. Geol. Soc. London **149**: 655–668.

Tucker, M.E. (2001) *Sedimentary Petrology: An Introduction to the Origin of Sedimentary Rocks*. Blackwell Science, Oxford, p. 260.

Tucker, M.E. and Marshall, J. (2004) Diagenesis and geochemistry of Upper Muschelkalk (Triassic) build-ups and associated facies in Catalonia (NE Spain): a paper dedicated to Francesc Calvet. Geologica Acta **2**: 257–269.

Tucker, M.E. and Wright, V.P. (1990) *Carbonate Sedimentology*. Blackwell Science, Oxford, p. 482.

Van Lith, Y., Warthmann, R., Vasconcelos, C., McKenzie, J.A. (2003) Microbial fossilization in carbonate sediments: a result of the bacterial surface involvement in dolomite precipitation. Sedimentology **50**: 237–245.

Vasconcelos, C., Warthmann, R., McKenzie, J.A., Visscher, P.T., Bittermann, A.G. and van Lith, Y. (2006) Lithifyng microbial mats in Lagoa Vermelha, Brazil: modern Precambrian relics? Sediment. Geol. **185**: 175–183.

PART 2:
PHANEROZOIC STROMATOLITES

Föllmi
Delamette
Ouwehand
Krajewski
Massari
Westphal
Tunis
Pugliese
Jurkovšek
Ogorelec
Drobne
Riccamboni
Tewari
Yamamoto
Lee
Isozaki

Biodata of **Karl B. Föllmi, Michel Delamette**, and **Pieter Jan Ouwehand,** authors of ***"Aptian to Cenomanian Deeper-Water Hiatal Stromatolites from the Northern Tethyan Margin"***

Karl B. Föllmi is of Swiss nationality and was born and raised in the Netherlands. He returned to Switzerland when he was 19 and absolved his undergraduate, graduate, and Ph.D. studies at the ETH Zurich. He stayed at UC Santa Cruz for his postdoc and returned to Switzerland, where he obtained his habilitation (at ETH Zurich). In the following, he became professor in the field of sedimentology and geochemistry at the University in Neuchatel and subsequently professor in sedimentology at the University of Lausanne. His fields of interest are centered around large-scale processes at present and the history of the Earth, which have relevance to present day environmental change. This includes research on environmental changes in carbonate platforms in the Cretaceous (present day analogs: coral reef plateaus), phosphorite deposits (carrying the essential nutrient phospahte), organic carbon-rich deposits (sinks for atmospheric CO_2), weathering processes on the continent (as a source for essential biophile elements, and paleoclimatic reconstructions.

E-mail: **karl.foellmi@unil.ch**

V.C. Tewari and J. Seckbach (eds.), *STROMATOLITES: Interaction of Microbes with Sediments*, Cellular Origin, Life in Extreme Habitats and Astrobiology 18, 159–186
DOI 10.1007/978-94-007-0397-1_8,

Michel Delamette, born in 1956 at Chamonix (High Savoy, France), Michel Delamette graduated at the University of Chambéry (Savoy) and then at the University of Lyon where he obtained a Ph.D. in Paleontology in 1982. From 1979 to 1984, he was geologist assistant at the University of Geneva where he received a Ph.D. in Earth Sciences in 1986. From 1987 to 1990, he became a research fellow of the Swiss National Foundation for scientific research at the University of Fribourg (Switzerland). During a stay of 6 years, he carried paleontological expertises for the Museum of Geology of Lausanne and the Natural History Museum of Geneva, participated in a project of the International Park of Mont Blanc Massif, and wrote a book describing geological excursions in the Mont Blanc area (1993 f. ed., 2002 2nd ed.). Since 1996, he has been scientific attaché and operation manager at the Regional Natural Park of Chartreuse near Grenoble (France). His research work focuses mainly on mid Cretaceous deposits of the Southeast France (condensed phosphatic series of the Delphino-helvetic Platform and expanded black shale of the Vocontian Basin) together with the ammonite stratigraphy and paleontology.

E-mail: **michel.delamette@parc-chartreuse.net**

Pieter Jan Ouwehand was born in 1957 in Nieuwkoop, the Netherlands. In 1961, he moved with his parents to Switzerland. He grew up in the Swiss "Mittelland" and completed the "Kantonsschule" in Chur, Graubünden. Subsequently, he studied geology at the ETH Zürich and concluded with a diploma thesis on Triassic austroalpine sediments and a Ph.D. thesis on "mid"-Cretaceous helvetic sediments, under the guidance of prof. Rudolf Trümpy. Since then, Pieter Jan Ouwehand works in the domain of applied geology, and since 2000, he is the director of his own geology company. He was council member and president of CHGEOL (Swiss Association of Geologists) between 1998 and 2007 and 1998 and 2001, respectively. His involvement in the public outreach of Earth Sciences is important to Pieter Jan Ouwehand, and with many other colleagues, he is particularly engaged in the project "Erlebnis-Geologie Schweiz."

E-mail: **ou@wanner-so.ch**

Michel Delamette

Pieter Jan Ouwehand

APTIAN TO CENOMANIAN DEEPER-WATER HIATAL STROMATOLITES FROM THE NORTHERN TETHYAN MARGIN

KARL B. FÖLLMI[1], MICHEL DELAMETTE[2], AND PIETER JAN OUWEHAND[3]
[1]Université de Lausanne, Anthropole, CH-1015 Lausanne, Switzerland
[2]Parc Naturel Régional de Chartreuse, 38380 Saint-Pierre-de Chartreuse, France
[3]Wanner AG, CH-4501 Solothurn, Switzerland

Abstract A suite of deeper-water hiatal (DWH) stromatolites has been identified in the phosphatic and glauconitic sediments of Aptian to Cenomanian age in the alpine Helvetic thrust-and-fold belt, which represents the former northern Tethyan margin. The most important occurrences date from the latest Early to Late Aptian, the late Early to early middle Albian, and the Early Cenomanian. They are invariably associated with condensed phosphatic beds and occur preferentially on top of hardgrounds or on reworked pebbles and fossils. The zone of optimal stromatolite growth and preservation coincides with the zone of maximal sedimentary condensation, in the deeper parts of phosphogenic areas. The DWH stromatolites show variable morphologies, ranging from isolated laminae ("films") to internally laminated columns and crusts. They reach thicknesses of maximal 10 cm and are either preserved in phosphate or micrite. In the latter case, they may show peripheral impregnations of phosphate or iron oxyhydroxides. The quasi-complete lack of macroscopic sessile organisms suggests that the DWH stromatolites grew close to the upper boundary of an oxygen-minimum zone. Electron-scanning microscopic images show that the Early Cenomanian examples preserved in micrite consist of filamentous structures, which form spaghetti-like assemblages. They are interpreted as the remains of poikiloaerobic, heterotrophic microbes.

Coeval DWH stromatolites are known from the entire European segment of the northern Tethyan margin, and shallow-water counterparts are commonplace on Tethyan carbonate platforms. This indicates that, in general, paleoceanographic and paleoenvironmental conditions were appropriate for stromatolite growth and preservation. The here-described DWH stromatolites proliferated especially in time windows, which followed upon the oceanic anoxic periods OAE 1a (Early Aptian), 1b (latest Aptian and earliest Albian), and 1d (latest Albian). They may represent pioneer ecosystems, which thrived during the recovery phases following the "mid"-Cretaceous OAEs.

Keywords Stromatolites • Phosphates • Aptian • Albian • Cenomanian • Northern Tethyan Margin • Upwelling • Helvetic Alps • Benthic foraminifera

1. Introduction

Stromatolites of deeper-water, aphotic origin are less commonly preserved in the sedimentary record than their shallow-water counterparts. Phanerozoic examples are often associated with omission surfaces, hardgrounds, and/or condensed sediments and have been fossilized by authigenic mineralization processes such as the precipitation of iron and manganese oxyhydroxides or phosphates (e.g., Krajewski, 1981, 1984; Krajewski et al., 1994, 2000; Böhm and Brachert, 1993; Martín-Algarra and Vera, 1994; Vera and Martín-Algarra, 1994; Martín-Algarra and Sánchez-Navas, 1995, 2000; George, 1999; Zanin et al., 2003). This combination of sedimentological and mineralogical characteristics places the formation and preservation of the so-called deeper-water hiatal (DWH) stromatolites in a context of current-dominated sedimentary regimes in outer-shelf, slope, and basinal environments on the one hand and in isolated, drowned, banktop environments on the other. A further, often invoked condition to preserve DWH stromatolites are low-oxygen levels in bottom waters, which are as such devoid of microbial-mat predating organisms; however, in some cases, rapid burial and biomineralization may provide an alternative and/or complementary mechanism of stromatolite preservation (e.g., Brett and Seilacher, 1991; Föllmi, 1989, 1996). Effectively, the occurrence of modern DWH microbial mats has been established in upwelling regimes with well-developed oxygen-minimum zones, notably along the coast of Peru and Chile (e.g., Gallardo, 1977; Williams and Reimers, 1983; Schulz et al., 1996; Arning et al., 2008).

An exceptionally well-preserved suite of DWH stromatolites or microbialites (Delamette, 1990) has been documented in phosphate- and glauconite-rich sediments of "mid"-Cretaceous age on the former northern Tethyan margin, which is presently locked up in the Helvetic thrust-and-fold belt of the central European Alps ("Grès verts helvétiques"; Delamette, 1981, 1985, 1988a, b, 1990, 1994; Delamette et al., 1997; "Garschella Formation"; Föllmi and Ouwehand, 1987; Föllmi, 1986, 1989, 1990; Ouwehand, 1987) (Figs. 1 and 2). They occur in different stratigraphic levels of Aptian, Albian, and Cenomanian age and are either phosphatic or preserved in micrite of pelagic origin. In this contribution, we provide a review of these occurrences and reconstruct the paleoceanographic and paleoenvironmental conditions leading to their preservation.

2. Occurrences in Austria (Vorarlberg)

Sediments of the Garschella Formation in the Helvetic Alps of Vorarlberg (western Austria) contain well-preserved DWH stromatolites in different lithostratigraphic units (Figs. 1 and 2; Föllmi, 1986, 1989). They are all associated with the occurrence of highly condensed phosphatic beds and are either fossilized in micrite or in phosphatized form.

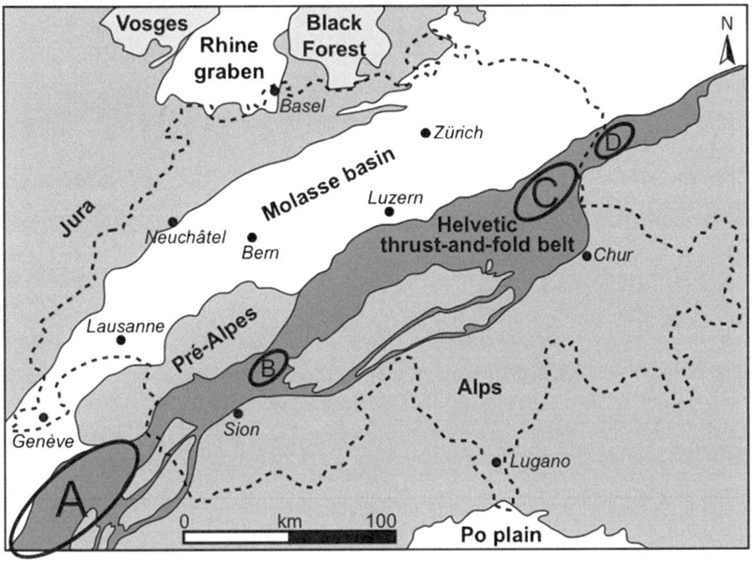

Figure 1. Simplified tectonic map of Switzerland and neighbor regions, with indication of the areas within the Helvetic thrust-and-fold belt, in which "mid"-Cretaceous DWH stromatolites have been discovered. (**A**) Haute Savoie (Delamette, 1981, 1985, 1988a, b, 1990, 1994; Delamette et al., 1997); (**B**) Col de la Plaine Morte (Gainon, 2001; Föllmi and Gainon, 2008); (**C**) Churfirsten-Säntis (Ouwehand, 1987); and (**D**) Vorarlberg (Föllmi, 1986, 1989, 1990).

2.1. APTIAN

The oldest DWH stromatolites are documented from highly condensed phosphatic sediments of latest Early to middle Late Aptian age (Luitere Bed; *furcata* to *nolani* Zones). They are restricted to distal occurrences, often in association with micritic, *Hedbergella*-containing carbonates present within the phosphatic sediments. Different associations related to different facies types are known. A first and frequently occurring association is in the form of mm-thin, phosphatic laminar stromatolites in phosphatic crusts and nodular layers (Fig. 3; e.g., Unter-Wäldle Alp near the Dornbirner Ache). The stromatolites occur attached to the phosphatic crusts or phosphatic nodules or to lithoclasts derived from the subjacent Urgonian Schrattenkalk Formation.

A second association occurs only locally: the stromatolites are preserved as columnar forms with a height of several cms, which are restricted to the infills of micritic, *Hedbergella*-containing carbonate in local pockets and fissures on top of the Urgonian Schrattenkalk Formation (Fig. 3a; Quarry Nofels, Schellenberg). The margins of the pockets and fissures are phosphatized, whereas the stromatolites themselves are conserved in micrite (Fig. 3a). They frequently include encrusting, agglutinated foraminifera (*Placopsilina* sp.).

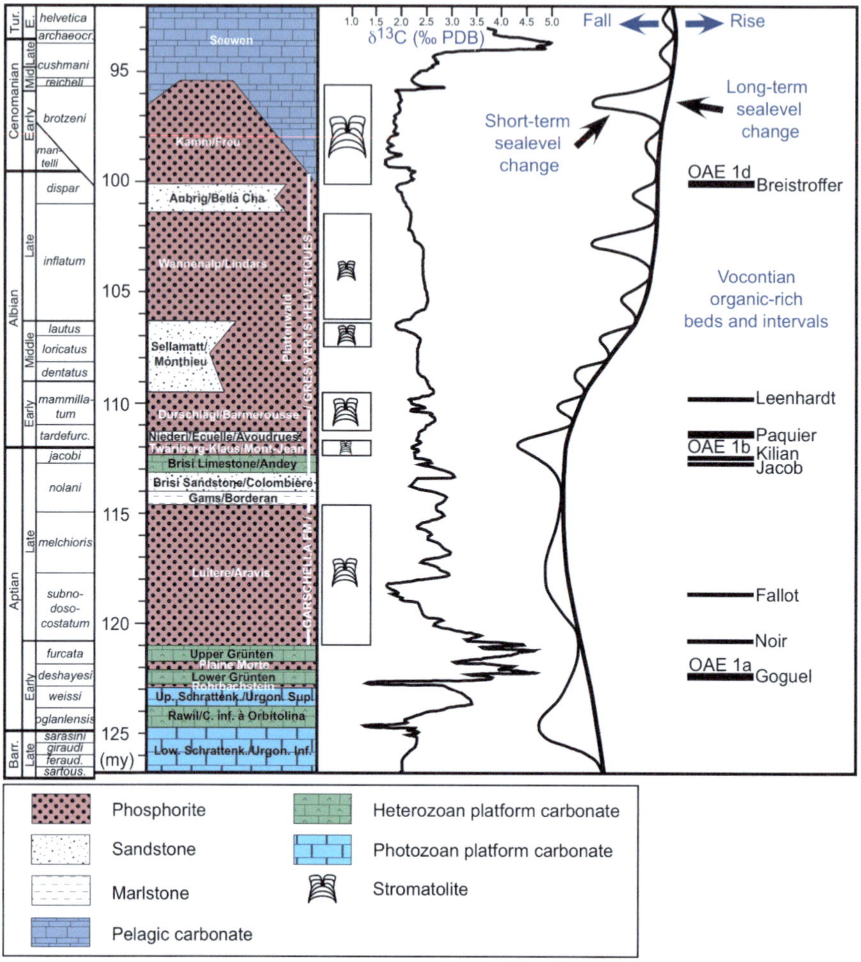

Figure 2. Highly simplified compilation of "mid"-Cretaceous stratigraphy of the Alpine Helvetic zone, with indication of the lithostratigraphic unit names in the germanophone Alps and their approximate equivalents in the francophone Alps. To the *right*, all stratigraphic levels are shown in which DWH stromatolites were identified. The most important ones are marked by *larger stromatolite symbols*. The whole-rock $\delta^{13}C$ curve is after Herrle et al. (2004) for the Aptian, Erbacher et al. (1996) and de Azevedo and Rodrigues (2000) for the Albian, and Jarvis et al. (2006) for the Cenomanian and Early Turonian. The eustatic sea-level curve is from Haq et al. (1987). The compilation of organic-rich intervals in the Vocontian Basin is from Bréhéret (1997).

2.2. APTIAN–ALBIAN BOUNDARY INTERVAL

Sporadic phosphatic stromatolite fragments occur attached to reworked and transported pebbles within a phosphatic resediment (Klaus Beds; latest Aptian to earliest Albian in age: *jacobi* to *tardefurcata* Zones: Dornbirner Ache, E of Ebnit).

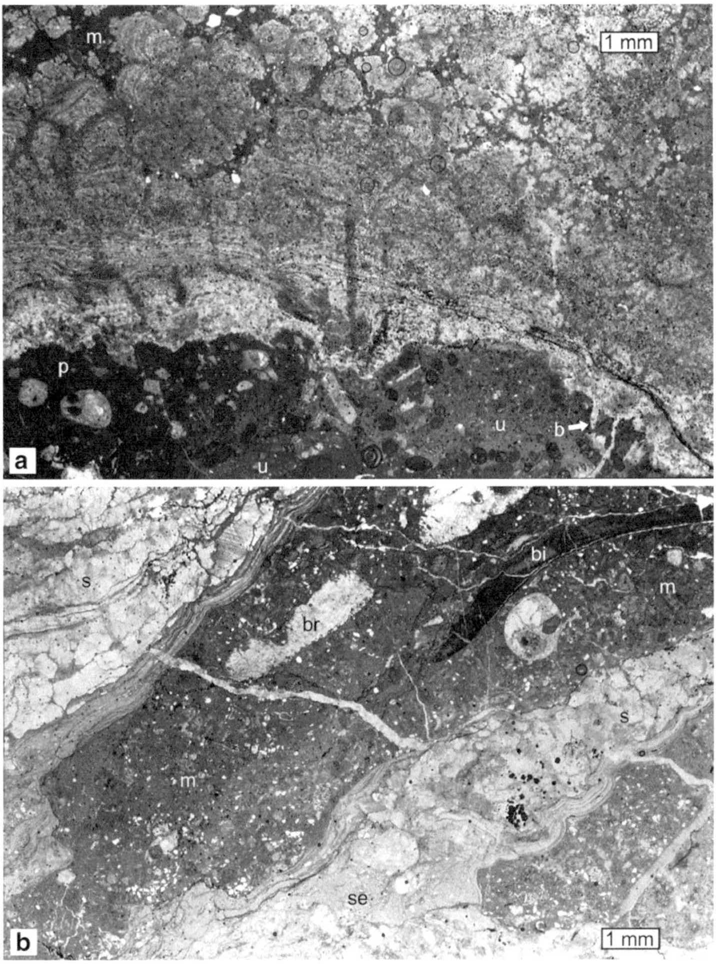

Figure 3. Occurrences of DWH stromatolites in the Luitere Bed (late Early to early Late Aptian) in Vorarlberg: (**a**) Thin-section photo of columnar stromatolites in a *Hedbergella*-bearing micritic infill (m) of a depression on top of the Urgonian Schrattenkalk Formation (u). The stromatolites are columnar and preserved in micrite; the top surface of the Urgonian limestone is bored (b) and peripherically partly phosphatized (p) (Nofels, Schellenberg); (**b**) Thin-section photo of phosphatic stromatolites (s) and sediment (se) within a phosphatic crust, enveloped by phosphatic laminations. Voids in the phosphatic crust are filled with partly phosphatized micrite (m), which contains bored fragments of brachiopod (br) and bivalve (bi) shells. From Föllmi (1986, 1989).

2.3. ALBIAN

A well-preserved occurrence of macroscopic stromatolites is observed on top of the Urgonian Schrattenkalk Formation in the quarry of Unter-Klien, in a succession of sediments attributed to the Hohenems Nappe (Wyssling, 1986). The stromatolites are phosphatic, columnar, up to 10 cm high, and associated with a nodular

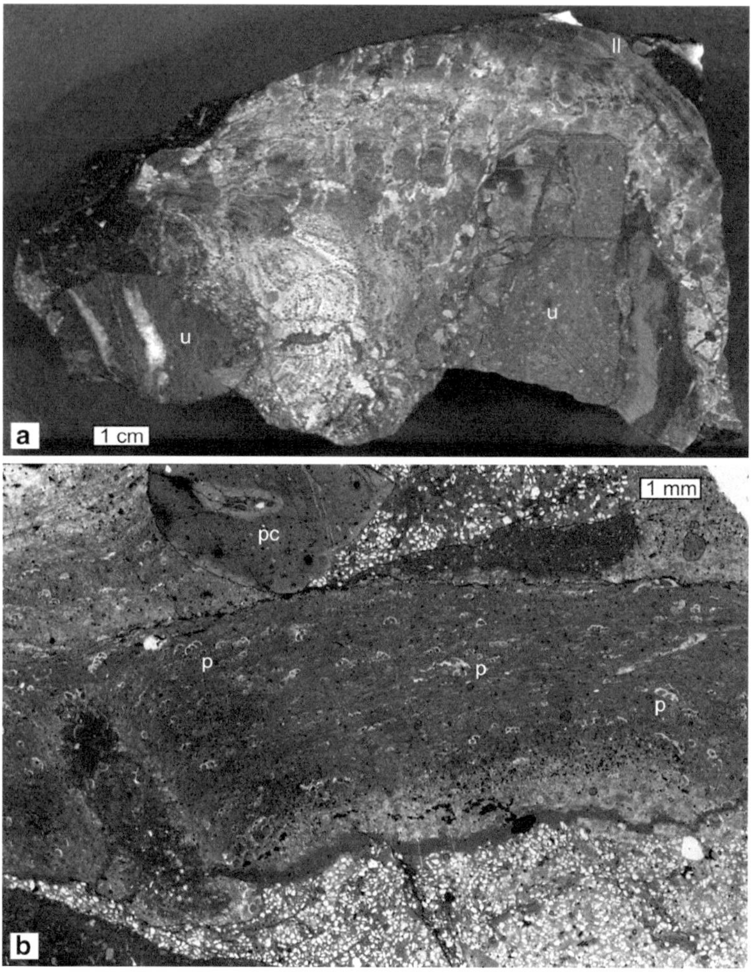

Figure 4. Occurrences of DWH stromatolites in a phosphatic bed on top of the Urgonian Schrattenkalk Formation, which is considered as a proximal equivalent of the Sellamatt Beds (late Middle Aptian) in Unterklien, Vorarlberg: (**a**) Slabbed sample showing phosphatic stromatolites attached to pebbles of Urgonian limestone (u). The colony on the *left side* of the image was tilted during stromatolite growth. The generation of stromatolites on top cover previous generations with an internally laminated layer (ll) without columns; (**b**) Thin-section photo of the same bed and locality showing parts of a phosphatized stromatolite covering a fine-grained glauconitic sandstone. The wrinkled stromatolite laminae are colonized by agglutinating benthic foraminifera (*Placopsilina* sp.: p). The particle in the upper part is a phosphatized clast (pc). From Föllmi (1986, 1989).

phosphatic layer at the base of sediments of the Garschella Formation (Fig. 4a). They are attached to the top surface of the Urgonian Schrattenkalk Formation, or on pebbles of the same formation, which are reworked into the phosphoritic layer. The basal phosphorite layer is dated from the late Middle Albian (Föllmi, 1986). The stromatolites show different orientations, which are probably related to the rotation of the pebbles during stromatolite growth (Fig. 4a). Agglutinating benthic foraminifera are frequently observed in the stromatolite laminations (*Placopsilina* sp.; Fig. 4b).

Further stromatolites are known from distal occurrences of the Albian phosphatic Plattenwald Bed, where they are present in the form of up to several cm thick phosphatic, laminated mats on top of the phosphorite bed (Fig. 5a–b;

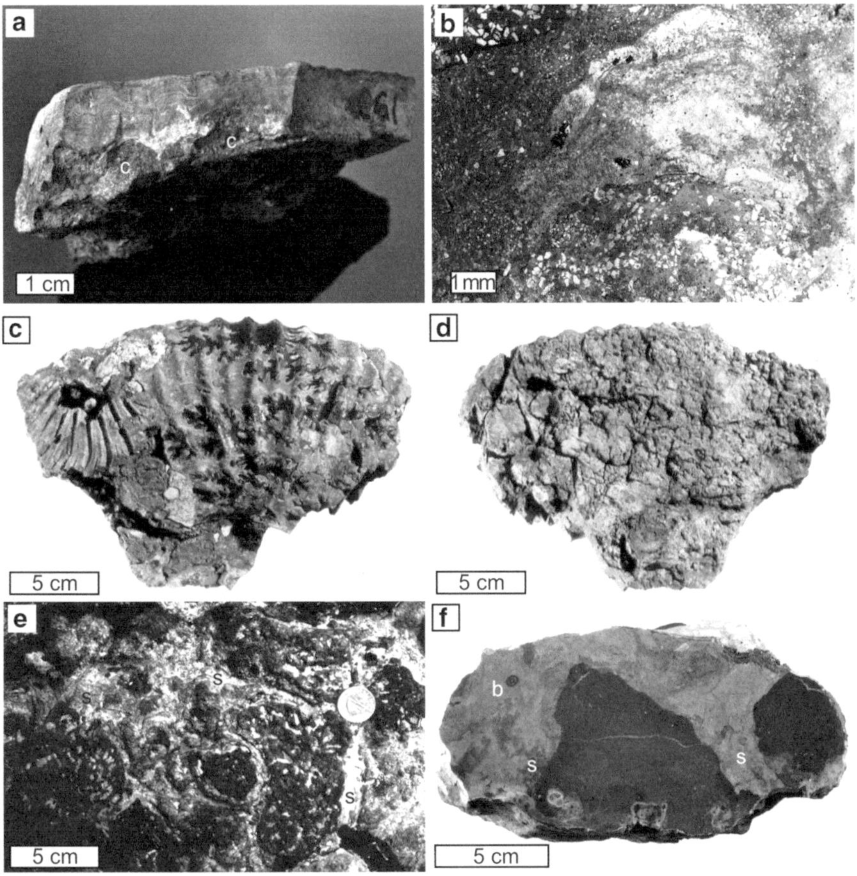

Figure 5. Occurrences of DWH stromatolites in the Plattenwald Bed (Albian) and on top of the Aubrig Beds (equivalent of the Kamm Bed; Early Cenomanian) in Vorarlberg: (**a**) Phosphatic stromatolite on top of a phosphatic crust (c) in the Plattenwald Bed (not younger than late Early Albian: Grosser Wald NW Vorder Schaner Alp); (**b**) Thin-section photo of a partly massive and partly crudely laminated, phosphatic stromatolite attached to a phosphate nodule-rich in quartz grains (probably Early Albian in age; W Ebnit); (**c** and **d**) *Lower* and *upper side* of a phosphatized ammonite mould (*Hypacanthoplites*) and a negative cast of another ammonite attached on the *left side*. The *upper side* shows a flat, eroded surface, covered by an irregular layer of phosphatic columnar stromatolites (S Simonsbach, SW Klaus); (**e**) Nodular top surface of the Aubrig Beds at Müselbach, with partly detached, peripherically phosphatized pebbles, which are bored and enveloped by stromatolites preserved in micrite (s; identical with the micrite of the overlying Seewen Formation); (**f**) Slabbed sample of the top surface of the Aubrig Beds at Müselbach showing two clasts of the Aubrig Beds, which are encrusted by micritic stromatolites. A belemnite is preserved in the **left part** (**b**). All images from Föllmi (1986, 1989), with the exception of (**e**).

e.g., in the area of Grosser Wald, NW of Schaner Alp) or on top of individual fossils and pebbles (Fig. 5c–d; Simonsbach, SW Klaus). Since the youngest phosphatic fossils in this area are not younger than late Early Albian, these occurrences are dated as Early Albian. They are also present in the form of up to 10 cm thick micritic and peripherically slightly phosphatized stromatolites, which are attached to pebbles and cobbles accumulated in former caves within the phosphatic Plattenwald bed, which are now filled in with micrite. The origin of these caves has been associated with the excavation of nonlithified sediments underneath phosphatic crusts by current activities (Föllmi, 1986, 1989). The micritic infill is as dated as Late Albian, based on the occurrence of planktonic foraminifera (Dornbirner Ache, Gunzmoos Alp, Simonsbach, E Feldkirch; Föllmi, 1986).

2.4. CENOMANIAN

Cm-thick, micritic stromatolitic crusts are found in fissures on top of a sandy, glauconitic, nodular limestone (Aubrig Beds in the Muselbach region; Fig. 5e, f). The micrite is rich in planktonic foraminifera and calcispheres and is identical with the micrite of the overlying Seewen Formation. The margins of the nodular surface are frequently bored and peripherically phosphatized. Rare phosphatic particles are trapped in the depressions between the nodules. This phase of phosphogenesis is correlated with the phosphatic Kamm Bed of more proximal areas in eastern Switzerland (Ouwehand, 1987). The foraminifera suggest an Early Cenomanian age (*brotzeni* Zone), and a temporal extension into the Middle Cenomanian (*reicheli* Zone) is not excluded (Föllmi, 1986).

3. Occurrences in Eastern Switzerland

3.1. APTIAN

The oldest occurrence of DWH stromatolites within the Garschella Formation is described from the basal phosphatic Luitere Bed of the early Late Aptian age, in the eastern part of the Churfirsten region (Figs. 1 and 2; Ouwehand, 1987). They occur either as phosphatic stromatolitic fragments (at the Garschella type locality and east of Sichelchamm) or as micritic stromatolotoid structures in up to 10 cm wide and 5 cm deep depressions on top of the Urgonian Schrattenkalk Formation, which are filled in by a *Hedbergella*-bearing micrite (east of Sichelchamm).

3.2. ALBIAN

The phosphatic Durschlägi Bed of late Early Albian age contains phosphatic stromatolite fragments in the region of the Zwinglipass (Säntis region; Ouwehand, 1987). A further occurrence of millimetric, phosphatic (and pyritized) stromatolite fragments was documented from the Albian phosphatic Plattenwald Bed in the region SW of Buchs (Rhine valley; Ouwehand, 1987). The basal phosphatic bed of the Garschella Formation near Säntis (Tierwis) also includes debris of phosphatic stromatolites. The age of this latter bed corresponds to the late Middle to early Late Albian (Ouwehand, 1987).

3.3. CENOMANIAN

An illustrative case of well-preserved stromatolites is known from distal occurrences of the partly phosphatic Kamm Bed (Figs. 6 and 7; Ouwehand, 1987).

Figure 6. Occurrence of DWH stromatolites in the Kamm Bed (Early Cenomanian), Tristencholben (**a**) and Sellamatt (**b**), Churfirsten area, eastern Switzerland: (**a**) Large sample showing peripherically phosphatized pebbles and cobbles (b) of the subjacent Aubrig Beds, which are colonized by stromatolites. The stromatolites are preserved in micrite and show peripherical impregnation by phosphate and iron oxyhydroxides. A first generation of stromatolites (1) is composed of well-developed columns, which may extend into laminated envelopes covering the lower flanks of the pebbles (s), whereas a second generation (2) consists of layered stromatolites. The matrix between the stromatolite-covered boulders consists in a pelagic micrite (m); (**b**) Sample showing the first-generation of columnar stromatolites directly on top of peripherically phosphatized pebbles of the Aubrig Beds. Note the upward widening of the columns. Compare Ouwehand (1987).

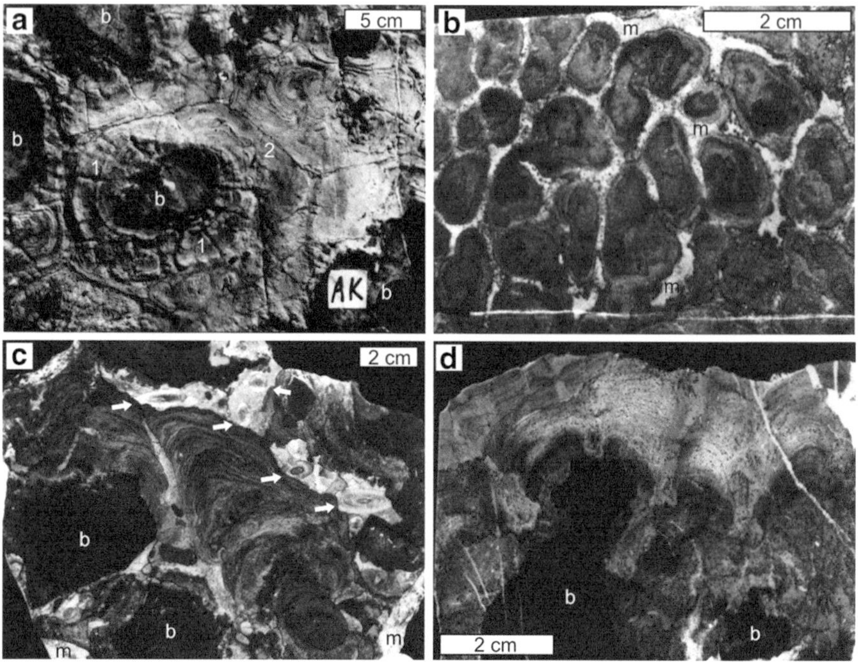

Figure 7. Occurrence of DWH stromatolites in the Kamm Bed (Early Cenomanian), Churfirsten area, eastern Switzerland: (**a**) Bedding plane on top of the Kamm Bed showing isolated, peripherically phosphatized nodules of the Aubrig Beds (b) enveloped by a first generation of columnar stromatolites (1) followed by a second generation consisting of layered stromatolites (2) (Selun, Churfirsten); (**b**) Slabbed and polished cut parallel to bedding on top of the Kamm Bed with irregularly formed stromatolite domes, which are embedded in pelagic micrite (m) (Platten, Churfirsten); (**c**) Cut-and-polished surface perpendicular to bedding from the upper part of the Kamm Bed. Irregularly laminated stromatolitic columns are attached on pebbles reworked from the underlying Aubrig Beds (b). The sediment in between the stromatolites consists of micrite (m) and contains numerous partly silicified belemnites (*arrows*) (Selun); (**d**) Cut-and-polished surface perpendicular to bedding from the upper part of the Kamm Bed. Stromatolites are attached to pebbles reworked from the Aubrig Beds (b) and show growth axes, which are inclined in different upward directions (Hinterrugg). All photos and descriptions from Ouwehand (1987).

They form a well distinguishable, up to 9 cm thick layer, which is laterally traced along a distance of approximately 10 km in a zone of approximately 6–9 km width throughout the Churfirsten region. The stromatolitic crust is composed of up to 3 cm broad, widening-upward columns and/or wrinkled and irregularly laminated, stratiform crusts (Figs. 6 and 7). They are preserved in micrite and may peripherically be impregnated by phosphate and iron oxyhydroxide. Their top surface shows small borings. At least two stromatolite generations are visible, which are distinguished by the different degree of peripheral mineralization and as such by color (Fig. 6a). The first-generation consist of columnar stromatolites, which colonize reworked and partly bored nodular clasts in the upper part of the Kamm Bed. They cover not only the higher parts of the nodules but appear also in thinner envelopes on the flanks and on lower parts of the nodules (Fig. 6a). They themselves are overgrown by a second generation, which forms a laterally continuous layer (Figs. 6a and 7a). The stromatolites are topped by a hiatal surface, which is covered by pelagic carbonates of the Seewen Formation. The pebbly substratum is dated as *appenninica* and early *brotzeni* zones (latest Albian), whereas the stromatolites themselves have a *brotzeni* zone age. The boring infills on top of the stromatolites have a *reicheli* zone age. The sediments of the pelagic Seewen Formation covering the stromatolites date from the *cushmani* zone (Ouwehand, 1987).

4. Occurrences in Central Switzerland

An occurrence of micritic and slightly phosphatic DWH stromatolites has recently been documented from the top of the Plattenwald Bed, in contact with the overlying Seewen Formation in the area of the Col de la Plaine Morte (S of Rawil; Fig. 1; Gainon, 2001; Föllmi and Gainon, 2008). They occur in association with micrite containing planktonic foraminifera of the Late Albian to the Middle Turonian, and are covered by pelagic sediments, which date from the Late Turonian (Fig. 8; Gainon, 2001). They are well comparable to the ones known from the Kamm Bed and their distal equivalents in eastern Switzerland and Vorarlberg.

Scanning-electron microscopy images reveal the presence of well-preserved, dense networks of elongated filaments, which are interpreted as the fossilized remains of microbes (Fig. 9; Gainon, 2001; Föllmi and Gainon, 2008).

DWH stromatolites have also been reported from different phosphatic beds of Albian and Early Cenomanian age at other sites in central Switzerland (Ouwehand, 1987); a detailed documentation is, however, still missing.

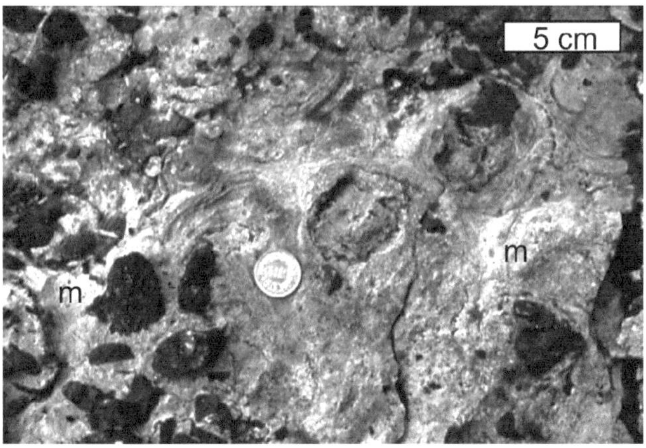

Figure 8. Occurrence of DWH stromatolites on top of the Plattenwald Bed in the region of Col de la Plaine Morte, central Switzerland. The stromatolites are preserved in micrite (m) and locally impregnated by iron oxyhydroxides and are attached to dark phosphatic clasts and fossils. The stromatolites form laterally continuous layers (from Gainon, 2001; compare also Föllmi and Gainon, 2008).

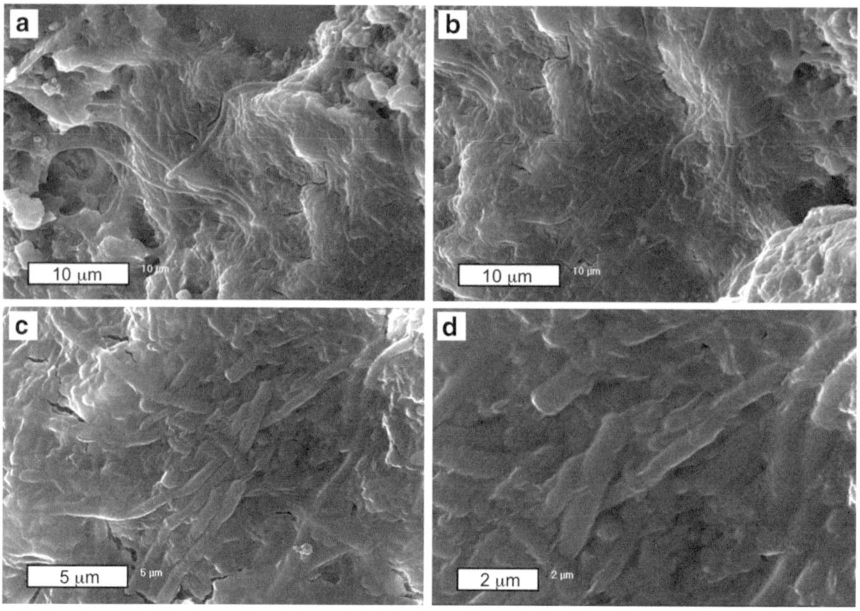

Figure 9. Occurrence of DWH stromatolites on top of the Plattenwald Bed in the region of Col de la Plaine Morte, central Switzerland: (**a**–**d**) Electron scanning microscopy images of a sample of the stromatolite occurrence shown in Fig. 8. Images (**a**) and (**b**) partly overlap and images (**c**) and (**d**) are magnifications of (**a**) and (**b**). All images show a dense network of filamentous corpuscules, which are interpreted as fossilized remains of poikiloaerobic, heterotrophic microbes (from Gainon, 2001; compare also Föllmi and Gainon, 2008).

5. Occurrences in Southeastern France

A detailed documentation of the presence of DWH stromatolites in the "Grès Verts Helvétiques" (Helvetic greensands), the western equivalent of the Garschella Formation, is given by Delamette (1981, 1988a, 1990, 1994) and Delamette et al. (1997) (Fig. 1).

5.1. APTIAN

The oldest stromatolites observed occur in the form of up to 2 cm thick, irregularly laminated crusts, which envelop parts of the walls of cavity incisions in a hardground on top of the "Calcaires Urgoniens" (Urgonian limestone: equivalent of the Schrattenkalk Formation) at Aujon (Massif de Platé, Haute Savoie). They are preserved in micrite and associated with serpulids. They are part of the Aujon Limestone and date from the early Late Aptian (Fig. 10a; Delamette, 1988a, 1990; Delamette et al., 1997).

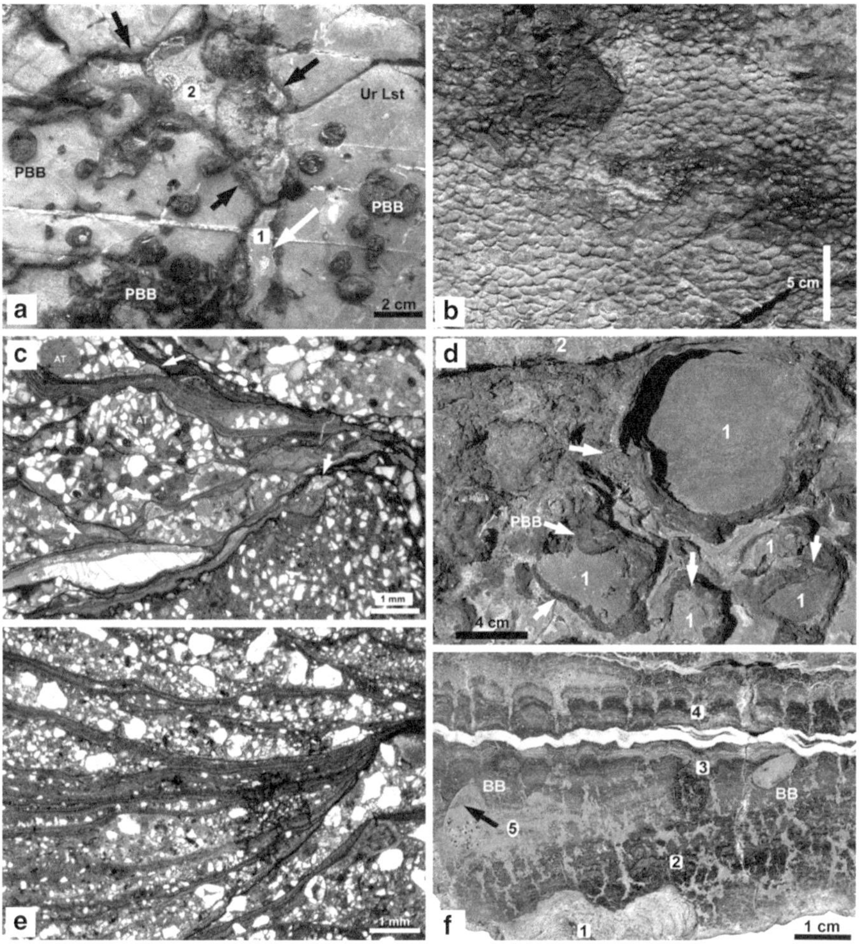

Figure 10. Occurrences of DWH stromatolites of Aptian and Albian age in southeastern France: (**a**) Hardground of early Late Aptian age on top of the "Calcaires Urgoniens" (Ur Lst) showing pockets with phosphatized rims (*black arrows*), bivalve borings (PBB), and stromatolites (1: *white arrow*), which is partly covered by bioclastic sediment (2) (Aujon, Platé Massif); (**b–f**) Albian stromatolites associated with phosphatic sediments (Arberroz phosphatic conglomerate of the Verds Beds) of Early to Middle Albian age of the Haute Savoie: (**b**) Phosphatic stromatolitic crust showing individual domes, which covers the bored surface of a sandy limestone on top of the Colombière Sandstone of Late Aptian age (Combe aux Puaires, Haut-Giffre Massif); (**c** and **e**) Thin-section photos of phosphatic laminae enveloping and separating phosphatized quartz-bearing sediments of the Early Albian age (Verds Beds). Note the presence of sessile foraminifera (*white arrows*) and agglutinated tubes (AT) of uncertain origin (foraminifera or bioturbation?) (**c**: Fenêtres à Grappins, Haut-Giffre Massif; **e**: Col du Freu, Bornes Massif); (**d**) Phosphatic stromatolitic envelopes (*arrows*) around bored pebbles (PBB), which were reworked from the subjacent Colombière Sandstone of the Late Aptian age (Combe aux Puaires, Haute-Giffre Massif); (**f**) Slabbed sample showing multiple stromatolite generations in Albian phosphatic conglomerate (Verds Beds): a sandy stromatolite (1) at the base is covered by dendritic phosphatic stromatolites ("thrombolites": 2), layered phosphatic stromatolites (3), and columnar phosphatic stromatolites (4). Note the presence of bivalve borings (BB: *Gastrochaenolites*), which are infilled by micritic stromatolites (5 with *black arrow*) (Combe aux Puaires, Haute-Giffre Massif). All images from Delamette (1988a, 1990).

5.2. ALBIAN

Numerous occurrences are known of phosphatic columnar stromatolites (up to 8 cm high and 2 cm thick), stromatolite crusts, and isolated stromatolite fragments in condensed phosphorite horizons (Fig. 10b–f; Arberroz phosphatic conglomerate of the Verds Beds; Arberroz and Nant d'Orlier: Massif de Platé; Chérente: Chaîne des Aravis; Oddaz and Fenêtres à Grappins: Massif du Haut-Giffre, all in Haute Savoie; Delamette, 1988a, 1990; Delamette et al., 1997). They have been dated as late Early to early Middle Albian in age. A lateral facies evolution from planar stromatolites to columnar stromatolites, phosphatic stromatolitic envelopes on reworked clasts of the underlying limestone, and glauconitic sandstone without stromatolites is observed (Fig. 11).

5.3. CENOMANIAN

Impressive occurrences of Cenomanian DWH stromatolites have been documented from numerous localities in the Haut Giffre, Platé, Bornes-Aravis, and Bauges Massifs of Haute Savoie (Freu Beds; Delamette, 1981, 1988a, 1990; Delamette et al., 1997). They are generally preserved in pelagic micrite and are partly phosphatic or impregnated by iron oxyhydroxides. They form up to 10 cm thick crusts, which envelop peripherically phosphatized sediment pebbles or fossils (e.g., ammonites; Figs. 12a and 13) and which may laterally pinch out by sub- and superjacent discontinuity surfaces (Fig. 10b). At least two generations have been identified, which may be separated by angular discontinuities (Figs. 12a and 13). The first generation is often formed of colonies of columnar stromatolites, whereas the second generation is more often part of laterally continuous crusts (Figs. 12a, c and 13).

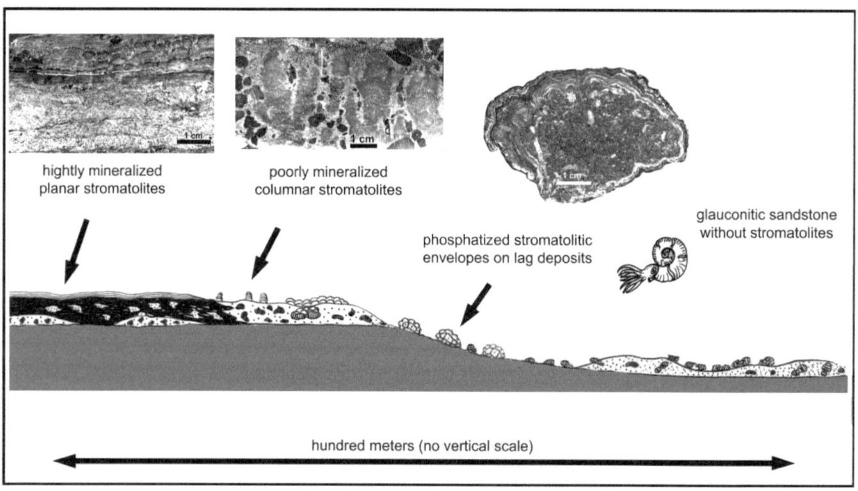

Figure 11. Lateral evolution of Albian stromatolites observed in the Haute Savoie, southeastern France.

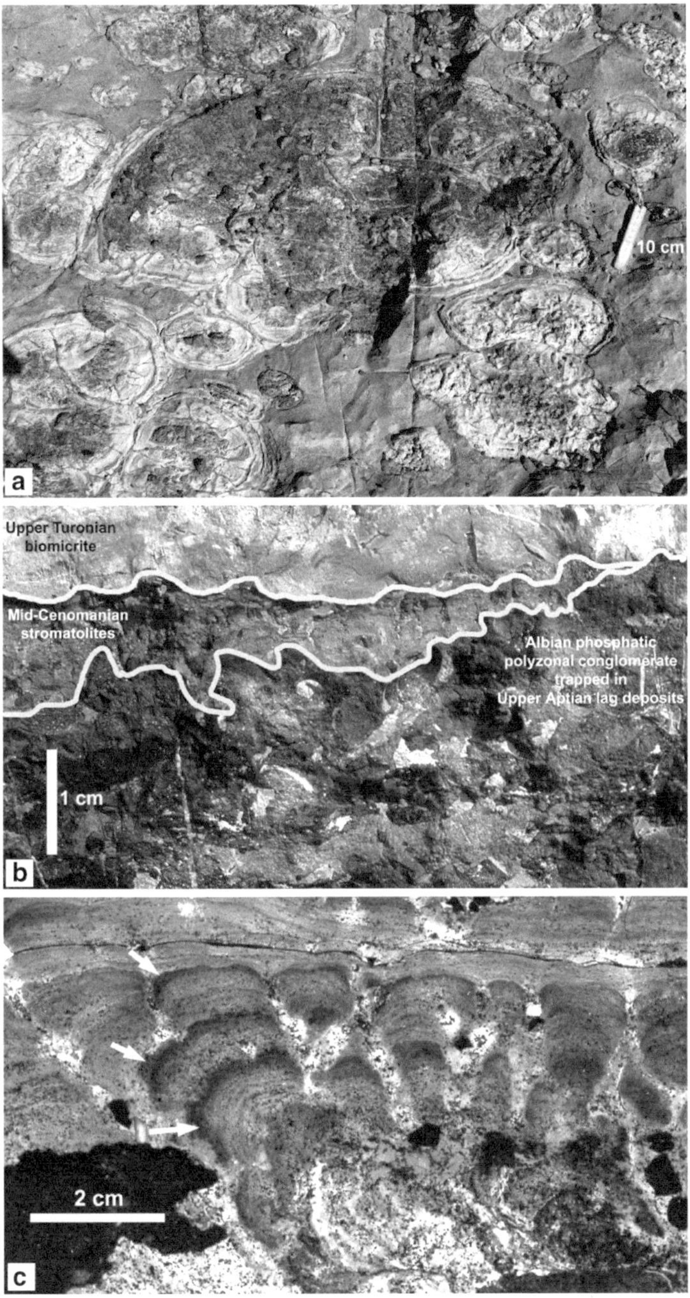

Figure 12. Occurrences of DWH stromatolites in the Freu Beds of Early to Middle Cenomanian age: (**a**) Stromatolitic envelopes ("oncoids") around phosphatized ammonite remains (Rochers des Fiz, Platé Massif): a first generation of columnar stromatolites is covered by a second generation of planar stromatolites (see Fig. 13 for more details); (**b**) Local occurrence of small columnar stromatolites at the Flaine oucrop (Platé Massif), which laterally disappear and make place for a hiatus between the Albian phosphatic bed and the Turonian biomicrite; (**c**) Polished slab from Chérente (Aravis Massif) showing multiple phases of phosphatization during the growth of the stromatolites (*white arrows*). Compare Delamette (1981, 1988a, 1990) for descriptions and similar images.

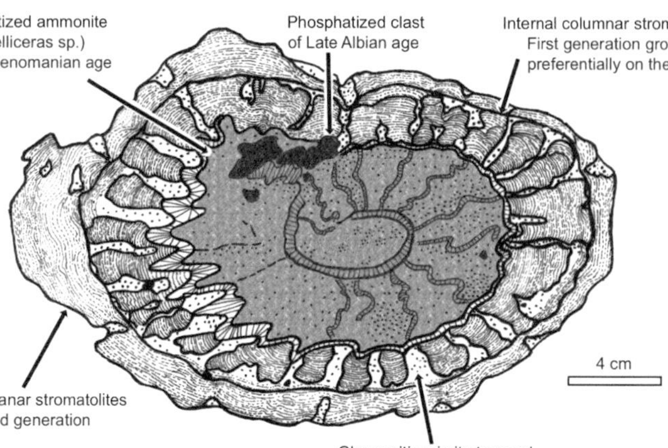

Figure 13. Two generations of stromatolites appear in the oncolitic structures around a phosphatized ammonite (*Mantelliceras* sp.) preserved in the Fiz Limestone of the Early to Middle Cenomanian age (Rochers des Fiz, Platé Massif). From Delamette (1988a).

6. Discussion

6.1. ASSOCIATION WITH DEEPER-WATER HIATAL SURFACES AND CONDENSED SEDIMENTS

The here-described Aptian, Albian, and Cenomanian DWH stromatolites are associated with (partly) phosphatized hardgrounds and hiatal surfaces, which are either isolated or allied with highly condensed phosphatic beds. The stromatolites may be present on top of a hardground (e.g., on top of the Urgonian limestone) and may also be topped by a hardground or hiatal surface (e.g., on top of the Early Cenomanian stromatolites). Stromatolitic remains also appear in the form of reworked fragments in condensed beds and in beds composed of allochthonous sediments. The stromatolites may directly be attached to the hardground but may also cover pebbles, cobbles, and fossils, which occur on top of the hardground or within depressions and caves associated with hardgrounds and hiatal surfaces. The hardgrounds and condensed phosphatic beds include up to several million years of formation as is indicated by ammonite and planktonic foraminiferal biostratigraphy (Fig. 2).

In all localities described here, the stromatolites occur together with open-marine faunal communities (e.g., ammonites and belemnites). In many localities, they are accompanied by or preserved in pelagic micrite containing planktonic foraminifera. In the Aptian, they occur on top of the drowned Urgonian shallow-water carbonate platform, and also there, the presence of the above-mentioned characteristics places them in a deepening-upward context.

The sum of these features, together with the absence of shallow-water facies or fauna, fossilized algae, and emersion surfaces, allows us to attribute the here-described stromatolites to a deeper-water environment. Their presence in depressions, fissures, and cave-like structures indicates that they were less dependent on light. We interpret these stromatolites as deeper-water forms, which were constructed by heterotrophic microbial communities in current-dominated regimes, where they were fossilized by authigenic mineralization processes (phosphate, iron oxyhydroxides, rarely pyrite), by trapped pelagic micrite, or by a combination of both. The term "deeper water" relates to the observation that the stromatolite occurrences are limited to the more distal part of the southern European shelf, which was at that time a passive margin lacking major tectonic activity (e.g., Pfiffner, 2009). Their depth did not surpass several hundreds of meters and cannot be considered as basinal.

6.2. MORPHOLOGY AND MINERALIZATION

The Helvetic DWH stromatolites range in size from a micrometric to a centimetric scale, with the largest forms reaching heights of approximately 10 cm. They appear in a continuum of morphologies, which span from laterally well-separated columnar forms ("dendroid" in Delamette, 1990) to closely spaced columnar forms ("colloform" in Delamette, 1990; Figs. 4a, 5b, 6, 7, and 10b, f), doming forms without clear lateral separation (Figs. 5a and 6a), and laterally continuous, planar layers lacking elevated structures (Fig. 12). They may entirely envelop clasts and fossils and represent oncoids in two-dimensional cuts (Fig. 7a). In some cases, stromatolites cover larger isolated pinnacle-like rock structures, which resisted erosion and which provide oncoid-like structures, when cut parallel to bedding in their higher parts (Ouwehand, 1987). In general, the stromatolitic structures range between millimeter-thin laminated layers and lenses separating different sediment generations and enveloping particles and fossils, which are often only visible in thin section ("films" in Delamette, 1990: Figs. 3b and 10c, e) and well-developed macroscopic forms. They are mostly clearly laminated, with lamination thicknesses ranging between several micrometers to millimeters. In thin section, laminations are not always obvious (Fig. 5b), and the stromatolites may appear in an irregular, dendritic way (Fig. 10f; "thrombolites" in Delamette, 1990) or only indicated by interspersed detrital minerals (Delamette, 1990).

They are preserved in phosphate and micrite, including different amounts of detrital particles (essentially quartz and glauconite). In the case of micrite, the presence of calcispheres, planktonic foraminifera, and other microfossils indicates that the sediment was of pelagic origin and trapped. In the case of fully phosphatic stromatolites, it is difficult to state if micrite was present as a precursor sedimentary phase or if the stromatolites were directly phosphatized, without intermittent sediment-trapping. In favor of the first hypothesis is the observation that micritic stromatolites often show a phosphatized periphery, indicating partial phosphatization following their formation with the help of micrite trapping (Fig. 12c). They may also be coated by iron oxyhydroxides. The local presence of borings and erosive discontinuities suggests early (peripheral) cementation of the stromatolites.

Delamette (1988a, 1990) observed a relationship between the morphology and mineralogy of the stromatolites, in that columnar forms are enriched in detrital material, whereas parallel-laminated, planar layers lack detrital material (Fig. 11). He assumed that in a sedimentary regime with detrital delivery, the microbial associations preferred columnar structures, in order to create bed-load bypass channels and avoid burial, whereas in a sedimentary regime without detrital advection, this type of growth strategy would not have been necessary. In addition to this, we observe an evolution within the same colonies from columnar forms to planar layers. In this case, stromatolite colonization appeared to have started in the form of numerous small columns, for example, around reworked and (semi-) lithified pebbles and cobbles, which formed themselves a substratum for the growth of laterally continuous planar layers and crusts (Figs. 4, 6, 12a, and 13).

6.3. ASSOCIATION WITH PHOSPHOGENESIS

The occurrence of the here-described DWH stromatolites from the northern Tethyan margin is essentially limited to phosphatized hardgrounds and condensed phosphatic beds. These horizons and beds have formed in a current-dominated sedimentary regime characterized by highly reduced sediment-accumulation rates, which resulted from a dynamic interplay between (sudden) sedimentation, erosion, and sediment reworking. Authigenesis of phosphate and glauconite, and in subordinate quantities of silica, pyrite, and iron oxyhydroxides, was an important process in the areas dominated by this sedimentary regime (Heim, 1934; Heim and Seitz, 1934; Schaub, 1936; Delamette, 1981, 1985, 1988a, b, 1990, 1994; Delamette et al., 1997; Föllmi, 1986, 1989, 1990, 1996; Föllmi and Delamette, 1991; Föllmi and Gainon, 2008; Ouwehand, 1987).

In many cases, the stromatolites appear on top of the phosphatic beds and in distal, deeper-water areas of the region of phosphogenesis, where sedimentary condensation processes were most intense. Since these beds were often formed in a regime of sea-level rise, the area of optimal stromatolite growth and preservation appears to be limited to the area of maximum condensation in the deeper part of the area of phosphogenesis.

A further observation is that the stromatolites often grew in and, in some cases, were restricted to depressions and cavities, which appear to have been created by differential erosion (in the case of the cavities, "tunneling erosion" by the local break-up of lithified surface crusts by currents and the excavation of underlying, nonlithified sediments; Föllmi, 1986, 1989; Delamette, 1988a). There they colonized the walls of the cavities and the exposed tops and flanks of pebbles and cobbles remaining on the cavity floors (Fig. 6a).

Many authors have established links between the presence of microbial colonies and the process of phosphogenesis, in that microbial activity induced chemical conditions appropriate for phosphogenesis (e.g., Lucas and Prévôt, 1984; Soudry and Lewy, 1988; Martín-Algarra and Sánchez-Navas, 1995, 2000; Schulz and Schulz, 2005). The association of DWH stromatolites and macroscopic phosphate accumulations in the Helvetic realm would indeed favor a link between the two; however, the bathymetric offset and the only partial overlap between the areas of maximal phosphogenesis and optimal stromatolite growth and preservation precludes a 1-to-1 relationship. Proximal areas of phosphogenesis often lack macroscopic and microscopic evidence for the presence of stromatolites, whereas in distal-most areas, stromatolites may be preserved in micrite, in spite of the presence of phosphate.

6.4. ENVIRONMENTAL CONDITIONS OF STROMATOLITE GROWTH

Whereas proximal occurrences of phosphate beds of Aptian to Cenomanian age in the Helvetic thrust-and-fold belt are rich in sessile benthic fossils, such as siliceous sponges, ahermatype corals, brachiopods, bryozoa, and crinoids, the distally occurring phosphatic beds containing stromatolites are generally devoid of benthic organisms, with the exception of the benthic microorganisms directly associated with the stromatolites (agglutinating benthic foraminifera, serpulids). Belemnites and rare ammonites are, however, present in the stromatolite beds. The absence of larger sessile benthic biota is taken as an indication of dysaerobic conditions prevailing in the bottom waters of the sites of stromatolite growth and preservation (Ouwehand, 1987; Delamette, 1988a, 1990). The presence of benthic ecosystems in coeval, more proximal occurrences suggests the presence of a bathymetric gradient in bottom-water oxygenation conditions that was probably related to the presence of an oxygen-minimum zone (OMZ) straddling the outer part of the northern Tethyan shelf and the slope. The area of optimal stromatolite growth and preservation would then indicate the zone, where the upper boundary of the OMZ interfaced with the outer shelf (Föllmi, 1989; Martín-Algarra and Vera, 1994).

The stromatolites of each time interval are well comparable throughout the entire Helvetic margin described here; the Cenomanian stromatolites, for example, show many common features for the area described here, such as their preservation in micrite, which may be partly phosphatized and impregnated by iron oxyhydroxides, the presence of several generations, and their attachment to lithoclasts and fossils. A distinction in the specific characteristics is, however, observed for each time window: the Aptian forms are mostly associated with *Hedbergella* and serpulids, whereas the Albian stromatolites show a firm association with benthic agglutinated foraminifera (*Placopsinella* sp.). The Cenomanian forms are almost entirely devoid of benthic foraminifera and serpulids but may include planktonic foraminifera instead. It should also be noted that the Aptian stromatolites appear rather locally, the Albian stromatolites are more widespread, and the Cenomanian forms occupy broad zones along the northern Tethyan margin (Delamette, 1990). This evolution through time away from the inclusion of benthos and toward their frequent occurrence may be the expression of the overall sea-level rise, which took place during the Albian and Cenomanian (Fig. 2) and a corresponding rise of the OMZ, leading to larger flat shelf areas intersecting with the upper boundary of the OMZ.

6.5. MICROBES

Microbial structures are very well preserved in the Cenomanian stromatolites of the Col de la Plaine Morte area in the central Swiss Helvetic Alps. They are filamentous, with a maximal length of approximately 20 μm and a diameter between 5 and 7 μm (Fig. 9). They are difficult to be interpreted with regards to their energy requirements and metabolic pathways. Since they are associated with deeper-water stromatolites, which also occur in cavities, they were most likely light independent and therefore heterotrophic. This would generally exclude cyanobacteria. Sulfur-oxydizing bacteria such as *Beggiatoa*, *Thioploca*, and *Thiomargarita* seem also unlikely since they usually have larger filament widths (up to several 100 μm; e.g., Schulz et al., 1996; Ahmad et al., 2006). The presence of sulfur-oxydizing microbes is not totally excluded, though, since sulfide minerals (pyrite) are present in sediments associated with the stromatolites. We exclude, however, the presence of methanotrophic microbes, based on a stable-isotope profile measured through a well-developed stromatolite of Early Cenomanian age in eastern Switzerland, where $\delta^{13}C$ values remain positive (>1.3‰; Ouwehand, 1987).

We interpret these Cenomanian forms as remains of poikiloaerobic, heterotrophic, filamentous microbes, which formed dense spaghetti-like colonies. The stromatolitic structures appear to be dominated by this type of microbes, and other morphologies (e.g., coccoid forms) were not detected.

6.6. REGIONAL PALEOCEANOGRAPHIC CONDITIONS

From the evidence discussed above, it appears that DWH stromatolite growth and preservation on the Helvetic shelf was favored by regional paleoceanographic conditions. A first mechanism is given by the presence of a current-dominated sedimentary regime, which was responsible for the formation of hardgrounds and hiatal surfaces, erosion, and sediment reworking and which maintained a system of sediment bypass to deeper parts of the shelf (Delamette, 1985, 1988a, b; Föllmi, 1986, 1989, 1990; Föllmi and Delamette, 1991; Ouwehand, 1987). As the probable cause of this sedimentary regime, the E-W-directed northern Tethyan current system was invoked, which was forced to rise onto the outer part of the northern Tethyan shelf, probably through its westward broadening. This may have led to an obstructional upwelling of deeper waters onto the outer part of the shelf, which were enriched in nutrients. As a consequence, primary productivity rates were high, and the partial decomposition of exported organic matter induced the installation of an OMZ. At the same time, the combination of a dynamic sedimentary regime with highly reduced accumulation rates and the delivery of phosphate onto the shelf by upwelling waters and the decomposition of organic matter were instrumental in the process of phosphogenesis. The DWH stromatolites found optimal conditions of growth and preservation in distal parts of the area of phosphogenesis, in the zone of maximal condensation along the axis of E-W-directed current, in the zone where the upper boundary of the OMZ intersected with the shelf. In this zone, oxygen levels in the bottom waters were sufficiently low to prevent the installation of epibenthos feeding upon microbial mats.

6.7. GLOBAL ENVIRONMENTAL CONDITIONS

In the Helvetic realm, important phases of optimal DWH-stromatolite growth and preservation occurred during the early Late Aptian, the late Early Albian, and the late Early Cenomanian (Fig. 2). These time windows follow upon periods of shallow-water carbonate platform drowning (during the late Early Aptian and the latest Aptian) and the take-up of pelagic carbonate deposition (latest Albian). They also follow upon periods of widespread dysaerobic to anaerobic conditions during the late Early Aptian (oceanic anoxic episode OAE 1a), the latest Aptian and earliest Albian (OAE 1b), and the latest Albian (OAE 1d) (Fig. 2).

The "mid"-Cretaceous time period seems generally favorable to stromatolite growth, both in deeper-water as well as in shallow-water environments. Occurrences of DWH stromatolites have been documented along the northern Tethyan margin from Poland to southern Spain (Krajewski, 1981, 1984; Krajewski et al., 2000; Gebhard, 1983; Martín-Algarra and Vera, 1994; Vera and Martín-Algarra, 1994; Martín-Algarra and Sánchez-Navas, 2000; Zitt et al., 2006). Coeval occurrences in or close to the Tethyan realm are known from the Briançonnais (Royant et al., 1970), the Austroalpine units representing the southern margin of the Tethys (Kindle et al., 1987), and northern Spain (Reitner et al., 1995).

During the same period, shallow-water stromatolites seem commonplace in the Tethyan realm. Aptian occurrences are, for instance, documented from shallow-water carbonates in eastern Spain, Tunisia, Turkey, and Oman (Riding and Tomas, 2005; Zaghbib-Turki, 2003; Yilmaz and Altiner, 2006: Immenhauser et al., 2005). Albian and Cenomanian examples are known from the Apennines (Borgomano, 2000), the Vasco-Cantabrian Basin in northern Spain (Neuweiler, 1993), and the Adriatic carbonate platform in Istria, Slovenia, and Croatia (Tišljar et al., 1998; Husinec and Jelaska, 2006; Palci et al., 2007).

It appears from this that, in addition to the regional paleoceanographic conditions favoring DWH stromatolite growth along the northern Tethyan margin, further factors are required to explain the general ubiquity of stromatolite build-ups throughout the Tethyan realm. We postulate a causal relationship between the periods of platform drowning and oceanic anoxic episodes, and the succeeding phases of intense stromatolite growth, and consider the stromatolites as a pioneering ecosystem, which became installed in the periods of recovery following the crises in oxygen conditions in ocean bottom waters and crises in shallow-water platform growth. The Early Cenomanian phase of stromatolite growth is somewhat different, in that it coincides with the onset of widespread pelagic carbonate deposition ("chalk seas"; e.g., Hay, 2008). This phase can also be seen as a phase of recovery, in that during this period, carbonate production took up again in a general way, after a long phase of reduced carbonate deposition in shallow and deeper waters along the northern Tethyan margin (Föllmi et al., 1994, 2006, 2007) and a general decline in reef growth and aragonitic reef builders (e.g., Kiessling et al., 2008).

7. Conclusions

A remarkable suite of phosphatic and micritic stromatolites occurs in association with Aptian to Cenomanian glauconitic and phosphatic sediments in the Helvetic thrust-and-fold belt of the central European Alps. These sediments were deposited on the outer part of the southern European shelf, which was part of the northern Tethyan margin. The stromatolites grew on top of hardgrounds, colonized reworked pebbles and cobbles, enveloped fossil remains, attached to the walls of cavities, and formed films on sediment surfaces. They are commonly associated with hiatal surfaces at the base, within, or on top of condensed phosphatic beds. They are preserved in phosphate and micrite, which may be partly phosphatized or impregnated by iron oxyhydroxides. They are present in the form of micro- to millimetric laminations, millimetric to 10 cm high laminated columns, or in laminated layers. Scanning electron microscopy reveals a dense network of filamentous bodies in the Cenomanian, micritic stromatolites, which are interpreted as the remains of poikiloaerobic, heterotrophic microbes.

The association with hiati, open-marine fauna, phosphates, and pelagic sediments and the lack of shallow-water sediments and fauna shows their deeper-water character. They occur in the zone of maximal condensation, in the deeper, more distal part of the area of phosphogenesis on the outer shelf of the southern European margin. The almost complete lack of macroscopic sessile fauna suggests that they were formed in a zone that interfaced with the upper limit of the oxygen-minimum zone.

The omnipresence of DWH stromatolites and their shallow-water counterparts in the Tethyan realm during the "mid"-Cretaceous suggests that the general environmental and oceanographic conditions were favorable to the growth and preservation of stromatolites. The occurrence of the here-described DWH stromatolites in periods following oceanic anoxic episodes let us speculate that they represent pioneering ecosystems, which were associated with the recovery phases following the stressful oceanic anoxic episodes.

8. Acknowledgments

We thank Vinod C. Tewari for his invitation to contribute a chapter to this book. Jacob Zopfi, Christina Glunk, and Eric Verrecchia of the University of Lausanne are acknowledged for their valuable insights on present-day microbial ecosystems and stromatolites in general. A first version of this contribution was carefully and constructively reviewed by Agustin Martín-Algarra and Krysztof Krajewski. Their comments greatly improved this paper. We are grateful to the Swiss National Science Foundation, which generously supported us during various stages of our career.

9. References

Ahmad, A., Kalanetra, K.M. and Nelson, D.C. (2006) Cultivated *Beggiatoa* spp. define the phylogenetic root of morphologically diverse, noncultured, vacuolated sulfur bacteria. Can. J. Microbiol. **52**: 591–598.

Arning, E.T., Brigel, D., Schulz-Vogt, H.N., Holmkvist, L., Jorgensen, B.B., Larson, A. and Peckmann, J. (2008) Lipid biomarker patterns of phosphogenic sediments from upwelling regions. Geomicrobiol. J. **25**: 69–82.

Böhm, F. and Brachert, T.C. (1993) Deep-water stromatolites and *Frutexites* Maslov from the Early and Middle Jurassic of S-Germany and Austria. Facies **28**: 145–168.

Borgomano, J.R.F. (2000) The upper cretaceous carbonates of the Gargano-Murge region, southern Italy: a model of platform-to-basin transition. AAPG Bull. **84**: 1561–1588.

Bréhéret, J.G. (1997) L'Aptien et l'Albien de la fosse vocontienne (des bordures au bassin). Evolution de la sédimentation et enseignements sur les événements anoxiques. Publ. Soc. Géol. du Nord **25**: 164

Brett, C.E. and Seilacher, A. (1991) Fossil Lagerstätten: a taphonomic consequence of event sedimentation, In: G. Einsele, W. Ricken and A. Seilacher (eds.) *Cycles and Events in Stratigraphy*. Springer-Verlag, Berlin, pp. 283–297.

De Azevedo, R.L.M. and Rodrigues, R. (2000) Towards $\delta^{13}C$ and $\delta^{18}O$ global curves for Albian carbonate sections. 31th Intern. Geol. Congress, Rio de Janeiro, CD abstract vol.

Delamette, M. (1981) Sur la découverte de stromatolites circalittoraux dans la partie moyenne du Crétacé nord-subalpin (Alpes occidentales françaises). C. Rend. Acad. Sci. **292**: 761–764.

Delamette, M. (1985) Phosphorites et paléocéanographie: l'exemple des phosphorites du Crétacé moyen delphino-helvétique. C. R. Acad. Sci. **300**: 1025–1028.

Delamette, M. (1988a) l'Evolution du domaine Helvétique (entre Bauges et Morcles) de l'Aptien supérieur au Turonien: Séries condensées, phosphorites et circulations océaniques. Publ. Dép. Géol. Paléont. Univ. Genève **5**: 316.

Delamette, M. (1988b) Relation between the condensed Albian deposits of the Helvetic domain and the oceanic current-influenced continental margin of the northern Tethys. Bull. Soc. Geol. France **8**: 739–745.

Delamette, M. (1990) Aptian, Albian and Cenomanian microbialites from the condensed phosphatic deposits of the Helvetic shelf, western Alps. Eclog. Geol. Helv. **83**: 99–121.

Delamette, M. (1994) Les séries à phosphorites aptiennes à cénomaniennes de la plate-forme helvétique en Haute-Savoie et en Valais. Publ. Dép. Géol. Paléont. Univ. Genève **14**: 101–135.

Delamette, M., Charollais, J., Decrouez, D. and Caron, M. (1997) Les grès verts helvétiques (Aptien moyen - Albien supérieur) de Haute-Savoie, Valais et Vaud (Alpes occidentales Franco-Suisses): Analyse stratigraphique et inventaire paléontologique. Publ. Dép. Géol. Paléont. Univ. Genève **23**: 400.

Erbacher, J., Thurow, J. and Littke, R. (1996) Evolution patterns of radiolaria and organic matter variations: a new approach to identify sea-level changes in mid-Cretaceous pelagic environments. Geology **24**: 499–502

Föllmi, K.B. (1986) Die Garschella- und Seewer Kalk-Formation (Aptian-Santonian) im Vorarlberger Helvetikum und Ultrahelvetikum. Mitt. Geo. Inst. ETH Univ. Zürich NF **262**: 391.

Föllmi, K.B. (1989) Evolution of the Mid Cretaceous triad: platform carbonates, phosphatic sediments, and pelagic carbonates along the northern Tethys margin. Lect. Notes Earth Sci. **23**: 153.

Föllmi, K.B. (1990) Condensation and phosphogenesis: example of the Helvetic Mid-Cretaceous (northern Tethyan Margin), In: A.J.G. Notholt and I. Jarvis (eds.) *Phosphorite Research and Development. Geol. Soc. London, Sp. Publ. 52*, pp. 237–252.

Föllmi, K.B. (1996) The phosphorus cycle, phosphogenesis, and marine phosphate-rich deposits. Earth-Sci. Rev. **40**: 55–124.

Föllmi, K.B. and Delamette, M. (1991) Comment on: model simulation of mid-cretaceous ocean circulation. Science **251**: 94.

Föllmi, K.B. and Gainon, F. (2008) Demise of the northern Tethyan Urgonian carbonate platform and subsequent transition towards pelagic conditions: the sedimentary record of the Col de la Plaine Morte area, central Switzerland. Sediment. Geol. **205**: 142–159.

Föllmi, K.B. and Ouwehand, P.J. (1987) Die Garschella-formation und Götzis-Schichten (Aptian-Coniacian): Neue stratigraphische Daten aus dem Helvetikum der Ostschweiz und des Vorarlbergs. Eclog. Geol. Helv. **80**: 141–191.

Föllmi, K.B., Weissert, H., Bisping, M. and Funk, H. (1994) Phosphogenesis, carbon-isotope stratigraphy, and carbonate-platform evolution along the Lower Cretaceous northern tethyan margin. GSA Bull. **106**: 729–746.

Föllmi, K.B., Godet, A., Bodin, S. and Linder, P. (2006) Interactions between environmental change and shallow-water carbonate build-up along the northern Tethyan margin and their impact on the early Cretaceous carbon-isotope record. Paleoceanogr. **21**: doi:10.1029/2006PA001313.

Föllmi, K.B., Bodin, S., Godet, A., Linder, P. and van de Schootbrugge, B. (2007) Unlocking paleo-environmental information from early Cretaceous shelf sediments in the Helvetic Alps: stratigraphy is the key! Swiss J. Geosci. **100**: 349–369.

Gainon, F. (2001) Etude géologique dans la région du Rawil: Cartographie et étude stratigraphique et sédimentologique des Couches à Orbitolines supérieures et de la Formation de Garschella. Unpublished diploma thesis, University de Neuchâtel, 90.

Gallardo, V.A. (1977) Large benthic microbial communities in sulphide biota under Peru-Chile subsurface countercurrent. Nature **268**: 331–332.

Gebhard, G. (1983) Stratigraphische Kondensation am Beispiel Mittelkretazischen Vorkommen im perialpinen Raum. Unpublished Ph.D. thesis, University of Tübingen, 145.

George, A.D. (1999) Deep-water stromatolites, Canning Basin, northwestern Australia. Palaios **14**: 493–505.

Haq, B.U., Hardenbol, J. and Vail, P.R. (1987) Chronology of fluctuating sea level since the Triassic. Science **235**: 1156–1167.

Hay, W.W. (2008) Evolving ideas about the Cretaceous climate and ocean circulation. Cret. Res. **29**: 725–753.

Heim, A. (1934) Stratigraphische Kondensation. Eclog. Geol. Helv. **27**: 372–383.

Heim, A. and Seitz, O. (1934) Die mittlere Kreide in den helvetischen Alpen von Rheintal und Vorarlberg und das Problem der Kondensation. Denkschr. Schweiz. Naturf. Ges. **69**: 185–310.

Herrle, J.O., Kößler, P., Friedrich, O., Erlenkeuser, H. and Hemleben, C. (2004) High-resolution carbon isotope records of the Aptian to Lower Albian from SE France and the Mazagan Plateau (DSDP Site 545): a stratigraphic tool for paleoceanographic and paleobiologic reconstruction, Earth Planet. Sci. Lett. **218**: 149–161.

Husinec, A. and Jelaska, V. (2006) Relative sea-level changes recorded on an isolated carbonate platform: Tithonian to Cenomanian succession, southern Croatia. J. Sediment. Res. **76**: 1120–1136.

Immenhauser, A., Hillgärtner, H. and van Bentum, E. (2005) Microbial-foraminiferal episodes in the early Aptian of the southern Tethyan margin: ecological significance and possible relation to oceanic anoxic event 1a. Sedimentology **52**: 77–99.

Jarvis, I., Gale, A.S., Jenkyns, H.C. and Pearce, M.A. (2006) Secular variation in the Late Cretaceous carbon isotopes: a new $\delta^{13}C$ carbonate reference curve for the Cenomanian-Campanian (99.6–70.6Ma). Geol. Mag. **143**: 561–608.
Kiessling, W., Aberhan, M. and Villier, L. (2008) Phanerozoic trends in skeletal mineralogy driven by mass extinctions. Nature Geosci. **1**: 527–530.
Kindle, P.J., Schindler, U., Ouwehand, P.J. and Weissert, H. (1987) Mid-Cretaceous phosphorites of austro-alpine nappes in Austria and Switzerland. Terra cogn. **7**: 284.
Krajewski, K.P. (1981) Phosphate microstromatolites in the High-Tatric Albian (Tatras Mountains). Ann. Soc. Geol. Polon. **19**: 175–183.
Krajewski, K.P. (1984) Early diagenetic phosphate cements in the Albian condensed glauconitic limestones of the Tatra Mountains, western Carpathians. Sedimentology **31**: 443–470.
Krajewski, K.P., Van Cappellen, P., Trichet, J., Kuhn, O., Lucas, J., Martín-Algarra, A., Prévôt, L., Tewari, V.C., Gaspar, L., Knight, R.I. and Lamboy, M. (1994) Biological processes and apatite formation in sedimentary processes, In: K.B. Föllmi (ed.) *Concepts and Controversies in Phosphogenesis. Eclogae. Geol. Helv. 87*, pp. 701–745.
Krajewski, K.P., Lesniak, P.M., Lacka, B. and Zawidzki, P. (2000) Origin of phosphatic stromatolites in the Upper Cretaceous condensed sequence of the Polish Jura Chain. Sediment. Geol. **136**: 89–112.
Lucas, J. and Prévôt, L. (1984) Synthèse d'apatite par voie bactérienne à partir de matière organique phosphaté et de diver carbonates de calcium dans les eaux douce et marine naturelle. Chem. Geol. **42**: 101–118.
Martín-Algarra, A. and Sánchez-Navas, A. (1995) Phosphate stromatolites from condensed cephalopod limestones, Upper Jurassic, southern Spain. Sedimentology **42**: 893–919.
Martín-Algarra, A. and Sánchez-Navas, A. (2000) Bacterially mediated authigenesis in Mesozoic stromatolites from condensed pelagic sediments (Betic Cordillera, Southern Spain), In: C.R. Glenn, J. Lucas and L. Prévôt-Lucas (eds.) *Marine Authigenesis: From Global to Microbial. SEPM Sp. Publ. 66*, pp. 499–525.
Martín-Algarra, A. and Vera, J.A. (1994) Mesozoic pelagic phosphate stromatolites from the Penibetic (Betic Cordillera, southern Spain), In: J. Betrand-Sarfati and C. Monty (eds.) *Phanerozoic Stromatolites II.* Kluwer, Dordrecht, pp. 345–391.
Neuweiler, F. (1993) Development of Albian microbialites and microbialite reefs at marginal platform areas of the Vasco-Cantabrian Basin (Soba reef area, Cantabria, N. Spain). Facies **29**: 231–249.
Ouwehand, P.J. (1987) Die Garschella-Formation ("Helvetischer Gault", Aptian-Cenomanian) der Churfirsten-Alvier Region (Ostschweiz). Mitt. Geol. Inst. ETH Univ. Zürich, NF **275**: 296.
Palci, A., Jurkovšek, B., Kolar-Jurkovšek, T. and Caldwell, M.W. (2007) New palaeoenvironmental model for the Komen (Slovenia) Cenomanian (Upper Cretaceous) fossil lagerstätte. Cret. Res. **29**: 316–328.
Pfiffner, O.A. (2009) *Geologie der Alpen.* Haupt Verlag, Bern, 359.
Reitner, J., Wilmsen, M. and Neuweiler, F. (1995) Cenomanian/Turonian sponge microbialite deep-water hardground community (Liencres, northern Spain). Facies **32**: 203–212.
Riding, R. and Tomas, S. (2005) Stromatolite reef crusts, Early Cretaceous, Spain: bacterial origin of in situ-precipitated peloid microspar? Sedimentology **53**: 23–34.
Royant, G., Rioult, M. and Lanteaune, M. (1970) Horizon stromatolithique à la base du Crétacé supérieur dans le Briançonnais ligure. Bull. Soc. Geol. France **7**: 372–374.
Schaub, H.P. (1936) Geologie des Rawilgebietes. Eclogae. Geol. Helv. **29**: 337–407.
Schulz, H.N. and Schulz, H.D. (2005) Large sulfur bacteria and the formation of phosphorite. Science **307**: 416–418.
Schulz, H.N., Jorgensen, B.B., Fossing, H.A. and Ramsing, N.B. (1996) Community structure of filamentous, sheat-building sulfur bacteria, Thioploca spp., off the coast of Chile. Appl. Environ. Microbiol. **62**: 1855–1862.
Soudry, D. and Lewy, Z. (1988) Microbially influenced formation of phosphate nodules and megafossil moulds (Negev, southern Israel). Palaeogeogr. Palaeoclimatol. Palaeoecol. **64**: 15–34.
Tišljar, J., Vlahović , I., Velić , I., Matiččec, D. and Robson, J. (1998) Carbonate facies evolution from the late Albian to Middle Cenomanian in Southern Istria (Croatia): influence of synsedimentary tectonics and extensive organic carbonate production. Facies **38**: 137–151.

Vera, J.A. and Martín-Algarra, A. (1994) Mesozoic stratigraphic breaks and pelagic stromatolites in the Betic Cordillera, southern Spain, In: J. Betrand-Sarfati and C. Monty (eds.) *Phanerozoic Stromatolites II*. Kluwer, Dordrecht, pp. 319–344.

Williams, L.A. and Reimers, C. (1983) Role of bacterial mats in oxygen-deficient marine basins and coastal upwelling regimes: preliminary report. Geology **11**: 267–269.

Wyssling, G.W. (1986) Der frühkretazische helvetische Schelf im Vorarlberg und Allgäu. Jb. Geol. BA **129**: 161–265.

Yilmaz, I.Ö. and Altiner, D. (2006) Cyclic palaeokarst surfaces in Aptian peritidal carbonate successions (Taurides, southwest Turkey): internal structure and response to mid-Aptian sea-level fall. Cret. Res. **27**: 814–827.

Zaghbib-Turki, D. (2003) Cretaceous coral-rudist formations in Tunisia, In: E. Gili, M. El Hédi Negra and P.W. Skelton (eds.) *North African Cretaceous Carbonate Platform Systems*. Kluwer, Dordrecht, pp. 83–110.

Zanin, Y.N., Luchimina, V.A., Levchuk, M.A. and Pisavera, G.M. (2003) Stromatolites and oncolites in Mesozoic deposits of the west Siberian plate. Russ. Geol. Geophys. **44**: 1348–1352.

Zitt, J., Vodrazka, R., Hradecka, L., Svobodova, M. and Zagorsek, K. (2006) Late Cretaceous environments and communities as recorded at Chrtníky (Bohemian Cretaceous Basin, Czech Republic). Bull. Geosci. **81**: 43–79.

Biodata of **Dr. Krzysztof P. Krajewski**, author of ***"Phosphatic Microbialites in the Triassic Phosphogenic Facies of Svalbard"***

Dr. Krzysztof P. Krajewski is at present an associate Professor at the Institute of Geological Sciences and Department of Antarctic Biology, Polish Academy of Sciences (PAS) in Warszawa, Poland. He obtained his Ph.D. degree in 1986, and continued his studies and research at the Polish Academy of Sciences, University of Oslo, and University of Oldenburg. He completed his DSc degree in 2001 in the fields of sedimentary petrology and geochemistry. His scientific interests focus on the interactions between biological activity and mineral deposition in carbonate, phosphate, and organic carbon-rich sedimentary systems. For more than 20 years, Krajewski has carried out geological research in the Arctic, where he concentrated on the Mesozoic phosphorites and petroleum source beds. He is currently involved in research on the evolution of Cenozoic sedimentary systems in Antarctica.

E-mail: **kpkraj@twarda.pan.pl**

V.C. Tewari and J. Seckbach (eds.), *STROMATOLITES: Interaction of Microbes with Sediments*, Cellular Origin, Life in Extreme Habitats and Astrobiology 18, 187–222
DOI 10.1007/978-94-007-0397-1_9,

PHOSPHATIC MICROBIALITES IN THE TRIASSIC PHOSPHOGENIC FACIES OF SVALBARD

KRZYSZTOF P. KRAJEWSKI[1,2]
[1]*Institute of Geological Sciences, Polish Academy of Sciences, Twarda 51/55, 00-818, Warszawa, Poland*
[2]*Department of Antarctic Biology, Polish Academy of Sciences, Ustrzycka 10/12, 02-141, Warszawa, Poland*

Abstract The Triassic organic carbon-rich, phosphogenic facies setting in Svalbard, NW Barents Sea Shelf provides an insight into ancient shelf depositional system that shows similarities to modern phosphogenic shelves reinforced by coastal upwellings. Enhanced deposition and diagenesis of marine organic matter on the Svalbard shelf promoted phosphogenesis in bottom sediments that resulted in the formation of various nodular and peloidal phosphate deposits as well as peculiar phosphatic accumulations related to benthic activity of microbial mats. The phosphatic microbialites are interpreted to have been dominated by mats of colorless filamentous sulfur bacteria, which proliferated on elevated sediment bodies and submarine slopes under conditions of highly reduced or halted sedimentation. They provided local depositional systems capable of precipitating and significantly concentrating carbonate fluorapatite (CFA) in the sediment due to sealing of the sediment surface to phosphate diffusion, narrowing chemical gradients essential for the phosphorus pumps, reactive phosphate buildup, and the mineral precipitation. The precipitation of apatite was a very early phenomenon that resulted in the formation of phosphatic matrix or cement filling the original pore space of biolaminated sediment sequences. The matrix formation was a complex process, involving pulses of apatite precipitation associated with fossilization of microbial communities, oxidation of organic matter, enrichment with rare earth elements, and pyritization of the phosphatic fabric. The formation of phosphatic matrix was limited to a narrow submat zone, which coincided with the upper limit of the sulfate reduction zone in organic-rich sediment. The horizons with well-defined phosphatic microbialites show high concentrations of CFA (up to 32 wt% P_2O_5), though their contribution to the phosphorus pool of the phosphogenic facies is subordinate. Nodular and peloidal accumulations dominate the phosphorite fraction of the Triassic sequence, but these phosphates usually give lower burial phosphorus concentrations in the sediment.

Keywords Phosphatic microbialites • Triassic • Phosphogenesis • Coastal upwelling • Benthic • Microbial mats • Sedimentation • Carbonate fluorapatite • Fossilization • Pyritization • Phosphorus • Svalbard

1. Introduction

A standard interpretation of phosphorite formation in modern upwelling settings is based on Baturin's model of phosphogenesis, which postulates apatite precipitation within the pore environment of organic carbon (OC)-rich sediment as a result of diagenetic decomposition of P-containing organic matter deposited in excess from productive water column (Baturin, 1978, 1982, 1983; Baturin and Bezrukov, 1979). Apatite deposition seems to be preferentially linked to the water/sediment interface and the near-surface sediment layer where the generation of phosphate ion is most intense (Jahnke et al., 1983; Froelich et al., 1988; McManus et al., 1997; Schuffert et al., 1998). However, supersaturation conditions with respect to apatite and its precipitation to form a major mineral phase require additional mechanisms that prevent dissolved phosphate from escaping to the overlying water (Burnett and Riggs, 1990; Föllmi, 1996; Trappe, 1998). Several such mechanisms have been proposed, including chemical and microbiological traps at the interface (Glenn et al., 1994; Jarvis et al., 1994; Krajewski et al., 1994). One attempt points to the role of benthic microbial mats in fixation of phosphate ion to form intracellular polyphosphate, and its release and deposition in the form of apatite in the subsurface environment (Reimers et al., 1990). This hypothesis is supported by a common occurrence of microbial mats dominated by colorless filamentous sulfur bacteria on modern OC-rich, phosphogenic seafloors (e.g., Gallardo, 1977; Soutar and Crill, 1977; Rosenberg et al., 1983; Henrichs and Farrington, 1984; Glenn and Arthur, 1988; Schulz et al., 1996, 1999). Schulz and Schulz (2005) documented that these mats on the Namibian shelf are capable of sufficiently concentrating dissolved phosphate in pore water to promote precipitation of hydroxyapatite. Recently, laminated phosphorite crusts related to the activity of bacterial mats have been described from phosphogenic sediments on the Peruvian shelf (Arning et al., 2009a, b).

This paper presents the Triassic OC-rich, phosphogenic facies setting in Svalbard, NW Barents Sea Shelf (Fig. 1) that provides an insight into ancient depositional system showing similarities to modern phosphogenic shelf systems reinforced by coastal upwellings. It is highly enriched in marine organic matter and encompasses a variety of types of phosphate deposits formed over a range of bottom environments, from coastal to deep shelf. Phosphatic microbialites revealed there are interpreted to have developed as a result of activity and phosphatization of microbial mats dominated by colorless filamentous sulfur bacteria. They are used to study their role in allocating phosphogenesis in ancient OC-rich sediments deposited under high-biological productivity conditions.

2. Geological Background

Svalbard is the emergent northwestern corner of the Barents Sea Shelf (Fig. 1), bordered on the north by passive continental margin toward the Arctic Ocean and on the west by spreading axis of the Knipovich Ridge (Talwani and Eldholm, 1977).

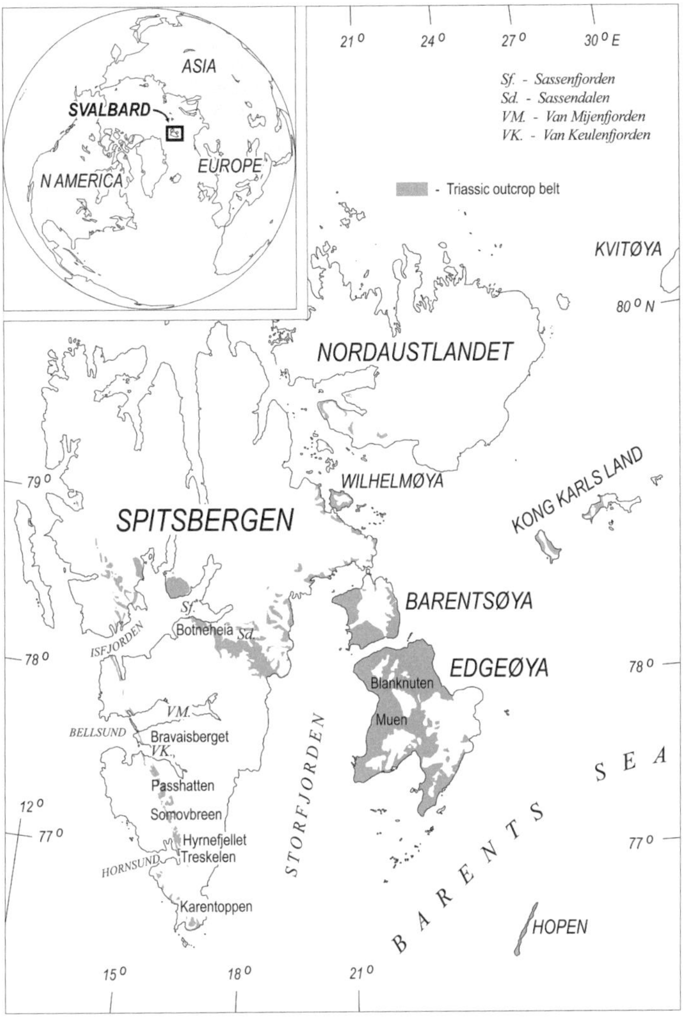

Figure 1. Map of Svalbard archipelago showing the Triassic outcrop belt (*shaded*).

The islands of Svalbard (including Spitsbergen, Nordaustlandet, Edgeøya, Barentsøya, Wilhelmøya, Kong Karls Land, Hopen, and others) were uplifted by Late Mesozoic and Cenozoic crustal movements, providing insight into geological structure of the region and the stratigraphic column from Paleoproterozoic to Paleogene (Harland, 1997; Dallmann et al., 2002). The Caledonian basement (Paleoproterozoic through Ordovician) hosting tectonic depressions filled by the Devonian Old Red Sandstone is covered by a single sedimentary complex of Late Paleozoic and Mesozoic platform strata (Dallmann, 1999). The opening of the Arctic and North Atlantic oceans in Paleogene, and the interdependent stages of separation of Svalbard and Greenland shaped the present tectonic structure of Svalbard. A thrust-and-fold belt stretches along western and southern Spitsbergen, owing its origin to dextral plate transform setting with strike-slip and convergent movements (Maher et al., 1986). The belt is bordered from the east by simultaneously formed Central Spitsbergen (Paleogene) Basin that passes eastwards into Eastern (Mesozoic) Svalbard Platform. The reverse uplift of the western Spitsbergen orogen as well as northern Spitsbergen and northern Nordaustlandet caused erosion of younger strata and exposition of the Caledonian basement and Devonian through Permian succession.

The Triassic sedimentary sequence crops out in southern, western, central, and eastern Spitsbergen, on Edgeøya, Barentsøya, Wilhelmøya, Kong Karls Land, and Hopen. It consists of the lower part dominated by marine facies, and the upper part dominated by deltaic facies, which are classified into the Sassendalen and Kapp Toscana groups, respectively (Mørk et al., 1999). The phosphogenic facies concentrate in the upper part of the Sassendalen Group (Fig. 2). They occur in the Middle Triassic (Anisian–Ladinian) OC-rich strata that are discerned as the Bravaisberget Formation in southern and western Spitsbergen, and as the Botneheia Formation in central Spitsbergen and eastern Svalbard (Krajewski et al., 2007; Krajewski, 2008). This phosphogenic horizon reflects regional transgressive pulse of the Panthalassa Ocean on the so-called western land area dominated by deltaic systems, which provided highly productive shelf environments to the NW Barents Sea Shelf (Krajewski, 2000a). The deposition area was likely a big embayment bordered on the west by shallow shelf and deltaic settings of western and southern Spitsbergen (Fig. 3), and on the east and southeast by prograding clastic bodies of the deltaic systems of the southern Barents Sea Shelf (Riis et al., 2008). This embayment had north connection with deep oceanic basins (Cocks and Torsvik, 2007).

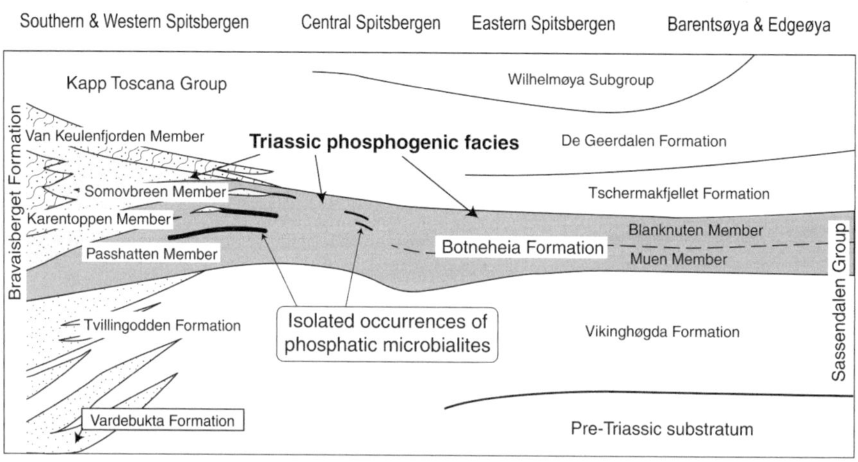

Figure 2. Lithostratigraphic scheme of the Triassic sedimentary formations in Svalbard showing position of the Middle Triassic phosphogenic facies (*shaded*) and the occurrences of phosphatic microbialites. Lithostratigraphy after Mørk et al. (1999), Krajewski (2000a, 2008), and Krajewski et al. (2007).

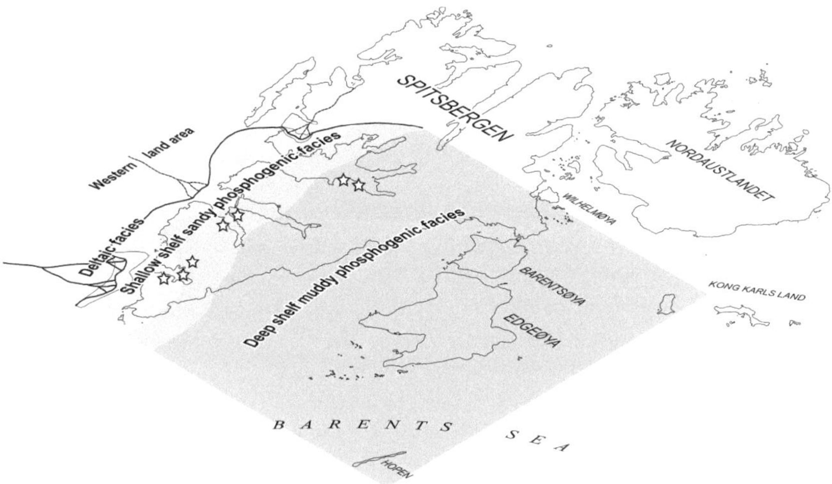

Figure 3. Simplified paleogeographic reconstruction of the Middle Triassic phosphogenic shelf in Svalbard. *White stars* indicate the occurrences of phosphatic microbialites. Adopted from Krajewski (2000b).

The reasons of high-biological productivity during the Middle Triassic in Svalbard are as yet poorly understood. There are, however, accumulating data that suggest upwelling from deeper oceanic water masses located further north (Krajewski, 2006). Calculations of phosphorus volume preserved in the Bravaisberget Formation in Spitsbergen suggest phosphorus burial rates comparable to those reported from modern continental margins reinforced by coastal upwellings (Krajewski, 2000e). During the Middle Triassic time, the Svalbard shelf was located at moderate paleolatitudes (40–50°N) and exposed toward the Panthalassa Ocean (Steel and Worsley, 1984; Harland, 1997), providing favorable framework for upwelling-forced phosphogenesis (Parrish and Curtis, 1982). In similar paleoceanographic situations, other Triassic OC-rich, phosphogenic sediments deposited at margins of the Arctic Ocean, including the Shublik Formation in northern Alaska (Detterman et al., 1975; Detterman, 1989; Parrish et al., 2001), the Murray Harbor Formation in the Sverdrup Basin (Embry, 1988; Mørk et al., 1989), and the New Siberian Islands deposits (Egorov and Baturin, 1987), making together a broad, high-latitude phosphogenic province (Krajewski, 2005).

3. Triassic Phosphogenic Facies

The phosphogenic facies in Svalbard can be divided into two major groups: (1) the shallow shelf sandy phosphogenic facies and (2) the deep shelf muddy phosphogenic facies (Fig. 3). Distribution of these facies reflects the distance from shore and from major delta outlets, the bathymetry and sea-level changes during the depositional history, as well as the sea bottom topography and dynamics of sediment bodies (Krajewski, 2000a, 2006).

The sandy phosphogenic facies occurs in the Passhatten and Somovbreen members of the Bravaisberget Formation (Fig. 2). It occupies shallower, inner part of the shelf basin, forming a NNW-SSE stretching belt in western and southern Spitsbergen (Fig. 3). It passes southwestwards into deltaic and brackish facies of the Karentoppen and Van Keulenfjorden members, which are devoid of phosphatic accumulations. The facies encompasses a variety of clastic sediment types, though fine-grained sandstones and silty shales are most common (Fig. 4a–c). Sedimentary sequences include shallow-water sandy sheets and bodies, sand bars, slopes with clastic slumps, and local depressions with wedges of OC-rich sediments. Most of the sediment shows traces of intense mechanical and biological reworking, suggesting well-aerated bottom environment (Mørk, 2006). Clastic components are represented by quartz grains, though debris of siliceous sponge spicules are common, and locally abundant, in the sand bar sequences in particular. The phosphorite fraction is represented by phosphate nodules, phosphatic fossil molds, coprolites, phosphatized fish and reptilian bones, phosphatized burrows, and subordinate peloids (Krajewski et al., 2007; Mørk and Bromley, 2008). Most of this fraction originated as a result of pulses of rapid precipitation of apatite (possibly preceded by an unstable calcium phosphate phase) at shallow burial depths. The apatite makes cement between detrital grains and fills original porosity of biogenic debris. Pristine phosphate nodules are commonly observed, though horizons of reworked and redeposited nodules typically occur in high-energy sedimentary sequences.

The muddy phosphogenic facies occurs in the Muen and Blanknuten members of the Botneheia Formation (Fig. 2). It covers vast areas of deeper, outer part of the shelf basin in central and eastern Spitsbergen, and in Edgeøya and Barentsøya (Fig. 3). The facies is dominated by black, OC-rich shales and mudstones (Fig. 4d–f). Sedimentary sequences include muddy sheets on shelving bottom, bottom current deposits, submarine slopes with gravity mass transport deposits, and extensive depressions filled with phosphatic mudrock. Sedimentary, petrographic, and geochemical data suggest wide variations in bottom ventilation, from fully oxic bottoms with burrowing infauna to stagnant, euxinic depressions (Krajewski, 2000b, 2008). The phosphorite fraction is dominated by nodules and peloids. The apatite cementation took place in the upper part of sediment column, though at much shallower depths than in the sandy phosphogenic sediments (Krajewski, 2000c). It resulted in the formation of nodular bodies and peloids preserving molds of radiolaria, thin-shelled pelecypods (*Daonella*), and *Tasmanites* algae. Pristine deposits are common, but reworked horizons occur repeatedly in some sequences, witnessing changing dynamics of the muddy bottom environment. In the sequences of euxinic depressions, reptilian and fish bone beds reflecting mass mortality events are observed.

Despite morphological and textural variations of the phosphorite fraction, all the discerned phosphate deposits in the Triassic facies are composed of one phosphate mineral, which is crypto- to microcrystalline carbonate fluorapatite (CFA) containing 2–4% carbon dioxide in the crystal lattice (Fig. 5).

Figure 4. (**a**) Shallow shelf sandy phosphogenic facies of the Somovbreen Member exposed in inner Hornsund, southern Spitsbergen is developed in the form of massive sequence of fine-grained, bioturbated sandstones containing numerous phosphate nodules. (**b**) Recurrent rows of pristine phosphate nodules (*arrow*) in sandy phosphogenic facies in the southernmost Spitsbergen. Hammer is 30 cm long. (**c**) Energy-dispersive map of phosphorus of matrix of a pristine sandy phosphate nodule. Apatite (*white*) occurs in the form of interparticle cement (a) as well as some peloids (p). Detrital components (*black*) are dominated by debris of siliceous sponge spicules (s) and quartz grains (q). (**d**) Deep shelf muddy phosphogenic facies of the Blanknuten Member exposed in southern Barentsøya is developed in the form of black, massive mudrock containing numerous phosphate nodules and peloids. (**e**) Recurrent rows of pristine phosphate nodules (*arrow*) in muddy phosphogenic facies in central Spitsbergen. Hammer head is 14 cm long. (**f**) Energy-dispersive map of phosphorus of matrix of a pristine muddy phosphate nodule. Apatite (*white*) occurs in the form of interparticle, cluster-like cement (a) as well as common molds of radiolaria (r). *Black* is late diagenetic calcite cement. Panels (**c**) and (**f**) are adopted from Krajewski (2000a).

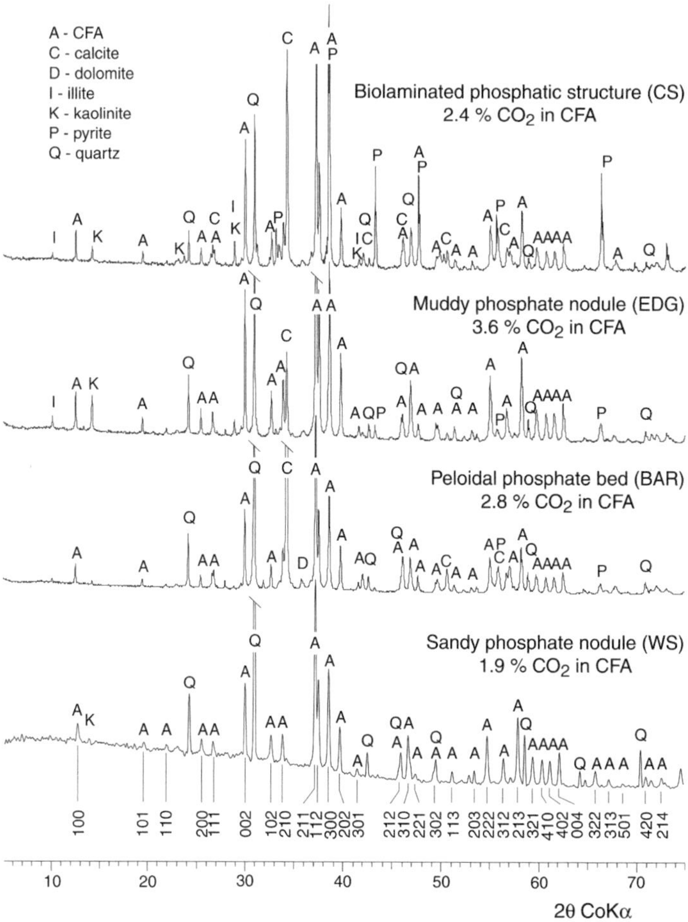

Figure 5. X-ray diffraction patterns of representative samples of phosphorite fractions and phosphatic sediments in the Triassic phosphogenic facies of Svalbard. The content of carbon dioxide in CFA was calculated using the equations of Gulbrandsen (1970) and Shuffert et al. (1990). *Numbers* indicate *hkl* reflections of CFA. *CS* central Spitsbergen, *WS* western Spitsbergen, *BAR* Barentsøya, *EDG* Edgeøya.

4. Phosphatic Microbialites

Phosphatic microbialites have been revealed at several locations close to the transition between the sandy and muddy phosphogenic facies (Fig. 3). These structures tend to occur in condensed sequences and/or on hiatal surfaces developed in OC-rich sediment, which were exposed at sea bottom for prolonged periods of time (Figs. 6 and 7). These sequences are interpreted to reflect either

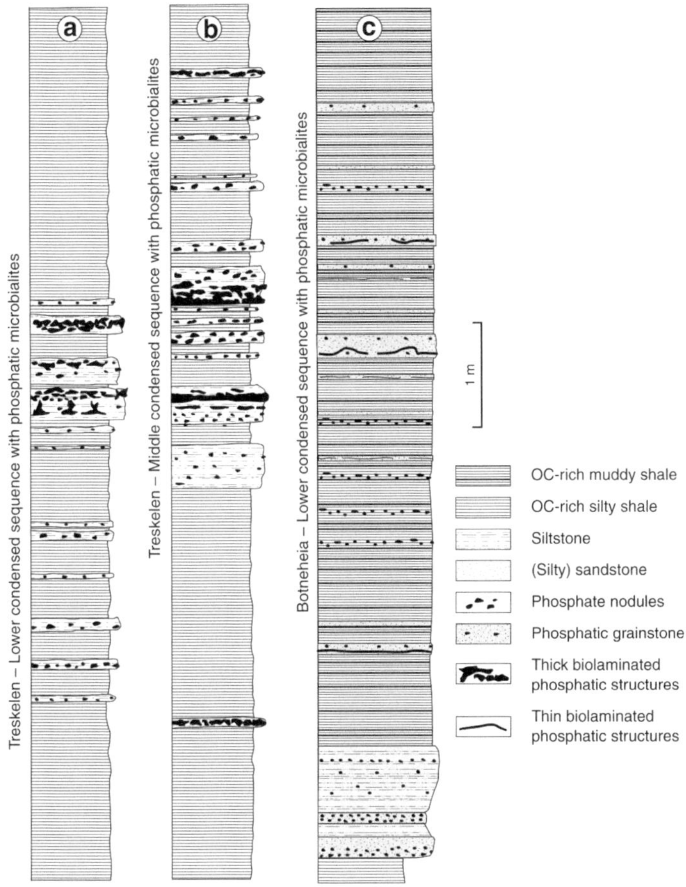

Figure 6. Examples of the occurrence of phosphatic microbialites in condensed sequences on siliceous sand bars (**a** and **b**) and on submarine muddy slope (**c**).

submarine slope environments or submarine bars and migrating sediment bodies surrounded by local depressions.

4.1. PETROGRAPHIC FEATURES

Phosphatic microbialites can be distinguished among other types of phosphorite bodies discerned in the Triassic phosphogenic facies by their accretional morphology, internal lamination, and the presence of mineral remnants of microbial mat community. These structures vary in size from discrete, microscopic laminations on hiatal surfaces and/or in winnowed beds to macroscopic phosphatic bodies showing generations of growth and outcrop-scale lateral extent. The best developed are

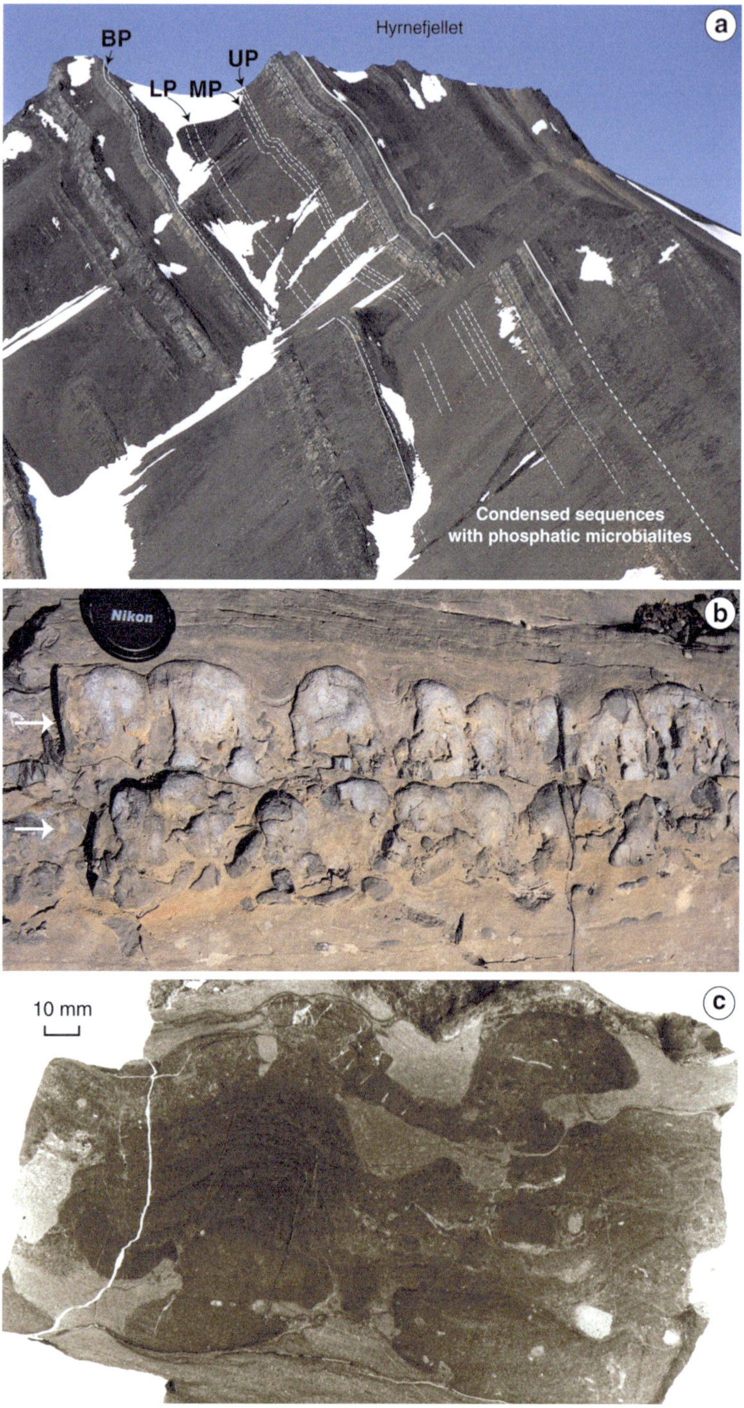

Figure 7. (**a**) Section through the Bravaisberget Formation in inner Hornsund (ca. 100 m thick), southern Spitsbergen showing recurrent condensed sequences with phosphatic microbialites. BP, LP, MP, and UP are basal, lower, middle, and upper condensed sequences, respectively. (**b**) Phosphatic

phosphatic microbialites associated with condensation at margins of submarine bars dominated by siliceous sponge. Their morphology seems to reflect the distribution of microbial mat at the sediment surface and stages of phosphatization of the mat-stabilized sediment. The phosphatic bodies tend to form horizontal bands within the sediment and/or juxtaposed columnar structures that grew upward (Fig. 7b). Flat-laminated phosphatic bands correlate with pronounced discontinuity surfaces in some condensed sequences. Incipient developments of these structures up to a few millimeters thick are found associated with some omission or winnowed surfaces. Intergrowths of columnar and flat-laminated structures are the dominant microbial mat deposits in the condensed horizons (Fig. 7c). The nature of their development implies a complex interplay between the lateral mat expansion and vertical mat growth, sediment accretion, and mat burial. The morphologies of these complex structures contrast with usually ovoid to ellipsoidal phosphate nodules. They are similar to biolaminated phosphorite crusts revealed in phosphogenic sediments on the Peruvian shelf (Arning et al., 2009b).

The internal lamination of phosphatic microbialites usually is difficult to notice, either on fresh or weathered surfaces. It becomes better pronounced on polished surfaces subjected to chemical treatment (Fig. 7c). The lamination consists of alternating thin laminae (0.1–1 mm) dominated by compact, crypto- to microcrystalline CFA, and thicker laminae (1–5 mm) composed of clastic and/or biogenic sediment, and cemented by CFA (Fig. 8). The nature of this lamination is independent of the petrographic and textural features of the host sediment. The lamination is best pronounced in flat-laminated structures and broad columns. These structures are inferred to mark prolonged nondeposition during the formation of the condensed horizons.

The internal sediment of phosphatic microbialites shows close affinities to the host sediment. The sediment in the columnar structures passes sideward into nonphosphatic sediment without any noticeable discontinuity surfaces (Fig. 8a, d). These transitions are, however, sharp and enhanced by different weathering resistance of apatite-cemented and the host sediments (Fig. 7b). In the marginal parts of the columns, thin laminae become irregular and dashed as they reach the border of phosphatization zone. This internal organization suggests that, despite the present-day columnar shapes, the biolaminated structures did not form elevated morphological units at the seafloor, but originated from punctuated microbial mat growth on a rather flat sediment surface. Unconformities noted between some sets of laminae correlate to scouring episodes in the neighboring sediment. They define generations of biolaminated structures, which

Figure 7. (continued) microbialite in condensed horizon developed in OC-rich sandy mudstone. Two generations of biolaminated columnar structures (*arrows*) rest on erosional surfaces. Winnowed phosphate nodules are incorporated in the microbialite. (**c**) Intergrowth of columnar and flat-laminated phosphatic structures developed as a result of microbial mat growth and sediment phosphatization in OC-rich sandy mudstone. Vertical section, polished surface after staining to obtain color contrast between phosphatic and nonphosphatic sediment.

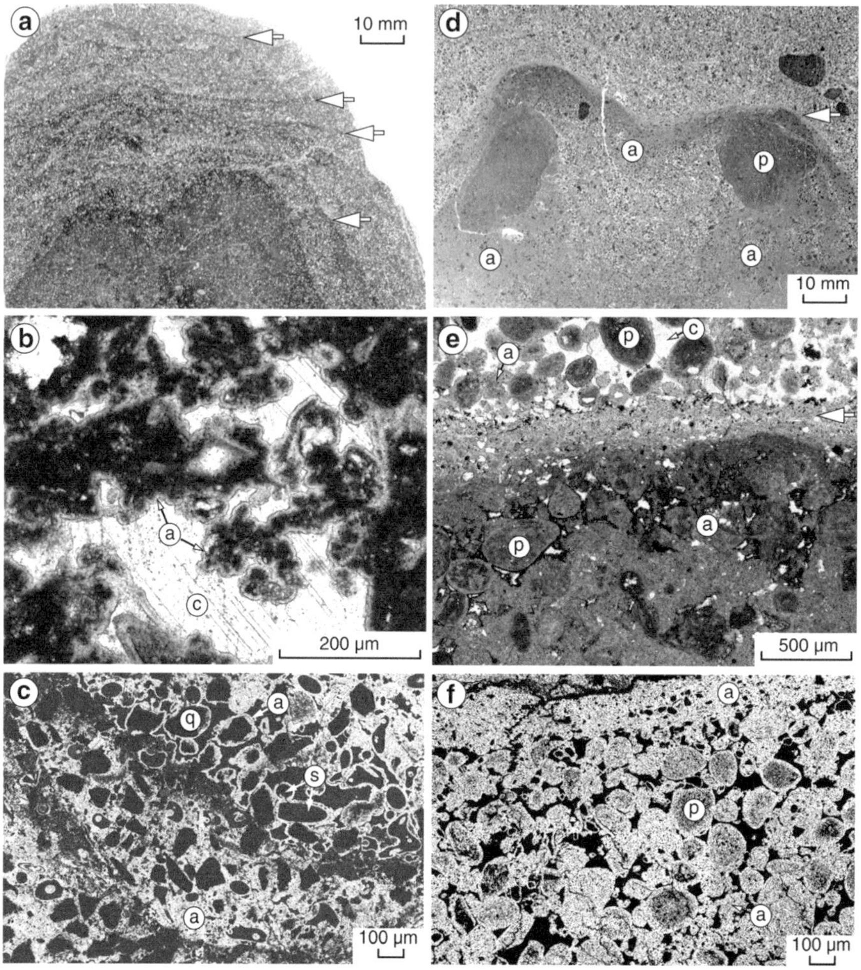

Figure 8. Microscopic features of phosphatic microbialites. (**a**) Irregular lamination of phosphatic microbialite (*arrows*) overgrowing phosphate nodule in a siliceous bar sequence. Thin section photograph. (**b**) Delicate meshwork of apatitic cement with numerous organic inclusions (a) making matrix of phosphatic microbialite shown in (**a**). The remaining pore space is replaced by diagenetic calcite cement (c). Transmitted light microscope, normal light. (**c**) Energy-dispersive map of phosphorus of matrix of phosphatic microbialite shown in (**a**). Apatite (a) occurs in the form of cloth-like cement, and rims around quartz grains (q) and debris of siliceous sponge spicules (s). (**d**) Thin, undulated apatitic lamina (*arrow*) covering discrete omission surface in phosphatic grainstone in a submarine muddy slope sequence. Larger dark bodies (p) are reworked phosphate nodules. Formation of the lamina was associated with the development of irregular apatite cementation zones (a). Thin section photograph. (**e**) Contact between two laminae (*arrow*) in phosphatic microbialite developed in winnowed phosphatic grainstone. The laminae are built up of phosphate peloids and coated grains (p) cemented by a meshwork of cluster-like apatite (a). The lower lamina shows more generations of apatitic cement than the upper one as well as strong pyritization (*black*). The remaining pore space is replaced by diagenetic calcite cement (c). Transmission light microscope, normal light. (**f**) Energy-dispersive map of phosphorus of matrix of phosphatic microbialite shown in (**d**). Apatite (*white*) occurs in the form of peloids (p) and cloth-like cement (a). Panels (**c**), (**d**), and (**f**) are adopted from Krajewski (2000a).

developed independently from one another, though they reflect the actual mat distribution on the sediment surface.

All the analyzed phosphatic microbialites show similar pattern of CFA emplacement into the mat-stabilized clastic and/or phosphatic sediment. The growth of a lamina started from isolated apatitic clusters that precipitated in the interparticle pore space and encapsulated numerous organic microinclusions (Fig. 8b, c). The clusters become bigger upward the lamina, and they join one another to form a compact phosphatic matrix, which terminates in a rim of purely apatitic accretionary material (Fig. 8e, f). The laminae could either be buried in the sediment due to rapid deposition of detrital grains, biogenic debris, or phosphatic peloids (Fig. 8d), or build out upwards repeating the mechanism outlined above (Fig. 8a).

Authigenic pyrite is intimately associated with the OC-rich, phosphogenic facies and all the types of phosphate deposits. This pyrite owes its origin to bacterial reduction of sulfate and reactive ferric iron in the upper part of sediment column (Berner, 1970, 1984; Canfield, 1989). Phosphatic microbialites are more pyritized than any other type of phosphate deposits in the Triassic phosphogenic facies (Fig. 9). The predominant portion of pyrite occurs in (poly)framboidal form (Fig. 9c), suggesting its formation at high supersaturation levels in the uppermost part of the sulfate reduction zone (Berner, 1985; Wilkin and Barnes, 1997). A part of early pyrite is developed in euhedral form (octahedral to cubic). This pyrite postdates the framboid formation, but predates reduction of the porosity (Fig. 9b, d), suggesting also an early formation, though at lower levels of saturation (Raiswell, 1982; Dill and Kemper, 1990). The enhanced pyritization of phosphatic microbialites is consistent with inferred low or nil sedimentation rates during formation of the condensed horizons, and with stabilization of the oxic/anoxic interface within the microbial mats. Thicker biolaminated structures show many generations of apatitic cement and a sequence of stages of its pyritization.

4.2. MICROBIAL FABRICS

Scanning electron microscopy observations of phosphatic microbialites reveal filamentous-like apatitic ultrastructures, which are likely to represent remnants of the original mat-forming community (Fig. 10a, b). These ultrastructures tend to concentrate within thin, apatite-rich laminae where they can be observed in incompletely cemented pore space. All these traces occur in the form of hollow tubes 1–3 μm wide and 30 and more micrometer long, though usually curved and disrupted in the pore space. The tubes are attached to clastic and/or biogenic sediment grains, and they also provide a delicate meshwork between the grains. The condensation of phosphatic matrix in the laminae leads to a consequent incorporation of this meshwork into a compact or globose apatitic groundmass, in which discrimination of individual tubes is no longer possible.

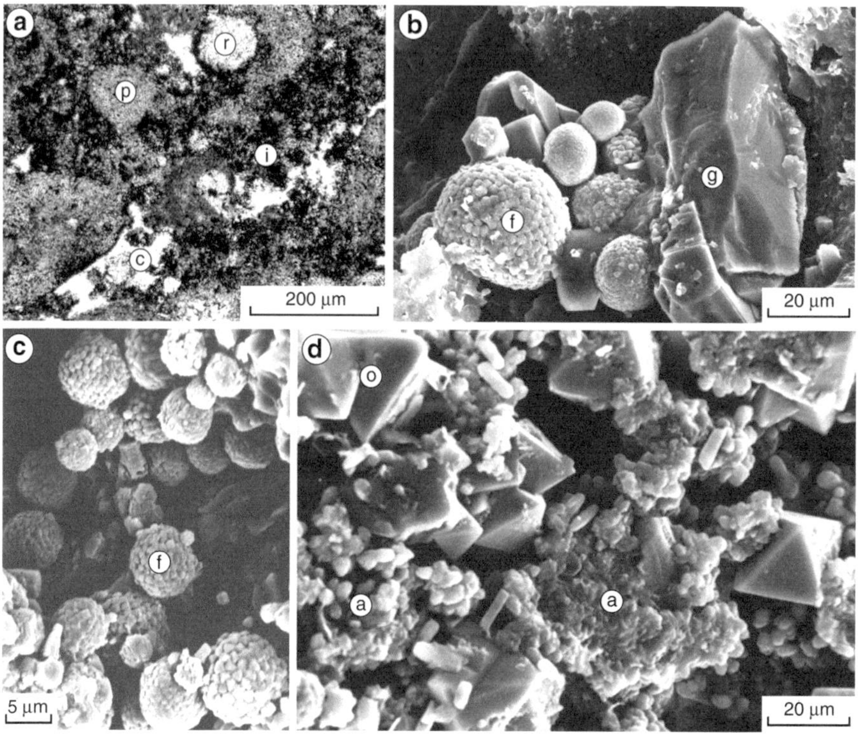

Figure 9. Pyritization of phosphatic microbialites. (**a**) Strong pyritization (*black*) of phosphatic matrix (*gray*). Pyrite occurs in the form of framboids and small crystals making irregular impregnation zones (i) and fringes around peloids (p) and radiolaria molds (r). The remaining pore space is replaced by diagenetic calcite cement (c). Transmission light microscope, normal light. (**b**) Mineral sequence of framboidal pyrite (f) followed by precipitation of microgranular pyrite (g) in a void in microbial lamina. (**c**) Polyframboidal pyrite (f) in a void in microbial lamina. (**d**) Octahedral pyrite (o) in porous, cluster-like apatitic cement (a) in microbial lamina. Note the minute CFA particles incorporated in pyrite crystals. (**b–d**): scanning electron microscope; preparations after dissolution of diagenetic calcite cement.

Tube-like fossilization of filamentous microorganisms was a common phenomenon in many ancient microbialites associated with marine phosphogenic systems, and it was also demonstrated in the course of laboratory experiments (Krajewski et al., 1994). This process implies a development of phosphatic overgrowths around individual filaments and their subsequent decomposition to provide central canals within the mineral incrustations (Soudry and Champetier, 1983). In phosphatic microbialites of Svalbard, the tubes show very irregular development of apatite overgrowths, which suggests an advanced decomposition of a filamentous microorganism at the time of CFA precipitation.

It is difficult to reconstruct the specific composition of microbial mat community on the basis of mineral ultrastructures preserved in phosphatic microbialites.

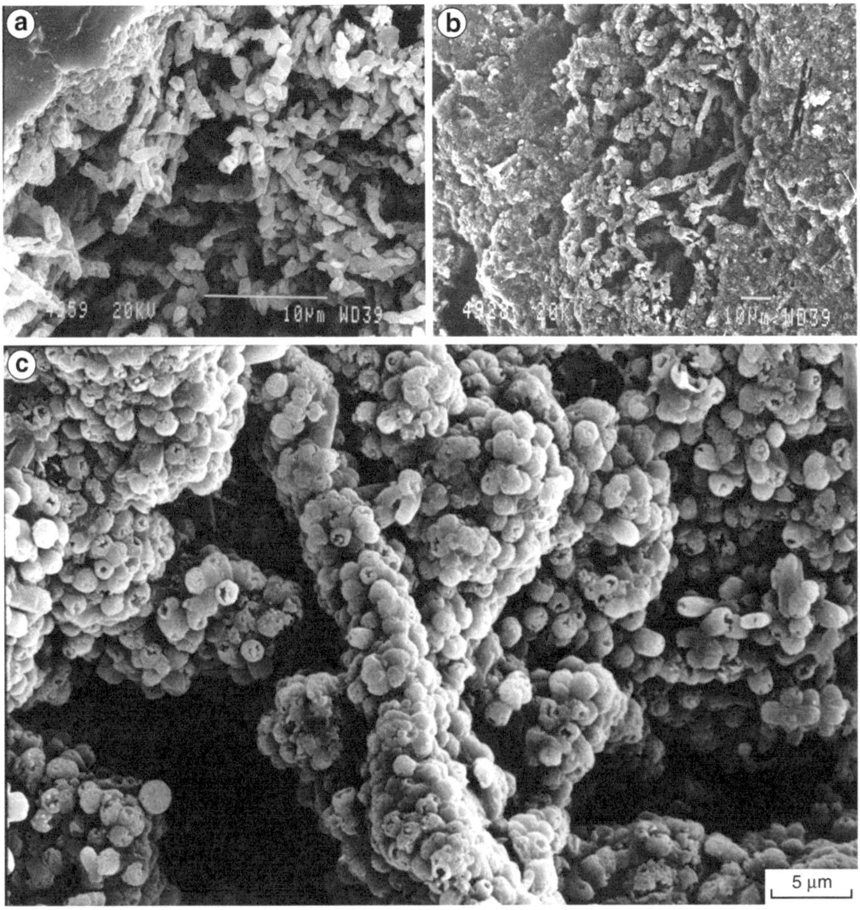

Figure 10. Microbial fabrics in phosphatic microbialites. (**a**, **b**) Tube-like microbial ultrastructures in matrix of phosphatic microbialite. (**c**) Densely packed coccoid microbial ultrastructures in matrix of phosphatic microbialite. The apatitic capsules show apertures exposed toward the open pore space. (**a–c**) The mineral incrustations are composed of cryptocrystalline CFA. Scanning electron microscope; preparations after dissolution of diagenetic calcite cement. Adopted from Krajewski (2000a).

However, it seems likely that similar to modern mats on phosphogenic shelves, the microbial mats in the Triassic of Svalbard were dominated by colorless filamentous sulfur bacteria. These bacteria are chemohetero- to chemoautotrophs (Morita et al., 1981; Krumbein, 1983; Nelson and Jannasch, 1983; Wiessner, 1981; Strohl, 1989; Jørgensen and Gallardo, 1999). Therefore, they are independent of the presence and intensity of bottom illumination. They can form gradient-type mat accumulations on OC-rich sediments even in deep shelf and slope environments (Williams, 1984). The studied microbialites occur both in the shallow and deep shelf facies where they exhibit similar structure organization and lamination. This lamination resembles those illustrated from modern OC-rich sediments stabilized by *Beggiatoa* mats in the Santa Barbara Basin (Soutar and Crill, 1977). Complex biolaminated structures in the condensed horizons show similar internal organization to the laminated phosphorite crusts on the Peruvian shelf, which are interpreted to represent phosphatized sulfur-oxidizing bacterial mats (Arning et al., 2009a, b). The tube-like microstructures preserved in the Svalbard microbialites are far smaller than those illustrated from shallow-water phosphatic microbialites, interpreted to be remnants of cyanobacteria (Soudry and Champetier, 1983; Soudry, 1987, 2000; Mitrov et al., 1987). The apparent lack of branching suggests that they are not fungi (Dahanayake and Krumbein, 1985; Bréhéret, 1991). Their dimensions correspond to thin filaments described from phosphatic biolaminites of the Monterey Formation (Williams and Reimers, 1983; Williams, 1984; Reimers et al., 1990) as well as to filaments from shale-hosted microbial mat structures in the Central Superbasin in Australia (Logan et al., 1999). These filaments are interpreted to be remnants of filamentous sulfur bacteria. It should be noted, however, that modern colorless filamentous sulfur bacteria embrace species of several different filament sizes (Wiessner, 1981; Schulz et al., 1996, 1999; Schulz and Schulz, 2005), some of which are considerably larger (even orders of magnitude) than the sizes elucidated on the basis of dimensions of mineral tube-like ultrastructures preserved in the Triassic microbialites. Sizes of the tube-like ultrastructures in Svalbard correspond to thin, not sheath-forming filaments of *Beggiatoa* (Wiessner, 1981; Strohl, 1989).

The development of phosphatic matrix in the microbialites was associated with a far better preservation of a coccoid microbial community (Fig. 10c), different from the mat-forming community, but similar, if not identical, to microbial community preserved throughout the pristine phosphate accumulations in the Triassic phosphogenic facies (Krajewski, 2000a). The contribution of these microbes as a substratum to matrix formation could be explained by their abundance in the OC-rich sediment, and by the overlapped rapid apatite precipitation. This may suggest that they represent remnants of a common decomposing community that proliferated in the uppermost part of OC-rich sediment column. The coccoid ultrastructures show apatitic capsules with common apertures exposed toward the pore space, which indicates that CFA precipitation took place on microbial cells showing metabolic activity (Krajewski et al., 2000). These ultrastructures resemble very closely to apatitic capsules developed around heterotrophic bacteria during laboratory phosphate precipitation experiments (Lucas and Prévôt, 1985).

All the analyzed traces of microorganisms in the Triassic phosphogenic facies developed as a result of extracellular precipitation of CFA, though complex histories of the infilling of apatitic incrustations suggest stages of the microbial matter decomposition and continuing precipitation.

4.3. GEOCHEMICAL FEATURES

4.3.1. Organic Carbon Versus Mineral Phosphorus

The pyrolytic and chemical characteristics of organic carbon in the sandy and muddy phosphogenic facies as well as in phosphatic microbialites point to a common marine source of sedimentary organic matter and similar pathways of its diagenetic transformation (Figs. 11 and 12). The organic matter is hydrogen-rich and oxygen-poor, represented by kerogen type I/II. The exception is the nonphosphogenic deltaic facies of southern Spitsbergen, which is dominated by land-derived organic fractions (kerogen type III). The observed differences in hydrogen

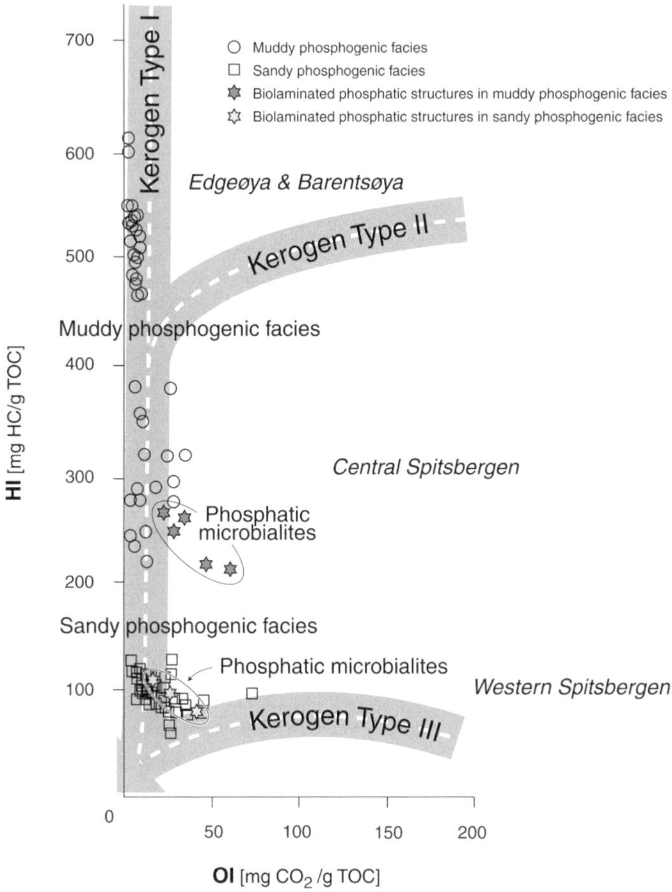

Figure 11. Plot of Rock-Eval pyrolytic hydrogen index (HI–mgHC/gTOC) and oxygen index (OI–mgCO_2/gTOC) values obtained from phosphorite fractions and phosphatic sediments in the Triassic phosphogenic facies of Svalbard against mean evolution paths of kerogen type I, II, and III.

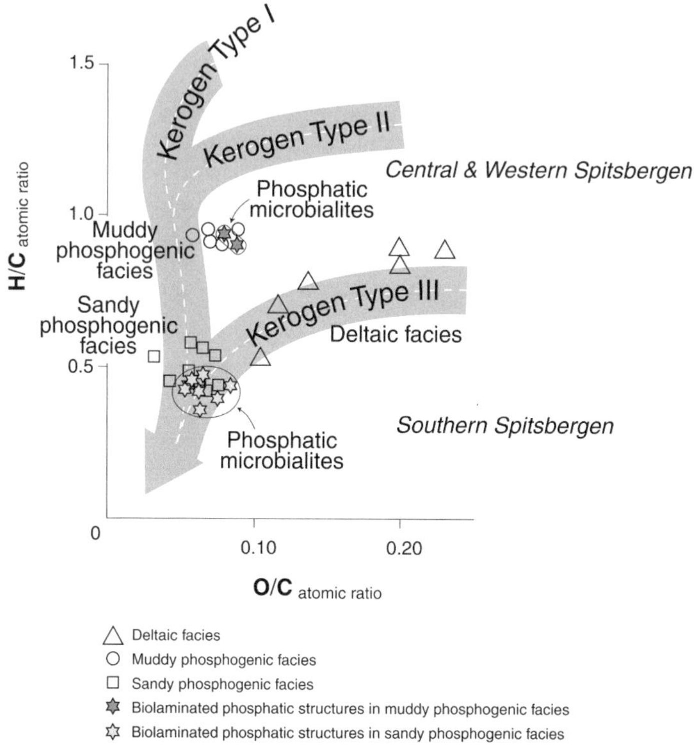

Figure 12. Plot of H/C and O/C atomic ratios of kerogen extracted from phosphorite fractions and phosphatic sediments in the Triassic phosphogenic facies of Svalbard against mean evolution paths of kerogen type I, II, and III. Adopted from Krajewski (2000e), with new data.

content in kerogen of the phosphogenic facies reflect different levels of organic metamorphism in major tectonic units of Svalbard, from early mature to mature sequences in the Eastern Svalbard Platform (vitrinite reflectance $R_o = 0.5–0.8\%$) through mature sequences in the Central Spitsbergen Basin ($R_o = 0.7–0.9\%$) to overmature sequences along the thrust-and-fold belt of western ($R_o = 1.0–1.4\%$) and southern Spitsbergen ($R_o \sim 2.0\%$) (Mørk and Bjorøy, 1984; Krajewski, 2000d, e).

A slight enrichment in oxygen is observed in organic matter of phosphatic microbialites, compared with neighboring nodular and peloidal phosphates. This can be demonstrated for the central and western Spitsbergen microbialites (Fig. 11), and even for microbialites occurring in overmature sequences of southern Spitsbergen (Fig. 12). Microscopic survey shows that kerogen is concentrated within organic-rich particles like aggregates, clusters, fecal pellets, and products of their disintegration, which point to preferential preservation of organic matter settled from the water column (Fig. 8). Within these particles, kerogen occurs in the form of amorphous fraction with common unicellular bodies interpreted to be remnants of planktonic microalgae (Krajewski, 2000d).

Microbial ultrastructures in phosphatic microbialites are widely devoid of preserved organic matter. This suggests that the oxygen increase in kerogen reflects enhanced oxidation of the settled organic matter in the microbial mat environment during its prolonged exposition at sea bottom.

There are huge variations in organic content in the muddy phosphogenic facies, resulting from changing depositional conditions and variations in bottom oxygenation across the Svalbard shelf (Fig. 13). The greatest values (up to 10 wt% organic carbon) reflect deposition under euxinic conditions in local depressions in the deep shelf environment and low levels of organic metamorphism in the Eastern Svalbard Platform (Krajewski, 2008). The sandy phosphogenic facies and the condensed sequences with phosphatic microbialites show smaller though comparable value ranges, reflecting enhanced regeneration of organic matter under dominating oxic conditions and in hiatal environments. The decrease in the content of organic carbon southwestwards in Spitsbergen is also a result of elevated levels of organic metamorphism that led to kerogen maturation and expulsion of bitumen during the post-Triassic geological evolution of the region (Figs. 11 and 12).

The phosphate deposits show wide variations in phosphorus content, from a few percent to more than 30% P_2O_5 (Fig. 13 and Table 1). They reflect variations in the concentration of phosphorite fraction in sediment as well as the advancement

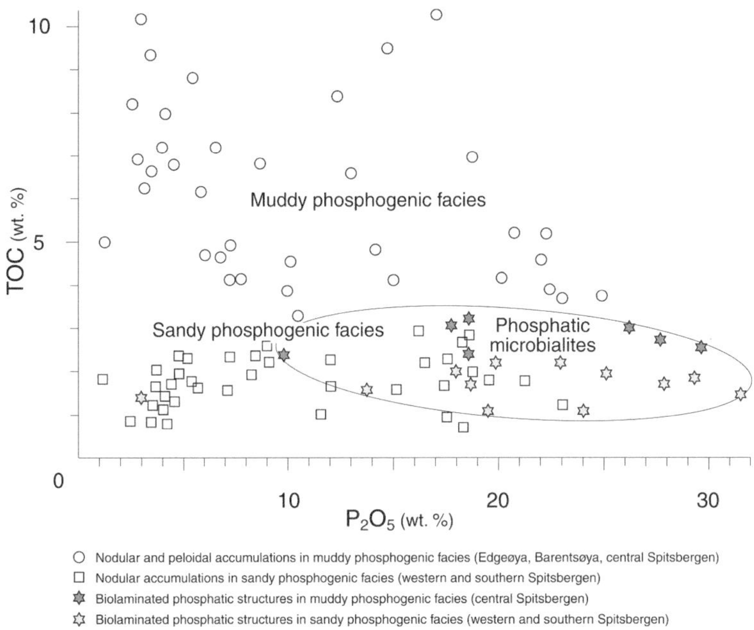

Figure 13. Plot of total organic carbon (TOC in wt%) and mineral phosphorus (P_2O_5 in wt%) values obtained from phosphorite fractions and phosphatic sediments in the Triassic phosphogenic facies of Svalbard. Adopted from Krajewski (2000a), with new data.

Table 1. Major element contents (wt%) in representative samples of compact phosphate nodules and phosphatic microbialites in the sandy and muddy phosphogenic facies of Svalbard, measured by ICP-ES.

Sample	SiO_2	Al_2O_3	Fe_2O_3	MgO	CaO	Na_2O	K_2O	TiO_2	P_2O_5	MnO	LOI[a]
Sandy phosphogenic facies											
Phosphate nodule (WS)	19.90	1.16	0.93	0.42	43.24	0.15	0.31	0.08	18.54	0.01	15.00
Phosphate nodule (SS)	22.64	2.06	1.07	1.08	40.16	0.32	0.58	0.13	17.63	0.04	14.10
Biolaminated structure (SS)	12.77	2.41	2.03	0.37	44.28	0.15	0.53	0.11	31.57	0.01	4.00
Biolaminated structure (WS)	7.78	1.98	1.96	1.28	45.97	0.11	0.29	0.04	25.19	0.03	14.80
Muddy phosphogenic facies											
Phosphate nodule (EDG)	17.46	3.64	1.09	0.66	40.55	0.26	0.56	0.10	23.06	0.08	11.80
Phosphate nodule (CS)	17.80	1.21	0.76	0.56	43.02	0.10	0.23	0.05	20.19	0.02	15.90
Biolaminated structure (CS)	12.29	2.68	4.19	0.49	42.92	0.29	0.54	0.11	29.83	0.01	6.00
Biolaminated structure (CS)	11.86	3.34	2.57	1.28	44.64	1.01	0.06	0.16	27.42	0.03	7.50

CS central Spitsbergen, *SS* southern Spitsbergen, *WS* western Spitsbergen, *EDG* Edgeøya
[a]Lost in ignition

of apatite cementation of its primary and secondary pore space. Indistinct cementation zones in both the sandy and muddy phosphogenic facies usually contain up to 15 wt% P_2O_5. Compact phosphate nodules and peloids show higher phosphorus concentrations (up to 25 wt% P_2O_5) that correspond to advanced cementation at isolated sites in sediment. Phosphatic microbialites show the highest concentrations of phosphorus (up to 32 wt% P_2O_5), suggesting that the bottom areas covered by microbial mats provided best conditions for precipitating and concentrating CFA on the Svalbard shelf. A similar association of microbial mat-generated phosphatization and burial phosphorus concentration was reported from the Cretaceous Mishash Formation of Israel (Soudry, 1987, 2000), though the two phosphogenic systems show pronounced differences in environmental controls and the nature of phosphate-focusing microbial mat communities.

4.3.2. Isotopic Composition of Apatitic Carbon and Sulfur

Isotopic compositions of carbon and sulfur in substituting carbonate and sulfate ions in the lattice of sedimentary apatite are important proxies to the chemistry of phosphogenic environments (McArthur et al., 1986; Jarvis et al., 1994). The CFA-forming phosphorite fraction in the Triassic phosphogenic facies has

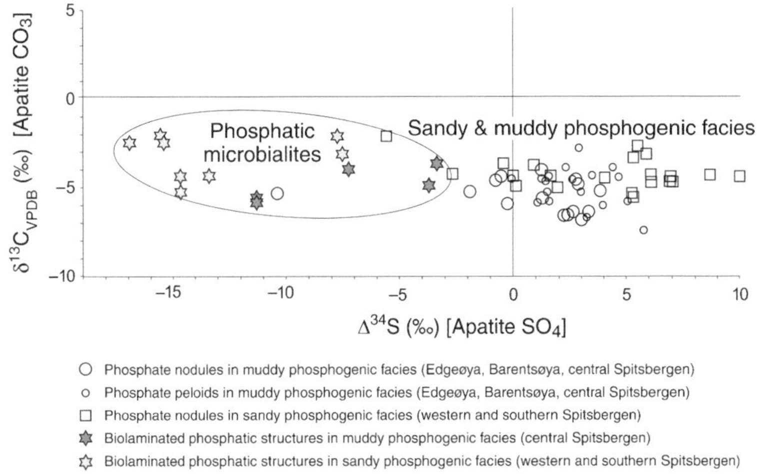

Figure 14. Plot of $\delta^{13}C$ (‰) and $\Delta^{34}S$ (‰) values of CFA obtained from phosphorite fractions and phosphatic sediments in the Triassic phosphogenic facies of Svalbard. $\Delta^{34}S$ was calculated using the following equation: $\Delta^{34}S_{apatite} = \delta^{34}S_{apatite} - \delta^{34}S_{coeval\ marine\ sulfate}$, where $\delta^{34}S_{coeval\ marine\ sulfate} = +14.5$‰ to +15.7‰, mean +15.1‰ (Fanlo and Ayora, 1998).

been found to contain enough substitutions for precise analysis (Krajewski, 2000b, c). Isotopic data already collected from Svalbard are shown in Fig. 14. All the analyzed samples show a rather narrow range of $\delta^{13}C_{VPDB}$ values (−7‰ to −1‰), with no noticeable difference between the discerned phosphorite types. These results suggest rather constant contribution of carbon dioxide resulted from oxidation of organic matter to carbonate pool in phosphogenic environments of the Svalbard shelf. They are consistent with the analysis of organic carbon presented above. However, a pronounced difference in isotopic composition of apatitic sulfur between phosphorite fractions of the sandy and muddy facies on one hand and the phosphatic microbialites on the other is observed. After correction for a likely value of coeval marine sulfate ($\delta^{34}S_{VCDT}$: +14.5‰ to +15.7‰; Fanlo and Ayora, 1998), most of phosphate deposits in the sandy and muddy phosphogenic facies falls in a range from the seawater values to the ones considerably heavier ($\Delta^{34}S$: −2‰ to +10‰). These values reflect changing position of the major zone of phosphogenesis in sediment, from near surface in the muddy facies to shallow burial in the sandy facies, depending on the degree of bottom oxygenation and the position of redoxcline (Krajewski, 2000b). They are comparable with those obtained from many other marine phosphorite deposits (e.g., Bliskovski et al., 1977; Benmore et al., 1983; McArthur et al., 1986; Piper and Kolodny, 1987; Mertz, 1989; Kolodny and Luz, 1992; Compton et al., 1993; Shen et al., 1998).

In contrast, phosphatic microbialites gave strikingly smaller values ($\Delta^{34}S$: −18‰ to −2‰), suggesting that sulfate sulfur trapped in the precipitating CFA was isotopically much lighter than the marine sulfate (Fig. 14). The plausible explanation of this composition is a contribution of sulfate originated as a result of oxidation of bacterially generated hydrogen sulfide. This strongly supports the hypothesis presented here on the colorless sulfur bacteria composition of the studied microbial mats. The mats dominated by colorless filamentous sulfur bacteria intensely oxidize hydrogen sulfide that is generated in anoxic sediment column and undergoes upward diffusion (Jørgensen, 1982; Nelson et al., 1986; Hagen and Nelson, 1997; Yamamoto-Ikemoto et al., 1998). Sulfide oxidation at a narrow oxic/anoxic interface stabilized by the mats may occur simultaneously by different pathways, both microbiological and abiological (Ferdelman et al., 1997). The oxidation of hydrogen sulfide by these bacteria occurs in a sequence of steps, with intermediate formation of intracellular granules of elemental sulfur (Wiessner, 1981; Strohl, 1989), though the end product is almost entirely sulfate ion generated to the surrounding medium (Otte et al., 1999). Abiological processes include autocatalytic oxidation of hydrogen sulfide by free oxygen and, to a lesser extent, by direct reaction with ferric (hydr)oxides. Dissimilatory sulfate reduction in the sediment is associated with pronounced isotopic fractionation of sulfur, with ^{32}S isotope being preferentially reduced to sulfide (Chambers and Trudinger, 1979; Habicht and Canfield, 1997). The sulfate ion eventually generated within the narrow environment of the mat also contains isotopically light sulfur (Toran and Harris, 1989; Otte et al., 1999), which is likely to be incorporated, in the form of SO_4 lattice substitutions, in the crystallizing CFA. Elevated content of isotopically light sulfate sulfur in apatite of the laminated phosphorite crusts off Peru is also interpreted to record oxidation of biogenic H_2S by sulfide-oxidizing bacteria (Arning et al., 2009a).

4.3.3. Isotopic Composition of Pyritic Sulfur

Isotopic composition of pyritic sulfur in the phosphorite fractions versus their content of iron is shown in Fig. 15. This composition is lighter than the assumed composition of coeval marine sulfate, confirming origin of authigenic pyrite as a result of dissimilatory bacterial reduction of sulfate in the phosphogenic sediment. The likely isotopic fractionation during this process ranges from −50‰ to −5‰. Similar fractionations associated with dissimilatory sulfate reduction are commonly observed in modern organic-rich marine sediments (Chambers and Trudinger, 1979; Habicht and Canfield, 1997), and strongly negative $\delta^{34}S$ values are noted in many biogenic pyrites (Strauss, 1997).

Phosphatic microbialites show more negative $\Delta^{34}S$ values (−40‰ to −20‰) than most nodular and peloidal phosphates of both the sandy and muddy phosphogenic facies. The difference may reflect increased contribution of sulfate originated from oxidation of hydrogen sulfide in the microbial mat environment compared with the OC-rich bottoms devoid of mat cover. Recurrent reduction of sulfate and oxidation/disproportionation of sulfide and intermediate sulfur compounds are likely processes leading to light isotopic composition of pyritic sulfur (Canfield and Thamdrup, 1994; Habicht et al., 1998). Phosphatic microbialites

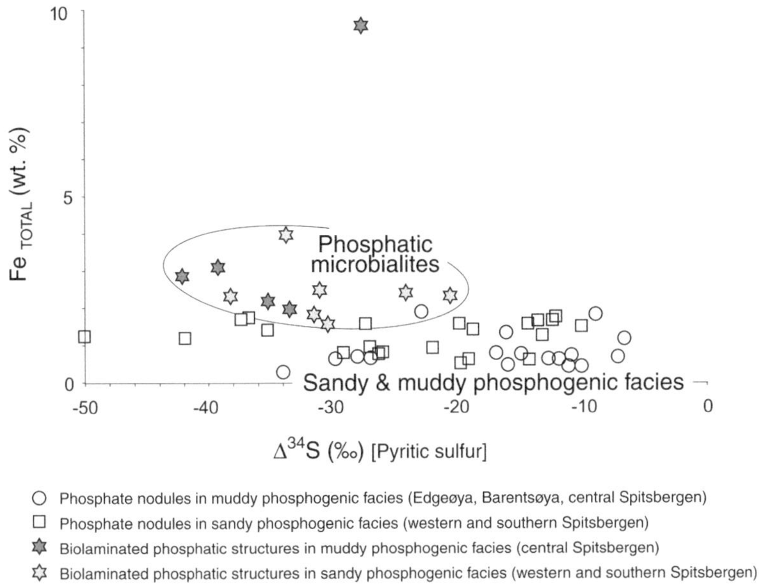

Figure 15. Plot of total content of iron (Fe_{TOTAL} in wt%) and $\Delta^{34}S$ (‰) values of pyritic sulfur obtained from phosphorite fractions and phosphatic sediments in the Triassic phosphogenic facies of Svalbard. $\Delta^{34}S$ was calculated using the following equation: $\Delta^{34}S_{pyrite} = \delta^{34}S_{pyrite} - \delta^{34}S_{coeval\ marine\ sulfate}$, where $\delta^{34}S_{coeval\ marine\ sulfate} = +14.5‰$ to $+15.7‰$, mean $= +15.1‰$ (Fanlo and Ayora, 1998).

show elevated content of iron, which is consistent with their enhanced pyritization revealed in microscopic survey. Minute intergrowths of sphalerite have also been noted, and the content of iron correlates with Zn (20–600 ppm in nodules and peloids; 600–2,100 ppm in phosphatic microbialites). This suggests that the microbial mats stabilized the oxic/anoxic interface at the sea bottom, and that the biolaminated sediments occurring below became collectors of reduced Fe, Zn, and S.

4.3.4. Rare Earth Elements

Rare earth elements (REEs) in the phosphorite fractions were normalized against post-Archean shales from Australia (PAAS; McLennan, 1989), and their distributions are shown in Fig. 16. Cerium anomalies were calculated as $Ce^* = \log[3Ce_n/(2La_n + Nd_n)]$ where n represents shale normalized contents (Wright et al., 1987). In all the analyzed samples, REEs are associated with CFA. The dominant cryptocrystalline nature of CFA and the absence of diagenetic recrystallization or neoformation features in the analyzed samples suggest that the REE concentrations reflect geochemical processes in the phosphogenic environments. Similar distribution patterns of REEs in the nodular and peloidal deposits in sequences showing different levels of organic metamorphism (R_o from 0.5% to 2.0%) suggest that the

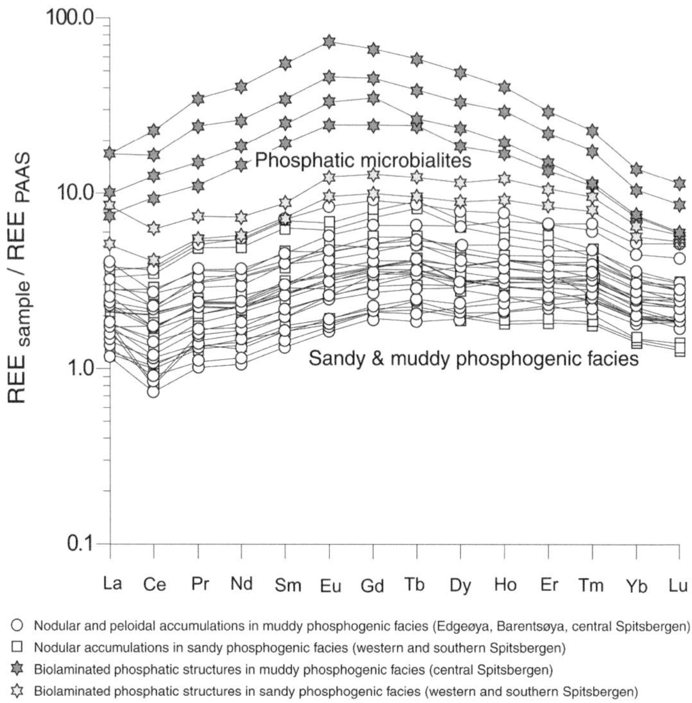

Figure 16. Rare earth element (REE) distributions in phosphorite fractions and phosphatic sediments in the Triassic phosphogenic facies of Svalbard. REE contents are normalized against post-Archean shales from Australia (PAAS).

maturation of kerogen and generation of bitumen in the phosphogenic facies did not affect the REE concentrations in CFA.

Most phosphatic nodules and peloids in the sandy and muddy phosphogenic facies contain between 300 and 600 ppm (ΣREE). This value range is comparable to the ones typically noted in unaltered sedimentary phosphorites (e.g., Altschuler, 1980; McArthur and Walsh, 1984; Piper et al., 1988). The REE distributions show modified seawater pattern (Elderfield and Greaves, 1982; Jarvis, 1992), with weak negative cerium anomalies ($Ce^* = -0.4$ to -0.05). A slight enrichment in intermediate REEs and similar depletion in heavy REEs are observed, suggesting that the incorporation mechanism was not a simple uptake of REEs from seawater. The content of REEs in pore waters of the phosphogenic environments was likely affected by an excess deposition and diagenesis of marine organic matter under anoxic sulfidic conditions.

Phosphatic microbialites show considerably higher concentrations of REEs than the nodular and peloidal accumulations (ΣREE = 1,100–5,400 ppm). The highest concentrations have been noted in biolaminated structures in the muddy phosphogenic facies of central Spitsbergen (2,300–5,400 ppm).

Lower concentrations noted in biolaminated structures in the sandy phosphogenic facies of western Spitsbergen (1,100–1,400 ppm) correspond to the highest concentrations in the nodular and peloidal accumulations in the muddy facies (ca. 1,000 ppm). The increase in the content of REEs in phosphatic microbialites is associated with the evolution of their distributions from modified seawater pattern showing weak negative cerium anomaly (Ce* ~ −0.1) to concave-down pattern without cerium anomaly. The latter are depleted in both the light and heavy REEs compared with intermediate REEs, with top enrichment centered on Eu and Gd (Fig. 16). Similar patterns have been recognized in several biogenic apatites (e.g., Wright et al., 1987; Gradjean and Albarède, 1989) and phosphate nodules and grains in OC-rich sequences (e.g., McArthur and Walsh, 1984; Tlig et al., 1987; Kidder and Eddy-Dilek, 1994), though their origin remains ambiguous (Jarvis et al., 1994). The REE distributions in phosphatic microbialites seem to have been shaped by their prolonged exposition close to the water/sediment interface, during which continued degradation of organic matter in the microbial mat-shielded environment affected both the element enrichment and abundance changes. Disappearance of the cerium anomaly in highly REE-enriched microbialites is consistent with the precipitation of CFA under cumulative deposition and reduction of ferric (hydr)oxides, which usually contain increased amounts of Ce owing to Ce^{4+} scavenging from seawater (Martín-Algarra and Sánchez-Navas, 1995).

5. Discussion

The Triassic sedimentary sequence in Svalbard provides an excellent example of OC-rich, phosphogenic shelf depositional system developed as a response to high-biological productivity event in the Mesozoic of the Barents Sea Shelf (Fig. 17). The overall development of the phosphogenic facies resembles sediments in modern OC-rich, phosphogenic environments on continental shelves reinforced by coastal upwellings (e.g., Burnett et al., 1980; Baturin, 1982; Glenn and Arthur, 1988;

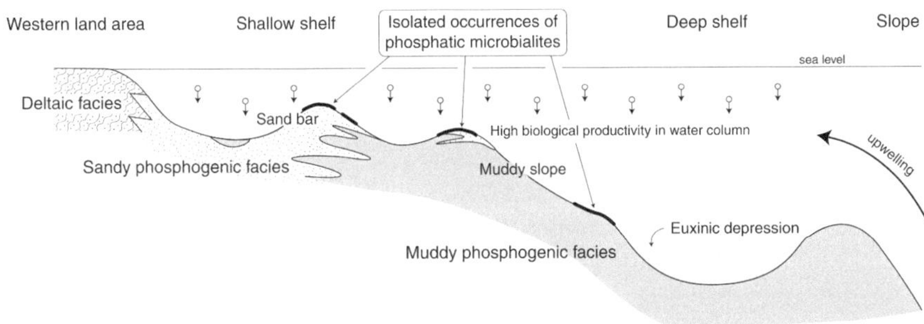

Figure 17. Paleoenvironmental synthesis of the Triassic phosphogenic facies in Svalbard showing locations of preferential development of phosphatic microbialites.

Garrison and Kastner, 1990). The common occurrence of microbial mats dominated by colorless filamentous sulfur bacteria on these shelves (e.g., Gallardo, 1977; Henrichs and Farrington, 1984; Schulz et al., 1996, 1999) has been suggested to drive phosphogenesis in bottom sediments due to interface-linked phosphorus transformations (Schulz and Schulz, 2005; Arning et al., 2009a). Phosphatic microbialites revealed in Svalbard might be considered fossil phosphatic sulfur bacteria mats, and therefore used to test their role in phosphorite formation in ancient OC-rich environments.

The Triassic phosphogenic facies in Svalbard embrace a variety of phosphate deposits, including nodules, peloids, phosphatic fossil molds, burrows, coprolites, and phosphatized skeletal remnants, originated due to rapid, punctuated precipitation of CFA in the uppermost parts of OC-rich sediment column (Krajewski, 2000a). Regional development of these facies on the Svalbard shelf was induced by excess deposition of fresh, marine organic matter from the water column, which by recycling in bottom environments provided a continuous source of organically bound phosphorus for apatite precipitation (Fig. 16). The vast majority of phosphate deposits is represented by nodular and peloidal accumulations originated over a range of bottom environments, from fully oxic shallow shelf sandy bottoms to suboxic/anoxic deep shelf muddy bottoms with local euxinic depressions (Krajewski, 2006). Among these deposits, the phosphatic microbialites seem to represent a peculiar, spatially restricted type of phosphate deposition that required a unique set of conditions to develop and preserve in the burial sequences. These conditions were associated with temporarily suppressed or halted sedimentation on local elevations, submarine slopes, and flanks of migrating sediment bodies, which provided stable bottoms for colonization and proliferation of microbial mats. The available evidence points to a subordinate role of this type of phosphate deposit as phosphorus sink in the Triassic phosphogenic facies. This does not preclude its role as local phosphorus collectors in parts of the shelf depositional system, which were especially prone to sedimentary condensation.

Despite their only local contribution to the phosphogenic facies, the phosphatic microbialites are shown to have provided a favorable place for early and intense CFA precipitation in OC-rich sediment. Their burial concentration of mineral phosphorus is clearly the highest among the discerned phosphate deposits in the Triassic sequence. The phosphatization zone of biolaminated structures has a lateral extent that imprints the original mat distribution at the surface. This association is very prominent throughout all the investigated microbialites. One way of explaining this association was presented by Williams and Reimers (1983), Reimers et al. (1990), and Schulz and Schulz (2005) who located a direct and immediate source for apatite precipitation in the mat itself. These authors have shown that sulfur bacteria are capable of transitorily fixating phosphate ion to form intracellular polyphosphate, which under anoxic conditions would be liberated to supersaturate pore fluids leading to precipitation of apatite (Gächter et al., 1988; Gächter and Meyer, 1993). However, this scenario does not explain the formation of the vast majority of phosphate deposits in Svalbard, which provide no evidence of growth close to microbial mat systems. All these deposits show abundant apatitic ultrastructures preserved during the formation of phosphatic matrices and grains, which are interpreted to be remnants of bacterial decomposing community in sediment. The apatite-pyrite association strongly suggests phosphogenesis at the upper margin of anoxic sulfidic zone, located either beneath the water/sediment interface in the sandy oxic environments or close to it in the muddy dysoxic to anoxic environments. Similar bacterial ultrastructures and apatite-pyrite mineral sequences are observed in phosphatic microbialites, along with highly degraded remnants of a filamentous community interpreted to be sulfide-oxidizing bacteria. This suggests that biogeochemical microenvironments of apatite precipitation in phosphatic microbialites and elsewhere in the phosphogenic facies were similar, though they showed different intensity, duration, and lateral extent of the precipitation process. There should therefore be other aspects of a microbial mat-shielded environment that are likely to explain extensive subsurface phosphatization.

1. *Sealing of the water/sediment interface to chemical diffusion.* The resistant framework of microbial mat stabilized the subsurface pore environment for an extended period of time during seafloor history of the condensed horizons. This environment would provide conditions for continuous precipitation of any authigenic mineral phase whose ionic counterparts actually were at the (super) saturation level (Krumbein, 1983; Cohen et al., 1984; Cohen and Rosenberg, 1989). Phosphate ion generated to the pore environment of biolaminated sediment, mostly from decomposition of P-containing organic matter but also from dissolution of skeletal hydroxyapatite and reduction of ferric (hydr)oxides, could precipitate in this stable environment to form a high-grade apatitic matrix. High concentrations of CFA and REEs observed in the phosphatic microbialites are consistent with sedimentologic observations suggesting nondepositional interludes over OC-rich bottoms and phosphogenesis in the microbial mat-shielded environments.
2. *Narrowing chemical gradients and pH shifts.* The filamentous sulfur bacteria mats are capable of narrowing and stabilizing the H_2S/O_2 interface at or close to the sediment surface (Jørgensen, 1982; Jørgensen and Revsbech, 1983; Nelson et al., 1986; SenGupta et al., 1997). The narrow interface is essential for these bacteria to outcompete autocatalytic oxidation of hydrogen sulfide by oxygen (Fenchel and Bernard, 1995). The intense metabolic sulfide oxidation at this interface (Heijs et al., 1999) results in local production of protons, which lower pH values of the microenvironment (Boudreau, 1991). Lowered pH values would favor a rapid precipitation of phosphate from the solution owing to a catalytic effect of hydrogen ions (Nathan and Sass, 1981; Van Cappellen, 1991; Van Cappellen and Berner, 1991). Isotopically light sulfate sulfur in CFA of the microbialites documents that the oxidation of H_2S was an important process during phosphogenesis in the microbial mat-shielded environments.

3. *Enhancing effects of the iron-related phosphorus pump mechanism.* A portion of phosphate ion generated in OC-rich sediment escapes to the overlying water, where it may be sorbed on iron (hydr)oxides settling to the bottom (Froelich et al., 1988). This phosphate is liberated during reduction of ferric iron in the anoxic sulfidic environment, providing surplus phosphate source for apatite precipitation (Glenn et al., 1994). The sorption–desorption cycles of phosphate associated with iron (hydr)oxides widely depend on small-scale shifts of the H_2S/O_2 interface (Ingall et al., 1993; Gunnars and Blomqvist, 1997). The narrow interface located within the microbial mat is a subject of oscillation reflecting changes of bottom water oxygen level, variations in productivity, and the rate of mat burial and vertical microbiota migration (Huettel et al., 1996; SenGupta et al., 1997; Jørgensen and Gallardo, 1999). The complex phosphate-pyrite sequences observed in thicker biolaminites suggest that vertical migration of the interface was a common phenomenon in the sedimentary mat systems, resulting in pulses of CFA precipitation followed by pyritization of the phosphatic matrix.

6. Conclusions

The Triassic organic carbon-rich, phosphogenic facies setting in Svalbard, NW Barents Sea Shelf developed in a broad spectrum of shelf depositional environments in response to a regional event of high biological productivity reinforced by upwelling from oceanic basin. Enhanced deposition and diagenesis of marine organic matter promoted shelf-wide phosphogenesis in bottom sediments that resulted in the formation of various nodular and peloidal phosphate deposits as well as peculiar, spatially restricted phosphatic accumulations related to benthic activity of microbial mats. The phosphatic microbialites developed under temporarily suppressed or halted sedimentation on local elevated sediment bodies and slopes that provided stable bottoms for colonization and proliferation of microbial mats. The microbial mats were most probably dominated by colorless filamentous sulfur bacteria, which, by growing on sulfidic gradients, provided depositional systems capable of precipitating and significantly concentrating CFA in the sediment due to focusing and intensifying the interface-linked processes of phosphorus transformation. Precipitation of apatite in the mat-stabilized sediments led to the formation of thin, high-grade phosphorite horizons in parts of the phosphogenic shelf area. These horizons show the highest concentration of phosphorus (up to 32 wt% P_2O_5) among the phosphate deposits preserved in the Triassic sequence. However, the spatial and temporal limitations of the growth and phosphatization of microbial mats confine the importance of phosphatic microbialites to local and restricted phosphorus collectors in the Svalbard shelf basin.

7. Acknowledgments

The author is grateful to Professor Vinod C. Tewari for his invitation to contribute to this book. Fieldwork in Svalbard was supported by research grant of the Ministry of Science and Higher Education No. IPY/279/2006, and carried out under the aegis of the IVth International Polar Year project KINNVIKA (IPY ID: 564; Internal Project No. 10 "*Geological processes in the formation of the Arctic phosphogenic province*"). The manuscript benefited from review by David Soudry.

8. References

Altschuler, Z.S. (1980) The geochemistry of trace metals in marine phosphorites. Part I. Characteristic abundances and enrichment, In: Y.K. Bentor (ed.) *Marine Phosphorites. Geochemistry, Occurrence, Genesis. SEPM Special Publication No. 29*, pp. 19–30.

Arning, E.T., Birgel, D., Brunner, B. and Peckmann, J. (2009a) Bacterial formation of phosphatic laminites off Peru. Geobiology **7**: 295–307.

Arning, E.T., Lückge, A., Breuer, C., Gussone, N., Birgel, D. and Peckmann, J. (2009b) Genesis of phosphorite crusts off Peru. Mar. Geol. **262**: 68–81.

Baturin, G.N. (1978) *Phosphorites in the Ocean*. Smirnov Institute of Oceanology. SSSR Academy of Sciences, Izdatelstvo Nauka, Moscow, 232 pp. (in Russian).

Baturin, G.N. (1982) *Phosphorites on the Sea Floor. Origin, Composition and Distribution. Developments in Sedimentology 33*. Elsevier, Amsterdam, 343 pp.

Baturin, G.N. (1983) Some unique sedimentological and geochemical features of deposits in coastal upwelling regions, In: J. Thiede and E. Suess (eds.) *Coastal Upwelling; Its Sediment Record. Part A*. Plenum Press, New York, pp. 11–27.

Baturin, G.N. and Bezrukov, P.L. (1979) Phosphorites on the sea floor and their origin. Mar. Geol. **31**: 317–332.

Benmore, R.A., Coleman, M.L. and McArthur, J.M. (1983) Origin of sedimentary francolite from its sulphur and carbon isotope composition. Nature **302**: 516–518.

Berner, R.A. (1970) Sedimentary pyrite formation. Am. J. Sci. **268**: 1–23.

Berner, R.A. (1984) Sedimentary pyrite formation: an update. Geochim. Cosmochim. Acta **48**: 605–615.

Berner, R.A. (1985) Sulphate reduction, organic matter decomposition and pyrite formation. Philos. Trans. R. Soc. Lond. **A315**: 25–38.

Bliskovski, V.Z., Grinienko, V.A. Migdisov, L.L. and Savina, L.I (1977) Isotopic composition of sulfur in minerals of phosphorite ores. Geochemistry **8**: 1208–1216 (in Russian).

Boudreau, B.P. (1991) Modelling the sulphide-oxygen reaction and associated pH gradients in porewaters. Geochim. Cosmochim. Acta **55**: 145–159.

Bréhéret, J.-G. (1991) Phosphatic concretions in black facies of the Aptian-Albian Marnes bleues Formation of the Vocontian basin (SE France), and at Site DSDP 369: evidence of benthic microbial activity. Cretaceous Res. **12**: 411–435.

Burnett, W.C. and Riggs, S.R. (eds.) (1990) *Phosphate Deposits of the Word. Vol. 3. Neogene to Modern Phosphorites.* Cambridge University Press, Cambridge, 464 pp.

Burnett, W.C., Veeh, H.H. and Soutar, A. (1980) U-series, oceanographic and sedimentary evidence in support of recent formation of phosphate nodules off Peru, In: Y.K. Bentor (ed.) *Marine Phosphorites. Geochemistry, Occurrence, Genesis. SEPM Special Publication No. 29*, pp. 61–71.

Canfield, D.E. (1989) Reactive iron in marine sediments. Geochim. Cosmochim. Acta **53**: 619–632.

Canfield, D.E. and Thamdrup, B. (1994) The production of ^{34}S-depleted sulfide during bacterial disproportionation of elemental sulfur. Science **266**: 1973–1975.

Chambers, L.A. and Trudinger, P.A. (1979) Microbiological fractionation of stable sulfur isotopes: a review and critique. Geomicrobiol. J. **1**: 249–293.

Cocks, L.R.M. and Torsvik, T.H. (2007) Siberia, the wandering northern terrane, and its changing geography through the Paleozoic. Earth-Sci. Rev. **82**: 29–74.

Cohen, Y. and Rosenberg, E. (eds.) (1989) *Microbial Mats. Physiological Ecology of Benthic Microbial Communities*. American Society for Microbiology, Washington, 494 pp.

Cohen, Y., Castenholz, R.W. and Halvorsen, H.O. (eds.) (1984) *Microbial Mats: Stromatolites. MBL Lectures in Biology 3*, 498 pp.

Compton, J.S., Hodell, D.A., Garrido, J.R. and Mallinson, D.J. (1993) Origin and age of phosphorite from the south-central Florida Platform: relation of phosphogenesis to sea-level fluctuations and $\delta^{13}C$ excursions. Geochim. Cosmochim. Acta **57**: 131–146.

Dahanayake, K. and Krumbein, W.E. (1985) Ultrastructure of a microbial mat-generated phosphorite. Mineralium Deposita **20**: 260–265.

Dallmann, W.K. (ed.) (1999) *Lithostratigraphic Lexicon of Svalbard. Upper Palaeozoic to Quaternary Bedrock. Review and Recommendations for Nomenclature Use*. Norsk Polarinstitutt, Tromsø, 318 pp.

Dallmann, W.K., Ohta, Y., Elvevold, S. and Blomeier, D. (eds.) (2002) *Bedrock Map of Svalbard and Jan Mayen*. Norsk Polarinstitutt Temakart 33.

Detterman, R.L. (1989) Triassic phosphate deposits, north-eastern Alaska, USA, In: A.J.G. Notholt, R.P. Sheldon and D.F. Davidson (eds.) *Phosphate Deposits of the World. Vol. 2. Phosphate Rock Resources*. Cambridge University Press, Cambridge, pp. 14–17.

Detterman, R.L., Reiser, H.N., Brosgé, W.P. and Dutro Jr., J.T., (1975) Post-Carboniferous stratigraphy, northern Alaska. U.S. Geological Survey Professional Paper **886**, 46 pp.

Dill, H. and Kemper, E. (1990) Crystallographic and chemical variations during pyritization in the upper Barremian and lower Aptian dark claystones from the Lower Saxonian Basin (NW Germany). Sedimentology **37**: 427–443.

Egorov, A.Yu. and Baturin, G.N. (1987) Phosphorites in Triassic deposits of the Novosibirsk Islands. Dokl. Akad. Nauk SSSR **297**: 921–925 (in Russian).

Elderfield, H. and Greaves, M.J. (1982) The rare-earth elements in sea-water. Nature **296**: 214–219.

Embry, A.F. (1988) Triassic sea-level changes: evidence from the Canadian Arctic Archipelago, In: C.K. Wilgus, B.S. Hastings, H. Posamentier, J. Van Wagoner, C.A. Ross and C.G.St.C. Kendall (eds.) *Sea-Level Changes: An Integrated Approach. SEPM Special Publication No. 42*, 249–259.

Fanlo, I. and Ayora, C. (1998) The evolution of the Lorraine evaporite basin: implications for the chemical and isotope composition of the Triassic ocean. Chem. Geol. **146**: 135–154.

Fenchel, T. and Bernard, C. (1995) Mats of colourless sulphur bacteria. 1. Major microbial processes. Mar. Ecol. Prog. Ser. **128**: 161–170.

Ferdelman, T.G., Lee, C., Pantija, S., Harder, J., Bebout, B.M. and Fossig, H. (1997) Sulfate reduction and methanogenesis in a *Thioploca*-dominated sediment off the coast of Chile. Geochim. Cosmochim. Acta **61**: 3065–3079.

Föllmi, K.B. (1996) The phosphorus cycle, phosphogenesis and marine phosphate-rich deposits. Earth Sci. Rev. **40**: 55–124.

Froelich, P.N., Arthur, M.A., Burnett, W.C., Deakin, M., Hensley, V., Jahnke, R., Kaul, L., Kim, K.-H., Roe, K., Soutar, A. and Vathakanon, C. (1988) Early diagenesis of organic matter in Peru continental margin sediments: phosphorite precipitation, In: W.C. Burnett and P.N. Froelich (eds.) *The Origin of Marine Phosphorite. The Results of the R. V. Robert D. Conrad Cruise 23-06 to the Peru Shelf*. Mar. Geol. **80**: pp. 309–343.

Gächter, R. and Meyer, J.S. (1993) The role of microorganisms in mobilization and fixation of phosphorus in sediments. Hydrobiologia **253**: 103–121.

Gächter, R., Meyer, J.S. and Mares, A. (1988) Contribution of bacteria to release and fixation of phosphorus in lake sediments. Limnol. Oceanogr. **33**: 1542–1558.

Gallardo, V.A. (1977) Large benthic microbial communities in sulphide biota under Peru-Chile Subsurface Countercurrent. Nature **268**: 331–332.

Garrison, R.E. and Kastner, M. (1990) Phosphatic sediments and rocks recovered from the Peru margin during ODP Leg 112. Proc. Ocean Drill Prog. Sci. Results **112**: 111–134.
Glenn, C.R. and Arthur, M.A. (1988) Petrology and major element geochemistry of Peru margin phosphorites and associated diagenetic minerals: authigenesis in modern organic-rich sediments, In: W.C. Burnett and P.N. Froelich (eds.) *The Origin of Marine Phosphorite. The Results of the R. V. Robert D. Conrad Cruise 23-06 to the Peru Shelf*. Mar. Geol. **80**: pp. 231–267.
Glenn, C.R., Föllmi, K.F., Riggs, S.R., Baturin, G.N., Grimm, K.A., Trappe, J., Abed, A.M., Galli-Olivier, C., Garrison, R.E., Ilyin, A.V., Jehl, C., Rohrlich, V., Sadaqah, R.M.Y., Schidlowski, M., Sheldon, R.E. and Siegmund, H. (1994) Phosphorus and phosphorites: sedimentology and environmental concerns. Eclogae Geol. Helv. **87**: 747–788.
Gradjean, P. and Albarède, F. (1989) Ion probe measurement of rare earth element in biogenic apatites. Geochim. Cosmochim. Acta **53**: 3179–3183.
Gulbrandsen, R.A. (1970) Relation of carbon dioxide content of apatite of the Phosphoria Formation to regional facies. U.S. Geol. Surv. Prof. Pap. **700B**, B9–B13.
Gunnars, A. and Blomqvist, S. (1997) Phosphate exchange across the sediment-water interface when shifting from anoxic to oxic conditions – an experimental comparison of freshwater and brackish-marine systems. Biogeochemistry **37**: 203–226.
Habicht, K.S. and Canfield, D.E. (1997) Sulphur isotope fractionation during bacterial sulphate reduction in organic-rich sediments. Geochim. Cosmochim. Acta **61**: 5351–5361.
Habicht, K.S., Canfield, D.E. and Rethmeier, J. (1998) Sulphur isotopic fractionation during bacterial reduction and disproportionation of thiosulphate and sulphite. Geochim. Cosmochim. Acta **62**: 2585–2595.
Hagen, K.D. and Nelson, D.C. (1997) Use of reduced sulfur compounds by *Beggiatoa* spp.: enzymology and physiology of marine and freshwater strains in homogeneous and gradient cultures. Appl. Environ. Microbiol. **63**: 3957–3964.
Harland, W.B. (1997) *The Geology of Svalbard. Geological Society London Memoir No. 17*, 521 pp.
Heijs, S.K., Jonkers, H.M., van Gemerden, H., Schaub, B.E.M. and Stal, L.J. (1999) The buffering capacity towards free sulphide in sediments of a coastal lagoon (Bassin d'Arcachon, France) – the relative importance of chemical and biological processes. Estuar. Coast Shelf Sci. **49**: 21–35.
Henrichs, S.M. and Farrington, J.W. (1984) Peru upwelling region sediments near 15°S. 1. Remineralization and accumulation of organic matter. Limnol. Oceanogr. **29**: 1–19.
Huettel, M., Forster, S., Kloser, S. and Fossing, H. (1996) Vertical migration in the sediment-dwelling sulfur bacteria *Thioploca* spp. in overcoming diffusion limitations. Appl. Environ. Microbiol. **62**: 1863–1872.
Ingall, E.D., Bustin, R.M. and Van Cappellen, P. (1993) Influence of water column anoxia on the burial and preservation of carbon and phosphorus in marine shales. Geochim. Cosmochim. Acta **57**: 303–316.
Jahnke, R.A., Emerson, S.R., Roe, K.K. and Burnett, W.C. (1983) The present-day formation of apatite in Mexican continental-margin sediments. Geochim. Cosmochim. Acta **47**: 259–266.
Jarvis, I. (1992) Sedimentology, geochemistry and origin of phosphatic chalks, the Upper Cretaceous deposits of NW Europe. Sedimentology **39**: 55–97.
Jarvis, I., Burnett, W.C., Nathan, Y., Almbaydin, F.S.M., Attia, A.K.M., Castro, L.N., Flicoteaux, R., Hilmy, M.E., Husain, V., Qutawnah, A.A., Serjani, A. and Zanin, Y.N. (1994) Phosphorite geochemistry: state-of-the art and environmental concerns. Eclogae Geol. Helv. **87**: 643–700.
Jørgensen, B.B. (1982) Ecology of the bacteria of the sulphur cycle with special reference to anoxic-oxic interface environments. Philos. Trans. R. Soc. Lond. **B298**: 543–561.
Jørgensen, B.B. and Gallardo, V.A. (1999) *Thioploca* spp.: filamentous sulfur bacteria with nitrate vacuoles. FEMS Microbiol. Ecol. **28**: 301–313.
Jørgensen, B.B. and Revsbech, N.P. (1983) Colorless sulfur bacteria, *Beggiatoa* spp., and *Thiovulum* spp., in O_2 and H_2S microgradients. Appl. Environ. Microbiol. **45**: 1261–1270.
Kidder, D.L. and Eddy-Dilek, C.A. (1994) Rare-earth element variation in phosphate nodules from mid-continent Pennsylvanian cyclothems. J. Sediment. Res. **A64**: 584–592.

Kolodny, Y. and Luz, B. (1992) Isotope signatures in phosphate deposits: formation and diagenetic history, In: N. Clauer and S. Chaudhuri (eds.) *Isotopic Signatures and Sedimentary Records*. Springer-Verlag, Berlin, pp. 69–121.

Krajewski, K.P. (2000a) Phosphogenic facies and processes in the Triassic of Svalbard. Stud. Geol. Pol. **116**: 7–84.

Krajewski, K.P. (2000b) Isotopic composition of apatite-bound sulphur in the Triassic phosphogenic facies of Svalbard. Stud. Geol. Pol. **116**: 85–109.

Krajewski, K.P. (2000c) Diagenetic recrystallization and neoformation of apatite in the Triassic phosphogenic facies of Svalbard. Stud. Geol. Pol. **116**: 111–137.

Krajewski, K.P. (2000d) Phosphorus concentration and organic carbon preservation in the Blanknuten Member (Botneheia Formation, Middle Triassic) in Sassenfjorden, Spitsbergen. Stud. Geol. Pol. **116**: 139–173.

Krajewski, K.P. (2000e) Phosphorus and organic carbon reservoirs in the Bravaisberget Formation (Middle Triassic) in Hornsund, Spitsbergen. Stud. Geol. Pol. **116**: 175–209.

Krajewski, K.P. (2005) The Arctic phosphogenic province. Polish Academy of Sciences, Annual Report 2005, pp. 83–85.

Krajewski, K.P. (2006) Phosphogenic facies in the Triassic of Svalbard. Abstr. Proc. Geol. Soc. Norway **3**: 89–90.

Krajewski, K.P. (2008) The Botneheia Formation (Middle Triassic) in Edgeøya and Barentsøya, Svalbard: lithostratigraphy, facies, phosphogenesis, paleoenvironment. Pol. Polar Res. **29**: 319–364.

Krajewski, K.P., Van Cappelen, P., Trichet, J., Kuhn, O., Lucas, J., Martín-Algarra, A., Prévôt, L., Tewari, V.C., Gaspar, L., Knight, R.I. and Lamboy, M. (1994) Biological processes and apatite formation in sedimentary environments. Eclogae Geol. Helv. **87**: 701–745.

Krajewski, K.P., Leśniak, P.M., Łącka, B. and Zawidzki, P. (2000) Origin of phosphatic stromatolites in the Upper Cretaceous condensed sequence of the Polish Jura Chain. Sediment. Geol. **136**: 89–112.

Krajewski, K.P., Karcz, P., Woźny, E. and Mørk, A. (2007) Type section of the Bravaisberget Formation (Middle Triassic) at Bravaisberget, western Nathorst Land, Spitsbergen, Svalbard. Pol. Polar Res. **28**: 79–122.

Krumbein, W.E. (1983) Stromatolites – the challenge of a term in space and time. Precambrian Res. **20**: 493–531.

Logan, G.A., Calver, C.R., Gorjan, P., Summons, R.E., Hayes, J.M. and Walter, M.R. (1999) Terminal Proterozoic mid-shelf benthic microbial mats in the Central Superbasin and their environmental significance. Geochim. Cosmochim. Acta **63**: 1345–1358.

Lucas, J. and Prévôt, L. (1985) The synthesis of apatite by bacteria activity: mechanism, In: J. Lucas and L. Prévôt (eds.) *Phosphorites*. Mémoire Sciences Géologiques, Strasbourg **77**: 83–92.

Maher, H.D., Craddock, H.D. and Maher, K.A. (1986) Kinematics of Tertiary structures in Upper Paleozoic and Mesozoic strata on Midterhuken, West Spitsbergen. Geol. Soc. Am. Bull. **97**: 1411–1421.

Martín-Algarra, A. and Sánchez-Navas, A. (1995) Phosphate stromatolites from condensed cephalopod limestones, Upper Jurassic, Southern Spain. Sedimentology **42**: 893–919.

McArthur, J.M. and Walsh, J.N. (1984) Rare-earth geochemistry of phosphorites. Chem. Geol. **47**: 191–220.

McArthur, J.M., Benmore, R.A., Coleman, M.L., Soldi, C., Yeh, H.-W. and O'Brien, G.W. (1986) Stable isotopic characterization of francolite formation. Earth Planet. Sci. Lett. **77**: 20–34.

McLennan, S.M. (1989) Rare earth elements in sedimentary rocks: influence of provenance and sedimentary processes. In. B.R. Lipin and G.A. McKay (eds.) *Geochemistry and Mineralogy of Rare Earth Elements. Review in Mineralogy 18*, pp. 169–200.

McManus, J., Berelson, W.M., Coale, K.H., Johnson, K.S. and Kilgore, T.E. (1997) Phosphorus regeneration in continental margin sediments. Geochim. Cosmochim. Acta **61**: 2891–2907.

Mertz Jr., K.A. (1989) Origin of hemipelagic source rocks during Early and Middle Miocene, Monterey Formation, Salinas Basin, California. Am. Assoc. Pet. Geol. Bull. **73**: 510–524.

Mitrov, Yu.V., Zanin, Yu.N., Krasilnikova, N.A., Gurevich, B.G., Krivoputskaya, L.M., Krasilnikova, I.G. and Sukhov, Yu.K.M. (1987) *Ultrastructures of Phosphorites (Atlas of Pictures)*. Izdatelstvo Nauka, Moscow, 233 pp. (in Russian).

Morita, R.Y., Hurriaga, R. and Gallardo, V.A. (1981) *Thioploca*: methylotroph and significance in the food chain. Kieler Meeresforschung **5**: 384–389.
Mørk, A. (2006) Trace fossils of the Middle Triassic regressive systems tract of Svalbard. Abstr. Proc. Geol. Soc. Norway **3**: 111–115.
Mørk, A. and Bjorøy, M. (1984) Mesozoic source rocks in Svalbard, In: A.M Spencer et al. (eds.) *Petroleum Geology of the North European Margin*. Norwegian Petroleum Society, Graham and Trotman, London, pp. 371–382.
Mørk, A. and Bromley, R.G. (2008) Ichnology of a marine regressive system tract: the Middle Triassic of Svalbard. Polar Res. **27**: 339–359.
Mørk, A., Embry, A.F. and Weitschat, W. (1989) Triassic transgressive-regressive cycles in the Sverdrup Basin, Svalbard, and the Barents Shelf, In: J.D. Collinson (ed.) *Correlation in Hydrocarbon Exploration*. Graham and Trotman, London, pp. 113–130.
Mørk, A., Dallmann, W.K., Dypvik, H., Johannessen, E.P., Larssen, G.B., Nagy, J., Nøttvedt, A., Olaussen, S., P elina, T.M. and Worsley, D. (1999) Mesozoic lithostratigraphy, In: W.K. Dallmann (ed.) *Lithostratigraphic Lexicon of Svalbard. Review and Recommendations for Nomenclature Use. Upper Palaeozoic to Quaternary Bedrock*. Norsk Polarinstitutt, Tromsø, pp. 127–214.
Nathan, Y. and Sass, E. (1981) Stability relation of apatites and calcium carbonates. Chem. Geol. **34**: 103–111.
Nelson, D.C. and Jannasch, H.W. (1983) Chemoautotrophic growth of marine *Beggiatoa* in sulphide-gradient cultures. Arch. Microbiol. **136**: 262–269.
Nelson, D.C., Revsbech, N.P. and Jørgensen, B.B. (1986) Microoxic-anoxic niche of *Beggiatoa* spp.: microelectrode survey of marine and freshwater strains. Appl. Environ. Microbiol. **52**: 161–168.
Otte, S., Kuenen, J.G., Nielsen, L.P., Pearl, H.W., Zopfi, J., Schulz, H.N., Teske, A., Strotmann, B., Gallardo, V.A. and Jørgensen, B.B. (1999) Nitrogen, carbon, and sulfur metabolism in natural *Thioploca* samples. Appl. Environ. Microbiol. **65**: 3148–3157.
Parrish, J.T. and Curtis, R.L. (1982) Atmospheric circulation, upwelling, and organic-rich rocks in the Mesozoic and Cenozoic eras. Palaeogeogr. Palaeoclimatol. Palaeoecol. **40**: 31–66.
Parrish, J.T., Droser, M.L. and Bottjer, D.J. (2001) A Triassic upwelling zone: the Shublik Formation, Arctic Alaska, U.S.A. J. Sediment. Res. **71**: 272–285.
Piper, D.Z. and Kolodny, Y. (1987) The stable isotopic composition of a phosphorite deposit: $\delta^{13}C$, $\delta^{34}S$, and $\delta^{18}O$. Deep Sea Res. **34**: 897–911.
Piper, D.Z., Baedecker, P.A., Crock, J.G., Burnett, W.C. and Loebner, B.J. (1988) Rare earth elements in the phosphatic-enriched sediments of the Peru Shelf, In: W.C. Burnett and P.N. Froleich (eds.) *The Origin of Marine Phosphorite. The Results of the R.V. Robert D. Conrad Cruise 23-06 to the Peru Shelf*. Mar. Geol. **80**: 269–285.
Raiswell, R. (1982) Pyrite texture, isotopic composition and the availability of iron. Am. J. Sci. **282**: 1244–1263.
Reimers, C.E., Kastner, M. and Garrison, R.E. (1990) The role of bacterial mats in phosphate mineralization with particular reference to Monterey Formation, In: W.C. Burnett and S.R. Riggs (eds.) *Phosphate Deposits of the World. Vol. 3. Neogene to Modern Phosphorites*. Cambridge University Press, Cambridge, pp. 300–311.
Riis, F., Lundschien, B.A., Høy, T., Mørk, A. and Mørk, M.B.E. (2008) Evolution of the Triassic shelf in the northern Barents Sea region. Polar Res. **27**: 318–338.
Rosenberg, R., Arnitz, W.E., De Flores, E.C., Flores, L.A., Carbajal, G., Finger, I. and Tatazona, J. (1983) Bentos biomass and oxygen deficiency in the upwelling system off Peru. J. Mar. Res. **41**: 263–279.
Schuffert, J.D., Kastner, M., Emanuele, G. and Jahnke, R.A. (1990) Carbonate-ion substitution in francolite: a new equation. Geochim. Cosmochim. Acta **54**: 2323–2328.
Schuffert, J.D., Kastner, M. and Jahnke, R.A. (1998) Carbon and phosphorus burial associated with modern phosphorite formation. Mar. Geol. **146**: 21–31.
Schulz, H.N. and Schulz, H.D (2005) Large sulfur bacteria and the formation of phosphorite. Science **307**: 416–418.

Schulz, H.N., Jørgensen, B.B., Fossing, H.A. and Ramsing, N.B. (1996) Community structure of filamentous, sheath-building sulfur bacteria, *Thioploca* spp. off the coast of Chile. Appl. Environ. Microbiol. **62**: 1855–1862.

Schulz, H.N., Brinkhoff, T., Ferdelman, T.G., Marine, M.H., Teske, A. and Jørgensen, B.B. (1999) Dense population of a giant sulfur bacterium in Namibian shelf sediments. Science **284**: 493–495.

SenGupta, B.K., Platon, E., Bernhard, J.M. and Aharon, P. (1997) Foraminiferal colonization of hydrocarbon-seep bacterial mats and underlying sediment, Gulf of Mexico slope. J. Foraminiferal Res. **27**: 292–300.

Shen, Y., Zhao, R., Chu, X. and Jiajin, L. (1998) The carbon and sulfur isotope signatures in the Precambrian-Cambrian transition series of the Yangtze Platform. Precambrian Res. **89**: 77–86.

Soudry, D. (1987) Ultra-fine structures and genesis of the Campanian Negev high-grade phosphorites (southern Israel). Sedimentology **34**: 641–660.

Soudry, D. (2000) Microbial phosphate sediments, In: R.E. Riding and S.M. Awramik (eds.) *Microbial Sediments*. Springer-Verlag, Berlin, pp. 127–136.

Soudry, D. and Champetier, Y. (1983) Microbial processes in the Negev phosphorites (southern Israel). Sedimentology **30**: 411–423.

Soutar, A. and Crill, P.A. (1977) Sedimentation and climatic patterns in the Santa Barbara Basin during the 19th and 20th centuries. Geol. Soc. Am. Bull. **88**: 1161–1172.

Steel, R.J. and Worsley, D. (1984) Svalbard's post-Caledonian strata – an atlas of sedimentational patterns and palaeogeographic evolution, In: A.M. Spencer et al. (eds.) *Petroleum Geology of the North European Margin*. Norwegian Petroleum Society, Graham and Trotman, London, pp. 109–135.

Strauss, H. (1997) The isotopic composition of sedimentary sulfur through time. Palaeogeogr. Palaeoclimatol. Palaeoecol. **132**: 97–118.

Strohl, W.R. (1989) Beggiatoaceae, In: P.M. Bryant, N. Pfening and J.G. Holt (eds.) *Bergey's Manual of Systematic Bacteriology*. Williams and Wilkins, Baltimore, pp. 2089–2106.

Talwani, M. and Eldholm, O. (1977) Evolution of the Norwegian-Greenland Sea. Geol. Soc. Am. Bull. **88**: 969–999.

Tlig, S., Sassi, A., Belayouni, H. and Machel, D. (1987) Distribution de l'uranium, du thorium, du zirconium, du hafnium et des terres rares (TR) dans les grains de phosphates sédimentaires. Chem. Geol. **62**: 209–221.

Toran, L. and Harris, R.F. (1989) Interpretation of sulfur and oxygen isotopes in biological and abiological oxidation. Geochim. Cosmochim. Acta **53**: 2341–2348.

Trappe, J. (1998) *Phanerozoic Phosphorite Depositional Systems. Lecture Notes in Earth Sciences 76*. Springer-Verlag, Berlin, 316 pp.

Van Cappellen, P. (1991) The Formation of Marine Apatite: A Kinetic Study. Ph.D. Thesis. Yale University.

Van Cappellen, P. and Berner, R.A. (1991) Fluorapatite crystal growth from modified seawater solutions. Geochim. Cosmochim. Acta **55**: 1219–1234.

Wiessner, W. (1981) The family Beggiatoaceae, In: M.P. Starr, H. Stolp, H.G. Trüper, A. Balows and H.G. Schlegel (eds.) *The Prokaryotes. A Handbook of Habitats, Isolation, and Identification of Bacteria*. Vol. I. Springer-Verlag, New York, pp. 380–389.

Wilkin, R.T. and Barnes, H.L. (1997) Formation processes of framboidal pyrite. Geochim. Cosmochim. Acta **61**: 323–339.

Williams, L.A. (1984) Subtidal stromatolites in Monterey Formation and other organic-rich rocks as suggested contributors to petroleum formation. Am. Assoc. Pet. Geol. Bull. **68**: 1879–1893.

Williams, L.A. and Reimers, C. (1983) Role of bacterial mats in oxygen-deficient marine basins and coastal upwelling regimes: preliminary report. Geology **11**: 267–269.

Wright, J., Schrader, H. and Holser, W.T. (1987) Paleoredox variations in ancient oceans recorded by rare earths in fossil apatite. Geochim. Cosmochim. Acta **51**: 631–644.

Yamamoto-Ikemoto, R., Matsui, S., Komori, T. and Bosque-Hamilton, E.K. (1998) Interactions between filamentous sulfur bacteria, sulfate reducing bacteria and poly-P accumulating bacteria in anaerobic-oxic activated sludge from a municipal plant. Water Sci. Technol. **37**: 599–603.

Biodata of **Professor Francesco Massari** and **Prof. Dr. Hildegard Westphal**, authors of ***"Microbialites in the Middle–Upper Jurassic Ammonitico Rosso of the Southern Alps (Italy)"***

Professor Francesco Massari took a degree in Geological Sciences at Padova University on 18 July 1963. Since the 13 December 1980 he holds the chair of Sedimentology at the Faculty of Sciences of the Padova University. He has been member of the Scientific Council of the research centre of Alpine Geodynamics having seat at the Department of Mineralogy and Petrology of the Padova University. Francecso Massari had until 2007 a chair of Sedimentology at Department of Geology, Paleontology and Geophysics of the Padova University. Since 2008 he is retired but maintains a research contract with the Padova University for a duration of 2 years. His scientific activity has concerned several stratigraphic-sedimentologic themes and is particularly focused on recent terrigenous and carbonate sediments of foredeeps and post-orogenic basins. During the years 1992–1995 he has been member of the Editorial board of Sedimentology. During last years he has worked on the sedimentology and sequence stratigraphy of Pleistocene terrigenous and carbonate deposits of the Bradanic Foredeep, Crotone Basin, Salento peninsula, and Sicily. He coordinated a number of national research programmes of the Italian Ministry of Education (MURST) on the cyclic sedimentation and climatic variability in the Plio-Pleistocene sequences of the southern Italy and Mediterranean area.

E-mail: **francesco.massari@unipd.it**

V.C. Tewari and J. Seckbach (eds.), *STROMATOLITES: Interaction of Microbes with Sediments*, Cellular Origin, Life in Extreme Habitats and Astrobiology 18, 223–250
DOI 10.1007/978-94-007-0397-1_10,

Prof. Dr. Hildegard Westphal has studied geology in Tübingen and Brisbane and got her Ph.D. in Kiel. Afterwards, she worked as carbonate sedimentologist at the Universities of Miami, Hannover, Erlangen, and Bremen, and is now director of the Leibniz Center for Marine Tropical Ecology in Bremen. She is interested in the role of environmental conditions on carbonate-producing sedimentary systems, in carbonate diagenesis, and reservoir properties.

E-mail: **Hildegard.westphal@uni-bremen.de**

MICROBIALITES IN THE MIDDLE–UPPER JURASSIC AMMONITICO ROSSO OF THE SOUTHERN ALPS (ITALY)

FRANCESCO MASSARI[1] AND HILDEGARD WESTPHAL[2]
[1]*Geoscience Department, Padova University, Via Giotto 1, Padova, Italy*
[2]*Leibniz Center for Marine Tropical Ecology, Fahrenheitstrasse 6, 28357 Bremen, Germany*

Abstract The Trento Plateau (South-Alpine domain of the western passive continental margin of the Apulian plate), during the Middle–Late Jurassic was the site of extended microbialite development in Ammonitico Rosso facies. The setting was a current-swept, pelagic plateau located far from sources of terrigenous supply, characterized by very low average sedimentation rate. Both stromatolites and leiolites occur, commonly associated and passing into one another. For the carbonate-dominated Ammonitico Rosso depositional setting, it may be inferred that microbialites grew in well-oxygenated waters of normal salinity, arguably characterized by very low nutrient availability and probably located in the photic zone. A decimetre-scale basic rhythm occurring in some intervals of the local succession consists of a slightly clayey nodular unit stabilized by a carbonate, microbial unit. Transition from the lower to the upper unit of the rhythm may imply increase in hydrodynamics and oxygenation at the bottom and a progressive reduction in trophic level and bioturbation rate by burrowing metazoans. These rhythms are tentatively attributed to a precessional control.

Keywords Ammonitico Rosso • Middle–Late Jurassic • Pelagic • Stromatolites • Leiolites • Nodular to microbial rhythm • Microbialites • Southern Alps Italy • Cyanobacteria • Trento Plateau

1. Introduction

Microbialites are in steady but punctuated decline throughout the geological time, especially after the Palaeozoic (Kiessling, 2002; Kiessling and Flügel, 2002) and are presently confined to settings generally prohibitive for the life of other benthic organisms. Nevertheless, they still played a role in Mesozoic marine sediments. Their progressive reduction in the Phanerozoic is generally attributed to the progressively increased biodiversity, which may have led to ecological restriction of microbial mats by higher organisms such as grazers and burrowing animals that destroy the delicate structure of the mats.

In the Phanerozoic, microbialites in marine subtidal settings are typically associated with condensed sedimentary sequences. They commonly display a bright-red colour due to the presence of finely dispersed hematite and have been early lithified. Microbialitic structures in the Jurassic Ammonitico Rosso sequences are exposed on the Trento Plateau, a submarine structural high of the South-Alpine domain of NE Italy (e.g., Sturani, 1964; Jenkyns, 1971; Bernoulli and Jenkyns, 1974; Massari, 1979, 1981; Clari et al., 1984; Zempolich, 1993). The palaeogeographic position of this plateau during the Middle–Late Jurassic was that of a current-swept plateau far from continental landmasses and from any terrigenous and platform-derived carbonate sources. Important factors contributing to the widespread occurrence of microbialites undoubtedly were a very low sedimentation rate and the availability of hard substrates (such as early-lithified nodules exhumed at the sediment–water interface, shells or shell fragments, and hardgrounds).

A long-standing controversy concerns the deep versus shallow palaeo-water depth of deposition of these sequences and associated microbialites. The spectrum of palaeo-water depths proposed ranges from the wave base down to the Aragonite Compensation Depth. Jenkyns (1971) favoured a maximum depositional depth of about 200 m but probably considerably shallower, as is the case of many modern and ancient seamount terraces. Deposition in the photic zone was assumed by Bernoulli and Jenkyns (1974). Sturani (1971) suggested a water depth below wave base but not deeper than the base of the photic zone (less than 150 m). Zempolich (1993) suggested that microbial mats were composed of cyanobacteria and/or bacteria operating in the deep photic zone, above deep storm wave base (110–200 m). Some authors argued that the depositional setting was sporadically in the reach of at least major storms (e.g., Massari, 1981; Zempolich, 1993). On the other hand, Winterer and Bosellini (1981), Ogg (1981) and Winterer (1998) argued for a drowning of the Trento Plateau below the photic zone, passing through the aragonite lysocline and aragonite compensation depth. Winterer (1998) argued that stromatolitic domes of the Jurassic Mediterranean Tethys formed at depths ranging from a minimum of 150 m to depths where there was significant or even total dissolution of aragonite. On the basis of a straight-line extrapolation of Pliensbachian–Aalenian subsidence rates for Bajocian to Tithonian times, he estimated for the Ammonitico Rosso of Trento Plateau a depth increase from 200 to 275 m in the Bathonian, up to 350–450 m in the Kimmeridgian. For the reddish cherty and marly limestones of late Callovian Oxfordian, known in the Trento Plateau as Fonzaso Fm, Winterer (1998) proposed a deposition below the ACD, which was estimated by the author at about 400 m at that time.

An unambiguous proof of the actual palaeo-water depth is so far missing. Here, we focus on the Ammonitico Rosso facies and associated microbialites exposed on the Trento Plateau (mostly in the area of Lessini Mountains, north of Verona), in order to contribute to assessing the palaeo-bathymetry of the depositional setting and the environment in which the microbialites developed.

Biomarker data, at present unavailable, would be of critical importance to document the involvement of phototropic cyanobacterial mats. These data will be the subject of a future paper.

2. Geologic and Stratigraphic Setting

From the Early Jurassic on, the subsiding carbonate platforms of the northwestern Tethys underwent rifting and transform faulting, resulting in a complex mosaic of basins, pelagic plateaux and shallow platforms. Swells and basins were subject to faulting along the western passive continental margin of the Apulian plate, which eventually led to the opening of an arm of the Tethys Ocean during the Middle Jurassic (Bernoulli and Jenkyns, 1974; Gaetani, 1975; Winterer and Bosellini, 1981).

The Middle and Late Jurassic generally was a time of condensed pelagic sedimentation on the continental margins of the Tethys situated far from continental areas (Bernoulli and Jenkyns, 1974). Paucity of planktonic microorganisms and very low sedimentation rate may partly account for the condensed nature of the deposits.

Although there are many different facies of red nodular Ammonitico Rosso deposited in variable depositional settings, the focus here is on the condensed facies occurring on the pelagic swell of the Jurassic South-Alpine domain known as Trento Plateau. Here, a condensed and discontinuous pelagic succession of Ammonitico Rosso facies, no more than 30 m thick, bears ammonite faunas of several biozones and stretches from the upper Bajocian to the Tithonian. The Trento Plateau was a horst block of the southern continental margin of the Tethys (Fig. 1) (Winterer and Bosellini, 1981; Ogg, 1981). It evolved from a strongly subsiding Hettangian to Aalenian Bahamian-style carbonate platform to a current-swept plateau with greatly reduced pelagic sedimentation. This submarine ridge was bounded by two basins, the Lombardian Basin to the west and the Belluno Basin to the east. Due to its isolated palaeogeographic position, it received a minimal supply of nutrients and terrigenous clastics when compared to epicontinental settings of the northern Eurasian area (Bernoulli and Jenkyns, 1974). The average sedimentation rate was exceedingly low, on the order of a few millimetre per kyr, and stratigraphic gaps are common.

The fault blocks generated by the rifting show an asymmetry in their stratigraphy, with a thinner sedimentary succession in the west-central part of the plateau that reflects an eastward tilt of the Trento Plateau (Bosellini, 1973). Thin and discontinuous Bajocian sequences of clean-washed *Bositra*-bearing coquinas and crinoidal calcarenites (the so-called *Posidonia alpina* beds auctt.) commonly underlie and in places interfinger with the lowermost part of the Ammonitico Rosso succession (Sturani, 1971). The latter constitutes a formation, the "Rosso Ammonitico Veronese", originally named and introduced in the stratigraphic lexicon by Dal Piaz (1956). (In the following, we use the abbreviation RAV for

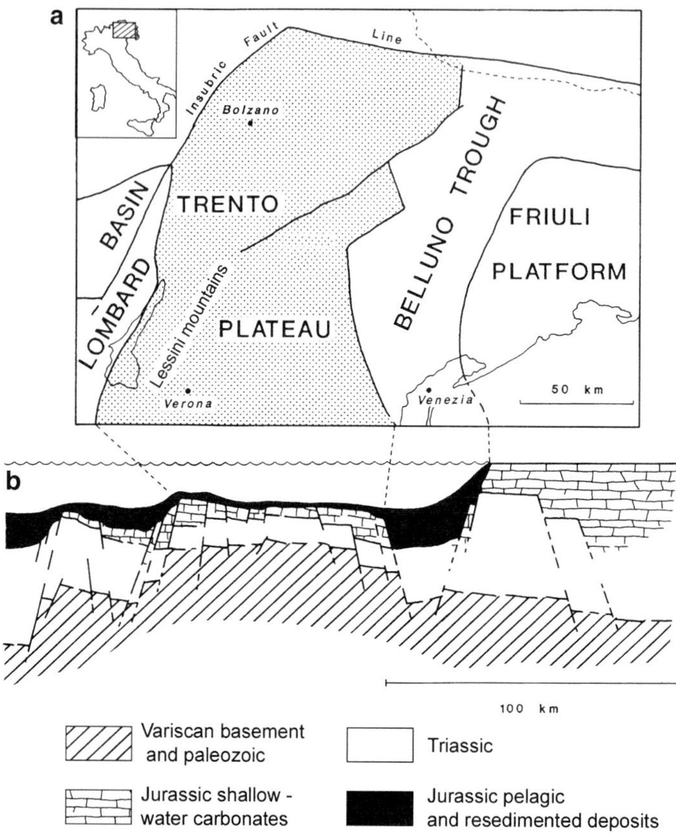

Figure 1. (**a**) Schematic reconstruction of the Jurassic palaeogeography [slightly modified after Martire (1992)]. (**b**) Schematic E-W palinspastic profile. This study is focused on Middle–Upper Jurassic microbialites occurring in the Ammonitico Rosso facies on the Trento Plateau, particularly in the Lessini Mountains area (N of Verona).

"Rosso Ammonitico Veronese" to refer to the studied succession). Red nodular pelagic limestones are also currently described as Ammonitico Rosso according to the Anglo-Saxon usage.

The pelagic limestones of the RAV are subdivided into three recently formalized units (Martire et al., 2006). Its lower unit (Rosso Ammonitico Inferiore, RAI, Aalenian to lowermost Upper Callovian) is generally quite massive, and commonly contains microbialites. The middle unit (Rosso Ammonitico Medio, RAM, Upper Callovian to Middle Oxfordian) is characterized by thin-bedded, planar-parallel- to flaser-bedded limestones locally accompanied by nodules and layers of red chert. Both the lower and middle units locally include sparse beds of clean-washed microcoquinas of thin-shelled bivalves. The Rosso Ammonitico Superiore (RAS, Middle Oxfordian to Upper Tithonian) mostly consists of nodular limestones and commonly contains microbialites in the lower part (Martire et al., 2006).

The locally observed transitional passage from the grainstones of the "*Posidonia alpina*" beds to the RAV (Massari, 1981) suggests a gradual change of depositional setting, with a decrease in environmental energy suggestive of slow, progressive deepening. The limited vertical facies changes within the RAV suggests that, at least on the Trento swell, subsidence significantly slowed at the passage from the thick Lower Jurassic platform carbonates to the condensed Middle–Upper Jurassic pelagic deposits (Martire, 1992). Therefore, a rapid drowning, due to a collapse of the carbonate platform forming the substrate of the RAV, as suggested by Winterer and Bosellini (1981), or a simple straight-line extrapolation of Pliensbachian-Aalenian subsidence rates for Bajocian to Tithonian times, as suggested by Winterer (1998), are unlikely. The fabric and compositional changes in the upper part of the RAS and at the transition to the white pelagic limestones of the Maiolica are probably a signal of increase in sedimentation rate due to the nannoconid bloom.

3. Methods

Microbialite facies were analysed in the field and on polished slabs. Drawings based on field photographs were used to analyse the spatial relationships. In addition, 229 thin sections were examined for microfacies analysis. For SEM analysis, samples were cut perpendicular to the bedding, mounted on SEM stubs and polished with corundum powder. Subsequently, they were cleaned in an ultrasonic bath and etched in 0.1 N hydrochloric acid for 20 s and afterwards coated with gold.

4. Characteristics of the Rosso Ammonitico Veronese

4.1. GENERAL ASPECTS AND FAUNAL CONTENT

The open-marine succession of the Rosso Ammonitico Veronese (RAV) is generally characterized by red colour due to the presence of diffuse hematite and is strongly condensed and typically discontinuous, with common hardgrounds. The oxidized (ferric) state of iron suggests an aerobic depositional setting (holoaerobic sediment). A skeletal micropeloidal microfacies is dominant and ranges from wackestone to packstone and grainstone. A packstone texture typically characterizes the brick-red matrix of the nodular facies, which is slightly argillaceous. Conspicuous elements of the fauna include whole and fragmented ammonites, commonly preserved as steinkerns, belemnite rostra, thin-shelled bivalves, gastropod protoconchs, brachiopods, aptychi, rhyncholites, rare small solitary corals, crinoidal ossicles, and rare echinoids. Microfossil assemblages include both planktonic and benthic elements. Small, thin-shelled bivalves, protoglobigerinids (Middle Oxfordian), and *Saccocoma* (latest Kimmeridgian to early Tithonian) climaxed successively. Benthic foraminifers (*Nodosariidae*, *Ataxophragmiidae*), radiolarians (calcite replaced), calcispheres (*Stomiosphaera*, *Cadosina*) and *Globochaete*, are present with variable abundance. Polyphasic bioclasts that have been subjected to repeated phases of burial, exhumation, and reworking appear commonly bored and iron-stained at the periphery, and sometimes show a coating of a black Fe–Mn oxide crust with small *Frutexites*-like stromatolites.

The role of bottom currents on the submarine plateau controlling the effectiveness of winnowing, bypass, and erosion was emphasized by Jenkyns (1971), Massari (1981) and Martire (1992, 1996). Evidence of bottom current activity includes (1) planed-off hardground surfaces, (2) Fe–Mn mineralization indicative of prolonged sediment bypass, (3) exhumation and local overturn of ammonite steinkerns on the sea floor, as evidenced by stromatolitic cupolas developed on both sides of tests, (4) local faunal reworking and mixing, and (5) differences in lithology between fossil infill and surrounding sediment.

The sedimentation likely took place under oxidizing conditions, as suggested by the great density of burrows, lack of trace fossils assemblages indicative of low levels of oxygenation, and the presence of iron generally as Fe^{3+} in the Fe-hydroxide iron staining and mineralization on lithoclasts and hardground surfaces. Iron sulphide in the form of dispersed framboidal pyrite occurs only in a grey-greenish variety known as "Verdello", that crops out in a restricted area surrounding the town of Trento. This variety is probably to be considered as a meta-aerobic sediment (Fischer et al., 2009), due to a particular depositional environment, leading to a combination of bottom aeration followed by an anaerobic chemical overprint. In this case, microbial activity arguably induces anaerobic conditions leading to the reduction of iron to the ferrous state.

The substratum of the microbial growth, as illustrated below, is represented by any type of hard substrate which stood above the sea floor, such as hardgrounds, early-lithified nodules, larger lithoclasts, lumps, and ammonite moulds (Fig. 2).

Hardgrounds range from planar to quite irregular, locally mineralized surfaces, and are commonly overlain by planar to domed stromatolites. Planar surfaces of hardgrounds truncate underlying stromatolitic domes, ammonite moulds, and other underlying features, and thus appear to result from a process of erosional planation. Erosion was most likely related to the grinding action of coarse sediment forming temporary mobile covers of bioclasts and lithoclasts. These sediment covers may be regarded as thin lag deposits that formed during the omission phase and acted as abrasives when moved by bottom currents as traction bedload across the surface (Goldring and Kaźmierczak, 1974; Fürsich, 1979). The erosional process may have been aided by the heavily destructive action exerted by borers. As argued by Hallam (1967) and Martire (1992, 1996) the hardgrounds are likely to represent stages of sea-level lowstand and minimal bathymetry, an assumption supported by the characteristics of associated sediments, which bear evidence of accelerated winnowing and highest energy level.

In the Trento Plateau whitish to pinkish micro-lumachelle layers are locally interbedded with the predominant microbial and nodular facies, especially in the RAI and RAM, occasionally capping hiatal surfaces. They range in thickness from a few centimetres to 20 cm and consist of clean-washed grainstone to packstone

Figure 2. Alternation of nodular and microbial facies. Drawing from a polished slab. Caprino Veronese quarry, Lessini Mountains (Bajocian).

dominated by thin-shelled bivalves, crinoid ossicles, and micropeloids. These layers locally show normal grading, and planar to hummocky cross stratification, and are locally capped by combined flow ripples or wave ripples with the characteristic internal structure of chevron-like laminae. These features suggest that at least in certain areas of the Trento Plateau, and in certain intervals of the succession characterized by a significant decrease in bathymetry, particularly during the Callovian, the sea bottom was occasionally in the reach of storm-generated flows, at least during most severe events (Massari, 1981; Zempolich, 1993). These occasional events, characterized by high hydrodynamic energy, appear to be superimposed on the background of gentle and more or less constant bottom currents.

Neptunian dykes, either concordant or discordant with respect to the bedding, with infills either passive or forcefully injected, are common in the RAV. They are generally referred to a Callovian phase of block faulting, which created an uneven topography and significant thickness variations and lateral facies changes in the RAM (Martire, 1992, 1996).

4.2. EARLY DIAGENETIC NODULES

Early diagenetic nodules are the most distinctive and widespread feature of the RAV. Although the term nodule has been used by some authors in a comprehensive way (e.g., Martire et al., 2006), we here limit its meaning to sub-spherical to ellipsoidal, or lobate, centimetre-sized, usually pink-coloured carbonate bodies with sharp to faded boundaries. A variably developed suturing and stylolitization at their boundaries occurred during later compactional diagenesis and commonly resulted in irregular shapes.

The nodules may be simple or compound. The former are 4–27 mm across (average 14 mm) and typically consist of wackestone, sometimes preserving relict bioturbation textures inferred to have formed in the soupy-soft stage. This bioturbation is muted and faint, due to a lack of contrast between burrow matrix and fill, and is sometimes recognizable as poorly distinct meniscate burrows related to unidentified deposit feeders of the meiofauna. Its sporadic preservation may have been favoured by the early hardening which froze a texture which otherwise might not survive. Compound nodules (lumps) result from the aggregation of several types of carbonate bodies, such as smaller nodules, small oncoids, and lithoclasts. Their shape ranges from sub-spherical or ellipsoidal to irregular, sometimes lobate, and maximum dimension from 15 to 60 mm.

The nodules usually are uncompacted and maintain a roughly sub-spherical shape, pointing to selective early lithification. The notion of early lithification is supported by a number of aspects: (1) biomoldic voids of, e.g., aragonitic shells that are preserved without deformation; (2) randomly oriented intra-nodule skeletal debris lacking any compaction-related preferred orientation, which is in strong contrast to the compacted fabrics of the matrix; (3) identical texture and bioclast packing compared to those of lithoclasts; and (4) isotopic data (Jenkyns, 1974; Massari, 1981; Clari and Martire, 1996).

Later compaction and pressure solution in the deep burial realm increased the textural contrast between early-cemented nodules and the soft matrix. The brick-red packstone matrix is strikingly different from the nodules in that it shows extensive compactional and dissolution features. Some components present in the nodules, such as radiolarians, protoglobigerinids, and gastropod protoconchs, are absent in the matrix and are thought to have been removed by selective dissolution in the matrix. The skeletal grains in the matrix are much more closely packed than in the nodules and have been subject to intergranular pressure dissolution (Clari and Martire, 1996; Martire, 1996). The absence of diagenetically metastable components in the matrix and the higher concentration of components in the matrix that are inert against early diagenesis demonstrate the early diagenetic differentiation of the sediment into cemented nodules and uncemented matrix (cf. Westphal et al., 2000). The concentration of hematite and indirect enrichment of insoluble argillaceous residue, particularly along solution seams, is clearly the effect of differential behaviour of the soft matrix with respect to the early-lithified parts during the burial diagenesis.

Traces of early bioturbation within the nodules, coupled with common nodule overgrowth by microbial domes, indicate that nodules formed some centimetres below the sediment–water interface and were later exhumed to the sediment–water interface. The selective nodule hardening near the sediment–water interface is regarded as a consequence of the alteration of sediment features by the bioturbation in the soupy/soft stage. It has long been known that endobenthic organisms profoundly alter the properties of the sediments they inhabit, by means of sediment mixing, by varying water content and grain size, by increasing permeability and modifying the consistency of the substrate, by introducing new reactive organic matter in the form of metabolic products, such as mucous secretions and dead organisms, and by influencing microbial populations that mediate chemical reactions, thus creating new biogeochemical microenvironments (Bromley, 1996). In this way, the bioturbation can greatly disturb chemical gradients, such as pH–Eh gradients, and can lead to the intensification of ion exchange in interstitial spaces, so that in a marine phreatic system dominated by current-induced flushing of sea water through the pores of sediments, the bioturbated areas become sites of high cementation potential (see also Kaźmierczak, 1974). The rounded shape suggests that the nodules grew by centrifugal accretion of cement (Bathurst, 1971), starting from the areas where minute burrowing was more intense. Cases of early-lithified concretions exhumed to the sediment–water interface are commonly reported in literature, and their hardening is commonly regarded as an outcome of the changes in sediment characteristics due to bioturbation (e.g., Bathurst, 1971; Kaźmierczak, 1974; Kennedy and Garrison, 1975; Dromart, 1989; Fürsich et al., 1992).

Marlier facies, more common in the RAS, is characterized by ellipsoidal nodules with sharp to faded boundaries. These nodules, which are the typical components of the "nodular facies" of Clari and Martire (1996), show no evidence of exhumation to the sediment–water interface. Although evidence of early cementation was shown by syntaxial cements, as documented by Clari and Martire (1996), the ellipsoidal shape of the nodules with the long axis parallel to bedding suggests that their degree of hardening was incomplete, allowing a somewhat plastic reaction during loading compaction, with some flattening and development of anastomosing dissolution seams. The incomplete lithification may relate to the presence of clay in the sediment. This is in agreement with the observation that clay contents in excess of only a few percent are sufficient to prevent early cementation (e.g., Kennedy and Garrison, 1975). Lateral transitions can be documented within the Trento swell between the marlier facies and the limestone facies subject to early hardening (e.g., Martire, 1996). These differences probably reflect different clay content and net sedimentation rates.

4.3. LITHOCLASTS

Lithoclasts appear as usually angular fragments, commonly with iron-stained and minutely bored periphery. Sometimes they are compound and polyphasic, suggesting an origin by multiple events of burial, exhumation, reworking, and mineralization. Lithoclasts coated by a Fe–Mn crust, sometimes consisting of *Frutexites*-type stromatolites, are particularly common in the most condensed facies. Lithoclasts are found in a range of textures: they may float in the matrix, or occur inside compound nodules or as nuclei of oncoids, or are trapped sometimes within microbial structures.

The common presence of lithoclasts attests to the detachment of fragments from already lithified substrates, such as interspaces between microbial domes, and is therefore one of the main evidence of synsedimentary lithification. A process of fracturing in situ (autobrecciation) of early-hardened substrates is suggested by the large size range of the intraclasts, the common mutually fitting boundaries of adjacent clasts, and local incipient fractures not yet resulting in clast detachment. The genetic mechanism was probably represented in most cases by the formation of early dilatational fractures (Pratt, 1982) due to the pressure exerted by the growing cement crystals.

Another process of lithoclast generation may have been the removal by bottom flows of fragments from early-lithified substrates subjected to heavy bioerosion by boring organisms. This mechanism was probably of subordinate importance and only effective at the expense of lithified hardground surfaces exposed at the sea floor for a long time.

4.4. MICROBIALITES

Originally indistinctly classified into the stromatolites, microbialites were later recognized to show a large spectrum of variations in internal morphology (Kennard and James, 1986; Dill, 1991; MacIntyre et al., 1996; Feldmann and McKenzie, 1998; Braga et al., 1995; Planavsky and Ginsburg, 2009). Following Shapiro (2000), the term macrostructure is used to indicate the gross forms of microbialite bodies, such as columns and domes; mesostructure for those elements visible at the hand-sample scale (centimetre scale), such as laminations in the case of stromatolites, and microstructure for features visible using a light microscope (millimetre scale). Following Planavsky and Ginsburg (2009), the term fabric is used to refer to the morphology and composition of a section of microbialite.

4.4.1. Oncoids

Oncoids are accretionary carbonate bodies with rounded shape, and dimensions comparable to those of nodules. Their core is a simple or compound nodule, a lithoclast, or a fragmented or whole shell. Laminated stromatolitic he mispheroids cover only the upper half of the core during the successive growth stages (Fig. 3a).

Intermittent mechanical perturbations, due to the motion of infaunal organisms or unsteady activity of currents, lead to rotation or upset of the partially coated oncoid, and every time it settles in a new position, a new accretionary stage may begin. This eventually leads to the multiphase coating of the nucleus (Massari, 1983). The variable complexity of the oncoids depends on the duration of their exposure on the sediment–water interface, and therefore, to a certain degree, it is a measure of the stratigraphic condensation.

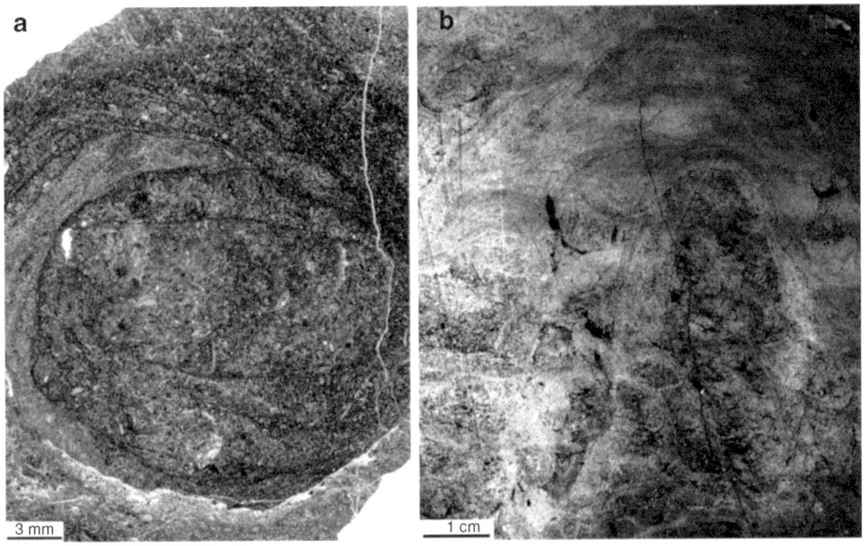

Figure 3. (**a**) Oncoid with asymmetric stromatolitic hemispheroids grown on different sides of a core represented by a bioeroded residual nodule (thin section, positive print). Covolo di Camposilvano section, Lessini Mountains (Callovian). (**b**) Leiolites capped by stromatolites. Note abundance of bioclasts and small vugs in the leiolites (peel, negative print). Rosso Ammonitico Veronese, unidentified quarry of the Lessini Mountains (Bathonian or Callovian).

4.4.2. Domal and Columnar Microbialites

The substratum of the microbial growth is represented by any type of hard substrate which stood above the sea floor, such as hardgrounds, exhumed early-lithified nodules, larger lithoclasts, lumps, ammonite moulds (cf. Dromart, 1989) (Fig. 2). Stromatolites grown on ammonite steinkerns are commonly developed as domes on truncated and planed-off upper shell surfaces. The presence of a hard substrate on which the mats could anchor is clearly a pre-requisite for the microbial mat growth.

Microbialites in the RAV of the Trento Plateau reveal large variations in internal fabrics, ranging from distinctly and regularly laminated to irregularly laminated and even structureless. On the basis of meso- and microstructure we distinguish stromatolites and leiolites. The term leiolite was introduced by Braga et al. (1995) to indicate a grain-dominated structureless microbialite lacking clear lamination or clots, containing in addition fenestrae and micrite, and was also drawn on by Dupraz and Strasser (2002) to indicate a microbialite with micropeloidal to dense microfabric. Here, we attribute to leiolites domed and columnar microbialites ranging from a few centimetres to 13 cm in height, characterized by non-laminated to patchily laminated micropeloidal and bioclastic, vuggy fabric. The contact with the surrounding deposits is usually sharp. Even though they are commonly non-laminated, they show variable microfabrics, sometimes with patchy, discontinuous, and irregular laminations, generally indistinct and locally anastomosing, which appear to have been disrupted by the bioturbation, and an abundance of small irregular vugs. The components include micropeloids and fine-grained to sand-sized bioclasts. The latter are commonly larger than those contained in the stromatolites and are sometimes associated with small nodules and lithoclasts. Bioturbation is recorded by subcircular to elliptic and sometimes elongate sections, sometimes accompanied by meniscate traces, with infill texturally and/or compositionally different from the surrounding sediment. Other evidence of bioturbation is provided by disruption (truncations, embayments) of the lamination. The microfabric apparently results from the disruptive influence of burrowing metazoans. Vugs may be depositional irregularities, or spaces formerly occupied by algae, or umbrella structures occurring below convex-up bivalve shells.

Leiolites and stromatolites occur either separately or, more commonly, as associated microbialites grading into one another. Although variable upward transitions into one another may be locally observed, the most common vertical transition is from leiolite to stromatolite (Fig. 3b), the latter occurring as a capping laminated drape on a leiolite column or dome. The frequent association of leiolites and stromatolites, common vertical transitions into one another, and presence of transitional terms between them, suggest that these microbial forms can develop in comparable environmental conditions. The leiolites probably differ from stromatolites because of a greater importance of microfabric disruption by burrowing metazoans. In this sense, we share the opinion of Planavsky and Ginsburg (2009) that a remodelling of a precursor fabric is involved, and that it would be preferable to use the general term microbialite, since the differences between the two forms are apparently linked to variations in the degree of metazoan overprinting, rather than to differences in the involved ecosystems.

Thrombolites were reported by Zempolich (1993) on the eastern margin of the Trento Plateau. However, typical thrombolites with the characteristic clotted micropeloidal microfabric (Kennard and James, 1986) were not identified.

The macrostructures of the stromatolites are variable. According to their growth pattern they can be differentiated into (1) unlinked, vertically stacked hemispheroids (SH, according to the acronyms suggested by Logan et al., 1964), (2) laterally linked gently convex domes (h/l of 0.25–0.3) (laterally linked hemispheroids, LLH), (3) columns, and (4) stratiform, planar-laminated structures (most commonly developed on planed-off hardgrounds) (Fig. 4).

Transitions in stromatolite macrostructures occur both vertically and laterally. Domes and columns are the most common types. The domes range in width from 2 cm (isolated domes above nodules) to 20 cm (domes developing on hardground surfaces) and are 2–8 cm in height. The columns have comparable height (exceptionally up to 13 cm), and are 2–6 cm wide at the base. This range of macrostructures would suggest growth in moderate to high current velocities (Jenkyns, 1971). In particular, domal hemispheroids growing separately probably formed where lateral linkage was inhibited by the energy level

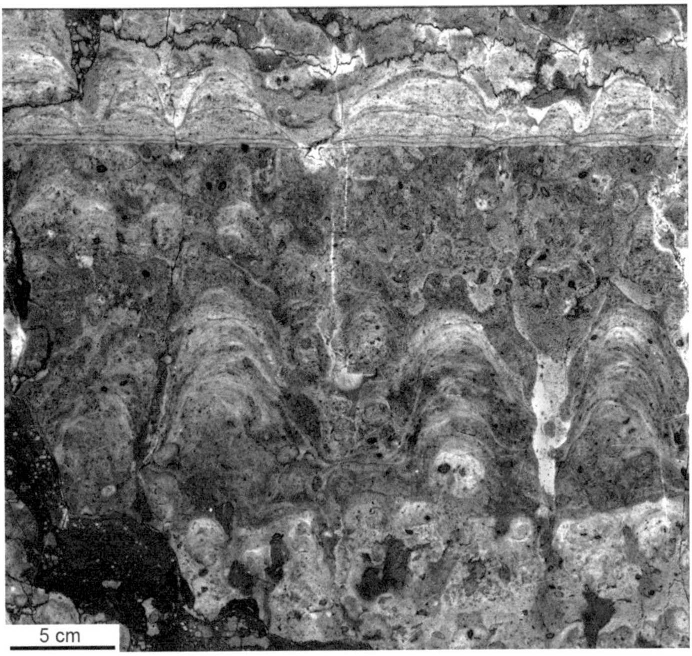

Figure 4. Stromatolite columns in the central-lower part are underlain by irregular leiolite columns. At the top a planar erosional hardground is covered by planar stromatolites grading upwards into domed stromatolites. Rosso Ammonitico Veronese, polished slab from an unidentified quarry of Lessini Mountains, probably S. Ambrogio Valpolicella area (Bathonian or Callovian).

of the environment, preventing growth in the interdomal depressions that were actively swept by currents.

Linked domes appear commonly grouped and coalesced to form clusters around 15–25 cm wide in plan view, showing lobate to polygonal boundaries, and locally forming small mounds with a relief of a few centimetres. Slabs cut parallel to the bedding surfaces at the level of microbial domes show a characteristic polygonal pattern of the interspaces between coalesced clusters, especially evident in the case of densely packed microbial structures.

Stromatolites overlying hardground surfaces show either domed or planar-laminated macrostructures, sometimes passing laterally or upwards into one another. The domes of hardground-covering stromatolites are significantly wider at their base (up to 15 cm, in exceptional cases up to 30 cm) in comparison to stromatolitic domes capping nodules or ammonite steinkerns in the stratigraphic intervals between hardgrounds. Moreover, the stromatolites overlying hardgrounds, show significantly thinner, finer-grained and more regular laminae, possibly reflecting a considerably lower net sedimentation rate (cf. Dupraz et al., 2006; Zamagni et al., 2009).

The basic stromatolitic microstructure is represented by alternating thin laminae of dense micrite (thickness 0.05–0.14 mm) and thicker laminae of granular

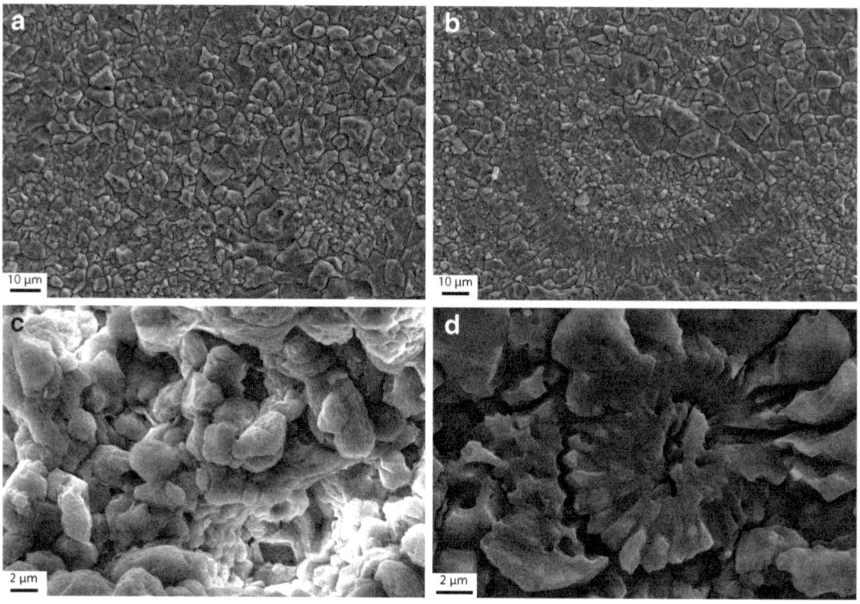

Figure 5. SEM micrographs of polished and etched surfaces of Rosso Ammonitico Veronese, Snell section, near Roverè, Lessini mountains (Bathonian). (**a**) Typical appearance of the microbialites. Heterogeneous distribution of micritic and microsparitic areas. The distribution of the micritic areas is reminiscent of a peloidal structure. (**b**) Microbialite with enclosed shell. Microspar cement is enclosing the shell and dominates the micrograph. (**c**) Micrite-sized carbonate grains. The *rounded shapes* are reminiscent of a microbial origin. (**d**) Coccoliths enclosed in the tightly cemented microbialite.

packstone to grainstone (thickness 0.3–2.6 mm). These packstone to grainstone laminae consist of a silt-sized peloids with well-defined contours, thin-shelled bivalves, protoglobigerinids, radiolarians, *Globochaete*, gastropod protoconchs, etc., in a micrite or microspar matrix. No preserved filaments or cells were observed. Scanning electron microscopy reveals that the stromatolites are micrites with most grains <4 µm in diameter. However, the distribution of crystal grain sizes is heterogeneous and locally microspar crystals >4 µm in diameter dominate (Fig. 5).

The fabric ranges from a tight mosaic, particularly in the microspar areas, to a loose mosaic in the micrite areas. Enclosed in this matrix are thin shells (bivalves and ostracods) and calcareous plankton (coccoliths). Cement crystals of micrite to microspar size grow on the skeletal material. The mosaic texture and the incorporation of grains points to continuing cementation of the stromatolites as is typical for modern microbialites (Westphal et al., 2010). Locally, nests of tiny round carbonate grains of 1 µm in diameter occur that are reminiscent of bacterial products. Other petrographic features that would directly reflect a microbial origin of the stromatolites, such as peloidal aggregates, cell chains, or filaments, have not been found. One reason for the lack of direct microstructural evidence of microbes could be that most microbes induce extracellular calcification through their metabolism or the extracellular substances they produce (Riding, 2000; Arp et al., 2001; Dupraz et al., 2004) instead of calcifying themselves. Continuing cementation further tends to obscure direct evidence. The heterogeneity, the micritic texture, and the incorporated components in the stromatolites are similar to those of modern microbialites (cf. Westphal et al., 2010) and are in agreement with, but nevertheless insufficient proof of, microbial activity.

The inferred mode of growth of the stromatolites was by trapping and binding of different kinds of particles into gravity-defying attitudes (agglutinated stromatolites, Riding, 2000). Trapped sediment particles are remarkably finer-grained and better-sorted than those accumulating in the interdome spaces, indicating that the process was selective: the microbial mucilages and meshes were able to hold only the finer fractions, as only particles entrained in suspension can be shifted to, and trapped by, the microbial mats. Tapering, wavy laminae and irregular lamination due to laterally variable thickness of the granular laminae are common (Fig. 6a).

Locally, larger particles such as lithoclasts with common peripheral iron staining, small nodules, or oncoids, are trapped within the granular laminae. The thickness, texture, and composition of the granular laminae probably vary as a function of the depositional rate and energy of the hydrodynamical agents.

Evidence of pauses in the growth of the stromatolitic structures is sometimes given by interlaminar micro-unconformities interrupting the growth sequence and truncating the underlying laminae; these internal discontinuity surfaces are commonly micro-bored and iron-stained.

Attached sessile micro-encrusters such as thecideacean brachiopods, serpulid tubes, and crinoid holdfasts rarely appear intergrown between the laminae and suggest competition for substrate space with the microbial mats; it is clear,

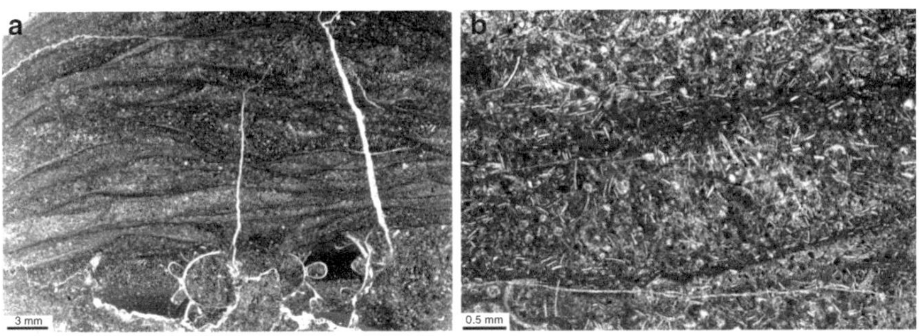

Figure 6. (**a**) Stromatolitic cap developed on an ammonite mould erosionally truncated on the upper side. The stromatolitic laminae show high lenticularity. Note the polyphasic internal sediments inside the ammonite mould, emplaced before its truncation. Rosso Ammonitico Veronese, Covolo di Camposilvano section, Lessini Mountains (Callovian). (**b**) Differences in the orientation of elongate particles within granular and micritic laminae of stromatolites. Random orientations of particles, commonly at high angles to the surface of the growing structure, and inverse grading occur in the granular laminae, whereas an orientation parallel and tangential to the mat surfaces predominates in the micritic laminae. Rosso Ammonitico Veronese, Chiusa section, Lessini Mountains (Oxfordian).

however, that the latter were ecologically more successful in the Ammonitico Rosso environment. The microbial mats were sometimes hosts to organisms that appear to have lived inside or between the microbial mats. These organisms have caused local embayments or truncation of the laminae. The flexibility of the laminae to these disruptions indicates that the penetration occurred prior to lithification. However, these effects of disruption are remarkably subordinate when compared to those displayed by the leiolites.

It is generally agreed that the formation of well-laminated stromatolites is induced by filamentous forms of microbes (Bathurst, 1971; Hoffman, 1976), which commonly occur as colonies of filaments densely arranged in mucilaginous films. The filaments consist of the cellular trichome in a thick mucilaginous sheath, to which sedimentary grains adhere. Because of their motility, they can glide up through a layer of sediment recently deposited on them and re-establish a new surface mat. In this way, laminated fabrics can result from concurrent algal growth and sediment deposition.

Granular and micritic laminae commonly show remarkable differences in the orientation of elongate particles, such as fragments of thin-shelled bivalves (Fig. 6b). These particles tend to arrange themselves with random orientations in the granular laminae, commonly at high angles to the mat surface, and sometimes also edgewise; moreover, the particles in these laminae may show an inverse grading. In contrast, in the micritic laminae the elongate particles commonly show an orientation parallel and tangential to the mat surfaces. A rhythmic alternation of trapping and binding mechanisms is suggested in the generation of the granular and micritic laminae, respectively, with a change in habit of the mat from erect filaments, with particles entangled in rapidly growing filaments, to a condition of a virtually smooth mat. In the first stage, possibly in response to phases of suspended sediment input, the growing mat incorporates sediment particles falling in between the erect bundles of the mesh of filaments; as a result, the particles rest without any preferred orientation or even edgewise, possibly with a selection of sizes leading to inverse grading. In the second stage, possibly characterized by reduced input of suspended sediment and low growth rates, the mat was presumably in a condition of smooth surface (prostrate filaments?), allowing elongate free-falling particles to be retained by the sticky surface of the mat. As a result, they are bound with an orientation predominantly parallel to the growth surface of the mat. It may be envisaged that this alternation results from a rhythmic growth pattern reflecting a balance between mat growth and sedimentation rate.

This inferred mode of growth is surprisingly similar to that reported by Noffke et al. (1997) in the modern siliciclastic tidal flats of North Sea. The authors describe small-scale, bipartite depositional units a few millimetres thick. These consist of a lower granular part, normally graded, containing randomly oriented quartz grains deposited from suspension by flood events of declining energy, and an upper organically dominated layer built by microbial mats

---

?indicates that it is possible but not proven that this is prostate filaments.

(cyanobacteria associated with benthic diatoms, sulphur bacteria, and several other bacterial groups). Following the decline of the hydrological energy, microphytobenthos colonizes the freshly sedimented layers, forming an organically dominated layer, onto which quartz grains settle down and become bound in the organic matrix with their long axes parallel to the bedding planes. There is no reworking and re-suspension of deposited grains during subsequent periods of increased flow, since the microbial mats increase the critical shear stress required to re-mobilize the grains, and therefore prevent erosion.

The growth rate of microbialites has been assumed by Martire (1992) and Martire and Clari (1994) to have been on average as low as a few millimetres per kyr. However, as noted by Jenkyns (1971) and Reid et al. (2000), favourable conditions for stromatolite growth may have occurred only for short periods of time, with phases of relatively high growth rates separated by long pauses. It may be speculated that fluctuating processes controlling the stromatolite growth are represented by changes in physico-chemical conditions of the bottom waters, planktonic productivity in the higher parts of the water column, bottom current action, sediment contribution, or nutrient supply.

It is likely that microbialites formed a hard, lithified structure in an early stage. This is suggested by the local presence of attached sessile micro-encrusters and the observation that without cementation the steep walls of the columns and domes would have collapsed. The microbial structures do not generally exhibit evidence of boring, possibly because of the protective living algal mat cover. In addition, microborers appear not to prefer microbialites as substrates, possibly because of the highly porous structure of the substrate (Heindel et al., 2009). It is suggested that the inherent ability of the microbial community to induce precipitation of a hard fabric (carbonate cement) could greatly reduce the disruptive effect of burrowers and grazers.

Locally, particles in the granular laminae of stromatolites show a re-organization and geopetal arrangement, suggesting that the cementation was differential, affecting the micritic laminae, while leaving, at least for a certain time, the granular laminae in an incoherent state. This suggests that a temporary condition of alternating lithified and non- or poorly lithified laminae existed, with laminae of dense micrite possibly representing biofilm calcification (Riding, 2000). Despite the rigidity of the structure, it may be envisaged that the stromatolites could develop high porosities and permeabilities, due to the removal of organic material through decay and oxidation. As a result, even weak currents can move through the structure, and the process may lead to rearrangement of the particles of the granular laminae (cf. Bathurst, 1971).

Early cementation of the microbial structures was probably induced by particular micro-environmental geochemical or biochemical conditions. It was probably a result of the action of anaerobic bacteria in the lower reducing zones of the mat (Bathurst, 1971), most probably sulphate-reducing bacteria (Morse and MacKenzie, 1990). This seems to be supported by the local occurrence of framboidal pyrite in the so-called Verdello variety of Ammonitico Rosso.

A combination of heterotrophic bacterial activities and degradation of biofilms rich in extracellular polymeric substance (EPS), releasing Ca^{2+}, has been proposed (Visscher et al., 2000) to lead to typically micritic or slightly coarser precipitation (Riding, 2000). Cementation probably occurred mainly during hiatal intervals that may be expressed by the omission surfaces regularly occurring at the top of the microbial structures (see below). In modern marine stromatolites, Reid et al. (2000) observed that development of surface films of exopolymer and subsequent heterotrophic bacterial decomposition forming thin crusts of microcrystalline carbonate mainly occur during discontinuities in sedimentation, and that further modification of the sediment, with formation of thicker lithified laminae by climax communities including endolithic coccoid cyanobacteria, occur during prolonged hiatal periods.

5. Basic Rhythms

In some intervals of the RAV the microbial structures are arranged in small-scale rhythms that form characteristic couplets ranging in thickness from 6 to 25 cm (Massari, 1979, 1981; Martire, 1996). A first stage, corresponding to the development of the lower unit of the couplet, is characterized by the formation of a pavement mostly consisting of nodules, to which other carbonate bodies of variable nature may be associated, such as lumps, lithoclasts, ammonite moulds, and oncoids. In the second stage, this pavement is stabilized by a microbial cover, which represents the upper unit of the couplet (Figs. 2 and 7).

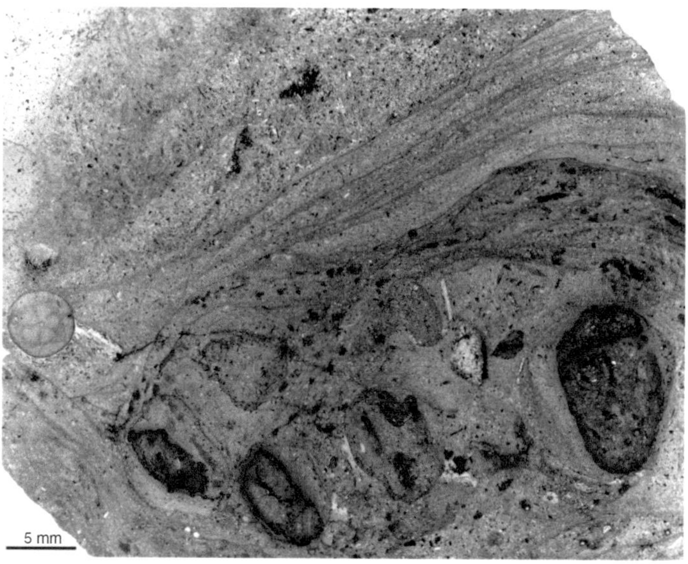

Figure 7. Basic rhythm showing a stromatolite drape "stabilizing" a pavement formed by several types of carbonate particles, among which are nodules, oncoids, and lithoclasts. Rosso Ammonitico Veronese, Piccola Mantova section, near Boscochiesanuova, Lessini Mountains (Callovian).

These rhythms are not ubiquitous and commonly obscured by burial compaction and stylolitization; they are most common and eye-catching in the lower part of the RAS, where microbialites occur as laterally linked domes growing on nodular (*lato sensu*) pavements. The rhythmic pattern is apparently missing or non-apparent (1) where microbial domes are smaller and unlinked or discontinuously linked, and (2) in the most condensed intervals that consist entirely of stromatolites.

The followings steps can be reconstructed in the generation of the basic rhythms (Fig. 8).

In the initial, soupy/soft state, the sediment areas burrowed by small, mobile deposit feeders of the meiofauna are inferred to be the preferential sites of gradual early lithification of the nodules some centimetres below the sediment–water interface (Fig. 8a).

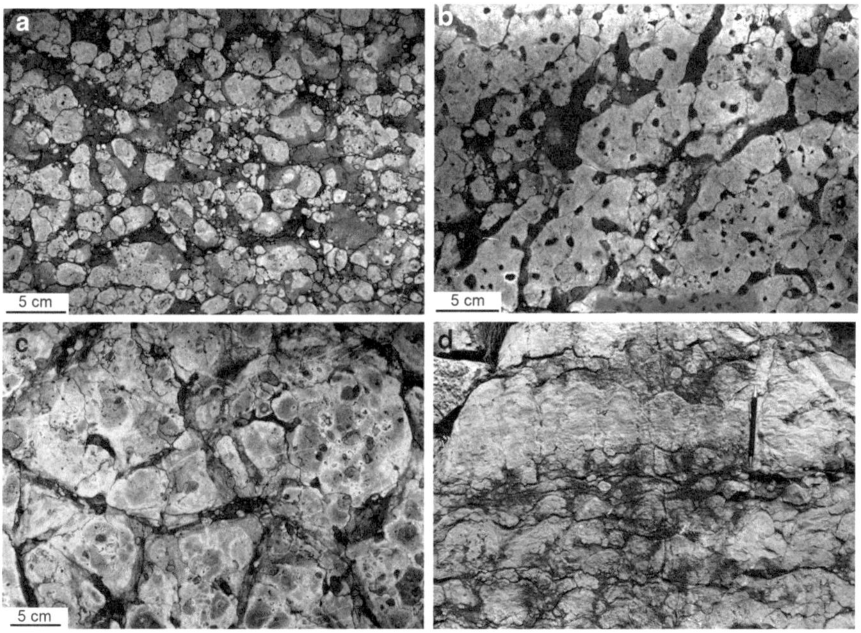

Figure 8. The structure of the basic rhythm is illustrated by three sections cut parallel to the bedding surfaces (**a–c**) at progressively higher levels of the rhythm. (**a**) Loose nodular facies. (**b**) First stage of the mat growth, with vertical *Thalassinoides* shafts (visible as rounded sections) mostly forced to localize in the interspaces between early-lithified parts. (**c**) Final stage of mat growth, with clusters of microbial domes bounded by a polygonal pattern of interspaces and sealing most of *Thalassinoides* shafts. Rosso Ammonitico Veronese, polished slabs from an unidentified quarry of Lessini Mountains. (**d**) Basic rhythms in section perpendicular to bedding. Erbezzo section, Lessini Mountains (Kimmeridgian) (pencil 14 cm long for scale).

The following step is the exhumation of the nodules, together with other hardened particles, on the sediment–water interface, probably due to current-induced winnowing of the interstitial unconsolidated sediment and/or turnover of the sediment by the activity of burrowing organisms. Bromley (1996) supports McCave's (1988) observation that diffusive bioturbation of fine-grained sediment is able to displace gravel-sized clasts in an upward direction, due to reduced shear strength and confining pressure upwards.

Once the nodules and locally associated hardened particles are exhumed, microbial growth on their top surfaces can take place, producing either oncoids through periodic displacements, or simple, upward growing domes in the case of stationary cores. At this stage, the bottom consistency is that of a firm ground and *Thalassinoides* burrows, 1–2 cm in diameter, probably produced by dwelling, filter-feeding thalassinidean Crustacea, may begin to create a network (Fig. 8b). The course of the burrows is deflected around the already lithified nodules and at the buried hardened surfaces of earlier rhythms (stenomorphic burrows in Bromley, 1975).

As above noted, burrowing activity by metazoans could partly continue during the growth of microbialites, particularly during the development of leiolite-type microbialites. Also *Thalassinoides* systems survived during the first stages of the growth of the mats. However, freedom of movement of their trace-makers was progressively restricted by the progressing lithification, the vertical shafts being mostly forced to localize in the interspaces between single microbial domes or clustered domes, although still mostly connected at their top with the sediment–water interface. With the development of laterally linked domes in the last stages of mat growth, the interspaces of smaller domes are bridged due to the tendency of the microbial films to extend over larger surfaces (cf. Olivier et al., 2004). The commonly observed upward-transition from leiolites to stromatolites is thought to record a progressive reduction and eventual cessation of bioturbation by metazoans in the upper unit of the basic rhythm. The final stage of mat growth is commonly characterized by clusters of microbial domes bounded by a polygonal pattern of interspaces (Fig. 8c). Vertical shafts of *Thalassinoides* are mostly sealed by the growing mat, with only a few persisting residual shafts still in contact with the sea water. Sealed shafts may appear incompletely infilled with sediment, with one or more voids left at the top, later occluded by clear, blocky calcite cement. These voids probably result from post-depositional volume reduction of the sediment fill by dewatering and particle settlement, not by burial compaction, prevented here by the rigidity of early-lithified walls. The appearance of the basic rhythms in field outcrops is shown in Fig. 8d.

It is suggested that the transition from the lower to the upper layer of the rhythm is accompanied by a change from soupy/soft ground to firm ground and eventually hardened bottom in the stage of microbial mat cementation, the latter marking a temporary interruption of the sedimentation, i.e. an omission stage. The concept of alternating softground and hardground stages was already illustrated by Hollmann (1961).

Moreover, the hypothesis of regularly fluctuating hydrodynamic conditions as one of the basic factors responsible of the small-scale rhythms of the RAV deserves consideration. The importance of the current activity as a primary controlling factor in the genesis of the Ammonitico Rosso facies has been stressed by Jenkyns (1971, 1974), Massari (1979), Ogg (1981), Comas et al. (1981), Clari et al. (1984), and Martire (1992). Due to the Bernoulli effect, constriction of bottom (tidal?) currents over the shallower floor of the Trento Plateau laterally bounded by two basins presumably resulted in an increase in velocity of the flow. This situation ensured a certain continuity of bottom currents during the rhythm development. The evidence provided by the structure of the basic rhythm suggests that bottom currents were subject to fluctuating energy. The lowermost energy level was probably experienced in the first stage, when the sediment was still in a soupy/soft state and bioturbated by minute deposit feeders. The limited influx of terrigenous fines and associated input of nutrients in this stage probably resulted in a relative increase of the trophic level, although presumably at a low degree.

It may be speculated that in the later stage, the settling of the *Thalassinoides* network and the subsequent bottom stabilization by microbial structures was accompanied by enhancement of the bottom current velocity and possibly a certain increase in bottom oxygenation. Bathurst (1971) found that the subtidal mats on the Bahamas can withstand current velocities at least twice as high and in some cases five times as high, as those that eroded the unbound sediment. The largest values of current velocity may be reached during the omission stage, during which the non-deposition was conceivably accompanied by a process of bypass or erosion.

Condensed sequences of the Trento Plateau are almost pure limestones, whereas coeval basinal deposits are much marlier, suggesting sediment fractionation with clay minerals swept into troughs (Jenkyns, 1971), where marl-limestone rhythms typically formed (e.g., the Sebino trough of the Lombard basin in the Kimmeridgian-Tithonian, Winterer and Bosellini, 1981). Ogg (1981) reported the presence in the RAS of rhythms with couplets of 20–30 cm thickness, with a marlier lower layer with sparse small nodules, grading upwards into a layer with larger nodules separated by marl seams, and finally into a limy, stylolitized layer. These findings indicate that the nodular-microbial rhythms typical of the shallower areas of the Trento Plateau, particularly in the lower part of the RAS, passed to marl-limestone rhythms in deeper settings and basinal areas. Marl-limestone rhythms, occurring in settings deeper than those of coeval nodular limestone lithofacies, are reported from the Upper Jurassic of the French Subalpine Basin (Dromart, 1989), where the marl interbeds contain oncolites and isolated microbial knobs whereas the limestone beds contain columnar stromatolites. Based on compositional difference of clay minerals between limestone beds and marl interbeds, the author attributes the rhythm to climatic changes that influenced the quantity and quality of the clay influx from the land with a periodicity controlled by the precession.

? indicates that the currents are possibly tidal but there is no final proof.

6. Discussion

Modern microbialites do not seem to be limited by the low nutrient concentrations occurring in their natural habitats (Pringault et al., 2004). In the carbonate-dominated Ammonitico Rosso depositional setting, located far from sources of terrigenous supply, and arguably characterized by very low nutrient availability, very low average sedimentation rates, and normal salinity, as faunal evidence clearly indicates, it is suggested that microbialites grew in oligotrophic, well-oxygenated waters of normal salinity, still located in the photic zone.

An environment not excessively deep is suggested by the fact that the sea bottom was occasionally in the reach of major storm-generated flows (Massari, 1981; Zempolich, 1993). Similar shallow depth and photic zone conditions were inferred for the Upper Jurassic red nodular limestones in Rosso Ammonitico facies of the Subbetic (S Spain) (Comas et al., 1981).

It may be speculated that the fluctuations recorded by basic nodular to microbialitic rhythms are controlled by the precession (cf. Dromart, 1989; Olivier et al., 2004). Martire and Clari (1994) calculated a sedimentation rate of 3.2 mm/kyr for the Rosso Ammonitico Veronese of the Trento Plateau, after evaluating both effects of compaction and major sedimentary gaps, i.e. those with duration quantifiable on a biostratigraphical basis. In the hypothesis of precessional control on the basic rhythmicity, this figure of sedimentation rate would imply an accretion rate of the sea floor by about 6 cm during a single rhythm. Since the thickness of basic rhythms in the RAV ranges 6–25 cm, one should conclude (1) that in the evaluation of the sedimentation rate, the time represented by the countless omission surfaces bounding the basic rhythms, although of non-quantifiable duration, should be taken into account, in addition of the time represented by more important gaps; (2) that, in this perspective, the sedimentation rate during the formation of basic rhythms might have been significantly higher than the value calculated by Martire and Clari (1994). This would imply a surprising scenario of relatively fast growth of microbialites followed by long omission periods, a possibility already considered by Jenkyns (1971) and Reid et al. (2000).

7. Conclusions

The most important factors allowing for the widespread occurrence of microbialites in the Rosso Ammonitico Veronese is thought to have been the presence of hard substrata and a very low average sedimentation rate. It is suggested that microbialites grew in oligotrophic, well-oxygenated waters of normal salinity, still located in the photic zone.

Concerning the factors involved in the genesis of the rhythmic nodular-microbial couplets, it may be inferred that the transition from the lower slightly clayey, nodular unit, to the upper carbonate, microbial unit may imply increase in hydrodynamics and oxygenation at the bottom, and progressive reduction in trophic level and burrowing by metazoans. A precessional control is assumed, resulting in short-term climatic variations that influenced the hydrodynamic regime of currents and the nature and amount of sediment influx, although fluctuations in the input of fine-grained terrigenous sediments and associated nutrients were arguably very weak during the deposition of the carbonate-dominated Rosso Ammonitico Veronese. Previous evaluations of the sedimentation rate of this pelagic facies (Martire and Clari, 1994) did not take into consideration the time represented by the countless omission surfaces bounding the basic rhythms. In the hypothesis of a precessional control of the rhythms, this new perspective opens a surprising scenario of relatively fast growth of microbialites followed by long omission periods.

8. Acknowledgements

We are grateful to the referees L. Martier and N. Nofke who provided very helpful feedback and to the editors of the book, particularly to Professor Vinod C. Tewari for his continuous and patient guidance.

9. References

Arp, G., Reimer, A. and Reitner, J. (2001) Photosynthesis-induced biofilm calcification and calcium concentrations in Phanerozoic oceans. Science **192**: 1701–1704.

Bathurst, R.G.C. (1971) *Carbonate Sediments and Their Diagenesis. Developments in Sedimentology 12*. Elsevier Scientific Publishing Company, Amsterdam, 620 pp.

Bernoulli, D. and Jenkyns, H.C. (1974) Alpine, Mediterranean, and Central Atlantic Mesozoic facies in relation to the early evolution of the Tethys, In: R.H. Dorr and R.H. Shaver (eds.) *Modern and Ancient Geosynclinal Sedimentation. SEPM Special Publications 19*. Society for Sedimentary Geology, Tulsa, pp. 129–160.

Bosellini, A. (1973) Modello geodinamico e paleotettonico delle Alpi Meridionali durante il Giurassico-Cretacico. Sue possibili applicazioni agli Appennini, In: Accordi et al. (eds.) *Moderne vedute sulla geologia dell'Appennino. Accad. Naz. Lincei 183*, pp. 163–205.

Braga, J.C., Martin, J.M. and Riding, R. (1995) Controls on microbial dome fabric development along a carbonate-siliciclastic shelf-basin transect, Miocene, SE Spain. Palaios **10**: 347–361.

Bromley, R.G. (1975) Trace fossils at omission surfaces, In: R.W. Frey (ed.) *The Study of Trace Fossils*. Springer-Verlag, New York, pp. 399–428.

Bromley, R.G. (1996) *Trace Fossils – Biology, Taphonomy and Applications*. Chapman & Hall, London.

Clari, P.A. and Martire, L. (1996) Interplay of cementation, mechanical compaction, and chemical compaction in nodular limestones of the Rosso Ammonitico Veronese (Middle–Upper Jurassic, northeastern Italy). J. Sediment. Res. **66**: 447–458.

Clari, P.A., Marini, P., Pastorini, M. and Pavia, G. (1984) Il Rosso Ammonitico Inferiore (Baiociano-Calloviano) nei Monti Lessini Settentrionali (Verona). Riv. Ital. Paleont. Strat. **90**: 15–85.

Comas, M.C., Oloriz, F. and Tavera, J.M. (1981) The red nodular limestones (Ammonitico Rosso) and associated facies: a key for settling slopes or swell areas in the Subbetic Upper Jurassic submarine topography (southern Spain), In: A. Farinacci and S. Elmi (eds.) *Rosso Ammonitico Symposium Proceedings*. Edizioni Tecnoscienza, Roma, pp. 119–136.

Dal Piaz, G. (1956) "Rosso Ammonitico di Verona", In: Lexique Stratigraphique International, 1 Europe, (11 Italie) – 87, Congr. Géol. Int. Mexico, Paris.

Dill, R.F. (1991) Subtidal stromatolites, ooids and crusted-lime muds at the Great Bahama Bank margin, In: R.H. Osborne (ed.) *From Shoreline to Abyss*. SEPM, Society for Sedimentary Geology, Tulsa, pp. 147–171.

Dromart, G. (1989) Deposition of Upper Jurassic fine-grained limestones in the western Subalpine Basin, France. Palaeogeogr. Palaeoclimatol. Palaeoecol. **69**: 23–43.

Dupraz, C. and Strasser, A. (2002) Nutritional modes in coral-microbialite reefs (Jurassic, Oxfordian, Switzerland): evolution of trophic structure as a response to environmental change. Palaios **17**: 449–471.

Dupraz, C., Visscher, P.T., Baumgartner, L.K. and Reid, R.P. (2004) Microbe–mineral interactions: early carbonate precipitation in a hypersaline lake (Eleuthera Island, Bahamas). Sedimentology **51**: 745–765.

Dupraz, C., Pattisina, R. and Verrecchia, E.P. (2006) Translation of energy into morphology: simulation of stromatolite morphospace using a stochastic model. Sediment. Geol. **185**: 185–203.

Feldmann, M. and McKenzie, J.A. (1998) Stromatolite-thrombolite associations in a modern environment, Lee Stocking Island, Bahamas. Palaios **13**: 201–212.

Fischer, A.G., Hilgen, F.J. and Garrison, R.E. (2009) Mediterranean contributions to cyclostratigraphy and astrochronology. Sedimentology **56**: 63–94.

Fürsich, F.T. (1979) Genesis, environments and ecology of Jurassic hardgrounds. N. Jb. Geol. Palaont. Abh. **158**: 1–63.

Fürsich, F.T., Oschmann, W., Singh, I.B. and Jaitly, A.K. (1992) Hardgrounds, reworked concretion levels and condensed horizons in the Jurassic of western India: their significance for basin analysis. J. Geol. Soc. London **149**: 313–331.

Gaetani, M. (1975) Jurassic stratigraphy of the Southern Alps: a review, In: C. Squires (ed.) *Geology of Italy*. Earth Sciences Society of the Libyan Arab Republic, Tripoli, pp. 377–402.

Goldring, R. and Kaźmierczak, J. (1974) Ecological succession in intraformational hardground formation. Paleontology **17**: 949–962.

Hallam, A. (1967) Sedimentology and palaeogeographic significance of certain red limestones and associated beds in the Lias of the Alpine region. Scottish J. Geol. **3**: 195–220.

Heindel, K., Wisshak, M. and Westphal, H. (2009) Microbioerosion in Tahitian reefs: a record of environmental change during the last deglacial sea-level rise (IODP #310). Lethaia **42**: 322–340.

Hoffman, P. (1976) Stromatolite morphogenesis in Shark Bay, western Australia, In: M.R. Walter (ed.) *Stromatolites. Developments in Sedimentology 20*. Elsevier Scientific Publishing Company, Amsterdam, pp. 261–271.

Hollmann, R. (1961) Über Subsolution und die "Knollenkalke" des Calcare Ammonitico Rosso Superiore in Monte Baldo (Malm; Norditalien). Neues Jb. Geol. Paläont. Abh. **119**: 163–179.

Jenkyns, H.C. (1971) The genesis of condensed sequences in the Tethyan Jurassic. Lethaia **4**: 327–352.

Jenkyns, H.C. (1974) Origin of red nodular limestones (Ammonitico Rosso, Knollenkalke) in the Mediterranean Jurassic: a diagenetic model, In: K.J. Hsu and H.C. Jenkyns (eds.) *Pelagic Sediments: On Land and Under the Sea. International Association of Sedimentologists Special Publication 1*. International Association of Sedimentologists, pp. 249–271.

Kaźmierczak, J. (1974) Crustacean associated hiatus concretions and eogenetic cementation in the Upper Jurassic of central Poland. N. Jb. Geol. Paläont. Abh. **147**: 329–342.

Kennard, J.M. and James, N.P. (1986) Thrombolites and stromatolites: two distinct types of microbial structures. Palaios **1**: 492–503.

Kennedy, W.J. and Garrison, R.E. (1975) Morphology and genesis of nodular chalks and hardgrounds in the Upper Cretaceous of southern England. Sedimentology **22**: 311–386.

Kiessling, W. (2002) Secular variations in the Phanerozoic reef ecosystem, in Phanerozoic reef patterns, In: W. Kiessling, E. Flügel and J. Golonka (eds.) *Phanerozoic Reef Patterns. SEPM Special Publication 72*. Society for Sedimentary Geology, Tulsa, pp. 625–690.

Kiessling, W. and Flügel, E. (2002) Paleoreefs – a database on Phanerozoic reefs, In: W. Kiessling, E. Flügel and J. Golonka (eds.) *Phanerozoic Reef Patterns. SEPM Special Publication 72*. Society for Sedimentary Geology, Tulsa, pp. 77–92.

Logan, B.W., Rezak, R. and Ginsburg, R.N. (1964) Classification and environmental significance of algal stromatolites. J. Geol. **72**: 68–83.

MacIntyre, I.G., Reid, R.P. and Steneck, R.S. (1996) Growth history of stromatolites in a Holocene fringing reef, Stocking Island, Bahamas. J. Sediment. Res. **66**: 231–242.

Martire, L. (1992) Sequence stratigraphy and condensed pelagic sediments. An example from the Rosso Ammonitico Veronese, northeastern Italy. Palaeogeogr. Palaeoclimatol. Palaeoecol. **94**: 169–191.

Martire, L. (1996) Stratigraphy, facies and synsedimentary tectonics in the Jurassic Rosso Ammonitico Veronese (Altopiano di Asiago, NE Italy). Facies **35**: 209–236.

Martire, L. and Clari, P. (1994) Evaluation of sedimentation rates in Jurassic-Cretaceous pelagic facies of the Trento Plateau: relevance of discontinuities and compaction. Giorn. Geol. **56**: 193–209.

Martire, L., Clari, P., Lozar, F. and Pavia, G. (2006) The Rosso Ammonitico Veronese (Middle–Upper Jurassic of the Trento Plateau): a proposal of lithostratigraphic ordering and formalization. Riv. Ital. Paleont. Strat. **112**: 227–250.

Massari, F. (1979) Oncoliti e stromatoliti pelagiche nel Rosso Ammonitico Veneto. Mem. Sci. Geol. **32**: 21.

Massari, F. (1981) Cryptalgal fabrics in the Rosso Ammonitico sequences in the Venetian Alps, In: A. Farinacci and S. Elmi (eds.) *Rosso Ammonitico Symposium Proceedings*. Edizioni Tecnoscienza, Roma, pp. 435–469.

Massari, F. (1983) Oncoids and stromatolites in the Rosso Ammonitico sequences (Middle–Upper Jurassic) of the Venetian Alps, Italy, In: T.M. Peryt (ed.) *Coated Grains*. Springer-Verlag, Berlin, pp. 358–366.

McCave, I.N. (1988) Biological pumping upwards of the coarse fraction of deep-sea sediments. J. Sediment. Petrol. **58**: 148–158.

Morse, J. and MacKenzie, F. (1990) *Geochemistry of Sedimentary Carbonates*. Elsevier, Amsterdam, 707 p.

Noffke, N., Gerdes, G., Klenke, T. and Krumbein, W.E. (1997) A microscopic sedimentary succession of graded sand and microbial mats in modern siliciclastic tidal flats. Sediment. Geol. **110**: 1–6.

Ogg, J.G. (1981) Middle and Upper Jurassic sedimentation history of the Trento Plateau (northern Italy), In: A. Farinacci and S. Elmi (eds.) *Rosso Ammonitico Symposium Proceedings*. Edizioni Tecnoscienza, Roma, pp. 479–503.

Olivier, N., Pittet, B. and Mattioli, E. (2004) Palaeoenvironmental control on sponge-microbialite reefs and contemporaneous deep-shelf marl-limestone deposition (Late Oxfordian, southern Germany). Palaeogeogr. Palaeoclimatol. Palaeoecol. **212**: 233–263.

Planavsky, N. and Ginsburg, R.N. (2009) Taphonomy of modern marine Bahamian microbialites. Palaios **24**: 5–17.

Pratt, B.R. (1982) Stromatolitic framework of carbonate mud-mounds. J. Sediment. Petrol. **52**: 1203–1227.

Pringault, O., de Wit, R. and Camoin, G. (2004) Functioning of modern marine microbialites built by benthic cyanobacteria. In: G. Pandalai (ed), *Recent Research Developments in Microbiology, Vol. VIII*, Research Signpost Publishers, Kerala, India, pp. 41–56.

Reid, R.P., Visscher, P.T., Decho, A.W., Stolz, J.F., Bebout, B.M., Dupraz, C., Macintyre, I.G., Paerl, H.W., Pinckney, J.L., Prufert-Bebout, L., Steppe, T.F. and DesMarais, D.J. (2000) The role of microbes in accretion, lamination and early lithification of mtodern marine stromatolites. Nature **406**: 989–992.

Riding, R. (2000) Microbial carbonates: the geological record of calcified bacterial-algal mats and biofilms. Sedimentology **47**: 179–214.

Shapiro, R.S. (2000) A comment on the systematic confusion of thrombolites. Palaios **15**: 166–169.

Sturani, C. (1964) La successione delle faune ad Ammoniti nelle formazioni medio-giurassiche delle Prealpi Venete occidentali. Mem. Inst. Geol. Min. Univ. Padova **24**: 63.

Sturani, C. (1971) Ammonites and stratigraphy of the "*Posidonia alpina*" beds of the Venetian Alps (Middle Jurassic). Mem. Inst. Geol. Min. Univ. Padova **28**: 190.

Visscher, P.T., Reid, R.P. and Bebout, B.M. (2000) Microscale observations of sulfate reduction: correlation of microbial activity with lithified micritic laminae in modern marine stromatolites. Geology **28**: 919–922.

Westphal, H., Head, M.J., Munnecke, A. (2000) Differential diagenesis of rhythmic limestone alternations supported by palynologic evidence. J. Sediment. Res. **70**: 715–725.

Westphal, H., Heindel, K., Brandano, M., Peckmann, J. and Cabioch, G. (2010) Microbialites as contemporaneous framework componets of coral reefs – the trophic paradox (deglacial of Tahiti, IODP 310). Facies **56**: 337–352.

Winterer, E.L. (1998) Paleobathymetry of Mediterranean Tethyan Jurassic pelagic sediments. Mem. Soc. Geol. Ital. **53**: 97–131.

Winterer, E.L. and Bosellini, A. (1981) Subsidence and sedimentation on a Jurassic Passive Continental Margin, Southern Alps, Italy. Am. Assoc. Pet. Geol. Bull. **65**: 394–421.

Zamagni, J., Košir, A. and Mutti, M. (2009) The first microbialite – coral mounds in the Cenozoic (Uppermost Paleocene) from the Northern Tethys (Slovenia): Environmentally-triggered phase shifts preceding the PETM? Palaeogeogr. Palaeoclimatol. Palaeoecol. **274**: 1–17.

Zempolich, W.G. (1993) The drowning succession in Jurassic carbonates of the Venetian Alps, Italy: a record of supercontinent breakup, gradual eustatic rise, and eutrophication of shallow-water environments, In: R.G. Loucks and J.F. Sarg (eds.) *Carbonate Sequence Stratigraphy – Recent Developments and Applications. AAPG Memoir 57*. AAPG, Tulsa, pp. 63–105.

Biodata of **Giorgio Tunis, Nevio Pugliese, Bogdan Jurkovšek, Bojan Ogorelec, Katica Drobne, Rodolfo Riccamboni**, and **Vinod Chandra Tewari**, authors of ***"Stromatolites as Markers of Biotic and Abiotic Events in the Karst Region"***

Professor Giorgio Tunis is researcher in Geology since 1980 and full professor since 2000 at the Faculty of Sciences of the University of Trieste, where he lectured in the following courses: Sedimentology, Facies Analysis and Depositional Systems, Geology of Oil Reservoirs, Basin Analysis, Geology, and Stratigraphy. The main field of interest concerns stratigraphy, facies analysis, and basin analysis of the Cretaceous-Tertiary clastic successions cropping out in the Southern Julian Prealps and, in part, in the Southern Carnian Prealps, in the mountains of the Southwestern Slovenia and in the hilly region of Central Istria (Croatia). He studied also the Cretaceous and Tertiary carbonate deposits of the Friuli Platform, Karst region, and Istria Platform. On this subject, he carried out stratigraphic, sedimentological, and paleoenvironmental researches on the existence and survival of dinosaurs on the Adriatic Carbonate Platform. At present, his main interest concerns biostratigraphy, paleontology, sedimentology, and taphonomy of Late Cretaceous rudist-bearing carbonate facies cropping out in Friuli (NE Italy), Karst and Southern Istria. G. Tunis is author/co-author of about 85 papers, some of them on international journals. He also participated to numerous international and national Congresses, to which he contributed with more than 50 oral communications and posters.

E-mail: **tunis@univ.trieste.it**

V.C. Tewari and J. Seckbach (eds.), *STROMATOLITES: Interaction of Microbes with Sediments*, Cellular Origin, Life in Extreme Habitats and Astrobiology 18, 251–272
DOI 10.1007/978-94-007-0397-1_11, © Springer Science+Business Media B.V. 2011

Professor Nevio Pugliese is a Full Professor in Palaeontology (University of Trieste) since 2000. He is currently Director of the National Antarctic Museum (Trieste University) since 2007. His scientific activity concerns the geological and paleontological topics focusing on the marine, brackish water, and freshwater microfaunae (mainly ostracods) to evidence their role of environmental, climatic, geographic, and biostratigraphic markers from Mesozoic to Quaternary.

He has participated in 80 national and international congresses. He is author and co-author of 140 scientific papers in national and international journals.

E-mail: **pugliese@univ.trieste.it**

Dr. Bogdan Jurkovšek (born 1952) is scientific councilor, employed at the Geological Survey of Slovenia. He graduated in 1976 at the University of Ljubljana (Faculty of Natural Sciences and Engineering) with his diploma work on biostratigraphic studies of the Carnian strata in the Mežica mine (Eastern Alps). He earned his Ph.D. in 1981 at the same university with a thesis on biostratigraphy of the Middle and Upper Triassic strata in Slovenia. Since 1977, he has been active at the Geological Survey of Slovenia as a mapping geologist. He is leader of the research-scientific program on Regional Geology of Slovenia with its main goal to elaborate geological maps. He is author and co-author of many scientific, technical and popular articles, and monographs predominantly from the field of biostratigraphy, paleontology, and geological cartography. In the last 15 years, his research work was focused on geology of the Trieste-Komen Plateau and resulted in the elaboration of two geological maps.

E-mail: **bogdan.jurkovsek@geo-zs.sl**

Nevio Pugliese

Bogdan Jurkovšek

Prof. Dr. Bojan Ogorelec is a scientific councilor (1984–2009) employed as Head of the Department of Sedimentology and Petrology at the Geological Survey of Slovenia from 2000 to 2006. He was director of the Geological Survey of Slovenia and retired in 2009. His specialization is in the field of carbonate rocks and recent marine and lacustrine sediments. He worked at the Heidelberg University, Germany in 1988 for thesis on microfacies and diagenesis of Upper Triassic Main dolomite and Dachstein limestone in SW Slovenia. He is also teaching at the Ljubljana University the courses on regional geology and sedimentology of carbonate rocks. Dr. Bojan is author and co-author of about 200 scientific papers in Slovenian and international geological journals. He was chief-editor of the Slovenian scientific journal *Geologija* from 1988 to 2009.

E-mail: **bojan.ogorelec@geo-zs.si**

Professor Katica Drobne received her Ph.D. in 1972 from the University of Ljubljana. It was published as monograph in Mém. suisse de Paléontologie (1977, Basel), in Ljubljana rewarded by state award. The subject of it was systematic description of Paleogene alveolinids and their stratigraphic and regional distribution on the NW part of Adriatic carbonate platform. Later, her research interest was enlarged besides the alveolinids on larger foraminifera such as agglutinated conical forams and complex miliolids. The latter with its correlation on the region from Tethys Australia (PNG), Oman, Mediterranean region, Pyrennean to Carribean realm. She found out different paleoenvironmental and paleocommunities of foraminiferal associations from the Upper Cretaceous to Priabonian (Paris 2009). Some period from 1985 year she is directed to detail micropaleontological research of K/T boundary on the NW part of Adriatic CP, more or less continuous development from Maastrichtian to Danian and upward in the Paleocene and the Eocene. In this research area, a large group of researchers for the biotic and abiotic topics by extensive numbers of publications is now active. Her recent interest is dedicated to the study, among the group named W.O.L.F., to connect the possible data of Paleogene microfossil associations of small and specially of larger foraminifera about the index species as the key-tool for the philogeny, biostratigraphy and paleobiography on the Periadriatic and surrounding countries enlarged to the east into Alpine–Himalayan region. She was co-organizer of several I.G.C.P. Paleogene projects 274, 286, 393 for many years, also on K/T boundary 552 and 555. She is working as scientific councilor (external collaborator) at Ivan Rakovec Institute of Paleontology, Scientific Research Centre of Slovenian Academy of Sciences and Arts, Ljubljana (Slovenia).

E-mail: **katica@zrc-sazu.si**

Dr. Rodolfo Riccamboni obtained his Bachelor's Degree in Geology from Trieste University, Italy, in 2006. Currently, he collaborates with the Geomorphology, Applied Geology, Informatic Cartography Group (GGACI) in the Department of Geosciences at Trieste University. His research topics include the micropaleontology and carbonate stratigraphy of the Trieste Karst area but recently he has focused the attention on the creation of contents and software architecture for multimedia and multi-theme guides for mobile devices.

E-mail: **rriccamboni@units.it**

Katica Drobne

Rodolfo Riccamboni

Professor Vinod C. Tewari is currently the Head of the Sedimentology Group at Wadia Institute of Himalayan Geology, Dehradun and a Senior Associate of International Centre for Theoretical Physics, Trieste, Italy. He obtained his Ph.D. from the University of Lucknow in *Geology* in 1986 and continued his research in Wadia Institute. Dr. Tewari taught Geology at Kumaon Univerisity, Nainital, Uttarakhand (U.K.), India as Professor of Geology. Professor Tewari's scientific interests are in the areas of Precambrian stromatolites, sedimentation, carbon isotope chemostratigraphy, genesis, early evolution and diversification of life and its astrobiological significance. He has been associated with the International Geological Correlation Program (I.G.C.P.) Projects on Stromatolites, Biosedimentology of Microbial Buildups and currently on *The Rise and Fall of Vendian Biota*. He has eighty research papers published to his credit, and edited several volumes of Himalayan Geology, India and Journal of Nepal Geological Society, Kathmandu, Nepal. Professor Tewari has visited UCLA, CSEOL, USA in 2007 and 2008 to work on Laser Raman Spectroscopy with Professor J.W. Schopf. He organized first *Indo-Soviet Symposium on Stromatolites and Stromatolitic Deposits* and other IGCP meetings in India. He has been one of the members of the organizing committee of the World Summit on Ancient Microscopic Fossils organized by Professor J.W. Schopf, held in University of California, Los Angeles, USA in 2008. He is a member of the Editorial Board of the international journal Astrobiology.

E-mail: **vtewari@wihg.res.in**

MICROBIALITES AS MARKERS OF BIOTIC AND ABIOTIC EVENTS IN THE KARST DISTRICT, SLOVENIA AND ITALY

GIORGIO TUNIS[1], NEVIO PUGLIESE[1], BOGDAN JURKOVŠEK[2], BOJAN OGORELEC[2], KATICA DROBNE[3], RODOLFO RICCAMBONI[1], AND VINOD CHANDRA TEWARI[4]

[1]Department of Geosciences, University of Trieste, via Weiss, 2, 34127 Trieste, Italy

[2]Geological Survey of Slovenia, Dimičeva 14, 1000 Ljubljana, Slovenia

[3]Ivan Rakovec Institute of Paleontology, ZRC SAZU, Novi Trg 2, P.O. Box 306, 1001 Ljubljana, Slovenia

[4]Wadia Institute of Himalayan Geology, Dehradun 248001, Uttarakhand, India

Abstract Microbialite units are commonly found within the Mesozoic–Tertiary carbonate succession of the Karst District of Slovenia and Italy. They record environmental conditions, which are linked with both local and global events.

Local events are recorded by scattered occurrences and are mostly related to changes in the hydrodynamic regime of the platform and/or synsedimentary tectonics, which are responsible for rapid environmental changes. In addition, freshwater stromatolites indicate periodic continental influences. Global events are recorded by more or less continuous units of microbialite, which can be related to sea-level changes, as documented in the Turonian (regressive oncoid limestones) and the Maastrichtian, K/T boundary and basal Danian (peritidal limestones linked with Milankovich cycles). In general, the microbialites provide evidence of environments where the higher order organic community declined and disappeared, after having tolerated a drastic variability. Below the microbialite units, the community was usually poorly diversified and composed of opportunistic taxa, such as small foraminifers (agglutinated forms, miliolids, discorbids and nonionids), thin-shelled gastropods and ostracods. The same or similar benthic taxa appeared again above the microbialite units indicating their ability to reestablish in the environment especially after peritidal conditions. Microbialite deposits are more evident and thicker in the Slovenia Karst District (Dolenja Vas and Sopada sections) than the Italian part (Padriciano sections).

Keywords Stromatolites • Paleoecology • Paleoclimate • Global events • Cretaceous • Tertiary • Karst region • Adriatic platform • Trieste • Italy • Slovenia • Komen • Foraminifera • Algae • Meghalaya • India

1. Introduction

Microbialites comprising the well-laminated stromatolites and the more clotted, unlaminated thrombolites are known since the early history of the Earth. The oldest microbialites span the geological record as far back as 3.45 billion years ago (Hofmann et al., 1999; Schopf and Walter, 1983), and they were generally diverse and abundant from 2,800 to 1,000 Ma (Riding, 2000). Chela Flores et al. (2011) reported that well-preserved fossil bacteria have been found in black cherts associated with microbialitic carbonates during the Archean and Proterozoic. A high diversification of the microbiota occurred during Proterozoic as clearly demonstrated in Lesser Himalaya, India (Tewari, 2004, 2007; Shukla et al., 2006; Schopf et al., 2008). Thus, the microbialites represent one of the first examples of organic activity due to cyanobacteria. In shallow-water oxygenated environments, these microbes may be classified as aerobic phototropes: they decompose organic material to inorganic components by redox processes. This results in progressive depletion in O_2, SO_4^-, CO_2 (see Riding, 2000 and the references therein). Stromatolites are typically laminated benthic deposits (Riding, 1991) due to microbial interactions (Gebelein, 1976) and to cycling of microbial communities and lithyfication (Reid et al., 2000). Moreover, they were and are able to construct differently shaped buildups (e.g., domes, columns, reef, crusts) in marine subtidal and inter/supratidal low latitude settings, in lacustrine and fluvial systems and, also, close to hydrothermal springs (hot-spring travertine) and cold seeps. Microbialites are also well developed in the carbonate succession of the Karst District (NE Italy and SW Slovenia), where the Mesozoic and Paleogene strata are represented by carbonate rocks. In the paleogeographic sense, this area was part of the vast Adriatic–Dinaric Carbonate Platform and was located between the Belluno Basin to the northwest and the Slovenian Basin (Buser, 1989; Buser et al., 2008) to the northeast. The Trieste–Komen Platform and the Slovenian Basin continue into Adriatic platform as illustrated by Cati et al. (1987), Herak (1987) and Velić et al. (2003). Microbialites are found at several levels in the succession spanning from early Cretaceous to Danian. This work aims to highlight the role of microbialites in the paleoenvironmental evolution of the Karst region, their mode of occurrence and the environmental changes related to biotic, abiotic, and climatic changes.

2. Geological Framework

From a geographic point of view, the platform corresponds to the present day Southern Prealps and Karst District (Fig. 1), and the Slovenian Basin can be related to the Julian Prealps, Tolmin Mountains and Goriška Brda region. The Mesozoic and Paleogene carbonate platform succession consists of deposits ranging from Valanginian to early Eocene (Jurkovšek et al., 1996; Jurkovšek, 2008). The early Cretaceous succession consists mainly of bedded limestones, dolomitic

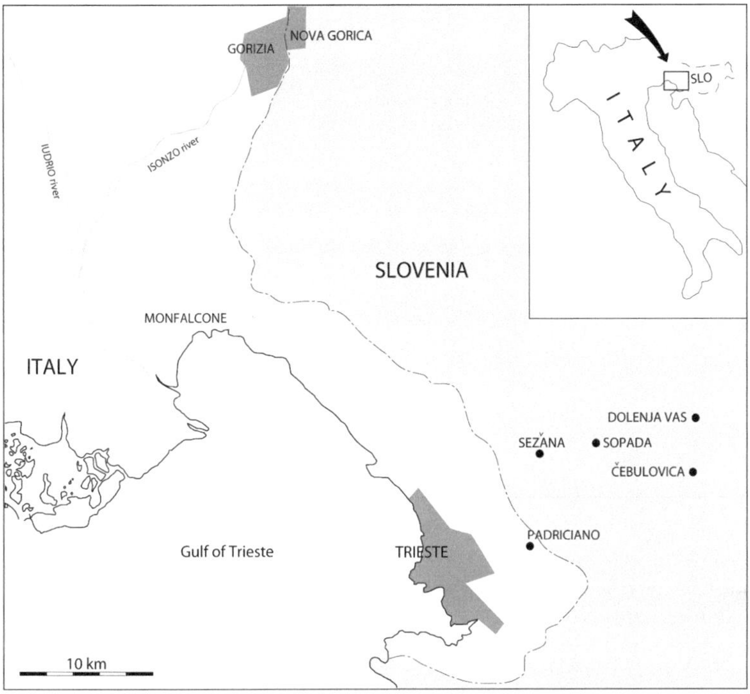

Figure 1. Location map of the Adriatic Platform in Trieste–Slovenia region.

lenses and intercalations of dolomitic breccias. With the exception of the regional platform drowning began at the end of the Cenomanian ("pelagic limestone" with Calcisphaerulidae), the Cretaceous section is characterized by prevailing shallow-water deposits with evident emergent phases at the Aptian–Albian boundary, during the late Albian-early Cenomanian, the late Campanian-early Maastrichtian (Jurkovšek et al., 1996; Jurkovšek, 2008) and the late Maastrichtian (Drobne et al., 1996; Jurkovšek et al., 1996; Jurkovšek, 2008; Pugliese et al., 1995, 2000). Paleocene beds, when identified, begin with peritidal deposits (basal Danian), followed by and successively with deposits bearing evidence of brackish or even freshwater environments, which alternate with shallow marine and low-energy environments of protected shelf and lagoon. Predominant marine deposits (i.e., miliolid limestone, coral-algal limestone and alveolinid-nummulitid limestone) continue up to the Middle Eocene recording an increasing depth from shoreline to open shelf followed by turbidites of the Flysch. A reference stratigraphic section is presented in Fig. 2. Microbialites are reported from early Cretaceous to Danian strata. Thus, our review mainly focuses only this time span.

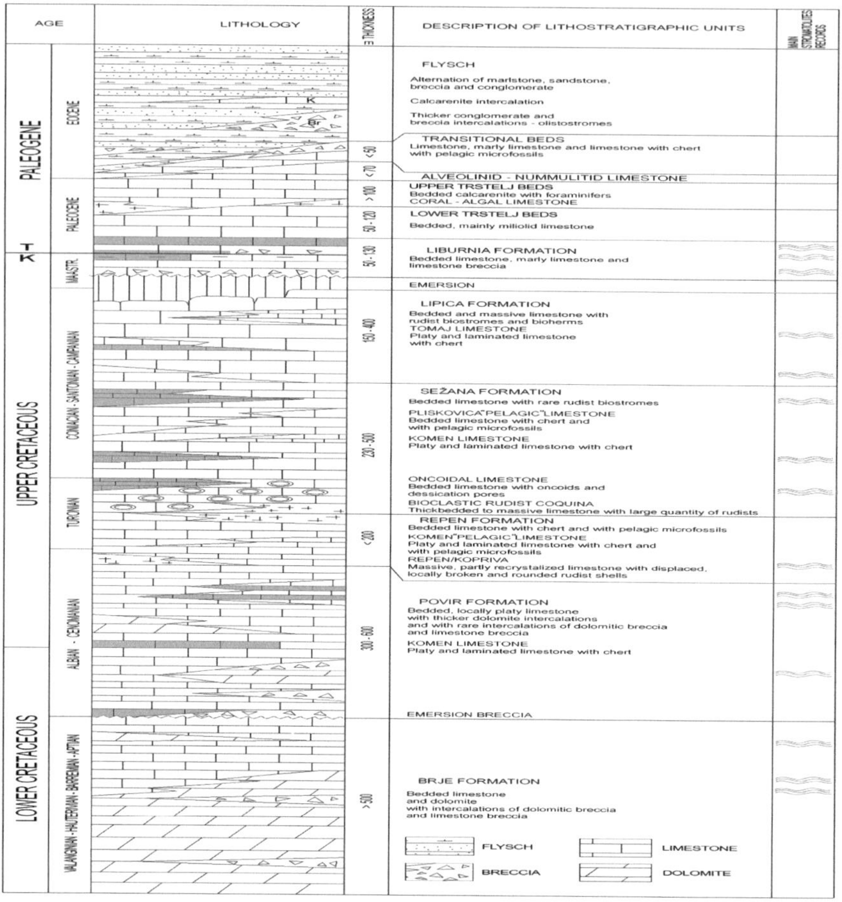

Figure 2. Lithocolumn of the stratigraphic section, Adriatic Platform.

3. Microbialites Through Time in the Karst District Region: Stratigraphy and Paleoenvironments

3.1. EARLY CRETACEOUS

The early Cretaceous generally consists of blackish grainy dolomites and dark limestone beds with dolomite packages and lenses (Jurkovšek et al., 1996; Jurkovšek, 2008). Paleosoils, dolomitic breccias and limestone breccias are present in the mid-lower part (Aptian/Albian) and mid-upper part (Albian) of the succession, respectively. The middle part shows thin concentrations of requienids. This interval records the first anoxic episode of the carbonate platform responsible for the deposition of black limestones. The early Cretaceous succession is terminated by a dolomitic breccia.

The outcropping succession in the Italian sector of the Karst ranges between the upper part of the Early Aptian and the Upper Albian (Tentor et al., 2000), whereas in the neighboring Slovenian region the oldest rocks are be Valanginan (Jurkovšek, 2008) and, possibly, Berriasian (Tentor et al., 2002).

3.2. HAUTERIVIAN–BARREMIAN–APTIAN

Prevailing dolomites and well-bedded limestones characterize this part of the succession, which is interpreted as deposited in a tidal flat setting as indicated by the dominant micritic matrix and fossil assemblages (Jurkovšek et al., 1996). Loferite laminae with shrinkage pores are mainly observed in the Hauterivian–Barremian interval, where local, short-lasting emersion phases of the platform are documented by thin levels of intraformational breccia and veins/small pockets of bauxitic clay. Microbialites are not very common, due to the late diagenetic dolomitization that obliterates most of the structures and textures typical of supratidal conditions. Other successions similar to this part of the Trieste–Komen Karst succession are found on proximal sectors of inner platform such as Trnovo forest (Koch et al., 1989) and generally on most of the Friuli area such as Cansiglio Plateau, Val Cellina, Bernadia Mountains, Iudrio valley (Sartorio et al., 1997). This indicates a general situation related to the maximum lateral extent of the platform peritidal facies, which also affected the most marginal sectors of the platform in the Friuli region.

3.3. ALBIAN

In the Italian Karst District, the Microbialites are sporadically present in thin laminated layers (microbial laminae), recording local extreme environmental conditions for the biota. In particular, the biofacies are characterized by oligotypical and opportunistic assemblages (small miliolids and nubeculariids, thin-shelled ostracods and gastropods). The Microbialites represent a critical interval, where faunal diversity was reduced and waited for the restoration of an aquatic environment.

Interesting data came out from the recent study of a drilling carried out close to Iamiano (Cucchi, personal communication). A thick dolomite-limestone breccia bed due to post-sedimentary tectonics and complex diagenetic processes is to be found between the Upper Albian and the Middle Cenomanian deposits of the Karst District (Tentor et al., 1994; Jurkovšek et al., 1996). This breccia bed crops out also in the neighboring Western Croatia, e.g., Čičarija Mountains close to Rijeka and in the zones of Cres and Krk islands (Vlahović et al., 2002). In the cores from the Iamiano drilling, the breccia bed overlies Upper Albian deposits with very poor microfaunal assemblages. The lithologies consist of limestones, dolomitized limestones, both late diagenetic and early-diagenetic dolomites and of a peculiar, 4-m thick layer of unlithified quartz silt. Shallowing-upward peritidal parasequences are rather common in the upper part of the section and these are classic such LLH stromatolites with wavy laminae. The inferred environment is a shallow lagoon, likely under-arid or semi-arid climatic conditions.

3.4. LATE CRETACEOUS

3.4.1. Late Cenomanian

The oldest localities with fossil fishes in the Karst District are attributed to the first interval of the Komen limestone of late Cenomanian (Cavin et al., 2000; Jurkovšek and Kolar-Jurkovšek, 2007). The available evidence in the Komen surrounding suggests that the depositional environment of the Komen limestone was an intraplatform basin, in the proximity of an exposed area such us tidal flat probably other supratidal environments were populated by conifers (Dobruskina et al., 1999) and where karstification could occur (Palci et al., 2008). The examined micro- and macrofaunal elements from the Moschenizze section (Monfalcone) suggest a stressed lagoonal environment, probably due to paleosalinity changes (Tentor and Tentor, 2007). Stromatolites and microbial laminae (mm-scale laminations), which are locally disrupted to form flat pebble conglomerates. The occurrence of chert nodules and sheets, dark gray and black, is noteworthy: an intraformational sponge specule silica source for chert nodules is suggested by the presence of calcitized sponge spicules in the host limestone. The interval in question extends laterally but discontinuously between Škrbina, Tomačevica, Volčji Grad (Slovenia) and Monfalcone (Italy). Global sea level rise during the late Cenomanian has significantly affected the carbonate sedimentation over a large part of the Adriatic–Dinaric platform system. The flooding is documented by a unit rich in Calcisphaerulidae and planktonic foraminifers (Komen Pelagic Limestone according to Jurkovšek et al., 1996; Jurkovšek, 2008). The drowning of the platform is traditionally connected with the Cenomanian–Turonian oceanic anoxic event and to the coeval sea level rise (Jenkyns, 1991; Gušić and Jelaska, 1990; Jurkovšek and Kolar-Jurkovšek, 2007). Bioclastic facies consisting of alternations of floatstone, rudstone, and grainstone with broken, more or less abraded but locally intact shells sometimes of Caprinidae (*Caprinula boissy*) and *Chondrodonta* abruptly overlie the pelagic limestone.

3.4.2. Upper Turonian (?)–Coniacian

Mudstone/wackestone containing *Decastronema kotori*, *Thaumatoporella* and scarce benthic foraminifers, often with desiccation structures and vadose crystal silt abruptly overlie the bioclastic limestones outcropping in the Northern Karst District (Mount S. Michele-Gorizia, Italy). A level of oncoidal limestones overlying the pelagic and bioclastic limestones of the Repen Formation (Jurkovšek et al., 1996; Jurkovšek, 2008) is present in the Slovenian Karst District and only sporadically in the Italian sector. The oncoidal level is evident in a quarry close to Slivia (Trieste, Italy). Caffau (1997) described the stratigraphic succession including this level, which may be referred to the Upper Turonian oncolitic horizon originally found on the Brač Island (Gušić and Jelaska, 1990) and mapped by Jurkovšek (2008) in the Slovenian Karst. According to the aforementioned Croatian and Slovenian workers, this level records the sudden drop of the eustatic sea-level (Haq et al., 1987a, b).

The following succession (Upper Turonian?–Coniacian) is characterized by prevalent mud-rich sediments at the base which are interpreted as shallow, low-energy lagoon deposits. Herein laminar, fine-grained intertidal stromatolite-bearing layers are often observed both on the Italian and on the Slovenian region of the Karst District. Thin intercalations of platy, stromatolitic and laminated limestones with fossil fish are reported by Cavin et al. (2000), close to the hamlet of Kobjeglava, 15 km far from Sežana (Slovenia).

3.4.3. Santonian

Common lithofacies in this part of the Karst District are represented by peloidal mudstone/wackestone with *Decastronema kotori* and wackestone/packstone with abundant benthic foraminifers alternating with rudist floatstone. Large rudist-rich bodies, which laterally pass into bioclastic grainstone and packstone characterize this unit on the whole. Microbial laminites are scarce and can be observed mainly in the lower part of the interval, where fenestrae and dissolution cavities, bioturbation and pedogenic alteration are also present. A gradual change during Santonian from peritidal/shallow subtidal cycles to predominant subtidal conditions, which reflect a ramp-like open shelf environment has been postulated by Tentor (2007). Two Santonian levels with fossil fish, conifers and other rare fossils are well known in literature: i.e., Polazzo (Fogliano-Redipuglia, Italy) and Skopo (Slovenia). Close to Polazzo locality (D'Erasmo, 1952; Dalla Vecchia et al., 2001), a thin succession, 2 m thick, embedded within rudist-bearing limestones is made of laminites and stromatolites. The inferred environment is a very shallow, low-energy lagoon, where terrestrial influxes are clear. The 4-m thick Skopo succession (Ogorelec et al., 1987; Cavin et al., 2000) is represented by alternating beds of dark fossiliferous wackestone/packstone, laminites, flat pebble conglomerates and stromatolite-bearing layers. Chert nodules and sheets also occur.

3.4.4. Campanian

On the Italian side of the Karst District, extensive emergence of the platform with important erosional phenomena occurred during the Campanian p.p. (*pro parte*), whereas on the Slovenian side, there were large drowned areas (10 km north of Sežana), which led to deposition of limestones with ammonites (Jurkovšek et al., 1996; Summesberger et al., 1996, 1999a, b) and other pelagic fauna (Jurkovšek and Kolar-Jurkovšek, 2007). A karstic depression, a few meters deep and some some hundreds of meters long, outcropping close to Cotici-Gorizia (Italy) has been recently examined by Venturini et al. (2008). The bed rock is made of bioclastic limestones with *Keramosphaerina tergestina* and is successively by altered bioclastic limestone with rudist shell fragments and clearly visible paleokarstic phenomena. Karstic limestone breccia at the base of the depression shows a micropaleontological assemblage represented by *Scandonea samnitica* and *Scandonea mediterranea*. The karstic depression is filled from the base to the top with the following: (1) oolitic bauxite, (2) oncolite rudstone with thrombolite (thromboids), oncolitic stromatolites and stromatolites (Fig. 3), (3) well-bedded mudstone/wackestone with ostracods and charophyte remains at the base and a level rich of freshwater gastropods (*Viviparus* sp.), (4) a few dark mudstone levels with brackish facies containing discorbids and more or less abundant molluscs and ostracods and characterized by desiccation structures and thin plasticlast breccia levels. The depression is covered by light fossiliferous wackestone with *Murciella cuvillieri*, *Moncharmontia apenninica*, *Rotorbinella scarsellai*. This sequence suggests a fluctuating internal transgression according to the well-known "blue hole" model (Rasmussen and Neumann, 1988). Similar successions to this one were recognized in several localities from the Adriatic-Dinaric Carbonate Platform (ADCP) such as the Hauterivian–Barremian site of Kolone, Western Croatia (Dini et al., 1998) and many Liburnian sections of the Kvarner region and Istria,

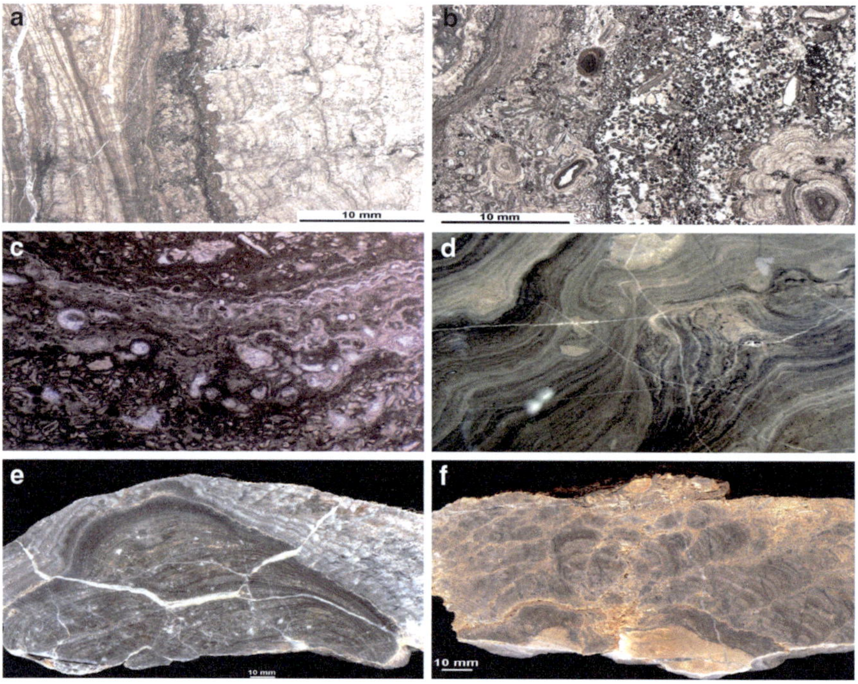

Figure 3. (**a**) Growth laminae of fibrous calcite and homogeneous micrite. Cotici (**b**) oncolitic rudstone with frombolite and sparse, abundant bauxitic oolites. Cotici (**c**) stromatolitic laminae interspersed with *Microcodium* prisms in Dolenja Vas East section. Dolenja Vas. (**d**) Stromatolitic bioherm just above the K/T boundary, Basovizza. (**e**) Dome-shaped stromatolite from lacustrine deposits of Cotici. Cotici (**f**) lacustrine cerebroid-type stromatolites.

Western Croatia (Bignot, 1972; Ćosović et al., 1994; Drobne et al., 1991). The presence of fossil charophytes and specimens of *Viviparus* suggest a shallow lacustrine depositional environment for the levels (2) and (3). The stromatolites which are 15–20 cm thick are mostly domal in shape and well preserved. They generally show, at the mesostructural scale, calcite/micrite couplets of growth laminae (Fig. 3). Textural and geochemical analysis of stromatolites is currently in way.

3.4.5. Maastrichtian

During the Maastrichtian, repeated emergence and local floodings are known to have occurred over the whole Karst region. The Maastrichtian succession is characterized by alternating freshwater, brackish and marine facies, namely the Liburnian Formation. The last deposits are typical of a shallow protected lagoonal environment (Ogorelec et al., 2001): with stromatolites mostly in the marine facies. Wackestone with *Rhapydionina liburnica*, *Fleuryana adriatica*, *Cuneolina ketini*, rare lithosomes with rudists (*Bournonia*) and more common rudist shell concentrations (*Gyropleura*, *Apricardia*), laminites, stromatolite-bearing layers (Fig. 3), mudstone with charophyte and ostracods, wackestone with discorbids, molluscs, ostracods and *Anomia*, coal lenses with *Stomatopsis*, *Cosinia*, etc., veins/small pockets of bauxite and breccias with plasticlasts have been recognized along the classical profiles of Čebulovica, Divača-Kozina, Dolenja Vas, Sopada (Slovenia), and Padriciano (Italy) and the stratigraphic section were carefully described by various authors, among them Stache (1889), Pavlovec (1963), Bignot (1972), Cousin (1981), Drobne et al. (1988, 1989), Pugliese et al. (1995), Jurkovšek et al. (1996), Gregorič et al. (1998), Pugliese et al. (2000), Ogorelec et al. (2001), etc. The Maastrichtian, particularly the latest Maastrichtian, K/T and Danian beds contain the most beautiful examples of the peritidal stromatolites of the Karst District. The beds of the interval in question form a thicker, better exposed section in Slovenia (Dolenja Vas) than in Italy. However, peritidal cycles including the stromatolites seem to be more numerous in the Italian region rather than in the Slovenian one. Geochemical analysis on the peritidal cycles highlight the fact that the stromatolites often record evident negative shifts of $\delta^{13}C$; in particular, the most negative values occur close to high concentration of disarticulated fragments of *Microcodium* entrapped by cyanobacterial laminae.

4. K/T Boundary

Close to Padriciano village (Trieste), Pugliese et al. (1995) identified the K/T boundary within a peritidal succession. The finding followed those of other localities in Slovenia (Drobne et al., 1988, 1989; Delvalle and Buser, 1990; Ogorelec et al., 1995, 2007). The occurrence of the K/T boundary is documented by paleontological data (definitive disappearance of the above reported Maastrichtian taxa and subsequent appearance of Paleocene ones – see Pugliese et al., 1995, 2000; Drobne et al., 1996, 2007), paleomagnetical data with identification of Ch29r (Marton et al., 1995), iridium anomaly (Hansen et al., 1995) and a negative shift of $\delta^{13}C$ (Ogorelec et al., 1995, 2007; Dolenec et al., 1995; Riccamboni, 2005; Tewari et al., 2007).

5. Paleogene

The Danian section is mostly constituted by peritidal carbonates and rarely also of shallow open marine facies with the dasycladaceans and coral reefs and patches (Barattolo, 1998; Turnšek and Drobne, 1998). The lower part of the Selandian interval points out also a lagoon environment characterized by not very common passages from the subtidal to the intertidal/supratidal zone. The lower part of the Selandian section indicates a moderately high energy open lagoon/coastal marine settings (Caffau et al., 2002). Marine coastal settings characterize the Thanetian. Stromatolites are widely distributed throughout the Danian but become progressively rare through the Paleocene section. The Paleocene–Eocene algal–foraminiferal Lakadong Limestone of the Shillong Plateau, Meghalaya, northeastern India is a shallow tidal deposit and eastern extension of the Tethys ocean (Tewari et al., 2007, 2010).

6. Danian

Stromatolites are widely distributed in the Danian interval, which is characterized by various lithofacies (Drobne et al., 1988; Ogorelec et al., 2001; Caffau et al., 2002). Fossiliferous mudstone and wackestone are documented by *Decastronema barattoloi* (De Castro) and *Bangiana hanseni* (Drobne) (Riccamboni, 2005; Ćosović et al., 2008; Drobne et al., 2007). These limestones are the most common lithologies alternating with levels indicating short-lasting exposures of the platform. They are characterized by flat pebble conglomerates, shrinkage pores, stromatolites, emergence breccias and diffuse rhizogenic structures (Ogorelec et al., 2001; Košir, 2004). Most sediments have been deposited as mud in a very shallow and low-energy environment (e.g., lagoon). Supratidal breccias and paleosoil layers indicate emergence in the area while characeans and lagynophores indicate episodic paleosalinity fluctuations and the changing of marine to brackish or even a freshwater regime (Ogorelec et al., 2001).

7. Discussion

Stromatolites are relatively abundant in the peritidal successions of the Karst District in supratidal/intertidal settings, whereas they are absent in subtidal normal marine and in high-energy platform or carbonate ramp environments. Common occurrences of microbialite are present in Hauterivian/Barremian, Albian, Upper Turonian (?)–Coniacian, and Maastrichtian–Danian strata of the Karst District of the Adriatic-Dinaric Carbonate Platform. This area was affected by synsedimentary tectonic movements close to the Aptian/Albian, Albian/Cenomanian and Campanian/Maastrichtian boundaries. As a result of this, large zones of the Karst District were affected by tectonic uplift and relatively long-lasting phases of subaerial exposure. Subsequently, the peritidal environments were reestablished over the area in question. Thus, microbialites reflect the maximum lateral extension of peritidal facies over the Karst Platform. Also, temporary flooding of the Platform occurred during the latest Cenomanian and this strongly affected the sedimentation of the internal platform areas of the Karst District. Successively, low-energy peritidal facies reappeared across the area.

The stromatolite-bearing layers have been rather precisely situated by us within the framework of the stratigraphic section of the Karst District, but, often, their lateral extension is limited and discontinuous: this is the case of the late Cenomanian localities of Škrbina, Tomačevica etc. and of the Santonian sites of Polazzo and Skopo. It is supposed that local tectonics controlled the relationships between different areas and the formation of related accommodation spaces. Moreover, the presence of a tectonically induced more or less irregular topography caused the co-existence of areas characterized by different distribution patterns of the subtidal, flat intertidal, pond and supratidal facies.

So far as the climatic control is concerned, it is speculated that the microbially laminated limestones and/or dolostones with dessication cracks, paleokarst structures and paleosoils of late Albian age (e.g., Iamiano drilling) may be referred to a semi-arid to arid climatic regime (evaporitic environment?), while the Maastrichtian/Danian stromatolites may be attributed to wetter conditions (ongoing research). Besides, marine stromatolites are recorded from the succession of the Trieste–Komen plateau, a recent study documents the presence of well-preserved lacustrine stromatolites of late Campanian/Maastrichtian age. Oncolites, oncolitic stromatolites and stromatolites overlying an oolitic bauxite level filled one karstic depression during an important emergence phase of the platform. These stromatolites with well-preserved growth fabrics, textures, and probably unaltered geochemical signatures may provide interesting insights regarding the physico-chemical conditions of the environment and paleoclimatic information from stable carbon and oxygen isotope values (Andrews et al., 1997; Nehza et al., 2009). The whole Karst section reveals the continuing importance of microbial carbonates throughout much of the Cretaceous until the Danian. Successively, microbial carbonates seem to have been absent or very scarce in the remaining part of Paleocene, and this decline of the stromatolites from the Cretaceous onwards is a global event and previous workers have also mentioned about it (Webb, 1996; Riding, 2000).

8. Conclusions

Stromatolites are commonly found along the carbonate sequence of the Karst District. They record environmental conditions, which are linked with both local and global events. Local events are supported by scattered findings and are mostly related to changes of the hydrodynamic regime of the platform and/or synsedimentary tectonics, which are responsible for rapid environmental changes. Moreover, the presence of freshwater stromatolites may be included into settings characterized by continental influence. Global events are indicated by more or less continuous horizons of stromatolites, which can be related to sea-level changes. This is the case of the oncoid limestones and the peritidal cycles of Maastrichtian, K/T boundary and basal Danian. These levels can be related to marine regressions. Moreover, the peritidal cycles of Maastrichtian, K/T boundary and basal Danian may be related to other causes, including those Milankovich rhythms. Within the cycles, stromatolites locally capped by *Microcodium*, are common and indicate inter-supratidal settings. From the paleobiological point of view, the stromatolites provide evidence of environments where the metazoan community declined and disappeared, after having tolerated a drastic variability. Below the stromatolitic layers, the community was usually poorly diversified and constituted by opportunistic taxa, such as small foraminifers (agglutinated forms, miliolids, discorbids and nonionids), thin-shelled gastropods and ostracods. The same or similar taxa appeared again above the stromatolitic layers indicating ability to reenter the depositional environment, so evidencing their capability to enter the settings recording a renewal of benthic life.

9. Acknowledgments

We are grateful to Professor M.E. Tucker, Durham, U.K. for kindly reviewing the chapter and providing valuable suggestions for improving it.

10. References

Andrews, J.E., Riding, R. and Dennis, P.F. (1997) The stable isotope record of environmental and climatic signals in modern terrestrial microbial carbonates from Europe. Palaeogeogr. Palaeoclimatol. Palaeoecol. **129**: 171–189.

Barattolo, F. (1998) Dasycladacean green algae and microproblematica of the uppermost Cretaceous - Paleocene in the Karst area (NE Italy and Slovenia), In: L. Hottinger and K. Drobne (eds.) *Paleogene Shallow Benthos of the Tethys*. Opera DELA SAZU 34(2). Ljubljana, pp. 65–127.

Bignot, G. (1972) Recherches stratigraphiques sur les calcaires du Cretacé supérieur et de l'Eocène d'Istrie et des régions voisines. Essai de révision du Liburnien. Travaux Lab. Micropaléontologie, Université Paris, ParisVI 2, pp. 1–353.

Buser, S. (1989) Development of the Dinaric and the Julian Carbonate Platforms and of the Intermediate Slovenian Basin (NW Yugoslavia). Mem. Soc. Geol. Ital. **40**(1987): 313–320.

Buser, S., Kolar-Jurkovšek, T. and Jurkovšek, B. (2008) The Slovenian Basin during the Triassic in the Light of Conodont data. Boll. Soc. Geol. Ital. (Ital. J. Geosci.) **127**(2): 257–263.

Caffau, M. (1997) Palaeontological and stratigraphic description of a rudist deposit of the upper Turonian in Slivia, Trieste Karst, Italy. Ann. Istrian Mediterr. Stud. **11**: 141–160.

Caffau, M., Drobne, K. and Pugliese, N. (2002) Stop 4 Padriciano section, In: J. Hohenneger, R. Melis, P. Pervesler and N. Pugliese (eds.) *The Third International Congress EMMM 2002, Field Excursion Guide*, pp. 29–34.

Cati, A., Sartorio, D. and Venturini, S. (1987) Carbonate platforms in the subsurface of the Northern Adriatic area. Mem. Soc. Geol. Ital. **40**: 295–308.

Cavin, L., Jurkovšek, B. and Kolar-Jurkovšek, T. (2000) Stratigraphic succession of the Upper Cretaceous fish assemblages of Kras (Slovenia). Geologija **43**(2): 165–195.

Chela-Flores, J., Montenegro, M.E., Pugliese, N., Tewari, V., and Tuniz, C. (2011) Evolution of plant-animal interaction. In: Z. Dubinsky and J. Seckbach (eds.) *Cellular Origin, Life in Extreme Habitats and Astrobiology, All Flesh is Grass*, **16**, Springer, Dordrecht, The Netherlands, pp. 9–34.

Ćosović, V., Koi, M., Marjanac, T., Moro, A., Gušić, I., Jelaska, V. (1994) Some paleontological evidence of the Eocene trasgression of the Adriatic Carbonate Platform. First International Meeting on Perimediterranean Carbonate Platforms, Marseille, Abstract 49–50.

Ćosović, V., Drobne, K., Ogorelec, B., Moro, A., Koič , M., Šoštarko, I., Tarlao, A. and Tunis, G. (2008) *Decastronema barattoloi* (DE CASTRO), characteristic fossil of the Paleocene and the Eocene peritidal sediments from the Adriatic carbonate platform. Geol. Croat. **61**(2–3): 321–332.

Cousin, M. (1981) *Les rapports Alpes-Dinarides; les confins de l'Italie et de la Yugoslavie*. Société Géologique du Nord 5. Villeneuve, 1042 pp.

D'Erasmo, G. (1952) Nuovi ittioliti cretacei del Carso Triestino. Atti Mus. Civ. Stor. Nat. Trieste **28**: 81–122.

Dalla Vecchia, F.M., Rigo, D., Tentor, M., Pacor, G. and Moratto, D. (2001) Il sito paleontologico di Polazzo (Gorizia): dati e prospettive, In: M.C. Perri (a cura di) *Giornate di Paleontologia 2001*, Giornale di Geologia, ser. 3, **62**(2000 Suppl). Bologna, pp. 151–156.

Delvalle, D. and Buser, S. (1990) Mikrofacies analysis of limestone from the Upper Cretaceous to the Lower Eocene of SW Slovenia (Yugoslavia). Geologija **31/32**: 351–394.

Dini, M., Tunis, G. and Venturini, S. (1998) Continental brackish and marine carbonates from the Lower Cretaceous of Kolone-Barbariga (Istria, Croatia): stratigraphy, sedimentology and geochemistry. Palaeogeogr. Palaeoclimatol. Palaeoecol. **140**: 245–269.

Dobruskina, I.A., Jurkovšek, B. and Kolar-Jurkovšek, T. (1999) Upper Cretaceous flora of Slovenia. Ann. Ser. Hist. Nat. **9**(2): 243–256.

Dolenec, T., Cucchi, F., Giacomich, R., Marton, E. and Ogorelec, B. (1995) Abiotic characteristics of carbonate rocks from the K/T boundary on the Karst area (isotops, geochemistry, geochronology and paleomagnetism), In: Fourth International Workshop ESF Sci. Network "Impact Cratering and Evolution of Planet Earth", Ancona May 1995, Abstracts and Field Trips, Ancona, pp. 68–69.

Drobne, K., Pavlovec, R. and Ćosović, V. (1991) Section Ragancini-Lišani, Lower and Middle Lutetian, In: *Introduction to the Paleogene SW Slovenia and Istria*. Field Trip Guidebook IGCP Project 286 "Early Paleogene Benthos" 2nd Meeting Postojna, Ljubljana, pp. 83–85.

Drobne, K., Ogorelec, B., Pleničar, M., Zucchi Stolfa, M.L. and Turnšek, D. (1988) Maastrichtian, Danian and Thanetian Beds in Dolenja Vas (NW Dinarides, Yugoslavia), Microfacies, Foraminifers, Rudists and Corals. Razprave 4. Razr. SAZU **29**: 147–224, 35 pls.

Drobne, K., Ogorelec, B., Pleničar, M., Barattolo, F., Turnšek, D. and Zucchi-Stolfa, M.L. (1989) The Dolenja Vas section, a transition from Cretaceous to Paleocene in the NW Dinarides, Yugoslavia. Mem. Soc. Geol. Ital. **40**(1987): 73–84.

Drobne, K., Ogorelec, B., Dolenec, T., Marton, E. and Palinkaš, L. (1996). Biota and abiota at the K/T boundary in the Dolenja Vas section, Slovenia, In: K. Drobne, Š. Goričan and B. Kotnik (eds.) *The Role of Impact Processes in the Geological and Biological Evolution of Planet Earth.* International Workshop, Geology of Western Slovenia Field-guide, Ljubljana, pp. 163–181.

Drobne, K., Ogorelec, B. and Riccamboni, R. (2007) *Bangiana hanseni* n.gen. n.sp. (Foraminifera), an index species of Danian age (Lower Paleocene) from the Adriatic carbonate platform (SW Slovenia, NE Italy, Herzegovina). Razprave 4. Razr. SAZU **48**(1): 5–71, 12 pls.

Gebelein, C.D. (1976) Open marine subtidal and intertidal stromatolites (Florida, the Bahamas and Bermuda), In: M.R. Walter (ed.) *Stromatolites. Developments in Sedimentology*. Elsevier, New York, pp. 381–388.

Gregorič, M., Caffau, M., Lenaz, D. and Demin, A. (1998) Late Maastricthian?-Paleocene unaltered glassy microspherules at Padriciano (Trieste Karst, NE Italy): a preliminary report. Razprave 4. Razr. SAZU **39**(6): 221–233.

Gušić, I. and Jelaska, V. (1990) Stratigrafija gornjokrednih naslaga otoka Brača u okviru geodinamske evolucije Jadranske karbonatne platforme (Upper Cretaceous Stratigraphy of the Island of Brač). Djela JAZU, Razr. Prir. Znan **69**: 160.

Hansen, H.J., Drobne, K. and Gwozdz, R. (1995) The K/T boundary in Slovenia: dating by magnetic susceptibility, Stratigraphy and iridium-anomaly in a debris flow, In: A. Montanari and R. Coccioni (eds.) ESF 4th International Workshop, Ancona, pp. 84–85.

Haq, B.U., Hardenbol, J. and Vail, P.R. (1987a). Mesozoic and Cenozoic chronostratigraphy and cycles of sea level changes: an integrated approach. Soc. Econ. Paleontol. Mineral. Spec. Publ. **42**: 71–108.

Haq, B.U., Hardenbol, J. and Vail, P.R. (1987b). Chronology of fluctuating sea levels since the Triassic. Science **23**: 1156–1167.

Herak, M. (1987) Relationship between Adriatic and Dinaric carbonate platforms. Mem. Soc. Geol. Ital. **40**: 289–293.

Hofmann, H.J., Grey, K., Hickman, A.H. and Thorpe, R.J. (1999) Origin of 3.45 G2 coniform stromatolites in Warrawoona Group, Western Australia. Geol. Soc. Am. Bull. **111**: 1256–1262.

Jenkyns, H.C. (1991) Impact of Cretaceous sea level rise and anoxic events in Mesozoic carbonate platform of Yugoslavia. Am. Assoc. Petrol. Geol. Bull. **75**(6): 1007–1017.

Jurkovšek, B. (2008) *Geological Map of the Northern Part of the Trieste-Komen plateau 1: 25.000*. Geological Survey of Slovenia, Ljubljana.

Jurkovšek, B. and Kolar-Jurkovšek, T. (2007) Fossil assemblages from the Upper Cretaceous Komen and Tomaj Limestones of Kras (Slovenia). N. Jb. Geol. Paläont. Abh. **245**(1): 83–92.

Jurkovšek, B., Toman, M., Ogorelec, B., Šribar, L., Šribar, L.J., Poljak, M. and Drobne, K. (1996) Formacijska geološka karta južnega dela Tržaško-komenske planote 1:50.000: kredne in paleogenske karbonatne kamnine (Geological map of the southern part of the Trieste-Komen Plateau: Cretaceous and Paleogene carbonate rocks). Inštitut za geologijo, geotehniko in geofiziko, Ljubljana, 143 pp.

Koch, R., Ogorelec, B. and Orehek, S. (1989) Microfacies and diagenesis of Lower and Middle Cretaceous carbonate rocks of NW-Yugoslavia (Slovenia, Trnovo Area). Facies **21**: 135–170.

Košir, A. (2004) *Microcodium* revisited: root calcification products of terrestrial plants on carbonate-rich substrates. J. Sediment. Res. **74**: 845–857.

Marton, E., Drobne, K., Cimerman, F., Ćosović , V. and Košir, A. (1995) *Paleomagnetism of Latest Maastrichtian Through Oligocene Rocks in Istria (Croatia), the Karst Region, and S of the Sava Fault (Slovenia)*. First Croatian Geological Congress, Proceedings, Opatija 2, pp. 355–360.

Nehza, O., Woo, K.S. and Lee, K.C. (2009) Combined textural and stable isotopic data as proxies for the mid-Cretaceous palaeoclimate: a case study of lacustrine stromatolites in the Gyeongsang Basin, SE Korea. Sediment. Geol. **214**: 85–99.

Ogorelec, B., Orehek, S., Buser, S. and Pleničar, M. (1987) Komen Beds – Skopo and Dutovlje, In: E. Colizza, R. Costa, F. Cucchi, N. Pugliese (eds.) *Guidebook Excursions of the International Symposium on the Evolution of the Karstic Carbonate Platform: Relation with other Periadriatic Carbonate Platforms*, pp. 61–66.

Ogorelec, B., Dolenec, T., Cucchi, F., Giacomich, R., Drobne, K. and Pugliese, N. (1995) Sedimentological and geochemical characteristics of carbonate rocks from the K/T boundary to Lower Eocene on the Karst area (NW Adriatic platform). First Croatian Geological Congress, Proceedings, Opatija 2, pp. 415–422.

Ogorelec, B., Drobne, K., Jurkovšek, B., Dolenec, T. and Toman, M. (2001) Paleocene beds of the Liburnia Formation in Čebulovica (Slovenia, NW Adriatic-Dinaric platform). Geologija **44**: 15–65, 11 pls.

Ogorelec, B., Dolenec, T. and Drobne, K. (2007) Cretaceous–Tertiary problem on shallow carbonate platform: carbon and oxygen excursions, biota and microfacies at the K/T boundary sections Dolenja Vas and Sopada in SW Slovenia, Adria CP. Palaeogeogr. Palaeoclimatol. Palaeoecol. **255**: 64–76.

Palci, A., Jurkovšek, B., Kolar-Jurkovšek, T. and Caldwell, W.M. (2008) New palaeonvironmental model for the Komen (Slovenia) Cenomanian (Upper Cretaceous) fossil *lagerstätte*. Cretaceous Res. **29**: 316–328.

Pavlovec, R. (1963). Stratigrafski razvoj starejšega paleogena v južnozahodni Sloveniji (Die stratigraphysche Entwicklung des ältern Paleogens im südwestlichen Teil Sloweniens). Razprave 4. Razr. SAZU, **7**, 421–556.

Pugliese, N., Drobne, K., Barattolo, F., Caffau, M., Galvani, R., Kedves, M., Montenegro, M.E., Pirini Radrizzani, C., Pleničar, M. and Turnšek, D. (1995) Micro and macrofossils from the K/T Boundary through Paleocene in the Northern Adriatic Platform. *1st Croatian Geological Congress Proceedings* 2, pp. 415–422.

Pugliese, N., Arbulla, D., Caffau, M. and Drobne, K. (2000) Strategie di vita nel biota daniano (SBZ1) del Carso Triestino (Italia), In: *Crisi biologiche, radiazioni adattative e dinamica delle piattaforme italiane*. Lettere ed Arti di Modena 21. Accademia nazionale delle Scienze, Collana Studi, pp. 215–220.

Rasmussen, K.A. and Neumann, A.C. (1988) Holocene overprint of Pleistocene paleokarst: bight of Abaco, Bahamas, In: N.P. James and P.W. Choquette (eds.) *Paleokarst*. Springer, New York, pp. 138–148.

Reid, R.P., Visscher, P.T., Dechu, A., Stolz, J.K., Bebout, B.M., Dupraz, C., Mac Intyre, J.G., Paerl, H.W., Pincheney, J.L., Aufert-Bebout, L., Steppe, T.F. and Des Marais, D.J. (2000) The role of microbes in accretion, lamination and early lithyfication of modern marine stromatolites. Nature **406**: 989–992.

Riccamboni, R. (2005) Evidenze del passaggio Cretacico-Terziario nel Carso triestino e sloveno. Unpublished thesis, University of Trieste, 205 pp.

Riding, R. (1991) Calcified cyanobacteria, In: R. Riding (ed.) *Calcareous Algae and Stromatolites*. Springer, Berlin, pp. 57–87.

Riding, R. (2000) Microbial carbonates: the geological record of calcified bacterial-algal mats and biofilms. Sedimentology, **47**(Suppl. 1): 179–214.

Sartorio, D., Tunis, G. and Venturini, S. (1997) The Iudriuo Valley section and the evolution of the northeastern margin of the Friuli platform (Iulian Prealps, NE Italy-W Slovenia). Mem. Soc. Geol. Ital. **49**: 163–193.

Schopf, J.W. and Walter, M.R. (1983) Archean microfossils: new evidence of ancient microbes, In: Schopt, J.W. (ed.) *Earth's Earliest Biosfere: Its Origin and Evolution* 1. Princeton University Press, Princeton, pp. 214–239.

Schopf, J.W., Tewari, V.C. and Kudryavtsev, A.B. (2008) Discovery of a new Chert permineralised microbiota in the Proterozoic Buxa Formation of the Ranjit Window, Sikkim, Northeast India, and its Astrobiological implications. Astrobiology **8**(4): 735–746.

Shukla, M., Tewari, V.C., Babu, R. and Sharma, A. (2006). Microfossils from the Neoproterozoic Buxa Dolomite, West Siang District, Arunachal Lesser Himalaya, India and their significance. J. Pal. Soc. India **51**: 57–73.

Stache, S.G. (1889) Die Liburnische Stufe und deren Grenzhorizonte. Abh. k. k. Geol. R. A. **13**(1): 1–170, 6 Taf.

Summesberger, H., Jurkovšek, B. and Kolar-Jurkovšek, T. (1996) Aptychi associated with ammonites from the Lipica-Formation (Upper Cretaceous, Slovenia). Ann. Nat. Hist. Mus. Wien, Ser. A Mineral. Petrogr. Geol. Paläontol. Anthropol. Prähist. **97A**: 1–19.

Summesberger, H., Jurkovšek, B. and Kolar-Jurkovšek, T. (1999a) Upper jaws of Placenticeratidae from the Karst Plateau (Upper Cretaceous, Slovenia). Ann. Nat. Hist. Mus. Wien, Ser. A Mineral. Petrogr. Geol. Paläontol. Anthropol. Prähist. **101A**: 119–122.

Summesberger, H., Jurkovšek, B. and Kolar-Jurkovšek, T. (1999b) Rollmarks of soft parts and a possible crop content of Late Cretaceous ammonites from the Slovenian karst, In: F. Olóriz and F.J. Rodríguez-Tovar (eds.) *Advancing Research on Living and Fossil Cephalopods: Development and Evolution: Form, Construction, and Function: Taphonomy, Palaeoecology, Palaeobiogeography, Biostratigraphy, and Basin Analysis*. Kluwer, New York, pp. 335–344.

Tentor, A. (2007) Osservazioni stratigrafiche sul Monte Brestovi (Carso goriziano). Nat. Nascosta **35**: 1–23.

Tentor, A and Tentor, M. (2007) Primo rinvenimento di Frenelopsis nel Cenomaniano superiore del Carso Isontino (Colle di Moschenizza, Monfalcone, Gorizia). Nat. Nascosta **34**: 6–15.

Tentor, M., Tunis, G. and Venturini, S. (1994) Schema stratigrafico-tettonico del Carso isontino. confine del Carso triestino. Nat. Nascosta **9**: 1–32.

Tentor, M., Tunis, G. and Venturini, S. (2000) I depositi infra-cretacici della zona di confine del Carso triestino. Nat. Nascosta **1**(21): 1–28.

Tentor, M., Tunis, G. and Venturini, S. (2002) Segnalazione di livelli valanginiano-berrasiani nella successione del Carso: Plateau di Trieste-Komen (Slovenia occidentale). Mem. Soc. Geol. Ital. **57**: 153–164.

Tewari, V.C. (2004) Microbial diversity in Meso-Neoproterozoic Formations, with particular reference to the Himalaya. In: J. Seckbach (ed.) *Origins, Cellular Origin, Life in Extreme Habitats and Astrobiology*, **6**, Kluwer, The Netherlands, pp. 515–528.

Tewari, V.C. (2007) The rise and decline of the Ediacaran biota: palaeobiological and stable isotopic evidence from the NW and NE Lesser Himalaya, India, In: P. Vickers-Rich and P. Komarower (eds.) *Rise and Fall of the Ediacaran Biota. Special Publication 286.* Geological Society of London, pp. 77–101.

Tewari, V.C., Stenni, B., Pugliese, N., Drobne, K., Riccamboni, R. and Dolenec, T. (2007) Peritidal sedimentary depositional dacies and carbon isotope variation across K/T boundary carbonates from NW Adriatic platform. Palaeogeogr. Palaeoclimatol. Palaeoecol. **255**: 77–86.

Tewari, V.C., Kumar, K., Lokho, K and Siddaiah, S. (2010) Lakadong Limestone: Paleocene–Eocene boundary carbonate sedimentation in Meghalaya, northeastern India. Curr. Sci. **98**(1): 88–94.

Turnšek, D. and Drobne, K. (1998) Paleocene corals from the northern Adriatic platform, In: L. Hottinger and K. Drobne (eds.) *Paleogene Shallow Benthos of the Tethys*. 2. DELA – Opera SAZU, Cl. 4, 34(2), 129–154, 10 Pls.

Velić, I., Tišljar, J., Vlahović, I., Matičec, D. and Bergant, S. (2003) Evolution of the Istrian part of the Adriatic Carbonate Platform from the Middle Jurassic to the Santonian and Formation of the flysch basin during the Eocene: main events and regional comparison, In: I. Vlahović and J. Tišljar (eds.) *Evolution of Depositional Environments from the Palaeozoic to the Quaternary in the Karst Dinarides and the Pannonian Basin.* Field Trip Guidebook 22nd IAS Meeting of Sedimentology, Opatija, September 17–19, 2003, Zagreb (2003), pp. 3–17.

Venturini, S., Tentor, M. and Tunis, G. (2008) Episodi continentali e dulcicoli ed eventi biostratigrafici nella sezione campaniano-maastrichtiana di Cotici (Mt. San Michele, Gorizia). Nat. Nascosta **36**: 6–23.

Vlahovič, I., Tišljar, J., Fuček, L., Oštric, N., Prtoljan, B., Velić, I. and Matičec, D. (2002) The origin and importance of the dolomite-limestone breccia between the Lower and Upper Cretaceous deposits of the Adriatic Carbonate Platform: an example from Čičarija Mt (Istria, Croatia). Geol. Croat. **55**(1): 45–55.

Webb, G.E. (1996) Was Phanerozoic reef history controlled by the distribution of non-enzymatically secreted reef carbonates (microbial carbonate and biologically induced cement?). Sedimentology **43**: 947–971.

Biodata of **Dr. Atsushi Yamamoto, Emeritus Professor Kwang-Choon Lee**, and **Professor Yukio Isozaki**, authors of ***"Lower Cretaceous Stromatolites in Far East Asia: Examples in Japan and Korea"***

Dr. Atsushi Yamamoto, currently at the Department of Life Science, Kinki University (Japan), obtained his Ph.D. from the University of Tokyo in 2009. His research focuses on the morphological evolution of fossil stromatolites with respect to recent cyanobacterial biomats, and on the *in situ* experiments of forming stromatolites from plane cyanobacterial biomats.

E-mail: **atsushi@life.kindai.ac.jp**

Emeritus professor Kwang-Choon Lee of Department of Mineral and Mining Engineering at Sangji University (Korea) is currently studying Cretaceous nonmarine stromatolites, and caves. Also he is concerned in conservation and sustainable development of geological heritage, i.e., designation of natural monument or geopark, etc.

E-mail: **kclee@sangji.ac.kr**

Atsushi Yamamoto

Kwang-Choon Lee

V.C. Tewari and J. Seckbach (eds.), *STROMATOLITES: Interaction of Microbes with Sediments*, Cellular Origin, Life in Extreme Habitats and Astrobiology 18, 273–287
DOI 10.1007/978-94-007-0397-1_12, © Springer Science+Business Media B.V. 2011

Professor Yukio Isozaki of the University of Tokyo is currently studying the mass extinction and relevant environmental changes across the Paleozoic–Mesozoic boundary. He and his colleagues clarified that the famous, oldest (Early Archean) "stromatolites" from the Pilbara craton (Western Australia) actually occur in deep-sea beds deposited around an ancient mid-oceanic ridge with hydrothermal activity, denying the long-believed photosynthetic cyanobacterial origin in shallow marine settings.

E-mail: **isozaki@ea.c.u-tokyo.ac.jp**

LOWER CRETACEOUS STROMATOLITES IN FAR EAST ASIA: EXAMPLES IN JAPAN AND KOREA

ATSUSHI YAMAMOTO[1,3], KWANG-CHOON LEE[2], AND YUKIO ISOZAKI[1]

[1]*Department of Earth Science and Astronomy, The University of Tokyo, Komaba, Meguro, Tokyo 153-8902, Japan*

[2]*Department of Mineral and Mining Engineering, Sangji University, Wonju-si, Gangwon-do 220-702, South Korea*

[3]*Department of Life Science, Kinki University, Kowakae, Osaka 577-8502, Japan*

Abstract The morphology, structure, and mode of occurrence of the Early Cretaceous nonmarine stromatolites from the Kanmon Group in northern Kyushu (SW Japan) and from the Gyeongsang Supergroup in South Korea are reviewed. These stromatolites were formed likely in intertidal–supratidal settings of a large intracontinental lake (the Kanmon–Gyeongsang Basin) that were occasionally exposed to arid condition. Seven macrostructures of stromatolite are recognized. In particular, three of them are dominant, i.e., flat type, columnar type, and domal type. The thickness of clastic layers of the stromatolites and the associated sedimentary facies suggest that the flux of terrigenous clastics into the depositional site may have controlled the morphological diversity of the stromatolites, i.e., the flat type occurred in very shallow nearshore (supratidal), while the domal type in relatively deeper, offshore (intertidal).

Keywords Lower Cretaceous • Japan • Korea • Lacustrine • Paleoenvironment • Stromatolite • Nonmarine • Intertidal

1. Introduction

With respect to the Precambrian world, stromatolites are relatively rare in the Phanerozoic except some specific intervals such as the Cretaceous (Awramik and Sprinkle, 1999; Riding, 2006). The occurrences of Cretaceous stromatolite were reported mainly from coral/sponge reefs in the circum-Atlantic regions, such as Spain (e.g., Neuweiler et al., 1999) and Mexico (e.g., Beraldi-Campesi et al., 2004). Most of them are commonly associated with other calcareous biota, thus do not form stromatolite-dominant reefs like the Precambrian examples (Leinfelder and Schmid, 2000). The Cretaceous stromatolites from Japan and Korea are unique in this regard, because they provide rare examples of monotonously stromatolite-dominant units within terrigenous clastic packages. Such a pure stromatolite reef might form only in nonmarine settings. Dominant reef builders of Early Cretaceous, such as corals and rudists, could not inhabit nonmarine environments. These Cretaceous stromatolites in East Asia are analog to the Precambrian stromatolites in morphology and probably also in ecology.

The occurrences of Lower Cretaceous stromatolites in East Asia were recorded already in the early twentieth century: first from the Jinju area in South Korea (Kodaira, 1922) and then from northern Kyushu in Japan (Kobayashi, 1936). Later, Ishijima (1977) and Seo et al. (1994) added descriptions of the stromatolites from northern Kyushu and discussed their fresh-water origin. Also in S. Korea, many stromatolite beds were described from the vicinity of Daegu by Lee et al. (1991), whose finding was followed up by later studies focused on morphology, internal structure, and carbon and oxygen isotopic ratios of the stromatolites (e.g., Woo et al., 2002, 2004; Nehza and Woo, 2006; Nehza et al., 2009). Recently, Yamamoto et al. (2009) clarified detailed mode of occurrence of the stromatolite from northern Kyushu, with special attention to their close association with evaporitic sedimentary features and to their branching pattern. By analyzing carbon/oxygen isotopic ratios in a lamina by lamina manner for the S. Korean specimens, Nehza et al. (2009) speculated that seasonal return of arid condition formed the rhythmic alternation of the stromatolites. In this short article, fundamental characteristics of these unique Cretaceous stromatolites in Japan and S. Korea are reviewed, and their geological implications are briefly discussed.

2. Stromatolite-Bearing Lower Cretaceous Strata

The Lower Cretaceous stromatolites occur in nonmarine strata in both N. Kyushu and S. Korea (Fig. 1). The Lower Cretaceous rocks in N. Kyushu and western Honshu (main island of Japan) are called the Kanmon Group (Matsumoto et al., 1982), while that in S. Korea the Gyeongsang Supergroup (Chang et al., 2003). The nonmarine Cretaceous strata of N. Kyushu and S. Korea, currently separated by the Tsushima strait, were likely deposited in the same large fresh-water basin (e.g., Kobayashi, 1941; Chough et al., 2000), which is tentatively called the Kanmon–Gyeongsang Basin in this article. The Cretaceous rocks in the two areas are correlated with each other on the basis of the lithologic similarity and molluscan fossils (Hase, 1960; Ota, 1960).

The Kanmon Group consists of two subgroups: the Wakino Subgroup (lower) and the Shimonoseki Subgroup (upper) (Fig. 2). The Kanmon Group overlies unconformably the basement composed of the Carboniferous high-P/T metamorphic rocks (Nishimura and Shibata, 1989), and is in turn covered unconformably by the mid-Cretaceous Yahata Formation (Ota and Yabumoto, 1998). The Wakino Subgroup is composed mainly of well-bedded shale, sandstone, and conglomerate (e.g., Ota, 1953). The Shimonoseki Subgroup is composed of sandstone, tuff, and andesite lava (e.g., Hase, 1960; Ota and Yabumoto, 1998). The depositional setting of the Kanmon Group was regarded as a mid-latitude,

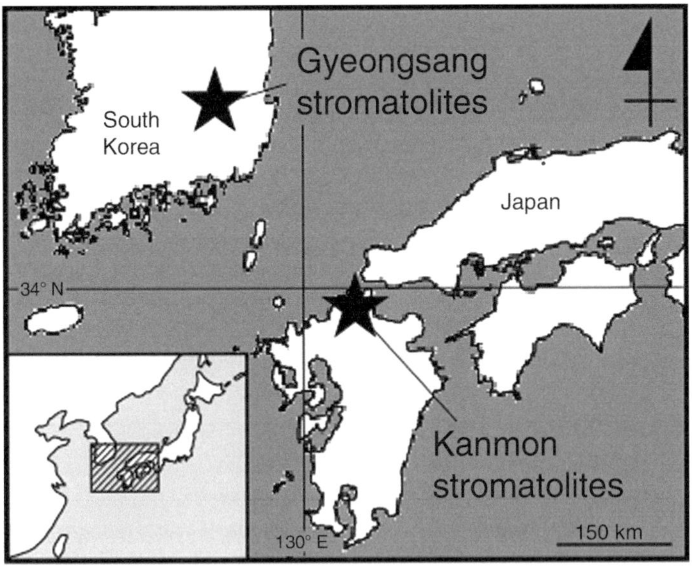

Figure 1. The Early Cretaceous stromatolite localities in N. Kyushu (Japan) and S. Korea.

Figure 2. Correlation of the Lower Cretaceous strata in N. Kyushu and S. Korea. Stromatolites occur from the shaded formations [compiled from Ishijima (1977), Lee et al. (1991), Lee and Woo (1996), Seo et al. (1994) and Yamamoto et al. (2009)].

intracontinental lacustrine environment under arid to semiarid climate (e.g., Ota and Yabumoto, 1998). The stromatolites occur in the Sengoku, Lower Wakamiya, and Upper Wakamiya formations of the Wakino Subgroup (Fig. 2; Ishijima, 1977; Yamamoto et al., 2009).

The Gyeongsang Supergroup in S. Korea is composed of the Sindong, Hayang, and Yucheon groups in ascending order (Fig. 2; e.g., Chang, 1975; Choi, 1985; Huh and Hayashi, 2001). The Sindong Group is composed of shale, sandstone, and conglomerate (Chang, 1975). The Hayang Group is composed mainly of shale, sandstone, and conglomerate, with minor amount of volcanic rocks that range from mafic to felsic in composition (Chang, 1975; Choi, 1985; Huh and Hayashi, 2001). The Yucheon Group is composed mainly of felsic lava and tuff with subordinate sedimentary rocks (Chang, 1975). The Gyeongsang Supergroup was regarded as a lacustrine sedimentary unit deposited in an intracontinental basin (e.g., Chough et al., 2000; Nehza et al., 2009) like the Kanmon Group in Japan. The stromatolites occur in the Jinju Formation of the Sindong Group (Lee and Woo, 1996), and also in the Banyawol, Hwasan, and Sinyangdong Formations of the Hayang Group (Fig. 2; Lee and Woo, 1996).

The Kanmon Group and the Gyeongsang Supergroup are mutually correlated (Fig. 2) on the basis of their lithologic similarity and molluscan fossils, such as Trigonioididae bivalves (Hase, 1960; Ota, 1960). The Early Cretaceous charophyte, *Clypeator jiuquanensis* (the Hauterivian to Early Barremian), from the Nakdong Formation and the radiometric ages of intrusive rocks (120–82 Ma) constrain that the Kanmon Group and the Gyeongsang Supergroup were deposited probably during the Hauterivian to Albian (Early Cretaceous) age (Chang, 1997; Chang et al., 1997; Matsumoto et al., 1982; Ota and Yabumoto, 1998). The Jinju Formation in S. Korea is correlated with the stromatolite-bearing Lower Wakamiya Formation in Kyushu.

3. Morphotypes of Stromatolites

3.1. STROMATOLITES FROM THE KANMON GROUP

All the Cretaceous stromatolites from N. Kyushu occur in the black shale (e.g., Fig. 4c). The stromatolite-bearing beds are intercalated in the conglomerate-dominant sections (Fig. 3a). The Lower Wakamiya Formation yields stromatolites in four horizons (Fig. 3a). Stromatolites of the Lower Wakamiya Formation have three macrostructures (Table 1): i.e., flat type (Fig. 4a), columnar type (Fig. 4b), and nodular type (Fig. 4c). In particular, the former two types are abundant. The flat-type stromatolite is composed of horizontal laminae of about 0.1 mm thick in average. The columnar-type stromatolite consists of composite columns of ca. 10 cm height. The nodular-type stromatolite is composed of composite spherical structures oriented randomly. The laminae of both columnar and nodular types are about 0.1 mm thick. Among the three, the columnar-type stromatolite is the biggest in size that reaches about 30 cm in height. The flat-type stromatolite consists of alternating clastic layers and micritic layers. The flat type is unique in

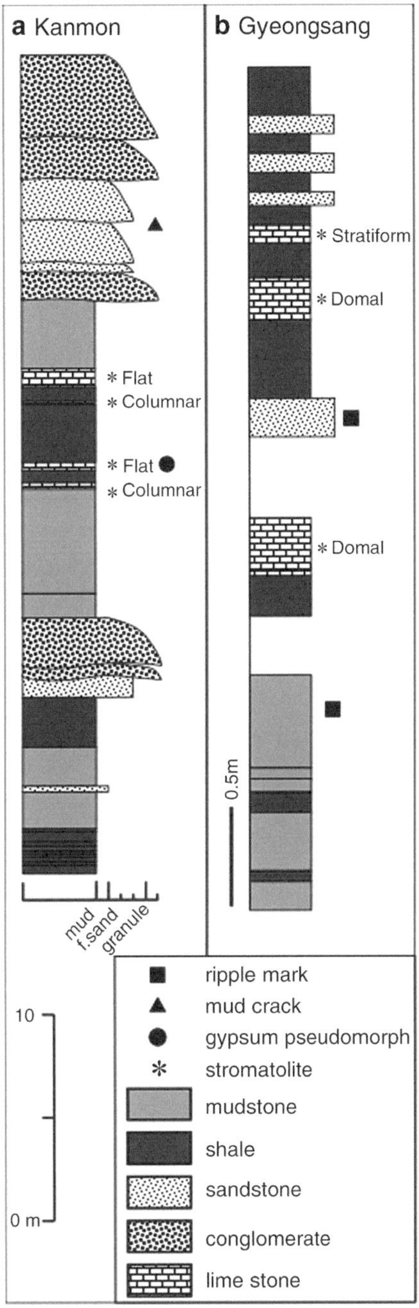

Figure 3. Columnar sections of typical stromatolite-bearing strata: (**a**) the Lower Wakamiya Formation [J apan, modified from Yamamoto et al. (2009)], (**b**) the Sinyangdong Formation [Korea, modified from Nehza and Woo (2006)].

Table 1. Seven macrostructures of the Kanmon and Gyeongsang stromatolites. Flat- and columnar-type stromatolites in Japan and domal type in S. Korea are dominant.

Locality	Macrostructure	Secondary structure	Alternation	Abundance
Japan	Flat type	Smooth lamina	Clastics/micrite	Abundant
	Columnar type	Column	Siliciclastic-rich micrite/ organic carbon-rich	Abundant
	Nodular type	Column	Siliciclastic-rich micrite/ organic carbon-rich	Rare
S. Korea	Stratiform type	Column/smooth lamina	Fibrous-calcite/micrite	Rare
	Domal type	Column/smooth lamina	Fibrous-calcite/micrite	Abundant
	Oncolitic type	Smooth lamina	Fibrous-calcite/micrite	Common
	Rod type	Column/smooth lamina	Fibrous-calcite/organic carbon	Rare

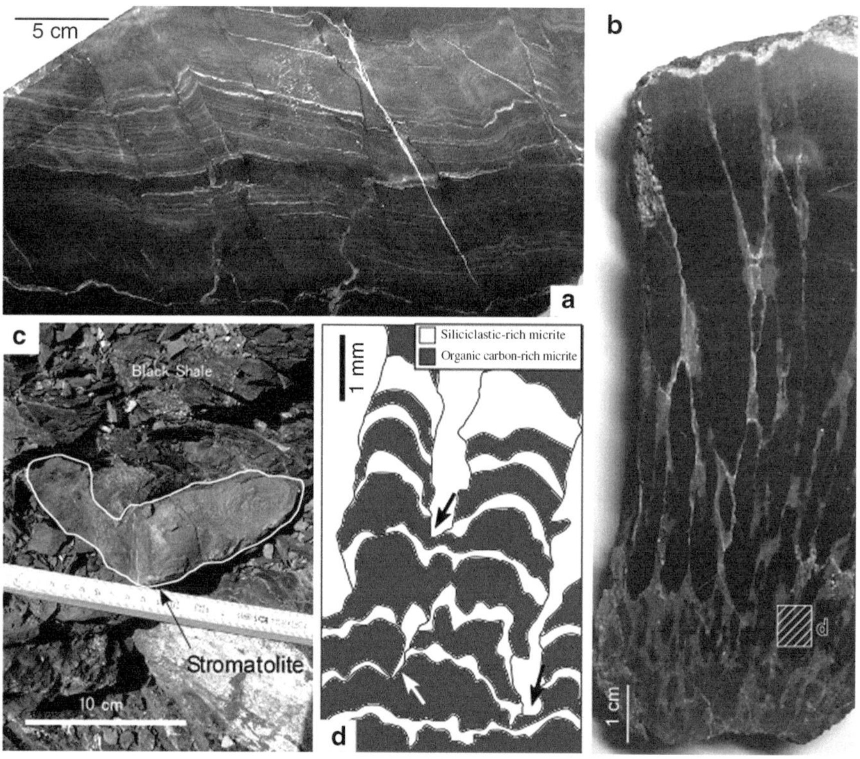

Figure 4. Morphological variation of Kanmon stromatolites: (**a**) flat type, (**b**) columnar type. A *white square* indicates the illustrated area of (**d**, **c**) nodular type born from the black shale, (**d**) branching observed in columnar type. *Arrows* show branching points.

intercalating occasionally siliciclastic laminae of over 0.3 mm thick, whereas the columnar- and nodular-type stromatolites lack the clastic layers. The internal laminations of the rest two types consist of alternating siliciclastic-rich micrite lamina and organic carbon-rich micrite lamina. On the basis of the random distribution of the internal spheres and the types of alternation, the nodular-type stromatolites are regarded as broken forms of the columnar type. The micritic laminae of all three types likely correspond to primary biomats (Yamamoto et al., 2009).

3.2. STROMATOLITES FROM THE GYEONGSANG SUPERGROUP

The stromatolites from the Gyeongsang Supergroup occur in the black shale just like those from the Kanmon Group in Japan (Fig. 3b; e.g., Lee et al., 1991). Nonetheless, in contrast to those of the Kanmon Group, the stromatolite-bearing shales of the Gyeongsang Supergroup are commonly associated with fine-grained sandstone, and never occur with coarse-grained sandstone or conglomerates (e.g., Lee et al., 1991). The stromatolites of the Gyeongsang Supergroup have four macrostructures (Table 1), i.e., domal type (Fig. 5a, b, e), stratiform type, oncolite type (Fig. 5c), and rod type (Fig. 5d). The Hayang Group of the Middle Gyeongsang Supergroup yields three of the four types: domal type, oncolite type, and stratiform type. The domal type is abundant and includes domes with smooth laminae (Fig. 5a) and/or domes with columnar laminae (Fig. 5b). The maximum diameter of stromatolites from the Hayang Group reaches 1 m or more (Fig. 5e).

All the stromatolites from the Hayang Group are composed of alternating micritic lamina and fibrous-calcite lamina (Woo et al., 2004; Nehza and Woo, 2006). The thickness of each micritic/fibrous-calcite couplet is about 2–5 mm (Nehza and Woo, 2006). The micrite laminae may have microbial origin, whereas the fibrous-calcite laminae might have been chemically precipitated (Woo et al., 2002). This common character in the alternation suggests that three types of the Hayang stromatolites were formed under almost the same environment. The stromatolites from the Hayang Group are composed of both micrite and fibrous-calcite laminae that include a minor amount of terrigenous clastics; however, clastic-dominant laminae that are common in the Kanmon Group are absent (Nehza et al., 2009; Nehza and Woo, 2006; Woo et al., 2002).

The stromatolites from the Jinju Formation (the Sindong Group) are of the rod type (Fig. 5d) and unique in having a cavity in the center (Lee and Woo, 1996). This cavity may have derived from a wood trunk around which the stromatolite columns/laminae grew, and the putative wood itself may have decayed out later to leave a central cavity (Lee and Woo, 1996). The internal columns/laminae of the rod-type stromatolites are very similar to the secondary columns/laminae of the columnar type of the Japanese stromatolites.

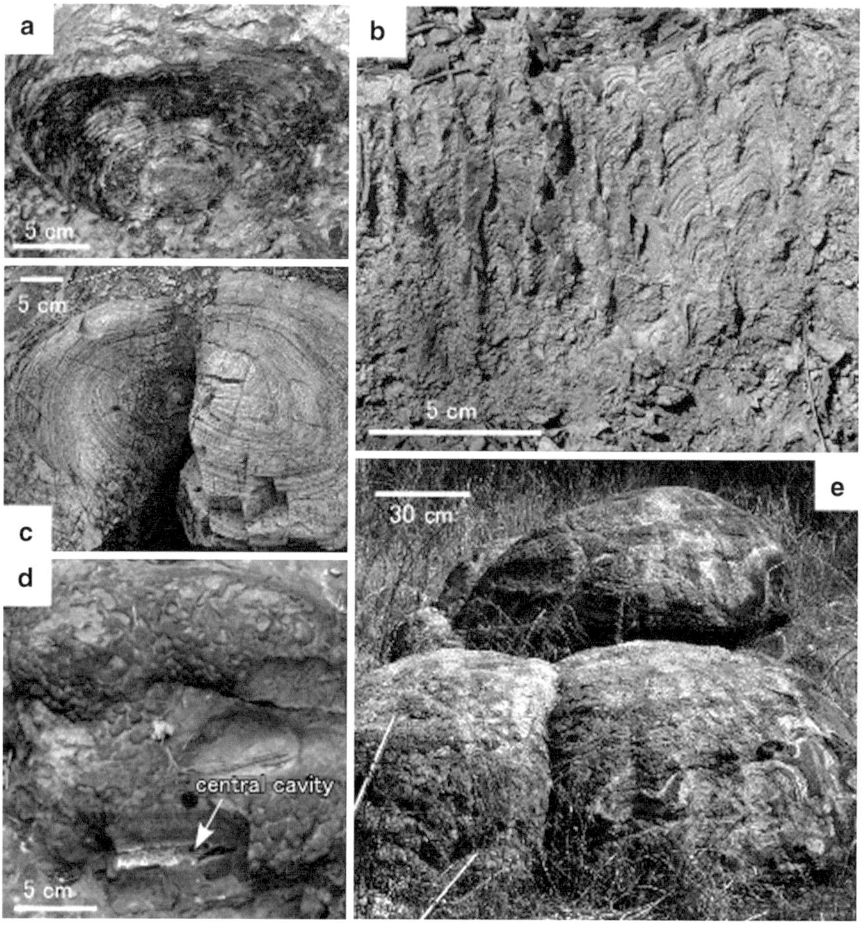

Figure 5. Morphological variation of Gyeongsang stromatolites: (**a**) domal stromatolites with smooth laminae (Hwasan Formation), (**b**) domal stromatolites with many columns (Sinyangdong Formation), (**c**) oncolitic stromatolite (Hwasan Formation), (**d**) rod-shaped Jinju stromatolites (overlook view), (**e**) huge domal stromatolites (Banyawol Formation).

4. Discussion

4.1. STROMATOLITE-FORMING ENVIRONMENTS

As discussed above, the Lower Cretaceous stromatolites from N. Kyushu and S. Korea have several similarities and dissimilarities. Despite the slight difference in age, all these stromatolites occur in black shales with mud cracks/gypsum pseudomorphs (e.g., Lee et al., 1991; Yamamoto et al., 2009). These sedimentary features indicate that the stromatolites have formed in a very shallow-water environment, such as intertidal to supratidal conditions where stromatolites were occasionally exposed to arid conditions (e.g., Lee et al., 1991; Nehza and Woo, 2006; Yamamoto et al., 2009).

The occurrence of apparent laminae composed dominantly by terrigenous clastics characterizes the stromatolites of the Kanmon Group, whereas those of the Gyeongsang Supergroup completely lack them (see Sect. 3). Among the three dominant types of the Kanmon and Gyeongsang stromatolites (Table 1), the flat type has thicker clastic laminae (sometimes >0.3 mm) than the columnar type, and the domal type lacks them. These differences in clastic content among the studied stromatolites suggest that morphological variations were controlled by the physical distance from the source of terrigenous clastics. The flat-type stromatolites likely formed in the nearest environment to the source with the highest clastic flux. The domal-type stromatolite probably formed in relatively offshore environments away from the source, because the clastic content is quite small. Thus, the successive form change in stromatolites, from the flat type to the columnar type and then to the domal type, probably represents a lateral transition of sedimentary facies from nearshore (supratidal) to offshore (intertidal) (Fig. 6). Accordingly, as to the stromatolite-bearing beds, the Kanmon Group was deposited in a proximal setting, whereas the Gyeongsang sections in distal setting with respect to the provenance of terrigenous clastics. The intimate association of the stromatolite-bearing shales and conglomerates in the Kanmon Group (Yamamoto et al., 2009) supports the above interpretation.

Previous studies (e.g., Hoffman, 1976) suggest that sedimentation rate and location within coastal settings are likely two major controlling factors that affect the morphology of modern stromatolites. The domal-type stromatolite commonly occurs in relatively offshore environments, whereas the flat-type stromatolites in nearshore environments in Shark Bay (Logan, 1961). The distribution of the columnar-type stromatolite is restricted between those of the domal type and the flat type. According to such relationships among morphology, formed environment and sedimentation rate recognized in modern stromatolites, the flat-type

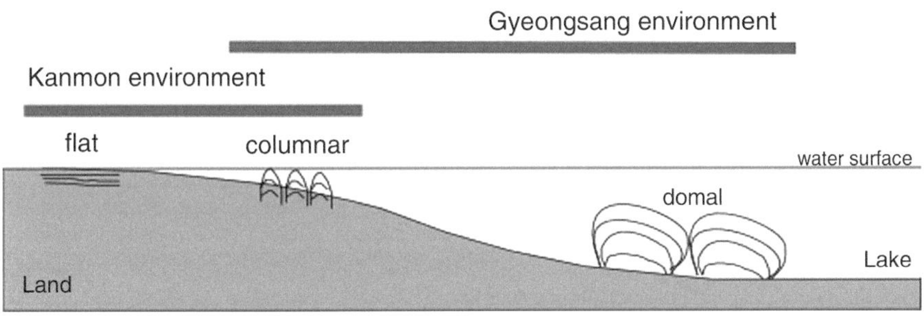

Figure 6. Estimated environment of abundant Kanmon and Gyeongsang stromatolites. The stromatolites formed in supratidal to intertidal environments. Gyeongsang stromatolites were formed more offshore than Kanmon stromatolites. Morphological change of the both stromatolites was probably suited to that of modern examples.

stromatolites of the Kanmon Group likely formed in a nearshore environment with higher sedimentation rate, whereas the domal-type stromatolites from the Gyeongsang Supergroup in an offshore environment with lower sedimentation rate (Fig. 6).

4.2. TRANSITION OF STROMATOLITE-FORMING AREA

In Early Cretaceous East Asia, the Kanmon Group in N. Kyushu and the Gyeongsang Supergroup in S. Korea likely formed in the southeastern and northwestern parts of the large nonmarine intracontinental Kanmon–Gyeongsang Basin, respectively (e.g., Chough et al., 2000). According to the paleogeographic reconstruction of Cretaceous East Asia (Fig. 7a; Chough et al., 2000; Maruyama et al., 1997), the Kanmon Group was deposited in the southeastern coast of the Kanmon–Gyeongsang Basin, whereas the Gyeongsang Supergroup in the northwestern coast. During the Wakino–Sindong time (Fig. 2), stromatolites formed probably all around the basin (Fig. 7b) including N. Kyushu and S. Korea. On the other hand, during the Shimonoseki–Hayang time (Fig. 2), stromatolites occurred only in S. Korea, particularly along the northwestern coast of the Kanmon–Gyeongsang Basin (Fig. 7c). This fact may indicate that the preexisting, stromatolite-forming stable environments have been modified considerably in the beginning of the Shimonoseki–Hayang time (ca. 100 Ma).

In the Early Cretaceous, a large-scale subduction-related magmatism occurred in East Asia to form the Ryoke granite batholith belt in SW Japan arc, as a result of the subduction of the Izanagi–Pacific mid-oceanic ridge beneath Asia (Isozaki and Maruyama, 1991; Isozaki, 1996; Maruyama et al., 1997). In N. Kyushu, the Early Cretaceous arc-related volcanic rocks (Kitahikoshima Formation)

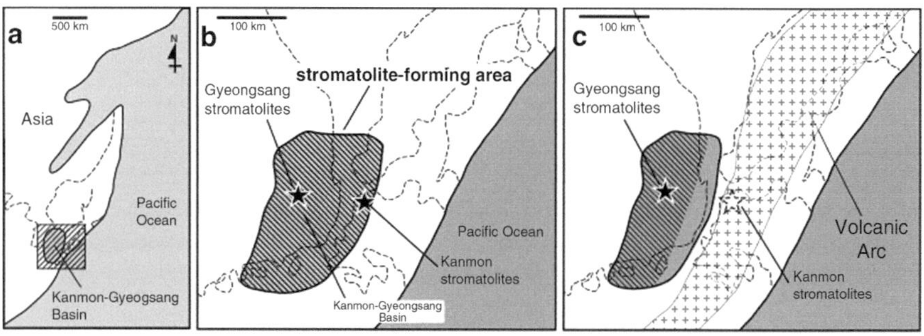

Figure 7. Change of stromatolite-forming area in the Kanmon–Gyeongsang Basin. Stromatolites were formed (**a**) the Kanmon–Gyeongsang Basin in the Early Cretaceous East Asia. The *shadowed square* indicates the area shown in (**b** and **c**), (**b**) Wakino–Sindong time. Stromatolites were formed all around the basin, (**c**) Shimonoseki–Hayang time. Stromatolites were formed only in the west side of the basin because of volcanic activities. *Dashed lines* show previous position of Japanese islands and Korean Peninsula [modified from Chough et al. (2000) and Maruyama et al. (1997)].

formed during the Shimonoseki time (Ota and Yabumoto, 1998). This intensive volcanism and associated uplift of the arc complex likely increased the sedimentary flux into the Kanmon–Gyeongsang Basin, in particular to its southeastern margin, to change the basin geometry. During the shrinkage of the basin on its eastern side (Fig. 7c), the stromatolite-forming environments were destroyed or migrated to the west.

5. Summary

1. The Early Cretaceous stromatolites of the Kanmon Group (SW Japan) and the Gyeongsang Supergroup (S. Korea) were formed in intertidal–supratidal environments, occasionally exposed to arid conditions, along the large-scale intracontinental lake margin in East Asia.
2. Three abundant types of stromatolites are recognized, i.e., flat type, columnar type, and domal type. The thickness of clastic layers of the stromatolites and the associated sedimentary facies suggest that the flux of terrigenous clastics into the depositional site may have controlled this morphological diversity of the Early Cretaceous nonmarine stromatolites. The flat type occurred in very shallow proximal settings, while the domal type in relatively deeper, distal ones.

6. Acknowledgments

We thank Professor Vinod C. Tewari for giving us a chance of writing this article. We thank Prof. Kazushige Tanabe for giving us many beneficial and critical advices about the Kanmon stromatolites. We also thank Dr. Atsushi Fujii for giving us valuable information about the locality of Kanmon stromatolites. This study was supported by Grant-in-Aid from Japan Society of Promotion of Sciences (no. 20224012 to Y.I.).

7. References

Awramik, S.M. and Sprinkle, J. (1999) Proterozoic stromatolites: the first marine evolutionary biota. Hist. Biol. **13**: 241–253.

Beraldi-Campesi, H., Cevallos-Ferriz, S.R.S. and Chacón-Baca, E. (2004) Microfossil algae associated with Cretaceous stromatolites in the Tarahumara Formation, Sonora, Mexico. Cretaceous Res. **25**: 249–265.

Chang, K.H. (1975) Cretaceous stratigraphy of southeast Korea. J. Geol. Soc. Korea **11**: 1–23.

Chang, K.H. (1997) Korean Peninsula, In: E.M. Moores and R.W. Fairbridge (eds.) *Encyclopedia of European and Asian Regional Geology*. Chapman and Halls, London, pp. 465–473.

Chang, K.H., Park, S.O. and Kim, H.S. (1997) Cretaceous stratigraphy and geologic history of medial Kyongsang Basin: tectonic and volcanism. Geosci. J. **1**: 2–9.

Chang, K.H., Suzuki, K., Park, S.O., Ishida, K. and Uno, K. (2003) Recent advance in the Cretaceous stratigraphy of Korea. J. Asian Earth Sci. **21**: 937–948.

Choi, H.I. (1985) Sedimentation and evolution of the Cretaceous Gyeongsang Basin, southeastern Korea. J. Geol. Soc. **143**: 29–40.

Chough, S.K., Kwon, S.T., Ree, J.H. and Choi, D.K. (2000) Tectonic and sedimentary evolution of the Korean peninsula: a review and new view. Earth Sci. Rev. **52**: 175–235.

Hase, A. (1960) The Late Mesozoic formations and their molluscan fossils in West Chugoku and North Kyushu, Japan. J. Sci. Hiroshima Univ. Ser. C **3**: 281–342.

Hoffman, P. (1976) Stromatolite morphogenesis in Shark Bay, Western Australia, In: M.R. Walter (ed.) *Stromatolites*. Developments in Sedimentology 20. Elsevier Scientific Publishing Company, Amsterdam, pp. 261–271.

Huh, M. and Hayashi, K. (2001) Nonmarine Cretaceous ostracoda and dinosaur sites in South Korea. ISO 2001, Field Excursion Guidebook, pp. 1–20.

Ishijima, W. (1977) Cretaceous algal stromatolites from Kokura, Kitakyushu City, Japan. Bull. Natl. Sci. Mus. Ser. C (Geol.) **3**: 175–182.

Isozaki, Y. (1996) Anatomy and genesis of a subduction-related orogen: a new view of geotectonic subdivision and evolution of the Japanese Islands. Island Arc **5**: 289–320.

Isozaki, Y. and Maruyama, S. (1991) Studies on orogeny based on plate tectonics in Japan and new geotectonic subdivision of the Japanese Islands. J. Geogr. **100**: 697–761 (in Japanese with English Abstract).

Kobayashi, T. (1936) The geologic structure of southwestern Japan and its Mesozoic palaeogeography (Pt. 4). J. Geol. Soc. Japan **43**: 531–541 (in Japanese).

Kobayashi, T. (1941) Sakawa orogenic cycle and its bearing on the Japanese Islands Chapter 8, The Cretaceous formations of the Tsushima Facies and meso-volcanism. J. Fac. Sci. Imp. Univ. Tokyo Sect. 2 **5**: 401–406.

Kodaira, R. (1922) *Geology of Chin-jyu district, Korea (MS)*. The graduation thesis of the Imperia. University of Tokyo.

Lee, K.C. and Woo, K.S. (1996) Lacustrine stromatolites and diagenetic history of carbonate rocks of Chinju Formation in Kunwi area, Kyongsangbukdo, Korea. J. Geol. Soc. Korea **32**: 351–365.

Lee, K.C., Woo, K.S., Paik, K.H. and Choi, S.J. (1991) Depositional environments and diagenetic history of the Panyawol, Hwasan, and Shinyangdong Formations, Kyongsang Supergroup, Korea. With emphasis on carbonate rocks. J. Geol. Soc. Korea **27**: 471–492.

Leinfelder, R.R. and Schmid, D.U. (2000) Mesozoic reefal thrombolites and other microbolites, In: R.E. Riding and S.M. Awramik (eds.) *Microbial Sediments*. Springer, Berlin, pp. 289–294.

Logan, B.W. (1961) *Cryptozoon* and associate stromatolites from the Recent, Shark Bay, Western Australia. J. Geol. **69**: 517–533.

Maruyama, S., Isozaki, Y., Kimura, G. and Terabayashi, M. (1997) Paleogeographic maps of the Japanese islands: plate tectonic synthesis from 750 Ma to the present. Island Arc **6**: 121–142.

Matsumoto, T., Obata, I., Tashiro, M., Ohta, Y., Tamura, M., Matsukawa, M. and Tanaka, H. (1982) Correlation of marine and nonmarine formations in the Cretaceous of Japan. Fossils **31**: 1–26 (in Japanese with English abstract).

Nehza, O. and Woo, K.S. (2006) The effect of subaerial exposure on the morphology and microstructure of stromatolites in the Cretaceous Sinyangdong Formation, Gyeongsang Supergroup, Korea. Sedimentology **53**: 1121–1133.

Nehza, O., Woo, K.S. and Lee, K.C. (2009) Combined textural and stable isotopic data as proxies for the mid-Cretaceous paleoclimate: a case study of lacustrine stromatolites in the Gyeongsang Basin, SE Korea. Sediment. Geol. **214**: 85–99.

Neuweiler, F., Gautrer, P., Thiel, V., Lange, R., Michaelis, W. and Reitner, J. (1999) Petrology of Lower Cretaceous carbonate mud mounds (Albian, N. Spain): insights into organomineralic deposits of the geological record. Sedimentology **46**: 837–859.

Nishimura, Y and Shibata, K. (1989) Mode of occurrence and K-Ar ages of metagabbroic rocks in the "Sangun metamorphic belt", Southwest Japan. Mem. Geol. Soc. Japan **33**: 343–357 (in Japanese with English abstract).

Ota, Y. (1953) Geological study on the late Mesozoic system in northern Kyushu (First). On the Mesozoic system in Mt. Kasagi district (so-called Wakino district), Kurate, Fukuoka. Bull. Fukuoka Gakugei Univ. **2**: 206–213.

Ota, Y. (1960) The zonal distribution of the non-marine fauna in the Upper Mesozoic Wakino Subgroup (studies of the molluscan fauna of the non-marine Upper Mesozoic Kwanmon Group. Part 5). Mem. Fac. Sci. Kyushu Univ. Ser. D Geol. **9**: 187–209.

Ota, M. and Yabumoto, Y. (1998) The Lower Cretaceous Kanmon Group. *Quadrangle series Fukuoka (14) No.34, Geology of the Kokura District*, pp. 32–40 (in Japanese).

Riding, R. (2006) Microbial carbonate abundance compared with fluctuations in metazoan diversity over geological time. Sedimentology **185**: 229–238.

Seo, S.G., Sakai, T. and Okada, H. (1994) Depositional environments of the Wakino Subgroup of the Lower Cretaceous Kanmon Group in the Kitakyushu Area, Japan. Mem. Fac. Sci. Kyushu Univ. Ser. D **18**: 41–60.

Woo, K.S., Yoon, H.S. and Lee, K.C. (2002) Paleoclimatic investigation from the stromatolite of the Cretaceous Sinyangdong Formation, Gyeongsang Supergroup, South Korea. The 1st and 2nd Symposiums on Geology of Korea (special publication) 1, pp. 21–35.

Woo, K.S., Khim, B.K., Yoon, H.S. and Lee, K.C. (2004) Cretaceous lacustrine stromatolites in the Gyeongsang Basin (Korea): records of cyclic change in paleohydrological condition. Geosci. J. **8**: 179–184.

Yamamoto, A., Tanabe, K. and Isozaki, Y. (2009) Lower Cretaceous fresh-water stromatolites from northern Kyushu, Japan. Paleontol. Res. **23**: 139–149.

PART 3:
MODERN STROMATOLITES
(MARINE, LACUSTRINE, HOTSPRINGS)

Browne
Chacón-Baca
Gómez
Montejano
Barrera
Sanchez-Ramos
Falicia Goh
Handley
Campbell
Foster
Green
Reid
Gaspar
Bowlin
Custals
Andres
Farías
Poiré
Arrouy
Albarracin

Biodata of **Dr. Kathleen M. Browne**, author of ***"Modern Marine Stromatolitic Structures: The Sediment Dilemma"***

Dr. Kathleen M. Browne is currently Assistant Provost and Director of the Science Education and Literacy Center (SELECT) at Rider University. She obtained her Ph.D. in marine geology and geophysics from the University of Miami, Rosenstiel School of Marine and Atmospheric Science and continued her work after joining the faculty in the Geological, Marine, and Environmental Sciences Department at Rider University. She expanded her work to include studies of New Jersey eutrophic lakes until moving into her present position. Dr. Browne's interests now also include improving the science and math literacy and teaching practices of K-16 instructors, guiding district staff to implement coherent science and mathematics curricula, and the assessment of student learning.

E-mail: **browne@rider.edu**

V.C. Tewari and J. Seckbach (eds.), *STROMATOLITES: Interaction of Microbes with Sediments*, Cellular Origin, Life in Extreme Habitats and Astrobiology 18, 291–312
DOI 10.1007/978-94-007-0397-1_13,

MODERN MARINE STROMATOLITIC STRUCTURES: THE SEDIMENT DILEMMA

KATHLEEN M. BROWNE
Rider University, 2083 Lawrenceville Road, Lawrenceville, NJ 08619, USA

Abstract Modern marine stromatolites have been considered for some time as "windows" into the past that can potentially reveal information about the circumstances in which ancient examples formed. Many studies have focused on lithified, laminated examples from Shark Bay Western Australia over the last 50 years and from the Bahamas over the last 30 years. The early works provided thorough studies of the environments and the stromatolitic structures and discussed many unknowns left to be explained. In the last 20 years, significant advancements in the study of microbial ecosystems have resulted from innovative methodologies and technologies. This paper is a brief summary of recent developments in the research on the formation of lithified modern normal marine stromatolitic structures from the Bahamas.

Keywords Alkaline lake • Bahamas • Cement • Cyanobacteria • Endolith • EPS • Exopolymer • Substances • Laminae • Lamination • Microbialite • Microbial mat • Precipitation • Schizothrix • Sediment • Sedimentation • Stromatolite • Stromatolitic • Sulfate-reducing bacteria

1. Bahamian Stromatolites and Microbialites

Lithified subtidal and intertidal structures resembling ancient stromatolites have been found in the Bahamas at numerous locations (Dravis, 1983; Dill et al., 1986; Dill, 1991; Reid and Browne, 1991; Reid et al., 1995; Ginsburg and Planavsky, 2008; Golubic and Focke, 1978). All are located nearby the edge of the eastern side of the Bahamian plateau. Subtidal settings are typically high energy sites influenced by tidal currents or wave energy in sandy embayments, and intertidal sites are sandy beach areas experiencing wave energy typical of the windward site of the platform (Reid et al., 1995). In many cases, well-formed cohesive microbial mats were noted on the tops of these structures. The mat on top of the bioherms and the structures in the hardened interiors at three localities have been studied closely: intertidal stromatolites on the backreef side of an algal ridge fringing reef complex at Stocking Island (Reid and Browne, 1991), very shallow subtidal examples in a similar complex at Highborne Cay (e.g., Reid et al., 2000; Andres and Reid, 2006), and deeper subtidal examples in Adderly Channel (Dill et al., 1986; Browne, 1993; Feldmanm and McKenzie, 1998; Planavsky and Ginsburg, 2009).

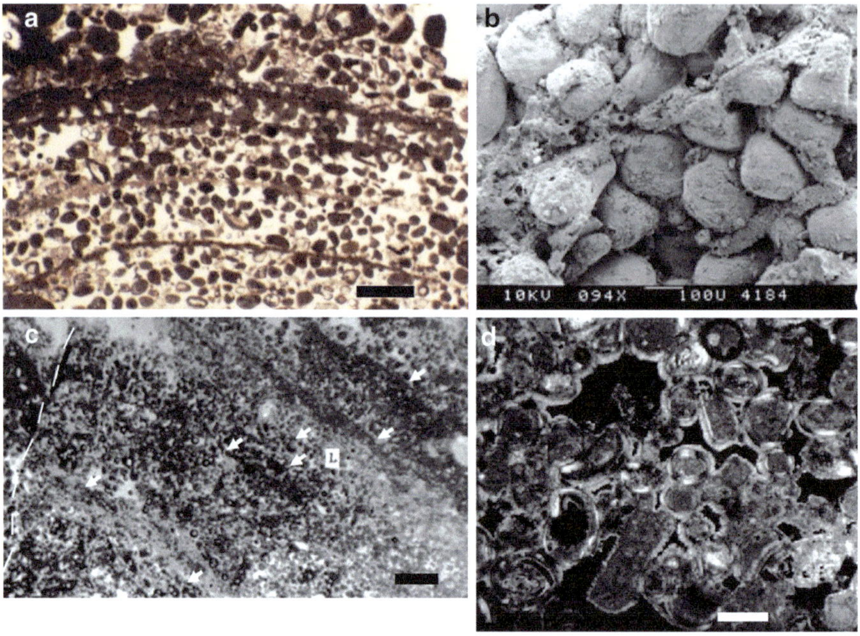

Figure 1. Photomicrograph of sections from Bahamian stromatolites. (scale). (**a**) Thin section of lithified layering in intertidal sample, Stocking Island (500 μm). (**b**) SEM view of section of lithified layering from subtidal stromatolite, Adderly Channel. Micritic cement crust runs from *top right* to *lower left* corners. Above and below the cement crust, micropeloidal and grain-coating cements occur (100 μm). (**c**) Thin section of faintly laminated sediment (L) adjacent to pocket of unlaminated (C), coarser sediment (separated by *dashed white line*). Adderly Channel subtidal stromatolites. White *arrows mark* cement crusts (1 mm). (**d**) This section of unlaminated sediment showing presumably inorganic cement coating grains (300 μm).

Lamination in the hardened parts of the bioherms consists of couplets of *cement crusts* (thin micritic crusts) and thicker *weakly lithified fine sand* (Fig. 1). There are two types of cement crusts: thin micritic crusts (Reid and Browne, 1991; Browne, 1993; Reid et al., 1995; Golubic and Browne, 1996) and fused grain layers (Reid et al., 2000, 2003, 2008). Micritic crusts range from 40 to 100 μm thick in living mat samples, and contain very little sediment, silt-sized (Reid and Browne, 1991; Browne, 1993). The fused grain layers in hardened parts of stromatolites can be 1–2 mm thick and more laterally continuous than micritic crusts, and thus much more substantial and stronger than the latter (Reid et al., 2008). Weakly lithified porous layers range in thickness from 500 to 3,000 μm; cements typically occur as coating grains, at grain contacts and pore-filling micropeloidal clusters. While called stromatolites when first described, closer examination indicates that some examples do not contain the millimeter-scale lamination associated with stromatolites as a predominant internal structure. For example, subtidal structures found at ~10 m and shallower depths in Adderly Channel contain a mix of laminated and unlamented fabrics and have been termed thrombolites (Feldmanm and McKenzie, 1998) and most recently microbialites, rather than stromatolites (Planavsky and Ginsburg, 2009).

2. Lamination Formation

To understand the origin of the laminated portions of these structures, studies have looked to the living community on tops of the bioherms. As would be expected in a normal marine setting, diverse, eukaryote-dominated communities can be found on some (e.g., Dravis, 1983; Dill et al., 1986). But laminated microbial mats with thin micritic crusts, dominated by prokaryotes have been found on the surfaces of several of these hardened structures (Reid and Browne, 1991; Browne, 1993; Reid et al., 1995; Golubic and Browne, 1996) (Fig. 2). The layering found within the mats resembles closely the lamination preserved in parts of the hardened portions of the structures. The motile, filamentous, multi-trichomed cyanobacterium, *Schizothrix gebleinii*, was found to be the dominant microbe in the mats. In a normal marine environment, the occurrence of a prokaryote-dominated community is typically indicative of environmental stress that prohibits most other organisms from occurring in abundance (Golubic, 1991). Field observations and sediment preserved in mats and lithified portions of stromatolites indicate that the primary stress in all the above settings is high sedimentation rates. The mats are periodically buried by migrating sand dunes in channels or near shore sand bodies, and when not buried are often subjected to high rates of sediment entrainment from sandy areas surrounding the stromatolites and microbialites. Examples in intertidal settings may also be influenced by environmental stresses typical in this environment, although it should be noted that eukaryote-dominated algal turf communities can be found in the nearby *Schizothrix* mats at equivalent or slightly higher elevations in the intertidal and sometimes directly adjacent to sediment sources (e.g., Reid and Browne, 1991; Reid et al., 1995).

Schizothrix-dominated mats have been studied extensively in recent years, particularly at the very accessible intertidal and shallow subtidal sites at Highborne Cay, because the internal layering in the surficial microbial mats is similar to the preserved lithified internal parts found beneath the mats. Based on the observations of mat structure and internal lithified lamination from the many localities of stromatolites in the Bahamas, the findings from the Highborne Cay examples seem to be applicable for all *Schizothrix*-dominated mats that produce stromatolitic structures in the Bahamas and as such have been used to make reasonable generalizations about lamination formation by this mat. A thorough comparison of the intertidal, shallow subtidal (1 m) and deeper subtidal (10 m; Adderly Channel) examples would be worthwhile to determine whether there are any differences detectable either in the mats or in the lithified products. A brief summary of what is now known about various aspects of the layering in mats is provided

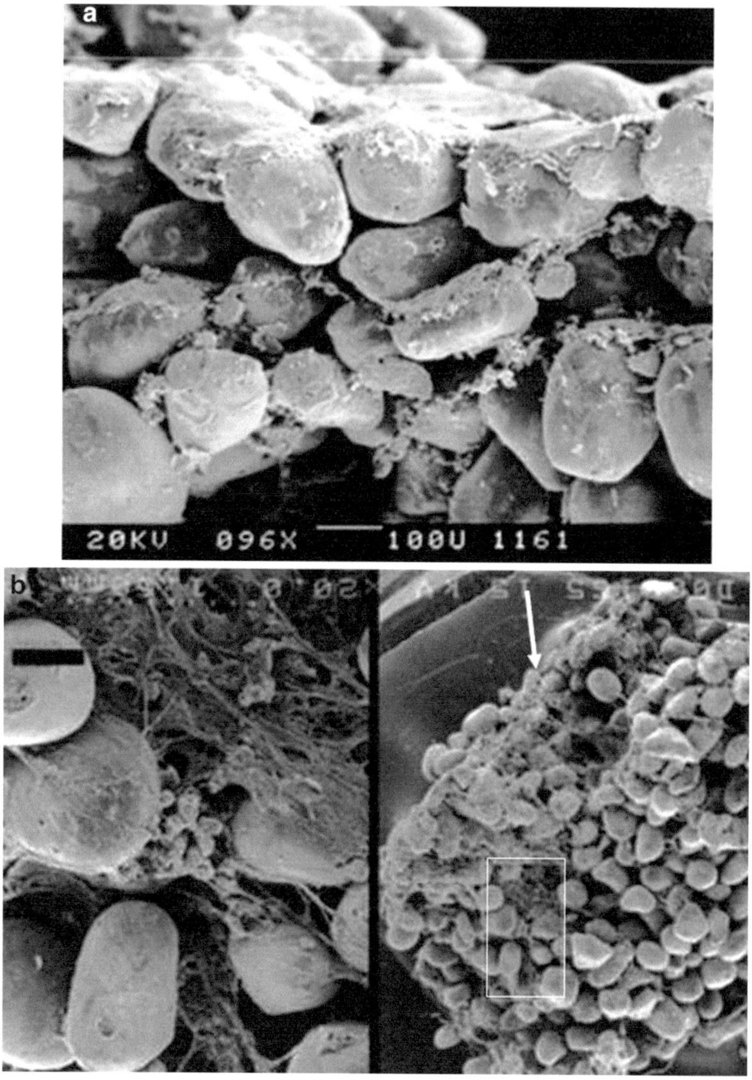

Figure 2. SEM photomicrographs of cement crust in living *Schizothrix* mats. (**a**) Side view of mat sample showing surficial organic film and underlying delicate cement crusts presumably formed by earlier biofilms (*arrows*) (scale = 100 μm). (**b**) *Right side*: Side view of mat the bulk of which is likely a Type 1 mat showing filamentous organics oriented vertically among abundant sand grains (top indicated by *arrow*) (scale = 1.5 mm). *Box* in *right photo* outlines area magnified on *left*. *Left photo*: micro-peloidal cements within organic matrix (scale = 100 μm).

here. Reid et al. (2000, 2003, 2008) report on three types of mats from the shallow subtidal stromatolites at Highborne Cay, which represent stages in the mat development relative to changes in sedimentation rates, as inferred by sediment grain size incorporated into the mats and microbe activity. Type 1 mats are sediment rich and dominated by a sparse population of *Schizothrix gebleinii* oriented vertically around the abundant sand-sized sediment grains (Fig. 2b). In this state, evidence indicates that sedimentation rates are high, grains adhere to the mat surface due to sticky EPS (extracellular polymeric substances) produced by *Schizothrix*, and the motile trichomes are continuously adjusting to the newly accreted sediment by moving upward around the grains, abandoning sheaths in their wake. The filaments, sheaths and EPS bind the sediment grains holding the mat together as the top surface continues to grow upward from additional sediment accretion. Several studies have noted that there are very few other organisms present in this stage of mat development (e.g., Golubic and Browne, 1996; Reid et al., 2000) indicating that sedimentation rates can very effectively eliminate not only eukaryotes but also most prokaryotes. Limited cement precipitation has been found in this mat type occurring as grain contact cements and "micropeloidal" clusters (Fig. 2b). Type 2 mats consist of a surficial "biofilm" of diverse heterotrophic microbes, copious amounts of amorphous EPS, aragonite crystals, and silt-sized grains. Immediately beneath this film is a sparse to moderately dense population of *Schizothrix*, with enmeshed filaments oriented randomly. The mats consist of a community that can exist when sedimentation rates decrease, as inferred by the incorporation of only silt-sized sediment. At such times, heterotrophic microbes become more prevalent and they and their products concentrate on the mat surface. In studies of the deeper subtidal examples, dense mats of horizontally oriented *Schizothrix* have been found at the mat surface (Browne, 1993, Fig. 3.4; Golubic and Browne, 1996, Fig. 6) topping the vertically oriented filaments (Type 1), which may represent a transition from Type 1 to Type 2 mats in Reid et al. (2000), or might be a mat structure more common in deeper subtidal mats. Golubic and Browne (1996) did not describe heterotroph biofilms in the deeper mats but it is likely that they were simply not detected in this earlier work. The heterotrophic biofilms contain micritic cement crusts resembling similar features found in lithified portions of the stromatolites (Reid et al., 2000, 2003, 2008) (Fig. 2a showing cements free of organics beneath mat surface) and may be where similar crusts form in the deeper subtidal mats as well (Browne, 1993; Golubic and Browne, 1996). Spherical aggregates 2–5 μm in diameter of 1 μm long aragonite needles have been observed embedded in the EPS matrix (Reid et al., 2000, 2003, 2008). Type 2 mats further develop into Type 3 mats with continued lower sedimentation rates and consist of abundant *Solentia* sp., an endolithic coccoid cyanobacteria; randomly oriented *Schizothrix* filaments occur below the calcified biofilm (Reid et al., 2000). This endolith bores into the carbonate grains and bores can quickly infill with carbonate precipitation; apparent microbe movement from one grain to another aids in cementing or "fusing" grains together (Reid et al., 2000; MacIntyre et al., 2000).

The various mat types noted in Highborne Cay samples point to mat community maturation or succession as the living surface adjusts to changes in sedimentation rates (see Reid et al., 2000, 2008 for summary diagram). For initial substrate colonization, some sort of stable substrate in high sedimentation zones appears to be necessary for the first steps in the growth of a stromatolitic mat

(Ginsburg and Planavsky, 2008). With substrates present, *Schizothrix* colonizes as a pioneer organism (Reid et al., 2000) and stabilizes sediment that adheres to the mat sufficiently to form cohesive sediment-rich mats. Numerous studies (Neumann et al., 1970; Stal, 1994; Yallop et al., 1994; Dade et al., 1990) have shown that sediment stabilization is accomplished by microbial communities, and cyanobacteria in particular are very effective in doing so through the combination of cyanobacterial entrapment of sediment and the EPS it adds to the mat. Paterson et al. (2008) have shown how effective microbes specifically from Highborne Cay shallow subtidal stromatolites are in stabilizing sediment. In addition, experimental aquarium studies of reconstituted stromatolite mat material have shown that *Schizothrix* can quickly recolonize the sediment surface (~2 days), increase sediment stabilization over time, and the vertical sequence of microbes in type 2 mats can be replicated. In these experiments, the thickness of the mats was significantly greater than thicknesses seen in the modern example studies cited in this review. And interestingly, the vertical sequence of microbes was recreated *without* replicating the variety of environmental factors known to occur in the real environments. These results indicate that certainly more work is needed to understand the factors that influence mat formation and community successions.

Once the pioneer *Schizothrix* mat is established on a stable substrate, the mat community appears to adjust as sedimentation rates vary. When sedimentation rates are high, grains adhere to the surface and the phototactic *Schizothrix* adjusts by moving upward and binding the grains. Browne (1993) and Golubic and Browne (1996) found in some mat samples *Schizothrix* concentrated on the sediment surface, producing mat layers with significantly more organic matter, minimal sediment and orientations of *Schizothrix* filaments that range from horizontally oriented to randomly oriented and enmeshed (for example, see Fig. 6 in Golubic and Browne, 1996). They proposed that decreases in sedimentation rates might be the reason for the development of this denser mat. Type 2 mats described in Reid et al. (2000) resemble this stage in mat development; however, their studies revealed bacterial biofilms above the *Schizothrix* dense layers. They described this mat type as "hiatal" or discontinuity surfaces, where the community has "matured" from the pioneer cyanobacterial community to one in which other microbes become more prominent on the surface of the mat community (Reid et al., 2000). It is in fact these discontinuity surfaces that define the lamination in uncemented mats and are the layers that contain micritic crusts. Type 3 mats represent a further maturation of the mat community as sedimentation rates continue to be low (Reid et al., 2000). The endoliths are able to concentrate and their activity alters sediment and cement crusts already incorporated into the mats.

There is only limited information about lateral variations in surface community at any one time. On a small scale with just the cohesive laminated microbial mat, Reid et al. (2000) found that at Highborne Cay, Type 1 mats are significantly more common than Type 2 or 3, which makes sense in a high-energy setting, if high depositional rates limit growth of other organisms. If sedimentation rates are important, then occurrence of the three mat types should somehow correlate with variations in sedimentation rates with Type 1 being more abundant closer to the sediment source, and Type 2 and 3 further away from the sediment. But if the mat surface is not smooth, or flow is variable over the mats then distribution of mat types might be complex. Further complicating the issue may be seasonal fluctuations in Type 2 and 3 mat types at least in intertidal and shallow subtidal mats (Foster et al., 2009). Additional research is needed to tie the proposed sedimentation rate controls to mat type occurrences.

Finally, much progress has been made in identifying the organisms in the microbial mats responsible for the creation of stromatolitic bioherms. In earlier studies, organisms in these microbial mats were identified by morphological studies primarily. Through various methodologies recently developed (e.g., fluorescence in situ hybridization, combined cultivation and molecular techniques), it has been determined, not surprisingly, that there are more organisms in these cohesive mats than were originally identified (Baumgartner et al., 2006; Havemann and Foster, 2008; Foster et al., 2009). This is an exciting development in that it enables researchers to investigate the roles other organisms may play in stromatolite growth and preservation. At the moment, it is unclear how prevalent many of the newer species discovered in the mat are, and what, if any role they play in the various processes involved in lamination accretion. Interestingly in studies of the microbial mats described above by Foster et al. (2009), *Schizothrix gebeleinii* was not found in any of the cultivars possibly due to the difficulty in recovering cells encased in thick polysaccharide sheaths and protected by calcium-enriched EPS during the nucleic acid extractions (Foster et al., 2009). The authors point out that their findings should compliment earlier morphological studies until more is learned about this organism. Similar analyses to distinguish the mat communities in Type 1 mats versus Type 2 mats would also be quite interesting, although likely extremely difficult. It is possible that the filamentous cyanobacteria prevalent in these mats could be renamed at some point in time through more modern techniques. But whatever its name, it plays a key role in formation of these biosedimentary structures.

3. Microbial Mat Mineral Precipitation

Numerous studies have focused on mineral precipitation in the microbial mats for several reasons: lithification of the microbe-constructed structures is essential for preservation in the sedimentary record; the micritic crusts and fused grain layers actually define the layering in the bioherms; the mineralization phenomena in and of itself is worthy of research and explanation; and the sub-millimeter micritic laminae most closely resemble lamination in most ancient stromatolites. When cement crusts were first noted in subtidal living Bahamian *Schizothrix* mats from Adderly Channel, metabolic processes within the *Schizothrix* mats (meaning the community of organisms, not any one organism), such as photosynthesis or sulfate reduction, were hypothesized to be driving precipitation (Browne, 1993;

Golubic and Browne, 1996 and further described in Seong-Joo et al., 2000) given notable work cited in these studies. Researchers focusing on mineral precipitation suggested decomposition by sulfate-reducing bacteria (SRB) specifically as the cause of carbonate precipitation in microbial mats (see summary in Sprachta et al., 2001). However, cement precipitation in stromatolites has been very difficult to study for several reasons. The microbes, the vertical thickness of mat communities they create, and the areas where mineral precipitation has been observed are very tiny. Also undetermined in these earlier studies were the members of the microbial communities beyond what appeared to be dominant organisms. Without the use of fine-scale instrumentation and more knowledge about many of the organisms involved, studies of metabolic activity the chemical changes caused by metabolic processes were extremely difficult. Innovative methodologies and tools have been used since the work cited above to help with these efforts.

Through numerous recent efforts that have systematically addressed various aspects of the mat biology and chemistry, there is much evidence to show that SRB in the Type 2 biofilm and extracellular polymeric substances (EPS) do in fact play an important role not only in sediment accretion but also in mat mineralization (e.g., Reid et al., 2000, 2008; Stolz et al., 2001; Decho et al., 2005; Visscher and Stolz, 2005; Andres et al., 2006; Stolz, 2008; Dupraz et al., 2008; RIBS-http://www.stromatolites.info). This is another exciting development in understanding the mats involved and the structures they produce. A brief summary, not at all indicative of the work involved to come to these conclusions, is provided here; readers should consult Dupraz et al. (2008) and references cited within their work for a comprehensive summary. In Type 2 mats, which are characterized by the presence of micritic crusts within the living mat, it appears that degradation of EPS drives $CaCO_3$ precipitation in the mats. The micritic crusts occur, where copious amounts of EPS are found (Reid et al., 2000) and maximum SRB activity occurs (Visscher et al., 2000). EPS is secreted by *Schizothrix* to stabilize cells, create a chemically protective microenvironment for cells, and to bind calcium to protect cells from $CaCO_3$ precipitation (Decho and Kawaguchi, 2003, 2008). EPS can be produced and broken down very quickly. Initial degradation of EPS matrix by inorganic and/or organic processes can release low molecular weight organic carbon (LMWOC) along with calcium bound in EPS polymers; SRB break down the LMWOC increasing alkalinity. The combined increase in calcium and alkalinity increases the calcium carbonate saturation index and precipitation results. Isotopic analyses of the micritic crusts support this explanation; the crusts are slightly depleted in ^{13}C suggesting precipitation from heterotrophic processes (Andres et al., 2006). Most recently, Aloisi (2008) has argued through geochemical modeling that calcium binding by EPS has a minimal effect on prohibiting mineral precipitation during photosynthesis. Calculations include large calcium fluxes between the mat and surrounding waters. If this model accurately represents the real mat systems, then it is more likely that precipitation is driven by photosynthesis rather than sulfate reduction (Aloisi, 2008). Andres et al. (2006) argue, however, that the isotopic composition of the crusts along with chemical gradients found in the mats indicate the mat waters are not in equilibrium with seawater and isotopic signatures reflect local microbial processes within the mat. Additional research is needed to resolve this debate.

When endolithic organisms join the community and form Type 3 mats, their activity further enhances the layering by altering and welding together sediment grains originally trapped in the mat (see MacIntyre et al., 2000 for a summary of this process). Thus, the cyanobacteria play a critical role in accreting sediment. They trap sediment during high sedimentation events. When sedimentation rates decrease, the cyanobacteria adjust their activity but continue to add EPS to the mat and bind the sediment, and a film of SRB can establish itself on the stabilized sediment/mat surface, which plays a critical role in creating micritic crusts by breaking down the EPS. At this point in their development, the mats function as "geochemical bioreactors" with the combined community acting as an "alkalinity engine" by changing calcium and carbonate concentrations within the EPS matrix and providing mineral nucleation sites (Visscher and Stolz, 2005; Dupraz et al., 2008).

4. Preservation

Preservation of the laminated fabric in the Bahamian mats ultimately requires sufficient lithification for the structures to survive erosion as well as limited exposure to alteration processes. The cement crusts and fused grain layers described above are key features in defining lamination and preserving it. Micritic crusts and fused grain layers form laterally along a roughly horizontal plane within Type 2 and 3 mats. Each micritic cement crust is thought to be a record of a depositional hiatus but because there must be lateral variation in the maturity of the surface mat across the top of the stromatolite, there would also be lateral variation in cement crust formation, producing essentially a patchy occurrence. In hardened portions of the stromatolitic bioherms, this patchiness appears as laterally discontinuous crust and fused grain layers in thin sections of the layering. Cementation in between these crusts can also be important. Note the peloidal micrite between grains mentioned in Reid and Browne (1991) and in Fig. 2.

It is unclear how much of the various cements described above are needed to prevent the structure from crumbling apart once organic materials decay. But the very subtle definition to laminae in some thin sections of hardened parts of the bioherms, and the limited intragranular cements indicate that it does not take a tremendous amount. The fused grain layers when present are certainly significantly stronger than the micritic crusts. Intragranular cements must also be sufficient to preserve the spacing of grains; otherwise, when the organic materials decay away, the sediment trapped by the Type 1 mats would simply fall to the top of the underlying cement crust. Note, for example, that pore space in the thin section photographs of hardened parts of intertidal stromatolites (Reid and Browne, 1991; Fig. 5a and Reid et al., 1995; Plate 7, Fig. 1a, b), and in this paper (Fig. 1a, c), that some areas of sediment are very loosely packed with very limited grain to grain contacts.

There is some indication that a maximal thickness of the couplet of sediment-rich layers and capping dense microbial layer/biofilm (combined Type 1 and 2/3) for preservation to occur. Sections of hardened stromatolite lamination show a relatively consistent millimeter-scale lamination thickness. Field observations of sediment accretion and mat development on deeper subtidal stromatolites, however, suggest there could be greater variation in the thicknesses of the Type 1 and 2/3 sequence in the surficial mat than the same couplet preserved within the hardened parts of the bioherms, although more data would be needed to test this statistically (Browne, 1993; personal observation). Recent findings suggest, however, that the differences between layer thickness in mats and lithified lamination seen in earlier studies may be an artifact of observation techniques. In the intertidal and shallow subtidal mat samples, EPS in mats can prevent one from seeing subtly defined laminae but thin sections of the mat material reveal layering where it is not visible in vertical sections of the matt (Reid, personal communication). Alternatively, it may be that the layering differences noted deeper subtidal examples do occur in this deeper setting; these mats may need to be reexamined using updated techniques. Assuming the earlier observations of the deeper subtidal mats are accurate, thicker sediment-rich layers could be explained by prolonged periods of high sedimentation. Even with sufficient micritic crusts, the thicker layers may simply be too thick for the very limited micropeloidal cements in the sand-rich layers to hold them together. If thinner couplets are more likely preserved, then the hardened structures preferentially preserve layering produced under those circumstances, whatever they are. More work is definitely needed here.

Beyond the microbial mat cementation, the internal parts of the stromatolite can be further lithified through inorganic cementation (Planavsky and Ginsburg, 2009; Fig. 1d). In addition, large sections of the subtidal examples from Adderly Channel show that eukaryotes can modify the original structure effectively by destroying some of the original laminated structure (the microbialites of Planavsky and Ginsburg, 2009). Certainly, field observations of these subtidal microbialites show numerous examples of eukaryote-covered tops, with communities including a wide range of prokaryotic and eukaryotic microbes as well as macro plants and animals; the composition of this community has been described by Dill et al. (1986). Microbial communities still occur among macroeukaryotes but typically consist of a much more diverse population of eukaryotes and prokaryotes and as a result, are significantly less cohesive and not laminated (Browne et al., 2000). Essentially, they produce a loosely bound fluffy coating of organisms and sand-sized sediment. Small patches of cohesive mats typically coating high knobby points can still be found, however, and likely continue lamination accretion. Why these small cohesive mat patches occur among a more diverse community is also less certain. There is some indication that the fluffy diatom coatings might aid sediment accretion by trapping grains, and underlying cyanobacteria moving upward through that loosely trapped sediment to bind it, allowing thicker layers to accrete (Reid et al., 2008). Alternatively, they may simply be remnants of mats formed at a previous time when sedimentation rates limited colonization to the low diversity microbe communities. The eukaryote community might be considered a later stage in the succession of organisms on these hardened structures. Assuming sedimentation rates are the only control in the occurrence of well-formed microbial mats, if the bioherms are left exposed with their tops high above the sediment surrounding them for extended periods of time, eukaryotic plants and animals would have opportunity to colonize the surfaces. A diversity of microbes would dilute the cohesiveness of the community and prohibit the formation of laminated sediment. A diversity of plants and animals could disrupt sediment accumulation and some can contribute significantly to bioerosion of the structure they have colonized. Larger bioherms may be more altered by eukaryotes since their tops would stand higher above the sediment surface longer than would be the case for smaller bioherms. An extreme example of bioerosion was described by Dill et al. (1986) in a setting where sediment no longer surrounds "relict" bioherms. Further work is clearly needed to better understand the most diverse communities found on microbialite tops.

The model to explain variations in the communities that are found on the tops and sides of the Bahamian microbialites is a relatively simple model that relies on sedimentation rate changes. Field observations in the real setting reveal notable variation in the community covering the tops of stromatolites even when the tops are nearly buried by passing sand dunes. It is highly unlikely that the proposed community succession from microbial mats to the eukaryotic community is well synchronized with variations in sedimentation rates. Thus, any single observation would show the surface community in some instances not yet adjusted to newly established sedimentation conditions. Surface communities probably do not always have enough time to adjust to changes and may be out of synch with sedimentation change at times. More work monitoring the changes that individual bioherms experience relative to sedimentation rate changes is needed here as well.

5. Discussion

The progress that has been made in understanding stromatolite layering in modern marine examples over the last 2 decades is impressive. For many years, microbial mat chemistry has been suspected to be the driving force behind not just the lithification and preservation of modern marine stromatolitic fabric, but contributes significantly to the formation of lamination. Without a doubt, learning the details of this early stage cement precipitation will expand the abilities of geologists to understand ancient examples, perhaps from many environmental conditions. With all this progress, however, some of the finer details of when, where, how, and from what exactly the early cements form still require discovery and explanation. Ideally, catching the cementation in the act within a living microbial mat should reveal quite a bit assuming the technology available provides the necessary evidence to tie that precipitation to specific processes. Given the progress made, it may be that the tools to measure microscale biogeochemical phenomena exist now and the larger challenge may be in either replicating the natural surrounding environmental conditions for experimental studies, or collecting data in situ. Another challenge may be in applying techniques used in studies of more accessible intertidal and shallow subtidal Bahamian examples (e.g., Highborne Cay) to deeper, less accessible examples (e.g., Adderly Channel). Identifying distinctive mat characteristics and preserved evidence of different depositional environments, if they exist, would surely be useful to geologists studying ancient examples. It seems likely that such results will be forthcoming. And when they come, it is also likely that they will reveal possible connections to similar, or not so similar, systems where other microbialites form. There are some lingering questions beyond the lithification and layering origins of stromatolitic deposits that seem to warrant further investigation. What are the limits to *Schizothrix* mat's ability to create laminated structures? To what degree does sedimentation *really* drive the formation of the stromatolitic layering in modern normal marine settings? Given answers to these questions, what exactly can be applied to ancient examples?

5.1. WHAT ARE THE LIMITS TO SCHIZOTHRIX MAT'S ABILITY TO CREATE LAMINATED STRUCTURES?

Recent developments have shown that there are many more microbes in the *Schizothrix* mats from Bahamian stromatolites and that some, specifically SRB, play a critical role in driving cement crust precipitation. It remains to be determined what roles all the other microbes discovered play. All evidence thus far points to *Schizothrix* serving as a pioneer organism setting up the circumstances to allow sediment to be trapped and stabilized by the mat (with all its co-habitants), for the cohabitating SRB in more mature *Schizothrix* mats to break down the EPS added by *Schizothrix* and SRB, altering the mat chemistry to cause cement precipitation, and for an endolithic cohabitant, *Solentia*, in even more mature mats to contribute further to lithification and to alteration of the carbonate grains in the mats. *Schizothrix* is present because it can not only survive but also prosper in apparently high sedimentation-rate settings. But with other filamentous cyanobacteria recently identified in the mats, it may be time to reconsider whether some of these cohabitants also play a role in constructing and/or lithifying the sediment-rich structures.

Regardless of contributions made by each microbe present, we can certainly say that collectively organisms in the *Schizothrix* mat are very effective in making millimeter-scale microbial lamination, contributing to lithification of the layering and bioherms, and incorporating structures that closely resemble many ancient stromatolites in some ways. Since it appears that sedimentation rates are a controlling factor that allow these mats to form in a setting where a diverse population of microbes and macro plants and animals occur, it makes sense that even slight variations in sedimentation rates would cause shifts in the mat community.

In both intertidal and subtidal stromatolites settings, it appears likely that when sedimentation is not a controlling factor, many other organisms would dominate the setting, or at least grow in large enough numbers to prohibit *Schizothrix* from making laminated cohesive mats cement crusts and fused grain layers. *Schizothrix* is motile and produces abundant EPS making sticky mats. Sediment dropped by currents onto this mat get stuck and *Schizothrix* trichomes shaded by the grains move out of their sheaths around the grains into more sunlight. But are there any intrinsic factors in the mat that influence construction and preservation of these structures? How quickly, for how long, and when within a day/night and/or seasonal cycle can *Schizothrix* adjust to sediment? What sedimentation rate is *too* much for *Schizothrix* to keep up? Are there other factors or conditions that would cause it to change movement in response to sedimentation, or change the stickiness of the mat? What might be the reasons for the possible seasonal variations in mat types noted in Foster et al. (2009)? As reported in Dupraz et al. (2009), periodic burials may serve as selective pressure that favors the bacterial communities, in particular cyanobacteria (Andres and Reid, 2006; Kromkamp et al., 2007; Perkins et al., 2007). But what happens when burial does not occur, a stromatolite surface is covered by a eukaryotic community and then sedimentation rates increase? Are the eukaryotes systematically eliminated from the community as conditions worsen for them and *Schizothrix* recoat the surface, or is burial needed to catastrophically wipe out the eukaryotic community? When a *Schizothrix* mat is well formed, what happens over time in terms of community succession when sedimentation rates decrease for extended periods of time, or permanently? Do eukaryotes simply take over or does the *Schizothrix* mat somehow prohibit colonization by small or large eukaryotes making it difficult for colonization? Do community successions occur even during prolonged high sedimentation rates? Are there detectable differences in the mats from the intertidal, shallow subtidal, and deeper subtidal stromatolite mats? What are the impacts of prolonged burial on microbial mats? [Kromkamp et al., 2007 and Perkins et al., 2007 have addressed this area to some degree; their results are discussed in Reid et al. (this volume).] How long can a microbial mat-covered surface stay buried before the mat dies? What happens when/if a mat dies during burial; is cementation enhanced? Answers to some of these questions could have implications for the next questions posed here.

5.2. TO WHAT DEGREE DOES SEDIMENTATION REALLY DRIVE THE FORMATION OF MODERN MARINE STROMATOLITIC LAYERING?

For whatever the reasons, *Schizothrix*-created mats do not appear to form in settings with low sedimentation rates. So sedimentation rate *does* drive at least part of the process. In all studies of these mats, it has been assumed that denser mats (biofilms) form when sedimentation rates drop. There is, however, only limited evidence that shows a correlation between the occurrences of specific mat stages with higher sedimentation rates (thus, the reason for some of the questions above). Browne (1993) and Browne et al. (2000) proposed that higher sedimentation rates occur when crests of dunes and ripples superimposed on large sand dunes pass by microbialite structures in deeper subtidal settings. Similarly, Reid et al. (2000) suggest that *Schizothrix* is on the move leaving vertically oriented filaments and empty sheaths among abundant ooid sand (Type I mat) when the surrounding sediment surface is nearby the tops of structures coated with *Schizothrix* mats. The occurrence of *Schizothrix* mats on low relief stromatolites and macroalgae on high relief examples reported at many locations supports this idea as well (Reid et al., 1995). Reid et al. (this volume) report on variations of burial and exposure rates for stromatolites and thrombolites in the shallow water setting at Highborne Cay at numerous locations within the reef complex; their results make it clear that just the simple process of burial and unburial does not simply explain variations in internal structures of the microbialites. Finer-scale observations that look at small-scale (centimeter and smaller) variations in the mats relative to sediment dynamics are needed to tie the mat types to actual sedimentation. Beyond confirming the proposals made by above authors, more detailed studies may also reveal indirectly whether there are other factors or conditions that would cause the formation of the Type 2 and 3 mats. The decreased sedimentation rates are inferred from changes in size of grains incorporated in the mats; could there be some other reason that smaller and fewer grains are incorporated in the Type 2 mats? Reid et al. (2003) suggest that trapping of sediment could be biologically controlled through the stickiness of EPS (most likely), growth rates, motility and response to light, or chemical cues, or in other words, by biological intrinsic variables. Foster et al. (2009) indicate that Type II and III mats "incur fluctuations in their abundance throughout the year" suggesting seasonal influences on mat occurrence, thus even if there are intrinsic factors, extrinsic factors still appear to be important. If there are intrinsic factors, what are they and to what degree are they imprinted into the preserved layered fabric of the hardened structures? If it is the case that there are no other factors or conditions controlling mats other than sedimentation, then predominant sedimentary processes would be the driver controlling lamination formation. For example, if sediment is consistently and sufficiently supplied to the entire mat surface, then it might be possible that every spot would be in Stage I development. If sedimentation decreases consistently across the mat surface with an increase in distance from the sediment source, then Type 2 and 3 mats could conceivably develop and cement crusts form pervasively. Does this happen? If there is variation in flow across the surface and/or availability of sediment, then the mat surface could be in varied stages of development, along with the layering it forms, at different locations. Flow experiments conducted by the RIBS research group have shown that microbe distribution around a hemispherical shape changes with different flow velocities, and with*out* sediment stressing the mats (http://www.stromatolites.info, cyanobacterial ecology page in lab investigations section). If this is true in the natural setting, then flow without sediment should be considered here. If there are both sedimentary and biologic controls on the mat transitions from Type 1 to Type 2 and 3 mats, what are their respective contributions to the final laminated structures produced? And finally, if it turns out that the transitions from one mat to another are purely biologic, then is the sediment important only to exclude other organisms that would otherwise dilute the *Schizothrix*-dominated mats, and more mature variations of them?

5.3. IN COMPARING MODERN STROMATOLITES WITH ANCIENT EXAMPLES, WHAT DO WE DO WITH THE COARSE SEDIMENT?

Researchers of modern stromatolites argue that the mats producing stromatolites today are analogs for ancient stromatolites in that they enable us to study the role of the bacterial community in carbonate precipitation of micritic laminae (e.g., recently Dupraz et al., 2008; Paterson et al., 2008). Without a doubt there is much to be discovered still about the processes involved in stromatolite formation and lithification and applied to ancient examples. Numerous studies (e.g., as summarized in Vasconcelos et al., 2006) suggest that "many of the early biochemical reaction pathways, which are basic to life, such as anaerobic sulfur metabolism, can be considered to have been dynamically conserved in the evolutionary process over billions of years" and that "phylogenetically deep-branching microorganisms…are still important players in contemporary 'relic' ecosystems" (p. 180). So, what we see in modern stromatolites may provide a window into the past. But does the sediment fog up the window? Since laminae in most ancient examples consist of thin micritic crusts without apparent sediment grains, they are more likely the result of direct precipitation, inorganic or organically mediated, rather than trapping and binding of sediment (Altermann, 2004). Dupraz et al. (2008) argue that "the ecological model of lamina formation associated with the cycling of bacterial communities at the surface of the [modern] build-ups can possibly be transposed in the fossil record and provides an important conceptual model for the fossil counterparts." The challenge now is to apply the model to see what else we can learn about ancient stromatolites.

Micritic crusts do form in the Bahamian mats but sedimentation is a critical process in the lamination formation and a notable difference in the structures produced by the mats. Can we simply ignore the sediment in modern structures in comparing them to ancient examples? Interestingly, if one did just theoretically remove the sediment grains, the texture that would remain would roughly resemble ancient laminae: thin continuous micritic crusts alternating with more porous peloidal micrite (the rare micropeloidal precipitates in Type 1 mats). Perhaps sedimentation was a factor in Precambrian examples but the grains were simply not incorporated into the structures. Or perhaps sediment was incorporated in more examples than we have been able to tell but finer grains sizes, or diagenetic processes and/or recrystallization have hidden the detrital components (Allwood, 2008). Through modeling Dupraz et al. (2006) show that by adjusting intrinsic and extrinsic variables, and specifically sedimentation, morphologies and branching patterns similar to those found in Precambrian stromatolites can be produced. But their models do not yet address lamination. Maybe sedimentation controlled large scale morphologies but had no impact on the small-scale lamination. Or perhaps as Dupraz et al. (2008) suggest, other microbial processes could have played an important role in the Precambrian. If so, what do we know about modern examples that can help us better understand ancient examples?

The sediment dilemma seems significant enough that some Precambrian stromatolites researchers have begun to shift attention to modern laminated structures produced in alternative settings. Stromatolites can be found in modern highly alkaline lake settings that are thought to more closely resemble Precambrian seas (e.g., Kazmierczak et al., 2004). Microbialites found in such settings contain structures similar to those found in Precambrian deposits (Kazmierczak et al., 2004; Kazmierczak and Kempe, 2006). In the quasi-marine caldera lake on Satonda Island, Indonesia, for example, an apparently monospecific mat of coccoid pleurocapsalean cyanobacteria produces well-laminated sub-millimeter fabrics that contain no sediment (Kempe and Kazmierczak, 2007). Stromatolites described from the sodium carbonate-dominated, highly alkaline caldera lakes of Niuafo'ou Island, Tonga contain many internal structures similar to those found in Precambrian deposits, although the lake waters do not appear to support stromatolite formation at this point in time (Kazmierczak and Kempe, 2006). Alternatively, carbonate laminae from hypersaline lagoonal settings with seasonal fluctuations in water chemistry may help us understand some ancient examples. Vasconcelos et al. (2006) describe living mats from the hypersaline Lagoa Vermelha, Brazil that contain Ca–Mg carbonate laminae. They suggest that the anaerobic sulfur cycling organisms in these mats may offer a "unique natural laboratory in which to study early biochemical processes involved in carbonate formation." (p. 181).

Better explanations of ancient stromatolite examples may require researchers to look at the normal marine examples *combined with* high alkaline and/or high salinity examples as those settings have conditions similar to ancient seas that overlap to some degree. In normal marine settings where there are less chemical fluctuations, stromatolites occur that possess internal layering resembling ancient examples if the sediment was not present. In isolated settings, water bodies that more closely resemble alkaline Precambrian seas contain on their shorelines stromatolitic structures with sediment-free layering that more closely resembles some ancient examples. Perhaps we can learn something from both marine and alkaline settings that, combined, will help us understand ancient examples. Undoubtedly, the developments in each research effort will continue to inform each other. And with Precambrian geologists continuing to look for the best modern analogs as geomicrobiologists develop the methodologies and tools to study modern examples, many questions will be answered.

6. Conclusions

The review provided here makes it clear that significant progress has been made in understanding how modern normal-marine stromatolitic structures are created. We have a better idea of which organisms are involved, how successions in mat community members play a role in the formation of a laminated fabric, the likely microbial activity and biogeochemical processes that cause cementations and fusion of sediment grains that define layering, and a bit about proposed sedimentation variations that may in part cause mat community successions. This progress has been achieved despite the small scales involved in these systems (size of microbes and cements, mat thicknesses, changes in mat chemistry) and the complex interactions of stromatolite communities with varying environmental conditions. Evolving innovative methodologies and tools, and persistent studies have been critical to this progress. There are still, of course, many unanswered questions, with some posed above. Perhaps one of the most significant in terms of better understanding the geologic rock record is what these modern examples can tell us about ancient stromatolites. Learning about the differences in stromatolite formation in subtly different depositional environment as well as dramatically different depositional environments (intertidal vs. subtidal normal-marine and either of the normal marine settings vs. hypersaline or alkaline lake environments) may broaden our perspective and better enable us to see how each example can assist geologists in studying ancient examples.

7. Acknowledgments

The author thanks the editors of this volume, Professors Vinod C. Tewari and Joseph Seckbach, for their invitation to contribute, and Professor Tewari for his encouragement and guidance. I thank also an anonymous reviewer and Dr. R. Pamela Reid for their thorough critical reviews and suggestions, which improved this chapter significantly.

8. References

Allwood, A. (2008) Genesis of 3.43 billion year old stromoatlites: geochemical imaging, spectroscopy and microfacies analysis in recrystallized accretionary structures, In: Y. Watanabe, B. Kamber and A. Allwood (conveners) *Biological and Environmental Signatures in Archean Rocks*. Astrobiology **8**(2): 319–326.

Aloisi, G. (2008) The calcium carbonate saturation state in cyanobacterial mats throughout Earth's history. Geochim. Cosmochim. Acta **72**: 6037–6060, doi: 10.1016/j.gca.2008.10.007.

Altermann, W. (2004) Evolving life and its effect on Precambrian sedimentation, In: P.G. Eriksson, W. Altermanm, D.R. Nelson, W.U. Mueller and O. Catuneanu (eds.) *The Precambrian Earth: Tempos and Events, Development in Precambrian Geology 12*. Elsevier, Amsterdam, pp. 539–545.

Andres, M.S. and Reid, R.P. (2006) Growth morphologies of modern marine stromatolites: a case study from Highborne Cay, Bahamas. Sediment. Geol. **185**: 319–328, doi: 10.1016/j.sedgeo.2005.12.020.

Andres, M.S., Sumner, D.Y., Reid, R.P. and Swart, P.K. (2006) Isotopic fingerprints of microbial respiration in aragonite from Bahamian stromatolites. Geology **34**(11): 973–976, doi: 10.1130/G22859, A.1; 2 figures.

Baumgartner, L.K., Reid, R.P., Dupraz, C., Decho, A.W., Buckley, D.H., Spear, J.R., Przekop, K.M. and Visscher, P.T. (2006) Sulfate reducing bacteria in microbial mats: changing paradigms, new discoveries. Sediment. Geol. **185**: 131–145, doi: 10.1016/j.sedgeo.2005.12.008.

Browne, K.M. (1993) Lamination in recent Bahamian subtidal stromatolites: origin and lithification, unpublished Ph.D. dissertation, University of Miami, Coral Gables, 296 p.

Browne, K.M., Golubic, S. and Seong-Joo, L. (2000) Shallow marine microbial carbonate deposits, In: R.E. Riding and S.M. Awramik (eds.) *Microbial Sediments*. Springer-Verlag, Berlin, pp. 233–249.

Dade, W.B., Davis, J.D., Nichols, P.D., Nowell, A.R.M., Thistle, D., Trexler, M.B. and White, D. (1990) Effects of bacterial exopolymer adhesion on the entrainment of sand. Geomicrobiol. J. **8**: 1–16.

Decho, A.W. and Kawaguchi, T. (2003) Extracellular polymers (EPS) and calcification within modern marine stromatolites, In: W.E. Krumbein, D.M. Paterson and G.A. Zavarzin (eds.) *Fossil and Recent Biofilms – A Natural History of Life on Earth*. Kluwer, Dordrecht, pp. 227–240.

Decho, A.W. and Kawaguchi, T. (2008) Extracellular polymers (EPS), calcification within modern marine stromatolites, In: Y. Dilek, H. Fumes and K. Muehlenbachs (eds.) *Links Between Geological Processes, Microbial Activities and Evolution of Life: Microbe and Geology*. Springer, Netherlands, pp. 227–240.

Decho, A.W., Vusscher, P.T. and Reid, R.P. (2005) Production and cycling of natural microbial exopolymers (EPS) within a marine stromatolite. Palaeogeogr. Palaeoclimatol. Palaeoecol. **219**: 71–86, doi: 10.1016/j.palaeo.2004.10.015.

Dill, R.F. (1991) Subtidal stromatolites, ooids, and crusted-lime muds at the Grand Bahama Bank Margin, In: R.H. Osborne (ed.) *From Shoreline to Abyss, SEPM Special Publication*. Society for Sedimentary Geology, pp. 147–171.

Dill, R.F., Shinn, E.A., Jones, A.T., Kelly, K. and Steinen, R.P. (1986) Giant subtidal stromatolites forming in normal saline waters. Nature **324**: 55–58.

Dravis, J.J. (1983) Hardened subtidal stromatolites, Bahamas. Science **219**: 385–386.

Dupraz, C., Pattisina, R. and Verrecchia, E.P. (2006) Translation of energy into morphology: simulation of stromatolite morphospace using a stochastic model. Sediment. Geol. **185**: 185–203, doi: 10.1016/j.sedgeo.2005.12.012

Dupraz, C., Reid, R.P., Braissant, O., Decho, A.W. and Norman , R.S. (2009) Processes of carbonate precipitation in modern microbial mats, Earth Science Reviews **96**(3): 141–162, doi: 10.1016/j.earscirev.2008.10.005.

Feldmanm, M. and McKenzie, J.A. (1998) Stromatolite-thrombolite associations in a modern environment, Lee Stocking Isalnd, Bahamas. Palaios **13**: 201–212.

Foster, J.S., Green, S.J., Ahrendt, S.R., Golubic, S., Reid, R.P., Hetherington, K.L. and Bebout, L. (2009) Molecular and morphological characterization of cyanobacterial diversity in the stromatolites of Highborne Cay, Bahamas. ISME J. **3**: 1–15.

Ginsburg, R.N. and Planavsky, N.J. (2008) Diversity of Bahamian microbialite substrates, In: Y Dilek, et al. (eds.) *Links Between Geological Processes, Microbial Activities and Evolution of Life*. Springer, Netherlands, pp. 177–195.

Golubic, S. (1991) Modern stromatolites – a review, In: R. Riding (ed.) *Calcareous Algae and Stromatolites*. Springer, Berlin, pp. 541–561.

Golubic, S. and Browne, K.M. (1996) *Schizothrix gebeleienii* sp. nova builds subtidal stromatolites, Lee Stocking Island, Bahamas. Algol. Stud. **83**: 273–290.

Golubic, S. and Focke, J.W. (1978) *Phormidium hensersonii* Howe: identity and significance of a modern stromatolite building microorganism. J. Sediment. Petrol. **48**: 751–764.

Havemann, S.A. and Foster, J.S. (2008) Comparative characterization of the microbial diversities of an artificial microbialite model and a natural stromatolite. Appl. Environ. Microbiol. **74**(23): 7410–7421.

Kazmierczak, J. and Kempe, S. (2006) Genuine modern analogues of Precambrian stromatolites from caldera lakes of Niuafo'ou Island, Tonga. Naturwissenschaften **93**: 119–126, doi: 10.1007/s00114-005-0066-x

Kazmierczak, J., Kempe, S. and Altermann, W. (2004) Microbial origin of Precambrian carbonates: lessons from Modern analogues, In: P.G. Eriksson, W. Altermanm, D.R. Nelson, W.U. Mueller and O. Catuneanu (eds.) *The Precambrian Earth: Tempos and Events, Development in Precambrian Geology 12*. Elsevier, Amsterdam, pp. 545–564.

Kempe, S. and Kazmierczak, J. (2007) Hydrochemical key to the genesis of calcareous non-laminated and laminated cyanobacterial microbialites, In: J. Seckbach (ed.) *Algae and Cyanobacteria in Extreme Environments, Cellular Origin, Life in Extreme Habitats and Astrobiology 11*. Springer, Dordrecht, pp. 242–264.

Kromkamp, J.C., Perkins, R., Dijkman, N., Consalvey, M., Andres, M. and Reid, R.P. (2007) Resistance to burial of cyanobacteria in stromatolites. Aquat. Microb. Ecol. **48**: 123–130.

MacIntyre, I.G., Prufert-Bebout, L. and Reid, R.P. (2000) The role of endolithic cyanobacteria in the formation of lithified laminae in Bahamas stromatolites. Sedimentology **47**: 915–921.

Neumann, A.C., Gebelein, C.D. and Scoffin, T.P. (1970) The composition, structure and erodibility of subtidal mats, Abaco, Bahamas. J. Sediment. Petrol. **40**: 274–297.

Paterson, D.M., Aspden, R.J., Visscher, P.T., Consalvey, M., Andres, M.S., et al. (2008) Light-dependant biostabilisation of sediments by stromatolite assemblages. PLoS ONE **3**(9): e3176, doi: 10.1371/journal.pone.0003176.

Perkins, R.G., Kromkamp, J.C. and Reid, R.P. (2007) Importance of light and oxygen for photochemical reactivation in photosynthetic stromatolite communities after natural sand burial. Mar. Ecol. Prog. Ser. **349**: 23–32.

Planavsky, N. and Ginsburg, R.N. (2009) Taphonomy of modern marine Bahamian microbialites. Palaios **24**: 5–17, doi: 10.2110/palo.2008.p08-001r

Reid, R.P. and Browne, K.M. (1991) Interidal stromatolites in a fringing Holocene reef complex, Bahamas. Geology **19**: 15–18.

Reid, R.P., Macintyre, I.G., Browne, K.M., Steneck, R.S. and Miller, T. (1995) Modern marine stromatolites in the Exuma Cays, Bahamas: uncommonly common. Facies **33**: 1–18.

Reid, R.P., Visscher, P.T., Decho, A.W., Stolz, J.F., Bebout, B.M., Dupraz, C., MacIntyre, I.G., Paerl, H.W., Pinckney, J.L., Prufert-Bebout, L., Steppe, T.F. and DesMarais, D.J. (2000) The role of microbes in accretion, lamination and early lithification of modern marine stromatolites. Nature **406**: 989–992.

Reid, R.P., Dupraz, C., Visscher, P.T. and Sumner, D.Y. (2003) Microbial processes forming modern marine stromatolites: microbe-mineral interactions with a three-billion-year rock record, In: W.E. Krumbein, D.M. Paterson and G.A. Zavarzin (eds.) *Fossil and Recent Biofilms a Natural History of Life on Earth*. Kluwer, Dordrecht, pp. 103–118.

Reid, R.P., Bowlin, E., Gaspar, A.P., Andres, M., Macintyre, I.G. and Stolz, J.F. (2008) Two years in the life of a stromatolite, Highborne Cay Bahamas: Assessing the role of algal eukaryotes in stromatolite formation, In: *Geobiology of Stromatolites, International Kalkowsky Symposium*, Gottingen, Germany. October 4–11th, Abstracts and Program.

Seong-Joo, L., Browne, K.M. and Golubic, S. (2000) On stromotlite lamination, In: R.E. Riding and S.M. Awramik (eds.) *Microbial Sediments*. Springer-Verlag, Berlin, pp. 16–24.

Sprachta, S., Camoin, G., Golubic, S. and Le Campion, Th. (2001) Microbialites in a modern lagoonal environment: nature and distribution, Tikehau atoll (French Polynesia). Palaeogeogr. Palaeoclimatol. Palaeoecol. **175**: 103–124.

Stal, L.J. (1994) Microbial mats: ecophysiological interactions related to biogenic sediment stabilisation, In: W.E. Krumbein, D.M. Paterson and L.J. Stal (eds.) *Biostabilization of Sedim*ents. University of Oldenburg, Germany, pp. 41–53.

Stolz, J.F. (2008) Structure of marine biofilms: flat laminated mats and Modern marine stromatolites, In: Y. Dilek, H. Fumes and K. Muehlenbachs (eds.) *Links Between Geological Processes, Microbial Activities and Evolution of Life: Microbe and Geology*. Springer, Berlin, pp. 65–76.

Stolz, J.F., Feinstein, T.N., Salsi, J., Visscher, P.T. and Reid, R.P. (2001) TEM analysis of microbial mediated sedimentation and lithification in modern marine stromatolites. Am. Mineral. **86**: 826–833.

Vasconcelos, C., Warthmann, R., McKenzie, J.A., Visscher, P.T., Bittermann, A.G. and van Lith, Y. (2006) Lithifying microbial mats in Lagoa Vermelha, Brazil: Modern Precambrian relics?, Sedimentary Geology **185**(3–4): 175–183, doi: 10.1016/j.sedgeo.2005.12.022.

Visscher, P.T., Reid, R.P. and Bebout, B.M. (2000) Microscale observations of sulfate reduction: correlation of microbial activity with lithified micritic laminae in modern marine stromatolites, Geology **28**: 919–922.

Visscher, P.T. and Stolz, J.F. (2005) Microbial mats as bioreactors: populations, processes, and products. Palaeogeogr. Palaeoclimatol. Palaeoecol. **219**: 87–100, doi: 10.1016/j.palaeo.2004.10.016.

Yallop, M.L., de Winder, B., Paterson, D.M. and Stal, L.J. (1994) Comparative structure, primary production and biogenic stabilization of cohesive and non-cohesive marine sediments inhabited by microphytobenthos. Estuar. Coast Shelf. Sci. **39**: 565–582

Biodata of **Dr. Elizabeth Chacón B., Dr. Esther Berrendero Gómez, Dr. Marco A. Sanchez Ramos, Dr. Gustavo Montejano**, and **Dr. Juan M. Malda Barrera**, authors of ***"Are Cyanabacterial Mats Precursors Stromatolites?"***

Dr. Elizabeth Chacón B. is currently Research Professor at the Faculty of Earth Sciences, UANL, Mexico, since 2006. She obtained her Ph.D. from the Faculty of Sciences at the UNAM within an Exchange DAAD-Graduate Program between DAAD (in George-August University, Göttingen) and UNAM. She continued her postdoctoral research at the Louis Pasteur University (Strassbourg), and later she joined the academic team of Prof. F. García-Pichel at ASU. Her scientific interest in Origins of Life and Astrobiology since the early Bachelor years has conducted her to the study of biosignatures derived from fossil and recent microbialites.

E-mail: **elizach@fct.uanl.mx**

Dr. Esther Berrendero Gómez is currently post-graduate research of the Faculty of Sciences in the National Autonomous University of México (UNAM), Mexico, and the FCT (UANL) with a MAEC-AECID fellowship of Ministerio de Asuntos Exteriores y de Cooperación, Spain, and CONACyT. She obtained her Ph.D. from the Autonomous University of Madrid (UAM) in 2008; her general research area includes the diversity and distribution of cyanobacteria in natural communities. Her scientific interest lies in the morphological, molecular, evolution, and ecophysiological diversity of cyanobacteria from carbonate environments.

E-mail: **esther.berrendero@uam.es**

Elizabeth Chacón B.

Esther Berrendero Gómez

V.C. Tewari and J. Seckbach (eds.), *STROMATOLITES: Interaction of Microbes with Sediments*, Cellular Origin, Life in Extreme Habitats and Astrobiology 18, 313–341
DOI 10.1007/978-94-007-0397-1_14, © Springer Science+Business Media B.V. 2011

Dr. Marco A. Sanchez-Ramos is a full time Professor from the Facultad of Biotic Resources, in the Autonomous University of Queretaro (UAQ) in Queretaro, Mexico. Early in his career he joined the research group in the Institute of Cell Physiology from the National Autonomous University (UNAM) in the area of Neuroendocrinology, Brain Assimetry, and Neurophysiology, where he obtained his Ph.D. His professional experience includes projects in Biopharmacy and Neurophysiology; although his main stream of research is Physiology, his multiple and broad interests cover from Evolution through Human Behavior. Although he is normally devoted to research and teaching activities at the Bachelor and Graduate Level, he is currently engaged in several interdisciplinary academic as science diffusion projects as well.

E-mail: **masr@uaq.mx**

Dr. Gustavo Montejano is a full time Professor from the Biology Department at the Science Faculty at UNAM, Mexico City, where he does research and teaches Cyanoprokaryota Taxonomy, Botanics, and Ecology of Continental Algae, among other courses since several years. He obtained his Ph.D. in Biological Sciences from the National Autonomous University of Mexico (UNAM) and his main academic activities include the ecology and taxonomy of stream algae, and particularly, the biology and taxonomy of cyanobacteria, among other cultural and culinary interests as well. Over the years he has amounted a good collection of algae and cyanobacterial specimens doing extensive lab and field research work, including biological sampling from many diverse and sometimes 'unknown' sites around Mexico.

E-mail: **gmz@ciencias.unam.mx**

Marco A. Sánchez Ramos

Gustavo Montejano

Dr. Juan M. Malda Barrera is a fulltime Professor from the Facultad of Biotic Resources, in the Autonomous University of Queretaro (UAQ). He holds a Bch. Degree in Biology from the Autonomous Metropolitan University (UAM). Later he pursued a Master Degree in Philosophy (UAQ) and after, a Ph.D. in Philosophy from the Estado de Morelos, Mexico. He has been teaching since his early years as graduate student. Among his diverse courses are Evolution, Paleontology, Biology and Society, Neoevolution, Epistemology. Besides Evolution and Paleontology he has explored scientific novel writings and science diffusion as well.

E-mail: **jandlam@yahoo.com**

ARE CYANOBACTERIAL MATS PRECURSORS OF STROMATOLITES?

ELIZABETH CHACÓN B.[1], ESTHER BERRENDERO GÓMEZ[2], GUSTAVO MONTEJANO[2], JUAN M. MALDA BARRERA[3], AND MARCO A. SANCHEZ-RAMOS[3]

[1]*Facultad de Ciencias de la Tierra, Universidad Autónoma de Nuevo León, Carretera Cerro Prieto Km 8 Linares, Nuevo León 67700, Mexico*

[2]*Laboratorio de Ficología, Facultad de Ciencias, UNAM, Apdo. Postal 70-620, Ciudad Universitaria Copilco, Coyoacán 04510, Mexico*

[3]*Facultad de Ciencias Naturales, Universidad Autónoma de Querétaro, Avenida de las Ciencias S/N Col, Juriquilla, Querétaro 76230, Mexico*

Keywords Cyanobacteria • Stromatolites • Microbialites • Paleoenvironments • Geobiologica • Geobiology • Uniformitarism • Kalkowsky • Mats

1. Microbial Mats as Precursors of Stromatolites

Since James Hutton established the principle of uniformitarism in 1788, direct comparisons between ancient and present processes have been key elements in geological and paleontological observations. Lyell and Darwin successfully applied uniformitarism using reasoning and inference to obtain the missing data from incomplete evidence (Gould, 1983). This actualistic approach has also permeated the study of stromatolites and their living analogs, modern microbial mats. As early as 1908, Kalkowsky (1908), who coined the term *stromatolites*, recognized not only the organic nature of stromatolites, but also the participation of microbial life in their construction (Riding, 2008, and references therein), as he wrote: '*stromatolites have a fine more or less even layered fabric....and the participation of simple plants gave rise to limestone precipitation*' (translation from Prof. J. Paul, 2008). The structural similarity between algal mats and intertidal stromatolites was documented since the beginning of the twentieth century; Walcott (1914) even presumed the participation of cyanobacteria by then (Awramik and Grey, 2005). The early observations of Black (1933) also included the participation of cyanobacteria in the modern marine microbialites from the Bahamas. The precipitation of calcium carbonate, erosive boring, and binding of sediments were identified as main processes in these algal carbonates (Fritsch, 1945; Pettijohn,

1957; Sharp, 1970). Later Logan et al. (1964) proposed a popular general classification system to group the main morphologies found in 'algal stromatolites' when he analyzed recent stromatolites from Shark Bay in Australia.

Confronted with an incomplete fossil record, the approach normally applied to establish the relation between stromatolites and microbial mats is to use recent microbial mats to infer past processes and fossil stromatolites to reconstruct paleoenvironments. If microbial mats are the morphological precursors of stromatolites, their detailed analysis should provide the basis for understanding stromatolites, establishing the multiple relations between sedimentological and all other geobiological interactions.

The most popular definition states that stromatolites accrete by the continuous processes of trapping, binding and/or precipitation of sedimentary particles by microbial communities (Awramik, 1977) followed by cementation (Knoll, 2008). From these three accretion mechanisms, the most compelling to decipher in the sedimentary record is precipitation, and whether this precipitation is abiogenic or biogenic. Krumbein (1983) used a more flexible concept, defining stromatolites as organosedimentary laminated calcareous rocks (fossil or not) with an origin related to microbial life (but see review in Riding (2008)). Other types of microbial carbonate rocks are grouped together with stromatolites in the general definition of *microbialites* (Burne and Moore, 1987). According to Semikhatov et al., (1979), a stromatolite is an attached, laminated, lithified sedimentary growth structure accreting from a point of limited surface of initiation, where their diagnostic and unifying character is their laminated fabric, although general observations on how they grow and lithify are diverse (Grotzinger and Knoll, 1999). Paleontological evidence has shown that microbial behavior can be recorded in sediments and that variations in morphology may results from variation in both microbiota and depositional environments (Sumner, 2000); therefore, understanding stromatolites implies to understand the microbial community proliferating on a given substrate, their accretion mechanism, and their diagenetic history within an integrated context.

Classically stromatolites have been considered to be organosedimentary structures resulting from lithified microbial mats where cyanobacteria are main contributors; in fact, the few examples of extant stromatolites show well-developed cyanobacterial mats; most Proterozoic stromatolites probable accreted by microbial trapping and/or binding (Knoll, 2008); in contrast, some Archean stromatolites show a fabric where precipitation has been the main process, and therefore the analysis of each stromatolite occurrence should be individually analyzed (Knoll, 2008). Owing to their great morphological similarity, most described Precambrian microfossils have been interpreted as cyanobacteria (Golubic, 1991; see discussion in Golubic et al. (2000) and in Knoll (2008)). But microfossiliferous stromatolites are rather rare, and, in case of any microfossil occurrences in Archean stromatolites, the size of cyanobacterial microfossils would be difficult to reconcile with the very fine lamination in Archean stromatolites. On the other hand, cyanobacteria display a significant morphological and metabolic versatility (Stal, 1991), two attributes that confer them with a great adaptation capacity that

Figure 1. Examples of stromatolites from different strata. (**a**) Classical stromatolite from the Kalkowsky locality in Lower Saxony, Germany. (**b**) A close-up reveals that stromatolitic laminations are covered by regular-sized oolites. (**c–i**) Diverse macrostructures found within Cretaceous stromatolites from the Tarahumara Formation in Sonora, Mexico.

position them as key precursors of early ecosystems. Whether cyanobacteria were Archean builders remains questionable, but their geobiological impact on Earth history has been unique and crucial: at some point during the Precambrian, they released the poisonous and now vital oxygen, changing forever the course of geobiological evolution on Earth (Fig. 1).

2. The Persistent Cyanobacteria

Unlike any other biological group, cyanobacteria triggered major evolutionary events that shaped the biosphere that we see today: cyanobacteria were the first organisms to employ oxygenic photosynthesis, being responsible of the Earth's atmosphere transition from anoxic to oxic (Ehrlich, 1981); cyanobacteria are considered modern representatives of the evolutionary ancestors of chloroplasts in plants too (Olsen et al., 1994; Wang et al., 1999; Cavalier-Smith, 2003).

Paleoecologically they are important because of their role as primary producers and were likely the main builders of ancient ecosystems on Earth (Rothschild and Mancinelli, 1990).

2.1. CYANOBACTERIAL ATTRIBUTES

Cyanobacteria are photosynthetic prokaryotic organisms that taxonomically group within eubacteria. They use basic growth requirements such as light, CO_2, and H_2O for energy, carbon, and electron sources, respectively (Stal, 1995). Other species are facultative photoheterotrophs, utilizing light as energy source and organic compounds (glucose, sucrose or fructose) as carbon source (Castenholz, 2001). In darkness, they gain energy by respiring endogenous carbohydrates produced during light time (Stal, 1995).

Cyanobacteria are known to occur in oxic as well as anoxic environments (Castenholz, 1973) and only a relatively small number may perform anoxygenic photosynthesis using sulfide as electron donor (Schmidt, 1988). In a few cases, cyanobacteria are also capable of chemoheterotrophic growth in darkness, when both energy and carbon are derived from organic compounds (Smith, 1982). Some species use various nitrogen sources (NO_3, NH_4^+), although they have a preference for ammonium (Guerrero and Lara, 1987). They can use inorganic phosphorous available in the environment, and some of them posses phosphatase to utilize organic phosphorous (Grossman et al., 1994).

Cyanobacteria are a singular group since they have the ability to perform a photosynthesis similar to that found in eukaryotic algae and vascular plants (Whitton, 1992), but their structural organization and biochemistry correspond to Gram-negative bacteria (Stanier and Cohen-Bazire, 1977). As a group they exhibit a great diversity in morphology and organization; they may be found as unicells (e.g., *Chroococcus* in Fig. 2a) and filaments (Fig. 2b, c) that may present true or false branches (Fig. 2e, f, respectively). True-branching cyanobacteria are characterized by the presence of branching point cells in contact with three different neighboring cells, where longitudinal, oblique, and transverse cell divisions occur (Gugger and Hoffmann, 2004), giving rise to main branching types Y, T,

Figure 2. Morphological diversity in cyanobacteria from different localities around Mexico. (**a**) Example of unicellular cyanobacteria, *Chroococcus* sp. from sulfurous mats of Baño de San Ignacio [BSI], in Linares, NL, México; note the small hormogonium (*arrow*). (**b**) Simple uniseriate filament from [BSI] identified as *Johannes* sp., surrounded by a mucilaginous sheath (*arrow*). (**c**) *Phormidium* sp. derived from [BSI] showing carbonate precipitates along the filament. (**d**) *Hyella* sp., a typical cosmopolitan endolithic cyanobacteria (**e**) *Mastigocladus* sp., example of filamentous cyanobacteria with true branches (*arrow*) from Agua Salada, Morelos, México. (**f**) Example of false-branching in *Tolypothrix* (*arrow*). (**g**) Tapered filament from a marine cultivated *Calothrix*. (**h**) Typical trichome

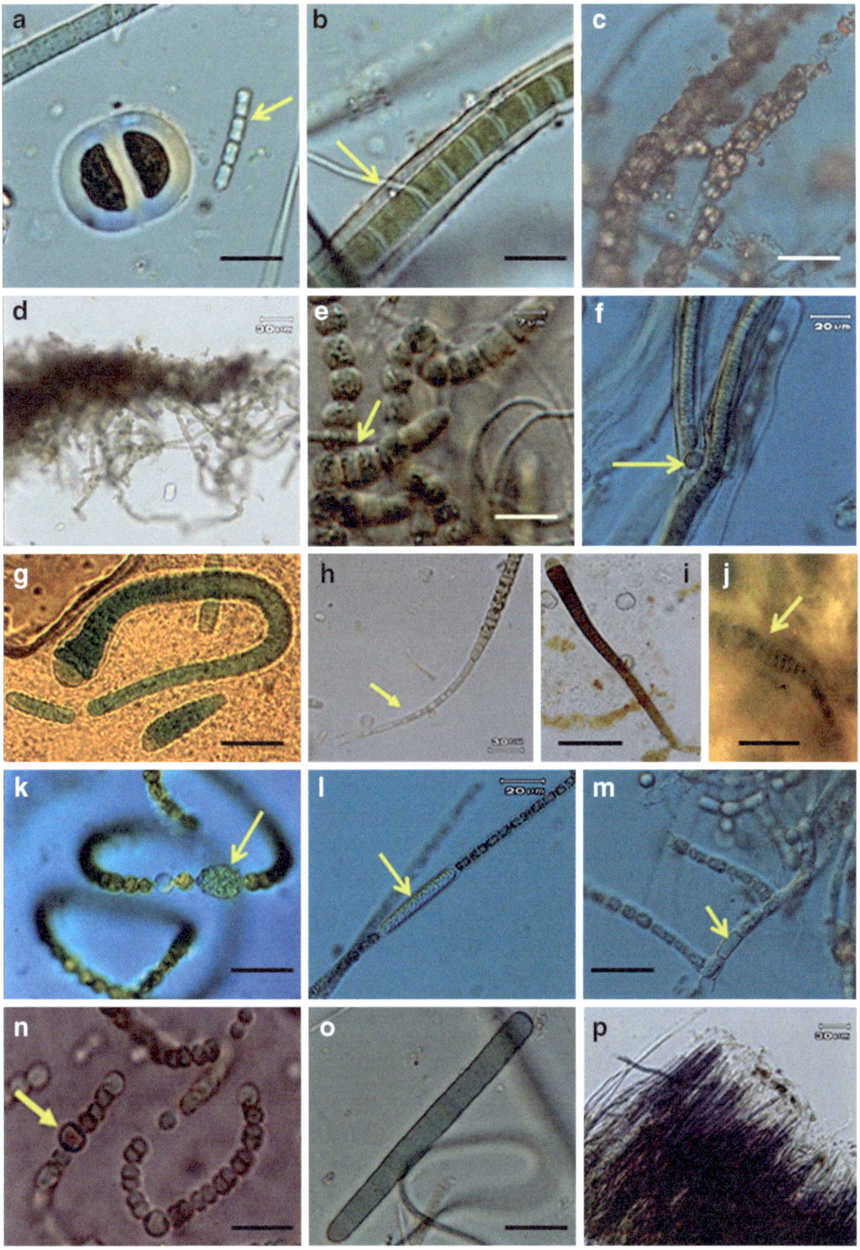

Figure 2. (continued) in *Rivularia* and (**i**) *Calothrix* sp. both derived from Laguna Atotonilco. (**j**) A fossil tapered filamentous remain from the Rivularaceae family preserved in the Huepac chert. (**k**) Heterocyst in *Anabaena* sp. (**l**) *Aphanizomenon* sp. showing an elongated akinete from Laguna de Atotonilco. (**m**) *Hapalosiphon* sp. showing an enlarged heterocyst (*arrow*). (**n**) Intercalary heterocysts developing in *Nostoc* sp. (**o**) Hormogonium from [BSI]. (**p**) Population of vertically-oriented filaments of *Heteroleiblenia* sp.

and V (Golubic et al., 1996). On the other hand, false-branching cyanobacteria show intervening cells (*intercalary heterocyst*) or cells produced by the fragmentation of the trichome within the sheath, followed by elongation of the newly generated trichomes ends and the protrusion of one of them through the sheath (Adams, 2000).

Certain species of filamentous cyanobacteria exhibit characteristic tapered filaments (Fig. 2g–j). In some cases, the apical cells become narrowed and colorless, forming multicellular hairs (Fig. 2h); typically the cells lose their chlorophyll and thylakoids split apart to form liquid-filled vacuoles. Apparently these apical hairs develop in response to phosphorous limitation, though shorter hairs can be produced experimentally under iron deficiency (Douglas et al., 1986). These are characteristic of environments with variable phosphorus levels and where phosphorous limitation is produced in intermittent periods (Pentecost and Whitton, 2000).

Within cyanobacteria a number of morphophysiological strategies evolved that protected nitrogenase from oxygen, including a temporal separation of oxygenic photosynthesis and nitrogen fixation, occurring during day and night, respectively (Bergman et al., 1997). Some filamentous cyanobacteria exhibit the differentiation of a specialized cell, the *heterocyst*, providing a suitable microaerobic environment for the nitrogenase activity (Giovannoni et al., 1988). Well-developed heterocysts are shown in Fig. 2f, k, m, n. The heterocyst differentiation is irreversible and they develop when a suitable source of combined nitrogen becomes limited (Whitton, 1992). Heterocysts, unlike vegetative cells, have a very thick cell wall that decreases gas diffusion, including oxygen (Cardemil and Wolk, 1981). Species of cyanobacteria inhabiting well-oxygenated environments are able to fix atmospheric nitrogen inside the heterocyst (Wolk et al., 1994); the most representative is *Nostoc* (Fig. 2n) where intercalary heterocysts are found along the filament. There are other arrangements, such as in *Rivularia* (Fig. 2i), where there is just a single terminal heterocyst. The shape, number, and position of heterocysts along a filament are basic taxonomic characters used in the traditional classification systems (Komárek and Anagnostidis, 1989). Many heterocystous cyanobacteria also form a second cell type known as *akinete*, which is a specialized spore-like cell formed under unfavorable environmental conditions, as showed in Fig. 2l, and thus, they are able to survive cold and desiccation much better than vegetative cells (Adams and Duggan, 1999). Under suitable conditions akinetes germinate to produce new filaments (Herdman, 1987; Adams, 1992). Those cyanobacterial species that form true filaments all form *hormogonia* (illustrated in Fig. 2g, o), which are modified filaments that play crucial roles in number of important physiological processes in cyanobacteria (Tandeau de Marsac, 1994). Their main function is reproduction (Fay, 1983) but they are capable of mobility too (Whitton, 1992). Hormogonia are generally differentiated in filamentous cyanobacteria; they form from one or more collapsed intercalary cells known as *necridium*, or by the separation between two adjacent cells; in both of cases their formation depends on environmental conditions (Wood et al., 1986).

2.2. ENVIRONMENTAL RANGE OF CYANOBACTERIA

Cyanobacteria, as an evolutionarily ancient group, have acquired multiple strategies that enable them to continuously adapt to the dynamic physical environment (Dvornyk and Nevo, 2003). Because of their versatile metabolism and their adaptation ability, they have colonized almost all habitats on Earth. They are found in terrestrial and aquatic environments of freshwater and marine waters, where they may be found as benthic, planctonic, and intertidal populations (Fay, 1983); they even occur in extreme environments such as hot desert soils or in polar regions (Stal, 2000).

Cyanobacteria frequently inhabit freshwaters and soils with higher pH values. Many cyanobacteria display a temperature-optimum for growth (Castenholz, 2001). Some species tolerate even higher temperatures, up to 74°C (Castenholz, 1984), and they proliferate in hot springs (Dillon and Castenholz, 2003; Steunou et al., 2006), terrestrial rocks, and hot desert soils (García-Pichel et al., 2003; Nagy et al., 2005), and are ubiquitous in polar ponds (Vincent et al., 1993).

In nature cyanobacteria may be found as single filaments and diverse type of mats, but generally tend to cluster forming characteristics colonies, such as those formed by *Nostoc* showed in Fig. 2m; *Nostoc* may equally be found as extensive soft mats, as illustrated in Fig. 3c. Cyanobacteria may also associate with other microorganisms in microbial mats (Fig. 3a, b) or form symbiotic associations with variable eukaryotic partners, including plants (Rasmussen and Syvenning, 2001), fungi and invertebrates (Wilkinson, and Fay, 1979), fuelling the cycling of nutrients in terrestrial ecosystems.

Many of the free-living cyanobacteria and those engaged in symbiotic associations with other organisms have contributed significantly to the productivity of aquatic and terrestrial habitats, including cultivated lands, particularly in the tropics (Whitton and Potts, 2000). Additionally cyanobacteria are of considerable importance in the natural environment as initial colonizers of arid land and as primary producers of organic matter (Fay, 1983). The abundant microbial life found in the Dry Valleys from Antarctica provide examples of how primary producers adopt different habits, including development of epilithic, edaphic, hypolithic, and endolithic communities, and microbial mats in wetter habitats, such as lake margins and in stream beds where cyanobacteria often dominate with eukaryotic algae or with other picoplancton components (Howes, 1989; Elster and Komarek, 2003; Ivanikova et al., 2007).

In general, cyanobacteria resist desiccation very well, playing a key role in the stability of the surface crusts of semi-deserts and in the fertility of soils used for farming in arid regions (Whitton, 2000). Their tolerance to high salinities and fluctuating conditions allows them to dominate in many hypersaline marine lagoons and inland saline lakes too (Javor, 1989; García-Pichel et al., 1998; López-Cortés et al., 2001).

Equally, adaptation to low light is an important factor in modern picocyanobacteria (Paerl, 1990) and in some lakes (Stockner et al., 2000). Furthermore,

Figure 3. *Examples of some extreme environments from Central Mexico.* (**a**) Locality of Los Azufres where thermal sulfurous springs bathe vivid cyanobacterial mats in Michoacan, Mexico. (**b**) A close-up of in situ-silica precipitating sites where cyanobacterial mats at temperatures around 67°C. (**c**) *Nostoc* sp. colonies from La Huasteca Potosina. (**d**) Volcanic maar of Rincon de Parangueo [RP] with stromatolitic mounds. (**e**) Domal stromatolitic mounds ranging from a few centimeters up to 15–20 cm in diameter occurring around the crater-lake. (**f**) Microbial mats associated to ephemeral lakes in [RP]. (**g**) Dissecation sedimentological pattern in [RP]. (**h**) The presence of a red ponds is often observed. (**i**) Internal section of saline crust covered by endoliths. (**j–k**) Thin section of the stromatolitic micro-fabric of domal bioherms. (**l**) Petrographic thin-section showing the alternate stromatolitic lamination of interpreted microbial origin. (**m–p**) SEM micrographs showing pinnate diatoms covered by mucilaginous material, cyanobacterial filaments, and coccoidal structures.

some cyanobacteria display a high tolerance to ultraviolet radiation types UVA, UVB and UVC (García-Pichel et al., 1992) ; Dillon and Castenholz, 1999; Dillon and Castenholz, 2003). This ability may have been a particularly important advantange during the early evolution of cyanobacteria (García-Pichel et al., 1992), back in Precambrian times, when the UVC levels were higher and before the oxygen blocker (Dillon and Castenholz, 1999).

Such a wide tolerance range may be accomplished by different cyanobacterial strategies: by synthesizing internal compounds such as mycosporine-like amino acids, by switching DNA reparation mechanisms, gene regulation, and by gliding motility in soft microbial mats to escape the high irradiance intensities (Castenholz, 2004).

3. Cyanobacterial Microbial Mats

As mats cyanobacteria are found in a broad range of environments too, some of which can be considered extreme, such as hypersaline ponds and lakes, thermal springs, dry and hot deserts, extremely cold environment of polar regions, and even in hypersaline and acidic pools (Stal, 2000). They develop as mat covers on solid surfaces of abiogenic and biogenic substrates, in aquatic or semi-aquatic habitats, and they can grow and expand on sediments of shallow waters with low energy and little grazing, their ecological tolerance of cyanobacteria being a key factor in the development of a mat.

Microbial mats are dense communities of vertically stratified microorganisms that proliferate on the surface and on the shallow subsurface of sediments (Stolz, 2000). These mats typically consist of a matrix of mucilage in which cyanobacterial trichomes and other algal cells are embedded together with other heterotrophic and chemoautotrophic microorganisms and diverse inorganic materials as sand grains. These microorganisms secrete extracellular polymeric substances (EPS), which enhance the stability of sediments (Gerdes et al., 2000; Sarkar et al., 2008). The morphology, structure, and color of microbial mats are determined by the dominant species, sediment features, and other environmental factors (Stal, 2000).

Cyanobacterial mats exhibit marked depth-related light and chemical gradients, where pH and concentrations of dissolved inorganic carbon (DIC) and O_2 (DO) daily oscillate, allowing a wide variety of metabolic diversity to be represented and stratify accordingly (Des Marias, 2003). In fact, the oxic zone reflects a dynamic balance between photosynthetic production and consumption of oxygen by a host of sulfide-oxidizing and heterotrophic bacteria (Des Marais, 2003). Together with other bacteria cyanobacteria degrade organic matter generating microchemical gradients concentrated at the sediment-mineral interface, leading to lithification in decaying mats (Stolz et al., 2001). In some cases, the precipitation of $CaCO_3$ is associated with the photosynthetic uptake of CO_2 (Pentecost and Bauld, 1988;

Ludwig et al., 2005); in others the carbonate-precipitation-promoting process has been more related to cyanobacterial degradation of EPS (Dupraz et al., 2006), to sulfate-reducing bacteria (Visscher et al., 2000; Dupraz and Visscher, 2005) and to the thermodynamics of the microenvironments in mats (Arp et al., 2003) and their interactions with cyanobacteria and heterotrophic bacteria (Altermann et al., 2006).

Microbial mats display special physical characteristics that confer them to behave as structural units: mechanical integrity, stability and cohesiveness and they are embedded within extracellular polymeric substances or EPS (Weckesser et al., 1979; Costerton et al., 1995), composed of highly organized polymers such as cellulose and of small molecules such as uronic acid and amino acids among others (Decho and Kawaguchi, 2003). This EPS accumulates outside the cells, forming an organic matrix to which microbes attach. They function as biosystems with a daily recycling of O, C and S, where photosynthesis, aerobic respiration, fermentation, sulfide oxidation, sulfate reduction, and methanogenesis overlap (Fenchel et al., 1998). Some recent studies have stressed the importance of light as a key factor in the generation of microbialites for the stabilization of a mat, arguing that photosynthesis was the key in the proliferation and persistence of Precambrian microbial mats (Andres and Reid, 2006; Paterson et al., 2008). In some cases, the cohesive nature of the mucopolysacharides of cyanobacterial sheaths is an essential factor to trap sedimentary particles (Stoodley et al., 2002). Microbial baffling and trapping concentrate sedimentary particles, stabilize them against erosion, increasing their preservation potential (Noffke and Paterson, 2008). Grotzinger and Knoll (1999) identified three laminae forming processes: mat growth, sedimentary deposition, and mineral precipitation, which played a significant role in the accretion of Precambrian stromatolites. Quantitative data from modern microbialites have also shown the importance of early lithification in these processes (Dupraz and Visscher, 2005). Among the main factors involved during mat development are the following: a founder community that starts the colonization of the substrate, a source of energy supply and a continuous supply of organic and inorganic metabolites. Once a surface is colonized, the development of the whole community continues to stabilization (Reid et al., 2000).

Microbial mats occur worldwide and those developing under extreme environments are particularly interesting because they may be more related to primitive environment analogs. Along the Mexican Volcanic Belt (MVB), diverse mesophilic and thermal environments occur; especially beautiful are the sulfurous hot springs from Los Azufres, Michoacan (Fig. 3a, b), where microbial mats developing under turbulent waters and silica sinter precipitation along a well-defined temperature gradient (from 67°C to 26°C) may be found. Most of these recent mats are formed by cyanobacteria (*Chroccoccus* and filamentous morphotypes) along with other eukaryotic microorganisms, such as chlorophytes and penate diatoms (Morales-Puente et al., 2002). In fact, the presence of green algae and diatoms in modern microbialites is rather common (Chacón-Baca, 2002).

The other ephemeral microbial mat formed under highly alkaline conditions is the maar lake of Rincon de Parangueo, in Valle de Santiago (Malda et al., 2002). Small microbialites occur around the former crater-lake (Fig. 3d–h), showing a coarse fabric and a 1-cm-thick-finely laminated-layer at the top with the typical stromatolitic fabric as seen in petrographic thin-section (Fig. 3i–l). Among the main biotic component in these microbialites are penate diatoms, filamentous and coccoidal cyanobacterial, and in minor proportions, other unidentified bacteria (Fig. 3m–p). A multidisciplinary research in this locality is currently under progress (Aranda et al., 2009); although the main analysis focus in cyanobacteria derived from this environment, it is clear that archaeal groups and other eubacteria are present as well.

3.1. WHO CAN BUILD STROMATOLITES?

The generalized view holds that cyanobacteria are major builders in modern microbial mats (Awramik, 1977; Jing et al., 2005; Roeselers et al., 2006; Lau et al., 2009), a paradigm that has been promoted by the orthodox methodologic approach, and by our familiarity with cyanobacterial morphotypes (Westall, 2005). Microbial communities of green and sulfur purple bacteria, including chemoorgano-heterotrophs, fermenters, chemoorganoautotrophos and iron and sulfate reducing bacteria, are important structural and functional components of modern microbial mats (Baumgartner et al., 2006).

Besides microscopic methods, modern tools currently employed in polyphasic approaches to study microbial mats include the following: (a) *Molecular methods*, such as terminal restriction fragments analysis and 16S rRNA gene sequencing, DGGE (denaturing gel electrophoresis), DNA sequencing analysis, and ARISA profiles (automated ribosomal intergenic spacer analyzer) (Havemann and Foster, 2008; Sompong et al., 2008; Miller et al., 2009; Dillon et al., 2009; Leuko et al., 2007; Lau et al., 2009; Allen et al., 2009), and (b) *Biochemical Methods*, including the analysis of phospholipid fatty acid (PLFA) profiles, isolation and culturing, nanoSIMS and CARD-FISH approach (catalyzed reporter deposition-fluorescence in situ hydridization; Fike et al., 2008; Shiraishi et al., 2008) as well as the distribution and isotopic composition of lipids derived from mats (Jahnke et al., 2001, 2008 and references therein; Orphan et al., 2008; Schouten et al., 2007), from different environmental samples (van der Meer et al., 2005), and the characterization of hopanoids biomarker that may represent a link to link stromatolites and modern microbial mats (Jahnke et al., 2004). Modern analytical tools for dissection of mats and biofilms are reviewed by de Beer and Kuhl (2001) and in the updated review on Molecular ecology and polyphasic approaches by García-Pichel (2008). During the last few years, modern molecular tools are increasingly showing that although cyanobacteria are indeed a major biomass component, principally in the first upper millimeter of some mats (Ley et al., 2006), they comprise a small proportion of the total diversity (Papineau et al., 2005). Also they have showed that a significant diversity and abundance of other noncyanobacterial populations exist at different depths of the vertical profile of mats, as probably they must have been in the past (Jørgensen, 1982; Ward et al., 1992). Recent works actively documenting the community composition from a wide range of environments include the microbial mats at the Black Sea, where the anaerobic oxidation of methane is a significant process (Krüger et al., 2005), the noncyanobacterial mat-forming microbial communities dominated by the anoxygenic phototrophs GNSB (green nonsulfur-bacteria) such as filamentous Chloroflexi (Ley et al., 2006), which is able to form laminated mats too (Boomer et al., 2009) and found in diverse hot springs (Schouten et al., 2007). α- and γ-proteobacteria were among the most abundant groups in thermophilic mats, followed by Chlorobi, and less frequently by other phototrophs such as Chloroflexi and Cyanobacteria (Lau et al., 2009).

On the other hand, estimations on the cyanobacterial isolates from Hamelin Pool in Shark Bay stromatolites include *Microcoleus*, *Euhalothece*, *Chroococcus*, *Spirulina*, *Lyngbya*, *Pleurocapsa*, *Stanieria*, *Chroococcidiopsis*, *Xenoccocus*, and *Halothece* among others (Brendan et al., 2004). A more recent molecular analysis of the hypersaline smooth and pustular microbial mats communities from the lagoon of Shark Bay indicated the predomination of Actinobacteria, Bacteroidetes, Chloroflexi, Cyanobacteria, Gemmatimonas, Planctomycetes, Alphaproteobacteria, Gammaproteobacteria, Deltaproteobacteria (Allen et al., 2009), while the analysis of intertidal columnar stromatolites (considered as analogs to the oldest stromatolites) revealed a high diversity of microbial populations being the most abundant sequences of alfa and gamma proteobacteria and cyanobacteria from the genera *Euhalothece*, *Gloeocapsa*, *Gloeothece*, *Chroococcidiopsis*, *Dermocarpella*, *Acaryochloris*, *Geitlerinema*, and *Schizothrix*, besides other noncyanobacterial communities characterized by sequences related to the halophilic archaea (Goh et al., 2009).

In spite of this increasing acknowledgment of the plurality of microbial populations in mats, cyanobacteria continue to be key factors in the development of microbialites, mainly during the colonization of the sedimentary substrate, the accretion of lithified mats, and because oxygenic photosynthesis is the main source of primary productivity (Papineau et al., 2005). Particularly interesting are microbial mats from the hypersaline basin of Guerrero Negro; previous microbiological studies showed that cyanobacteria dominated surface layers along with other five phyla: *Chloroflexi, Spirochaetes, Proteobacteria, Bacterioidetes, and Firmicutes* (Nübel et al., 1999); today at least 42 phyla have been reported (Ley et al., 2006). Molecular analysis in diverse hot springs revealed the following cyanobacterial species: *Synechococcus* spp., *Phormidium cf. boryanum*, and *Leptolyngbya* spp. in hot springs (Sompong et al., 2008).

The new awareness of microbial diversity and their daily biogeochemical oscillations in modern mats (Krüger et al., 2005; Allen et al., 2009; Lau et al., 2009; Schouten et al., 2007; Dillon et al., 2009) clearly raises the probability that cyanobacteria may have not been the first, nor the only one stromatolite builder back in the Early Precambrian (Olson, 2006). Sulfate reducing bacterial (SRB) communities also form microbial mats; they are abundant and active in the oxic zones of mats, and tolerate oxygen and in the lithified zones; they could also be considered as early stromatolites builders (Baumgartner et al., 2006); additionally, other paleontological evidence suggests their presence in the early Archean fossil record (Westall, 2008).

3.2. WHAT DOES IT TAKE TO BUILD A STROMATOLITE?

The generation of stromatolites involves complex and multiple factors and interactions over time and under constant changing environmental conditions. According to their synoptic relief, stromatolites may occur as *bioherms* (a mound-like domelike, lenslike, or reeflike rock developed in situ), or as *biostromes* (a distinctly bedded laterally extended rock). Some of the factors influencing their morphogenesis include the microbial community, geochemical environment, sedimentary input, current conditions, wave energy, underlying substrate, climate, depth of formation, light abundance and levels of nutrient availability (Planavsky and Grey, 2008). Stromatolites may accrete through the continuous trapping, binding and/or precipitation of sedimentary particles, or by the relative contribution of these processes, sometimes in parallel with dissolution by endolitic microorganisms that promote lithification by forming well-indurated stromatolitic layers (Macintyre et al., 2000), being calcification the most complex issue in stromatolite accretion. Microbial precipitation primarily results from two major processes: (1) 'biologically induced' precipitation, where microbial activities generate biogeochemical conditions that facilitate precipitation; and (2) 'biologically influenced' precipitation, where passive interactions of extracellular biopolymers and the geochemical environment drive precipitation (Frankel and Bazylinski, 2003). A common location for such biopolymers is the microbial 'biofilm' (i.e., cells surrounded within a matrix of extracellular polymeric substances or EPS; Douglas, 2005), a key factor for accretion. The EPS is a highly hydrated biopolymer that microorganisms secrete outside their cells embedding the organism into a biofilm or mat. Its composition includes high-molecular-weight mucous secretions from cells to the extracellular matrix as well as small organic and inorganic molecules (Kawaguchi and Decho, 2002). Among the macromolecules identified are proteins, carbohydrates, glycoproteins, and uronic acids. In this way, the EPS creates a microenvironment for sessile cells conditioned by the chemical nature of the EPS matrix (Flemming et al., 2007). The EPS production is crucial to development of the mat and it is not exclusive from cyanobacteria, other microorganisms may produce EPS. Sometimes EPS can be interchanged with terms such as *slime* or *mucilages* or even cyanobacterial sheaths, according to their association with cells; slimes never have a direct contact with cells, and it is used to describe a soluble or nonmembrane-bound looser binding to the bioaggregates in comparison to capsules; their description is according to the extraction procedure (Nielsen and Jahn, 1999), also including all polymeric material not directly anchored in cell wall. Sheaths are bound-cell components and *mucilages* is a general term applied mainly to the EPS of plants, but its composition is similar (95–97% carbohydrates and 5–3% aminoacids), with uronic and ferulic acids as well (Knee et al., 2001). A review of EPS is found in Wingender et al. (1999), its role in biofilms (Flemming et al., 2007) and in mats (Decho et al., 2005). Dupraz et al. (2006) have taken a holistic approach to detail the factors at the micro, meso, and macroscales. Among the extrinsic factors are substrate, sedimentation, alkalinity, salinity, nutrients, temperature, hydrodynamics, and competition, while the intrinsic factors involved are CO_2, $CaCO_3$, recycling of elements, EPS cohesion, metabolic pathways, and EPS, among others. Other important observations indicate that light enhances stromatolite cohesion and EPS production, stabilizing the macrostructure (Paterson et al., 2008). Besides these factors, it is clear that a copious production of EPS in microbial mats provides an ecological advantage to microbial communities that have to cope with UV irradiation, dehydration and even as defense against grazers in *Phormidium* (Pajdak et al., 2001). At the biochemical level, the EPS helps to maintain the structure and facilitates the extracellular enzymatic activity and chemical signaling (Xavier and Foster, 2007), specifically the process is known as *quorum sensing* (QS), which probably plays a significant role in microbial biofilms and mats (Decho et al., 2008, 2009). Briefly, QS refers to a density-dependant behavior as a response to the extracellular concentration of a small chemical molecule that induces a concerted or coordinated response (Gera and Srivastava, 2006); this social behavior displays a concerted response including changes in gene expression, levels of secondary messengers (metabolites) or protein modifications (Galperin and Gomelsky, 2005). This communication among cells relies on the production and release of small signaling molecules called *autoinducer*, which allows to respond to changes in growth conditions, most often in response to population density, a process called QS (Whitehead et al., 2001). Once an extracellular threshold concentration of the autoinducer molecule or compound has been reached, a set of genes are coordinately activated or repressed within the bacterial population (Dove et al., 2003). This unified response is an emergent property related to nutrient acquisition (Park et al., 2003) and to the regulation of phenotypes (Gera and Srivastava, 2006). QS also regulates surface topology (Park et al., 2003), microbial competition and motility (Dingding et al., 2006), and it is a general response displayed by many diverse microbial communities, including Gram-negative and Gram-positive bacteria (Gera and Srivastava, 2006). Decho et al. (2009) have extracted some basic autoinducers molecules from natural marine microbial mats, and have suggested that QS mechanisms may have evolved early in stromatolitic microbial mat communities, where periodicity is determined in part by diel geochemical gradients. The QS mechanism could also help to explain some macroscopic accretion patterns in stromatolites, for instance the change of hemispheroidal morphology to branching, the upward protrusion, and the change of accretion patterns over infinitely long-distance (in microns) among the building microbial populations during stromatolite morphogenesis. Notably well-developed branching patterns and multiple branching stages observed in ancient stromatolites are an open question (Planavsky and Grey, 2008); one might further speculate that these branching-patterns would require a well-concerted and coordinated response of ancient microbial communities to a specific environmental autoinducer in fluctuating concentrations of oxygen?

In recent stromatolites from Highborne Cay, Reid et al. (2000) identified at least three main components: *Schizothriz*, heterothrophic bacteria, and endolithic cyanobacteria, including *Solentia*. Cyanobacteria are the main producers of organic matter that will be consumed by sulfate-reducing bacteria that in turn will sustain a wide variety of phototrophic and chemotrophic bacteria in mats (Des Marais, 2003), and may be quantitatively important in terms of biomass. Nonetheless, cyanobacterial mats may not be the only potential morphological precursors for stromatolites. Theoretically, any microbial population able to precipitate and trap minerals facilitating nucleation could form microbialites (see examples in Douglas, 2005); for instance, microbial mats composed of green sulfur bacteria, which have the ability to operate as photoheterotrops and phototrophs (Schouten et al., 2007). The precursors of stromatolites would be structures like layered mats lithified at the base and soft at the top, such as those reported from Bahamas.

The transition from microbial mat to stromatolites would be represented by the lithified layered cyanobacterial microbial mats but other nonlaminated mat could be producing microbialites. In fact any microbial mat capable of trapping, binding and/or precipitating minerals, and able to undergo lithification and fossilization may produce a variety of carbonate precipitates (Benzerara et al., 2006). Within the mat the uppermost surficial layer (a few mm in thickness) is the biologically active accreting portion of the stromatolites that results from responding microbial communities (Awramik, 1992; Chacón-Baca, 2002). A further stage in stromatolite development would be the partially lithified microbial mats. Detailed studies on modern stromatolites from Highborne Cay have shown that they grow by the continuous trapping of oolitic sands, formation of micritic crust, and formation of fused-grain laminae (Stolz et al., 2001). Carbonate minerals such as calcite and aragonite commonly precipitate within the amorphous exopolymer matrix of EPS, rather than onto the surface of cyanobacterial sheaths (Reid et al., 2000). Another key observation has been the occurrence of lithified laminae in their texture (Visscher et al., 1998, 2000). Not only the formation but also the persistence of microbial mats is essential for stromatolite development; mats have been proven to be resistant to hydrodynamic forces, and they are stable through time by the precipitation of microbial $CaCO_3$ (Visscher and Stolz, 2005).

In clear contrast to the general model for stromatolites formation, other complex communities of algae and cyanobacteria occur, which do not form a typical microbial mat, but rather are responsible for the formation of stromatolitic structures in calcareous streams of central Mexico (Osorio-Santos, 2007). In such cases, the initial stages of these structures include the colonization of a round substrate (generally a pebble), one or two dominating cyanobacterial

Figure 4. *Microbial fabrics*. (**a**) Hand-sample of a Triassic stromatolite from Southern Germany. (**b**) Microstructure in the stromatolite shown in (**a**) showing nonlinking micritic columns with variable width. (**c**–**d**) Cretaceous stromatolites from Huepac in outcrop (**c**) and fine-grained irregular lamination as seen by SEM (**d**). (**e**) Petrographic thin-section of the Huepac stromatolites showing a more or less flat and even lamination with a small increase in convexity. (**f**) Sparry and fine-grained (*dark*) lamination as preserved in the associated chert. Note the similarity with lamination found in *Conophyton gargaricum* discussed by Riding (2008). (**g**) Recent thrombolite from Cuatro Cienegas showing a clotted fabric (**i**) and a regular convex lamination pattern of sparry calcite and a clotted fabric (**j**). (**h**) Oncolite from la Huasteca Potosina showing well-laminated layers dominated by two cyanobacterial species and an internal clotted fabric.

species belonging to the genera *Romeria* and *Leptolyngbya* that establish on the substrate and that have a high capacity of trapping and binding carbonates. After this initial colonization, this new, hard, and porous substrate is covered by a rich and complex community, which includes algae and cyanobacteria (Fig. 4). Nearly 30 species have been identified in mature communities, most of them correspond to cyanobacteria and diatoms among them *Hyella, Pseudanabaena, Phormidium, Homoeothrix, and Oscillatoria,* among other. Less commonly, rhodophytes may be found, and even when they are not very diverse they are abundant, occurring as juvenile stages named 'chantransia'. All preliminary results suggest that this community is responsible for the typical lamination of the oncolite without developing from a layered mat precursor (Osorio-Santos, 2007).

4. Concluding Remarks

Ever since Kalkowsky stromatolites research has continuously been expanded from different methodological perspectives; a century later stromatolites continue to be actively investigated to document the early history of life on Earth (Awramik, 2006; Awramik and Grey, 2005). Stromatolites provide an indirect and complex evidence of early life on Earth, and today they represent a crysol of biosignatures in which much of the modern research is focusing (Banfield et al., 2001; Cady et al., 2003; Brendan et al., 2004; Seckbach and Walsh, 2008). Microbial mats represent vast living reservoirs of diverse information and a potential window to the past, providing good models for the generation of stromatolites regarding the processes and mechanisms of accretion, as well as the variety of factors in play, according to a specific environmental setting. Their persistence through time suggests emergent properties in mats, since they respond and cope as an integral unit, and their study should be the first stages to decode stromatolites but with limitations, since the translation from mats to stromatolites is not direct. However, modern cyanobacterial mats may be more related to thrombolitic microbialites than to Archean stromatolites. Although their analysis has increased our general understanding about microscopic interactions between mineral particles and the extracellular matrix (EPS), their gross microstructure does not support a direct analog to the fine lamination observed in ancient stromatolites. Probably modern studies are just the tip of the iceberg in reference to what microbial mats diversity and dynamics imply. New and detailed biogeochemical information must reside inside other prokaryotic mats from anaerobic or harsh environments, such as those forming at Guerrero Negro or at the Black Sea, where anaerobic oxidation of methane prevails or microbial mats under continuous and desiccation cycles that resist most severe fluctuating conditions and show well-developed lamination of diverse taxonomic groups. In spite of the vast information about the geomicrobiological interactions of microorganisms through evolution and the greater resolution of analytical tools, stromatolites continue to be a puzzling and crucial piece in the map of life.

5. Acknowledgments

The authors gratefully acknowledge the expert and insightful and constructive comments from the reviewers (Alain Préat, Jesse Dillon, and an anonymous reviewer). The authors are especially grateful to the editor, Prof. Vinod C. Tewari for his kind invitation and his constant encouragement, kindness, and patience. This work was funded by project PROMEP-103-5/07/2523 (SEP) and by Research Project CONACyT-P2-83500-CB to E. Chacón, and a fellowship (to E. Berrendero) from the Ministerio de Asustos Exteriores y de Cooperación, Spain.

6. References

Adams, D.G. (1992) Multicellularity in cyanobacteria, In: S. Mohan, C. Dow and J. A. Cole (eds.) *Prokaryotic structure and function, a new perspective*. Society for General Microbiology Symposium, *Vol*. 47. Cambridge, UK, Cambridge University Press, pp. 341–384.

Adams, D.G. (2000) Symbiotic interactions, In: B.A. Whitton and M. Potts (eds.) *The Ecology of Cyanobacteria*. Kluwer Academic, Dordrecht, pp. 523–561.

Adams, D.G. and Duggan, P. S. (1999) Heterocyst and akinete differentiation in cyanobacteria, New Phytol. **144**: 3–33.

Allen, M.A., Goh, F., Burnns, B.P. and Neilan, B.A. (2009) Bacterial, archaeal and eukaryotic diversity of smooth and pustular microbial mat communities in the hypersaline lagoon of Shark Bay. Geobiology **7**: 82–96.

Altermann, W., Kazmierczak, J., Ohren, A. and Wright, D.T. (2006) Cyanobacterial calcification and its rock-building potential during 3.5 billion years of Earth history. Geobiology **4**: 147–166.

Andres, M.S. and Reid, R.P. (2006) Growth morphologies of modern marine stromatolites, a case study from Highborne Cay, Bahamas. Sediment. Geol. **185**: 319–328.

Aranda-Gómez, J.J., Chacón, E., Charles-Polo, M., Solorio-Munguía, J.G., Vega-González, M., Moreno-Arredondo, A. and Origel-Gutiérrez, G. (2009) Collapse structures at the bottom of a recently desiccated maar lake: Rincón de Parangueo maar, Valle de Santiago, México, IAVCEI – CVS – IAS 3IMC Proceedings, Malargüe, Argentina, 2 pp.

Arp, G., Reimer, A. and Reitner, J. (2003) Microbialite formation in seawater of increased alkalinity, Satonda Crater Lake, Indonesia. J. Sediment. Res. **73**: 105–127

Awramik, S.M. (1977) Stromatolites, In: C. Ponnamperuma (ed.) *Chemical Evolution of the Early Precambrian*. Academic Press, New York, pp. 111–123.

Awramik, S.M. (1992) The history and significance of stromatolites, In: M. Schidlowski (ed.) *Early Organic Evolution*. Springer-Verlag, Berlin, pp. 435–449.

Awramik, S.W. (2006) Respect for stromatolites. Nature **441**: 700–701.

Awramik, S.M. and Grey, K. (2005) Stromatolites: biogenicity, biosignatures, and bioconfusion. Proceed. SPIE **59060**: 1–9.

Banfield, J.F., Moreau, J.W., Chan, C.S., Welch, S.A. and Little, B. (2001) Mineralogical biosignatures and the search for life in Mars. Astrobiology **1**: 447–465.

Baumgartner, L.K., Reid, R.P., Dupraz, C., Decho, A.W., Buckley, D.H., Spear, J.R., Przekop, K.M. and Visscher, P.T. (2006) Sulphate reducing bacteria in microbial mats: changing paradigms, new discoveries. Sediment. Geol. **185**: 131–145.

Benzerara, K., Menguy, N., López-García, P., Yoon, T.H., Kazmierczak, J., Tyliszczak, T., Guyot, F. and Brown, G.E. (2006) Nanoscale detection of organic signatures in carbonate microbialites. Proc. Natl. Acad. Sci. **103**: 9440–9445.

Bergman, B., Gallon, J.R., Rai, A.N. and Stal, L.J. (1997) N_2-Fixation by non-heterocystous cyanobacteria. FEMS Microbiol. Ecol. **19**: 139–185.

Black, M. (1933) The algal sediments of Andros Island, Bahamas. Philos. Trans. R. Soc. Lond. B Biol. Sci. Ser. B. **122**: 165–192.

Boomer, S.M., Noll, K.L., Geesey, G., Dutton, E.B. (2009) Formation of Multilayered Photosynthetic Biofilms in an Alkaline Thermal Spring in Yellowstone National Park, Wyoming, Appl. Environ. Microbiol. **75**: 2464–2475.

Brendan, B.P., Goh, F., Allen, M. and Neilan, B.A. (2004) Microbial diversity of extant stromatolites in the hypersaline marine environment of Shark Bay, Australia. Environ. Microbiol. **6**: 1096–1101.

Burne, R.V. and Moore, L.S. (1987) Microbialithes, organosedimentary deposits of benthic microbial communities. Palaios **2**: 241.

Cady, S.L., Farmer, J.D., Grotzinger, J.P., Schopf, J.W. and Steele, A. (2003) Morphological biosignatures. Astrobiology **3**: 351–368.

Cardemil, L. and Wolk, L.P. (1981) Isolated heterocysts of *Anabaena variabilis* synthesize envelope polysaccharides. Biochem. Biophys. Acta **671**: 265–276.

Castenholz, R.W. (1973) Ecology of blue-green algae in hot springs, In: N.G. Carr and B.A. Whitton (eds.) *The Biology of Blue-Green Algae*. Blackwell, Oxford, pp. 379–414.
Castenholz, R.W. (1984) Composition of hot spring microbial mats, a summary, In: Y. Cohen, R.W. Castenholz and H.O. Halvorson (eds.) *Microbial Mats Stromatolites*. Alas R. Liss, New York, pp. 101–119.
Castenholz, R.W. (2001) General characteristics of the cyanobacteria, In: E. Garrity, D.R. Booner and R.W. Castenholz (eds.) *Bergey's Manual of Systematic Bacteriology 1*. Springer, New York, pp. 474–487.
Castenholz, R.W. (2004) Phototrophic bacteria under UV stress, In: J. Seckbach (ed.) *Origins, Evolution and Biodiversity of Microbial Life*. Kluwer Academic, Dordrecht, pp.445–461.
Cavalier-Smith, T. (2003) Genomic reduction and evolution of novel genetic membranes and protein targeting machinery in eukaryote-eukaryote chimaeras (meta-algae). Phil. Trans. R. Soc. Lond. **358B**: 109–134.
Chacón-Baca, E. (2002) *Biogeochemical Study of Stromatolites from the Tarahumara Formation in Sonora, Mexico (Estudio Biogeoquímico de los Estromatolitos de la Formación Tarahumara)*. UNAM, México, 138 pp.
Costerton, J.W., Lewandowski, Z., Caldwell, D.E., Korber, D.R. and Lappin-Scott, H.M. (1995) Microbial biofilms. Ann. Rev. Microbiol. **49**: 711–745.
de Beer, D. and Kühl, M. (2001) Interfacial processes and activities in biofilms and microbial mats, In: B.P. Boudreau and B.B. Jørgensen (eds.) *The Benthic Boundary Layer*. Oxford University Press, Oxford, pp. 374–394.
Decho, A.W. and Kawaguchi, T. (2003) Extracellular polymers (EPS) and calcification within modern marine stromatolites, In: W.E. Krumbein, D.M. Paterson and G.A. Zavarzin (eds.) *Fossil and Recent Biofilms*. Kluwer Academic, Dordrecht, pp. 227–240.
Decho, A.W., Visscher, P.T. and Reid, R.P. (2005) Production and cycling of natural microbial exopolymers (EPS) within a marine stromatolite. Palaeogeogr. Palaeoclimatol. Palaeoecol. **219**: 71–86.
Decho, A., Visscher, P.T., Braissant, O., Dupraz, C., Norman, R.S., Reid, P.R. and Stolz, J.F. (2008) An emerging framework for understanding marine stromatolites formation, In: J. Reitner, N.V. Queric and M. Reich (eds.) *Geobiology of Stromatolites. International Kalkowsky-Symposium, Göttingen Abstract Volume*. Universitätsdrucke Göttingen, Germany, pp. 21–22.
Decho, A.W., Visscher, P.T., Ferry, J., Kawaguchi, T., He, L., Przekop, K.M., Norman, R.S. and Reid, R.P. (2009) Autoinducers extracted from microbial mats reveal a diversity of N-acylhomoserine lactones (HSL) and abundance changes that may relate to diel pH. Environ. Microbiol. **11**: 409–420.
Des Marais, D.J. (2003) Biogeochemistry of Hypersaline Microbial Mats. Biol. Bull. **204**: 160–167.
Dillon, J. and Castenholz, R. (1999) Scytonemin, a cyanobacterial sheath pigment, protects against UVC radiation: implications for early photosynthetic life. J. Phycol. **35**: 673–681.
Dillon, J.G. and Castenholz, R.W. (2003) The synthesis of the UV-screening pigment, scytonemin, and photosynthetic performance in isolates from closely related natural populations of cyanobacteria (*Calothrix* sp.). Environ. Microbiol. **5**: 484–491.
Dillon, J.G., Miller, S., Bebout, B., Hulla, M., Pinel, N. and Stahl, D.A. (2009) Spatial and temporal variability in stratified hypersaline microbial mat community. FEMS Microbiol. Ecol. **68**: 46–58.
Dingding, A., Danhorn, T., Fuqua, C. and Parsek, M.R. (2006) Quorum sensing and motility mediate interactions between *Pseudomonas aeruginosa* and *Agrobacterium tumefaciens* in biofilm cocultures. Proc. Natl. Acad. Sci. U.S.A. **103**: 3828–3833.
Douglas, S. (2005) Mineralogical footprints of microbial life. Am. J. Sci. **305**: 503–525.
Douglas, D., Peat, A., Whitton, B.A. and Wood, P. (1986) Influence of iron status on structure of the cyanobacterium (blue–green alga) *Calothrix parietina*. Cytobios **47**: 155–165.
Dove, J.E., Yasukawa, K., Tinsley, C.R. and Nassif, X. (2003) Production of the signalling molecule, autoinducer-2, by Neisseria meningitidis, lack of evidence for a concerted transcriptional response. Microbiology **149**: 1859–1869.
Dupraz, C. and Visscher, P.T. (2005) Microbial lithification in marine stromatolites and hypersaline mats. Trends Microbiol. **13**: 429–438.

Dupraz, C., Patissina, R. and Verrecchia, E.P. (2006) Translation of energy into morphology, Simulation of stromatolite morphospace using a stochastic model. Sediment. Geol. **185**: 185–203.

Dvornyk, V. and Nevo, E. (2003) Genetic polymorphism of cyanobacteria under permanent natural stress, a lesson from the "Evolution Canyons". Res. Microbiol. **154**: 79–84.

Ehrlich, H.L. (1981) *Geomicrobiology*. Marcel Dekker, Inc., New York, 408 pp.

Elster, J. and Komarek, O. (2003) Ecology of periphyton in a meltwater stream ecosystem in the maritime Antarctic. Antarct. Sci. **15**: 189–201.

Fay, P. (1983) *The Blue-Greens (Cyanophyta-Cyanobacteria), The Institute of Biology's Studies in Biology no 160.* Edward Arnold, Baltimore, 88 pp.

Fenchel, T., King, G.M. and Blackburn, T.H. (1998) *Bacterial Biogeochemistry*. Academic Press, San Diego.

Fike, D.A., Gammon, C.L., Ziebis, W. and Orphan, V.J. (2008) Micron-scale mapping of sulfur cycling across the oxycline of a cyanobacterial mat: a paired nanoSIMS and CARD-FISH approach. ISME J. **2**: 749–759.

Flemming, H.C., Strathmann, M. and Leon Morales, C.F. (2007) Microbial effects, Chapter 9, In: B. Westrich and U. Forstner (eds.) *Sediment Dynamics and Pollutant Mobility in Rivers, an Interdisciplinary Approach*. Springer-Verlag, Berlin.

Frankel, R.B. and Bazylinski, D.A. (2003) Biologically-induced mineralization in bacteria, In: P.M. Dove, J.J. De Yoreo and S. Wiener (eds.) *Biomineralization*. Rev. Mineral. Geochem. **54**: 95–114.

Fritsch, F.E. (1945) *The Structure and Reproduction of the Algae*. Cambridge University Press, London, 940 pp.

Galperin, M.Y. and Gomelsky, M. (2005) Bacterial signal transduction modules: from genomics to biology. ASM News **71**: 326–333.

Garcia-Pichel, F., Sherry, N.D. and Castenholz, R.W. (1992), Evidence for an ultraviolet sunscreen role of the extracellular pigment scytonemin in the terrestrial cyanobacterium *Chlorogloeopsis* sp., Photochem. Photobiol **56**: 17–23.

García-Pichel, F., Nübel, U. and Muyzer, G. (1998) The phylogeny of unicellular, extremely halotolerant cyanobacteria, Arch Microbiol **169**: 469–482.

García-Pichel, F., Johnson, S.L., Youngkin, D. and Belnap, J. (2003) Small-scale vertical distribution of bacterial biomass and diversity in biological soil crusts from arid lands in the Colorado Plateau, Microb Ecol **46**: 312–321.

García-Pichel, F. (2008) Molecular ecology and environmental genomics, In: A. Herrero and E. Flores (eds.) *The Cyanobacteria, Molecular Biology, Genetics and Evolution*. Caister Academic Press, Norfolk, pp. 60–87.

Gera, Ch. and Srivastava, S. (2006) Quorum-sensing, the phenomenon of microbial communication. Curr. Sci. **90**: 666–676.

Gerdes, G., Klenke, T. and Noffke, N. (2000) Microbial signatures in peritidal siliciclastic sediments, a catalogue. Sedimentology **47**: 279–308.

Giovannoni, S.J., Turner, S., Olsen, G.J., Barns, S., Lane, D.J. and Pace, N.R. (1988) Evolutionary relationships among cyanobacteria and green chloroplasts. J. Bacteriol. **170**: 3584–3592.

Goh, F., Allen, M.A., Leuko, S., Kawaguchi, T., Decho, A.W., Burns, B.P., Neilan, B.A. (2009) Determining the specific microbial populations and their spatial distribution within the stromatolite ecosystem of Shark Bay. The ISME J. **3**: 383–396.

Golubic, S. (1991) Modern stromatolites – a review, In: R. Riding (ed.) *Calcareous Algae and Stromatolites*. Springer-Verlag, Berlin, pp. 541–561.

Golubic, S., Hernandez-Marine, M. and Hoffmann, L. (1996) Developmental aspects of branching in filamentous Cyanophyta/Cyanobacteria. Algol. Stud. **83**: 303–329.

Golubic, S., Seong-Joo, L. and Browne, K.M. (2000) Cyanobacteria: architects of sedimentary structures, In: R. Riding and S.M. Awramik (eds.) *Microbial Sediments*. Springer-Verlag, Berlin, pp. 57–67.

Gould, S.J. (1983) *Hen's Teeth and Horse's Toes*. W. W. Norton, New York.

Grossman, A.R., Schaefer, M.R., Chiang, G.G. and Collier, J.L. (1994) The responses of cyanobacteria to environmental conditions, light and nutrients, In: A. Bryant (ed.) *The Molecular Biology of Cyanobacteria*. Kluwer Academic, Dordrecht, pp. 641–675.

Grotzinger, J.P. and Knoll, A.H. (1999) Precambrian stromatolites, evolutionary milestones or environmental dipsticks? Ann. Rev. Earth Planet. Sci. **27**: 313–339.

Guerrero, M.G. and Lara, C. (1987) Assimilation of organic nitrogen, In: P. Fay and C. Van Baalen (eds.) *The Cyanobacteria*. Elsevier, Oxford, pp. 163–186.

Gugger, M.F. and Hoffmann, L. (2004) Polyphyly of true branching cyanobacteria (Stigonematales). Int. J. Syst. Evol. Microbiol. **54**: 349–357.

Havemann, S.A. and Foster, J.S. (2008) A comparative characterization of the microbial diversity in an artificial microbialite model and a natural stromatolite. Appl. Environ. Microbiol. **74**: 7410–7421.

Herdman, M. (1987) Akinetes, structure and function, In: P. Fay and C. Van Baalen (eds.) *The Cyanobacteria*. Elsevier, Amsterdam, pp. 227–250.

Howes, I. (1989) Filamentous green algae in freshwater streams on Signy Island, Antarctica. Hydrobiologia **172**: 1–18.

Ivanikova, N.V., Popels, L.C., Michael, R., McKay, L. and Bullerjahn, G.S. (2007) Lake superior supports novel clusters of cyanobacterial picoplankton. Appl. Environ. Microbiol. **73**: 4055–4065.

Jahnke, L.L., Eder, W., Huber, R., Hope, J.M., Hinrichs, K.-U., Hayes, J.M., Des Marais, D.J., Cady, S.L. and Summons, R.E. (2001) Signature lipids and stable carbon isotope analyses of Octopus Spring hyperthermophilic communities compared with those of *Aquificales* representatives. Appl. Environ. Microbiol. **67**: 5179–5189.

Jahnke, L., Tsegereda, E., Hope, J., Turk, K.A., van Zuilen, M., Des Marais, D., Farmer, J.D. and Summons, R.E. (2004) Lipid biomarker and carbon isotopic signatures for stromatolite-forming, microbial mat communities and *Phormidium*. Geobiology **2**: 31–47.

Jahnke, L.L., Orphan, V.J., Embaye, T., Turk, K.A., Kubo, M., Summons, R.E. and Des Marais, D.J. (2008) Lipid biomarker and phylogenetic analyses to reveal archaeal biodiversity and distribution in hypersaline microbial mat and underlying sediment. Geobiology **6**: 394–410.

Javor, B. (1989) *Hypersaline Environments, Microbiology and Geochemistry*. Springer-Verlag, Berlin, 328 pp.

Jing, H.M., Aitchison, J.C., Lacap, D.C., Peeraornpisal, Y., Sompong, U., Pointing, S.B. (2005) Community phylogenetic analysis of moderately thermophilic cyanobacterial mats from China, The Philippines and Thailand. Extremophiles **9**: 325–332.

Jørgensen, B.B. (1982) Mineralization of organic matter in the sea bed – the role of sulfate reduction. Nature **296**: 643–645.

Kalkowsky, E. (1908) Oolith und Stromatolith im norddeutschen Buntsandstein. Zeitschr. Deutsch. Geol. Gesellschf. **60**: 68–125.

Kawaguchi, T. and Decho, A.W. (2002) Isolation and biochemical characterization of extracellular polymeric secretions (EPS) from modern soft marine stromatolites (Bahamans) and its inhibitory effect on $CaCO_3$ precipitation. Prep. Biochem. Biotech. **32**: 51–63.

Knee, E.M., Gong, F.C., Gao, M., Tepliski, M., Jones, A.R., Foxworthy, A., Mort, A.J. and Bauer, W.D. (2001) Root mucilage from pea and its utilization by rhizosphere bacteria as a sole carbon source. Mol. Plant Microb. Interact. **14**: 775–784.

Knoll, A.H. (2008) Cyanobacteria and earth history, In: A. Herrero and E. Flores (eds.) *The Cyanobacteria, Molecular Biology, Genetics and Evolution*. Caister Academic Press, Norfolk, pp. 1–20.

Komárek, J. and Anagnostidis, K. (1989) Modern approach to the classification system of Cyanophytes 4 Nostocales. *Arch. Hydrobiol.* Suppl 82, 3 Algol. Stud. **36**: 247–345.

Krüger, M., Treude, T., Wolters, H., Nauhaus, K., Boetius, A. (2005) Microbial methane turnover in different marine habitats, Paleogeog. Paleoclimat. Palaeoecol. **227**: 6–17.

Krumbein, W.E. (1983) Stromatolites – the challenge of a term in time and space. Precamb. Res. **20**: 493–531.

Lau, M.C., Aitchison, J.C. and Pointing, S.B. (2009) Bacterial community composition in thermophilic microbial mats from the five hot springs in Central Tibet. Extremophiles **13**: 139–149

Leuko, S., Allen, M.A., Goh, F., Burns, B.P., Walter, M.R. and Neilan, B.A. (2007) Analysis of intergenic spacer region length polymorphisms to investigate the halophilic archaeal diversity of stromatolites and microbial mat. Extremophiles **11**: 203–210.

Ley, R.E., Harris, J.K., Wilcox, J., Spear, J.R., Miller, S.R., Bebout, B.M., Maresca, J.A., Bryant, D.A., Sogin, M.L. and Pace, N.R. (2006) Unexpected diversity and complexity of the Guerrero Negro hypersaline microbial mats. Appl. Environ. Microbiol. **72**: 3685–3695.

Logan, B.W., Rezak, R. and Ginsburg, R.N. (1964) Classification and environmental significance of algal stromatolites. J. Geol. **72**: 68–83.

López-Cortés, A., Garcia-Pichel, F., Nübel, U. and Vázquez-Juarez, R. (2001) Cyanobacterial diversity in extreme environments in Baja California, México, a polyphasic study. Int. Microbiol. **4**: 227–36.

Ludwig, R., Al-Horani, F.A., de Beer, D. and Jonkers, H.M. (2005) Photosynthesis-controlled calcification in a hypersaline microbial mat. Limnol. Oceanogr. **50**: 1836–1843.

Macintyre, I.G., Bebout, L.P. and Reid, P.A. (2000) The role of endolithic cyanobacteria in the formation of lithified laminae in bahamian stromatolites. Sedimentology **47**: 915–921.

Malda, J., López-Sauceda, J., Morales-Puente, P., Cienfuegos, E., Sánchez-Ramos, M. and Chacón, E. (2002) Microbialites from a highly saline crater lake in Rincon de Parangueo, Mexico. *ISSOL Abstracts. OLEB 36*. Kluwer Academic, Dordrecht, 404 pp.

Miller, S.R., Strong, A.L., Jones, K.L. and Ungerer, M.C. (2009) Bar-coded pyrosequencing reveals shared bacterial community properties along the temperature gradients of two alkaline hot springs in Yellowstone National Park. Appl. Environ. Microbiol. **75**: 4565–4572.

Morales-Puente, P., Cienfuegos, E., Sanchez-Ramos, M., Fragoso, D. and Chacon, E. (2002) *Carbon Isotopic Variations of Recent Microbial Mats Developed at Sulfurous Springs. ISSOL Abstracts. OLEB 36*. Kluwer Academic, Dordrecht, 403 pp.

Nagy, M.L., Pérez, A. and Garcia-Pichel, F. (2005) The prokaryotic diversity of biological soil crusts in the Sonoran Desert (Organ Pipe Cactus National Monument, AZ). FEMS Microbiol. Ecol. **54**: 233–245.

Nielsen, P.H. and Jahn A. (1999) Extraction of EPS, In: J. Wingender, T.R. New and H.C. Flemming (eds.) *Microbial Extracellular Polymeric Substances*. Springer-Verlag, Berlin, pp. 49–72.

Noffke, N. and Paterson, D. (2008) Microbial interactions with physical sediment dynamics, and their significance for the interpretation of Earth's biological history. Geobiology **6**: 1–4.

Nübel, U., Garcia-Pichel, F., Kühl, M. and Muyzer, G. (1999) Quantifying microbial diversity: morphotypes, 16S rRNA genes, and carotenoids of oxygenic phototrophs in microbial mats. Appl. Environ. Microbiol. **65**: 422–430

Olsen, G.J., Woese, C.R. and Overbeek, R. (1994) The winds of (evolutionary) change, breathing new life into microbiology. J. Bacteriol. **176**: 1–6.

Olson, J.M. (2006) Photosynthesis in the Archean era. Photosynth. Res. **88**: 109–17.

Orphan, V.J., Jahnke, L.L., Embaye, T., Turk, K.A., Pernthaler, A. and Des Marais, D.J. (2008) Characterization and spatial distribution of methanogens and methanogenic biosignatures in hypersaline microbial mats. Geobiology **6**: 376–393.

Osorio-Santos, K. (2007) Comunidades algales asociadas con la formación de oncolitos en corrientes calcáreas de la Huasteca Potosina. Bch. Thesis, Facultad de Ciencias, UNAM, Mexico, 59 pp.

Paerl, H.W. (1990) Physiological ecology and regulation of N_2 fixation in natural waters. Adv. Microb. Ecol. **8**: 305–344.

Pajdak-Stós, A., Fialkowska, E. and Fyda, J. (2001) *Phormidium autumnale* (Cyanobacteria) defense against three ciliate grazer species. Aquat. Microbiol. Ecol. **23**: 237–244.

Papineau, D., Walker, J.J., Mojzsis, S.J. and Pace, N.R. (2005) Composition and structure of microbial communities from stromatolites of hamelin pool in Shark Bay, Western Australia. Appl. Environ. Microbiol. **71**: 4822–4832.

Park, S., Wolanin, P.M., Yuzbashyan, E.A., Lin, H., Darnton, N.C., Stock, J.B., Silberzan, P. and Austin, R. (2003) Influence of topology on bacterial social interaction. Proc. Natl. Acad. Sci. U.S.A. **100**: 13910–13915.

Paterson, D.M., Aspden, R.J., Visscher, P.T., Consalvey, M., Andres, M.S., Decho, A.W., Stolz, J. and Reid, R.P. (2008) Light-dependant biostabilisation of sediments by stromatolite assemblages. PloS Biol. **3**: 3171–3176

Paul, J. (2008) Kalkowsky Stromatolites from the Lower Saxony. Excursion, In: J. Reitner, N.V. Queric and M. Reich (eds.) *Geobiology of Stromatolites. International Kalkowsky-Symposium October 4–11, 2008, Abstracts*, Gottingen, pp. 119–121.

Pentecost, A. and Bauld, J. (1988) Nucleation of calcite on the sheaths of cyanobacteria using a simple diffusion cell. Geomicrobiol. J. **6**: 129–135.

Pentecost, A. and Whitton, B.A. (2000) Limestones, In: B.A. Whitton and M. Potts (eds.) *The Ecology of Cyanobacteria, Their Diversity in Time and Space*. Kluwer Academic, Dordrecht, pp. 257–279.

Pettijohn, F.J. (1957) *Sedimentary Rocks*, 2nd edn. Harper and Row, New York, 718 pp.

Planavsky, N. and Grey, K. (2008) Stromatolite branching in the Neoproterozoic of the Centralian Superbasin, Australia: an investigation into sedimentary and microbial control of stromatolite morphology. Geobiology **6**: 33–45.

Rasmussen, U. and Syvenning, M.M. (2001) Characterization by genotypic methods of symbiotic *Nostoc* strains isolated from five species of Gunnera. Arch. Microbiol. **176**: 204–210.

Reid, R.P., Visscher, P., Decho, A., Stolz, J., Bebout, B., Dupraz, C., Macintyre, R.P., Paerl, H.W., Pinckney, J.L., Prufert-Bebout, L., Steppe, T.F. and Des Marais, D.J. (2000) The role of microbes in accretion, lamination and early lithification of modern marine stromatolites. Nature **406**: 989–992.

Riding, R. (2008) Abiogenic, microbial and hybrid authigenic carbonate crusts: components of Precambrian stromatolites. Geol. Croatica **61**: 73–103.

Roeselers, G., Norris, T.B., Castenholz, R.W., Rysgaard, S., Glud, R.N., Kuhl, M. and Muyzer, G. (2006) Diversity of phototrophic bacteria in microbial mats from Arctic hot springs (Greenland). Environ. Microbiol. **9**: 26–38.

Rothschild, L.J and Mancinelli, R.L. (1990) Model of carbon fixation in micobial mats from 3,500 Myr ago to the present. Nature **345**: 710–712.

Sarkar, S., Bose, P.K., Samanta, P., Sengupta, P. and Eriksson, P.G. (2008) Microbial mat mediated structures in the Ediacaran Sonia Sandstone, Rajasthan, India, and their implications for Proterozoic sedimentation. Precamb. Res. **162**: 248–263.

Schmidt, A. (1988) Sulfur metabolism in cyanobacteria, Meth Enzymol **167**: 572–583.

Schouten, S., van der Meer, M.T.J., Hopmans, E.C., Rijpstra, W.I.C., Reysenbach, A.-L., Ward, D.M. and Damste, J.S. (2007) Archaeal and bacterial glycerol dialkyl glycerol tetraether lipids in Hot Springs of Yellowstone National Park. Appl. Environ. Microbiol. **73**: 6181–6191.

Seckbach, J. and Walsh, M. (2008) *From Fossils to Astrobiology: Records of Life on Earth and the Search for Extraterrestrial Biosignatures, Cellular Origin, Life in Extreme Habitats and Astrobiology*. Springer, Dordrecht.

Semikhatov, S.M., Gebelein, C.D., Cloud, P., Awramik, S.M. and Benmore, W.C. (1979) Stromatolite morphogenesis-progress and problems. Can. J. Earth Sci. **16**: 992–1015.

Sharp, J.H. (1970) Distribution, morphology, and accretion rate of recent subtidal algal stromatolites, Bermuda. J. Sediment. Res. **40**: 568–578.

Shiraishi, F., Zippel, B., Neu, T.R. and Arp, G. (2008) In situ detection of bacteria in calcified biofilms using FISH and CARD–FISH. J. Microbiol. Methods **75**: 103–108.

Smith, A.J. (1982) Modes of cyanobacterial carbon metabolism. In: N.G. Carry and B.A. Whitton (eds.), *The Biology of Cyanobacteria*., vol. 19, Blackwell Scientific Publishers, Oxford, pp. 47–85.

Sompong, U., Anuntalabhochai, S., Cutler, R.W., Castenholz, R.W. and Peerapornpisal, Y. (2008) Morphological and phylogenetic diversity of cyanobacterial populations in six hot springs of Thailand. ScienceAsia **34**: 153–162.

Stal, L.J. (1991) The metabolic versatility of the mat-building cyanobacteria *Microcoleus chtlonoplastes* and *Oscillatoria limosa* and its ecological significance. Algol. Stud. **64**: 453–467.

Stal, L.J. (1995) Physiological ecology of cyanobacteria in microbial mats and other communities. New Phytol. **131**: 1–32.

Stal, L.J. (2000) Cyanobacterial mats and stromatolites, In: B.A. Whitton and M. Potts (eds.) *The Ecology of Cyanobacteria, Their Diversity in Time and Space*. Kluwer Academic, Dordrecht, pp. 61–120.

Stanier, R.Y. and Cohen-Bazire, G. (1977) Phototrophic prokaryotes, the cianobacteria. Ann. Rev. Microbiol. **31**, 225–274.

Steunou, A.S., Bhaya, D., Bateson, M.M., Melendrez, M.C., Ward, D.M., Brecht, E., Peters, J.W., Kuhl, M. and Grossman, A.R. (2006) In situ analysis of nitrogen fixation and metabolic switching in unicellular thermophilic cyanobacteria inhabiting hot spring microbial mats. Proc. Natl. Acad. Sci. U.S.A. **103**: 2398–2403.

Stockner, J., Callieri, C. and Cronberg, G. (2000) Picoplankton and other non-bloomforming cyanobacteria in lakes, In: B.A. Whitton and M. Potts (eds.) *The Ecology of Cyanobacteria, Their Diversity in Time and Space*. Kluwer Academic, Dordrecht, pp. 195–231.

Stolz, J.F. (2000) Structures of microbial mats and biofilms, In: R.E. Riding and S.M. Awramik (eds.) *Microbial Sediments*. Springer-Verlag, Berlin, pp. 1–8.

Stolz, J.F., Feinstein, T.N., Salsi, J., Visscher, P. and Reid, P.R. (2001) TEM analysis of microbial mediated sedimentation and lithification in modern marine stromatolites. Am. Mineral. **86**: 826–833.

Stoodley, P., Sauer, K., Davies, D.G. and Costerton, J.W. (2002) Biofilms as complex differentiated communities. Annu. Rev. Microbiol. **56**: 187–209.

Sumner, D.Y. (2000) Microbial versus environmental influences on the morphology of late Archean Fenestrate microbialites, In: R. Riding and S.M. Awramik (eds.) *Microbial Sediments*. Springer-Verlag, Berlin, pp. 307–314.

Tandeau de Marsac, N. (1994) Differentiation of hormogonia and relationships with other biological processes, In: D.A. Bryant (ed.) *The Molecular Biology of Cyanobacteria*. Kluwer Academic, The Netherlands, pp. 825–842.

van der Meer, M.T.J., Schouten, S., Bateson, M.M., Nubel, U., Wieland, A., Kuhl, M., de Leeuw, J.W., Sinninghe Damste, J.S. and Ward, D.M. (2005) Diel variations in carbon metabolism by green nonsulfur-like bacteria in alkaline siliceous hot spring microbial mats from Yellowstone National Park. Appl. Environ. Microbiol. **71**: 3978–3986.

Vincent, W., Castenholz, R.W., Downes, M.T. and Howard-Williams, C. (1993) Antarctic cyanobacteria: light, nutrients, and photosynthesis in the microbial mat environment. J. Phycol. **29**: 745–755.

Visscher, P.T. and Stolz, J.F. (2005) Microbial mats as bioreactors, populations, processes, and products. Palaeogeog. Palaeoclimatol. Palaeoecol. **219**: 87–100.

Visscher, P.T., Reid, P.R. and Bebout, B.M. (1998) Microscale observations of sulfate reduction: correlation of microbial activity with lithified micritic laminae in modern marine stromatolites. Am. Mineral. **83**: 1482–1493.

Visscher, P.T., Reid, R.P. and Bebout, B.M. (2000) Microscale observations of sulfate reduction, Correlation of microbial activity with lithified micritic laminae in modern marine stromatolites. Geology **28**: 919–922.

Walcott, C.D. (1914) Cambrian geology and Palentology III., Precambrian Algonkian algal flora. Smithsonian Miscel. Collect. **67**: 77–156.

Wang, D.Y.C., Kumar, S. and Hedges, S.B. (1999) Divergence time estimates for the early history of animal phyla and the origin of plants, animals and fungi. Proc. Biol. Sci. **266B**: 163–171.

Ward, D.M., Bateson, M.M., Weller, R. and Ruff-Roberts, A.L. (1992) Ribosomal RNA analysis of microorganisms as they occur in nature. Adv. Microb. Ecol. **12**: 219–286.

Weckesser, J., Drews, G. and Mayer, H. (1979) Lipopolysaccharides of photosynthetic prokaryotes. Annu. Rev. Microbiol. **33**: 215–239.

Westall, F. (2005) Life on the early earth: a sedimentary view. Science **308**: 366–367.

Westall, F. (2008) Morphological biosignatures in early terrestrial and extraterrestrial materials. Space. Sci. Rev. **135**: 95–114.

Whitehead, N.A., Barnard, A.M., Slater, H., Simpson, N.J. and Salmond, G.P. (2001) Quorum-sensing in Gram-negative bacteria. FEMS Microbiol. Rev. **25**: 365–404.

Whitton, B.A. (1992) Diversity, ecology and taxonomy of the cyanobacteria, In: N.H. Mann and N.G. Carr (eds.) *Photosynthetic Prokaryotes, Biotechnology Handbooks 6*. Plenum Press, London, pp. 1–51.

Whitton, B.A. (2000) Soils and rice-fields, In: B.A. Whitton and M. Potts (eds.) *The Ecology of Cyanobacteria, Their Diversity in Time and Space*. Kluwer Academic, Dordrecht, pp. 233–255.

Whitton, B.A. and Potts, M. (eds.) (2000) *The Ecology of Cyanobacteria, Their Diversity in Time and Space*. Kluwer Academic, Dordrecht, pp. 233–255.

Wilkinson, C.R. and Fay, P. (1979) Nitrogen-fixation in coral-reef sponges with symbiotic cyanobacteria. Nature **279**: 527–529.

Wingender, J., Neu, T.R. and Flemming, H.-C. (1999) What are bacterial extracellular polymeric substances?, In: J. Wingender, T.R. Neu and H.C. Flemming (eds.) *Microbial Extracellular Polymeric Substances*. Springer Verlag, Berlin, pp. 1–19.

Wolk, C.P., Ernst, A. and Elia, J. (1994) Heterocyst metabolism and development, In: A. Bryant (ed.) *The Biology of Cyanobacteria*. Kluwer Academic, Dordrecht, pp. 769–823.

Wood, P., Peat, A. and Whitton, B.A. (1986) Influence of phosphorous status on fine structure of the cyanobacterium (blue–green alga) *Calothrix parietina*. Cytobios **47**: 89–99.

Xavier, J.B. and Foster, K.R. (2007) Cooperation and conflict in microbial biofilms. Proc. Natl. Acad. Sci. U.S.A. **104**: 876–881.

Biodata of **Falicia Goh**, author of ***"Living Stromatolites of Shark Bay, Western Australia: Microbial Inhabitants"***

Dr. Falicia Goh is currently a Post-Doctoral Fellow with the Genome Institute of Singapore. Previously, she was the senior research scientist at Department of Discovery Biology in Merlion Pharmaceuticals Pte Ltd. She obtained her Ph.D. from the University of New South Wales in 2008. Dr. Goh scientific interests are in the areas of: taxonomy and phylogeny of cyanobacterial and halophilic archaea, biology and geobiology of biofilms, geomicrobiology, and biomineralisation, osmoadaptation of microorganisms, and discovery of natural products from the environment.

E-mail: **silef78@yahoo.com**

V.C. Tewari and J. Seckbach (eds.), *STROMATOLITES: Interaction of Microbes with Sediments*, Cellular Origin, Life in Extreme Habitats and Astrobiology 18, 343–358
DOI 10.1007/978-94-007-0397-1_15, © Springer Science+Business Media B.V. 2011

LIVING STROMATOLITES OF SHARK BAY, WESTERN AUSTRALIA: MICROBIAL INHABITANTS

FALICIA GOH[1,2]
[1]*Genome Institute of Singapore,* 60 Biopolis Street, *Singapore 138672, Singapore*
[2]*School of Biotechnology and Biomolecular Sciences, University of New South Wales, Sydney, NSW 2052, Australia*

Keywords Living stromatolites • Microbes • Abiogenic • Metabolic • Microbial inhabitants • Modern stromatolites • Hamelin Pool • Shark Bay • Western Australia • Intertidal

1. Modern Living Stromatolites

An abundance of stromatolites throughout the Earth's geological records has been widely documented. The oldest examples of these preserved stromatolites are dated to more than three billion years of age (Byerly et al., 1986; Lowe, 1980; Walter et al., 1980). It is under debate whether they are of abiogenic or biogenic origin (Schopf et al., 2002), since geological processes of sedimentation can mimic the layering caused by microbes (Grotzinger and Knoll, 1999). However, recent studies have also reflected a growing acceptance of the oldest stromatolites from the Pilbara as biogenic (Allwood et al., 2006). This revealed that through the examination of different stromatolite morphologies and their similarities to modern microbially mediated carbonates, these formations were not purely the result of an abiogenic phenomenon. However, evidences of metabolic processes and biogeochemical cycles occurring in stromatolites are rarely preserved in these fossilised stromatolites. Modern stromatolites can thus be viewed as analogues to their fossilised counterparts for studying early life on Earth (McNamara and Awramik, 1992). In Western Australia, both living and fossil stromatolites have been identified (Fig. 1). It is also one of the locations where some of the oldest examples of fossil stromatolites are found and proven to be of biogenic origin (Allwood et al., 2006; Byerly et al., 1986; Lowe, 1980; Summons et al., 1999; Walter et al., 1980). Until the discovery of small lithified subtidal columnar stromatolites in the Bahamas (Steppe et al., 2001), the only subtidal marine examples known to be living and undergoing lithification were those in Hamelin Pool at Shark Bay, Western Australia. Hamelin Pool in Western Australia spans an area of ca. 1,220 km^2 with an average tidal range of ca. 60 cm (Logan and Cebulski, 1970). High net evaporation rates and average water temperatures throughout the year between 17°C and 27°C create a hypersaline environment with at least twice the

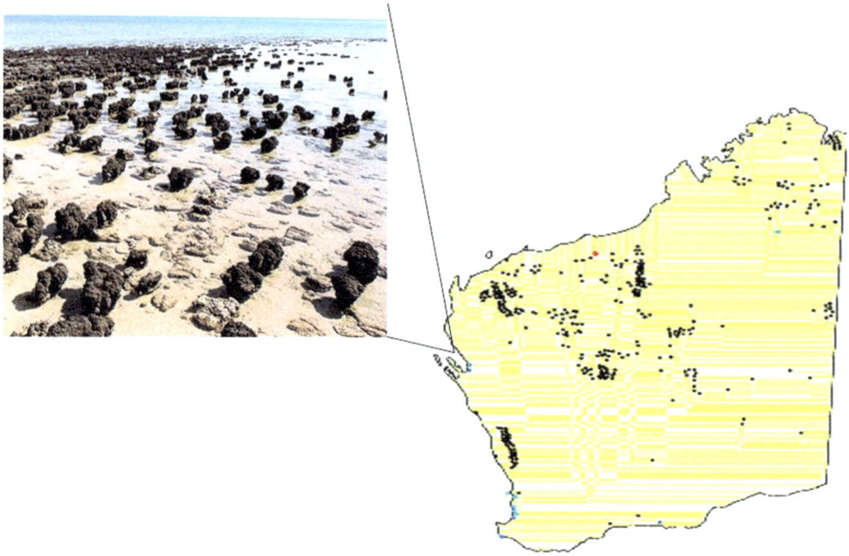

Figure 1. Location of stromatolites in Western Australia (Picture obtained from Australian Geological Survey). *Blue dots* are the location of living stromatolites, *red dot* is the location of 3.45 billion-year-old egg-carton stromatolites and *black dots* are the location of fossil stromatolites. *Inset*: stromatolites from Hamelin Pool, Shark Bay Western Australia.

salinity of seawater (Arp et al., 2001). The stromatolites and microbial mats are widespread in intertidal and inner sublittoral platform areas. However, in most areas the living stromatolites are also accompanied by dead forms in the supratidal zone (Playford and Cockbain, 1976). Stromatolites in Hamelin Pool have a morphological diversity and abundance comparable to that of Proterozoic and early Paleozoic stromatolites. These factors make them one of the best examples of living stromatolites to study. There are two major types of marine stromatolites: (1) coarse-grained eualgal–cyanobacterial stromatolites, and (2) fine-grained cyanobacterial stromatolites (Awramik and Riding, 1988). The latter is found in intertidal columns, and early studies have suggested they are built by two communities of cyanobacteria, *Entophysalis* and *Schizothrix*, along with numerous other cyanobacteria, while the eualgal-cyanobacterial stromatolites are found in the subtidal columns (Awramik and Riding, 1988).

2. Assessment and Comparative Analyses of Microbial Populations of Shark Bay Stromatolites and Surrounding Seawater

A number of studies have been conducted on the living stromatolites and microbial mats found in open marine waters in the Bahamas, revealing a great deal of information on processes involved in the formation of these geobiologically significant formations (Decho et al., 2005; Dupraz et al., 2004; Dupraz and Visscher, 2005; Paerl et al., 2001; Pinckney et al., 1995; Reid et al., 2000). Studies have shown that carbonate precipitation occurring in the Bahamian stromatolites are due to a combination of microbial metabolic processes such as photosynthesis, sulphate reduction, and anaerobic sulphide oxidation (Visscher et al., 1998). The association of cyanobacteria, heterotrophic bacteria, and sulphate-reducing bacteria (SRB) with the lithification of the Bahamian stromatolites has been documented (Reid et al., 2000; Visscher, 2000), and the roles of these microorganisms are related to grain or sediment trapping and/or carbonate dissolution and carbonate precipitation.

These studies also revealed that in the Bahamas, stromatolite formation was balanced between sedimentation and intermittent lithification by several community types (Reid et al., 2000). Filamentous *Schizothrix* sp. was involved in the binding and trapping of sand grains, and, subsequently, a heterotrophic bacterial biofilm with abundance of extracellular polymeric substances (EPSs) was formed. In this step, calcium carbonate precipitation and dissolution was closely regulated by sulphate reduction, sulphate oxidation, EPS production, and EPS degradation (Decho et al., 2005; Kawaguchi and Decho, 2002; Visscher, 2000). In the biofilms, cyanobacterial photosynthesis also creates diel fluctuations of geochemical gradients (Jørgensen, 2001). High concentrations of oxygen produced during the day that results in large amounts of organic carbon (Revsbech et al., 1983). This carbon source is then utilised by the sulphate-reducers. Several other cyanobacteria, such as *Leptolyngbya*, *Symploca*, *Phormidium*, *Gloeocapsa,* and *Trichodesmium* sp. have also been cultivated from the Bahamian stromatolites and are likely to be involved in such a mechanism (Foster et al., 2009). Microboring activities of coccoid endolithic cyanobacteria, e.g., *Solentia* sp., have also resulted in the formation of micritised grains, which assisted the growth of stromatolites and subsequently stabilised and preserved the structure (Macintyre et al., 2000). Information on mat lithification and stromatolite formation from the Bahamas has thus provided an important knowledge base for subsequent studies on the Shark Bay stromatolites.

The geology and environment of the Shark Bay stromatolites has been characterised in some detail, and studies on microbial composition as determined by microbial isolation and molecular techniques have also been undertaken (Neilan et al., 2002; Burns et al., 2004; Papineau et al., 2005; Goh et al., 2009). Although there is evidence of some flagellates in Shark Bay (Al-Qassab et al., 2002), much of the work on both the Bahamian and the Shark Bay stromatolites has shown that eukaryotes are relatively rare (Reid et al., 2000; Burns et al., 2004; Papineau et al., 2005). The higher salinity in Shark Bay may lead to an exclusion of most of the higher organisms, particularly those that may graze on stromatolites. Secondary metabolic production has also been shown to occur in cyanobacteria isolated from these Shark Bay stromatolites (Burns et al., 2005). Many organisms are known to produce bioactive secondary metabolites in response to environmental stresses (Baker et al., 1995), and the stromatolitic microorganisms thriving in the hypersaline reaches of Shark Bay may have developed similar defense mechanisms.

Such analyses provide invaluable information to enhance our understanding of this system and help in the conservation of this World Heritage Site.

Early studies suggested the Shark Bay intertidal stromatolites were dominated by only few microorganisms, mainly cyanobacteria such as *Entophysalis* sp. and *Schizothrix* sp. (Awramik and Riding, 1988). Recent studies have characterised in-depth the Shark Bay communities using both molecular and microbial culturing approaches (Neilan et al., 2002; Burns et al., 2004; Papineau et al., 2005; Goh et al., 2009). These demonstrated that the microbial communities present in these stromatolites are not simply comprised of cyanobacteria but include a range of phylogenetically distinct bacteria and archaea.

2.1. CYANOBACTERIAL COMMUNITIES

By using molecular nucleic acid tools, the Shark Bay stromatolite cyanobacterial 16S rDNA sequences were shown to be affiliated to the orders Pleurocapsales, Chroococcales, and Oscillatoriales. Sequences from seawater surrounding the stromatolites were also reported recently, and these sequences were mainly related to Chroococcales and Oscillatoriales (Goh et al., 2009). No sequences from the Nostocales order was identified in any of the studies conducted. Recently, similar molecular studies were conducted on the Bahamian stromatolites and the same orders of cyanobacteria (Pleurocapsales, Chroococcales, and Oscillatoriales) were identified and the absence of Nostocales (known nitrogen fixers) sequences or isolates were confirmed (Foster et al., 2009). These authors have suggested that unicellular pleurocapsales cyanobacteria such as *Stanieria* sp. might be involved in nitrogen fixation and iron sequestration.

The intertidal Shark Bay stromatolites were characterised by *Chroococcidiopsis*, *Dermocarpella,* and *Pleurocapsa* from Pleurocapsales (Goh et al., 2009). Under the order Chroococcales, *Gloeocapsa*, *Gloeothece*, *Cyanothece,* and *Euhalothece* were identified, and the presence of these mirrors the composition of other hypersaline ecosystems (Garcia-Pichel et al., 1998; Nübel et al., 2000). Within the Oscillatoriales, the study by Burns et al. (2004) revealed that most sequences of the filamentous cyanobacteria belonged to *Symploca*, *Plectonema* and a novel clade of filamentous cyanobacteria. A recent study by Goh et al. (2009) showed that oscillatoriales sequences were from the genera *Leptolyngbya*, *Schizothrix,* and *Geitlerinema*, while Papineau et al. (2005) identified *Microcoleus* sp. from the same environment. These studies all described different oscillatoriales species identified from these Shark Bay stromatolites. Thus, the diversity of cyanobacteria in this environment is greater and far more complex than previously described and results from these various studies are best combined to facilitate a better understanding of the overall diversity present (Table 1).

As shown by confocal microscopy (Fig. 2a), unicellular cyanobacterial species were located in the uppermost layer of the Shark Bay stromatolites and displayed endolithic properties (Goh et al., 2009). Although the endolithic unicellular cyanobacterium *Solentia* sp. involved in the formation of the Bahamian marine

Table 1. Representative cyanobacterial 16S rRNA gene sequences identified from clones or isolates of the Shark Bay stromatolites and the seawater collected from surrounding the stromatolites.

Samples		Analysis of 16S DNA sequences Nearest relative in GenBank	Accession no.	References
Stromatolites	*Clones*	*Cyanothece* sp. PCC 7424	AF132932	Goh et al. (2009)
		Cyanothece sp. ATCC 51142	AF132771	
		Gloeocapsa sp. KO38CU6	AB067575	
		Synechococccus sp. HOG	AF448075	
		Acaryochloris sp. Awaji-1	AB112435	
		Euhalothece sp. strain MPI95AH13	AJ000710	
		Cyanothece sp. ATCC 51142	AF132771	
		Gloeocapsa sp. PCC 73106	AF132784	
		Geitlerinema sp. PCC 7105	AF132771	
		Xenococcus PCC 7305	AF132783	
		Lyngbya hieronymusii	AF337650	
		Chroococcidiopsis sp. PCC 6712	AJ344557	
		Dermocarpella incrassata	AJ344559	
		Gloeothece sp. KO11DG	AB067577	
	Clones	*Microcoleus* sp. PCC 7420	X70770	Papineau et al. (2005)
		Pleurocapsa sp.	X78081	
	Clones	*Prochloron* sp.	X63141	Burns et al. (2004)
		Cyanothece sp. SKT4126	AB067581	
		Euhalothece sp.	AJ000713	
		Chroococcidiopsis sp. PCC 6712	AJ344557	
		Symploca sp. PCC 8002	AB039021	
		Gloeocapsa sp. PCC 73106	AF132784	
		Plectonema sp. F3	AF091110	
		Stanieria cyanosphaera	AF132931	
		Phormidium sp. E18	AF263339	
		Microcoleus sp.	X70770	
	Isolates	*Symploca* sp. PCC 8002	AB039021	
		Uncultured cyanobacterium FL06705	AF446264	
		Leptolyngbya sp. PCC 7104	AB039012	
		Plectonema sp. F3	AF091110	
		Cyanothece sp. SKTU136	AB067581	
		Pleurocapsa minor	AJ344564	
		Synechococcus sp. PCC 8807	AF448076	
		Xenococcus sp. PCC 7305	AF132783	
		Microcoleus sp.	X70770	
		Nostoc sp. PCC 7120	AP003598	
Seawater	*Clones*	Uncultured *Chroococcus* sp.	DQ058856	Goh et al. (2009)
		Euhalothece sp. strain MPI96N304	AJ000713	
		Chroococcidiopsis sp.	DQ914863	
		Uncultured *Gloeobacter* sp.	DQ058873	
		Gloeocapsa sp. KO38CU6	AB067575	
		Uncultured cyanobacterium clone	AM177414	
		Thermosynechoccoccus elogatus BP-1	AP005376	

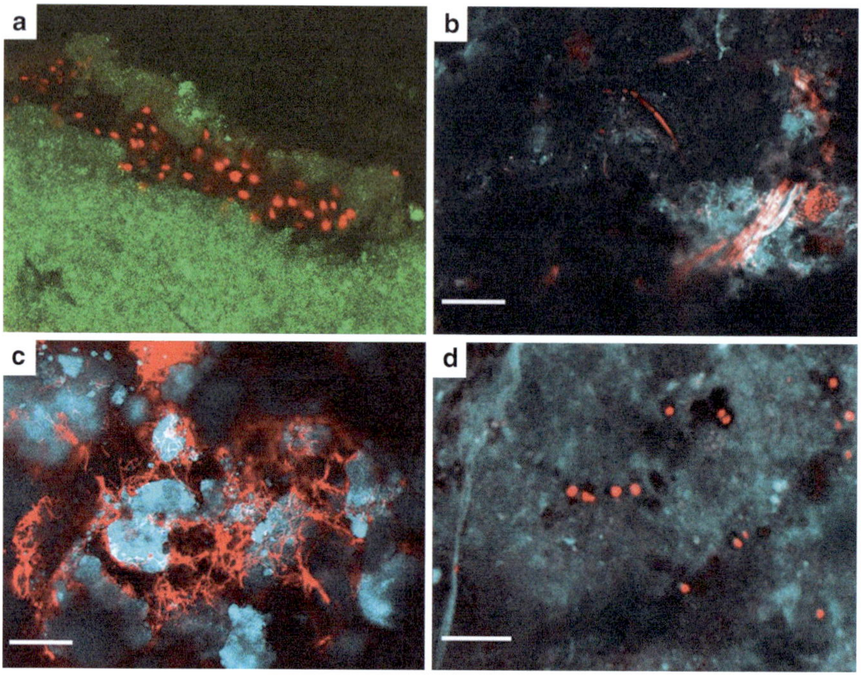

Figure 2. Krypton–Argon confocal images of smooth mat and stromatolites from Hamelin Pool, Shark Bay, Western Australia. (**a**) Stromatolite vertical surface dominated by unicellular cyanobacteria. Scale = 10 mm, (**b**) Top layer of mat (scale = 20 μm), (**c**) cross-section of mat showing binding and trapping sand grains (scale = 20 μm) and (**d**) black layer of the mat consisting of endolithic cyanobacteria (scale = 10 μm).

stromatolites (Macintyre et al., 2000) was not identified in Shark Bay, the unicellular cyanobacteria detected here appeared to display similar traits. Endoliths in the Bahamian marine stromatolite type-3 mats (Reid et al., 2000) contributed to the fusion of sand grains, and confocal microscopy images of smooth mats occurring in close proximity to the intertidal Shark Bay stromatolites revealed the same process present in these mats (Fig. 2b–d). Scanning electron microscopy also revealed boring activities in the Shark Bay intertidal stromatolites (Goh et al., 2009). The uppermost layer of the Shark Bay stromatolites had a distinctive black colouration and pigment analysis identified the presence of scytonemin (Goh et al., 2009). This layer has previously been linked with *Gloecapsa* sp., a unicellular cyanobacterium also cultivated and identified from the Shark Bay stromatolites (Dupraz et al., 2004). Scytonemin quenches light and has a protective effect against ultraviolet (UV) radiation (Lewin, 2006), and this may be significant in protecting community members in the high UV environment of Shark Bay (Palmisano et al., 1989). The differences in microbial diversity observed between various studies on the Shark Bay stromatolites

(Burns et al., 2004; Papineau et al., 2005; Goh et al., 2009) may be due to various environmental challenges such as fluctuations in the salinity of the water in Hamelin Pool and therefore lead to microbial succession. Microbial succession has also been shown to occur in the Bahamian stromatolites (Reid et al., 2000). Whether this is an underlying mechanism in the Shark Bay stromatolites, however, remains to be established.

2.2. HETEROTROPHIC BACTERIAL COMMUNITIES

Based on the molecular analysis of 16S rRNA genes from total genomic DNA extracted from Shark Bay stromatolites, the most abundant sequences were novel proteobacteria (Papineau et al., 2005). In an analysis by Burns et al. (2004), culturable bacteria from this environment were dominated by *Bacillus* sp. (Low G + C Gram-positive bacteria). Recently, Barabesi et al. (2007) also identified a gene cluster involved in calcium carbonate precipitation in *Bacillus subtilis*, and this group could therefore also potentially contribute to carbonate precipitation in Shark Bay. The data from molecular analysis of bacterial 16S rDNA sequences also revealed the presence of 15 other groups belonging to α-proteobacteria, γ-proteobacteria, δ-proteobacteria, acidobacteria, Bacteriodetes/chlorobi, two unknown clades, candidate division OP-11, firmicutes, candidate division TM6, planctomycetes, actinobacteria, chloroflexi, *Deinococcus,* and thermotogales (Goh et al., 2009). Interestingly, it is known that *Deinococcus* are able to withstand ionising radiation and thus survive prolonged desiccation (Mattimore and Battista, 1996), and the stromatolites in Shark Bay are often exposed to extended periods of desiccation. The Shark Bay intertidal stromatolite sequences related to proteobacteria/purple bacteria, firmicutes/Gram-positive bacteria, cyanobacteria, and green sulphur bacteria have also been identified from other living stromatolites (Bauld et al., 1979; Visscher et al., 1999). The α and γ-proteobacteria, Bacteriodetes/chlorobi, planctomycetes, chloroflexi, and cyanobacteria were also identified from the seawater surrounding the Shark Bay stromatolites (Goh et al., 2009). The dispersal of microorganisms from biofilms is a common phenomenon in the environment (Characklis, 1990; Brading et al., 1995), and these results suggest this may potentially be occurring in Shark Bay.

SRB, such as *Desulfovibrio* sp. and *Desulfobacter* sp. (Visscher et al., 1998), and facultative phototrophs such as *Rhodospirillium* and *Rhodobacter* (Steppe et al., 2001), have been associated with precipitation processes in stromatolites. Both of these groups were also identified from the Shark Bay stromatolites. The process of sulphate reduction by the SRB can mediate calcium carbonate precipitation (Visscher et al., 1998). Fluorescence in situ hybridisation (FISH) has also been used on the Bahamian stromatolites to show that these SRB are located at the oxic zone of the stromatolite surface and that they intermingle with the phototrophic cyanobacteria (Baumgartner et al., 2006), also shown in the Shark Bay stromatolites (Goh et al., 2009). In addition to SRB, sulphide-oxidising bacteria (green sulphur bacteria) were also identified. Studies on the Bahamian stromatolites have shown that hydrogen sulphide is reoxidised by the sulphide oxidising bacteria resulting in calcium carbonate dissolution (Visscher et al., 1998). It has also been shown by Decho et al. (2005) that uronic acid monomers of cyanobacterial exopolymeric substances (EPSs) were mineralised by the same group of bacteria. Thus, the close interactions between these two groups of microorganisms indicate that they have strong metabolic associations in these ecosystems.

In addition to the phototrophic bacteria, stromatolite formation is also dependent on both aerobic and anaerobic heterotrophic bacteria. As all aerobic heterotrophic bacteria respire, they use the oxygen provided by the phototrophic bacteria during photosynthesis and in turn provide the carbon dioxide to be used by the phototrophic bacteria (Visscher et al., 1998). This is limited to the day as photosynthesis can only occur in the presence of light, and therefore oxygen levels are insufficient for heterotrophic respiration during the night. It has therefore been suggested that sulphate reduction is another mechanism to respire in the absence of oxygen (Visscher et al., 1998). Sulphate reduction can also occur in the presence of oxygen but at a lower rate (Canfied and Des Marais, 1991).

Pinckney et al. (1995) proposed that calcium carbonate precipitation occurred through cyanobacterial oxygenic photosynthesis and subsequently dissolved during aerobic heterotrophic respiration. Thus, photosynthesis and aerobic respiration have opposite effects on calcium carbonate precipitation. No net lithification occurs when the rates of photosynthesis and aerobic respiration were the same. However, since photosynthesis supports both aerobic and anaerobic respiration, some net carbonate precipitation could occur (Visscher et al., 1998). As photosynthesis occurs during the day, there is typically no oxygen evolved from this process at night for aerobic heterotrophic respiration (Visscher et al., 1998). In addition to this geochemical cycling, EPS produced by cyanobacteria and other microbes under certain conditions also play a crucial role in mineral precipitation either as a site for nucleation or through binding of calcium ions (Arp et al., 1999; Decho and Kawaguchi, 1999). This inhibits calcium carbonate precipitation. However, degradation of the EPS by heterotrophic bacteria or UV light releases calcium ions, which then facilitates calcium carbonate precipitation. EPS is likely to play a crucial role in stromatolite formation, though the extent is currently unknown.

2.3. ARCHAEAL COMMUNITIES

A diverse community of archaea, particularly halophiles, were also documented in the Shark Bay stromatolites (Burns et al., 2004; Papineau et al., 2005; Goh et al., 2009). Sequences related to Crenarchaeota were identified across these studies. However, the roles of these archaea in the stromatolites remained unknown. Michaelis et al. (2002) have shown that the association of SRB with archaea can result in carbonate precipitation due to the coupling of methane oxidation with sulphate reduction. Some halophilic archaea are also capable of fixing carbon dioxide (Javor, 1988). At this moment, we can only speculate on the possible role of archaea in carbonate precipitation. Images from FISH conducted on a cross-section of Shark Bay stromatolites also showed a close interaction between the archaea and cyanobacteria (Goh et al., 2009). A novel halophilic archaeon, *Halococcus hamelinensis*, has also been isolated from this environment (Goh et al., 2006). Further studies on this microorganism could potentially reveal an important role for archaea in these geobiological structures. The presence of halophilic archaea could also be explained by the high salinity and drastic fluctuations in salinity of Hamelin Pool. It has been shown that in hypersaline environments, increase of biodiversity favoured the growth of halophilic microorganisms (Abed et al., 2007; Wieland and Kuhl, 2006). Consequently, mechanisms of adaptation to a hypersaline environment are also likely to be important functions of both stromatolite archaea and other microorganisms.

3. Adaptation of Stromatolite Microorganisms to a Hypersaline Environment

Microorganisms develop various mechanisms for their survival under extreme environmental conditions related to temperature, pressure, salinity, pH, radiation, desiccation, oxygen, and other extreme circumstances (Rothschild and Mancinelli, 2001). These mechanisms allow tight biological control of the microenvironment in stromatolites, thus ensuring the survival of the microbial communities (Paerl et al., 2000). Environmental fluctuations such as diel, tidal, and seasonal changes influence light intensity, salinity, temperature, and nutrient availability, and therefore will affect the microbial communities present. Thus, the macromorphology of stromatolites is affected by current velocity, sediment grain size distribution, mat community diversity, and carbonate saturation. An excellent example of this is seen in the studies conducted on the Bahamian stromatolites whereby the macromorphology was observed to be influenced by various physical environmental factors (Andres and Reid, 2006). Dupraz and Visscher (2005) showed that the calcium carbonate saturation indices also change the carbonate alkalinity and pH, thus affecting the microbial lithification of the stromatolites. The buffering capacity of alkaline water of an environment also prevents carbonate dissolution (Arp et al., 2003; Dupraz et al., 2004). Thus, the maintenance and stability of stromatolites are dependent on a balance between the microbial diversity, geochemical cycling and the prevailing environmental factors.

Hypersalinity, in particular, is likely to be a crucial factor in determining the dominant physiologies that may be present in the Shark Bay stromatolites. Salinity gradients are formed due to tidal flows and the evaporation of seawater, leading to the presence of diverse microbial species adapted to different ranges of salinity. Previously, low diversity of bacterial and archaeal communities at higher salinities has been observed in microbial mats in other locations previously (Benlloch et al., 2002; Jungblut et al., 2005). However, recently, it has been shown that Guerrero

Negro hypersaline mats contained a high diversity of both bacteria and archaea, of which a high percentage of these rRNA gene sequences were unrelated to known organisms (Ley et al., 2006; Robertson et al., 2009). This high diversity of bacteria and archaea were also observed in the Shark Bay stromatolites. As mentioned earlier, EPS produced by cyanobacteria or other bacteria serves as matrices in facilitating or inhibiting calcium carbonate precipitation, trapping nutrients, preventing desiccation, protecting against UV radiation, and providing water channels for the transport of metabolites and signal molecules (Decho, 2000). However, when EPS-producing bacteria were exposed to stress conditions, the concentration of uronic acid in the cells and the amount of EPS increased (Uhlinger and White, 1983). This subsequently affected calcium carbonate precipitation or dissolution. Cyanobacterial photosynthesis is also known to be impaired by salinity, due to the high accumulation of oxygen (Garcia-Pichel et al., 1999; Wieland and Kuhl, 2006). However, such inhibition is removed in microbial mats from hypersaline conditions that tend to be slimy, less compact, and less densely colonised. These conditions lead to lower accumulations of oxygen, thus creating conditions more favourable for photosynthesis to occur. This inhibition of photosynthesis can be reduced by adding the osmolytes sucrose or EPS to cultures of cyanobacteria (Chen et al., 2003). Thus, although salinity has a strong limiting effect on microbial activity, the physiological adaptation of resident microorganisms can overcome these stresses. This is an area of on-going research in Shark Bay.

In the Shark Bay stromatolites, the production of osmolytes such as sucrose and trehalose by the archaea, cyanobacteria, or bacteria could counteract environmental metabolic inhibition by the hypersaline environment. The presence of these compatible solutes is likely to allow organisms to survive and thrive under conditions of high salinity and desiccation. Cyanobacteria are also known to accumulate carbohydrates such as dissacharides under osmotic stress (Stal and Reed, 1987; Warr et al., 1985). Osmolytes such as sucrose, trehalose, glycine betaine, and various oligosaacharides (pentasaccharides and trisaccharides) were identified from the Shark Bay stromatolite community (F. Goh, unpublished data). Disaccharides such as trehalose and sucrose are protectants against desiccation by maintaining membrane integrity during cycles of drying and wetting (Leslie et al., 1995). These oligosaccharides can also be an important carbon source for other microorganisms (Sakugawa et al., 1990). Oligosaccharides synthesised by cyanobacteria and utilised by other organisms in a close niche environment such as the stromatolites may represent an example of mutualism.

Few studies have investigated the metabolic processes occurring in Shark Bay stromatolites or the changes that may occur in community composition over time. Studies on the Shark Bay stromatolites have indicated the presence of cyanobacteria, aerobic heterotrophs, anoxygenic phototrophs, sulphate reducers, sulphide oxidisers, and halophilic archaea. The likely roles of these groups of microorganisms in geochemical cycling in the Shark Bay stromatolites include carbon and nitrogen fixation and sulphur cycling. Furthermore, depending on the species of aerobic heterotrophs present, links with processes of fermentation, respiration, and the cycling of other nutrients can also be made. The functional groups identified here are similar to those dominating many hypersaline microbial mats. In addition to the effects of microbial populations and salinity on stromatolite formation, other environmental parameters should be considered. The microbial population can sustain itself by possessing specific individual strategies or having different but complementary metabolic processes in the various groups of microorganisms present. The effects of these metabolisms are reflected in the geochemical cycle. Hence, it is a challenge to understand the events occurring in stromatolite ecosystems, and the presence of diverse metabolic groups of microorganisms strongly suggests a high level of biological complexity in this ecosystem.

4. Acknowledgements

The author acknowledges and thanks the following people: the editors Professor Vinod C. Tewari and Professor Joseph Seckbach for invitation to contribute in this book volume and much of the coordination and administration of the publication and the two anonymous referees for their suggestions and comments towards the improvement of this book chapter. Finally, the author thanks both Professor Brett Neilan and Dr. Brendan Burns for their guidance, suggestion, and advice throughout this project.

5. References

Abed, R.M.M., Kohls, K. and de Beer, D. (2007) Effect of salinity changes on the bacterial diversity, photosynthesis and oxygen consumption of cyanobacterial mats from an intertidal flat of the Arabian Gulf. Environ. Microbiol. **9**: 1384–1392.

Allwood, A.C., Walter, M.R., Kamber, B.S., Marshall, C.P. and Burch, I.W. (2006) Stromatolite reef from the Early Archaean era of Australia. Nature **441**: 714–718.

Al-Qassab, S., Lee, W.J., Murray, S., Simpson, A.G.B. and Patterson, D.J. (2002) Flagellates from stromatolites and surrounding sediments in Shark Bay, Western Australia. Acta Protozool. **41**: 91–144.

Andres, M.S. and Reid, R.P. (2006) Growth morphologies of modern marine stromatolites: a case study from Highborne Cay, Bahamas. Sediment. Geol. **185**: 319–328.

Arp, G., Reimer, A. and Reitner, J. (1999) Calcification in cyanobacterial biofilms of alkaline salt lakes. Eur. J. Phycol. **34**: 393–403.

Arp, G., Reimer, A. and Reitner, J. (2001) Photosynthesis-induced biofilm calcification and calcium concentrations in Phanerozoic oceans. Science **292**: 1701–1704.

Arp, G., Reimer, A. and Reitner, J. (2003) Microbialite formation in seawater of increased alkalinity, Satonda Crater Lake, Indonesia. J. Sediment. Res. **73**: 105–127.

Awramik, S.M. and Riding, R. (1988) Role of algal eukaryotes in subtidal columnar stromatolite formation. Proc. Natl. Acad. Sci. U.S.A. **85**: 1327–1329.

Baker, B.J., Kopitzke, R.W., Yoshida, W.Y. and McClintock, J.B. (1995) Chemical and ecological studies of the antarctic sponge *Dendrilla membranosa*. J. Nat. Prod. **58**: 1459–1462.

Barabesi, C., Galizzi, A., Mastromei, G., Rossi, M., Tamburini, E. and Peritol, B. (2007) *Bacillus subtilis* gene cluster involved in calcium carbonate biomineralization. J. Bacteriol. **189**: 228–235.

Bauld, J., Chambers, L.A. and Skyring, G.W. (1979) Primary productivity, sulfate reduction and sulfur isotope fractionation in algal mats and sediments of Hamelin Pool, Shark Bay, W.A. Aust. J. Mar. Freshw. Res. **30**: 753–764.

Baumgartner, L.K., Reid, R.P., Dupraz, C., Decho, A.W., Buckley, D.H., Spear, J.R., Przekop, K.M. and Visscher, P.T. (2006) Sulfate reducing bacteria in microbial mats: changing paradigms, new discoveries. Sediment. Geol. **185**: 131–145.

Benlloch, S., López-López, A., Casamayor, E.O., et al. (2002) Prokaryotic genetic diversity throughout the salinity gradient of a coastal solar saltern. Environ. Microbiol. **4**: 349–360.

Brading, M., Boyle, J. and Lappin-Scott, H. (1995) Biofilm formation in laminar flow using Pseudomonas fluorescens EX 101. J. Ind. Microbiol. **15**: 297–304.

Burns, B.P., Goh, F., Allen, M. and Neilan, B.A. (2004) Microbial diversity of extant stromatolites in the hypersaline marine environment of Shark Bay, Australia. Environ. Microbiol. **6**: 1096–1101.

Burns, B.P., Seifert, A., Goh, F., Pomati, F., Jungblut, A.-D., Serhat, A. and Neilan, B.A. (2005) Genetic potential for secondary metabolite production in stromatolite communities. FEMS Microbiol. Lett. **243**: 293–301.

Byerly, G.R., Lower, D.R. and Walsh, M.M. (1986) Stromatolites from the 3,300–3,500-Myr Swaziland supergroup, Barberton Mountain Land, South Africa. Nature **319**: 489–491.

Canfied, D.E. and Des Marais, D.J. (1991) Aerobic sulfate reduction in microbial mats. Science **251**: 1471–1473.

Characklis, W. (1990) Process analysis, In: W. Characklis and K. Marshall (eds.) *Biofilms*. Wiley, New York, pp. 50–51

Chen, L., Li, D. and Liu, Y. (2003) Salt tolerance of *Microcoleus vaginatus Gom.*, a cyanobacterium isolated from desert algal crust, was enhanced by exogenous carbohydrates. J. Arid Environ. **55**: 645–656.

Decho, A.W. (2000) Microbial biofilms in intertidal system: an overview. Cont. Shelf Res. **20**: 1257–1273.

Decho, A.W. and Kawaguchi, T. (1999) Confocal imaging of in situ natural microbial communities and their extracellular polymeric secretions (EPS) using nanoplast resin. Biotechniques **27**: 1246–1251.

Decho, A.W., Visscher, P.T. and Reid, R.P. (2005) Cycling and turnover of natural expolymers from a marine stromatolite. Palaeogeogr. Palaeoclimatol. Palaeoecol. **219**: 71–86.

Dupraz, C. and Visscher, P.T. (2005) Microbial lithification in marine stromatolites and hypersaline mats. Trends Microbiol. **13**: 429–438.

Dupraz, C., Visscher, P.T., Baumgartner, K. and Reid, R.P. (2004) Microbe–mineral interactions: early carbonate precipitation in a hypersaline lake (Eleuthera Island, Bahamas). Sedimentology **51**: 745–765.

Foster, J.S., Green, S.J., Ahrendt, S.R., Golubic, S., Reid, R.P., Hetherington, K.L. and Bebout, L. (2009) Molecular and morphological characterization of cyanobacterial diversity in the stromatolites of Highborne Cay, Bahamas. ISME J. **3**: 573–587.

Garcia-Pichel, F., Nubel, U. and Muyzer, G. (1998) The phylogeny of unicellular, extremely halotolerant cyanobacteria. Arch Microbiol. **169**: 469–482.

Garcia-Pichel, F., Kuhl, M., Nubel, U. and Muyzer, G. (1999) Salinity-dependent limitation of photosynthesis and oxygen exchange in microbial mats. J. Phycol. **35**: 227–238.

Goh, F., Leuko, S., Allen, M.A., Bowman, J.P., Kamekura, M., Neilan, B.A. and Burns, B.P. (2006) Halococcus hamelinensis sp. nov., a novel halophilic archaeon isolated from stromatolites in Shark Bay, Australia. Int. J. Syst. Evol. Microbiol. **56**: 1323–1329.

Goh, F., Allen, M.A., Leuko, S., Kawaguchi, T., Decho, A.W., Burns, B.P., Neilan, B.A (2009) Determining the specific microbial populations and their spatial distribution within the stromatolite ecosystem of Shark Bay. ISME J. **3**: 383–396.

Grotzinger, J.P. and Knoll, A.H. (1999) Stromatolites in Precambrian carbonates: evolutionary mileposts or environmental dipsticks? Annu. Rev. Earth. Planet Sci. **27**: 313–358.

Javor, B.J. (1988) CO_2 fixation in halobacteria. Arch. Microbiol. **149**: 433–440.

Jørgensen, B.B. (2001) Space for hydrogen. Nature **412**: 286–289.

Jungblut, A.-D., Hawes, I., Mountfort, D., Hitzfeld, B., Dietrich, D.R., Burns, B.P. and Neilan, B.A. (2005) Diversity within cyanobacterial mat communities in variable salinity meltwater ponds of McMurdo Ice Shelf, Antarctica. Environ. Microbiol. **7**: 519–529.

Kawaguchi, T. and Decho, A.W. (2002) A laboratory investigation of cyanobacterial extracellular polymeric secretions (EPS) in influencing CaCO3 polymorphism. J. Cryst. Growth **240**: 230–235.

Leslie, S., Israeli, E., Lighthart, B., Crowe, J. and Crowe, L. (1995) Trehalose and sucrose protect both membranes and proteins in intact bacteria during drying. Appl. Environ. Microbiol. **61**: 3592–3597.

Lewin, R.A. (2006) Black algae. J. Appl. Phys. **18**: 699–702.

Ley, R.E., Harris, J.K., Wilcox, J., Spear, J.R., Miller, S.R., Bebout, B.M., Maresca, J.A., Bryant, D.A., Sogin, M.L. and Pace, N.R. (2006) Unexpected diversity and complexity of the Guerrero Negro hypersaline microbial mat. Appl. Environ. Microbiol. **72**: 3685–3695.

Logan, B.W. and Cebulski, D.E. (1970) Sedimentary environments of Shark Bay, Western Australia, In: B.W. Logan, G.R. Davies, J.F. Read and D.E. Cebulski (eds.) *Carbonate Sedimentation and Environments, Shark Bay, Western Australia*. American Association of Petroleum Geologists, Tulsa, Okla, pp. 1–37.

Lowe, D.R. (1980) Stromatolites 3,400-Myr old from the Archean of Western Australia. Nature **284**: 441–443.

Macintyre, I.G., Prufert-Bebout, L. and Reid, R.P. (2000) The role of endolithic cyanobacteria in the formation of lithified laminae in Bahamian Stromalites. Sedimentology **47**: 915–921.

Mattimore, V. and Battista, J.R. (1996) Radioresistance of *Deinococcus radiodurans*: functions necessary to survive ionizing radiation are also necessary to survive prolonged desiccation. J. Bacteriol. **178**: 633–637.

McNamara, K.J. and Awramik, S.M. (1992) Stromatolites: a key to understanding the early evolution of life. Sci. Prog. **76**: 345–364.

Michaelis, W., Seifert, R., Nauhaus, K., et al. (2002) Microbial reefs in the Black Sea fueled by anaerobic oxidation of methane. Science **297**: 1013–1015.

Neilan, B.A., Burns, B.P., Relman, D. and Lowe, D.R. (2002) Molecular identification of cyanobacteria associated with stromatolites from distinct geographical locations. Astrobiology **2**: 271–280.

Nübel, U., Garcia-Pichel, F., Clavero, E. and Muyzer, G. (2000) Matching molecular diversity and ecophysiology of benthic cyanobacteria and diatoms in communities along a salinity gradient. Environ. Microbiol. **2**: 217–226.

Paerl, H.W., Pinckney, J.L. and Steppe, T.F. (2000) Cyanobacterial-bacterial mat consortia: examining the functional unit of microbial survival and growth in extreme environments. Environ. Microbiol. **2**: 11–26.

Paerl, H.W., Steppe, T.F. and Reid, R.P. (2001) Bacterially mediated precipitation in marine stromatolites. Environ. Microbiol. **3**: 123–130.

Palmisano, A.C., Summons, R.E. and Cronin, S.E. (1989) Lipophilic pigments from cyanobacterial (blue-green algal) and diatom mats in Hamelin Pool, Shark Bay, Western Australia. J. Phycol. **25**: 655–661.

Papineau, D., Walker, J.J., Mojzsis, S.J. and Pace, N.R. (2005) Composition and structure of microbial communities from stromatolites of Hamelin Pool in Shark Bay, Western Australia. Appl. Environ. Microbiol. **71**: 4822–4832.

Pinckney, J., Paerl, H.W., Reid, R.P. and Bebout, B.M (1995) Ecophysiology of stromatolite mats, Stocking Island, Exuma Cays, Bahamas. Microb. Ecol. **29**: 19–37.

Playford, P.E. and Cockbain, A.E. (1976) Modern algal stromatolites at Hamelin Pool, a hypersaline barred basin in Shark Bay, Western Australia, In: M.R. Walter (ed.) *Stromatolites*. Elsevier, Amsterdam, pp. 101.

Reid, R.P., Visscher, P.T., Decho, A.W., et al. (2000) The role of microbes in accretion, lamination and early lithification of modern marine stromatolites. Nature **406**: 989–992.

Revsbech, N.P., Jørgensen, B.B., Blackburn, T.H. and Cohen, Y. (1983) Microelectrode studies of the photosynthesis, O_2, H_2S and pH profiles of a microbial mat. Limnol. Oceanogr. **28**: 1062–1074.

Robertson, C.E., Spear, J.R., Harris, K.J. and Pace, N.R. (2009) Diversity and stratification of archaea in a hypersaline microbial mat. Appl. Environ. Microbiol. **75**: 1801–1810.

Rothschild, L.J. and Mancinelli, R.L. (2001) Life in extreme environments. Nature **409**: 1092–1101.
Sakugawa, H., Handa, N. and Yagi, K. (1990) Distribution of glycosylglycerols and oligosaccharides in the marine environment and their ecoloigcal significance in the deep sea. Mar. Biol. **106**: 309–313.
Schopf, J.W., Kuryavtsev, A.B., Agresti, D.G., Wdowiak, T.J. and Czaja, A.D. (2002) Laser-Raman imagery of Earth's earliest fossils. Nature **416**: 73–76.
Stal, L.J. and Reed, R.H. (1987) Low-molecular mass carbohydrate accumulation in cyanobacteria from a marine microbial mat in response to salt. FEMS Microbiol. Ecol. **45**: 305–312.
Steppe, T.F., Pinckney, J.L., Dyble, J. and Paerl, H.W. (2001) Diazotrophy in modern marine Bahamian stromatolites. Microb. Ecol. **41**: 36–44.
Summons, R.E., Jahnke, L.L., Logan, G.A. and Hope, J.M. (1999) 2-Methylhopanoids as biomarkers for cyanobacterial oxygenic photosynthesis. Nature **398**: 554–557.
Uhlinger, D.J. and White, D.C. (1983) Relationship between physiological status and formation of extracellular polysaccharide glycocalyx in *Pseudomonas altantica*. Appl. Environ. Microbiol. **45**: 64–70.
Visscher, P.T. (2000) Microscale observations of sulfate reduction: correlation of microbial activity with lithified micritic laminae in modern marine stromatolites. Geology **28**: 919–922.
Visscher, P.T., Reid, R.P., Bebout, B.M., Hoeft, S.E., Macintyre, I.G. and Junior, J.A.T. (1998) Formation of lithified micritic laminae in modern marine stromatolites (Bahamas): the role of sulfur cycling. Am. Mineral. **83**: 1482–1493.
Visscher, P.T., Gritzer, R.F. and Leadbetter, E.R. (1999) Low-molecular-weight sulfonates, a major substrate for sulfate reducers in marine microbial mats. Appl. Environ. Microbiol. **65**: 3272–3278.
Walter, M.R., Buick, R. and Dunlop, J.S.R. (1980) Stromatolites 3,400–3,500 Myr old from the North Pole area, Western Australia. Nature **284**: 443–445.
Warr, S.R.C., Reed, R.H. and Stewart, W.D.P. (1985) Carbohydrate accumulation in osmotically stressed cyanobacteria (blue-green algae): interactions of temperature and salinity. New Phytol. **100**: 285–292.
Wieland, A. and Kuhl, M. (2006) Regulation of photosynthesis and oxygen consumption in a hypersaline cyanobacterial mat (Camargue, France) by irradiance, temperature and salinity. FEMS Microbiol. Ecol. **55**: 195–210.

Biodata of **Dr. Kim M. Handley** and **Professor Kathleen A. Campbell**, authors of ***"Character, Analysis and Preservation of Biogenicity in Terrestrial Siliceous Stromatolites from Geothermal Settings"***

Dr. Kim M. Handley is currently a postdoctoral research fellow in the Department of Earth and Planetary Science, University of California, Berkeley, U.S.A. She obtained her Ph.D. from The University of Manchester in 2008. Her research interests include biogeochemical cycling in marine hydrothermal systems; proteogenomics of microbial communities in contaminated sediments; and stromatolite formation and methods of microfossil identification in terrestrial hot springs.

E-mail: **kim.handley@berkeley.edu**

Associate Professor Kathleen A. Campbell currently lectures in paleoecology and paleoenvironments in the School of Environment at the University of Auckland. She completed her Ph.D. at the University of Southern California in 1995 and a post-doctoral research associateship at NASA Ames Research Center in 1997. Her research interests include life in extreme environments; paleoecologic analysis on active margins; trace fossils as high-resolution paleoenvironmental indicators; sedimentology; and carbonate and silica diagenesis.

E-mail: **ka.campbell@auckland.ac.nz**

Kim M. Handley

Kathleen A. Campbell

V.C. Tewari and J. Seckbach (eds.), *STROMATOLITES: Interaction of Microbes with Sediments*, Cellular Origin, Life in Extreme Habitats and Astrobiology 18, 359–381
DOI 10.1007/978-94-007-0397-1_16, © Springer Science+Business Media B.V. 2011

CHARACTER, ANALYSIS, AND PRESERVATION OF BIOGENICITY IN TERRESTRIAL SILICEOUS STROMATOLITES FROM GEOTHERMAL SETTINGS

KIM M. HANDLEY[1] AND KATHLEEN A. CAMPBELL[2]
[1]*Department of Earth and Planetary Science, University of California at Berkeley, Berkeley, CA, USA*
[2]*Geology, School of Environment, University of Auckland, Private Bag 92019, Auckland 1142, New Zealand*

Keywords Hot spring • Stromatolite • Exopolymeric substances • Biofilm • Microbial mat • Archaea • Bacteria • Thermophilic • Cyanobacteria • Silica • Sinter • Diagenesis • Physicochemistry • Hydrodynamics • Preservation • Microscopy • Molecular phylogenetic techniques • Lipids • Astrobiology

1. Introduction

Siliceous stromatolites embody important records of past biological activity. The biogenic–mineralogical laminated structures form within a range of geothermal settings and physicochemical, hydrodynamic, and biological regimes. Silica provides an excellent medium for cellular preservation, due to its capacity to permineralize or encase microbial cells with an X-ray amorphous cement (Francis et al., 1978). As such, it has rendered some of the best microfossils in the geologic record. In particular, the Early Devonian, hot spring-related Rhynie and Windyfield cherts (Scotland) contain exceptionally clear and comprehensive examples of fossilized early terrestrial plant, animal (arthropod), and microbial life, including detailed preservation of internal structures, in addition to paleoenvironmental information (Trewin, 1994; Fayers and Trewin, 2004). Although not displaying the same level and extent of exemplary preservation, the internally laminated structures and/or filamentous character of Devonian-aged Drummond Basin (Australia) siliceous hot spring deposits (sinters) do exhibit a remarkable resemblance to microfacies commonly found in contemporary hot spring environments, e.g., high-temperature columnar and spicular and lower-temperature stratiform, streamer, and pseudo-columnar stromatolites (Walter et al., 1996, 1998). However, the interpretation of ancient siliceous deposits becomes increasingly uncertain with the destructive effects of extensive diagenetic overprinting on microfossils (e.g., Walter et al., 1980) and is generally biased toward the more enduring microfossil preservation in stromatolitic sinters formed in lower-temperature systems, where relatively large, sheathed cyanobacteria or diatoms dominate today. This review explores methods for detecting biogenic participation in modern or ancient siliceous stromatolites that are applicable to sinters formed over a range of environmental regimes and includes a brief overview of the physicochemical, hydrodynamic, and biogenic factors involved in siliceous stromatolite formation and diagenesis.

2. Mechanisms of Formation and Diagenesis

2.1. ENVIRONMENTAL EFFECTS ON SILICA CHEMISTRY/ MINERALOGY

Silica precipitates from hot spring waters as X-ray amorphous opal-A. The deposition of soluble silica (or monomeric or polymeric silica) forms a dense, glass-like "vitreous" sinter fabric, common in near-vent settings or acidic fluids (e.g., Handley et al., 2005; Schinteie et al., 2007); whereas, porous granular sinter textures form where fast rates of polymerization and/or aging are sufficient for colloid formation and deposition (White et al., 1956; Rothbaum and Rohde, 1979; Mroczek et al., 2000). Higher silica saturations promote the formation of smaller, more numerous colloids (Makrides et al., 1980). In mixed solutions, dense, cemented granular sinter develops by the precipitation of soluble silica at the point where colloids join, where the negative radius of curvature has a low interfacial energy (Iler, 1979; e.g., Rodgers et al., 2002). Gels can form and settle in hot spring waters by aggregation of colloids into networks either at alkaline pHs in the presence of salts or at acidic pHs where the ionic charge of colloids is small (Iler, 1979; e.g., White et al., 1956).

Precipitation occurs owing to gravitational settling of large or aggregated colloids, silica supersaturation, or the availability of substrates with small interfacial energies that lower the activation energy required for nucleation. Aggregation of colloids may occur through Brownian motion and London Forces (Hunter, 1993). Both precipitation and polymerization of monomeric silica to form colloids also are driven by evaporation and cooling in menisci, subaerially wicked water, and other subaerial water films or pooled water influenced by hydrodynamic activity, such as splash or spray (Hinman and Lindstrom, 1996; Lowe et al., 2001; Mountain et al., 2003). The rate of silica polymerization is pH dependant and is inhibited at pHs ≤ 5 and severely inhibited at pHs ≥ 9 (e.g., Brown and McDowell, 1983). Polymerization is enhanced by the presence of silica or heterogeneous nuclei in solution (White et al., 1956; Gallup, 1998). The rate and shorter induction periods before the onset of polymerization are also favored at large silica saturation ratios (silica concentration/equilibrium concentration) (Krauskopf, 1956; White et al., 1956; Makrides et al., 1980), and this is influenced little by the saturation-independent effect of temperature on polymerization rates (Rothbaum and Rohde, 1979; Makrides et al., 1980). Saturation increases rapidly with decreasing temperatures [e.g., Eq. (1), Fournier and Rowe, 1977; T is absolute temperature].

$$\text{Equilibrium concentration}\left(\text{ppm}\right) = 10^{-731/T+4.52} \tag{1}$$

2.2. ROLE OF BIOLOGICAL TEMPLATES AND FOSSILIZATION

Surfaces wetted by hot spring waters are invariably coated either by visible and typically colorful, mesophilic, and thermophilic photosynthetic mats at temperatures below ~73°C, or by microscopic thermophilic and hyperthermophilic biofilms above this temperature (e.g., Fig. 1; Brock, 1978; Cady and Farmer, 1996). Microbial biofilms are pivotal in both stromatolite laminae accretion and macrostructure development.

Microscale stromatolitic laminations are formed by changes in the rates or nature of microbial growth and silicification, while macroscale stromatolite morphologies are controlled to a large extent by microbial colonization patterns (e.g., Walter et al., 1972; Brock, 1978; Jones et al., 1998; Handley et al., 2005; Schinteie, 2005). In the latter case, the volume and spatial distribution of silica deposited is influenced by the high surface area of cell scaffolds upon and within which silicification occurs, and the cluster patterns of cells. The voluminous gelatinous matrix of exopolymeric substances (EPS) that biofilm- and mat-forming microbes secrete around themselves (Hall-Stoodley et al., 2004) is also prone to fossilization (e.g., Fig. 2; Weed, 1889; Westall et al., 2000; Handley et al., 2008).

The affinity silica exhibits for biofilms appears to be passively induced, owing to either smaller interfacial energies than the solution or protonated functional sites on cell walls (or sheaths) and within EPS (Urrutia and Beveridge, 1993; Westall et al., 1995). Even so, a synchrotron-based Fourier-transform infrared study of cyanobacteria undergoing silicification demonstrated a physiological response to mineral encrustation, involving a thickening of the polysaccharide sheath (Benning et al., 2004a, b). Higher rates of silicification can arise during cellular lysis, where cell wall deterioration increases the availability of functional groups as cytoplasmic material is released (Ferris et al., 1988). Silicification of biofilm components can result in the long-term preservation of either physical microfossils by encrustation, permineralization, or replacement of cell walls and cytoplasmic material (Westall et al., 1995; Toporski et al., 2002), or chemical or biological signatures (e.g., lipids).

2.3. PHYSICOCHEMICAL AND HYDRODYNAMIC EFFECTS ON STROMATOLITES

Siliceous stromatolites in hot springs form distinct structures according to physicochemical and hydrodynamic regime (Mountain et al., 2003; Lowe et al., 2001), and corresponding changes in microbiota, which can shift abruptly as conditions alter from subaqueous to subaerial or with distance from vent, such as water cooling along outflow channels (e.g., Figs. 1 and 3; Cassie and Cooper, 1989; Walter et al., 1998; Lowe et al., 2001; Fernandez-Turiel et al., 2005; Childs et al., 2008). Neutral and alkaline pH waters tend to be dominated by photosynthetic or nonphotosynthetic prokaryotes, while acidic waters (<pH ~3) tend to be inhabited by

Figure 1. Examples of vent-to-marsh biota-sinter gradients in discharge aprons of nearly neutral, alkali chloride hot springs from Yellowstone National Park (YNP, Wyoming, USA) and the Taupo Volcanic Zone (TVZ, New Zealand). (**a**) Sinter mound build-ups in Sentinel Meadows, Lower Geyser Basin, YNP. Top of Steep Cone in foreground hosts a circular, gently surging, boiling spring with

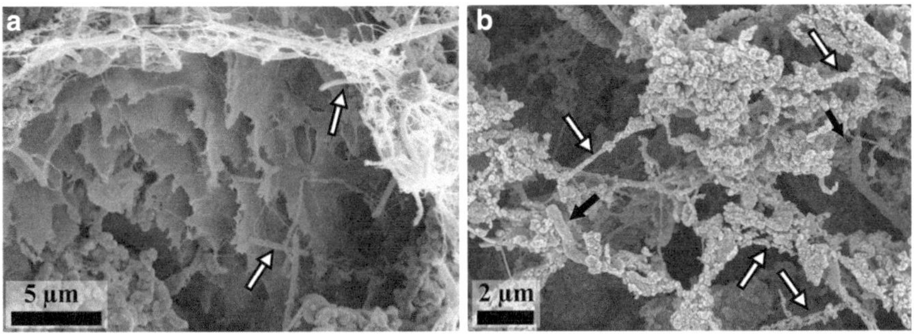

Figure 2. Scanning electron microscope (SEM) images of biofilms from a high-temperature stromatolitic sinter, Champagne Pool, Waiotapu, New Zealand. (**a**) Unsilicified cells (*arrows*) and desiccated sheet-like (*center*) and fibrillar (*across top*) EPS overlying sinter. Silica spherules of exposed sinter are visible along the bottom. (**b**) Silicified fibrillar EPS (*fine filaments, arrows*) and cells (*thick, small black arrows*).

acidophilic algae, such as *Cyanidium caldarium*, and acidophilic prokaryotes (e.g., thermophilic *Alicyclobacillus*, *Sulfobacillus*, and *Acidimicrobium* bacteria) (Cassie and Cooper, 1989; Jones et al., 2000; Johnson et al., 2003; Jones and Renaut, 2006; Schinteie et al., 2007). Stromatolites formed from acidic waters are characteristically finely laminated with dense, vitreous sinter (Jones et al., 2000; Mountain et al., 2003; Jones and Renaut, 2006; Schinteie et al., 2007).

Commonly observed associations among stromatolite structures, microbiota, and temperature are defined by changes in photosynthetic bacteria at ≤73°C (for neutral-alkaline pHs; cf. Walter et al., 1972, 1998; Brock, 1978; Walter, 1976a; Cady and Farmer, 1996; Jones et al., 2002; Lynne and Campbell, 2004; Fernandez-Turiel et al., 2005). Low-temperature (<45°C) waters and distal (from vent) marsh settings form clotted to peloidal sinter fabrics rich in oncoids, plant matter, animals, and diatom frustules (e.g., Figs. 1f, g and 3j, k; Walter, 1976a; Cassie and Cooper, 1989; Renaut et al., 1996; Campbell et al., 2001; Schinteie et al., 2007; Owen et al., 2008). Coarsely filamentous *Calothrix*-dominated cyanobacterial mats (brown to gray-green; Fig. 1c, e, f) also commonly form stratiform palisade (Fig. 4a, d) or shrubby structures (Fig. 3g, h) in distal sinter aprons. At mid-temperatures (~35–59°C), finely-filamentous *Phormidium*-dominated cyanobacterial

---

Figure 1. (continued) narrow, steep geyserite rim (cf. Braunstein and Lowe, 2001). (**b**) Detail of nodular and spicular geyserite rim at boiling pool margin, Sentinel Meadows. (**c**) Sinter apron on fault scarp at Orakei Korako, TVZ, showing discharge channels with mid-temperature (~50°C, *orange*) and low-T (~35°C, *brown*) cyanobacterial mat communities. (**d**) Mid-T pool with thick summertime overgrowth of cyanobacterial mats on Rainbow Terrace, Orakei Korako. *Arrow* over subaqueous columnar stromatolite. (**e**) Low-T (~35°C) distal sinter apron with brown-coated terracettes of *Calothrix* cyanobacterial mats at Grand Prismatic Spring, Midway Geyser Basin, YNP. (**f**) Low-T (~30°C to ambient) distal discharge area on Golden Fleece Terrace, Orakei Korako, with *Calothrix* mats and marsh plants adjacent to Golden Fleece Fault (*scarp in background*). (**g**) Dry Golden Fleece Terrace surface with loose sinter fragments and silica-cemented oncoids and plant stems.

mats (yellow orange to greenish; Fig. 1c, d) tend to prevail and are associated with the formation of columns (Fig. 1d, arrow), conical tufted (Fig. 3e, f), fenestral "bubble mat" (Fig. 3c, d), or palisade stromatolites (Fig. 4b, c). At moderately high-temperatures (~60–73°C), laminated stratiform or pustular structures typically develop from mats populated by *Chloroflexus* and *Synechococcus* (e.g., Fig. 1a yellow discharge channel adjacent to vent). Nonphotosynthetic biofilms at high-temperatures (>73°C) form knobby, columnar, or spicular structures (Figs. 1a, b and 3a), otherwise known as "geyserite," which were thought to be abiogenic (Castenholz, 1969; Walter, 1976b) until more recent work revealed microscopic biofilms (e.g., Cady and Farmer, 1996). The lamina thickness between mat- and biofilm-forming stromatolites differs between visually recognizable in the former (i.e., <73°C) to micron-scale in the latter (i.e., >73°C), and sinter is more likely to be dense and vitreous when precipitated at high temperatures (Weed, 1889; Walter, 1976b; Cady and Farmer, 1996; Braunstein and Lowe, 2001; Handley et al., 2005).

Stromatolite type is further controlled to a significant degree by a wide range of potential hydrodynamic settings (Lowe et al., 2001). Splash, spray, or fluctuating water levels (surging or even gentle ripples), for example, are most frequently associated with proximal vent sites, including geyser vents and spicular or columnar "geyserite" (e.g., Fig. 1a, b; Walter, 1976a; Jones et al., 1998; Braunstein and Lowe, 2001; Mountain et al., 2003). Free-forming pisoliths, ooids, and oncoids are a common subaqueous feature of turbulent geyser/vent pools, generated by multidirectional rolling motion (Walter, 1976a; Jones and Renaut, 1997). Oncoids also form under low-temperature wetting and drying conditions on some distal aprons (e.g., Renaut et al., 1996; Campbell et al., 2004), forming concentric silica laminae around plant fragments, invertebrates, or microbial filaments (e.g., Fig. 3j). In outflow channels, stromatolites are affected by flow rates. Mid-temperature columnar stromatolites will form at relatively deep water depths (e.g., Fig. 1d, arrow; Walter et al., 1972). *Calothrix* mats are apt to form coarsely filamentous palisades with thin sheet flow (Figs. 1e and 4a) but develop pustular mats with shrub-like structures in deeper water (Fig. 3g; Walter et al., 1998). On the other hand, at a range of temperatures, higher velocity flow can result in the formation of substrate-attached, streamer fabrics that comprise massive collections of silicified filamentous forms elongated in the direction of flow (Fig. 3i; Walter et al., 1998; Smith et al., 2003).

Figure 3. (continued) channel at Orakei Korako, trapping and silicifying bubbles from photosynthetic outgassing. Photo by Sue Turner. (**d**) Several fossilized bubbles (*white ovate areas*) partially filled in and surrounded by silicified, *gray-green* filamentous microbes from Tahunaatara sinter (~15 kyr BP, TVZ). Photo by Tim Buddle. (**e**) "*Phormidium*" cyanobacterial tufts silicifying in a drying pool, Sentinel Meadows, YNP. (**f**) Narrow, steep conical tufts (*arrows*) in Miocene Kohuamari sinter (Coromandel). (**g**) *Calothrix* "puff balls" (*arrows*) in still, low-T terrace pool (~35°C), Fountain Paint Pots, YNP. (**h**) Shrubby, fossilized puff balls (*arrows*); Broken Hills sinter. (**i**) Well-developed streamer fabric in high velocity outflow channel of Steady Geyser, Firehole Lake Drive, YNP. (**j**) Gastropod encased in filamentous oncoid (*arrows*), within clotted, peloidal fabric typical of geothermally influenced marsh facies; HB2 sinter core (<60 years BP), Tokaanu, TVZ. Photo by Kirsty Nicholson. (**k**) Pinnate diatoms (*small, gray rods*) set among large, brown amygdaloidal peloids in marsh facies, Tokaanu. Photo by Kirsty Nicholson.

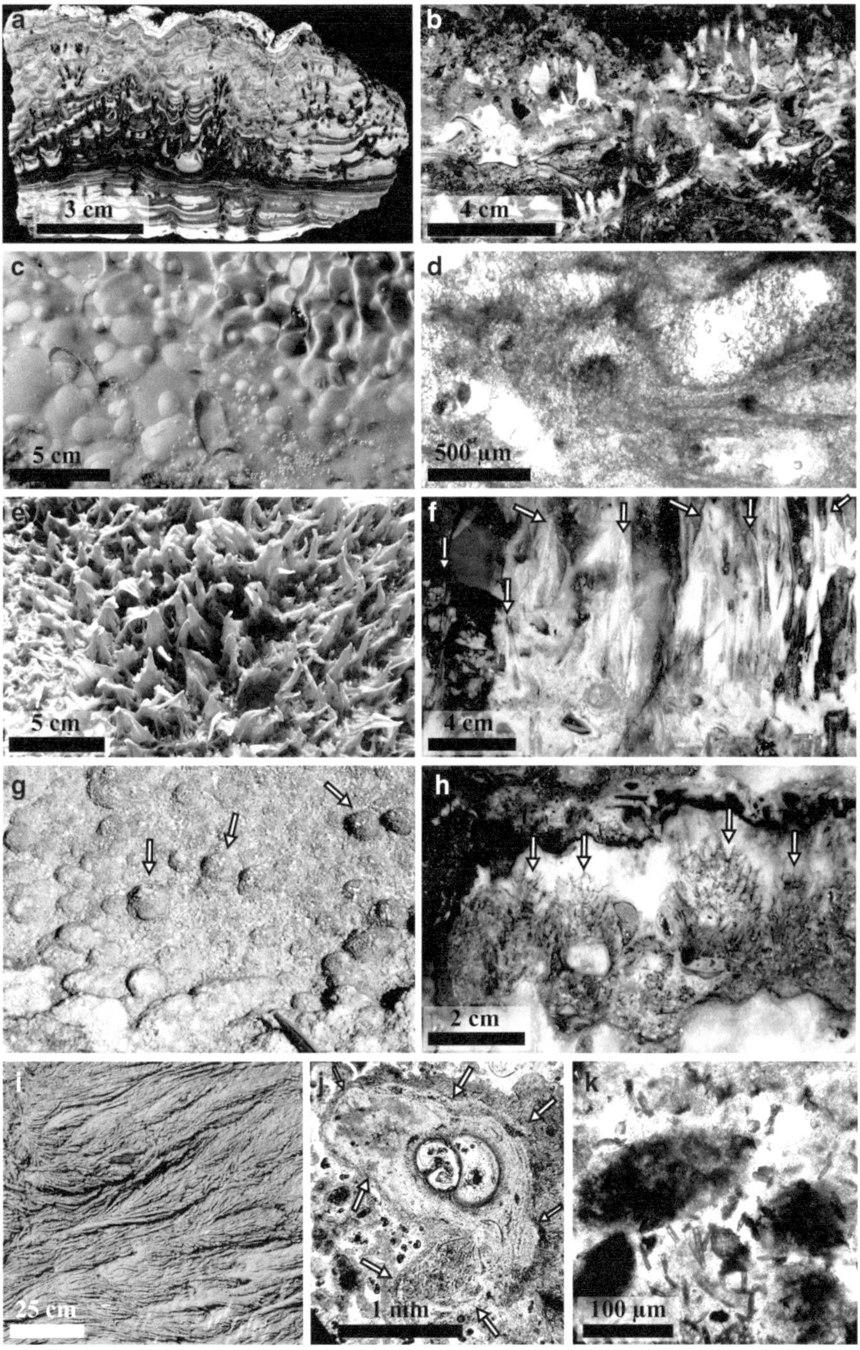

Figure 3. Detail of typical siliceous sinter fabrics from modern and fossil, nearly neutral, alkali chloride hot springs in Yellowstone and New Zealand. (**a**) Spicular geyserite cross-section (Holocene, Te Kopia, TVZ); Cady and Farmer (1996) illustrated filamentous biofilms forming upon similar high-T substrates from YNP. (**b**) Digitate microstromatolites (*white*) growing atop plant debris; Broken Hills sinter, Miocene Coromandel volcanic arc, New Zealand. (**c**) Mid-T cyanobacterial mat in discharge

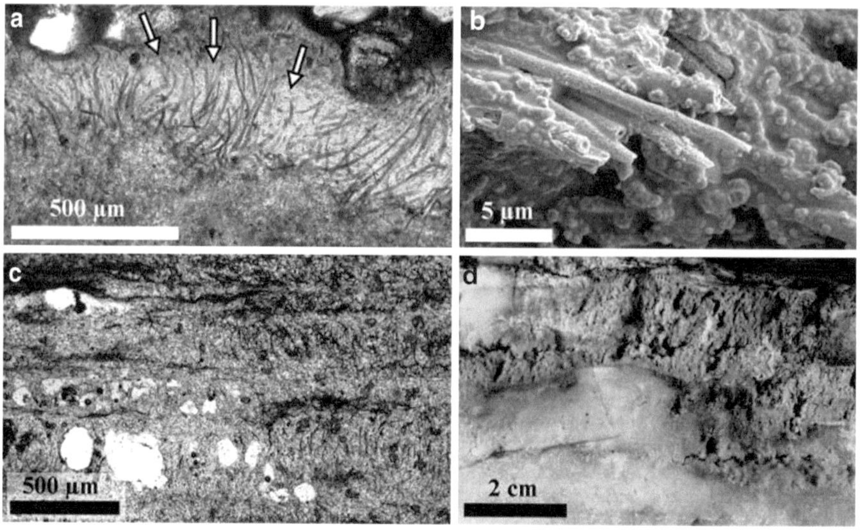

Figure 4. Silica encrustation and patchy diagenesis of filamentous palisade fabrics from Holocene sinters of the Taupo Volcanic Zone, New Zealand. (**a**) Recently silicified, coarsely filamentous mat (<60 years old) from HB2 spring, Tokaanu. Filaments (*arrows*) still preserve orange-brown organic pigmentation despite being silicified and covered by detritus. Photo by Kristy Nicholson. (**b**) SEM image of finely filamentous mat encrusted by opal-A microspheres. HB2 spring, Tokaanu. Image by Kristy Nicholson. (**c**) Finely filamentous palisade fabric with trapped gas bubbles (*white*) and erect filament horizons separated by prostrate filament laminae. Tahunaatara (~15 kyrs). Photo by Tim Buddle. (**d**) Patchy quartzose diagenesis (*white solid areas*) of coarsely filamentous thick palisade laminae from the Holocene (~4,700 years BP) Umukuri sinter (Campbell et al., 2001).

2.4. DIAGENESIS

Similar to wood petrification and diagenesis of siliceous marine sediments (e.g., Wise and Weaver, 1974; Oehler, 1975; Stein, 1982), siliceous sinter undergoes a series of mineralogic and morphologic changes once deposited (Herdianita et al., 2000a). Fresh sinter comprises noncrystalline microspheres of opal-A, which transform to noncrystalline, aligned nanospheres-upon-microspheres of opal-A/CT, then to paracrystalline lepispheres or dumbbells of opal-CT, various types of nanostructures of opal-C, and eventually to blocky, microcrystalline quartz (e.g., Herdianita et al., 2000a, b; Campbell et al., 2001; Lynne and Campbell, 2004; Lynne et al., 2007; Jones and Renaut, 2007). Silica residues, also found in some acidic geothermal areas, are destructional rather than constructional in origin and form smaller, irregular microspheres of opal-A, owing to acidification and remobilization of silica from surrounding country rock (Rodgers et al., 2002, 2004). Using electron backscatter diffractometry, Lynne et al. (2007) have shown that crystallographic restructuring of sinter occurs before the silica mineral phase transformations, which, in turn, precede nano- to micron-scale morphological changes within the sinter matrix. In other words, crystallographic axes realign at a fine-scale in the direction of the next most mineralogically mature silica phase prior to any sign (by XRD or SEM imaging) that the deposit is transitioning to the next phase. The spectrum of opal to quartz steps that occur during diagenesis is now understood to proceed at variable rates depending on environmental conditions, including changes in the water table, pH, heat, and composition (e.g., organic content, carbonate, iron, etc.) (Herdianita et al., 2000a, b; Guidry and Chafetz, 2003; Lynne et al., 2006, 2007).

Diagenesis affects microbial fabrics in at least two main ways. First, if silicification is early, and subsequent diagenetic modification not too severe, rapidly mineralizing hot spring deposits can serve as excellent archives for biological signatures in the geologic record and even form lagerstätte (e.g., Trewin, 1996; Guido et al., in press). Second, recrystallization and reprecipitation of silica phase minerals from opal to quartz can easily obscure and even destroy biofabrics (e.g., Cady and Farmer, 1996; Walter et al., 1996; Campbell et al., 2001; Lynne and Campbell, 2003; Jones et al., 2004). Trichome and sheath diameters of cyanobacteria, for example, fill in or become encrusted with silica minerals, such that fine filaments from mid-temperature sinter aprons are difficult to distinguish from coarse filaments of low-temperature apron areas. However, many sinter deposits do not undergo diagenesis uniformly (e.g., Campbell et al., 2001; Lynne et al., 2008). "Patches" of original macrofabrics are good targets to search for microscale preservation of biosignals in sinters, including microbial microtufts, laminae, filaments, and microbubbles from photosynthetic degassing within stromatolitic macrostructures (e.g., Guido and Campbell, 2009; Guido et al., in press). Current research is focused on identifying the (micro)environmental factors that contribute to the rate at which diagenesis proceeds and the causes of spatial variability in the quality of sinter fabric preservation in the geologic record (e.g., Campbell and Lynne, 2006).

3. Analytical Techniques

3.1. SAMPLING AND PRESERVING FRESH SINTER DEPOSITS

A large part of the ability to interpret stromatolites preserved in the geologic record derives from understanding modern systems. Numerous sedimentological and microbiological studies document the geochemical and hydrodynamic conditions affecting growth in relation to the mineralogical and microbiological textures formed (Walter et al., 1972; Walter, 1976a; Cady et al., 1996; Hinman and Lindstrom, 1996; Jones and Renaut, 1997; Jones et al., 1998), aiding interpretations of relict deposits (e.g., White et al., 1989; Walter et al., 1996, 1998; Campbell et al., 2001). The approach to understanding actively forming siliceous stromatolites includes a small subset of experimental growth studies designed to examine the transitional phases in stromatolite accretion. Doemel and Brock (1974, 1977) conducted light filtration experiments and added silicon carbide to the surfaces of nodular mats (alkaline hot spring waters, 46–70°C, Yellowstone National Park) to demonstrate the mechanism of stromatolite formation in siliceous hot springs. Rapid lamina accretion occurred owing to upward migration of phototrophic *Chloroflexus aurantiacus*. Hinman and Lindstrom (1996) used silicon carbide markers to demonstrate seasonal effects on the silicification of photosynthetic bubble mats. Other studies have employed artificial substrates to gauge stromatolite growth rates (e.g., Mountain et al., 2003; Handley et al., 2005). Furthermore, multiple time point sampling enabled tracking of the formation of microstromatolitic spicular sinter from Champagne Pool (75°C; New Zealand), from its origins as microcolonies of filamentous thermophilic prokaryotes, and its development through recurrent recolonization (Handley et al., 2005). By examining varying effects of water level and hydrodynamics (ripples) on substrates, the study also demonstrated the subtle balance between silica deposition and microbial growth rates required for spicule formation.

An important consideration for field sampling is the ability to capture the transient nature of stromatolite surfaces, and hence the different laminae-forming events, which typically alternate between porous (microbe-rich) and nonporous (evidently microbe-poor/abiotic) sinter development (e.g., Cady and Farmer, 1996; Mountain et al., 2003). The outcome of studies in which freshly deposited sinters are imaged also depends strongly on the choice of method used for sample preservation and analysis. Simple air drying of sinter samples post-collection and without fixation is sufficient to preserve large numbers of (silicified) microbial cells at sinter surfaces (e.g., Jones et al., 1997). Occasionally, localized films of desiccated EPS are also apparent (e.g., Jones et al., 1997). However, these biofilms are generally present in an altered state, owing to cellular decay, dehydration, and potential microbial contamination (e.g., fungal overgrowths) such that it is not possible to fully and accurately interpret their impact on stromatolite formation. The choice of biological fixative also has an effect on overall biofilm integrity. Garland et al. (1979) found that a combination of fixative and mucopolysaccharide stain resulted in better preservation of EPS and hence also retention of cells bound by the EPS, although the choice of stain/fixative clearly influenced the polysaccharide morphology.

3.2. ANALYSIS OF FRESHLY DEPOSITED SINTER

Multiple imaging methods have been employed to examine (sub)micron-scale textures and microbial–mineral associations in siliceous stromatolites. For freshly deposited sinter, the effect of the analytical technique on biofilm integrity and the nature of the target data (i.e., textural versus compositional) are important considerations. Scanning electron microscopy (SEM) is the core tool for examining textures and fabrics at a high resolution. Many, although not all, biological features are clearly distinguishable from the mineral matrix. The vacuum under which samples are placed in traditional SEM causes the collapse of poorly or unsilicified microbial cells, owing to the evaporation of water. This problem can be overcome by first substituting water for a fluid with a low surface tension prior to drying, such as liquid CO_2 via the critical-point drying method or hexamethyldisilazane (Anderson, 1951, 1952; Fratesi et al., 2004). Critical-point drying has been used to preserve the three-dimensional character of microbial cells on freshly deposited sinters from a range of physicochemical regimes, notably assisting in cell identification and interpretation of their roles in forming high-temperature and acidic microstromatolites (Cady and Farmer, 1996; Handley et al., 2005; Schinteie et al., 2007).

Other techniques are better suited to imaging the biofilms in their natural or near-natural form. Both cryo-SEM and environmental SEM (ESEM) enable analyses of cells and EPS in a hydrous three-dimensional state, through either flash-freezing biofilms and maintaining them at liquid nitrogen temperatures or analyzing them under a pressurized and humid atmosphere of air. In cryo-SEM studies of partially silicified cyanobacterial mats from Orakei Korako, New Zealand, morphologically well-preserved cells were shown not only to be bound together in a tight fibrillar mesh of EPS but also to contain fibrillar cytoplasmic material (Fig. 5a, b; Lynne and Campbell, 2003). In ESEM, the gelatinous character of EPS is readily apparent (e.g., estuarine biofilm, Little et al., 1991; various terrestrial biofilms, Douglas, 2005; sinter surface biofilm, Handley et al., 2008). Cryo-SEM tends to illustrate the polysaccharide backbone of EPS and can demonstrate the succession of structures formed throughout matrix desiccation and contraction, i.e., (1) the three-dimensional matrix, (2) two-dimensional films, and finally, (3) fibrillar (e.g., Handley et al., 2008). Real-time dehydration studies of the EPS matrix also can be undertaken by sublimation of ice in cryo-SEM (Fig. 5c, d) or by evaporation through decreasing the atmospheric pressure in ESEM (Handley et al., 2008, Fig. 5a–d). In this manner, effects of natural and post-collection dehydration can be simulated, and features typically obscured by EPS may then be exposed.

Transmission electron microscopy (TEM) permits visualization of the ultrastructural level detail of extra- and intracellular components. In sinter studies, TEM has been performed at ambient temperatures following alcohol dehydration, embedding, and ultrathin-sectioning (although forms of cryo-TEM are also possible) to examine the nature of cellular-level silicification and the process of microfossil formation. Results have indicated that the pattern of microfossil formation is dependent on the microbial community membership and thereby also the geothermal setting. For example, cells within a *Chloroflexus* mat, which had been undergoing silicification in a hot spring in Strokkur, Iceland, were encrusted extracellularly by silica spherules and within the cytoplasm only when cells appeared to have lysed (Konhauser and Ferris, 1996). In a similar TEM study, sheaths of the cyanobacterium, *Calothrix*, were shown to effectively exclude silica, resulting in extracellular silicification (Phoenix et al., 2000). In comparison, Handley et al. (2005) showed small, unsheathed filamentous cells from a thermophilic hot spring

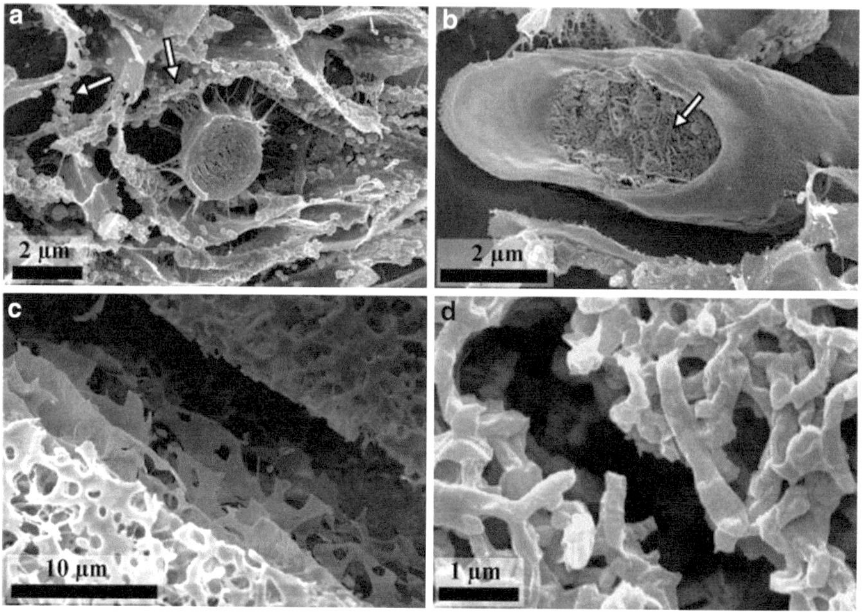

Figure 5. Cryo-SEM images of bacteria and associated EPS. (**a**) Cyanobacterial filament (*center*) from a mid-temperature mat at Orakei Korako, TVZ, surrounded by fibrillar and sheet-like EPS partially silicified by opal-A spherules (*arrows*). (**b**) Cross-section of a cyanobacterial filament (Mid-T, Orakei Korako), clearly depicting as yet unsilicified cytoplasmic material (*arrow*) encased in a sheath. (**c**) Cells (not visible) of the marine bacterium *Marinobacter santoriniensis* embedded in partially dehydrated sheet-like EPS. (**d**) Rod-shaped cell morphology of *M. santoriniensis* evident after further ice sublimation.

biofilm tended to be silicified both intra and extracellularly. Both these silicified cellular fractions remained clearly distinct from one another. Nonetheless, Lalonde et al. (2005) revealed a curious lack of cell-associated silicification in experiments with the thermophilic *Aquificales* bacterium *Sulfurihydrogenibium azorense*. Silica aggregated within the protein-rich EPS that was expressed in response to environmental silica but was not associated with cell surfaces.

In addition to textural characterizations are methods that enable clear identification of mineralogical (discussed below) and biofilm compositions. Confocal laser scanning microscopy (CLSM) combines two- or three-dimensional imagining with target-specific fluorescent staining (e.g., DNA, RNA, EPS) or autofluorescence (e.g., silica). One recent study of siliceous stromatolites employed CLSM to obtain compositional data and to examine spatial associations of sinter and biofilm components in their hydrated state (Handley et al., 2008). Use of CLSM illustrated the prevalence and wide distribution of biofilm EPS, supporting observations from SEM analyses that mineralization of EPS creates silicified films at the sinter surface or fibrillar structures that contribute to stromatolite formation and textures.

Molecular phylogenetic techniques have largely replaced phenotypic analyses in order to investigate the complexity of bacterial and archaeal community compositions and identify uncultured organisms. The use of both techniques has revealed correlations between phylogeny and geothermal regime, and hence stromatolite type, and tentatively imply function where closely related cultivated representatives exist. This is particularly important for diverse nonphotosynthetic communities, where phylogeny cannot be determined by light microscopy. Nonphotosynthetic, thermophilic communities identified on sinters deposited from several thermal springs in Yellowstone National Park, Iceland, and New Zealand, based on 16S rRNA gene analyses, for example, were found to be abundant in unclassified bacteria (i.e., with no near relatives in culture); organisms related to *Thermotogales*, *Thermus* spp., *Saccharomonospora glauca*, gammaproteobacteria, and the *Aquificales* bacteria *Thermocrinis ruber* and *Aquifex pyrophilus* (H_2, $S_2O_3^{2-}$ and/or S^0 oxidizing); and the archaea *Sulfolobales*, *Thermoplasmatales*, *Thermoproteales*, and *Desulfurococcales* (Reyesenbach et al., 1994; Huber et al., 1998; Inagaki et al., 2001; Blank et al., 2002; Meyer-Dombard et al., 2005; Benning and Tobler, 2008; Childs, 2008). Smith et al. (2003) used molecular phylogenetic techniques to identify bacteria responsible for forming large swaths of siliceous streamer fabric in a geothermal power station effluent drain. The community was found not to be dominated by a single organism but to contain a variety of bacteria, including relatives of the thermophilic betaproteobacterium, *Hydrogenophilus* sp. (H_2-oxidizing), *Chloroflexi*, and cyanobacteria. Bacterial communities have also successfully been identified from within sinter deposited years previously using 16S rRNA gene analysis (e.g., Neilan et al., 2002).

In environmental microbial ecology, there is also increasing accessibility and application of metagenomics, transcriptomics, and proteomics, as well as specific gene expression studies to community analysis, permitting detailed assessment of community structure and gene and protein expression (Maron et al., 2007; Bertin et al., 2008). Research groups have already begun applying these powerful techniques to microbial mats in hot spring systems. In a study by Bhaya et al. (2007), genomic sequencing of two dominant but ecologically dissimilar *Synechococcus* strains (i.e., ecotypes) from hot springs in Yellowstone National Park revealed differences in phosphate and nitrogen pathways, indicating that the strains had developed different nutrient requirements. In comparing these sequences to metagenomic data from mats, the authors also found greater strain-level variation in the *Synechococcus* populations at lower rather than at higher temperatures and identified a distinct population with genes for Fe(II) uptake and assimilation. In another study, the expression of oxygenic versus anoxygenic photosynthesis and nitrogen fixation genes was examined in a thermophilic microbial mat (Tibet) (Lau and Pointing, 2009). Results showed a clear vertical transition from sequences that indicate phototrophic and nitrogen-fixing chemolithoautotrophic bacteria at the surface to those indicating anoxygenic phototrophs with depth. In a similar study, Steunou et al. (2006) determined gene sequences specific to

Synechococcus ecotypes in a microbial mat (Octopus Spring, Yellowstone National Park). From these, they tracked the expression of the *Synechococcus* genes through day and night, depicting, for example, a decline in photosynthesis and respiration in the evening and an increase in nitrogen fixation toward the end of the day as photosynthesis declined and anoxia encroached in the mats.

3.3. ANALYSIS OF MODERN TO FOSSIL TERRESTRIAL HOT SPRING DEPOSITS

Certain analytical techniques are equally applicable to both modern and ancient sinters, such as mineralogical analyses of sinter and textural characterizations of sinter and microfossils. X-ray diffraction, used primarily in this context to differentiate between silica polymorphs, is complemented by Raman spectroscopy and thermal analysis in tracking sinter crystallinity and maturity (Herdianita et al., 2000a, b; Lynne et al., 2005, 2008). As for freshly deposited sinter, SEM underpins textural characterizations of fossil stromatolitic laminae and microbes. Optical light microscopy of thin sections of (micro)stromatolite cross sections is important for obtaining a well-contrasted overview of microstromatolitic laminations and microfossils – particularly laminae-forming microfossils – from low- or mid-temperature deposits (e.g., Braunstein, 1999; Jones et al., 2000; Campbell et al., 2001; Lynne and Campbell, 2003; Handley et al., 2005; Guido and Campbell, 2009; Guido et al., in press) and to reduce misidentifications of what may, in fact, be pseudofossils (cf. Buick, 1991). Microfossils can also be detected in high-temperature deposits by light microscopy where cells have been enlarged by silica encrustation but only to a very limited extent (Handley et al., 2005). SEM is particularly important when imaging the small diameter cells typical of (hyper)thermophiles (e.g., Cady and Farmer, 1996).

Refractory lipids provide a long-surviving source of information on members of eukaryotic, bacterial, and archaeal microbial assemblages as opposed to DNA, which is more readily degraded. Studies of modern and recently relict sinters prove the efficacy of the method for interpreting lipid signatures in ancient sinters and exploring differences in lipid preservation. Up to genera-level resolution of bacteria has been inferred from lipid compositions in modern hot spring settings, based on lipid profiles of pure cultures. Analyses of various refractory polar lipids (e.g., monoglycosyl, diglycosyl, and sulfoquinosovyl diglycerides, and phosphatidyl glycerol, 2β-methylbacteriohopanepolyols, and methylalkanes) have been used to distinguish among bacterial genera present in a modern thermophilic cyanobacterial mat at Octopus Spring, Yellowstone National Park (Ward et al., 1994; Jahnke et al., 2004). Meanwhile, lipids indicative of novel organisms, thermophiles, archaea, bacteria, and the bacterial genus *Roseiflexus* and order *Aquificales* were identified from silicified microbial communities within several high-temperature sinters from the Taupo Volcanic Zone, New Zealand (Pancost et al., 2005). Kaur et al. (2008) examined concentrations of bacterial and archaeal diether lipids of a similar chemical structure and expected level of preservation to determine temporal changes in community structures in a sinter accumulated over 900 years (Champagne Pool). Their results suggested a dramatic transition in the relative abundances of bacterial and archaeal communities.

4. Summary and Significance of Hot Springs and Their Deposits

The attraction of hot spring settings is both in the extreme microbial life they host and in their capacity to act as time capsules of paleoecological and paleoenvironmental information. Hot springs lure microbiologists to the discovery of novel enzymes, notably those that are thermostable (e.g., Turner et al., 2007). Arguably, the most famous case is that of the isolation of thermophilic *Thermus aquaticus* from a hot spring in Yellowstone National Park, followed by the isolation of thermostable Taq polymerase from *T. aquaticus* – a DNA polymerase widely used in polymerase chain reactions for DNA amplification (Brock and Freeze, 1969; Chien et al., 1976). In addition, the sinter deposited in many thermal springs is valued for preserving traces of microbial life deep into the geological record, and as such rendering insights into early forms of life on Earth (e.g., Trewin et al., 1994; Walter et al., 1998). However, there is also much interest in the application of terrestrial stromatolite studies to the search for fossilized life on Mars (e.g., Cady et al., 2003; Des Marais et al., 2008). Convincing evidence now exists for siliceous hot spring deposits on Mars. The Mars Reconaissance Orbiter (MRO) High Resolution Imaging Science Experiment (HiRISE) detected geomorphic features closely resembling terrestrial thermal features on Earth, including a large fracture system and aligned spring-like elliptical mounds with central vents, terracing, and concentric zoning of color (≈ composition) (Allen and Oehler, 2008). Meanwhile, silica-rich deposits were also discovered on Mars by Spirit rover, and they were found in conjunction with volcanic deposits (e.g., basalt, tephra, and hydrated ferric sulfates), giving strong indication for the siliceous material to have formed under (low-pH acid-sulfate) hydrothermal conditions (Squyres et al., 2008).

Future research on terrestrial hot springs will likely involve increased emphasis on genomic, transcriptomic, and proteomic studies to establish the intricacies of microbial communities and their metabolisms and better understand the ecology of these systems as analogs for early forms of life. Even so, detecting and interpreting biosignatures present in stromatolites necessitates that data from manifold approaches be taken into account, i.e., studies of modern sinters to identify relationships between environmental conditions and microbes that affect the overall morphological and internally laminated construction of stromatolites; microbial community information (phylogenetic and functional); and the underlying causes of the differential effects of diagenesis. This also includes examining the processes of biofilm silicification and fossilization, which can involve both cells and EPS, and can also be fraught by inconsistencies, such as species-specific fossilization biases (e.g., Westall, 1997; Westall et al., 2000;

Handley et al., 2008). The wider physical context of sinters, namely stromatolitic patterns in relict deposits that match microfacies in modern hot springs settings, and other potential nonmorphological evidence such as lipid biomarkers, are especially important for avoiding potential misapplication of a biological origin to abiotic structures, e.g., filament-like mineral inclusions (Hofmann and Farmer, 2000).

5. Acknowledgments

K.M. Handley acknowledges the Foundation for Research, Science and Technology's Geothermal for the New Millennium NSOF Extremophiles Programme and the Freeston Licensing Trust for funding. K.A. Campbell acknowledges the Royal Society of New Zealand's Marsden Fund and its Charles Fleming Senior Scientist Award for funding. We also thank B.Y. Lynne, P.R.L. Browne, B.W. Mountain, and D.M. Guido for scientific discussions; J.D. Farmer for introducing K.A.C. to hot springs at Yellowstone National Park; R. Renaut and S. Cady for helpful comments; and V. Tewari for the invitation to contribute.

6. References

Allen, C.C. and Oehler, D.Z. (2008) A case for ancient springs in Arabia Terra, Mars. Astrobiology **8**: 1093–1112.

Anderson, T.F. (1951) Techniques for the preservation of three-dimensional structure in preparing specimens for the electron microscope. Trans. N.Y. Acad. Sci. **13**: 130–134.

Anderson, T.F. (1952) Stereoscopic studies of cells and viruses in the electron microscope. Am. Nat. **86**: 91–100.

Benning, L.G. and Tobler, D.J. (2008) The metagenomics of biosilicification: causes and effects. Mineral. Mag. **72**: 221–225.

Benning, L.G., Phoenix, V.R., Yee, N. and Konhauser, K.O. (2004a) The dynamics of cyanobacterial silicification: an infrared micro-spectroscopic investigation. Geochim. Cosmochim. Acta **68**: 743–757.

Benning, L.G., Phoenix, V.R., Yee, N. and Tobin, M.J. (2004b) Molecular characterization of cyanobacterial silicification using synchrotron infrared micro-spectroscopy. Geochim. Cosmochim. Acta **68**: 729–741.

Bertin, P.N., Medigue, C. and Normand, P. (2008) Advances in environmental genomics: towards an integrated view of micro-organisms and ecosystems. Microbiology **154**: 347–359.

Bhaya, D., Grossman, A.R., Steunou, A.-S., Khuri, N., Cohan, F.M., Hamamura, N. Melendrez, M.C., Bateson, M.M., Ward, D.M. and Heidelberg, J.F. (2007) Population level functional diversity in a microbial community revealed by comparative genomic metagenomic analyses. ISME J. **1**: 703–713.

Blank, C.E., Cady, S.L. and Pace, N.R. (2002) Microbial composition of near-boiling silica-depositing thermal springs throughout Yellowstone National Park. Appl. Environ. Microbiol. **68**: 5123–5135.

Braunstein, D. (1999) The role of hydrodynamics in the structuring and growth of high temperature (73°C) siliceous sinter at neutral to alkaline hot springs and geysers, Yellowstone National Park. Unpublished Ph.D. thesis, Stanford University, pp. 163.

Braunstein, D. and Lowe, D.R. (2001) Relationship between spring and geyser activity and the deposition and morphology of high temperature (>73°C) siliceous sinter, Yellowstone National Park, Wyoming, U.S.A. J. Sediment. Res. **71**: 747–763.

Brock, T.D. (1978) *Thermophilic Microorganisms and Life at High Temperatures*. Springer-Verlag: New York, 465 pp.

Brock, T.D. and Freeze, H. (1969) *Thermus aquaticus* gen. n. and sp. n., a nonsporulating extreme thermophile. J. Bacteriol. **98**: 289–297.

Brown, K.L. and McDowell, G.D. (1983) pH control of silica scaling. Proceedings of the 5th New Zealand Geothermal Workshop, University of Auckland, Auckland, pp. 1–5.

Buick, R. (1991) Microfossil recognition in Archean rocks: an appraisal of spheroids and filaments from a 3500 m.y. old chert-barite unit at North Pole, Western Australia. Palaios **5**: 441–459.

Cady, S.L. and Farmer, J.D. (1996) Fossilization processes in siliceous thermal springs: trends in preservation along thermal gradients, In: G.R. Bock and J.A. Goode (eds.) *Evolution of Hydrothermal Ecosystems on Earth (and Mars?)*. John Wiley, Chichester, pp. 150–173.

Cady, S.L., Farmer, J.D., Grotzinger, J.P., Schopf, J.W. and Steele, A. (2003) Morphological biosignatures and the search for life on Mars. Astrobiology **3**: 351–368.

Campbell, K.A., Buddle, T.F. and Browne, P.R.L. (2004) Late Pleistocene siliceous sinter associated with fluvial, lacustrine, volcaniclastic and landslide deposits at Tahunaatara, Taupo Volcanic Zone, New Zealand. Trans. R. Soc. Edinburgh: Earth Sci. **94**: 485–501.

Campbell, K.A. and Lynne, B.Y. (2006) Diagenesis and dissolution at Sinter Island (456 yrs BP). Taupo Volcanic Zone: silica stars and the birth of quartz. Proceedings of the 28th New Zealand Geothermal Workshop, University of Auckland, Auckland, 7 pp.

Campbell, K.A., Sannazzaro, K., Rodgers, K.A., Herdianita, N.R. and Browne, P.R.L. (2001) Sedimentary facies and mineralogy of the late Pleistocene Umukuri silica sinter, Taupo Volcanic Zone, New Zealand. J. Sediment. Res. **71**: 727–746.

Cassie, V. and Cooper, R.C. (1989) Algae of New Zealand thermal areas. Bibl. Phycol. **78**: 1–159.

Castenholz, R.W. (1969) Thermophilic blue-green algae and the thermal environment. Bacteriol. Rev. **33**: 476–504.

Chien, A., Edgar, D.B. and Trela, J.M. (1976) Deoxyribonucleic acid polymerase from the extreme thermophile *Thermus aquaticus*. J. Bacteriol. **127**: 1550–1557.

Childs, A., Mountain, B.W., O'Toole, R. and Stott, M.B. (2008) Relating microbial community and physicochemical parameters of a hot spring: Champagne Pool, Wai-o-tapu, New Zealand. Geomicrobiol. J. **25**: 441–453.

Des Marais, D.J., Nuth III, J.A., Allamandola, L.J., Boss, A.P., Farmer, J.D., Hoehler, T.M., Jakosky, B.M., Meadows, V.S., Pohorille, A., Runnegar, B. and Spormann, A.M. (2008) The NASA astrobiology roadmap. Astrobiology **8**: 715–730.

Doemel, W.N. and Brock, T.D. (1974) Bacterial stromatolites: origin of laminations. Science **184**: 1083–1085.

Doemel, W.N. and Brock, T.D. (1977) Structure, growth, and decomposition of laminated algal-bacterial mats in alkaline hot springs. Appl. Environ. Microbiol. **34**: 433–452.

Douglas, S. (2005) Mineralogical footprints of microbial life. Am. J. Sci. **305**: 503–525.

Fayers, S.R. and Trewin, N.H. (2004) A review of the palaeoenvironments and biota of the Windyfield chert. Trans. R. Soc. Edinb. Earth Sci. **94**: 325–339.

Fernandez-Turiel, J.L., Garcia-Valles, M., Gimeno-Torrente, D., Saavedra-Alonso, J. and Martinez-Manent, S. (2005) The hot spring and geyser sinters of El Tatio, Northern Chile. Sediment. Geol. **180**: 125–147.

Ferris, F.G., Fyfe, W.S. and Beveridge, T.J. (1988) Metallic ion binding by *Bacillus subtilis*: implications for the fossilization of microorganisms. Geology **16**: 149–152.

Fournier, R.O. and Rowe, J.J. (1977) The solubility of amorphous silica in water at high temperatures and high pressures. Am. Mineral. **62**: 1052–1056.

Francis, S., Margulis, L. and Barghoon, E.S. (1978) On the experimental silicification of microorganisms. II. On the time of appearance of eukaryotic organisms in the fossil record. Precambrian Res. **6**: 65–100.

Fratesi, S.E., Lynch, F.L., Kirkland, B.L. and Brown, L.R. (2004) Effects of SEM preparation techniques on the appearance of bacteria and biofilms in the Carter Sandstone. J. Sediment. Res. **74**: 858–867.

Gallup, D.L. (1998) Aluminum silicate scale formation and inhibition (2): scale solubilities and laboratory and field inhibition tests. Geothermics **27**: 485–501.

Garland, C.D., Lee, A. and Dickson, M.R. (1979) The preservation of surface-associated micro-organisms prepared for scanning electron microscopy. J. Microsc. **116**: 227–242.

Guido, D.M. and Campbell, K.A. (2009) Jurassic hot-spring activity in a fluvial setting at La Marciana, Patagonia, Argentina. Geol. Mag. **146**: 617–622.

Guido, D.M., Channing, A., Campbell, K.A. and Zamuner, A. (in press) Jurassic geothermal landscapes and fossil ecosystems at San Agustín, Patagonia, Argentina. J. Geol. Soc. London **167**: 11–20.

Guidry, S.A. and Chafetz, H.S. (2003) Anatomy of siliceous hot springs: examples from Yellowstone National Park, Wyoming, USA. Sediment. Geol. **157**: 71–106.

Hall-Stoodley, L., Costerton, J.W. and Stoodley, P. (2004) Bacterial biofilms: from the natural environment to infectious diseases. Nat. Rev. Microbiol. **2**: 95–108.

Handley, K.M., Campbell, K.A., Mountain, B.W. and Browne, P.R.L. (2005) Abiotic–biotic controls on the origin and development of spicular sinter: *in situ* growth experiments, Champagne Pool, Waiotapu, New Zealand. Geobiology **3**: 93–114.

Handley, K.M., Turner, S.J., Campbell, K.A. and Mountain, B.W. (2008) Silicifying biofilm exopolymers on a hot-spring microstromatolite: templating nanometer-thick laminae. Astrobiology **8**: 747–770.

Herdianita, N.R., Browne, P.R.L., Rodgers, K.A. and Campbell, K.A. (2000a) Mineralogical and textural changes accompanying ageing of silica sinter. Miner. Deposita **35**: 48–62.

Herdianita, N.R., Rodgers, K.A. and Browne, P.R.L. (2000b) Routine instrumental procedures to characterise the mineralogy of modern and ancient silica sinters. Geothermics **29**: 65–81.

Hinman, N. and Lindstrom, R.F. (1996) Seasonal changes in silica deposition in hot spring systems. Chem. Geol. **132**: 237–246.

Hofmann, B.A. and Farmer, J.D. (2000) Filamentous fabrics in low-temperature mineral assemblages: are they fossil biomarkers? Implications for the search for a subsurface fossil record on the early. Plant. Space Sci. **48**: 1077–1086.

Huber, R., Eder, W., Heldwein, S., Wanner, G., Huber, H., Reinhard, R. and Stetter, K.O. (1998) *Thermocrinis ruber* gen. nov., sp. nov., a pink-filament-forming hyperthermophilic bacterium isolated from Yellowstone National Park. Appl. Environ. Microbiol. **64**: 3576–3583.

Hunter, R.J. (1993) *Introduction to Modern Colloid Science*. Oxford University Press, New York, 338 pp.

Iler, R.K. (1979) *The Chemistry of Silica: Solubility, Polymerization, Colloid and Surface Properties, and Biochemistry*. Wiley, New York, 866 pp.

Inagaki, F., Motomura, Y., Doi, K., Taguchi, S., Izawa, E., Lowe, D.R. and Ogata, S. (2001) Silicified microbial community at Steep Cone hot spring, Yellowstone National Park. Microbes Environ. **16**: 125–130.

Jahnke, L.L., Embaye, T., Hope, J., Turk, K.A., Zuilen, M.V., Des Marais, D.J., Farmer, J.D. and Summons, R.E. (2004) Lipid biomarker and carbon isotopic signatures for stromatolite-forming, microbial mat communities and *Phormidium* cultures from Yellowstone National Park. Geobiology **2**: 31–47.

Johnson, D.B., Okibe, N. and Roberto, F.F. (2003) Novel thermo-acidophilic bacteria isolated from geothermal sites in Yellowstone National Park: physiological and phylogenetic characteristics. Arch. Microbiol. **180**: 60–68.

Jones, B. and Renaut, R.W. (1997) Formation of silica oncoids around geysers and hot springs at El Tatio, northern Chile. Sedimentology **44**: 287–304.

Jones, B. and Renaut, R.W. (2006) Growth of siliceous spicules in acidic hot springs, Waiotapu geothermal area, North Island, New Zealand. Palaios **21**: 406–423.

Jones, B. and Renaut, R.W. (2007) Microstructural changes accompanying the opal-A to opal-CT transition; new evidence from the siliceous sinters of Geysir, Haukadalur, Iceland. Sedimentology **54**: 921–948.

Jones, B., Renaut, R.W. and Rosen, M.R. (1997) Vertical zonation of biota in microstromatolites associated with hot springs, North Island, New Zealand. Palaios **12**: 220–236.

Jones, B., Renaut, R.W. and Rosen, M.R. (1998) Microbial biofacies in hot-spring sinters: a model based on Ohaaki Pool, North Island, New Zealand. J. Sediment. Res. **68**: 413–434.

Jones, B., Renaut, R.W. and Rosen, M.R. (2000) Stromatolites forming in acidic hot-spring waters, North Island, New Zealand. Palaios **15**: 450–475.

Jones, B., Renaut, R.W., Rosen, M.R. and Ansdell, K.M. (2002) Coniform stromatolites from geothermal systems, North Island, New Zealand. Palaios **17**: 84–103.

Jones, B., Konhauser, K.O., Renaut, R.W. and Wheeler, R.S. (2004) Microbial silicification in Iodine Pool, Waimangu geothermal area, North Island, New Zealand: implications for recognition and identification of ancient silicified microbes. J. Geol. Soc. London **161**: 983–993.

Kaur, G., Mountain, B.W. and Pancost, R.D. (2008) Microbial membrane lipids in active and inactive sinters from Champagne Pool, New Zealand: elucidating past geothermal chemistry and microbiology. Org. Geochem. **39**: 1024–1028.

Konhauser, K.O. and Ferris, F.G. (1996) Diversity of iron and silica precipitation by microbial mats in hydrothermal waters, Iceland: implications for Precambrian iron formations. Geology **24**: 323–326.

Krauskopf, K.B. (1956) Dissolution and precipitation of silica at low temperatures. Geochim. Cosmochim. Acta **10**: 1–26.

Lalonde, S.V., Konhauser, K.O., Reyesenbach, A.-L. and Ferris, F.G. (2005) The experimental silicification of Aquificales and their role in hot spring sinter formation. Geobiology **3**: 41–52.

Lau, M.C.Y. and Pointing, S.B. (2009) Vertical partitioning and expression of primary metabolic genes in a thermophilic microbial mat. Extremophiles **13**: 533–540.

Little, B., Wagner, P., Ray, R., Pope, R. and Scheetz, R. (1991) Biofilms: an ESEM evaluation of artifacts introduced during SEM preparation. J. Ind. Microbiol. **8**: 213–222.

Lowe, D.R., Anderson, K.S. and Braunstein, D. (2001) The zonation and structuring of siliceous sinter around hot springs, Yellowstone National Park, and the role of thermophilic bacteria in its deposition, In: A.-L. Reyesenbach, M. Voytek and R. Mancinelli (eds.) *Thermophiles: Biodiversity, Ecology, and Evolution*. Kluwer Academic/Plenum Publishers, New York, pp. 143–166.

Lynne, B.Y. and Campbell, K.A. (2003) Diagenetic transformations (opal-A to quartz) of low- and mid-temperature microbial textures in siliceous hot-spring deposits, Taupo Vocanic Zone, New Zealand. Can. J. Earth Sci. **40**: 1679–1696.

Lynne, B.Y. and Campbell, K.A. (2004) Morphologic and mineralogic transitions from opal-A to opal-CT in low-temperature siliceous sinter diagenesis, Taupo Volcanic Zone, New Zealand. J. Sediment. Res. **74**: 561–579.

Lynne, B.Y., Campbell, K.A., Moore, J.N. and Browne, P.R.L. (2005) Diagenesis of siliceous sinter (opal-A to quartz) within 1900 years at Opal Mound Roosevelt Hot Springs, Utah, U.S.A. Sediment. Geol. **179**: 249–278.

Lynne, B.Y., Campbell, K.A., Perry, R.S., Browne, P.R.L. and Moore, J.N. (2006) Acceleration of sinter diagenesis in an active fumarole, Taupo volcanic zone, New Zealand. Geology **34**: 749–752.

Lynne, B.Y., Campbell, K.A., James, B.J., Browne, P.R.L. and Moore, J. (2007) Tracking crystallinity in siliceous hot-spring deposits. Am. J. Sci. **307**: 612–641.

Lynne, B.Y., Campbell, K.A., Moore, J. and Browne, P.R.L. (2008) Origin and evolution of the Steamboat Springs siliceous sinter deposit, Nevada, U.S.A. Sediment. Geol. **210**: 111–131.

Makrides, A.C., Turner, M. and Slaughter, J. (1980) Condensation of silica from supersaturated silicic acid solutions. J. Colloid Interface Sci. **73**: 345–367.

Maron, P.-A., Ranjard, L., Mougel, C. and Lemanceau, P. (2007) Metaproteomics: a new approach for studying functional microbial ecology. Microb. Ecol. **53**: 486–493.

Meyer-Dombard, D.R., Shock, E.L. and Amend, J.P. (2005) Archaeal and bacterial communities in geochemically diverse hot springs of Yellowstone National Park, USA. Geobiology **3**: 211–227.

Mountain, B.W., Benning, L.G. and Boerema, J.A. (2003) Experimental studies on New Zealand hot spring sinters: rates of growth and textural development. Can. J. Earth Sci. **40**: 1643–1667.

Mroczek, E.K., White, S.P. and Graham, D.J. (2000) Deposition of amorphous silica in porous packed beds – predicting the lifetime of reinjection aquifers. Geothermics **29**: 737–757.

Neilan, B.A., Burns, B.P., Relman, D.A. and Lowe, D.R. (2002) Molecular identification of cyanobacteria associated with stromatolites from distinct geographical locations. Astrobiology **2**: 271–280.

Oehler, J.H. (1975) Origin and distribution of silica lepispheres in porcelanite from the Monterey Formation of California. J. Sediment. Petrol. **45**: 252–257.

Owen, R.B., Renaut, R.W. and Jones, B. (2008) Geothermal diatoms: a comparative study of floras in hot spring systems of Iceland, New Zealand, and Kenya. Hydrobiologia **610**: 175–192.

Pancost, R.D., Pressley, S., Coleman, J.M., Benning, L.G. and Mountain, B.W. (2005) Lipid biomolecules in silica sinters: indicators of microbial biodiversity. Environ. Microbiol. **7**: 66–77.

Phoenix, V.R., Adams, D.G. and Konhauser, K.O. (2000) Cyanobacterial viability during hydrothermal biomineralization. Chem. Geol. **169**: 329–338.

Renaut, R.W., Jones, B. and Rosen, M.R. (1996) Primary silica oncoids from Orakeikorako hot springs, North Island, New Zealand. Palaios **11**: 446–458.

Reyesenbach, A.-L., Wickham, G.S. and Pace, N.R. (1994) Phylogenetic analysis of the hyperthermophilic pink filament community in Octopus Spring, Yellowstone National Park. Appl. Environ. Microbiol. **60**: 2113–2119.

Rodgers, K.A., Cook, K.L., Browne, P.R.L. and Campbell, K.A. (2002) The mineralogy, texture and significance of silica derived from alteration by steam condensate in three New Zealand geothermal fields. Clay Miner. **37**: 299–322.

Rodgers, K.A., Browne, P.R.L., Buddle, T.F., Cook, K.L., Greatrex, R.A., Hampton, W.A., Herdianita, N.R., Holland, G.R., Lynne, B.Y., Martin, R., Newton, Z., Pastars, D., Sannazarro, K.L. and Teece, C.I.A. (2004) Silica phases in sinters and residues from geothermal fields of New Zealand. Earth Sci. Rev. **66**: 1–61.

Rothbaum, H.P. and Rohde, A.G. (1979) Kinetics of silica polymerization and deposition from dilute solutions between 5 and 180°C. J. Colloid Interface Sci. **71**: 533–559.

Schinteie, R. (2005) Siliceous sinter facies and microbial mats from acid-sulphate-chloride springs, Parariki Stream, Rotokawa Geothermal Field, Taupo Volcanic Zone, New Zealand. Unpublished thesis, University of Auckland, pp. 144.

Schinteie, R., Campbell, K.A. and Browne, P.R.L. (2007) Microfacies of stromatolitic sinter from acid-sulphate-chloride springs at Parariki Stream, Rotokawa geothermal field, New Zealand. Palaeontol. Electronica **10**, 4A, 33, http://palaeo-electronica.org/2007_1/sinter/index.html.

Smith, B.Y., Turner, S.J. and Rodgers, K.A. (2003) Opal-A and associated microbes from Wairakei, New Zealand: the first 300 days. Mineral. Mag. **67**: 563–579.

Squyres, S.W., Arvidson, R.E., Ruff, S., Gellert, R., Morris, R.V., Ming, D.W., Crumpler, L., Farmer, J.D., Des Marais, D.J., Yen, A., McLennan, S.M., Calvin, W., Bell III, J.F., Clark, B.C., Wang, A., McCoy, T.J., Schmidt, M.E., de Souze Jr., P.A. (2008) Detection of silica-rich deposits on Mars. Science **320**: 1063–1067.

Stein, C.L. (1982) Silica recrystallisation in petrified wood. J. Sediment. Petrol. **52**: 1277–1282.

Steunou, A.-S., Bhaya, D., Bateson, M.M., Melendrez, M.C., Ward, D.M., Brecht, E., Peters, J.W., Kühl, M. and Grossman, A.R. (2006) *In situ* analysis of nitrogen fixation and metabolic switching in unicellular thermophilic cyanobacteria inhabiting hot spring microbial mats. Proc. Natl. Acad. Sci. U.S.A. **103**: 2398–2403.

Toporski, J.K.W., Steele, A., Westall, F., Thomas-Keprta, K.L. and McKay, D.S. (2002) The simulated silicification of bacteria – new clues to the modes and timing of bacterial preservation and implications for the search for extraterrestrial microfossils. Astrobiology **2**: 1–26.

Trewin, N.H. (1994) Depositional environment and preservation of biota in the Lower Devonian hot-springs of Rhynie, Aberdeenshire, Scotland. Trans. R. Soc. Edinburgh. Earth Sci. **84**: 433–442.

Trewin, N.H. (1996) The Rhynie Cherts: an Early Devonian ecosystem preserved by hydrothermal activity, In: G.R. Brock and J.A. Goode (eds.) *Evolution of Hydrothermal Ecosystems on Earth (and Mars?)*. John Wiley, Chichester, pp. 131–149.

Turner, P., Mamo, G., Karlsson, E.N. (2007) Potential and utilization of thermophiles and thermostable enzymes in biorefining. Microb. Cell Fact. **6**: 33, http://www.microbialcellfactories.com/content/6/1/9.

Urrutia, M.M. and Beveridge, T.J. (1993) Mechanism of silicate binding to the bacterial cell wall in *Bacillus subtilis*. J. Bacteriol. **175**: 1936–1945.

Walter, M.R. (1976a) Geyserites of Yellowstone National Park: an example of abiogenic "stromatolites", In: M.R. Walter (ed.) *Stromatolites. Developments in Sedimentology 20*. Elsevier, Amsterdam, pp. 87–112.

Walter, M.R. (1976b) Hot-spring sediments in Yellowstone National Park, In: M.R. Walter (ed.) *Stromatolites. Developments in Sedimentology 20*. Elsevier, Amsterdam, pp. 489–498.

Walter, M.R., Bauld, J. and Brock, T.D. (1972) Siliceous algal and bacterial stromatolites in hot spring and geyser effluents of Yellowstone National Park. Science **178**: 402–405.

Walter, M.R., Buick, R. and Dunlop, J.S.R. (1980) Stromatolites 3,400–3,500 Myr old from the North Pole area, Western Australia. Nature **284**: 443–445.

Walter, M.R., Des Marais, D., Farmer, J.D. and Hinman, N.W. (1996) Lithofacies and biofacies of mid-Paleozoic thermal spring deposits in the Drummond Basin, Queensland, Australia. Palaios **11**: 497–518.

Walter, M., Mcloughlin, S., Drinnan, A. and Farmer, J. (1998) Palaeontology of Devonian thermal spring deposits, Drummond Basin, Australia. Alcheringa **22**: 285–314.

Ward, D.M., Panke, S., Klöppel, K.-D., Christ, R. and Fredrickson, H. (1994) Complex polar lipids of a hot spring cyanobacterial mat and its cultivated inhabitants. Appl. Environ. Microbiol. **60**: 3358–3367.

Weed, W.H. (1889) Formation of travertine and siliceous sinter by the vegetation of hot springs. United States Geological Survey, 9th Annual Report, pp. 613–676.

Westall, F. (1997) Influence of cell wall composition on the fossilisation of bacteria and the implications for the search for early life forms, In: C.B. Cosmovici, S. Bowyer and D. Werthimer (eds.) *Astronomical and Biochemical Origins and the Search for Life in the Universe*. Editrice Compositori, Bologna, pp. 491–504.

Westall, F., Boni, L. and Guerzoni, E. (1995) The experimental silicification of microorganisms. Palaeontology **38**: 495–528.

Westall, F., Steele, A., Toporski, J., Walsh, M., Allen, C., Guidry, S., Mckay, D., Gibson, E. and Chafetz, H. (2000) Polymeric substances and biofilms as biomarkers in terrestrial materials: implications for extraterrestral samples. J. Geophys. Res. **105**: 24511–24527.

White, D.E., Brannock, W.W. and Murata, K.J. (1956) Silica in hot-spring waters. Geochim. Cosmochim. Acta **10**: 27–59.

White, N.C., Wood, D.G. and Lee, M.C. (1989) Epithermal sinters of Paleozoic age in north Queensland, Australia. Geology **17**: 718–722.

Wise, S.W. and Weaver, F.M. (1974) Cherification of oceanic sediments, In: K.J. Hsu and H.C. Jenkyns (eds.) *Pelagic Sediments: On Land and Under the Sea*. International Association of Sedimentologists, Oxford, pp. 301–326.

Biodata of **Jamie S. Foster** and **Stefan J. Green**, authors of ***"Microbial Diversity in Modern Stromatolites"***

Dr. Jamie S. Foster is currently an Assistant Professor at the University of Florida in the Department of Microbiology and Cell Science. Her research lab is located in the Space Life Science Lab at the Kennedy Space Center in Florida. Upon receiving her Ph.D. from the Universit y of Hawaii in 2000, Dr. Foster continued her research into complex microbial communities at the National Institutes of Health, Washington, DC and then later at the NASA Ames Research Center, CA. Her scientific research interests include evolution of complex symbioses, molecular interactions between microbes and their surrounding environment, and biological mechanisms associated with microbial carbon sequestration.

E-mail: **jfoster@ufl.edu**

Dr. Stefan J. Green is currently director of the DNA Services Facility in the Research Resources Center at the University of Illinois at Chicago. He obtained his Ph.D. from the Hebrew University of Jerusalem in 2004, and continued his studies and research in the Exobiology Branch of the NASA – Ames Research Center in Moffett Field, California, USA and subsequently in the Department of Oceanography at Florida State University, Tallahassee, FL. Dr. Green's scientific interests are in the area of molecular biology and environmental microbial ecology. His research has focused on three disparate systems: hypersaline microbial mats, compost-soil-plant interactions, and bioremediation of contaminated subsurface environments.

E-mail: **GreenDNA@uic.edu**

Jamie S. Foster

Stefan J. Green

V.C. Tewari and J. Seckbach (eds.), *STROMATOLITES: Interaction of Microbes with Sediments*, Cellular Origin, Life in Extreme Habitats and Astrobiology 18, 383–405
DOI 10.1007/978-94-007-0397-1_17,

MICROBIAL DIVERSITY IN MODERN STROMATOLITES

JAMIE S. FOSTER[1] AND STEFAN J. GREEN[2]
[1]*Department of Microbiology and Cell Science, Space Life Sciences Lab, University of Florida, Kennedy Space Center, Gainesville, FL 32899, USA*
[2]*Research Resources Center, University of Illinois at Chicago, Chicago, IL 60612, USA*

Abstract Poised at the biosphere–lithosphere interface, the microbial consortia associated with stromatolites have a profound impact on the evolution of Earth's environment. In this chapter, we review the current state of knowledge of microbial diversity in extant stromatolites by examining data generated using cultivation-independent molecular techniques. Specifically, we compare natural stromatolitic mat systems of three distinctive habitats: the hypersaline pools of Shark Bay, Australia; the open ocean stromatolites of Highborne Cay, Bahamas; and the lacustrine lagoons of Ruidera Pools, Spain. We compare these natural systems to an experimental artificial microbialite, looking for fundamental differences and similarities within the microbial communities. Of the 21 bacterial phyla or sub-phyla detected in the various stromatolite ecosystems, only *Cyanobacteria* were found dominant in all habitats. Within the phylum, few cyanobacterial ecotypes were common to all ecosystems. The marine and hypersaline stromatolite ecosystems had significantly higher bacterial diversity than did the artificial microbialite or the freshwater stromatolite, and the diversity approached that observed in non-lithifying hypersaline microbial mats. Finally, we consider the ecological insights provided by the acquisition of metagenomic sequence data for understanding stromatolite diversity and function. These high-throughput metagenomic sequencing approaches have been applied to modern stromatolitic and microbialitic mat communities and have facilitated a higher resolution characterization of microbial diversity at the molecular-level, thus providing an initial glimpse into the functional complexity of these dynamic ecosystems.

Keywords Stromatolites • Microbial diversity • Cyanobacteria • Highborne Cay • Shark Bay • Artificial microbialites • Ruidera Pool • Alphaproteobacteria • Metagenomics • Functional genetics

1. Introduction

For much of the Earth's history, stromatolites dominated the ancient landscape (Awramik, 1984). With fossils dating back over three billion years (Byerly et al., 1986; Grotzinger and Knoll, 1999), the geologic record of stromatolites clearly shows the extensive role microbes have played in the evolution of Earth's environment (Kasting, 2001). Stromatolite morphology and structure are the direct result of microbial activities and provide key insight into the nature of these complex ecosystems (Burne and Moore, 1987).

To examine the specific role that microbes have played in the formation and accretion of stromatolites, it is necessary to delineate the community composition of the microbial mats associated with these structures. Microbial mats are multilayered communities that drive the biogeochemical cycling of key elements within the stromatolite ecosystem. All "living" stromatolites are the products of microbial mat metabolic activity and geochemical cycling resulting in the accretion and precipitation of carbonate structures (Reid et al., 2000). However, not all microbial mats are conducive to the formation of stromatolitic structures (e.g., Dupraz and Visscher, 2005). Microbial mats capable of forming stromatolitic structures are dependent on the surrounding geochemical conditions, dominant metabolisms, and community composition (Havemann and Foster, 2008).

Prior to the molecular age, stromatolitic microbial mats were characterized primarily by morphology of the most dominant functional group (e.g., cyanobacteria), as well as the meso and macrostructure of the mineral structures (Papineau et al., 2005; Dupraz and Visscher, 2005). Now, with the advent of molecular high-throughput sequencing, ecotypes within the mat community can be delineated and compared among different stromatolite communities. The likelihood that the specific ecotypes found in modern stromatolitic mats can serve as exact proxies for Precambrian microbes is low; however, modern ecotypes may share with their ancient counterparts many basic fundamental metabolisms. Although many specific metabolic functions have been conserved over time, the exchange and rapid evolution of microbes via lateral gene transfer and interaction with viruses have made understanding the evolution of the microbial communities in stromatolites difficult. Despite such caveats, these modern analogs to Earth's microbial past offer an insightful view into the genetic and biochemical interactions between microorganisms and the ambient environment.

In this chapter, we produce a current census of stromatolite microorganisms. We compare the microbial diversities in the three prominent extant ecosystems – the hypersaline stromatolites of Shark Bay, Australia; the open ocean stromatolites of Highborne Cay, Bahamas; and the freshwater stromatolites of Ruidera Pool, Spain. We compare these natural ecosystems to an in vitro artificial microbialite model to ascertain if there are fundamental ecotypes associated with stromatolite formation. Finally, we discuss the emerging field of metagenomics and how this high-throughput sequencing approach has and will continue to rapidly expand our understanding of microbial diversity and functional complexity in modern stromatolites.

2. The Modern Stromatolite Habitats

Although once dominant on the early Earth, modern analogs of ancient stromatolites are far more limited in their global distribution. Although the frequency of their occurrence is low, the variety of habitats that stromatolites occupy is diverse.

In this section, we examine three aquatic habitats where microbe-dominated communities form laminated carbonate scaffolding via interconnected metabolic and geochemical processes. For our microbial census, we chose stromatolite ecosystems that have been analyzed by way of cultivation-independent cloning and sequencing of the small subunit ribosomal RNA gene (i.e., 16S rRNA). The habitats, locations, and GenBank accession numbers used in this comparative analysis are listed in Table 1. Only those sequences generated via domain-level clone libraries were used in phylogenetic and diversity analyses. As a result of variations in sequence length and position, not all sequences were appropriate for diversity analyses. Each stromatolite sample was processed by different researchers and methodologies, thus the direct comparison of these samples should be considered in this context. Likewise, we note that the statistical analyses of the sequence data can be sensitive to the number of clones analyzed for each sample (e.g., Hughes et al., 2001).

Table 1. Overview of modern stromatolite and microbial mat community 16S rRNA gene sequences.

Habitat	Location	Number of sequences	GenBank accession number	Used in analyses[a]	References
Hypersaline	Shark Bay	29	AY429113–AY429141	Yes	Burns et al. (2004)
Hypersaline	Shark Bay	63	AY430099–AY430161	Yes	Burns et al. (2004)
Hypersaline	Shark Bay	18	AY433816–AY433833	Yes	Burns et al. (2004)
Hypersaline	Shark Bay	34	AY435178–AY435211	Yes	Burns et al. (2004)
Hypersaline	Shark Bay	136	EF150675–EF150810	Yes	Goh et al. (2009)
Hypersaline	Shark Bay	478	EU851765–EU852242	Yes	Papineau et al. (2005)
Hypersaline	Shark Bay	34	AY604655–AY604688	No	Burns et al. (2005)
Marine	HBC[b]	174	EU917948–EU918121	Yes	Havemann and Foster (2008)
Marine	HBC	859	FJ911975–FJ912833	Yes	Baumgartner et al. (2009)
Marine	HBC	5	DQ822785–DQ822789	No	Unpublished
Marine	HBC	164	EU248965–EU249128	No	Foster et al. (2009)
Marine	Artificial[c]	408	EU917540–EU917947	Yes	Havemann and Foster (2008)
Freshwater	Ruidera	2	EU780449–EU780450	Yes	Santos et al. (2010)
Freshwater	Ruidera	80	EU753608–EU753687	Yes	Santos et al. (2010)
Freshwater	Ruidera	66	AY566310–AY566375	Yes	Santos et al. (2010)
Freshwater	Cuatros Ciénegas	13	AY541045–AY541057	No	Elser et al. (2005a)
Hypersaline	Guerrero Negro[d]	1655	DQ329539–DQ331020 DQ397339–DQ397511	Yes	Ley et al. (2006)

[a]Only sequences generated from domain-level clone libraries were used for the phylogenetic and diversity analyses. Where necessary, short sequences were excluded from analyses. Libraries generated from cyanobacterial primers or sequences isolated from DGGE were excluded from the phylogenetic analyses

[b]Sequences were derived from Type 2 stromatolites from Highborne Cay, Bahamas (HBC)

[c]Artificial microbialites were derived from stromatolites isolated from Highborne Cay, Bahamas

[d]Hypersaline microbial mats from Guerrero Negro, Mexico were used as a non-lithifying mat reference community

Nonetheless, the combined analysis of multiple ecosystems is powerful, and reveals that each ecosystem possesses different environmental challenges, influencing the adaptation and selection pressures on the stromatolite microbial communities.

Diversity indices for all four of the different communities were generated from raw sequence data downloaded from the GenBank sequence database (Table 2). Bacterial 16S rRNA gene sequences were aligned using the online software package Greengenes (DeSantis et al., 2006b), and these alignments were imported into the phylogenetic software package MEGA (Kumar et al., 2008). A distance matrix was generated from the alignments using only positions shared by all sequences in each dataset, and using the maximum composite likelihood substitution model. The distance matrix was processed using the software package DOTUR (Schloss and Handelsman, 2005), with a 97% sequence similarity used as the threshold for inclusion in an operational taxonomic unit (OTU). Sequence characterization was also performed using the aligned sequences by implementing the online Greengenes DNA maximum likelihood (DNAML) classification tool (Fig. 1). For the generation of phylogenetic trees, the aligned sequences were imported into the software package ARB (Ludwig et al., 2004) and inserted into a 16S rRNA gene tree by implementing the parsimony option while keeping the tree topology fixed using a bacterial 50% conservation filter. This approach allows the simultaneous insertion of sequences of varying length and position into a single phylogenetic tree produced from near full-length sequences. A compressed phylogenetic tree that depicts the relationship between all examined stromatolites habitats is shown in Fig. 2.

2.1. DIVERSITY IN MARINE STROMATOLITES

Geologic evidence has indicated that the salt content of the Precambrian ocean was at least twice that of the modern ocean (Knauth, 1998; Arp et al., 1999) and that the first stromatolitic microbial ecosystems most likely formed in these hypersaline or intertidal conditions (Monty, 1977; Arp et al., 2001). Characterizing the diversity in microbial mats that form laminated stromatolitic structures in modern marine habitats, specifically hypersaline environments, may facilitate our understanding of these ancient ecosystems, as salinity has been shown to be a major driving factor in environment-specific evolution (Lozupone and Knight, 2007).

2.1.1. Hypersaline Stromatolites in Shark Bay

The hypersaline stromatolites of Hamelin Pool, in Shark Bay, Australia represent one of the most prominent and abundant examples of living accreting stromatolites on Earth (Burns et al., 2004; Neilan et al., 2002; Papineau et al., 2005). First discovered in 1954 (Playford and Cockbain, 1976), the stromatolites of Shark Bay have had a significant impact on stromatolite research, as they were the first described microbe-induced build-ups analogous in size and shape to fossilized

Table 2. Bacterial 16S rRNA gene library diversity analyses of clone libraries.

	Shark Bay	HBC total[a]	HBC Type 1	HBC Type 2	HBC Type 3	Artificial Model[b]	Ruidera Pool	Guerrero Negro[b]
Sequences	412	811	190	271	299	201	72	464
OTUs[c]	161	269	101	128	139	88	39	204
Singlets[c]	98	151	61	79	91	58	28	117
Doublets[c]	23	45	26	24	18	18	4	39
OTUs[d] >5%(Sum)	3 (17%)	0 (0%)	1 (10%)	1 (9%)	1 (6%)	3 (30%)	5 (42%)	0 (0%)
Shannon (confidence)	4.54 (0.12)	4.96 (0.09)	4.29 (0.15)	4.43 (0.14)	4.51 (0.13)	3.87 (0.18)	3.36 (0.22)	4.95 (0.09)
Evenness	0.89	0.89	0.93	0.91	0.91	0.86	0.92	0.93
Chao1 (confidence)[e]	359 (273/509)	515 (427/651)	169 (136/323)	251 (196/353)	355 (257/533)	175 (132/261)	115 (66/252)	374 (308/481)
% Coverage[f]	76	79	68	71	70	71	61	75
Base pairs[g]	274	344	457	411	396	474	1,056	328

[a]HBC total represents the pooled sequences of the three major mat types (Type 1–3) of Highborne Cay, Bahamas

[b]The libraries of the artificial microbialites and the Guerrero Negro hypersaline mats were split due to the absence of overlapping regions. Only results examining the forward sequences are reported. The hypersaline Guerrero Negro was included as a non-lithifying reference mat community

[c]The values were calculated based on a 97% similarity threshold

[d]Number of OTUs at 97% similarity threshold or higher that represent greater than 5% of the population. The percentage of the total community these OTUs represent are in *parentheses*

[e]Values in *parentheses* represent the lower and upper 95% confidence intervals associated with Chao1 non-parametric estimator

[f]Percent coverage of the clone library using the formula (1 − [*N*-singlets]/[*N*])

[g]Number of base pairs that were used in the comparisons for each community

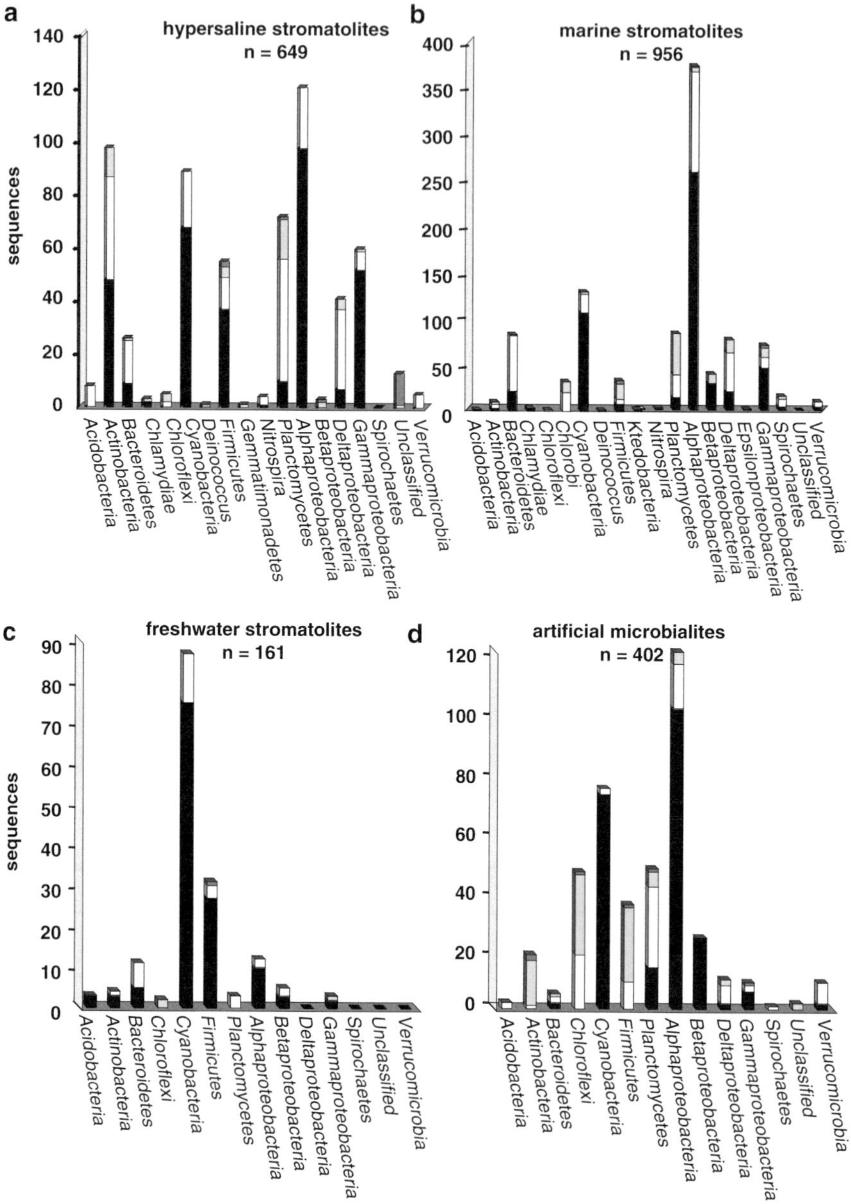

Figure 1. Overview of microbial diversity in modern stromatolite communities. *Bars* represent the total number of bacterial sequences (*n*) recovered from each phylum and separated based on their associated DNA maximum likelihood values (*black bars*, >0.90; *open bars*, 0.80–0.90; *light gray bars*, 0.70–0.80; and *dark gray*, <0.70). Graphs are also separated based on habitat and represent those sequences recovered from stromatolitic mats derived the (**a**) hypersaline waters of Shark Bay; (**b**) normal marine stromatolites of Highborne Cay; (**c**) freshwater lacustrine system of Ruidera Pool; and (**d**) laboratory cultivated microbialitic mats derived from Type 2 Highborne Cay stromatolites.

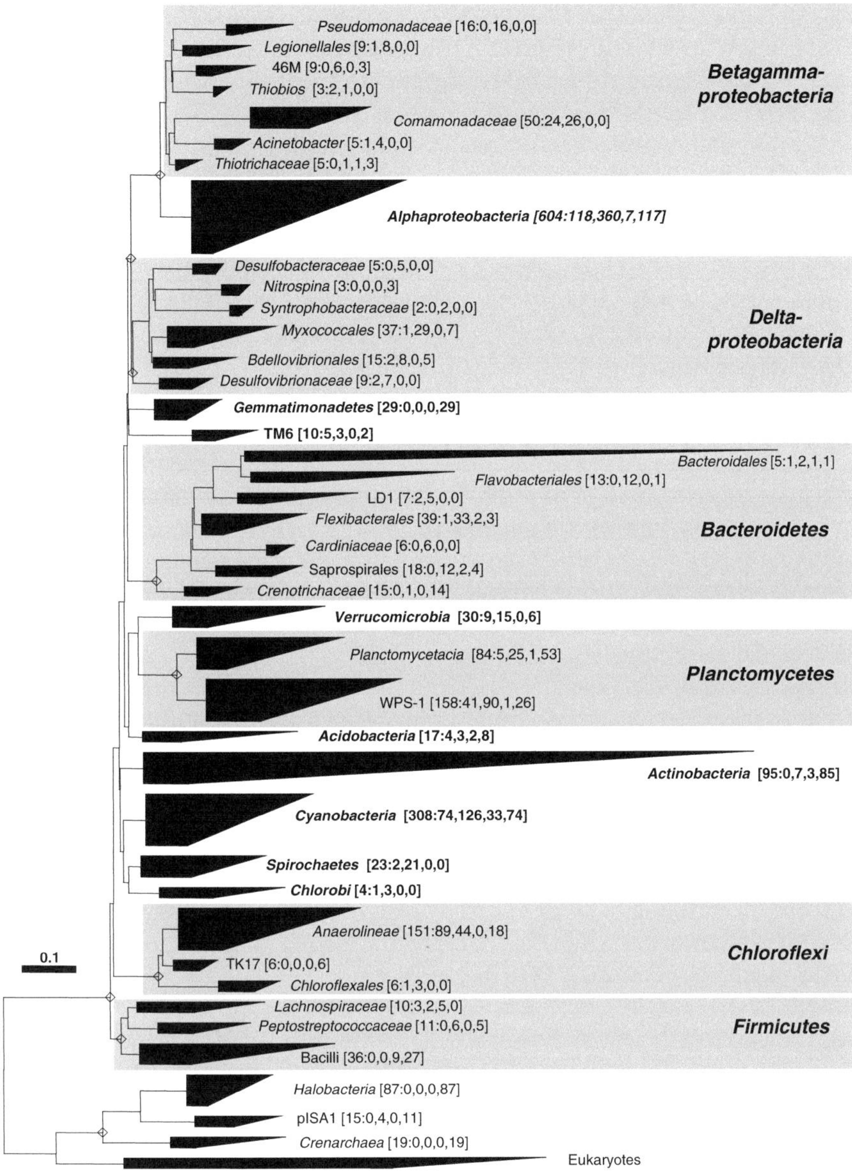

Figure 2. Phylogenetic tree of 16S ribosomal RNA (rRNA) gene sequences recovered from freshwater, marine and hypersaline stromatolites. Sequences were downloaded from GenBank, aligned using the software package Greengenes, and inserted into tree within the software package ARB using the parsimony tool and a bacterial 50% conservation filter. Major clades are compressed, and the total number of stromatolite sequences generated from clone libraries of artificial and natural stromatolite samples within each clade is indicated in *brackets*. Phyla or sub-phyla for which multiple taxa are shown (e.g., family- or order-level clades) are highlighted in *gray*. The *scale bar* represents 10% sequence divergence.

Precambrian stromatolites (Riding, 1991; Reid et al., 2003). The microfabrics of the Shark Bay stromatolites form via two major processes, the trapping and binding of organosedimentary material as well as the micritization of carbonates (Reid et al., 2003). Both processes require the involvement of microbes. Analyses of the 16S rRNA gene sequences recovered from the Shark Bay stromatolites ($n = 605$) clearly show a diverse and novel community of both archaea and bacteria (Burns et al., 2004; Papineau et al., 2005). Although archaea are important in modern stromatolites, this chapter focuses only on bacterial diversity. A comparison of all Shark Bay 16S rRNA gene sequences available in GenBank detected 17 different bacterial phyla or sub-phyla (Fig. 1a). Based on these 16S rRNA gene sequences, the Shark Bay stromatolites appear to be dominated by *Alphaproteobacteria* (20%), *Actinobacteria* (16%), and *Cyanobacteria* (15%). Note, however, only those sequences generated via domain-level primers (Lane, 1991; DeLong, 1992; Table 1) were used in these clone library comparisons. Sequences derived from denaturing gradient gel electrophoresis (DGGE), functional group-specific primers (e.g., cyanobacterial), or intergenic spacer region sequencing were not included in any of the phylum-level comparisons of this analysis. The other dominant phyla or sub-phyla in the Shark Bay stromatolites included *Planctomycetes* (11%), *Gammaproteobacteria* (10%), and the *Firmicutes* (9%). The high level of phylum-level diversity expanded at narrower taxonomic levels (Table 2). The Shannon index of the Shark Bay stromatolites was high (4.54), and a Chao1 estimate of 359 phylotypes at a 97% sequence identity threshold. This level of diversity was similar to that detected in the marine stromatolite from Highborne Cay. Both the marine stromatolite and the hypersaline stromatolite diversity indices approached the diversity level detected in a highly diverse non-lithifying microbial mat from Guerrero Negro, Baja, Mexico (Table 2; Ley et al., 2006). The microbial community in the Shark Bay stromatolite samples also exhibits a high degree of sequence novelty in several key taxa (Fig. 1a). For example, more than half the recovered sequences from the *Actinobacteria*, *Deltaproteobacteria*, and *Planctomycetes* had DNAML values below 0.90 (Fig. 1a). These DNAML values were calculated by comparing the target sequences to the reference sequences in Greengenes database (i.e., ARB). The closer these values are to 1.0, the higher the matching probability with sequences in the curated database (DeSantis et al., 2006a).

2.1.2. Open Ocean Stromatolites

The well-laminated carbonate build-ups along the margins of Exuma Sound, Bahamas, represent the only known example of extant stromatolites in normal marine salinity (Dravis, 1983; Dill et al., 1986; Reid et al., 1995). One location in particular, the island of Highborne Cay, has been the subject of several recent molecular, biogeochemical, and geological investigations into the formation and development of these modern accreting stromatolites (Reid et al., 2000; Decho et al., 2005; Perkins et al., 2007; Foster et al., 2009). There are three major mat types (Type 1, 2, and 3) found in the intertidal and subtidal zones of Highborne Cay.

Each mat type is classified based on the extent of carbonate precipitation and taxonomic richness (Reid et al., 2000; Baumgartner et al., 2009; Foster et al., 2009). Comparative analysis of the available 16S rRNA genes recovered from Highborne Cay stromatolites revealed that when pooled, these three stromatolitic mat types contain the highest level of diversity found in any modern stromatolite (Table 2). Together, there were 20 different phyla detected in the Highborne Cay stromatolitic mats with the *Alphaproteobacteria* (38%) and *Cyanobacteria* (18%) dominating (Fig. 1b). The diversity of each mat type was also high even when they were examined separately (Table 2). For example, in samples of Type 2 stromatolites, which are considered to be an intermediate stage of Highborne Cay stromatolite development (Reid et al., 2000), 18 different phyla or sub-phyla were detected. These samples were similarly dominated by the *Alphaproteobacteria* (20%) and *Cyanobacteria* (18%) (Havemann and Foster, 2008). Other taxa that were detected in high abundance were *Bacteroidetes* (9%), *Planctomycetes* (9%), *Deltaproteobacteria* (8%), and *Gammaproteobacteria* (7%; Fig. 1b). DNAML analyses of the combined Highborne Cay sequences revealed the highest level of sequence novelty in the *Chloroflexi*, *Planctomycetes*, and *Deltaproteobacteria* with the majority of sequences below 0.89 (Fig. 1b). These results may suggest that several of these ecotypes may be geographically localized to the stromatolites of Highborne Cay.

2.2. DIVERSITY IN FRESHWATER STROMATOLITES

In addition to the marine habitats, stromatolitic carbonate build-ups have also been reported in freshwater ecosystems (Pedley, 2000). Many of these freshwater stromatolitic communities undergo periodic desiccation events (Pedley, 2000) and some, such as the fresh water microbialites of Cuatros Ciénegas Basin, Mexico, have biological signatures that suggest that these freshwater pools once had a marine origin (Souza et al., 2006; Desnues et al., 2008).

2.2.1. Ruidera Pools, Spain

The only two modern freshwater microbialite ecosystems that have been comprehensively examined at the molecular level are the fluvio-lacustrine system in Ruidera Pools National Park in Central Spain, and the phosphorus-poor pools of the Cuatros Ciénegas Basin in Mexico. The latter ecosystem has been examined via high-throughput metagenomic sequencing and will be discussed elsewhere in this chapter. However, the stromatolitic tufas of Ruidera Pools represent one of the most prominent sites of carbon sequestration and precipitation in Europe (Ordóñez et al., 2005; Santos et al., 2009). Currently available 16S rRNA sequences mined from GenBank reveal that 54% of the sequences recovered from the Ruidera stromatolites are *Cyanobacteria* (Fig. 1c). The other dominant phylum from the Ruidera Pools was the *Firmicutes* (19%); however, quantitative PCR results indicated that these were not active members of the community (Santos et al., 2009). As these microbial communities are susceptible to the periodic desiccation and flooding events typical of the Ruidera Pools, the prominence of desiccation-resistant microbes such as the spore forming *Firmicutes* might not indicate a specific function essential for stromatolite formation, but merely a reflection of environmental adaptation. However, the taxonomic analyses of these freshwater microorganisms suggest that there are very few novel ecotypes in this system as most sequences had DNAML values above 0.90. In contrast to the marine and hypersaline stromatolites, the *Alphaproteobacteria* were relatively minor constituents (7%) of the total microbial community. Diversity indices reveal that the microbial consortia of the Ruidera Pool stromatolites are the lowest of all the communities compared and this may in part reflect the tremendous dominance of *Cyanobacteria* in the clone library (Table 2). Nonetheless, the clone library coverage was significantly lower (61%) for this freshwater ecosystem than the others. Additional sequencing will need to be performed to definitively demonstrate lower community diversity.

2.3. DIVERSITY IN ARTIFICIAL MICROBIALITES

To complement studies of the natural ecosystems, several researchers have developed laboratory analogs to model the initiation and development of modern stromatolites and microbialites (Fenchel and Kühl, 2000; Dupraz et al., 2006; Havemann and Foster, 2008; Foster and Mobberley, 2010). These artificial models can serve as proxies for the microbes and metabolic activity associated with microbialite formation. Comparisons between these artificial and natural communities have revealed key organisms associated with the carbon sequestration and precipitation processes (Havemann and Foster, 2008). Experimentally lithifying microbialites derived from the marine stromatolites of Highborne Cay show a decreased diversity relative to the natural stromatolites (Table 2; Fig. 1d). The decrease in diversity appears to represent a loss of diversity, as the microbial community in the experimental ecosystems was predominantly a subset of that found in the natural ecosystem (Havemann and Foster, 2008; Fig. 1b, d). The lower diversity levels can assist in developing laboratory-based models to isolate those specific ecotype and processes associated with carbonate precipitation in modern microbialites. Likewise, a lower diversity may also simplify the identification and isolation of those ecotypes essential for stromatolite formation. As in the Highborne Cay stromatolites, the artificial microbialites are dominated by *Alphaproteobacteria* (30%) and *Cyanobacteria* (18%; Fig. 1d). The representation of the other major phyla mirrors that of those in the natural system (Fig. 1d), clearly indicating that even after 1.5 years in cultivation the natural stromatolitic mat diversity was maintained under artificial conditions (Havemann and Foster, 2008). These results suggest that these artificial microbialites may serve as useful analogs that can be experimentally manipulated under simulated environmental conditions.

3. Dominant Phyla in Modern Stromatolites

Bacteria from the phylum *Cyanobacteria* are the predominant organisms present in all the stromatolites, and bacteria affiliated with the sub-phylum *Alphaproteobacteria* are the most abundant bacteria detected in the marine and hypersaline stromatolites. We discuss these organisms in greater detail later. In addition, we note the abundance of organisms affiliated with the phylum *Planctomycetes* and class *Anaerolineae* of the phylum *Chloroflexi* in the marine and hypersaline stromatolites. Although a subset of *Planctomycetes* are capable of anaerobic ammonia oxidation (ANAMMOX), many of the detected sequences in this study were not affiliated with these ANAMMOX bacteria (e.g., "Wittenberg polluted soil" or WPS bacteria of unknown physiology; Nogales et al., 2001). Likewise, bacteria from the class *Anaerolineae* are filamentous chemo-organotrophic anaerobes (Yamada et al., 2006) in a phylum known for green non-sulfur photosynthetic bacteria. We note that sequences affiliated with the *Chloroflexi* were the most abundant in the deeper, perpetually anaerobic zone of hypersaline microbial mats from Guerrero Negro, Baja, Mexico (Ley et al., 2006). Sequences affiliated with these taxa were observed to have low DNAML values (Fig. 1). This reflects a lack of isolated representatives and concurrently, a general ignorance of the role of the respective organisms in natural ecosystems. One surprising finding from the clone libraries of the marine and hypersaline systems was the absence of a substantial number of sequences affiliated with sulfate-reducing bacteria (SRB) of the sub-phylum *Deltaproteobacteria* (Fig. 2). These organisms have been shown to be abundant in non-lithifying microbial mat ecosystems (e.g., Minz et al., 1999) as well as lithifying stromatolites (e.g., Baumgartner et al., 2006), even in aerobic locations. Further work will be required to demonstrate that this is not an artifact of the sampling and analytical approach.

3.1. THE *CYANOBACTERIA*

Of all the phyla detected in modern stromatolites, none is more prominent than the *Cyanobacteria*. Often considered the driving metabolic force behind stromatolitic mat metabolism and early lithification (Pinckney and Reid, 1997; Dupraz et al., 2004; Visscher and Stolz, 2005), cyanobacteria represent a morphologically and molecularly diverse functional group (Golubic and Browne, 1996; Neilan et al., 2002; Reid et al., 2000; Stolz et al., 2001; Burns et al., 2004; Foster et al., 2009). Comparisons between the four communities revealed pronounced differences and similarities between the cyanobacterial populations (Fig. 3). One of the most surprising observations was the presence of *Leptolyngbya*-like cyanobacteria in all three natural communities (Fig. 3, Clusters 1–6). Interestingly most of the *Leptolyngbya*-like clusters contained sequences recovered from both Shark Bay and Ruidera Pool stromatolites (Fig. 3; Clusters 2–6). These ecotypes were

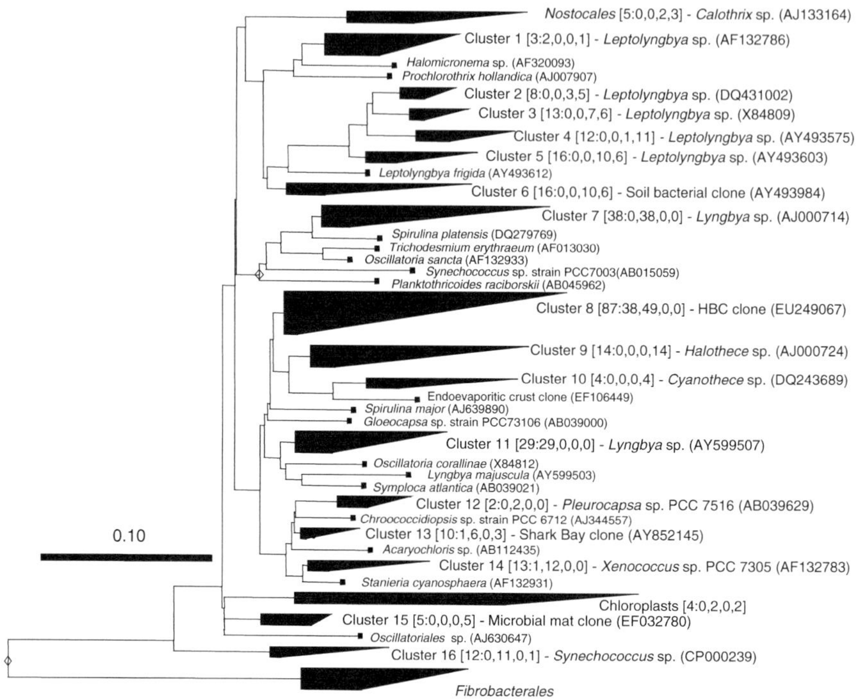

Figure 3. Phylogenetic tree of *Cyanobacteria* 16S rRNA gene sequences from clone libraries produced using general bacterial primers. Major clusters are compressed, and the total number of stromatolite sequences generated from clone libraries is indicated in brackets. The number of sequences from each of four types of stromatolites (artificial microbialite from Highborne Cay, natural stromatolite from Highborne Cay, natural stromatolite from Ruidera Pool, and natural stromatolite from Shark Bay) are indicated after a *semicolon* and separated by a *comma*. Representative sequences of each cluster are indicated. The *scale bar* represents 10% sequence divergence.

not detected in either the natural or artificial marine stromatolites, but together with several sequences from the *Nostocales*, they represented all the Ruidera Pool *Cyanobacteria*. Only *Leptolyngbya*-like organisms identified in Cluster 1 were common to Shark Bay and Highborne Cay. *Leptolyngbya* ecotypes have filamentous morphotypes and have been shown to exhibit key physiological characteristics associated with stromatolite formation (Foster et al., 2009). Despite some shared *Leptolyngbya* ecotypes, in general, there was limited overlap in cyanobacterial community composition between Shark Bay and Highborne Cay stromatolites. The dominant Shark Bay cyanobacteria included unicellular diazotrophs belonging to the *Cyanothece-Halothece* cluster (Fig. 3; Clusters 9 and 10). Such organisms have been found in several hypersaline non-lithifying microbial mat communities such as Guerrero Negro, Mexico (Omoregie et al., 2004; Ley et al., 2006; Green et al., 2008). Highborne Cay stromatolitic mats contained a highly distinctive cyanobacterial community, including representatives of the unicellular

Synechococcus spp. (Cluster 13) and *Xenococcus* spp. (Cluster 14) as well as filamentous *Lyngbya* spp. (Clusters 7 and 11). The recovery of 16S rRNA gene sequences from these putatively diazotrophic cyanobacteria in Highborne Cay stromatolites correlates well with previously described nitrogenous biochemical activities and *nifH* gene profiles of these communities (Steppe et al., 2001; Paerl et al., 2001). Interestingly, heterocystous cyanobacteria were essentially absent from these stromatolite communities. A few sequences affiliated with the order *Nostocales* (e.g., *Calothrix* sp.) were detected in the Ruidera Pool and Shark Bay stromatolites, but represent only a small fraction of the total cyanobacterial community (Fig. 3). Prior studies have shown that heterocystous cyanobacteria are absent from submerged hypersaline microbial mats, even under lowered salinity and sulfate conditions (e.g., Stal, 1995; Bebout et al., 2004; Green et al., 2008). This absence appears to extend to lithifying microbial mat communities as well, with an exception described later in this chapter.

3.2. THE *ALPHAPROTEOBACTERIA*

The *Alphaproteobacteria* represent one of the most metabolically diverse bacterial sub-phyla and are found in most stromatolite and microbial mat communities (Ward et al., 1998; Ley et al., 2006; Havemann and Foster, 2008). The stromatolitic mats of Highborne Cay and Shark Bay are dominated by this sub-phylum with particularly high representation by photoheterotrophic purple non-sulfur bacteria of the orders *Rhodobacterales* and *Rhizobiales*. Heterotrophic *Alphaproteobacteria*, such as those from the genus *Parvularcula* and prosthecate bacteria of the genera *Hyphomonas* and *Hyphomicrobium*, are also abundant in the HBC stromatolites. By far, phototrophic and heterotrophic *Alphaproteobacteria* were the most abundant and diverse in the natural Highborne Cay stromatolites, with representative sequences from half of the defined clusters (Fig. 4; Clusters 1, 4, 5, 9–14, 20). The artificial microbialite exhibited lower diversity of *Alphaproteobacteria* than the native system, though one salient shift in the *Alphaproteobacteria* community composition was the dramatic increase in relative abundance of the organisms affiliated with Cluster 13 (Fig. 4). In the artificial microbialite, these organisms represented approximately half of the total alphaproteobacterial community, and less than 10% of the native community. No specific physiology can be ascribed to these organisms, though their sequence similarity to other *Rhodobacterales* such as *Rhodopseudomonas* spp. may indicate a photosynthetic lifestyle. Although a number of genera of freshwater anoxygenic phototrophic bacteria are known (e.g., Yurkov and Beatty, 1998), these organisms, and *Alphaproteobacteria* in general, were essentially undetected in the freshwater stromatolites from Ruidera Pool (Santos et al., 2009). As with the total bacterial community, the alphaproteobacterial community diverged significantly between the marine HBC stromatolites and the hypersaline Shark Bay stromatolites. Although both systems contained organisms from the same clusters, the proportion of these organisms diverged

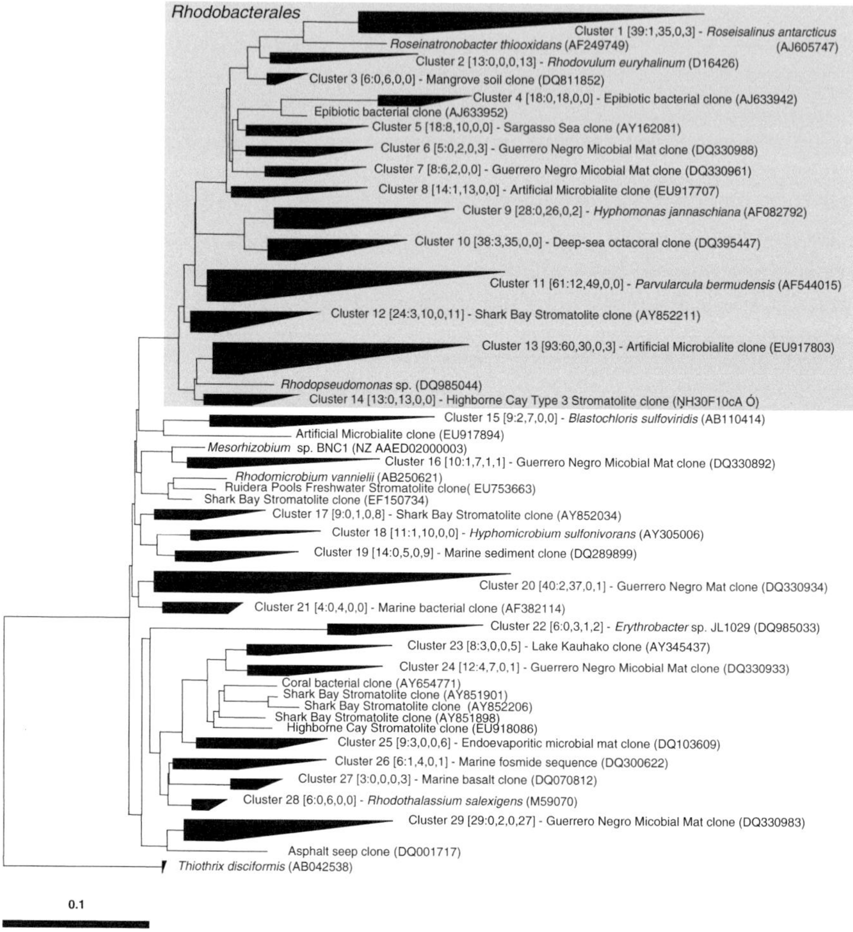

Figure 4. Phylogenetic tree of *Alphaproteobacteria* 16S RNA gene sequences from clone libraries produced using general bacterial primers. Major clusters are compressed, and the total number of stromatolite sequences generated from clone libraries is indicated in *brackets*. The number of sequences from each of four of stromatolites (artificial microbialite, natural stromatolite from Highborne Cay, natural stromatolite from Ruidera Pool, and natural stromatolite from Shark Bay) are indicated after a *semicolon* and separated by a *comma*. Representative sequences of each cluster are indicated. Sequences belonging to the order *Rhodobacterales* are highlighted in *gray*. The *scale bar* represents 10% sequence divergence.

significantly between the systems (e.g., Fig. 4; Clusters 1, 9, 13, 20, 29). Sequences affiliated with bacteria from the genus *Rhodovulum* (marine purple non-sulfur phototrophic bacteria; Fig. 4; Cluster 2; Hiraishi and Ueda, 1994) were the most abundant alphaproteobacterial sequence of the order *Rhodobacterales* detected in the hypersaline stromatolite. The other dominant *Alphaproteobacteria* detected in the Shark Bay stromatolite were unaffiliated with any described species, and have close representatives from clone libraries of submerged and endoevaporitic hypersaline microbial mats (Fig. 4; Clusters 25 and 29; Ley et al., 2006; Sørensen et al., 2005). A number of the defined alphaproteobacterial clusters in this study had no affiliation with described or isolated bacterial species, and these groups are in general composed largely of sequences recovered from marine or hypersaline microbial mats and stromatolites (Fig. 4; Clusters 6–8, 12–14, 20, 29). The physiology of these ecotypes has yet to be explored.

4. A New Beginning: The Future of Microbial Diversity Studies in Stromatolites

With the emergence of high-throughput pyrosequencing and the capacity for genome–genome comparisons, the reign of the 16S rRNA gene as the sole target for meaningful phylogenetic comparisons is coming to an end. The use of this gene for future phylogenetic comparisons will have to be taken in context with the rest of the genome and perhaps one day, the metagenome. Although this day has yet to arrive, new data have emerged that have rapidly increased our understanding of the microbial diversity and functional complexity of modern stromatolites.

4.1. METAGENOMICS IN MICROBIALITIC MATS

Comparing the collective genomes of microbes within complex communities via metagenomics (Handelsman, 2004) has facilitated the analysis of microbialitic mats at the molecular level. The first microbialitic mat communities to be examined via metagenomic sequencing were the freshwater microbialites of Cuatro Ciénegas Basin, Mexico, and the marine stromatolites of Highborne Cay, Bahamas (Breitbart et al., 2008; Desnues et al., 2008). Although to date neither of these microbialite ecosystems has been sequenced to saturation, these initial analyses have greatly expanded our understanding of the microbial ecotype composition and metabolic potential within the microbialite communities. The high diversity of these ecosystems, as observed in a metagenome analysis of the non-lithifying microbial mats from Guerrero Negro, Baja, Mexico (Kunin et al., 2008), will present a challenge for full characterization of these ecosystems.

Both the taxonomic richness and the functional gene diversity of these microbialite communities were determined by comparing the sequences to SEED, the preeminent database for metagenomic data sets (Overbeek et al., 2005). Sequences acquired via 454 pyrosequencing were compared against the SEED non-redundant database by BLASTX and mapped on to subsystems through the MG-RAST system (Meta Genome Rapid Annotation using Subsystem Technology; Overbeek et al., 2005). This method has been used successfully to identify the taxonomic diversity and the metabolic potential of bacteria from metagenomic sequences in a wide variety of microbial mat ecosystems including deep mine biofilms (Edwards et al., 2006), thermophilic hot springs (Bhaya et al., 2007), and non-lithifying hypersaline mat communities (Kunin et al., 2008).

In the Cuatros Ciénegas Basin, microbialites are found in a series of oligotrophic freshwater pools with the lowest levels of phosphorus (<0.1 μM) known in continental waters (Minckley, 1969; Elser et al., 2005a, b; Breitbart et al., 2008). Taxonomic profiles generated for the freshwater microbialites sampled from two pools of the Cuatros Ciénegas Basin, Pozas Azules (domal thrombolites) and Rio Mesquites (spherical oncolites), revealed that both were dominated by *Bacteria* (87% Pozas Azules; 95% Rio Mesquites; Breitbart et al., 2008; Desnues et al., 2008). Archaeal sequences were present in low abundance in both freshwater communities, with 1.8% in Pozas Azules and 0.7% in Rio Mesquites. Of the recovered archaeal sequences, roughly 86% were homologous to sequences of *Euryarchaeota*; few *Crenarchaeota* sequences were recovered. Eukaryotic diversity in these communities is also low, with only 10.6% of the recovered Pozas Azules metagenome sequences and 4.4% of the Rio Mesquites sequences sharing homology to eukaryotic organisms (Breitbart et al., 2008). The depleted phosphorus level appears to inhibit eukaryotic grazing, resulting in prokaryotic dominated communities (Elser et al., 2005a, b; Breitbart et al., 2008).

Interestingly, the taxonomic profile generated via metagenomics in the Rio Mesquites pool correlates to the 16S rRNA gene profiles generated for the freshwater stromatolites in Ruidera Pools, Spain. In both cases, the *Cyanobacteria* are dominant, comprising 74% and 54%, respectively, of 16S rRNA gene clone libraries. The Rio Mesquites microbialites contain a substantially higher proportion of *Cyanobacteria* from the order *Nostocales* relative to the Ruidera Pools stromatolite (44–21%, respectively) of the cyanobacterial community. These metagenomic analyses coupled with rRNA gene sequencing suggest that freshwater microbialites are dominated by cyanobacteria capable of nitrogen fixation, and can include heterocystous forms.

4.2. CORRELATING DIVERSITY WITH FUNCTION

Once the dominant ecotypes of the stromatolitic mats have been identified by rRNA gene sequence and through metagenomic analysis, the next challenge is to identify the function of each key organism within the community. Metagenomic sequencing from the Cuatros Ciénegas Basin microbialites has shown a significant environmental adaptation to the low phosphorus pools. SEED analyses of the recovered sequences showed a shift toward high affinity phosphate transporters, phosphatases, as well as phosphate sensing and regulation genes (Breitbart et al., 2008). Other key subsystems have been identified in these metagenomic sequencing efforts including genes associated with biofilm formation (e.g., tight adherence macromolecule transporters), EPS production (e.g., alginate and colanic acid production), and interspecies communication, such as the S-adenosylmethionine synthetases gene known to facilitate production of autoinducer-2 (Breitbart et al., 2008).

Such metagenomic data substantiate "common knowledge" about complex ecosystems, but also can provide surprising evidence for microbial processes that were not previously considered. For example, in a metagenomic analysis of a Guerrero Negro microbial mat, genes involved in photosynthesis were unsurprisingly heavily represented in the surface 2 mm (Kunin et al., 2008). However, this study showed an increase in the relative abundance of genes encoding for enzymes involved in sugar degradation with depth. Such data are at the first stage of analysis. By coupling the specific ecotypes to the expression of functional genes within the community (i.e., metatranscriptomics), isotopic signatures and geochemical cycling, we can potentially delineate the evolutionary adaptations of stromatolitic microbial mat community.

With the emergence of comparative metagenomics, a revolution of the species concept in microbes is underway (Ward et al., 2008). Organisms that share high levels of sequence similarity in 16S rRNA genes can have extensive differences in genome size and organization. One of the most compelling examples is of cyanobacteria in the thermophilic hot springs mats of Yellowstone. The complete genomes of closely related species of the genus *Synechococcus* (OS-A and OS-B', sharing >95% sequence similarity in the 16S rRNA gene) were recently analyzed and found to share little gene synteny (Bhaya et al., 2007). Such studies clearly show that despite morphological and 16S rRNA gene similarities, functional metagenomic data not only can identify these ecotypes, but also ascertain their physiological capacities and ecological significance within the community.

Such disparity within the genomes of organisms that share similar 16S rRNA gene sequences could be the result of genetic exchange with other cells and viral phages. Viruses are the most abundant biological entities in aquatic ecosystems (Suttle, 2005) and often outnumber prokaryotes 10:1 (Desnues et al., 2008). The role of phages in the marine stromatolites of Highborne Cay, Bahamas was recently examined. The vast majority of recovered phage sequences (>95%), acquired via metagenomic sequencing, were novel, suggesting biogeographical isolation of the Bahamian stromatolite community (Desnues et al., 2008). Phages, along with geochemical constraints on the microbial community, are likely driving forces for the evolution of new species within the stromatolitic microbial mat communities. Only through multi-lateral approaches of metagenomic sequencing coupled with biogeochemical and mineralogical analyses will the comprehensive microbial diversity of stromatolite ecosystems be delineated and understood.

5. References

Arp, G., Reimer, A. and Reitner, J. (1999) Calcification in cyanobacterial biofilms of alkaline salt lakes. Eur. J. Phycol. **34**: 393–403.

Arp, G., Reimer, A. and Reitner, J. (2001) Photosynthesis-induced biofilm calcification and calcium concentrations in Phanerozoic oceans. Science **292**: 1701–1704.

Awramik, S.A. (1984) Ancient stromatolites and microbial mats, In: Y. Cohen, R.W. Castenholz and H.O. Halvorson (eds.) *Microbial Mats: Stromatolites 3*. Alan R Liss, New York, pp. 1–22.

Baumgartner, L.K., Reid, R.P., Dupraz, C., Decho, A.W., Buckley, D.H., Spear, J.R., Przekop, K.M. and Visscher, P.T. (2006) Sulfate reducing bacteria in microbial mats: changing paradigms, new discoveries. Sediment. Geol. **185**: 131–145.

Baumgartner, L.K., Spear, J.R., Buckley, D.H., Pace, N.R., Reid, R.P., Dupraz, C. and Visscher, P.T. (2009) Microbial diversity in modern marine stromatolites, Highborne Cay, Bahamas. Environ. Microbiol. **11**: 2710–2719.

Bebout, B.M., Hoehler, T.M., Thamdrup, B., Albert, D., Carpenter, S.P., Hogan, M., Turk, K.A. and Des Marais, D.J. (2004) Methane production by microbial mats under low sulphate concentrations. Geobiology **2**: 87–96.

Bhaya, D., Grossman, A.R., Steunou, A.S., Khuri, N., Cohan, F.M., Hamamura, N., Melendrez, M.C., Bateson, M.M., Ward, D.M. and Heidelberg, J.F. (2007) Population level functional diversity in a microbial community revealed by comparative genomic and metagenomic analyses. ISME J. **1**: 703–713.

Breitbart, M., Hoare, A., Nitti, A., Siefert, J., Haynes, M., Dinsdale, E., Edwards, R., Souza, V., Rohwer, F. and Hollander, D. (2008) Metagenomic and stable isotopic analyses of modern freshwater microbialites in Cuatro Ciénegas, Mexico. Environ. Microbiol. **11**: 16–34.

Burne, R.V. and Moore, L.S. (1987) Microbialites: organosedimentary deposits of benthic microbial communities. Palaios **2**: 241–254.

Burns, B.P., Goh, F., Allen, M. and Neilan, B.A. (2004) Microbial diversity of extant stromatolites in the hypersaline marine environment of Shark Bay, Australia. Environ. Microbiol. **6**: 1096–1101.

Burns, B.P., Seifert, A., Goh, F., Pomati, F., Jungblut, A.D., Serhat, A. and Neilan, B.A. (2005) Genetic potential for secondary metabolite production in stromatolite communities. FEMS Microbiol. Lett. **243**: 293–301.

Byerly, G.R., Lowe, L.S. and Walsh, M.M. (1986) Stromatolites from 3300–3500 Myr Swaziland Supergroup, Barbeton Mountain Land, South Africa. Nature **319**: 489–491.

Decho, A.W., Visscher, P.T. and Reid, R.P. (2005) Production and cycling of natural microbial exopolymers (EPS) within a marine stromatolites. Palaeogeogr. Palaeoclimatol. Palaeoecol. **219**: 71–86.

DeLong, E.F. (1992) Archaea in coastal marine environments. Proc. Natl. Acad. Sci. U.S.A. **89**: 5685–5689.

DeSantis, T.Z., Hugenholtz, P., Larsen, N., Rojas, M., Brodie, E.L., Keller, K., Huber, T., Dalevi, D., Hu, P. and Andersen, G.L. (2006a) Greengenes, a chimera-checked 16S rRNA gene database and workbench compatible with ARB. Appl. Environ. Microbiol. **72**: 5069–5072.

DeSantis Jr., T.Z., Hugenholtz, P., Keller, K., Brodie, E.L., Larsen, N., Piceno, Y.M., Phan, R. and Andersen, G.L. (2006b) NAST: a multiple sequence alignment server for comparative analysis of 16S rRNA genes. Nucleic Acids Res. **34**: W394–399.

Desnues, C.G., Rodriguez-Brito, B., Rayhawk, S., Kelley, S., Tran, T., Haynes, M., Lui, H., Hall, D., Angly, F.E., Edwards, R.A., Thurber, R.V., Reid, R.P., Siefert, J., Souza, V., Valentine, D., Swan, B., Breitbart, M. and Rohwer, F. (2008) Biodiversity and biogeography of phages in modern stromatolites and thrombolites. Nature **452**: 340–345.

Dill, R.F., Shinn, E.A., Jones, A.T., Kelly, K. and Steinen, R.P. (1986) Giant subtidal stromatolites forming in normal salinity water. Nature **324**: 55–58.

Dravis, J.J. (1983) Hardened subtidal stromatolites, Bahamas. Science **219**: 385–386.

Dupraz, C. and Visscher, P.T. (2005) Microbial lithification in marine stromatolites and hypersaline mats. Trends Microbiol. **13**: 429–438.

Dupraz, C., Visscher, P.T., Baumgartner, L.K. and Reid, R.P. (2004) Microbe-mineral interactions: early carbonate precipitation in a hypersaline lake (Eleuthera Island, Bahamas). Sedimentology **51**: 745–765.

Dupraz, C., Pattisina, R. and Verrecchia, E.P. (2006) Translation of energy into morphology: simulation of stromatolite morphospace using a stochastic model. Sediment. Geol. **185**: 185–203.

Edwards, R.A., Rodriguez-Brito, B., Wegley, L., Haynes, M., Breitbart, M., Peterson, D.M., Saar, M.O., Alexander, S., Alexander, E.C. and Rohwer, F. (2006) Using pyrosequencing to shed light on deep mine microbial ecology. BMC Genomics **7**: 57.

Elser, J.J., Schampel, J.H., Garcia-Pichel, F., Wade, B.D., Souza, V., Eguiarte, L.E., Escalante, A. and Farmer, J.D. (2005a) Effects of phosphorus enrichment and grazing snails on modern stromatolitic microbial communities. Freshw. Biol. **50**: 1808–1825.

Elser, J.J., Schampel, J.H., Kyle, M., Watts, J., Carson, E.W., Dowling, T.E., Tang, C. and Roopnarine, P.D. (2005b) Response of grazing snails to phosphorus enrichment of modern stromatolitic microbial communities. Freshw. Biol. **50**: 1826–1835.

Fenchel, T. and Kühl, M. (2000) Artificial cyanobacterial mats: growth, structure, and vertical zonation patterns. Microb. Ecol. **40**: 85–93.

Foster, J.S. and Mobberley, J.M. (2010) Past, present, and future: microbial mats as models for astrobiological research. In: J. Seckbach and A. Oren (eds.) *Cellular Origin, Life in Extreme Habitats and Astrobiology: Microbial Mats: Modern and Ancient Microorganisms in Stratified Systems*. Springer, pp. 563–582.

Foster, J.S., Green, S.J., Ahrendt, S.R., Hetherington, K.L., Golubic, S., Reid, R.P. and Bebout, L. (2009) Molecular and morphological characterization of cyanobacterial diversity in the marine stromatolites of Highborne Cay, Bahamas. ISME J. **3**: 573–587.

Goh, F., Allen, M.A., Leuko, S., Kawaguchi, T., Decho, A.W., Burns, B.P. and Neilan, B.A. (2009) Determining the specific microbial populations and their spatial distribution within the stromatolite ecosystem of Shark Bay. ISME J. **3**: 383–396.

Golubic, S. and Browne, K.M. (1996) *Schizothrix gebeleinii* sp. nova builds subtidal stromatolites, Lee Stocking Island. Algol. Stud. **83**: 273–290.

Green, S.J., Blackford, C., Bucki, P. and Jahnke, L.L. (2008) A salinity and sulfate manipulation of hypersaline microbial mats reveals stasis in the cyanobacterial community structure. ISME J. **2**: 457–470.

Grotzinger, J.P. and Knoll, A.H. (1999) Stromatolites in Precambrian carbonates: evolutionary mileposts or environmental dipsticks? Annu. Rev. Earth Planet. Sci. **27**: 313–358.

Handelsman, J. (2004) Metagenomics: application of genomics to uncultured microorganisms. Microbiol. Mol. Biol. Rev. **68**: 669–685.

Havemann, S.A. and Foster, J.S. (2008) Comparative characterization of the microbial diversities of an artificial microbialite model and a natural stromatolite. Appl. Environ. Microbiol. **74**: 7410–7421.

Hiraishi, A. and Ueda, Y. (1994) Intrageneric structure of the genus *Rhodobacter*: transfer of *Rhodobacter suljidophilus* and related marine species to the genus *Rhodovulum* gen. nov. Int. J. Syst. Bacteriol. **44**: 15–23.

Hughes, J.B., Hellmann, J.J., Ricketts, T.H. and Bohannan, B.J. (2001) Counting the uncountable: statistical approaches to estimating microbial diversity. Appl. Environ. Microbiol. **67**: 4399–4406.

Kasting, J.F. (2001) Earth history. The rise of atmospheric oxygen. Science **293**: 819–820.

Knauth, L.P. (1998) Salinity history of the Earth's early ocean. Nature **395**: 554.

Kumar, S., Nei, M., Dudley, J. and Tamura, K. (2008) MEGA: a biologist-centric software for evolutionary analysis of DNA and protein sequences. Brief. Bioinform. **9**: 299–306.

Kunin, V., Raes, J., Harris, J.K., Spear, J.R., Walker, J.J., Ivanova, N., von Mering, C., Bebout, B.M., Pace, N.R., Bork, P. and Hugenholtz, P. (2008) Millimeter-scale genetic gradients and community-level molecular convergence in a hypersaline microbial mat. Mol. Syst. Biol. **4**: 198.

Lane, D.J. (1991) 16S/23S rRNA sequencing, In: E. Stackebrandt and M. Goodfellow (eds.) *Nucleic Acid Techniques in Bacterial Systematics*. Wiley, Chichester, pp. 115–175.

Ley, R.E., Harris, J.K., Wilcox, J., Spear, J.R., Miller, S.R., Bebout, B.M., Maresca, J.A., Bryant, D.A., Sogin, M.L. and Pace, N.R. (2006) Unexpected diversity and complexity of the Guerrero Negro hypersaline microbial mat. Appl. Environ. Microbiol. **72**: 3685–3695.

Lozupone, C.A. and Knight, R. (2007) Global patterns in bacterial diversity. Proc. Natl. Acad. Sci. U.S.A. **104**: 11436–11440.

Ludwig, W., Strunk, O., Westram, R., Richter, L., Meier, H., Yadhukumar, Buchner, A., Lai, T., Steppi, S., Jobb, G., Forster, W., Brettske, I., Gerber, S., Ginhart, A.W., Gross, O., Grumann, S., Hermann, S., Jost, R., Konig, A., Liss, T., Lussmann, R., May, M., Nonhoff, B., Reichel, B.,

Strehlow, R., Stamatakis, A., Stuckmann, N., Vilbig, A., Lenke, M., Ludwig, T., Bode, A. and Schleifer, K.H. (2004) ARB: a software environment for sequence data. Nucleic Acids Res. **32**: 1363–1371.

Minckley, W. (1969) *Environments of the Bolson of Cuatro Cienegas, Cuahuila, Mexico, with Special Reference to the Aquatic Biota. University of Texas El Paso Science Series 2*, Texas Western Press, El Paso, TX, pp. 1–65.

Minz, D., Fishbain, S., Green, S.J., Muyzer, G., Cohen, Y., Rittmann, B.E. and Stahl, D.A. (1999) Unexpected population distribution in a microbial mat community: sulfate-reducing bacteria localized to the highly oxic chemocline in contrast to a eukaryotic preference for anoxia. Appl. Environ. Microbiol. **65**: 4659–4665.

Monty, C. (1977) Evolving concepts on the nature and the ecological significance of stromatolites, In: E. Flügel (ed.) *Fossil Algae, Recent Results and Developments*. Springer-Verlag, Berlin, pp. 15–35.

Neilan, B.A., Burns, B.P., Relman, D.A. and Lowe, D.R. (2002) Molecular identification of cyanobacteria associated with stromatolites from distinct geographical locations. Astrobiology **2**: 271–280.

Nogales, B., Moore, E.R., Llobet-Brossa, E., Rossello-Mora, R., Amann, R. and Timmis, K.N. (2001) Combined use of 16S ribosomal DNA and 16S rRNA to study the bacterial community of polychlorinated biphenyl-polluted soil. Appl. Environ. Microbiol. **67**: 1874–1884.

Omoregie, E.O., Crumbliss, L.L., Bebout, B.M. and Zehr, J.P. (2004) Determination of nitrogen-fixing phylotypes in *Lyngbya* sp. and *Microcoleus chthonoplastes* cyanobacterial mats from Guerrero Negro, Baja California, Mexico. Appl. Environ. Microbiol. **70**: 2119–2128.

Ordóñez, S., González-Martin, J.A., García del Cura, M.Á. and Pedley, H.M. (2005) Temperate and semi-arid tufas in the Pleistocene to recent fluvial barrage system in the Mediterranean area: the Ruidera Lakes National Park (Central Spain). Geomorphology **69**: 332–350.

Overbeek, R., Begley, T., Butler, R.M., Choudhuri, J.V., Chuang, H.Y., Cohoon, M., de Crecy-Lagard, V., Diaz, N., Disz, T., Edwards, R., Fonstein, M., Frank, E.D., Gerdes, S., Glass, E.M., Goesmann, A., Hanson, A., Iwata-Reuyl, D., Jensen, R., Jamshidi, N., Krause, L., Kubal, M., Larsen, N., Linke, B., McHardy, A.C., Meyer, F., Neuweger, H., Olsen, G., Olson, R., Osterman, A., Portnoy, V., Pusch, G.D., Rodionov, D.A., Ruckert, C., Steiner, J., Stevens, R., Thiele, I., Vassieva, O., Ye, Y., Zagnitko, O. and Vonstein, V. (2005) The subsystems approach to genome annotation and its use in the project to annotate 1000 genomes. Nucleic Acids Res. **33**: 5691–5702.

Paerl, H.W., Steppe, T.F. and Reid, R.P. (2001) Bacterially mediated precipitation in marine stromatolites. Environ. Microbiol. **3**: 123–130.

Papineau, D., Walker, J.J., Mojzsis, S.J. and Pace, N.R. (2005) Composition and structure of microbial communities from stromatolites of Hamelin Pool in Shark Bay, Western Australia. Appl. Environ. Microbiol. **71**: 4822–4832.

Pedley, M. (2000) Ambient temperature freshwater microbial tufas, In: R. Riding and S.M. Awramik (eds.) *Microbial Sediments*. Springer, Heidelberg, pp. 179–186.

Perkins, R., Kromkamp, J.C. and Reid, R.P. (2007) Importance of light and oxygen for photochemical reactivation in photosynthetic stromatolite communities after natural sand burial. Mar. Ecol. Prog. Ser. **349**: 23–32.

Pinckney, J.L. and Reid, R.P. (1997) Productivity and community composition of stromatolitic microbial mats in the Exuma Cays, Bahamas. Facies **36**: 204–207.

Playford, P.E. and Cockbain, A.E. (1976) Modern algal stromatolites at Hamelin Pool, a hypersaline barred basin in Shark Bay, Western Australia, In: M.R. Walter (eds.) *Stromatolites. Developments in Sedimentology 20*. Elsevier, Amsterdam, pp. 389–411.

Reid, R.P., Macintyre, I.G., Browne, K.M., Steneck, R.S. and Miller, T. (1995) Modern marine stromatolites in the Exuma Cays, Bahamas – uncommonly common. Facies **33**: 1–17.

Reid, R.P., Visscher, P.T., Decho, A.W., Stolz, J.F., Bebout, B.M., Dupraz, C., Macintyre, I.G., Paerl, H.W., Pinckney, J.L., Prufert-Bebout, L., Steppe, T.F. and DesMarais, D.J. (2000) The role of microbes in accretion, lamination and early lithification of modern marine stromatolites. Nature **406**: 989–992.

Reid, R.P., James, N.P., Macintyre, I.G. and Dupraz, C.P. (2003) Shark Bay stromatolites: microfabrics and reinterpretation of origins. Facies **49**: 299–324.

Riding, R. (1991) Microbial carbonates: the geological record of calcified bacterial-algal mats and biofilms. Sedimentology **47**: 179–214.

Santos, F., Peña, A., Nogales, B., Soria, E., Garcia del Cura, M.Á., González-Martin, J.A. and Antón, J. (2010) Bacterial diversity in modern freshwater stromatolites from Ruidera Pools Natural Park, Spain. System. Appl. Microbiol. **33**: 209–221.

Schloss, P.D. and Handelsman, J. (2005) Introducing DOTUR, a computer program for defining operational taxonomic units and estimating species richness. Appl. Environ. Microbiol. **71**: 1501–1506.

Sørensen, K.B., Canfield, D.E., Teske, A.P. and Oren, A. (2005) Community composition of a hypersaline endoevaporitic microbial mat. Appl. Environ. Microbiol. **71**: 7352–7365.

Souza, V., Espinosa-Asuar, L., Escalante, A.E., Equiarte, L.E., Farmer, J., Forney, L., Lloret, L., Rodriguez-Martinez, J.M., Soveron, X., Dirzo, R. and Elser, J.J. (2006) An endangered oasis of aquatic microbial biodiversity in the Chihuahuan desert. Proc. Natl. Acad. Sci. U.S.A. **103**: 6565–6570.

Stal, L.J. (1995) Physiological ecology of cyanobacteria in microbial mats and other communities. New Phytol. **131**: 1–32.

Steppe, T.F., Pinckney, J.L., Dyble, J. and Paerl, H.W. (2001) Diazotrophy in modern marine Bahamian stromatolites. Microb. Ecol. **41**: 36–44.

Stolz, J.F., Feinstein, T.N., Salsi, J., Visscher, P.T. and Reid, R.P. (2001) TEM analysis of microbial mediated sedimentation and lithification in modern marine stromatolites. Am. Mineral. **86**: 826–833.

Suttle, C.A. (2005) Viruses in the sea. Nature **437**: 356–361.

Visscher, P.T. and Stolz, J.F. (2005) Microbial mats as bioreactors: populations, processes and products. Palaeogeogr. Palaeoclimatol. Palaeoecol. **219**: 87–100.

Ward, D.M., Ferris, M.J., Nold, S.C. and Bateson, M.M. (1998) A natural view of microbial biodiversity within hot spring cyanobacterial mat communities. Microbiol. Mol. Biol. Rev. **62**: 1353–1370.

Ward, D.M., Cohan, F.M., Bhaya, D., Heidelberg, J.F., Kühl, M. and Grossman, A. (2008) Genomics, environmental genomics and the issue of microbial species. Heredity **100**: 207–219.

Yamada, T., Sekiguchi, Y., Hanada, S., Imachi, H., Ohashi, A., Harada, H. and Kamagata, Y. (2006) *Anaerolinea thermolimosa* sp. nov., *Levilinea saccharolytica* gen. nov., sp. nov. and *Leptolinea tardivitalis* gen. nov., sp. nov., novel filamentous anaerobes, and description of the new classes *Anaerolineae* classis nov. and *Caldilineae* classis nov. in the bacterial phylum *Chloroflexi*. Int. J. Syst. Evol. Microbiol. **56**: 1331–1340.

Yurkov, V.V. and Beatty, J.T. (1998) Aerobic anoxygenic phototrophic bacteria. Microbiol. Mol. Biol. Rev. **62**: 695–724.

Biodata of **R. Pamela Reid, A. Patricia Lopes Gaspar, Emily M. Bowlin, Lilian Custals**, and **Dr. Miriam S. Andres**, authors of ***"Microbialites and Sediments: A 2-Year Record of Burial and Exposure of Stromatolites and Thrombolites at Highborne Cay Bahamas"***

Dr. R. Pamela Reid is Associate Professor in the Division of Marine Geology and Geophysics at the Rosenstiel School of Marine and Atmospheric Science, University of Miami, USA. She obtained her Ph.D. from the University of Miami in 1985 and continued her research at the Smithsonian Institution. Her scientific interests are carbonate sedimentation and diagenesis, geomicrobiology and biomineralization, biogenesis of stromatolites, and shallow-water alteration of carbonate skeletons.

E-mail: **preid@rsmas.miami.edu**

A. Patricia Lopes Gaspar is in the maritime cartography division of CARIS, Eindhoven Area, Netherlands, where she works as a cartographic and GIS specialist. She obtained her Masters degree in Marine Affairs from the University of Miami in 2007, with an internship involving sedimentation on a microbial reef.

E-mail: **pgaspar@caris.nl**

R. Pamela Reid

A. Patricia Lopes Gaspar

V.C. Tewari and J. Seckbach (eds.), *STROMATOLITES: Interaction of Microbes with Sediments*, Cellular Origin, Life in Extreme Habitats and Astrobiology 18, 407–425
DOI 10.1007/978-94-007-0397-1_18, © Springer Science+Business Media B.V. 2011

Emily M. Bowlin is a graduate student in the final stages of her Ph.D. in the division of Marine Geology and Geophysics at the Rosenstiel School of Marine and Atmospheric Science, University of Miami, USA. She obtained her BS from the University of Miami in 2003. Her scientific interests are carbonate sedimentology, geobiology, and stromatolite morphogenesis.

E-mail: **ebowlin@rsmas.miami.edu**

Lillian Custals is currently in the Divisions of Marine and Atmospheric Chemistry at the Rosenstiel School of Marine and Atmospheric Science, University of Miami, USA. She obtained her BS from the Universidad Nacional Pedro Henriquez Urena, Dominican Republic in 1981. Her scientific interests are transport, size distribution, and chemical composition of Aerosols and Mineral Dust in the atmosphere. She became involved in statistical analysis and data processing of the environmental data for this study.

E-mail: **lcustals@rsmas.miami.edu**

Emily M. Bowlin

Lillian Custals

Dr. Miriam S. Andres is currently a research scientist with the Carbonate Research and Development team of Chevron Energy Technology Company in San Ramon, California. She obtained her Ph.D. from the Swiss Federal Institute of Technology (ETH) – Zurich in 2002 and continued her research at the University of Miami. Her scientific interests include the evolution of microbialites through time and space. More specifically, the biogeochemistry of microbe–mineral interactions, isotopic fingerprinting of microbial pathways, as well as the role of biofilms in lithification processes.

E-mail: **miriam.andres@chevron.com**

MICROBIALITES AND SEDIMENTS: A 2-YEAR RECORD OF BURIAL AND EXPOSURE OF STROMATOLITES AND THROMBOLITES AT HIGHBORNE CAY BAHAMAS

R. PAMELA REID[1], A. PATRICIA LOPES GASPAR[2], EMILY M. BOWLIN[1], LILLIAN CUSTALS[1], AND MIRIAM S. ANDRES[3]
[1]*University of Miami/RSMAS-MGG, 4600 Rickenbacker Causeway, Miami, FL 33149, USA*
[2]*Division of Hydrography, Port of Lisbon Authority, Lisbon, Portugal*
[3]*Chevron Energy Technology Company, San Ramon, CA, USA*

Abstract An intensive 2-year field effort has characterized burial and exposure patterns for stromatolites and thrombolites currently forming in a fringing reef complex at Highborne Cay, Bahamas. Migrating sand waves in the nearshore environment bury and expose the microbialites for periods of days, weeks, or months. Intertidal, nearshore thrombolites and shallow subtidal stromatolites are buried for 30–70% of the year. Duration of exposure intervals is highly variable based on position along the reef tract. These distinctive burial and exposure patterns form a basis for understanding the distribution of microbial communities within the reef complex and the resultant microbialite accretion.

Keywords Stromatolites • Thrombolites • Sediments • Burial • Exposure • Microbial mats • Bahamas • Modern • Marine • Accretion • Bed load transport • Cyanobacteria • Photosynthesis • *Schizothrix* sp. • *Dicothrix* sp.

1. Introduction

Despite the ubiquity of microbialites in the Precambrian world, modern analogs to these ancient structures are found in only a few remote locations (Grotzinger and Knoll, 1999). One such site is Highborne Cay, on the western margin of Exuma Sound, Bahamas. There, both stromatolitic (laminated) and thrombolitic (unlaminated, with clotted structures) microbialites are actively forming in a fringing reef in open marine conditions (Reid et al., 1999; Littler et al., 2005; Myshrall et al., 2010). The proximity of both microbialite types makes the Highborne reef an ideal site for studying contrasting styles of microbial deposition.

As the only known modern analog to Precambrian stromatolites that formed in open marine environments, Exuma stromatolites have been the subject of intense investigation for the past decade (e.g., Desnues et al., 2008; Eckman et al., 2008; Dupraz and Visscher, 2005; Reid et al., 2000; Visscher et al., 2000). Integrated studies relating stromatolite microstructure to microbial populations and processes have provided a model for lamination formation (Reid et al., 2000). The stromatolites are formed by lithifying microbial mats that consist of distinctly layered microbial populations, with each layer representing a community that was, at one time, at the stromatolite surface. Three "mat types" each characterized by a distinct surface community are recognized: a filamentous cyanobacterial community, a biofilm community, and a coccoid cyanobacterial community. Each surface community is associated with the accretion of a distinct type of mineral deposit: trapped and bound sand grains, micritic laminae, and cemented layers of micritized grains. Subsurface lamination thus represents a chronology of former surface mats.

In 2003, the Research Initiative on Bahamian Stromatolites (RIBS) project initiated a comprehensive 5-year program to investigate intrinsic and extrinsic factors causing cycling of the surface microbial communities forming the stromatolites at Highborne Cay, Bahamas. An important component of this study was an intensive field monitoring effort, which included "A Year in the Life of a Stromatolite." The "year" evolved into 2 years of almost continuous monitoring of the field site from Jan 2005 to Dec 2006. During the field campaign, environmental and biological parameters influencing microbialite formation were closely monitored, and morphology and spatial distribution of microbialites were mapped. For a complete listing of papers published to date in the RIBS Contribution Series, see http://www.stromatolites.info.

Previous authors have speculated that growth of modern stromatolites in open ocean conditions requires periodic burial by sand to prevent overgrowth by epilithic algae and grazing/burrowing invertebrates (e.g., Dravis, 1983; Dill et al., 1986; Macintyre et al., 1996). Initial quantitative data on gross sedimentation in the Stocking Island reef complex, at the south end of the Exuma chain, showed that stromatolites were buried to depths of 0.5 m for periods of up to 100 days, limiting recruitment and colonization by macroalgae and corals (Steneck et al., 1998).

A rigorous analysis of the wave and sediment dynamics along the Highborne Cay reef complex suggests that stromatolite growth may be restricted to a relatively narrow range of flow energy and sediment transport conditions (Eckman et al., 2008). Nearshore wave dynamics are tightly linked to local wind conditions, and stromatolites experience gross rates of deposition of suspended sediment of 7–10 $\mathrm{g\,cm^{-2}\,day^{-1}}$ (Eckman et al., 2008). Bedload transport, responsible for the migration of nearshore sand waves, was not described by Eckman et al. (2008) and is the focus of the study reported here.

In the present study, we document burial and exposure of the Highborne Cay reef complex by migrating sand waves. This report represents an initial phase in the compilation of a comprehensive data set pertaining to stromatolite accretion. Forthcoming papers by Bowlin et al. (in preparation) will build on this sediment data to show how burial and exposure determine microbial mat composition, which in turn controls stromatolite accretion.

1.1. STUDY SITE

Highborne Cay (24°43′N; 76°49′W) at the north end of the Exuma chain of islands is approximately 4 km long and 2 km wide (Fig. 1). An unusual microbialite-algal ridge fringing reef complex stretches for 2.5 km along a sandy beach on the eastern shore of the island (Reid et al., 1999; Andres and Reid, 2006). The reef is best developed in the southern 1 km of the beach, where it was subdivided into ten research sites. Sites 1 and 2, in the south, where the beach is the most concave, are 20–50 m wide, with reef width decreasing to <10 m north of Site 2. North of Site 10, reef patches occur only as small isolated patches. Sediments in the swash zone and shallow subtidal portions of the reef at Sites 1–10 are oolitic sands with a median grain size of 175–275 μm; grain size coarsens significantly north of Site 10 (Eckman et al., 2008). Tides are semidiurnal with a maximum range of approximately 1 m.

Three distinct morphological zones are recognized within the reef (Fig. 2). From seaward to shoreward, these zones are described below.

Reef crest. This zone corresponds to the seaward edge of a coralline algal ridge formed by the branching coralline alga, *Neogoniolithon strictum*. Heads of the sediment tolerant *N. strictum*, up to 1 m in diameter, may be exposed at low tide. Coralline algal ridges were unknown in the Bahamas prior to reports of their discovery at Highborne Cay and Stocking Island by Steneck et al., 1997. All ten sites at the Highborne Cay reef, except Site 7, have a distinct reef crest.

Reef flat. This zone corresponds to the flat area of the algal ridge, shoreward of the reef crest. The reef flat is typically colonized by macroalgae, mainly *Chondria littoralis*, *Laurencia obtuse*, and *Batophora occidentalis* (Littler et al., 2005); microbial mats may also be present on the reef flat (Andres and Reid, 2006).

Back reef lagoon. A shallow lagoon extends from the reef flat to the beach. The dominant features in this back reef lagoon are stromatolites and thrombolites (Reid et al., 1999). The stromatolites form columnar heads and ridges in the subtidal areas of the back reef lagoon. The thrombolites form meter-sized platforms in the intertidal zone; thrombolites were further subdivided into nearshore thrombolites and seaward thrombolites, based on the difference in microbial mat communities of these structures (Myshrall et al., 2010).

To conduct an intensive field-monitoring program of the Highborne Cay reef, a small research station was constructed on the island in 2003. The initial plan was to monitor the reef 1 week per month. However, sediment movement along the beach proved far too dynamic for a 1-week per month sampling scheme to be effective; therefore the program evolved to almost continuous monitoring for a 2-year period from Jan 2005 to Dec 2006. Sporadic data sets are available from

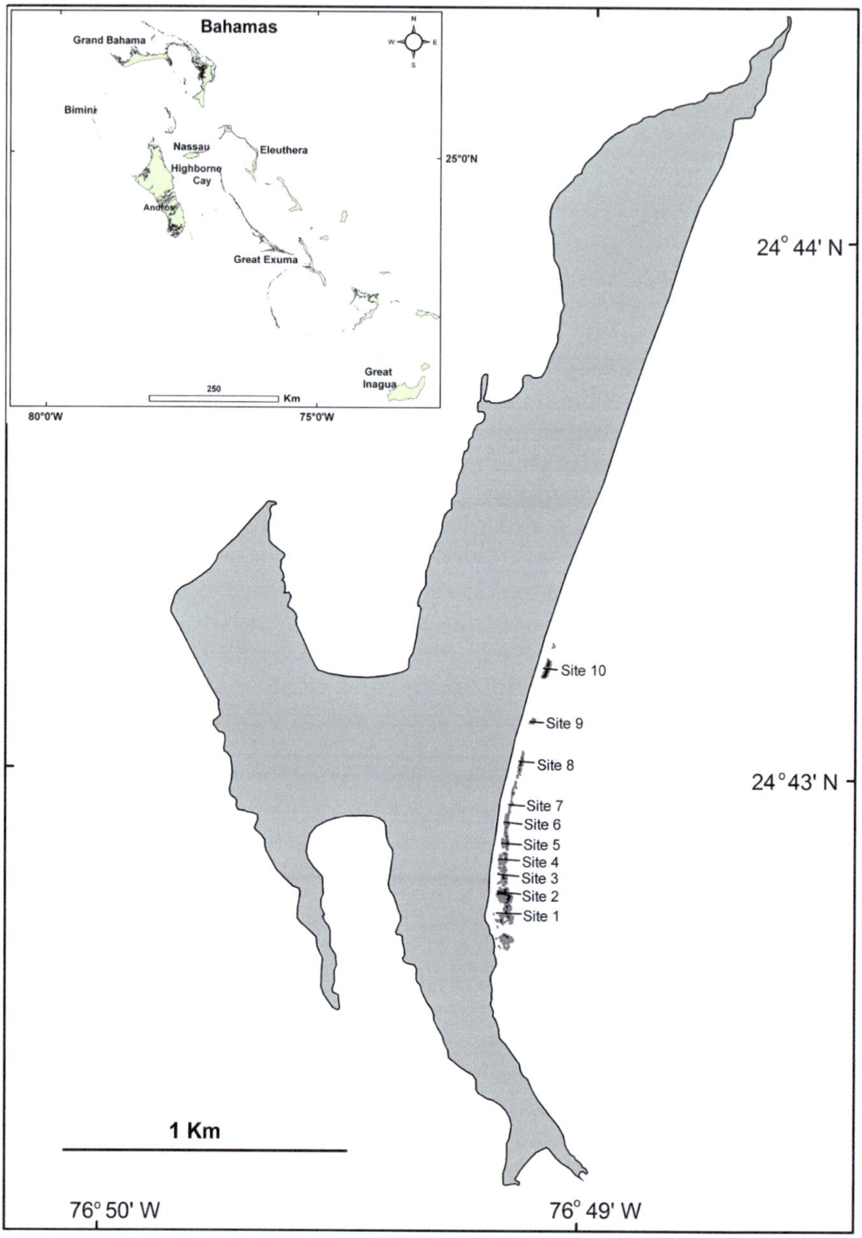

Figure 1. Map showing the location of the ten research sites established along the Highborne Cay stromatolite-algal ridge fringing reef complex. The study area extends from Site 1 in the south to Site 10 in the central area of the eastern shore of Highborne Cay.

Figure 2. Photographs showing the distinct reef zones at Site 4. (**a**) A panoramic view of Site 4 with thrombolites in the foreground and the reef crest in the background. (**b**) Stromatolites in back reef lagoon. (**c**) Reef flat covered with the macroalgae *Laurencia obtuse*. (**d**) Reef crest with heads of the coralline alga, *Neogoniolithon strictum* at the seaward edge of the reef platform.

June 2003 to Dec 2007, with almost continuous records for the period of 2005–2006. Methods and results pertaining to the investigation of burial and exposure patterns within the Highborne Cay reef are presented below.

2. METHODS

2.1. DATA COLLECTION

A total 84 markers (galvanized stakes and nails) were monitored within the ten research sites along the reef track. The stakes and nails were hammered into the top surfaces of thrombolites, stromatolites, reef flat, and reef crest. Visual observations at daily to weekly intervals, describing the stakes and nails as "buried," "exposed," or "sand level", were recorded on field sheets and entered into a database. "Buried" corresponds to greater than 2 cm of sediment covering the monitored surface, based on data showing that the mats are photosynthetically inactive at depths of 3 cm (Perkins et al., 2007). Actual depths of burial, as determined by exposed heights on the stakes or nails, were recorded in the database, but for the purposes of analysis in this study, these records are classified simply as buried. "Exposed" denotes a monitored surface free of sediment accumulation. "Sand level" denotes a surface at the level of the sand, or buried to a depth of less than or equal to 2 cm, and corresponds to a transition between burial and exposure events. Surfaces at sand level may be buried and unburied with shifting tides.

In addition to the visual surveys, photographs were taken at weekly to monthly intervals to document changes in surface mat communities at each stake and to contribute to the overall understanding of sediment movement along the reef complex; an example showing burial and exposure of a stromatolite nail at Site 1 is shown in Fig. 3. A total of 13,000 photographs for 2005 and 2006 are compiled in an image database.

Figure 3. A photographic time series showing different states of burial and exposure at nail N1.2, Site 1 (nail height is 11 cm). The nail was hammered into the top of a columnar stromatolite. (**a**) June 19, 2005, only the head of the nail is exposed, sand load is almost 11 cm. (**b**) October 10, 2005, stromatolite column is totally exposed. (**c**) November 11, 2005, the top surface of the stromatolite column is at sand level. (**d**) March 13, 2006, the stromatolite is buried, nail is partially buried; the stromatolite surface is buried by approximately 3 cm of sand.

2.2. DATA ANALYSIS

2.2.1. Burial and Exposure

2.2.1.1. Sedimentation Patterns for Individual Stakes

Each entry recorded in the sedimentation database was categorized as exposed (0 cm), sand level (0–2 cm), or buried (>2 cm). The time series of sedimentation data for each marker was then compiled into a bar graph that spanned the 2-year monitoring period, with each category represented by a distinctive color block. Missing data intervals were left blank, but subsequently some of the missing intervals were interpolated, with a "best guess" as to burial or exposure state, based on our accumulated knowledge of the dynamics of each site after 5 years of field observations.

The burial and exposure data for all stakes and nails within a single reef zone at each site were used to define average sedimentation patterns of different reef zones at each site, and average duration of exposure and burial events experienced by the microbialites, as presented below.

2.2.1.2. Sedimentation Patterns by Reef Zone

The exposure/burial visual survey data for the 84 markers were grouped by site and classified by reef zone (nearshore thrombolites, seaward thrombolites, stromatolites, reef flat, and reef crest) within each site. Each day for 2005–2006 was classified as either buried or exposed; sand level was grouped with exposed. Days lacking recorded data were handled as follows: if the marker showed the same burial or exposure state on both ends of the data gap, the days within the gap were assigned this same value; if burial state changed during the missing time interval, burial state was changed at the mid-point of the gap. The number of days each marker was exposed or buried in 2005 and 2006 was then calculated as a percent of the 2-year time span. The percentages for all markers at a given site within a specific reef zone were then averaged. The averages of percent of time buried vs. percent of time exposed for each site within each reef zone were graphed as histograms.

2.2.1.3. Duration of Microbialite Exposure and Burial Events

The average burial and exposure data for thrombolites and stromatolites at each site, as calculated in Sect. 2.2.1.3, were classified into event types according to the duration of the burial and exposure periods. Burial or exposure episodes lasting less than 7 days were classified as daily events; burial or exposure episodes of 7–30 days were classified as week-long events, and burial or exposure episodes of >30 days were classified as month-long events. The frequency of each event type was then calculated as a percent of total time exposed or buried, and averaged for all markers from a single site within each zone. These averages were graphed as histograms showing percent of each event type (i.e., month-long, week-long, or daily burial as a percent of total time buried; and month-long, week-long, or daily exposure as a percent of total time exposed) for each site within each of the three zones (nearshore thrombolites, seaward thrombolites, and stromatolites).

3. Results

3.1. SEDIMENTATION PATTERNS AT INDIVIDUAL STAKES

Burial and exposure data for each of the 84 individual stakes and nails for the 2005 and 2006 monitoring period were plotted as "bar" graphs; examples for stromatolite nails N1.3 at Site 1 and N8.3 at Site 8 are shown in Fig. 4. Blue on these figures represents periods of exposure, black represents buried, and red represents burial depths of less than 2 cm, termed "sand level," where the surface of the microbialite is typically buried and unburied with each tidal cycle. Interpolated (i.e., "best guess") intervals of burial and exposure are indicated by horizontal white bars overlying the blue and black; intervals of missing data that we are not comfortable interpolating were left white.

Stromatolites at Sites 1 and 8 experienced very different sedimentation patterns, as shown on Fig. 4. Site 1 stromatolites were typically exposed (blue) for long periods at a time, often several months in duration, with occasional sand level intervals (red). In contrast, Site 8 stromatolites experienced relatively short periods of exposure, typically lasting for days to weeks. Burial episodes at Site 1 are typically week-to-month long events, with burial at Site 8 being day, week, or month long in duration.

3.2. SEDIMENTATION PATTERNS BY REEF ZONE

Histograms in Fig. 5 show sedimentation patterns for each reef zone, by site, for the combined 2005 and 2006 monitoring period. Color codes of Fig. 5 indicate percent of total time buried (black) vs. exposed (blue). Data are averages of stakes

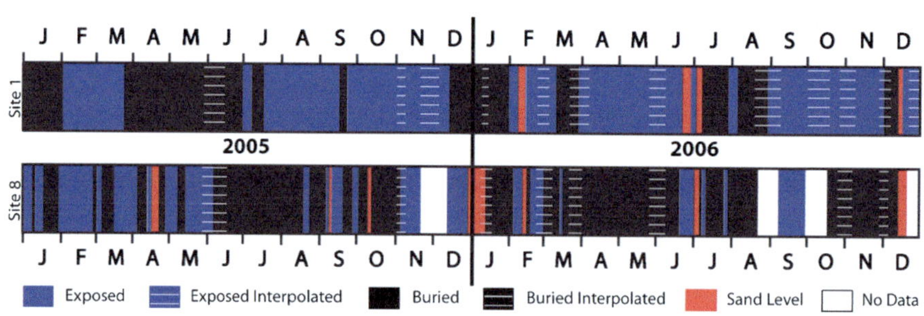

Figure 4. Bar graphs showing burial and exposure data for stromatolite nails N1.3 at Site 1 and N8.3 at Site 8. *Blue* represents periods of exposure, *black* represents buried, and *red* represents burial depths of less than 2 cm, termed "sand level," where the surface of the microbialite is often buried and unburied with each tidal cycle. Interpolated intervals are indicated by *white* horizontal bars; areas of missing data that cannot be interpolated are left *white*. Note that Site 1 stromatolites are typically buried and exposed for periods of 1 month or longer, whereas Site 8 stromatolites are typically exposed for periods of weeks to days.

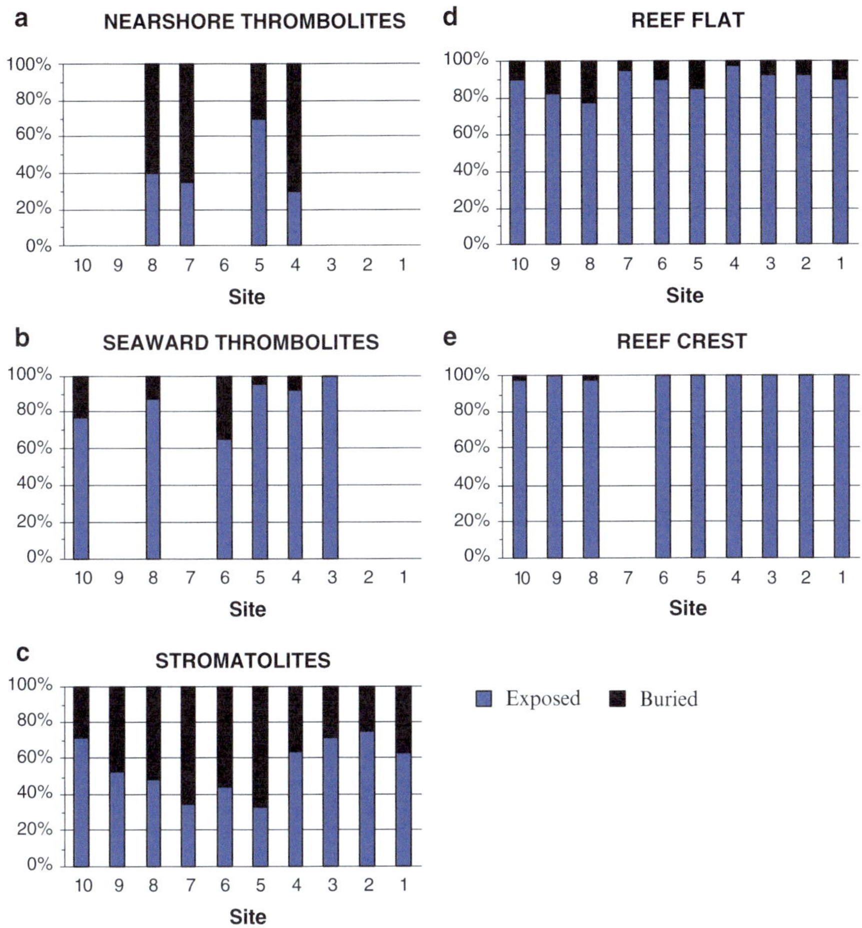

Figure 5. Histograms showing percentage of time the various reef zones at each site were exposed (*blue*) or buried (*black*) for the combined 2005 and 2006 observation period. Nearshore thrombolites (**a**) and stromatolites (**b** and **c**) in the back reef were exposed for much less than the reef flat (**d**) and reef crest (**e**).

from each zone at each site. Results show that the nearshore thrombolites and stromatolites in the back reef were much more buried than the reef flat and reef crest. Nearshore thrombolites (Fig. 5a) and stromatolites (Fig. 5c) were typically buried for 25–70% of the total observation period, contrasting with 5–25% burial of the reef flat (Fig. 5d) and almost continuous exposure of the reef crest zone (Fig. 5e). At sites with both nearshore and seaward thrombolites (Sites 4, 5, and 8), the seaward thrombolites (Fig. 5b) were much less buried (5–15%) than the nearshore equivalents (30–70% buried; Fig. 5a).

Sedimentation within the back reef also shows distinct patterns along the length of the reef. Stromatolites at the south and north ends of the study area,

Sites 1–4 and Site 10, were more exposed (65–75% of the year) than stromatolites in the reef's middle section at Sites 5–9 (30–50% exposed; Fig. 5c).

Results for 2005 and 2006 plotted as separate years show the same overall trends as the summaries for the combined data sets, although some individual zones at specific sites exhibit distinct yearly variability. For example, stromatolites at Site 8, which average 50:50 exposure vs. burial for the combined 2-year record (Fig. 5c), showed 60:40 exposure vs. burial in 2005, and 40:60 exposure vs. burial in 2006. Similarly stromatolites at Site 1, which averaged about 60:40 exposure vs. burial for the 2-year record, showed 50:50 exposure vs. burial in 2005 and 70:30 exposure vs. burial in 2006.

3.3. DURATION OF MICROBIALITE EXPOSURE AND BURIAL EVENTS

Figure 6 shows the duration of exposure and burial events for thrombolites and stromatolites in the back reef zone, defined as month-long (>30 days), week-long (7–30 days), and daily (<7 days) events. Since the reef flat and reef crest were almost continuously exposed, exposure duration is close to 100% month-long events. The following results emerged from analysis of the more complex burial and exposure patterns of the thrombolites and stromatolites.

Nearshore thrombolites were generally exposed for about 30–40% of the year, except for Site 5, which is exposed about 70% of the time (Fig. 5a). Exposure episodes were mainly 7–30 day (week-long) events, except for Site 5, where >30 day (month-long) long exposure was common (Fig. 6). Burial of nearshore thrombolites (60–70% of the year, 30% at Site 5; Fig. 5a) was dominated by month-long events (55–75%; Fig. 6d).

Seaward thrombolites were exposed for 65–95% of the 2-year monitoring record (Fig. 5b). Exposure events are mainly >30 days (month-long; Fig. 6b). The few burial events (5–35% of the year) were mainly 7–30 days (week-long) in duration (Fig. 6e).

As in overall patterns of burial vs. exposure, stromatolites show distinct differences in duration of exposure and burial events along the reef. Exposure ofa stromatolites at Sites 1–4 and Site 10 (65–75% of the year; Fig. 5c) was dominated by month-long episodes, whereas exposure episodes at sites 5–9 (35–50% of the year; Fig. 6b) were week- and month-long, with significant daily exposure events at Sites 5, 7, 8, and 9 (Fig. 6c). In addition, burial events at Sites 1 through 8 were dominantly month-long, whereas burial at Sites 9–10 was mainly week-long (Fig. 6f).

It should be noted that in all cases daily burial and exposure events might be underestimated because the typical time between observation periods was several days to a week. In addition, as in total burial vs. exposure, individual records of duration events in years 2005 and 2006 show overall trends that are

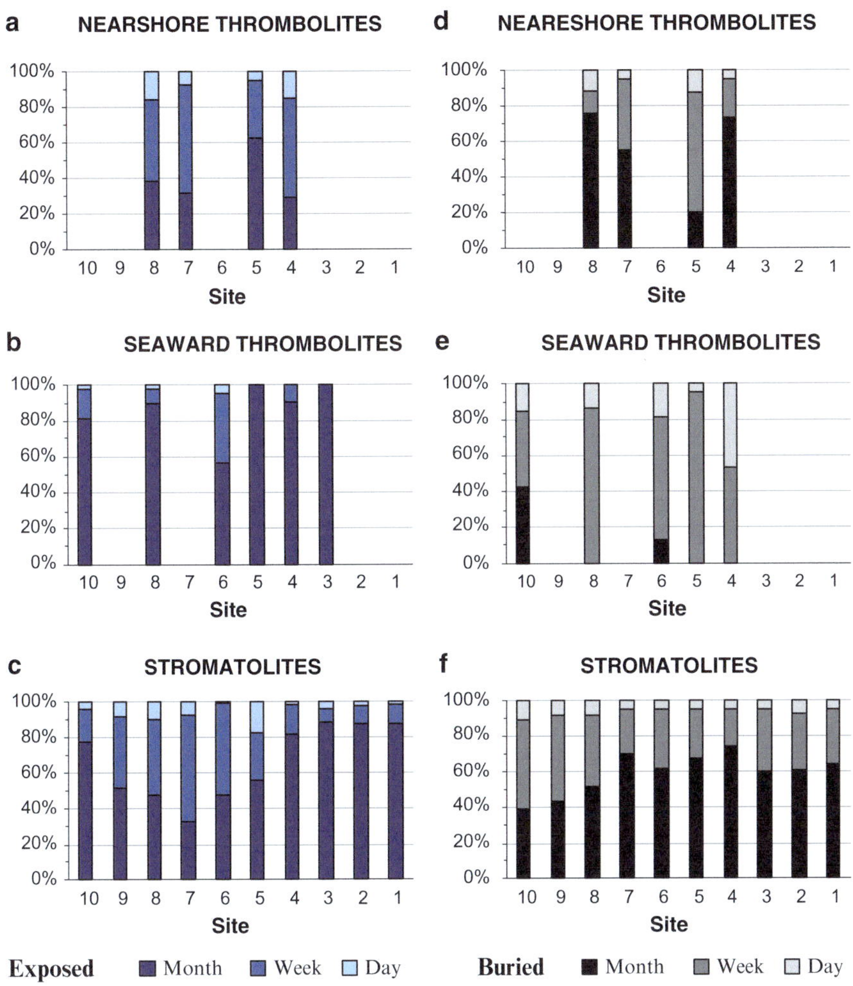

Figure 6. Histograms showing duration of exposure (**a, b, c**) and burial events (**d, e, f**) for thrombolites (**a, b, d, e**) and stromatolites (**c, f**) by site. Events are classified as month-long (>30 days), week-long (7–30 days), and daily (<7 days).

similar to those of the 2-year record, but specific patterns for any individual sites may vary from year to year. For example, for Site 8 stromatolites, approximately 50% of exposure in 2005 plus 2006 consisted of month-long events (Fig. 6c), but month-long events comprise 60% of the 2005 record, and only 30% of the 2006 record. Similarly, month-long burial of Site 1 stromatolites averaged 60% of total burial for the 2-year record (Fig. 6f), but was 85% of total burial in 2005 and 35% in 2006.

4. Discussion

The burial and exposure patterns described earlier reflect migrating sand waves in the nearshore region of the Highborne Cay beach. These sand waves rarely encroach on the reef crest, the highest portion of the reef; sand waves sometimes blanket the reef flat, and they regularly bury and unbury microbialites in the back reef lagoon. Like suspended sediment in Highborne reef (Eckman et al., 2008), bedload transport is linked to local wind conditions, but to date, we have been unable to predict burial and exposure at individual sites, based on the records of wind strength, direction, and duration. Sediment movement along the reef will also be affected by the complicated and variable morphologies of each individual site. Further analysis involving consideration of the physics of bedload transport in areas with high rugosity may help to predict burial and exposure patterns, but is beyond the scope of the present study, which is aimed at understanding affects of burial and exposure on microbialite accretion.

Alternating burial and exposure of the stromatolites and thrombolites in back reef lagoon of the Highborne Cay reef profoundly affects the communities that inhabit this reef zone. Of particular importance for stromatolite biogenesis is the ability of cyanobacteria to survive burial (Kromkamp et al., 2007; Perkins et al., 2007). Studies of cyanobacterial photophysiology using variable fluorescence showed that artificially buried samples of stromatolite cyanobacterial communities inactivated their photosynthetic electron transport when oxygen concentrations in the darkened sample decreased to less than 30% saturation. Post burial, the cyanobacterial community reactivated its photosynthetic activity within 1–2 h, but reactivation required both light and oxygen (Kromkamp et al., 2007). Based on these observations, Kromkamp and coauthors speculated that the redox state of the plastoquinone pool determines inactivation/reactivation process.

Results of photosynthesis in artificially buried stromatolite mats (i.e., Kromkamp et al., 2007) are supported by studies of natural burial processes (Perkins et al., 2007). Prokaroyotic cyanobacterial mats reactivated their quantum efficiency after 7–16 days of burial, although reactivation was slower with longer burial. In contrast, mats with surface communities dominated by diatoms did not have high tolerance for burial, with two out of three samples showing no recovery, and the third sample only low recovery after 7 days of burial (Perkins et al., 2007). Thus burial periods of longer than 1 week are likely to kill diatom communities on stromatolite surfaces, but cyanobacterial communities will be largely unaffected by burial.

In contrast to burial, which kills diatoms, diatom communities flourish on stromatolite surfaces during exposure events of several weeks to months, with different diatoms dominating seasonally (unpublished observations). Thus diatoms can become well established on the surfaces of Site 1 stromatolites, which experience characteristic exposure events with durations of >30 days. In contrast, diatoms are insignificant colonizers of stromatolites at Site 8, where exposure is typically limited to day- or week-long events. The balance of burial and exposure events is thus likely to be an important determinant in the net accretion of modern marine stromatolites.

Similarly, differences in burial patterns of nearshore and seaward thrombolites may account for differences in the microbial communities on the surfaces of these structures. Seaward thrombolites, which experience little to no burial, are colonized by "button mats," dominated by Cyanobacteria (mainly filamentous *Dicthrix* sp.; Planavsky et al., in press) and Alphaproteobacteria (Myshrall et al., 2010). In contrast, nearshore thrombolites, which are typically buried for up to 60–70% of the year, have surfaces colonized by "pink mat," so named because of an abundance of pigmented spore-forming cells that dominate the upper mm of the mat. The pink thrombolitc mats demonstrate a significantly higher level of diversity than the button mats, with the dominant phyla being Alphaproteobacteria and Deltaproteobacteria (Myshrall et al., 2010). As in the stromatolites, these differences in microbial surface mats are likely to influence accretion and subsurface microstructure of the microbialite.

5. Summary and Conclusions

Analysis of burial and exposure data for the Highborne Cay reef complex has identified characteristic sedimentation patterns for the various sites and reef zones of the microbialite-algal ridge reef system. Our findings are as follows:

- Stromatolites and thrombolites in back reef lagoon are frequently buried for periods ranging from days to months. The reef flat is occasionally blanketed by sediment, and the reef crest is almost never buried.
- At sites with both nearshore and seaward thrombolites (Sites 4, 5, 8), nearshore thrombolites were buried more (30–70% burial) than the seaward equivalents (5–15% burial).
- Nearshore thrombolites experienced frequent burial events longer than 30 days (month-long). Exposure of the nearshore thrombolites was mainly for 7–30 days (week-long), except for Site 5, where exposure events were commonly month-long.
- Seaward thrombolites were exposed mainly for periods greater than 30 days (month-long). The few burial events were mainly 7–30 days (week-long) in duration. At Site 10 burial events lasting 7–30 days (week-long) and greater than 30 days (month-long) were equally common.
- Stromatolites at the south and north ends of the reef (Sites 1–4 and Site 10) were more exposed than stromatolites in the reef's middle section (Sites 5–9).
- Exposure of stromatolites at Sites 1–4 and Site 10 (65–75% exposure) was dominated by month-long episodes, whereas exposure at Sites 5–9 (35–50% exposure) was characterized by week- and month-long episodes.
- Stromatolite at Sites 5, 7, 8, and 9 experienced significant daily exposure events.

These burial and exposure patterns form a basis for understanding the distribution of microbial communities within the reef complex, resulting in microbialite accretion. As shown by previous studies, cyanobacterial communities can survive periods of prolonged burial, whereas diatom communities cannot. Diatoms become established during prolonged exposure events. On the basis of the burial and exposure patterns described in this study, it is likely that diatoms play a role in stromatolite accretion at Site 1, but not at Site 8; these predictions will be verified by studies of microbial mats and accretion in forthcoming papers by Bowlin et al. (in preparation). Similarly, increased burial of nearshore thrombolites compared with seaward thrombolites may account for the differences in composition of the surface microbial mats on these structures, with seaward thrombolites dominated by Cyanobacteria and Alphaproteobacteria, and nearshore thrombolites dominated by Alphaproteobacteria and Deltaproteobacteria (Myshrall et al., 2010). In summary, bedload sediment transport in the form of migrating sand waves results in distinct patterns of burial and exposure for the microbialites, which are likely to be important determinant of microbial mat composition, and, in turn microbialite accretion.

6. Acknowledgments

We thank members of the RIBS team, who provided lively discussion and support for all aspects of the monitoring program. The crew of the RV Walton Smith and the management of Highborne Cay Plantations are also gratefully acknowledged for logistical support. Research was supported by National Science Foundation grant EAR-0221796. This is RIBS Contribution Number 56.

7. References

Andres, M.S. and Reid, P.R. (2006) Growth morphologies of modern marine stromatolites: a case study from Highborne Cay, Bahamas. Sediment. Geol. **185**: 319–328.

Bowlin, E.M., Reid, R.P., Gaspar, A.P. and Custals, L. (in preparation) Environmental controls on community cycling in modern marine stromatolites: insights into lamination formation. J. Sediment. Res.

Desnues, C., Rodrigues-Brito, B., Rayhawk, S., Kelley, S., Tran, T., Haynes, M., Liu, H., Furlan, M., Wegley, L., Chau, B., Ruan, Y., Hall, D., Angly, F.E., Edwards, R.A., Li, L., Vega Thurber, R., Reid, R.P., Siefert, J., Souza, V., Valentine, D.L., Swan, B.K., Breitbart, M. and Rohwer, F. (2008) Biodiversity and biogeography of phages in modern stromatolites and thrombolites. Nature **452**: 340–343.

Dill, R.F., Shinn, E.A., Jones, A.T., Kelly, K. and Steinen, R.P. (1986) Giant subtidal stromatolites forming in normal salinity waters. Nature **324**: 55–58.

Dravis, J.J. (1983) Hardened subtidal stromatolites, Bahamas. Science **219**: 385–386.

Dupraz, C. and Visscher, P.T. (2005) Microbial lithification in marine stromatolites and hypersaline mats. Trends Microbiol. **13**: 429–438.

Eckman, J.E., Andres, M.S., Marinelli, R.L., Bowlin, E.M., Reid, R.P., Aspden, R.J. and Paterson, D.M. (2008) Wave and sediment dynamics along a shallow subtidal sandy beach inhabited by modern stromatolites. Geobiology **6**: 21–32.

Grotzinger, J.P. and Knoll, A.H. (1999) Stromatolites in Precambrian carbonates: evolutionary mile-posts or environmental dipsticks? Annu. Rev. Earth Planet. Sci. **27**: 313–358.

Kromkamp, J.C., Perkins, R.G., Kijkman, N., Consalvey, M., Andres, M.S. and Reid, R.P. (2007) Resistance to burial of cyanobacteria in stromatolites. Aquat. Microb. Ecol. **48**: 123–130

Littler, D.S., Littler, M.M., Macintyre, I.G., Bowlin, E.M., Andres, M.S. and Reid, R.P. (2005) Guide to the dominant macroalgae of the stromatolite fringing reef complex, Highborne Cay, Bahamas. Atoll Res. Bull. 532: 67–91.

Macintyre, I.G., Reid, R.P. and Steneck, R.S. (1996) Growth history of stromatolites in a Holocene fringing reef, Stocking Island, Bahamas: Journal of Sedimentary Research, v. 66, p. 231–242.

Myshrall, K.L., Green, S.J., Mobberley, J.M., Havemann, S.A., Visscher, P.T., Reid, R.P. and Foster, J.S. (2010) Biogeochemical cycling and microbial diversity in the modern marine thrombolites of Highborne Cay, Bahamas. Geobiology v. 8, p. 337–354.

Perkins, R.G., Kromkamp, J.C. and Reid, R.P. (2007) Importance of light and oxygen for photo-chemical reactivation in photosynthetic stromatolites communities after natural sand burial. Mar. Ecol. Prog. Ser. **349**: 23–32.

Planavsky, N., Reid, R.P., Andres, M., Visscher, P.T., Myshrall, K.L. and Lyons, T.W. (2009) Formation and diagenesis of modern marine calcified cyanobacteria and implications for carbonate saturation in the ancient ocean. Geobiology v. 7, p. 566–576

Reid, R.P., Macintyre, I.G. and Steneck, R.S. (1999) A microbialites/algal ridge fringing reef complex, Highborne Cay, Bahamas. Atoll Res. Bull. **465**: 1–18.

Reid, R.P., Visscher, P.T., Decho, A.W., Stolz, J.F., Bebout, B.M., Dupraz, C., Macintyre, I.G., Paerl, H.W., Pickney, J.L., Prufert-Bebout, L., Steppe, T.F. and DesMarais, D.J. (2000) The role of microbes in accretion, lamination and early lithification of modern marine stromatolites. Nature **406**: 989–992.

Steneck, R.S., Macintyre, I.G. and Reid, R.P. (1997) Unique algal ridge systems of Exuma Cays, Bahamas. Coral Reefs **16**: 28–37.

Steneck, R.S., Miller, T.E., Reid, R.P. and Macintyre, I.G. (1998) Ecological controls on stromatolite development in a modern reef environment: a test of the ecological refuge paradigm. Carbon. Evap. **13**: 48–65.

Visscher, P.T., Reid, R.P. and Bebout, B.M. (2000) Microscale observations of sulfate reduction: correlation of microbial activity with lithified micritic laminae in modern marine stromatolites. Geology **28**: 919–922.

Biodata of **Dr. Maria Eugenia Farías, Professor Daniel Gustavo Poiré, María Julia Arrouy**, and **Professor Virginia Helena Albarracin**, authors of ***"Modern Stromatolite Ecosystems at Alkaline and Hypersaline High-Altitude Lakes in the Argentinean Puna"***

Dr. Maria Eugenia Farías was born in Córdoba, Argentina, and she is 41 years old and having three kids. She lives at San Miguel de Tucumán, Tucumán. She has got her Honors Degree in Biological Sciences (1992) and her Ph.D. in Microbiology (1996) both, from the National University of Tucumán. She also has a postdoctoral experience at Madrid-España (CSIC-CIB) and came back to Argentina under repatriated scientific CONICET program. At the moment, she is Associate Researcher at the National Research Council of Argentina (CONICET) and works at CCT-PROIMI in Tucumán. She has created and is the group leader of the Laboratory of Microbial Research on Andean Lakes (LIMLA) of the Pilot Plant of Microbial Industrial Processes (PROIMI, http:\\www.limla.com.ar). Her group worked on microbial diversity and survival mechanisms of the microbial communities living at the lakes of the Andean Puna since 2004. She undertook many field excursions to these environments located at the Northwest of Argentina: Laguna Vilama (Jujuy: 4,600 m), Salar de la Laguna Verde (Catamarca: 4,400 m), Laguna Socompa (Salta: 4,000 m) Lluillallaco (Salta: 6,000 m), Tolar Grande (Salta: 3,600 m), Laguna Brava (La Rioja: 4,000 m). She has made more than 30 scientific contributions in peer-reviewed journals and attended many conferences as invited lecturer. She is currently supervising seven Ph.D. thesis. She obtained funding for her research mainly from National Agency of Scientific Promotion, Antorchas Foundation and BBVA Foundation.

E-mail: **mefarias@proimi.org.ar**

V.C. Tewari and J. Seckbach (eds.), *STROMATOLITES: Interaction of Microbes with Sediments*, Cellular Origin, Life in Extreme Habitats and Astrobiology 18, 427–441
DOI 10.1007/978-94-007-0397-1_19, © Springer Science+Business Media B.V. 2011

Daniel Gustavo Poiré was born in La Plata, Argentina, in 1953. He obtained Degree in Geology in 1979 and Ph.D. in Natural Sciences (Geology) in 1987, both from the National University of La Plata. He obtained postdoctoral experience at The University of Liverpool, UK. Now he is Full Professor at the National University of La Plata (UNLP) and Independent Researcher at the National Research Council of Argentina (CONICET) at the Centro de Investigaciones Geológicas (UNLP-CONICET). He has lectured several postgraduate courses on sedimentological significance of trace fossils and stromatolites in Argentinean universities and in the University of Barcelona (Spain).

E-mail: **poiré@cig.museo.unlp.edu.ar**

María Julia Arrouy was born in Azul, Argentina, in 1981. She has got Degree in Geology in 2008 from the National University of La Plata (UNLP). She is a Ph.D. student in the UNLP and receiving fellowship of the National Research Council of Argentina (CONICET) and Cementos Avellaneda SA, at the Centro de Investigaciones Geológicas (UNLP-CONICET). She is also Teaching Assistant of Sedimentology courses at the UNLP.

E-mail: **mja@ol.cavellaneda.com.ar**

Daniel Gustavo Poiré

María Julia Arrouy

Virginia Helena Albarracin was born on August 14, 1978. She is at present an Assistant Researcher of the National Research Council (CONICET) in Argentina, working at the Laboratory of Microbial Research on Andean Lakes (LIMLA) of the Pilot Plant of Microbial Industrial Processes (PROIMI). Her areas of expertise are microbiology, molecular biology, microbial ecology, and biotechnology. She has got Honors Degree in Natural Sciences (2001) and her Ph.D. in Biochemistry (2007), both from the National University of Tucumán. She was awarded on several occasions by the University and the local governments for her **excellent academic records as a talented young woman**. At present, she is also an Assistant Professor in the University of Tucumán and the University of San Pablo-Tucumán. She has acquired professional experience at international scientific centers such as University of Jena, Germany (2004), Center of Marine Biotechnology, University of Maryland, USA (2007–2008), University of Mississippi, USA (2008), University of Puerto Rico, Puerto Rico (2008), Max-Planck Institute of Bioinorganic Chemistry, Mülheim, Germany (2008–2010) and has been awarded DAAD Visiting Professorship (2009) and Fulbright Visiting Scholar (2007). She has also presented many research papers in international conferences and published many papers in important peer-reviewed journals.

E-mail: **viralbarracin@gmail.com**

MODERN STROMATOLITE ECOSYSTEMS AT ALKALINE AND HYPERSALINE HIGH-ALTITUDE LAKES IN THE ARGENTINEAN PUNA

MARIA EUGENIA FARÍAS[1], DANIEL GUSTAVO POIRÉ[2], MARIA JULIA ARROUY[2], AND VIRGINIA HELENA ALBARRACIN[1]

[1]LIMLA, Planta Piloto de Procesos Industriales y Microbiológicos (PROIMI)CCT, CONICET, Av. Belgrano y Pasaje Caseros, 4000 Tucumán, Argentina

[2]Centro de Investigaciones Geológicas (CIG), Universidad Nacional de La Plata-CONICET, Calle 1 n 644, 1900 La Plata, Argentina

Keywords Stromatolites • Argentina • Sokompa Lake • High altitude • Andes • Biodiversity • Extremophilic • Hypersaline • Lake Clifton • Anaerobic autotrophic • Heterotrophic microorganisms

1. Introduction

High-altitude Lakes at the northwest of Argentina in the Puna and Andean regions (HAAL) are considered extreme environments of biotechnological interest (Fig. 1; Seufferheld et al., 2008). The HAAL ecosystems are almost unexplored systems of shallow lakes formed during the tertiary geological period, distributed in the geographical area called the Puna at altitudes from 3,000 to 6,000 m above sea level. Most of these wetlands are completely isolated, experience a wide daily range in temperatures (40°C), are slight saline to hypersaline, and are subject to low phosphate availability and high intensity of solar ultraviolet-B (UV-B) radiation. The HAAL ecosystems are unique not only for their geographical characteristics and broad range of extreme environments but also for their abundant biodiversity. The microbial communities that have evolved within these high-altitude aquatic ecosystems tolerate chemical and physical stresses such as wide fluctuations in daily temperatures, hypersalinity, and variable pH and have proved to be adapted to high levels of UV radiation, a low level of nutrient availability, and high concentrations of heavy metals, especially arsenic (Demergasso et al., 2004; Ferrero et al., 2004; Escudero et al., 2007; Fernandez Zenoff et al., 2006; Zenoff et al., 2006; Dib et al., 2008, 2010a, b; Ordoñez et al., 2009; Flores et al., 2009; Farías et al., 2009). The high altitude and low latitude geographical position of the HAAL expose the indigenous extremophilic communities to high solar irradiance (i.e., 165% higher than at sea level) with instantaneous UV-B flux reaching 17 Wm^{-2}; UV-flux is twice the amount of present-day equatorial Mars, while UV-B is half the amount on Mars. Hence, these environments have been proposed for future studies on astrobiology (Farías et al., 2009).

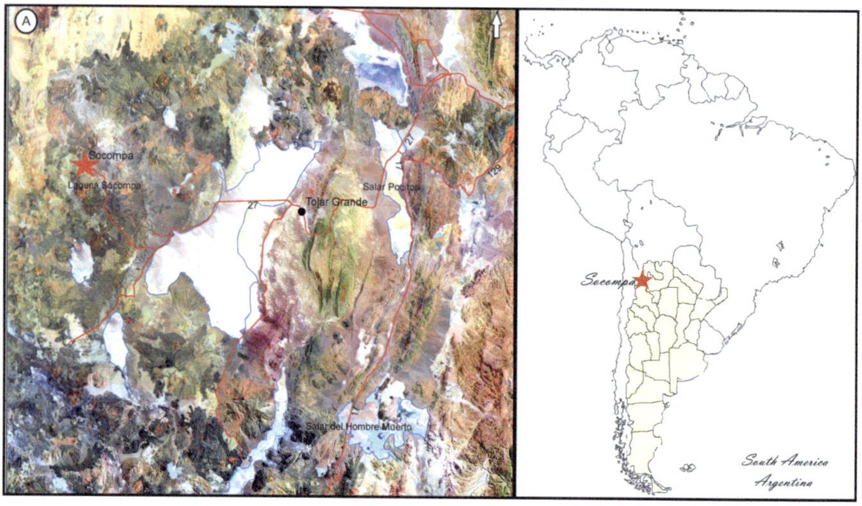

Figure 1. Location map of Socompa Lake.

Figure 2. Exposure conditions of stromatolites from Laguna Socompa during winter (August) (**a**), spring (October) (**b**), and summer (February) (**c**).

So far these outstanding microbial diversity and resistance mechanisms to extreme factors have been described in most Lakes at these locations, i.e., Laguna Azul, Laguna Verde, Laguna Negra, Laguna Vilama, Laguna Aparejos, Laguna Chaxas (Fernandez Zenoff et al., 2006; Zenoff et al., 2006; Dib et al., 2008, 2009, 2010a, b; Ordoñez et al., 2009; Flores et al., 2009; Farías et al., 2009). Nevertheless, there have been no studies or descriptions of these two interesting hypersaline lakes located near the border with Chile: i.e., Laguna Socompa and "Sea Eyes" at Tolar Grande (Fig. 2). Yet, these environments due to their high salinity content (Table 1) present a great potential to harbor modern stromatolite-like ecosystems and deserve further studies.

Stromatolites – internally laminated, macroscopic sedimentary structures, commonly of biological origin – form the dominant part of Earth's early fossil

Table 1. Physicochemical analyses of water and stromatolites from Laguna Socompa.

	Stromatolite	Water
Arsenic (mg/L)	32.1	33.8
Nitrite (mg/L)	0.08	0.08
Nitrate (mg/L)	42.4	58.1
Dissolved sodium (mg/L)	9.59	44.85
Phosphorous (mg/L)	232.8	34.96
Chlorure (mg/L)	16.38	68.03
Nitrogen (mg/L)	23.8	39.34
Total hardness (mg/L)	1,327	
Alkalinity (mg/L)	28.14	
Calcium (mg/L)	161	
Magnesium (mg/L)	64	
Sulfur (mg/L)	0.2	
Organic carbon % (g/100 g)	1.23	
Silicium soluble (mg/L)	450	
pH (upH 1:2)	8.5	9.0
Conductivity (mS/cm)	17.48	50.3
Organic matter % (g/100 g)	18.71	
Sulfate (g/L)	12.93	

Observations: *AA* Atomic absorption spectrometry, *EFM* spectrophotometry, *FLL* flame photometry, *SM* standard methods

record and so provide a potentially important source of information about early life. The oldest examples of these preserved formations are more than three billion years old and are found mainly in Western Australia and South Africa (Lowe, 1980; Walter et al., 1980; Byerly et al., 1986). Recent studies have reflected a widespread and growing acceptance of the oldest stromatolites from the Pilbara region of Western Australia as biogenic (Allwood et al., 2006) . However, stromatolites are shaped by a complex interaction of physical, chemical, and biological processes, and identifying unambiguous signatures of life from the preserved morphology of the structures can be extremely difficult (Poiré et al., 2005; Gaucher and Poiré, 2009). Modern stromatolites pave the way for studying all these process in an "in vivo" approach and, in this sense, are of utmost importance to give insights into the ecology and biogeochemistry of their Precambrian counterparts and to understand how Earth and its early biosphere may have co-evolved (Goh et al., 2009; Baumgartner et al., 2009).

Modern stromatolites have been so far recorded as (1) an hypersaline region of Hamelin Pool, Shark Bay in Western Australia (Goh et al., 2009), (2) shallow subtidal regions at the margin of Exuma Sound in the Bahamas (Foster et al., 2009), (3) freshwater areas at the Cuatro Ciénagas basin in Mexico (Desnues et al., 2008), and (4) Yellowstone Hot Spring (Lau et al., 2005). All of these locations are situated at the sea level where microorganisms cope with little or no stress conditions. In turn, in the desertic region of Salta, Northwestern Argentina, near the border of Chile, we have found characteristic stromatolite-like ecosystems lying and developing in shallow hypersaline lakes located above 4,000 m, under the pressure of harsh conditions, very similar to the ones that were present in the early Earth atmosphere (Belluscio, 2009). The aim of this work is to record the first description of extreme living stromatolites surviving above 4,000 m at Laguna Socompa. The environment, morphological description, and preliminary biological and mineralogical composition of these new modern stromatolites are presented in this study.

2. Study Area

Laguna Socompa (S 24°35′34″ W 68°12′42″) is located at the northern part of Salta Province in Argentina, at the base of the still active Socompa volcano (6,031 m above sea level), near the border with Chile, with incipient access roads, far from any urban population (Fig. 3). In a Tertiary and Cuaternary volcanic and volcaniclastic geological setting, the climate is dominated by arid and desertic conditions with daily temperatures ranging from 20 to −10°C in summer and 10 to −40°C in winter. The lagoon is shallow, the coast depth being not more than 60 cm. This lagoon presents high arsenic content (32 mg/L), a salinity gradient that ranges from 3 mS in the west side to 125 mS in the east side. Stromatolites are found in this area. They are located around the border of the lagoon. During the summer (Fig. 2c), they are partially exposed depending on the tide and hydrological regime, but they are completely submerged during winter and spring seasons (Fig. 2a, b). Hydrothermal water is introduced to Laguna Socompa as part of modern Andean volcanic system.

3. Environmental Physicochemical Conditions

Physicochemical properties of the water as well as the stromatolites are presented in Table 1 for comparison of water and stromatolite composition. As can be observed, high arsenic content and low nitrite and high nitrate content are present both in water and in stromatolites. Dissolved sodium is higher in water compared to stromatolites, phosphorous content is higher in stromatolites, and alkalinity and conductivity are lower in stromatolites; all these results demonstrate that environmental conditions for microbial development is less extreme in stromatolites than in water. As was largely established, microbial mats improve environmental condition for the development of bacteria and algae, which is a successful system for living in extreme environments. Microbial communities of Socompa's stromatolites seem not to be the exception. In addition, stromatolites in Socompa possess a high calcium and magnesium and low sulfur content, which correspond with the mineralogical composition that will be described below; alkalinity (8.5) is also a typical chemical characteristic that favors calcium precipitation. The high content of silica is directly related to the presence of diatom algae.

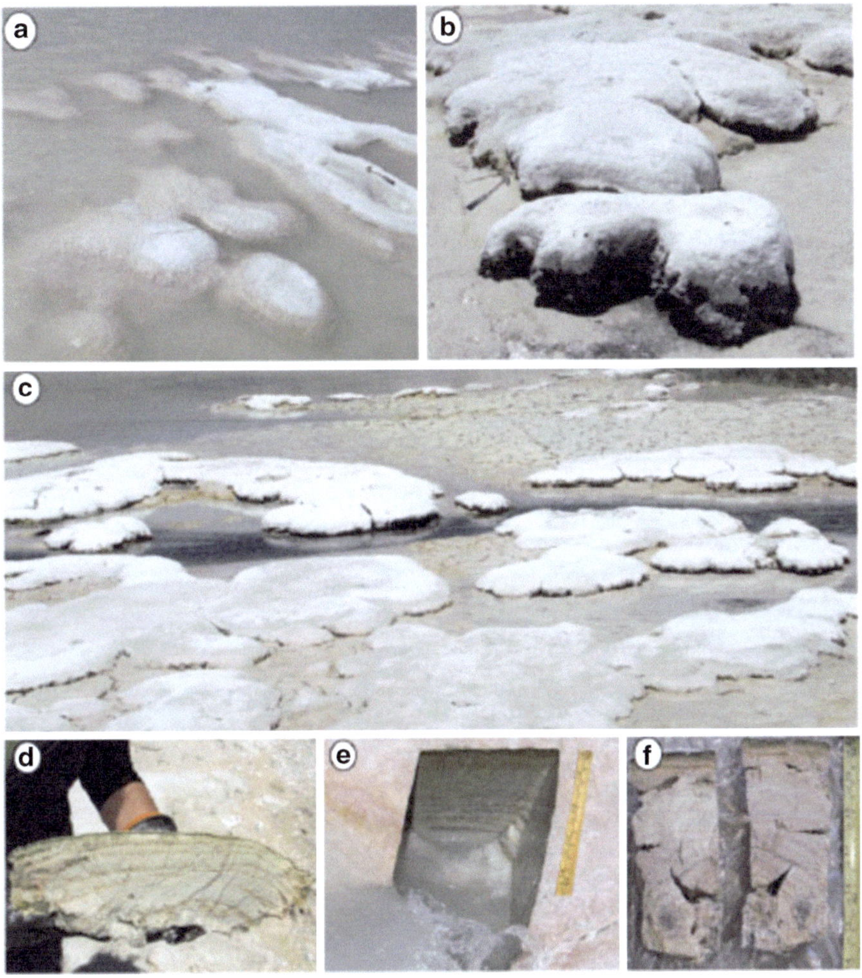

Figure 3. Socompa Lake stromatolites. (**a**) *Rounded* and *elongated*, low-domed bioherms on the lake coast. (**b**) Close-up of the domed bioherms. (**c**) *General view* of the domed bioherms and biostromes formed by coalescence of the bioconstruction. (**d**) *Vertical view* of the gently convex stromatolites. (**e**) Wavy laminar type in a *vertical view*. (**f**) Smooth laminar type and upward variations in the stromatolite laminar profile.

4. Morphological Description of Stromatolites

The stromatolites of the hypersaline–alkaline Socompa Lake are mainly broad, rounded, low-domed, up to 24 cm high, and 80 cm wide bioherms (Fig. 3a, b). However, elongated bioherms with the axe perpendicular to the coastal line are observed (Fig. 3a), while the coalescence of the domes is forming bigger domed biostromes (Fig. 3c). They are simple, unbranched, uniform, crustose, hemispherical, columnar stromatolites with flat to gently convex (3D), wavy (Fig. 3e) and smooth (Fig. 3f) laminar profile and rounded to oblong plant outline. By the way, some vertical section (Fig. 3f) shows variations in the laminar profile from the base to the top of the stromatolite. The lower part is steeply convex, the middle part is gently convex, and the upper one is flat. In the same way, some stromatolites show lateral and vertical lamina-morphology variations representing pseudocolumnar stromatolite (Fig. 4a).

In thin section, two kinds of microstructures have been recognized: film and vermiform microstructures. The more common one is the film microstructure (Fig. 4c) conformed by the alternation of lighter inorganic layer and darker

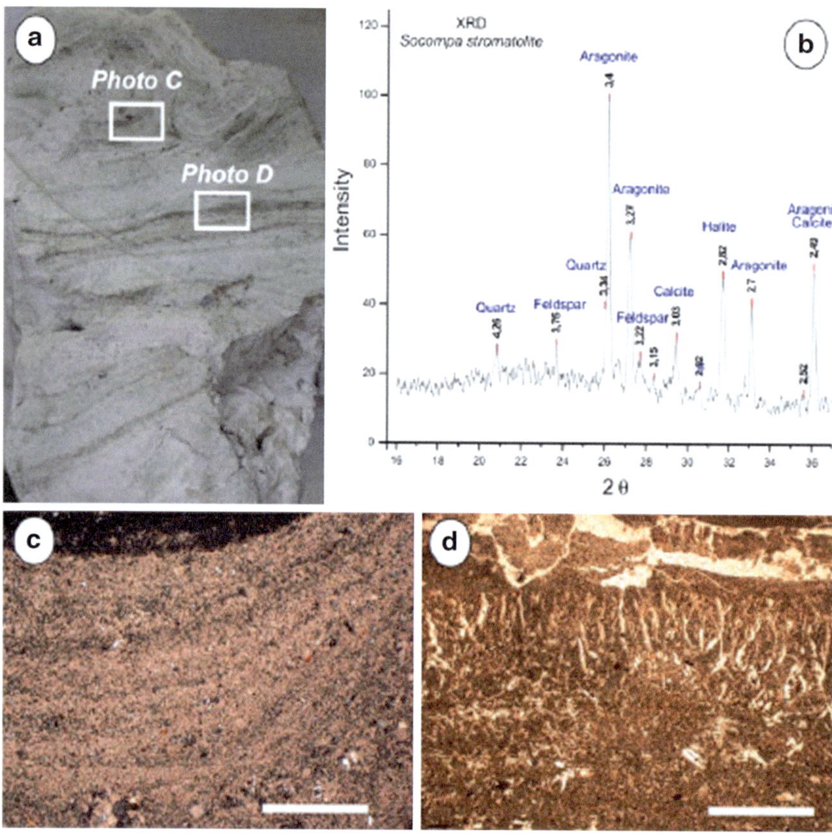

Figure 4. Socompa stromatolite mineral composition. (**a**) *Vertical view* showing the location of the studied thin sections. (**b**) X-ray diffractogram showing very abundant aragonite, with scarce calcite, halite, quartz, and feldspars. (**c**) Thin section photomicrograph showing film microstructure conformed by dark and light micrite laminas (cross-polarized light, *white bar* = 100 mμm). (**d**) Vermiform microstructure (plane polarized light, *white bar* = 100 mμm).

organic layer. The vermiform microstructure is observed in some microcrystalline organic layers characterized by abundant, vertical filament molds (Fig. 4d).

5. Mineral and Chemical Composition of Stromatolites

Mineral composition of the Socompa stromatolites (Fig. 4a) has been studied by both petrography and X-ray diffraction (XRD) analysis, while their texture and microstructure were observed in thin section by a transmitted-light Nikon microscopy. XRD analysis was carried out using a PANalytical X′Pert Pro series diffractometer equipped with a cooper-target tube, operated at 40 kV and 40 mA.

Random orientation XRD analysis indicates that the stromatolites are predominately composed of aragonite with small quantities of calcite, halite, quartz, and feldspars (Fig. 4b). Thin section analysis shows a typical laminar stromatolite microstructure (Bertrand-Sarfaty, 1976) formed by the alternations of light- and dark-colored laminas (Fig. 4c). The light micrite laminas are 30–70 mμm thick and the dark ones are up to 10 mμm thick. Few layers with vermiform microstructure have been observed, which are characterized by small, vertical tubes representing organic filament molds (Fig. 4d).

6. Biological Composition of Stromatolites

Our preliminary results showed that Socompa stromatolites are exposed to basic conditions (pH 8.5) and high arsenic content. Great quantities of silicium (450 mg/kg) and chlorophyll (70 μg/L) were present, which agreed with the fact that diatoms and chlorophyll were abundant in the samples.

Among diatoms, we could recognize different genera using both, optic and electron microscopy, i.e.: *Cymbella* sp., *Navicula* sp., *Hantzschia* sp., *Nitzschia* sp., *Synedra* sp., *Surirella* sp., *Rhopalodia* sp. and *Pinnularia* sp. (Fig. 5). Filamentous and nonfilamentous cyanobacteria were represented by: *Scytonema* sp, *Microcoleus* sp. and *Phormidium* sp. among others (Fig. 5).

Bacteria with diverse morphotypes were also observed in the stromatolite: cocci, diplococci, deinococci, rod, and filamentous bacteria were visible using electron microscopy (Fig. 6a). It is important to remark that bacteria were profusely developing in the stromatolites through all the layers examined. Moreover, bacteria seemed to "feed" or "colonize" on destroyed diatoms, probably suggesting the evolution in the development of the stromatolite, i.e., initial algae growth followed by their burring by sediments, colonization by bacteria on the burried layers, till complete mineralization and sedimentation of the organic matter (Fig. 6b).

A metagenomic approach was used to taxonomically identify the stromatolite microbial community. DGGE bands of the 16s rDNA gene indicated the affiliation of uncultured bacteria to Alpha, Beta, Gamma, and Delta Proteobacteria (Fig. 6c).

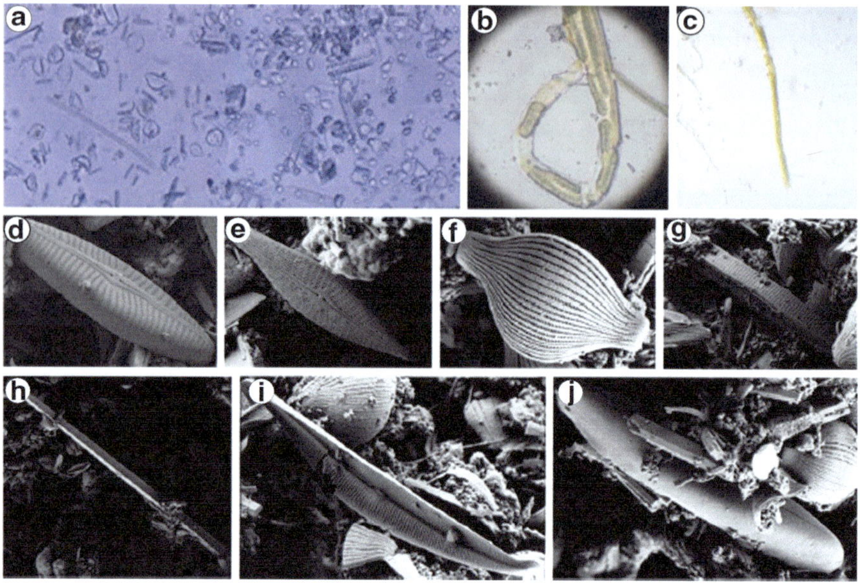

Figure 5. Optical (**a**–**c**) and scanning electron microscopy (**c**–**g**) of algae in stromatolites from Socompa Lake.

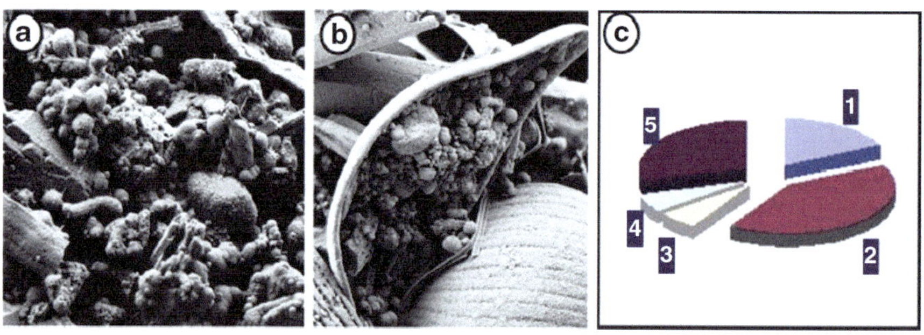

Figure 6. Scanning electron microscopy of stromatolites showing (**a**) bacteria with diverse morphotypes; (**b**) the close association of bacteria and destroyed diatoms; (**c**) distribution of bacterial diversity of stromatolites within the Proteobacteria group. 1. Beta-proteobacteria. 2. Gamma-proteobacteria. 3. Alpha-proteobacteria. 4. Delta-proteobacteria. 5. Non-culture clones, with no significant match with the database.

A great proportion of the bands showed identity with uncultured, nonspecified clone reported as marine origin. This fact may suggest that these stromatolites contain yet unknown bacteria from novel species or genus and current work is developed to unravel this issue. Isolation efforts are also being performed indicating that other bacterial groups are also represented in the stromatolites, such as actinobacteria and other high-GC% bacteria.

Macroscopic eukayotes (e.g., crustaceans or insects) have not been recorded in this lake; this is important because otherwise they would graze on the microbial mats and not allowed the formation of the laminated structures as has been seen before (Goh et al., 2009).

7. Conclusion

Stromatolites are the oldest known fossils, dating back more than three billion years. They were the dominant life-form on Earth for more than two billion years. Today, they are nearly extinct, living a precarious existence in only a few localities worldwide. These modern stromatolites are studied as models of the ancient earth. Stromatolites found in Socompa Lake are the first stromatolites reported at high altitude (4,000 m above sea level). Because of the altitude, the UV radiation is highly intense, the O_2 pressure is low, and the temperatures and winds are extreme. All these conditions found in Andes mountains recreate the ancient environment on earth, when ozone layer was not created. Stromatolites and O_2 are still not so abundant in the atmosphere. In that way, we propose that stromatolites found in Socompa are developing in the most similar conditions to the primitive earth comparing all modern stromatolites in literature. Thus, they are the best model for studying the ancient stromatolites development and their influence on geochemical cycle of earth.

Mineralogical composition of Socompa stromatolites is dominated mainly by aragonite, which is a less stable mineralogical form of calcium carbonate. Stromatolites are usually composed of calcite (Walter, 1977). However, aragonitic modern stromatolites have been recognized in some places such as the Lake Clifton, the Lake Thetis, in Western Australia (McNamara, 2009), and in Highborne Cay, in Bahamas (Andres et al., 2006).

Biological composition demonstrates typical stromatolites diversity formed by aerobic and anaerobic autotrophic and heterotrophic microorganisms; however, diversity of bacteria and archaea demonstrated that Socompa's stromatolites are a source of nonidentified new organisms. Detailed studies are underway to characterize this microbiota.

8. Acknowledgments

The authors thank the Municipio de Tolar Grande, Salta, and Gendamaria Nacional Argentina for logistic assistance in the field work. The authors are grateful to Professor Vinod C. Tewari, Wadia Institute of Himalayan Geology, Dehradun, India, for inviting them to contribute this chapter. One anonymous referee and Professor Tewari are also thanked for the review comments to improve the chapter. Professor Joseph Seckbach is thanked for editing the text. This work was partially supported by PICT 1221 and PIP Conicet 6237/05.

9. References

Allwood, A.C., Walter, M.R., Kamber, B.S., Marshall, C.P. and Burch, I.W. (2006) Stromatolite reef from the Early Archaean era of Australia. Nature **441**(7094): 714–718.

Andres, M., Sumner, D., Reid, R. and Swart, P. (2006) Isotopic fingerprints of microbial respiration in aragonite from Bahamian stromatolites. Geology **34**(11): 973–976.

Baumgartner, L.K., Spear, J.R., Buckley, D.H., Pace, N.R., Reid, R.P., Dupraz, C. and Visscher, P.T. (2009) Microbial diversity in modern marine stromatolites, Highborne Cay, Bahamas. Environ. Microbiol. **11**(10): 2710–2719.

Belluscio, A. (2009) High window on the past. Nature Digest **6**: 34–36. doi: 10.1038/ndigest.2009.091134.

Bertrand-Sarfati, J. (1976) An attempt to classify Late Precambrian stromatolite microstructure, In: M.R. Walter (ed.) *Stromatolites: Amsterdam*, Elsevier, pp. 251–259.

Byerly, G.R., Lower, D.R. and Walsh, M.M. (1986) Stromatolites from the 3,300–3,500-Myr Swaziland Supergroup, Barberton Mountain Land, South Africa. Nature **319**: 489–491.

Demergasso, C., Casamayor, E.O., Chong, G., Galleguillos, P., Escudero, L. and Pedros-Alio, C. (2004) Distribution of prokaryotic genetic diversity in athalassohaline lakes of the Atacama Desert, northern Chile. FEMS Microbiol. Ecol. **48**: 57–69.

Desnues, C., Rodriguez-Brito, B., Rayhawk, S., Kelley, S., Tran, T., Haynes, M., Liu, H., Furlan, M., Wegley, L., Chau, B., Ruan, Y., Hall, D., Angly, F.E., Edwards, R.A., Li, L., Thurber, R.V., Reid, R.P., Siefert, J., Souza, V., Valentine, D.L., Swan, B.K., Breitbart, M. and Rohwer, F. (2008) Biodiversity and biogeography of phages in modern stromatolites and thrombolites. Nature **452**(7185): 340–343.

Dib, J.R., Motok, J., Fernández Zenoff, V., Ordoñez, O. and Farías, M.E. (2008) Occurrence of resistance to antibiotics, UV-B and arsenic in bacteria isolated from extreme environments in high altitude (above 4,400 m) Andean wetlands. Curr. Microbiol. **56**: 510–517.

Dib, J.R., Weiss, A., Neumann, A., Ordoñez, O., Estévez, M.C. and Farías, M.E. (2009) Isolation of bacteria from remote high altitude Andean lakes able to grow in the presence of antibiotics. Recent Pat. Antiinfect. Drug Discov. **4**(1): 66–76.

Dib, J.R., Wagenknecht, M., Hill, R.T., Farías, M.E. and Meinhardt, F. (2010a) First report of linear megaplasmids in the genus Micrococcus. Plasmid **63**(1): 40–45 (Epub October 19, 2009).

Dib, J.R., Wagenknecht, M., Hill, R.T., Farías, M.E. and Meinhardt, F. (2010b) Novel linear megaplasmid from Brevibacterium sp. isolated from extreme environment. J. Basic Microbiol. **50**(3): 280–284.

Escudero, L., Chong, G., Demergasso, C., Farías, M.E., Cabrol, N.A., Grin, E., Edwin, M. and Yu, J.Y. (2007) Investigating microbial diversity and UV radiation impact at the high-altitude Lake Aguas Calientes, Chile. Abstr. Proc. SPIE Int. Soc. Opt. Eng. **1**. paper 6694-33: 1–11.

Farías, M.E., Fernandez-Zennof, V., Flores, R., Ordoñez, O. and Estevez, C. (2009) Impact of solar radiation on bacterioplankton in Laguna Vilama, an hypersaline Andean Saline Lake (4,650 m). Special issue on the high lakes project and other high altitude lakes. J. Geophys. Res. doi: 10.1029/2008JG000784.

Fernández Zenoff, V., Siñeriz, F. and Farías, M.E. (2006) Diverse responses to UV-B radiation and repair mechanisms of bacteria isolated from high-altitude aquatic environments. Appl. Environ. Microbiol. **72**(12): 7857–7863.

Ferrero, M., Farías, M.E. and Siñeriz, F. (2004) Preliminary characterization of microbial communities in high altitude wetlands of northwestern Argentina by determining terminal restriction fragment length polymorphisms. Rev. Latinoam. Microbiol. **46**(3–4): 72–80.

Flores, M.R., Ordoñezl, O.F. and Farías, M.E. (2009) Isolation of UV-B resistant bacteria from two Andean wetlands (4,400 m) with saline and hypersaline conditions. J. Appl. Gen. Microbiol. **55**: 447–458.

Foster, J.S., Green, S.J., Ahrendt, S.R., Golubic, S., Reid, R.P., Hetherington, K.L. and Bebout, L. (2009) Molecular and morphological characterization of cyanobacterial diversity in the stromatolites of Highborne Cay, Bahamas. ISME J. **3**(5): 573–587.

Gaucher, C. and Poiré, D.G. (2009) Biostratigraphy. Neoproterozoic-Cambrian evolution of the Río de la Plata Palaeocontinent. In: C. Gaucher, A.N. Sial, G.P. Halverson and H.E. Frimmel (eds.) *Neoproterozoic-Cambrian tectonics, global change and evolution: a focus on southwestern Gondwana*. Developments in Precambrian Geology **16**: 103–114, Elsevier. (ISBN-13: 978-0-444-53249-7).

Goh, F., Allen, M.A., Leuko, S., Kawaguchi, T., Decho, A.W., Burns, B.P. and Neilan, B.A. (2009) Determining the specific microbial populations and their spatial distribution within the stromatolite ecosystem of Shark Bay. ISME J. **3**(4): 383–396.

Lau, E., Nash, C.Z., Vogler, D.R. and Cullings, K.W. (2005) Molecular diversity of cyanobacteria inhabiting coniform structures and surrounding mat in a Yellowstone hot spring. Astrobiology **5**(1): 83–92.

Lowe, D.R. (1980) Stromatolites 3,400-Myr old from the Archean of Western Australia. Nature **284**: 441–443.

McNamara, K. (2009) Investigates the slime that ruled the earth for billons of years. Geoscientist **19**(12): 16–22.

Ordoñez, O.F., Flores, M.R., Dib, J.R., Paz, A. and Farías, M.E. (2009) Extremophile culture collection from Andean lakes: extreme pristine environments that host a wide diversity of microorganisms with tolerance to UV radiation. Microb. Ecol. **58**(3): 461–473.

Poiré, D.G., González, P.D., Canalicchio, J.M., García Repeto, F. and Canessa, N.D. (2005) Estratigrafía del Grupo Mina Verdún, Precámbrico de Minas, Uruguay. Latin American Journal of Sedimentology and Basin Analysis **12**(2): 125–143 (ISSN 1669-7316).

Seufferheld, M.J., Alvarez, H.M. and Farías, M.E. (2008) Role of polyphosphates in microbial adaptation to extreme environments. Appl. Environ. Microbiol. **74**(19): 5867–5874.

Walter, M.R. (1977) Interpreting stromatolites. Am. Sci. **65**: 563–571.

Walter, M.R., Buick, R. and Dunlop, J.S.R. (1980) Stromatolites 3,400–3,500 Myr old from the North Pole area, Western Australia. Nature **284**: 443–445.

Zenoff, V.F., Heredia, J., Ferrero, M., Siñeriz, F. and Farías, M.E. (2006) Diverse UV-B resistance of culturable bacterial community from high-altitude wetland water. Curr. Microbiol. **52**(5): 359–362.

PART 4: MODERN INSTRUMENTAL TECHNIQUES FOR THE STUDY OF STROMATOLITES AND MICROBIOTA

Igisu
Ueno
Nakashima
Awramik
Maruyama
Kilburn
Wacey
Tewari

Biodata of **Dr. Motoko Igisu, Dr. Yuichiro Ueno, Dr. Satoru Nakashima, Professor Stanley M. Awramik**, and **Dr. Shigenori Maruyama**, authors of ***"Micro-FTIR Spectroscopic Imaging of ~1,900 Ma Stromatolitic Chert"***

Dr. Motoko Igisu is currently Post-doctoral researcher of the Department of Earth Science and Astronomy in the University of Tokyo, Japan since 2008. She obtained her Ph.D. from Tokyo Institute of Technology in 2008 and continued her research at Tokyo Institute of Technology. Dr. Igisu's scientific interests are in the area of geochemistry of prokaryotic fossils and evolution of life.

E-mail: **igisu@ea.c.u-tokyo.ac.jp**

Dr. Yuichiro Ueno is currently the Assistant Professor of the Global Edge Institute in the Tokyo Institute of Technology, Japan. He obtained his Ph.D. from the Dept. EPS in the Tokyo Institute of Technology in 2002 and continued his studies and research at the University of Tokyo. Dr. Ueno's scientific interests are in the areas of: Archean biosphere, geochemistry of prokaryotic fossils, stable isotope biogeochemistry, and atmospheric evolution.

E-mail: **ueno.y.ac@m.titech.ac.jp**

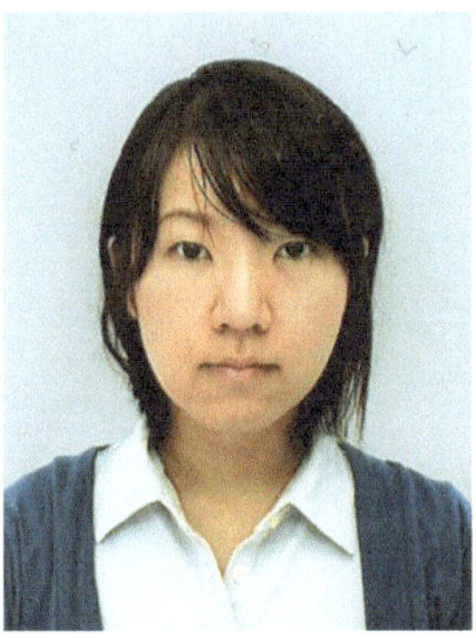

Motoko Igisu

Yuichiro Ueno

V.C. Tewari and J. Seckbach (eds.), *STROMATOLITES: Interaction of Microbes with Sediments*, Cellular Origin, Life in Extreme Habitats and Astrobiology 18, 445–461
DOI 10.1007/978-94-007-0397-1_20, © Springer Science+Business Media B.V. 2011

Dr. Satoru Nakashima is currently Professor at the Department of Earth and Space Science, Osaka University, Japan since 2005. He obtained his Ph.D. (Doctorat d'Etat es Science) from the University of Orleans, France in 1984 on the kinetic and thermodynamic studies of uranium reduction by sedimentary organic matter. He then worked at Japan Atomic Energy Research Institute for the geological disposal of radioactive wastes. He continued his research at Akita University, the University of Tokyo, Hokkaido University, Tokyo Institute of Technology before arriving at Osaka. He has been developing IR, Visible and Raman micro-spectroscopic methods for characterizing water, inorganic and organic phases in earth and planetary materials. By using these spectroscopic methods, Professor Nakashima studies quantitatively organic–inorganic and water–rock interactions in diverse areas ranging from astrobiology, origin and evolution of life, volcanology, seismology, rheology to earth's resource and environment.

E-mail: **satoru@ess.sci.osaka-u.ac.jp**

Dr. Stanley M. Awramik is Professor of Biogeology in the Department of Earth Science, University of California, Santa Barbara, California, USA. He obtained his Ph.D. from Harvard University in 1973. Professor Awramik's scientific interests are in the early history of life on Earth and his research focuses on microbial fossils and stromatolites.

E-mail: **awramik@geol.ucsb.edu**

Satoru Nakashima

Stanley M. Awramik

Dr. Shigenori Maruyama is currently Professor of the Department of Earth and Planetary Sciences in Tokyo Institute of Technology, Japan since 1993. He obtained his Ph.D. from Nagoya University in 1977 and continued his research at Toyama University, Stanford University, and the University of Tokyo. Professor Maruyama scientific interests are in the area of: system evolution of earth history, earth's dynamics, and evolution of life.

E-mail: **smaruyam@geo.titech.ac.jp**

MICRO-FTIR SPECTROSCOPIC IMAGING OF ~1,900 MA STROMATOLITIC CHERT

MOTOKO IGISU[1,2,3], YUICHIRO UENO[4], SATORU NAKASHIMA[5], STANLEY M. AWRAMIK[5], AND SHIGENORI MARUYAMA[1]

[1]*Department of Earth and Planetary Sciences, Tokyo Institute of Technology, O-okayama 2-12-1, Meguro-ku, Tokyo 152-8551, Japan*

[2]*Department of Earth and Space Science, Osaka University, 1-1 Machikaneyama-cho, Toyonaka-shi, Osaka 560-0043, Japan*

[3]*Department of Earth Science and Astronomy, The University of Tokyo, Komaba, Meguro-ku, Tokyo 153-8902, Japan*

[4]*Global Edge Institute, Tokyo Institute of Technology, Post No. I2-21, Meguro-ku, Tokyo 152-8551, Japan*

[5]*Department of Earth Science, University of California, Santa Barbara, CA 93106, USA*

Abstract Micro-Fourier transform infrared (FTIR) spectroscopic imaging analyses nondestructively revealed micrometer to millimeter-scale distributions of organic and inorganic functional groups in Proterozoic stromatolitic chert containing prokaryotic fossils from ~1,900 Ma Gunflint Formation. CH_3/CH_2 absorbance ratios ($R_{3/2}$) indicate a bacterial origin, but not Archaea, of most carbonaceous matter in the chert as well as in the microfossils themselves. However, the characterizations of the stromatolitic chert also show that $R_{3/2}$ value of carbonaceous matter existing with carbonates could be overestimated or underestimated. This technique is useful for searching and characterizing rapidly the organic matter in terrestrial and extraterrestrial samples at the micrometer to millimeter scale, and may provide useful information on the affinities of microfossils in the chert.

Keywords FTIR • Imaging • Functional group • Prokaryotic fossils • Gunflint Formation • Bacteria • Archaea • Aliphatic CH moiety • Chert • Carbonate • Carbonaceous matter

1. Introduction

The Precambrian is primarily a microbial world. Microbial fossils show carbonaceous cellular structure, which often resemble cyanobacteria and other prokaryotes (e.g., Schopf, 1992; Knoll, 2003). But morphological analysis of these microbial fossils is often not enough to determine the biology of such fossils and to determine their precise phylogenetic positions. In situ analytical techniques are now being applied to obtain chemical signatures of the micrometer-sized fossils. For example, micro-Raman spectroscopy is being used to demonstrate carbonaceous composition of putative organic-walled microfossils and to determine their degree of post-depositional alteration (Ueno et al., 2001, 2006; Kudryavtsev et al., 2001; Schopf et al., 2002, 2005; Brasier et al., 2002, 2004; Pasteris and Wopenka, 2002, 2003; Arouri et al., 1999, 2000; Marshall et al., 2005; Igisu et al., 2006). Raman spectra of Precambrian microfossils, however, cannot provide useful information for their taxonomy because Raman spectra of carbonaceous matter generally show a characteristic feature of amorphous carbon matter (e.g., Pasteris and Wopenka, 2003), which is not taxonomically specific. On the other hand, infrared (IR) spectroscopy can detect many polar bonds in carbonaceous matter (e.g., Rouxhet et al., 1980), and thus is potentially useful to detect taxon-specific chemical signature (e.g., Arouri et al., 1999, 2000; Marshall et al., 2005; Igisu et al., 2006). Our previous study showed that the aliphatic CH moieties (chain hydrocarbon parts) using CH_3/CH_2 absorbance ratios ($R_{3/2}$) can be useful for domain-level classification of prokaryotes (Bacteria or Archaea) (Igisu et al., 2009). However, even well-preserved Neoproterozoic prokaryotic fossils have quite small amounts of aliphatic CH moieties. If we search for chemical signatures in older and thus more challenging samples, more advanced micro-Fourier transform infrared (FTIR) technique will be helpful for searching and characterizing the organic signatures in microfossils.

Here, we report micro-FTIR spectroscopic imaging of stromatolitic chert from the ~1,900 million years old (Ma) Gunflint Formation of Canada. IR imaging was conducted on a doubly polished petrographic thin section at the millimeter scale to search for functional groups related to organic and inorganic matter. The distributions of functional groups were compared with the morphology of stromatolites. Based on these results, we discuss the $R_{3/2}$ of microfossils in the stromatolite.

2. Regional Setting, Stratigraphy, Age, and Stromatolites

The Gunflint Formation (Animikie Group) is a chemical-clastic sedimentary rock succession, 100–180 m thick (averages 122 m) that crops out in an ENE to WSW trend over a distance of about 268 km, from Gunflint Lake (Minnesota, USA) in the west to Schreiber Beach (north shore of Lake Superior, Ontario, Canada) in the east (Goodwin, 1956; Pufahl et al., 2000; Fig. 1). The Animikie Group, in ascending order, consists of the Kakabeka Conglomerate, Gunflint Formation, and Rove Formation, which have conformable relationships with one another (Floran and Papike 1975). The Rove Formation is unconformably overlain by the Mesoproterozoic Sibley Group (Morey and Ojakangas, 1982) and the Kakabeka Conglomerate rests unconformably on Archean metavolcanics (Pufahl et al., 2000).

Subdivision of the Gunflint Formation into members has changed over the years. By Fralick and Barrett (1995), two members were defined: a lower member

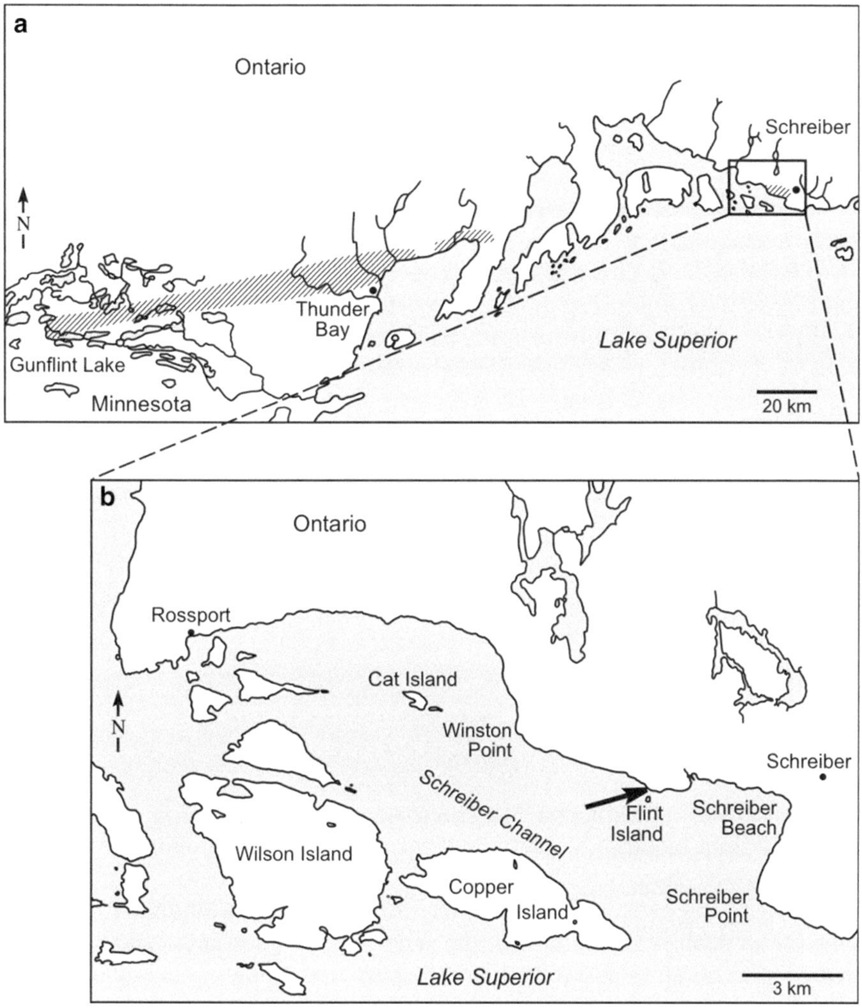

Figure 1. Map of northwestern region of Lake Superior showing (**a**) outcrop distribution (*diagonal line pattern*) of the Gunflint Formation [modified from Awramik and Barghoorn (1977)], and (**b**) location (*arrow*) where sample (GF 74) was collected, which today is in the Schreiber Channel Provincial Nature Reserve (http://www.ontarioparks.com/English/schr.html).

consisting of stromatolitic, ooidal, and granular cherts, and an upper member similar to the lower member. Each member represents a transgression with a shallowing phase separating the members (see Pufahl et al., 2000).

The age of the Gunflint Formation was imprecisely known until recently. In 2002, Fralick et al. (2002) reported a U–Pb age of 1,878.3 ± 1.3 Ma from a population of euhedral zircons recovered from re-worked volcaniclastic sediments from the middle part of the upper member. They interpreted this to be the depositional age of the formation.

Stromatolites from the Gunflint Formation have been a subject in a number of publications (e.g., Hofmann, 1969; Cloud and Semikhatov, 1969; Awramik and Semikhatov, 1979; Sommers et al., 2000). They occur in the basal portions of both members and are associated with the initial phases of each transgression. The stromatolites are interpreted to have formed in a wave-agitated, foreshore setting with a moderate shoreline gradient (see Hofmann, 1969; Pufahl et al., 2000).

The most comprehensive descriptions of the different types of stromatolites found in the formation were provided by Hofmann (1969); however, he did not give them binomials. Only one stromatolite has been treated taxonomically, *Gruneria biwabikia*, which was formally described by Cloud and Semikhatov (1969). Stromatolite morphologies include layered or stratiform, domical, columnar and columnar branching, compound (domical with layered and columnar branching morphologies, for example), and oncoids. Bioherms composed of many of these types are a common feature of the basal portions of the members (see Cloud, 1965, his Fig. 1 of bioherms at a locality 4 km west of Schreiber Beach; Fig. 1). Bioherms are up to about 1 m in diameter. No coniform stromatolites are known.

3. Materials and Methods

3.1. SAMPLE

The sample used in the micro-FTIR spectroscopy is a stromatolitic chert that was collected from a bioherm at the base of member 1 of the Gunflint Formation at a locality on the shore of Lake Superior, ~4 km west of Schreiber Beach (Fig. 1). The stromatolites here consist principally of layered and columnar stromatolites that grew on pebbles and cobbles of the underlying Kakabeka Conglomerate producing bioherms up to ~0.5 m in diameter. The stromatolites in the sample GF74, (a similar stromatolite is illustrated in Fig. 2) have parallel to subparallel stubby columns under 1 cm wide that infrequently branch, but frequently coalesce. These stromatolites resemble Form F of Hofmann (1969). At column margins, laminae characteristically overlap previous laminae, at times producing a wall. Laminae are thin, and usually <100 μm thick, boundaries between laminae are somewhat diffuse, in particular going from light to dark laminae. The laminae are composed mainly of concentrations of amorphous, at times granular, carbonaceous matter and microbial fossils (Fig. 3). These microfossils are carbonaceous and are often well preserved in the chert matrix (cryptocrystalline quartz). Based on the size, and their filamentous and coccoidal morphology, most of the microfossils are interpreted as cyanobacteria (Barghoorn and Tyler, 1965; Awramik and Barghoorn, 1977).

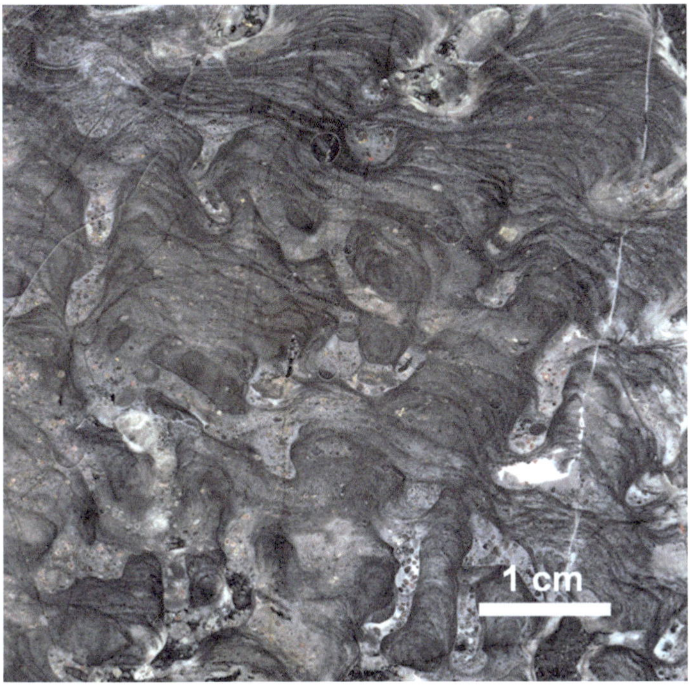

Figure 2. Stromatolite similar to the examples in the small sample (GF 74) used in this study. The sample illustrated here comes from bioherms at Winston Point (see Hofmann 1969), about 5 km WNW from where sample GF 74 was collected (Fig. 1). Yale Peabody Museum sample numbered 1,076.

The sample was cut with a rock saw, and a doubly polished petrographic thin section about 40 μm thick was prepared for petrographic study, obtaining the IR spectra as described previously (Igisu et al., 2006). The doubly polished thin section was immersed in acetone to dissolve used adhesive and clean up modern contamination on its surface. Based on the petrographic investigation of the thin section (Fig. 3), the stromatolitic chert is composed mainly of cryptocrystalline quartz (SiO_2), and contains brown to dark amber colored amorphous carbonaceous matter, microbial fossils (the same color as carbonaceous matter), pyrite, Fe-oxides, and carbonates.

3.2. MICRO-FTIR SPECTROSCOPIC IMAGING ANALYSIS

Micro-FTIR imaging spectroscopic analysis was conducted on the doubly polished petrographic thin section using an automated X–Y stage set in IR Multi-channel Viewer (Jasco FTIR6200 + IMV-4000) with a 16-channel linear array mercury

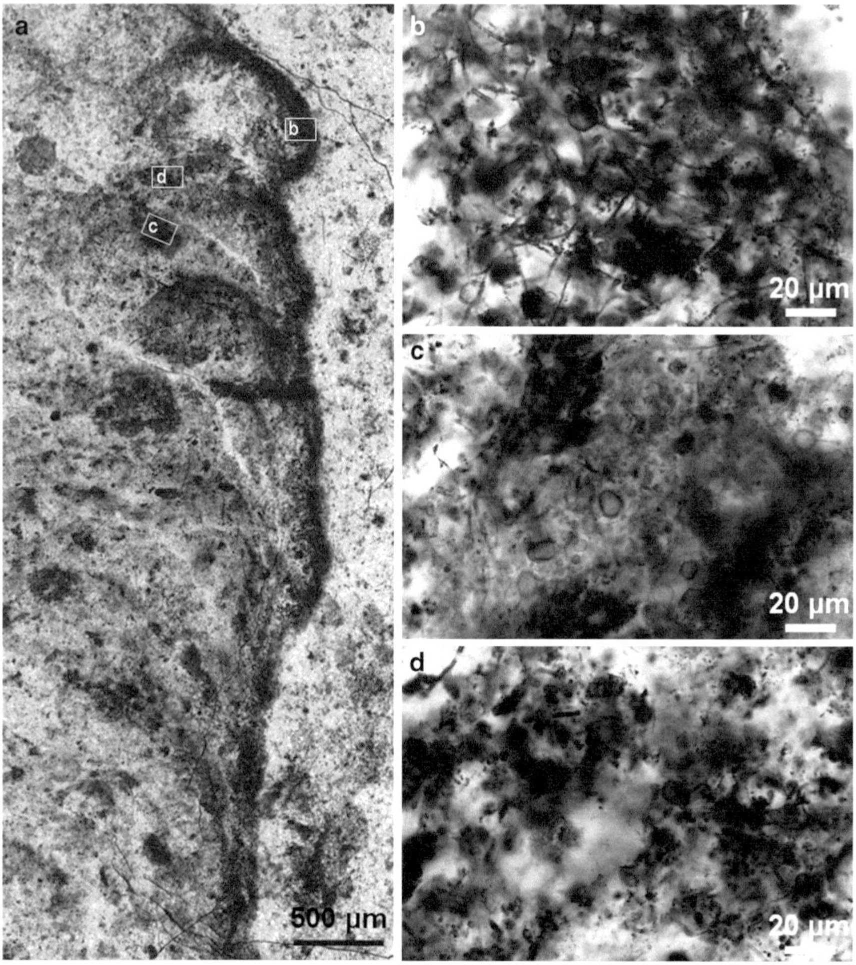

Figure 3. Optical photomicrographs of the stromatolitic chert GF74. (**a**) Prokaryotic fossil-bearing laminae. (**b**–**d**) Aggregated filamentous and coccoidal microfossils and amorphous organic matter.

cadmium telluride (MCT) detector. The thin section was placed over a hole in a sample holder; care was taken to ensure that a stromatolite column was positioned over the hole, and measured in the following manner. IR spectra for imaging analysis were obtained by collecting 16 scans (~2 s in total) in the spectral range from 4,000 to 1,000 cm^{-1} at an 8 cm^{-1} resolution. The spatial resolution was 12.5×12.5 μm^2 corresponding to a pixel area of the IR linear array detector. The total measured area was 1.2×5.0 mm^2, and 38400 IR spectra were obtained. The total analysis time is about 1.5 h. Analytical error was determined by duplicate measurements in the same position as the background spectrum and was better than 0.005 absorbance unit (AU) in the 3,000–1,300 cm^{-1} range, except in the CO_2 absorption region.

4. Results

Typical IR transmission spectra of carbonaceous matter in the thin section of the stromatolite are shown in Fig. 4. Aggregations of filamentous and coccoidal fossils as well as amorphous organic matter show IR absorption bands of aliphatic CH moieties (~2,960, 2,925, 2,850 cm^{-1}) in the range of >2,500 cm^{-1} (Fig. 4). The bands around 2,960 and 2,925 cm^{-1} are due to asymmetric stretching of end-methyl aliphatic CH_3 and methylene-chain CH_2, respectively (Bellamy, 1954). A weak band from symmetric stretching of aliphatic CH_2 is also seen around 2,850 cm^{-1} (Bellamy, 1954). Similar results were obtained in our previous study of cyanobacteria-like fossils in ~850 Ma stromatolitic cherts from the Bitter Springs Formation (Igisu et al., 2006), while symmetric stretching band of aliphatic CH_3 (~2,870 cm^{-1}) was not observed in this study.

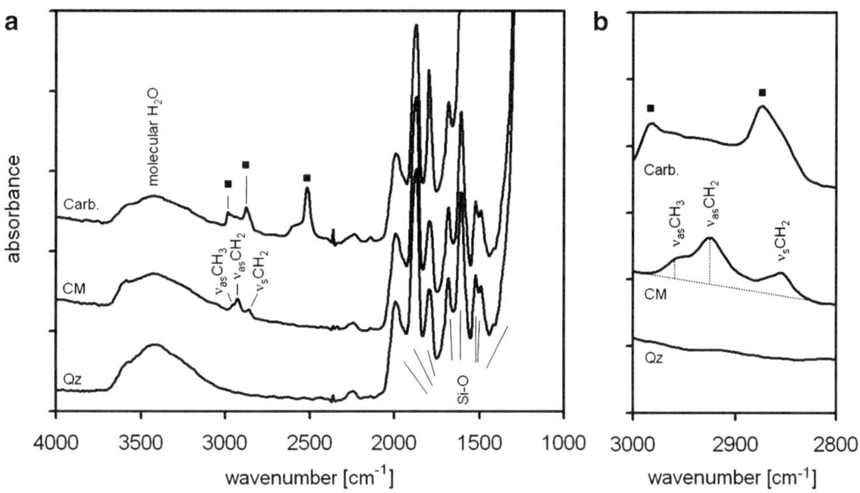

Figure 4. Representative IR transmission spectra of the microfossils (CM), matrix quartz (Qz), and carbonates (Carb.) in the stromatolitic chert. (**a**) Stacked IR spectra. IR band at 2,960 cm^{-1} is due to asymmetric aliphatic CH_3 ($\nu_{as}CH_3$), and those at 2,925 and 2,850 cm^{-1} are due to asymmetric and symmetric aliphatic CH_2 ($\nu_{as}CH_2$ and ν_sCH_2). IR bands at 2,990, 2,870 and 2,515 cm^{-1} are characteristics of carbonate (*square*). Broad band around 3,400 cm^{-1} is due to molecular water probably within the grain boundary of micro-quartz. Saturated bands (1,300–1,000 cm^{-1}) and seven bands (1,995, 1,870, 1,793, 1,684, 1,610, 1,525, and 1,492 cm^{-1}) are due to Si–O bonds of quartz. One scale unit for vertical axis corresponds to an absorbance value of 0.2. (**b**) Enlarged view of IR spectra of (**a**) in the range of 3,000–2,800 cm^{-1}. An example of baseline correction for elucidation of peak height is described as *dotted line*. One scale unit for the vertical axis corresponds to an absorbance value of 0.05.

The band around 3,400 cm^{-1} is due to O–H stretching vibrations. The broad 3,400 cm^{-1} band is generally considered to be due to liquid-like molecular water (H_2O) at grain boundaries of cryptocrystalline quartz (Nakashima et al., 1995; Ito and Nakashima, 2002; Igisu et al., 2006). Seven absorption bands at 1,995, 1,870, 1,793, 1,684, 1,610, 1,525, and 1,492 cm^{-1} are due to overtones and combinations of Si–O vibrations, and are characteristic of quartz (Nakashima et al., 1995; Ito and Nakashima, 2002; Igisu et al., 2006). The range 1,300–1,000 cm^{-1} (Si–O stretch) is saturated due to the thickness (~40 μm) of the thin section. The feature around 2,350 cm^{-1} is due to variation of CO_2 in air.

The carbonates among the minerals in the stromatolitic chert measured here also show IR bands at around 2,980, 2,875, and 2,515 cm^{-1} in the >2,500 cm^{-1} region together with more saturated bands in the <1,500 cm^{-1} region as shown in Fig. 4. Fe-oxide and pyrite do not show IR bands in >2,500 cm^{-1} region (Komada, 1985; RRUFF Database; http://rruff.info/).

The spatial distributions of the above components in the stromatolitic chert are shown in Fig. 5. The distributions of the bands (peak heights) at 2,960 and 2,925 cm^{-1} after baseline correction in raw IR transmission spectra roughly agree with thin section images of microfossil-bearing laminae (Fig. 5). The distribution of 2,515 cm^{-1} band agrees with those of carbonates, and the intense parts of 2,515 cm^{-1} band overlap with those of the 2,925 and 2,960 cm^{-1} bands (Fig. 5). This suggests the distributions at 2,960 and 2,925 cm^{-1} due primarily to aliphatic CH moieties might include absorptions of carbonates.

5. Discussion: Characterizing Aliphatic CH Moieties ($R_{3/2}$)

In order to evaluate a domain-level taxonomic origin of carbonaceous matter, we have introduced the aliphatic CH_3/CH_2 absorbance ratio ($R_{3/2}$) in our previous study as follows:

$$R_{3/2} = [\nu_{as}CH_3] / [\nu_{as}CH_2] \quad (1)$$

where $[\nu_{as}CH_3]$ and $[\nu_{as}CH_2]$ represent peak heights of asymmetric stretching bands for aliphatic CH_3 (end-methyl; ~2,960 cm^{-1}) and CH_2 (methylene-chain; ~2,925 cm^{-1}), respectively (Igisu et al., 2009). However, it should be noted that carbonates also show IR bands in the range of 3,000–2,800 cm^{-1}. This indicates that $R_{3/2}$ values of carbonaceous matter existing with carbonates can be overestimated or underestimated. Relationship of $R_{3/2}$ value to the content of carbonate is shown in Fig. 6. For better $R_{3/2}$ precision, we used only the data with 2,960 and 2,925 cm^{-1} absorbance more than three times of analytical error (0.005). Although the $R_{3/2}$ values with carbonate (absorbance at 2,515 cm^{-1} > 0.005) are about 0.82 ± 0.10 ($n = 359$), the $R_{3/2}$ values without carbonate are about 0.49 ± 0.08 ($n = 18$) (Fig. 6). In this study, the presence of carbonates tends to result in overestimation of $R_{3/2}$ values of carbonaceous matter.

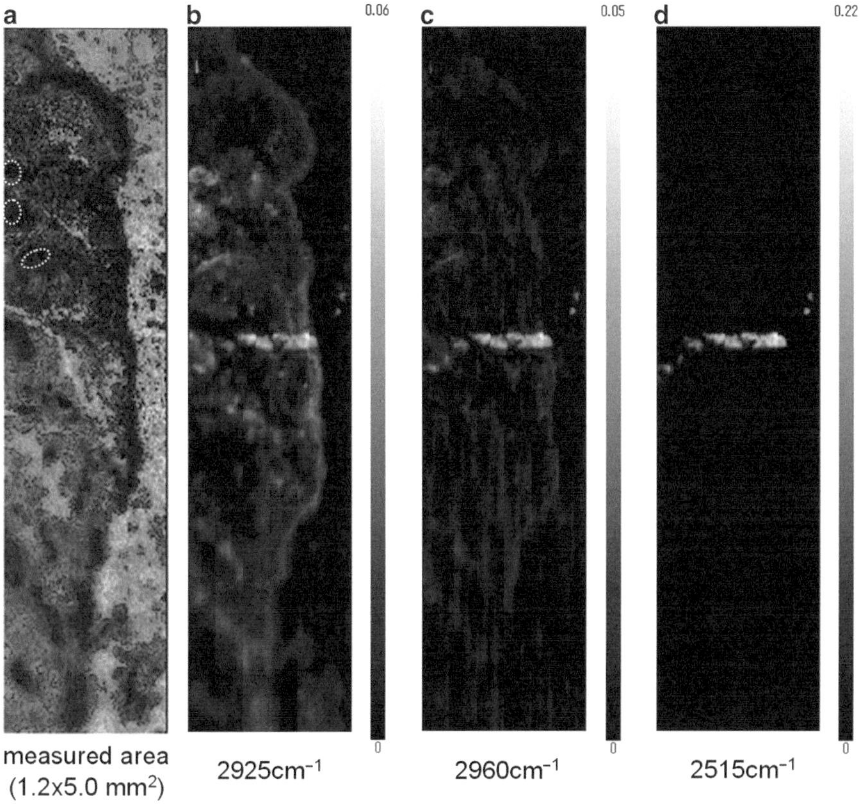

Figure 5. IR imaging results by the raw IR transmission spectra of the microfossil-bearing chert GF74. (**a**) Optical photomicrograph of the measured area (1.2×5.0 mm^2). *Dotted white circles* represent the areas, which include all the spot analyses used for $R_{3/2}$ calculation in Fig. 6c. (**b**–**d**) The spatial distribution maps for the peak heights at (**b**) 2,925, (**c**) 2,960, (**d**) 2,515 cm^{-1} after baseline correction. The color scale represents the higher peak height (*white*) and the lower one (*black*) in the absorbance unit (AU). Analytical errors in absorbance are 0.005 AU in (**b**–**d**).

Although HCl acid treatment can dissolve carbonates, it will be difficult to collect organic residues with prokaryotic cellular structures. It seems to be better to select carbonaceous matter without carbonates under the IR microscope for more precise $R_{3/2}$ evaluation. IR spectra of carbonate such as calcite ($CaCO_3$) and dolomite ($MgCO_3$) generally have an intense band around 1,450–1,410 cm^{-1} and ~2,515 cm^{-1} band together with smaller ~2,990 and ~2,875 cm^{-1} bands (Komada, 1985). Therefore, it should be checked in the $R_{3/2}$ evaluation of carbonaceous matter that a ~2,515 cm^{-1} band is not observed. Hence, we focus further discussions on IR bands that do not show the 2,515 cm^{-1} signal.

The above carbonaceous matter without carbonate is divided into aggregated filamentous and coccoidal fossils and amorphous organic matter (Fig. 3).

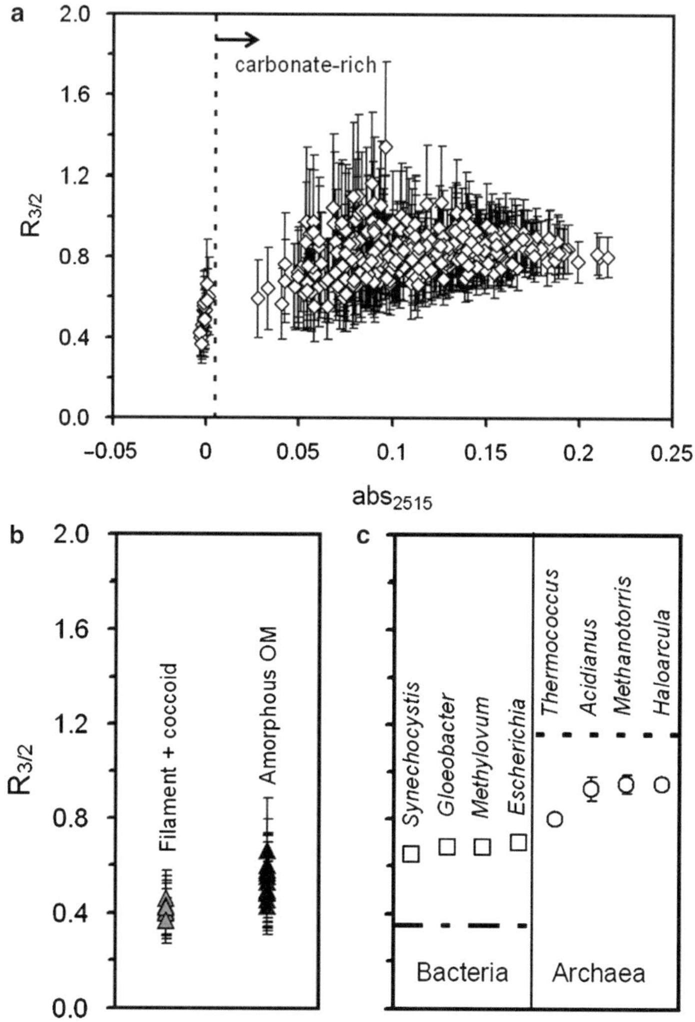

Figure 6. $R_{3/2}$ values of carbonaceous matter existing with and without carbonates, compared with whole cells and predicted lipids of extant prokaryotes. (**a**) Relationship between $R_{3/2}$ (IR absorbance ratio) and the content of carbonates (absorbance at 2,515 cm^{-1}). *Dotted line* represents upper limit of analytical error of absorbance at 2,515 cm^{-1} (0.005 AU). Individual error bars of $R_{3/2}$ values are calculated based on the duplicate analysis of air background spectra. All the data were obtained from GF74. (**b**) $R_{3/2}$ values of aggregated filamentous and coccoidal fossils and amorphous organic matter from GF74. Individual error bars of $R_{3/2}$ values are calculated based on the duplicate analysis of air background spectra. (**c**) $R_{3/2}$ values of extant Bacteria (*square*) and Archaea (*circle*) and predicted $R_{3/2}$ values of lipid for each domain (*dotted line*) [referred from Igisu et al. (2009)]. Positive correlation between $R_{3/2}$ value and ratio CH_3/CH_2 for *n*-alkane standard samples ($R_{3/2} = 2.56$ [CH_3/CH_2], $r^2 = 0.64$) can be used for estimating $R_{3/2}$ values of these domains' representative lipid.

The $R_{3/2}$ values are 0.41 ± 0.03 (n = 6) for aggregated filamentous and coccoidal microbial fossils, and 0.53 ± 0.07 (n = 12) for amorphous organic matter (Fig. 6b). The values are consistent with our previous study, and are roughly similar to the $R_{3/2}$ values of the predicted extant bacterial lipids ($R_{3/2}$ = 0.32–0.37) rather than archaeal lipids ($R_{3/2}$ = ~1.16) (Igisu et al., 2009) (Fig. 6b, c). This indicates that most microfossils and amorphous organic matter in this study belong to Bacteria. These results also show that this rapid imaging technique provides $R_{3/2}$ values of carbonaceous matter consistent with the previous spot analysis (Igisu et al., 2009).

In summary, micro-FTIR spectroscopic imaging analysis can be successfully applied to ~1,900 Ma stromatolites from the Gunflint Formation. The microfossil-bearing laminae in the stromatolitic chert contain immature carbonaceous matter with aliphatic CH_2 and CH_3. $R_{3/2}$ values indicate that most carbonaceous matter including prokaryotic fossils belong to Bacteria. However, it should be noted that carbonates also show IR signature in the similar range to aliphatic CH moieties. This method would be applicable for searching for chemical signatures in Earth's rock samples as well as extraterrestrial materials, and for chemically characterizing organic matter.

6. Conclusions

Micro-FTIR spectroscopic imaging analysis of a stromatolitic chert of the ~1,900 million-year-old Gunflint Formation containing extremely well-preserved microfossils provides the following conclusions:

(1) IR spectra of microfossil-bearing laminae show the 2,925 and 2,960 cm^{-1} bands due to aliphatic CH moieties. The distributions of these bands roughly agree with the morphology of the microfossil-bearing laminae, but the distributions are affected by the presence of carbonates, whose IR spectra show bands at around 2,980 and 2,875 cm^{-1}.
(2) It should be noted that $R_{3/2}$ values of carbonaceous matter occurring with some minerals such as carbonates can be both overestimated and/or underestimated, but these over and underestimations can be removed by careful evaluation of other, more intense bands (e.g., ~2,515 cm^{-1} band for carbonate).
(3) $R_{3/2}$ values of microfossils and amorphous organic matter without the presence of carbonate indicate that most microfossils and amorphous organic matter belong to Bacteria rather than Archaea. No signatures suggestive of eukaryotes were detected.

Micro-FTIR imaging analysis is useful for rapidly searching and characterizing organic signatures in the stromatolitic chert. Further detailed IR measurements, combined with traditional morphological analyses, may provide new approaches in the determination of the domain-level, taxonomic affinity of microfossils in the chert.

7. Acknowledgments

The major part of this study was conducted at Department of Earth and Space Science, Osaka University. We thank Profs. Hiroyuki Ohta, Satoshi Nakamura, and Dr. Mie Shimojima of Tokyo Institute of Technology and Dr. Ken Takai of JAMSTEC for providing bacteria and Archaea samples, and Jasco Co. Ltd. for their technical supports of micro-FTIR imaging analyses. We also thank Dr. David Wacey and an anonymous reviewer for helpful comments, Prof. Vinod C. Tewari who mainly handled the chapter for editing, review etc., and Prof. Joseph Seckbach for further handling. This research was partially supported by the twenty-first century COE program "How to build habitable planets," Tokyo Institute of Technology, sponsored by the Ministry of Education Culture, Sports, Technology and Science, Japan. M.I. is grateful to Yoshida Scholarship Foundation for the scholarship program "Doctor21". Y.U. is supported by MEXT's program "Promotion of Environmental Improvement for Independence of Young Researchers" under the Special Coordination Funds for Promoting Science and Technology and by KAKENHI (19740310).

8. References

Arouri, K., Greenwood, P.F. and Walter, M.R. (1999) A possible chlorophycean affinity of some Neoproterozoic acritarchs. Org. Geochem. **30**: 1323–1337.

Arouri, K., Greenwood, P.F. and Walter, M.R. (2000) Biological affinities of Neoproterozoic acritarchs from Australia: microscopic and chemical characterisation. Org. Geochem. **31**: 75–89.

Awramik, S.M. and Barghoorn, E.S. (1977) The Gunflint microbiota. Precambrian Res. **5**: 121–142.

Awramik, S.M. and Semikhatov, M.A. (1979) The relationship between morphology, microstructure and microbiota in Gunflint stromatolites. Can. J. Earth Sci. **16**: 484–495.

Barghoorn, E.S. and Tyler, S.A. (1965) Microorganisms from the Gunflint Chert. Science **147**: 563–577.

Bellamy, L.J. (1954) *The Infrared Spectra of Complex Molecules*. Wiley, New York, pp. 13–56.

Brasier, M.D., Green, O.R., Jephocoat, A.P., Kleppe, A.K., Van Kranendonk, M.J., Lindsay, J.F., Steele, A. and Grassineau, N.V. (2002) Questioning the evidence for Earth's oldest fossils. Nature **416**: 76–81.

Brasier, M.D., Green, O., Lindsay, J.F. and Steele, A. (2004) Earth's oldest (~3.5 Ga) fossils and the 'Early Eden hypothesis': questioning the evidence. Orig. Life Evol. Biosph. **34**: 257–269.

Cloud, P.E. (1965) Significance of the Gunflint (Precambrian) Microflora. Science **148**: 27–35.

Cloud, P.E. and Semikhatov, M.A. (1969) Proterozoic stromatolite zonation. Am. J. Sci. **267**: 1017–1061.

Floran, R.J. and Papike, J.J. (1975) Petrology of the low-grade rocks of the Gunflint Iron-Formation, Ontario-Minnesota. Geol. Soc. Am. Bull. **86**: 1169–1190.

Fralick, P.W. and Barrett, T.J. (1995) Depositional controls on iron formation associations in Canada, In: A.G. Plint (ed.) *Sedimentary Facies Analysis. International Association of Sedimentologists, Special Publication No. 22*. Blackwell Science, Oxford, pp. 137–156.

Fralick, P., Davis, D.W. and Kissin, S.A. (2002) The age of the Gunflint Formation, Ontario, Canada: single zircon U-Pb age determinations from reworked volcanic ash. Can. J. Earth Sci. **39**: 1085–1091.

Goodwin, A.M. (1956) Facies relations in the Gunflint Iron Formation. Econ. Geol. **51**: 565–595.

Hofmann, H.J. (1969) Stromatolites from the Proterozoic Animikie and Sibley groups. Geological Survey of Canada, Paper 68–69, 77 pp.

Igisu, M., Nakashima, S., Ueno, Y., Awramik, S.M. and Maruyama, S. (2006) In situ infrared microspectroscopy of ~850 million-year-old prokaryotic fossils. Appl. Spectrosc. **60**: 1111–1120.

Igisu, M., Ueno, Y., Shimojima, M., Nakashima, S., Awramik, S.M. Ohta, H. and Maruyama, S. (2009) Micro-FTIR spectroscopic signatures in bacterial lipids in Proterozoic microfossils. Precambrian Res. **173**: 19–26.

Ito, Y. and Nakashima, S. (2002) Water distribution in low-grade siliceous metamorphic rocks by micro-FTIR and its relation to grain size: a case from the Kanto Mountain region, Japan. Chem. Geol. **189**: 1–18.

Knoll, A.H. (2003) *Life on a Young Planet: The First Three Billion Years of Evolution on Earth.* Princeton University Press, Princeton, 277 pp.

Komada, H. (1985) *Infrared Spectra of Minerals Reference Guide to Identification and Characterization of Minerals for the Study of Soils.* Research Branch Agriculture, Canada.

Kudryavtsev, A.B., Schopf, J.W., Agresti, D.G. and Wdowiak, T.J. (2001) In situ laser-Raman imagery of Precambrian microscopic fossils. Proc. Natl. Acad. Sci. U.S.A. **98**: 823–826.

Marshall, C.P., Javaux, E.J., Knoll, A.H. and Walter, M.R. (2005) Combined micro-Fourier transform infrared (FTIR) spectroscopy and micro-Raman spectroscopy of Proterozoic acritarchs: a new approach to palaeobiology. Precambrian Res. **138**: 208–224.

Morey, G.B. and Ojakangas, R.W. (1982) Keweenawan sedimentary rocks of eastern Minnesota and northwestern Wisconsin. Geol. Soc. Am. Memoir **156**: 135–146.

Nakashima, S., Matayoshi, H., Yuko, T., Michibayashi, K., Masuda, T., Kuroki, N., Yamagishi, H., Ito, Y. and Nakamura, A. (1995) Infrared microspectroscopy analysis of water distribution in deformed and metamorphosed rocks. Tectonophysics **245**: 263–276.

Pasteris, J.D. and Wopenka, B. (2002) Images of the Earth's earliest fossils? Nature **420**: 476–477.

Pasteris, J.D. and Wopenka, B. (2003) Necessary, but not sufficient: Raman identification of disordered carbon as a signature of ancient life. Astrobiology **3**: 727–738.

Pufahl, P., Fralick, P. and Scott, J. (2000) Geology of the Paleoproterozoic Gunflint Formation. Institute on Lake Superior, Geology Proceedings, 46th Annual Meeting, Thunder Bay, Ontario, **46**(2), 44 pp.

Rouxhet, P.G., Robin, P.L. and Nicaise, G. (1980) Characterization of kerogens and of their evolution by infrared spectroscopy, In: B. Durand (ed.) *Kerogen.* Technip, Paris, pp. 163–175.

RRUFF Database, http://rruff.info/.

Schopf, J.W. (1992) Times of origin and earliest evidence of major biologic groups, In: J.W. Schopf and C. Klein (eds.) *The Proterozoic Biosphere, A Multidisciplinary Study.* Cambridge University Press, New York, pp. 587–593.

Schopf, J.W., Kudryavtsev, A.B., Agresti, D.G., Wdowiak, T.J. and Czaja, A.D. (2002) Laser-Raman imagery of Earth's earliest fossil. Nature **416**: 73–76.

Schopf, J.W., Kudryavtsev, A.B., Agresti, D.G., Czaja, A.D. and Wdowiak, T.J. (2005) Raman imagery: a new approach to assess the geochemical maturity and biogenicity of permineralized Precambrian fossils. Astrobiology **5**: 333–371.

Schreiber Channel Nature Reserve, http://www.ontarioparks.com/English/schr.html.

Sommers, M.G., Awramik, S.M. and Woo, K.S. (2000) Evidence for initial calcite-aragonite composition of Lower Algal Chert Member ooids and stromatolites, Paleoproterozoic Gunflint Formation, Ontario, Canada. Can. J. Earth Sci. **37**: 1229–1243.

Ueno, Y., Isozaki, Y., Yurimoto, H. and Maruyama, S. (2001) Carbon isotopic signatures of individual Archean microfossils (?) from Western Australia. Int. Geol. Rev. **43**: 196–212.

Ueno, Y., Isozaki, Y. and McNamara, K.J. (2006) Coccoid-like microstructures in a 3.0 Ga chert from Western Australia. Int. Geol. Rev. **48**: 78–88.

Biodata of **Matt R. Kilburn** and **David Wacey**, authors of ***"Elemental and Isotopic Analysis by NanoSIMS: Insights for the Study of Stromatolites and Early Life on Earth"***

Dr. Matt R. Kilburn is manager of the Ion Probe Facility at The University of Western Australia. His lab at UWA is the only lab in the world to house both the Cameca NanoSIMS 50 and the IMS 1280 large-geometry ion probes. He read Planetary Science at University College London, and obtained a Ph.D. on the partitioning of elements during planetary core formation from the University of Bristol. Dr Kilburn became interested in the study of early life while running the NanoSIMS lab at Oxford University, and he now spends an increasing amount of time using SIMS to test the biogenicity of local stromatolites, both ancient and modern.

E-mail: **Matt.Kilburn@uwa.edu.au**

Dr. David Wacey is a biogeochemist at the University of Western Australia. He obtained his D.Phil. from Oxford University studying the effects of SRB on biomineralization. His current research focuses on morphological, chemical and isotopic tracers of primitive Archean life, using cutting edge techniques such as NanoSIMS and Transmission Electron Microscopy. He is the author of a new introductory textbook on Archean life entitled "Early Life on Earth: A Practical Guide", released in 2009.

E-mail: **David.Wacey@uwa.edu.au**

Matt R. Kilburn

David Wacey

V.C. Tewari and J. Seckbach (eds.), *STROMATOLITES: Interaction of Microbes with Sediments*, Cellular Origin, Life in Extreme Habitats and Astrobiology 18, 463–493
DOI 10.1007/978-94-007-0397-1_21, © Springer Science+Business Media B.V. 2011

ELEMENTAL AND ISOTOPIC ANALYSIS BY NANOSIMS: INSIGHTS FOR THE STUDY OF STROMATOLITES AND EARLY LIFE ON EARTH

MATT R. KILBURN[1] **AND DAVID WACEY**[1,2]
[1]Centre for Microscopy, Characterisation and Analysis, The University of Western Australia, 35 Stirling Highway, Crawley, WA 6009, Australia
[2]School of Earth and Environment, The University of Western Australia, 35 Stirling Highway, Crawley, WA 6009, Australia

Abstract The ability to unambiguously identify and measure biogenic signals is the key to providing robust evidence for Earth's earliest life. Evidence for life on the early Earth has traditionally been founded on morphological studies, as this is the most obvious and tangible approach. Yet, many morphological features that have been interpreted as biogenic in origin can, in fact, equally be attributed to inorganic processes. Matching chemical signals, for example the presence of carbon and nitrogen, with morphology provides a more robust argument; the ability to correlate in situ isotopic signatures indicative of biological processing with morphological features would provide even more compelling evidence.

Morphological expressions of Earth's earliest life (e.g. microfossils, trace fossils, biominerals or microbial remains within stromatolites) tend to be very small and difficult to analyse in situ. The nano-secondary ion mass spectrometer (NanoSIMS) is a new generation of ion microprobe capable of chemically imaging sub-micron-sized features, providing high-resolution maps of biologically relevant elements such as C, N, P and S. Furthermore, sulphur and carbon isotopic analyses can be performed on micron-sized features with ~1–4‰ precision. This chapter summarises our recent work developing NanoSIMS protocols for isotopic and elemental analyses of candidate biological structures from both modern and ancient rocks.

Keywords Archean • Biogenicity • Biomineral • Carbon • Isotope • Early Life • Microfossil • NanoSIMS • Nitrogen • Pilbara • Stromatolite • Sulphur • Western Australia

1. Introduction

We know that planet Earth is about 4.5 billion years (Ga) old but what is less clear is when it first became home to life. It is unlikely that life gained a foothold before about 3.8 Ga because the first 700 million years of Earth history saw frequent and catastrophic meteorite impacts, which would have repeatedly sterilised the planet.

In contrast, there is abundant evidence for life in rocks younger than 3 Ga. Hence, much work has been carried out over the last ~30 years on the few isolated packets of rocks from the 3.0 to 3.8 Ga time period to study the transition from a non-biological to biological world, and to collect evidence for the first signs of biology that can be used in the search for extraterrestrial life.

Claims for the earliest life on Earth have been numerous (see Wacey, 2009 for a review), but there is still an intense debate about the validity of many of these reports. Ideally, any robust claim for life during this period needs to demonstrate a well-constrained geological context, evidence for biology-like morphology and chemical evidence for metabolic cycling, plus falsification of plausible non-biological origins (Brasier et al., 2006). Unfortunately, this is rarely possible given the great age, minute size and poor preservation of any putative signs of >3 Ga life. Our existing knowledge of >3 Ga life is based largely upon putative microfossils, fractionations of the isotopes of carbon and sulphur and features interpreted as fossilised microbial mats or stromatolites.[1] Each of these, at present, suffers from a degree of uncertainty in their interpretation. Here, we show how geochemical data obtained using nano-secondary ion mass spectrometry (NanoSIMS) can add to our knowledge base.

1.1. ELEMENTAL DISTRIBUTIONS IN STROMATOLITES

It is widely agreed that modern stromatolites are formed by the trapping, binding and cementing of sedimentary grains, as a result of the growth and metabolic activity of micro-organisms (e.g. Walter, 1976). For ancient examples, the formation mechanism is less straightforward because morphologically similar structures can be formed chemically without the aid of biology (e.g. McLoughlin et al., 2008). At present, therefore, stromatolites are a rather controversial indicator of early life on Earth.

The best examples of early Archean stromatolites come from the Pilbara Craton of Western Australia. Here, nodular, wavy-laminated and coniform stromatolites have been described from chert units within the ~3.49 Ga Dresser and ~3.43 Ga Strelley Pool Formations (Lowe, 1980; Walter et al., 1980; Hofmann et al., 1999; Allwood et al., 2006). They have been interpreted as biogenic, based largely upon simple macro-morphological comparisons with rare modern-day

[1] The term "stromatolite" can be used simply as a descriptive term without any indication of biological involvement, defined as "attached, laminated, lithified sedimentary growth structures, accretionary away from a point or limited surface of initiation" (Semikhatov et al., 1979). More commonly, however, the involvement of biology is assumed when a structure is termed a "stromatolite" – "accretionary sedimentary structures, commonly thinly layered, megascopic and calcareous, interpreted to have been produced by the activities of mat-building communities of mucilage-secreting micro-organisms, mainly photoautotrophic prokaryotes" (Schopf, 2006). We adopt the latter terminology in this chapter.

stromatolites, and have been used as a major line of evidence for the emergence of life on the early Earth (see Schopf, 2006 for a review). The problem here is that neither convincing microfossils, mat fabrics, nor elemental patterns indicative of biology have been found in these Archean structures and their biogenicity has been questioned on numerous occasions (Lowe, 1994; Grotzinger and Rothman, 1996; Brasier et al., 2006; McLoughlin et al., 2008). Those who favour a biological mechanism have dismissed the lack of microfossils and biogeochemical signals in such structures as merely due to poor preservation. However, this preservation potential of stromatolites has never been tested in detail, partly due to the lack of suitable technology, and partly due to a lack of coherence between studies of modern microbial mats/stromatolites and fossilised examples from the rock record.

Here, we outline the NanoSIMS methodology for high-resolution elemental imaging of both Phanerozoic and Precambrian stromatolite samples. We present data pertaining to micro- to nano-scale distributions of biologically important elements (e.g. C, N, S) and their spatial relationships within carbonate and silica rock matrices. Characteristic elemental distributions are observed, which may act as a baseline for future studies of Precambrian stromatolites.

1.2. SULPHUR ISOTOPES

Sulphur isotope data ($\delta^{34}S$; where $\delta^{34}S_{V\text{-}CDT}$ (‰) = $[(^{34}S/^{32}S_{sample}) - (^{34}S/^{32}S_{standard})/(^{34}S/^{32}S_{standard})] \times 1{,}000$) from early Archean sedimentary sulphides and sulphates can be informative both about levels of sulphate in the early Archean ocean and about the emergence of sulphate-reducing bacteria (SRB), and possibly, when combined with $\delta^{33}S$ and $\delta^{36}S$ data, the emergence of sulphur-oxidising and disproportionating microbes. Sulphur-processing microbes are often a major component of the modern microbial mat ecosystem (e.g. Baumgartner et al., 2006). Hence, spatially resolved $\delta^{34}S$, $\delta^{33}S$ and $\delta^{36}S$ data may be useful for constraining some of the microbial metabolisms operating in ancient stromatolites.

One problem with sulphur isotope studies has been the limited spatial resolution of conventional isotope measurement techniques. Standard combustion techniques require the isolation of a relatively large amount of sulphide (usually pyrite) from the host rock. Laser ablation inductively coupled plasma mass spectrometry (LA-ICP-MS) allows in situ analysis of sulphur isotopes but still requires sulphides of either ~30 μm or greater in size, or combined analysis of numerous smaller grains. Both of these techniques artificially homogenise the $\delta^{34}S$ signal, and have tended to provide data that is consistent with microbial mediation (Ohmoto et al., 1993; Shen et al., 2001) but also consistent with non-biological formation mechanisms (Runnegar et al., 2001).

A new generation of SIMS instruments (NanoSIMS) has both the spatial resolution and analytical precision to measure individual micro-sulphide grains and help to fill this gap in our knowledge. Here, we outline a methodology for achieving reproducible $\delta^{34}S$ data using NanoSIMS. Although there is still some work to be done to cross-calibrate this data with more conventional techniques, we demonstrate the ability to measure $\delta^{34}S$ with a precision of 1–4‰ from early Archean pyrites of <5 μm in diameter.

1.3. CARBON ISOTOPES

Carbon isotopes ($\delta^{13}C$) are widely used to argue for the biogenicity of Archean carbonaceous matter. Previous $\delta^{13}C$ data has appeared consistent with biological fractionation (i.e. $\delta^{13}C_{V-PDB}$ ~ −20 to −40‰) (e.g. Tice and Lowe, 2004; Walsh and Lowe, 1999). However, technological limitations have meant that most isotopic measurements have not been performed directly on the carbonaceous structures of interest (e.g. an individual microfossil, lining of a trace fossil or carbonaceous lamination). This casts some doubt upon carbon isotope reliability because such bulk rock isotopic measurements introduce the possibility of confusion with other carbon-containing species within the rock or from younger contamination. While standard ion microprobes do allow in situ measurement (e.g. Ueno et al., 2001), they lack the spatial resolution needed to accurately characterise ~1–20 μm-sized objects, sizes typical of putative signs of life in the early Archean. A second problem is that carbon isotopic evidence alone is not necessarily diagnostic for biology because non-biological processes such as "Fischer-Tropsch Type synthesis" can induce $\delta^{13}C$ fractionations of similar magnitudes (McCollum and Seewald, 2006). Therefore, carbon isotope data ideally needs to be accompanied both by an understanding of the geological context of the samples and by high-resolution, in situ mapping of coincident biologically significant elements (e.g. N, S and P) within and around the putative biological structures.

NanoSIMS affords the opportunity to obtain meaningful and reliable $\delta^{13}C$ data from very small objects. This is demonstrated using kerogen found within detrital sandstone grains from the 3.43 Ga Strelley Pool Formation, Western Australia. Isotopic precision of ~3‰ is demonstrated for kerogen linings only 2–5 μm in diameter. For larger carbonaceous objects, >5 μm in diameter, precision of ~1‰ has been achieved (Rasmussen et al., 2008).

2. NanoSIMS Analysis

Secondary ion mass spectrometry uses a high-energy primary ion beam (Cs^+ or O^-) to displace ions from the surface of a sample (secondary ions). These secondary ions are extracted to a mass spectrometer where they are separated by mass and counted. NanoSIMS uses a rastered primary ion beam to produce images much in the same way as a scanning electron microscope (SEM). In this case, however, the images are made up from the sputtered secondary ions, producing chemical maps of the rastered sample surface. NanoSIMS is equipped with a multicollection capability in which a static magnetic field deflects up to five ions or species into moveable detectors positioned at the appropriate position in the focal plane of the magnet (Hillion et al., 1994). Thus, it is possible to detect and map, for example C, O, N, Si and S simultaneously from the same scanned area. This is particularly important where delicate organic material may be rapidly eroded away by the primary ion beam. The mass spectrometer has been designed to allow high transmission at high mass resolution, that is, to achieve mass resolution that enables the separation of molecular or isotopic interferences on a particular mass (through the use of slits) without drastically reducing the signal. This results in the ability to measure isotope ratios simultaneously from the same sample volume. The precision of isotope measurements depends on the number of counts recorded from the lower abundance isotope. This, together with factors affecting the accuracy of NanoSIMS isotope analyses, will be discussed in more detail below.

2.1. ELEMENTAL MAPPING IN STROMATOLITES

Somewhat fortuitously, the elements commonly representative of organic material, C, N, O, S, have relatively high negative secondary ion yields when sputtered with a Cs^+ primary ion beam. This is due to their high electron affinity. Although N does not readily ionise during SIMS sputtering, it does have a particularly strong emission when coupled to C as the CN^- ion on mass 26. The Cs^+ primary ion beam also has the advantage that it can be focussed to less than 100 nm diameter. This results in the ability to map light elements at high spatial resolution and high sensitivity.

Ions from the transition elements (e.g. Fe, Co, Ni, Cu, Zn) are also commonly associated with/necessary for biological processing, acting as enzyme cofactors and/or potential energy sources for microbial redox reactions (Frausto da Silva and Williams, 2001). In SIMS, these elements yield positively charged secondary ions and are more efficiently produced by sputtering with an O^- primary beam. The O^- beam has a larger diameter than the Cs^+ beam, resulting in a slightly reduced lateral resolution of about 200 nm.

For chemical mapping of the relatively coarse features in stromatolites, secondary ion intensities were more important than high spatial resolution. Aberrations in the spectrometer limit the field of view to a maximum of about 80–100 μm. Therefore, images were typically acquired over a 50–60 μm field of view with an image resolution of 256 × 256 pixels. In a 50 μm field of view, each pixel measures 195 nm, so for images produced using the Cs^+ primary beam the maximum resolution is defined by the pixel size (as the primary ion beam is smaller than the pixel), but for the images acquired with the O^- primary beam, the maximum resolution is determined by the diameter of the beam.

The secondary ion yields from any given ion species vary depending on the primary ion species, the ionisation potential or electron affinity of the secondary ion, and the nature of the sample matrix. This makes quantification of multi-component systems extremely difficult with SIMS, and necessitates the use of well-known standards. Ion intensities are not directly comparable with one another, or with themselves in different matrices. This problem is further exacerbated in NanoSIMS where small-scale features of a particular matrix may dominate in only a few pixels of the image, giving rise to ion yield variability at the pixel-to-pixel scale.

As SIMS uses high voltages to accelerate the primary ions, the sample must be conductive. It is therefore necessary to coat non-conducting samples, such as minerals, glasses and ceramics, with a conductive metal film. Typically, a thin (5 nm) gold (Au) coat is used as Au generally does not react with the sample. During sputtering, the Au coat is eroded away exposing the underlying sample surface, which becomes, once again, non-conductive. The ionisation efficiency, and hence the secondary ion yield from the sample, increases with increasing primary ion implantation. As the ion beam erodes the Au coat away, there is an optimum time when the signal is strongest – shortly before the Au coat is entirely removed. After this stage, an electron flood gun may be used to dissipate charge build-up along the edges of the sputter crater, and allow secondary ions to leave the sample surface. This can be problematic when parts of the sample conduct – as the carbon-rich organic layers in stromatolites do.

2.1.1. Stromatolite Samples

Three samples were chosen to demonstrate NanoSIMS elemental mapping capability in stromatolites:

1. A modern lacustrine stromatolite from Lake Thetis, Western Australia (Figs. 1 and 2). Lake Thetis lies approximately 250 km north of Perth, Western Australia and is a permanent saline, alkaline and nutrient poor lake. These conditions exclude most macro-fauna and make the lake ideal for the growth of microbial communities. Grey et al. (1990) described the geomorphology of Lake Thetis. They showed the Lake Thetis basin formed more recently than the Middle Holocene (most likely 3,000–4,500 years B.P.), while radiocarbon dating of lithified stromatolites yielded ages of <1,500 years B.P. A number of terraces are observed surrounding the lake marking abandoned shorelines and recording a relative sea level fall from a high-stand at ~5,000 years B.P. (Brown, 1983). A ~0.5 m fluctuation of lake level between wet and dry periods affects a ~5 m wide strip of the foreshore that undergoes alternating periods of submergence and emergence (Grey et al., 1990).

   Five types of living microbial mats have been identified from Lake Thetis (crenulated, nodular, filamentous, diatomaceous and flocculent), each dominating different, roughly concentric zones in the lake and its surrounds (for details see Grey et al., 1990). Lithified stromatolite platforms and domes (Fig. 1a) occur on three terraces around the lake margin. Stromatolite domes are composed mostly of aragonite, are typically 50–100 cm in diameter and occur on the lowest terrace, defining the lake margin. Crenulate and nodular

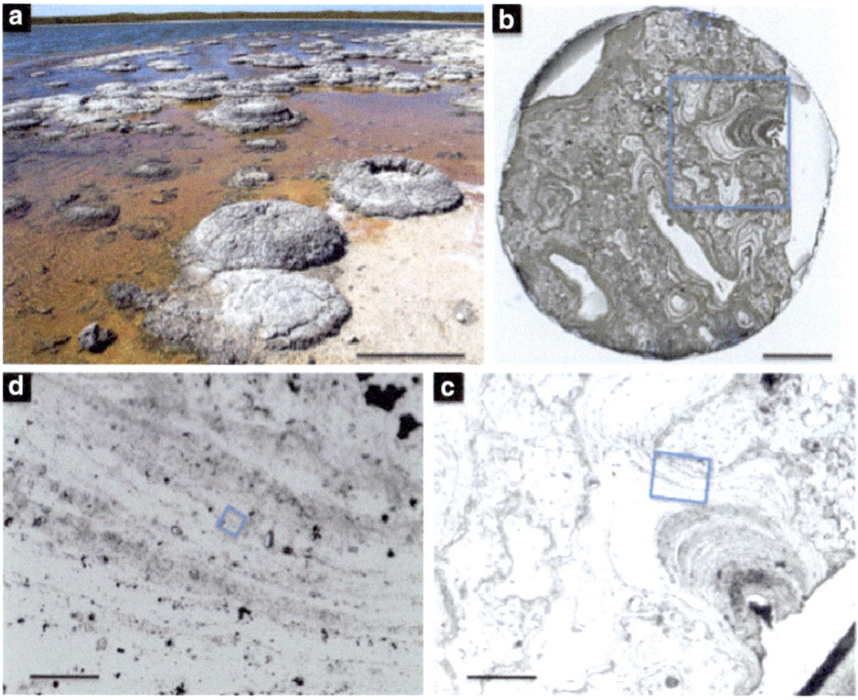

Figure 1. Modern lacustrine stromatolites from Lake Thetis, Western Australia. (**a**) Field photograph of <1,500-year-old lithified stromatolite domes associated with living nodular and diatomaceous microbial mats and pink microbial sediment, Lake Thetis shoreline; (**b**) scanned thin section of the outer part of a lithified stromatolite dome from Lake Thetis shoreline. The thin section has been cut into a 10 mm diameter circle for insertion into the NanoSIMS and a sub-area containing a micro-stromatolite has been selected for study (boxed area enlarged in **c**); (**c** and **d**) photomicrographs at increasing magnification of boxed area in (**b**) showing how individual laminae or groups of laminae can be targeted by NanoSIMS. Scale bar is 1 m for (**a**), 2 mm for (**b**), 600 μm for (**c**), 80 μm for (**d**).

living mats are intimately associated with these lithified domes but it is not clear whether they are now actively contributing to the domal stromatolite growth. In cross-section, the domes show variable textures, ranging from clotted (thrombolytic) in the dome centres to crudely laminated to distinctly laminated and rare branching columns at the outer margins. Lake Thetis stromatolites are interpreted to have formed by biologically induced carbonate precipitation (Grey et al., 1990).

The branching columns observed in some Lake Thetis stromatolites are of particular relevance to studies of ancient life. Branching is common in Precambrian stromatolites but is rarely observed in Phanerozoic examples. This is thought to be a consequence of increased bioturbation and grazing in the Phanerozoic inhibiting the development of mucilaginous layers that protect

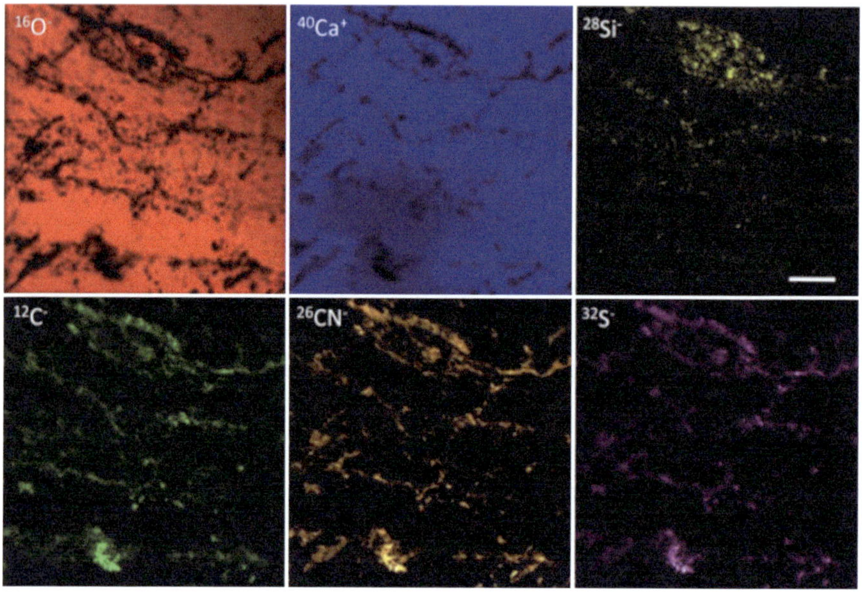

Figure 2. NanoSIMS ion images of laminations within a Lake Thetis stromatolite. The laminae run from left to right across the field of view and are highlighted by bright areas in the $^{12}C^-$, $^{26}CN^-$ and $^{32}S^-$ images, and black areas in the $^{16}O^-$ and $^{40}Ca^+$ images. The carbon image has been corrected for background carbon from calcium carbonate. $^{12}C^-$, $^{26}CN^-$ and $^{32}S^-$ correlate with one another spatially and in terms of signal intensity. $^{16}O^-$ and $^{40}Ca^+$ correlate with one another but anti-correlate with $^{12}C^-$, $^{26}CN^-$ and $^{32}S^-$, confirming that the observed $^{12}C^-$, $^{26}CN^-$ and $^{32}S^-$ signals come from organic material not from any carbonate or sulphate mineral phases. $^{28}Si^-$ shows some correlation with $^{12}C^-$, $^{26}CN^-$ and $^{32}S^-$, this is discussed in the text. Increased colour intensity reflects increased relative abundance of each ion. The *darker blue* colour in the lower half of the $^{40}Ca^+$ image is due to a decrease in ion signal when the conductive gold coat was sputtered away; scale bar is 8 μm.

the growing stromatolite. The Lake Thetis stromatolites therefore give a rare opportunity to study a potential true modern analogue for the most ancient stromatolites at the micro- to nano-scale. Our sample for NanoSIMS study comes from the outer margin of a domal stromatolite that exhibits laminations and micro-stromatolitic texture (Fig. 1b–d).

2. A modern marine stromatolite from Shark Bay, Western Australia (Fig. 3). Shark Bay, approximately 700 km north of Perth is home to diverse and abundant stromatolites and microbial mats. These occur in the intertidal and subtidal zones of a hypersaline embayment (Hamelin Pool), along almost 100 km of shoreline (Playford, 1990). Nine types of microbial mats have been identified in the bay, based on their surface morphology (Logan et al., 1974; Playford, 1990), along with domal and columnar stromatolites <4,000 years old. Some stromatolites are still actively growing today and a vertical accretion rate of ~0.4 mm/year is suggested by radiocarbon dating (Reid et al., 2003). Most Hamelin Pool stromatolites are unlaminated or weakly laminated with irregular

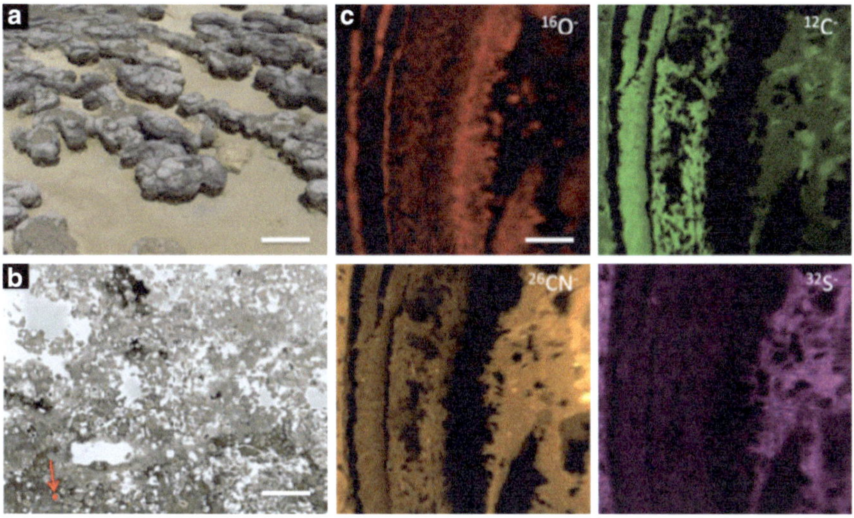

Figure 3. Modern marine stromatolites from Shark Bay, Western Australia. (**a**) Field photograph of living stromatolites from Hamelin Pool, Shark Bay; (**b**) scanned thin section of part of a lithified stromatolite collected from the shoreline of Hamelin Pool. This exhibits a rather clotted texture, similar to the "unlaminated calcarenite" described by Reid et al. (2003); (**c**) NanoSIMS ion images of oxygen, carbon, nitrogen (measured as CN^-) and sulphur from the area arrowed in (**b**). Increased colour intensity reflects increased relative abundance of each ion. The carbon image has been corrected for the background contribution from $CaCO_3$. $^{12}C^-$, $^{26}CN^-$ and ^{32}S correlate with one another but anticorrelate with $^{16}O^-$ just as they do in the Lake Thetis example. Scale bar is 50 cm for (**a**), 2.5 mm for (**b**), 10 μm for (**c**).

fenestral networks, quantities of trapped sediment and irregular outer surfaces, and are thought to be constructed mostly by pustular and colloform mat communities. More rarely, smooth mat communities construct well-laminated columns and mounds in the intertidal to shallowest subtidal zone (Playford, 1990). A number of recent studies have documented the microbial makeup of the stromatolites (e.g. Burns et al., 2004; Papineau et al., 2005) and some of the nine types of microbial mats (Allen et al., 2009) to begin to decipher the role of individual microbial groups in stromatolite construction at Shark Bay.

Our sample for NanoSIMS comes from the mid-intertidal zone. The fabric approximates that of the "unlaminated calcarenite" described by Reid et al. (2003) (compare our Fig. 3b with Reid et al., 2003, Plate 49, 1b and 1c), consisting of a mottled network of limestone (micrite, carbonate sand, peloids and oolites) and fenestrae (seen as holes in thin section). Occasional bivalve fragments and foraminifera are also cemented by micrite. This type of stromatolite was interpreted by Reid et al. (2003) as being formed by pustular microbial mats whose irregular surface morphology has inhibited laminae development. Our NanoSIMS data shows some of the chemical signals that are retained during the transformation from a living microbial mat to a stromatolite that can be preserved in the rock record.

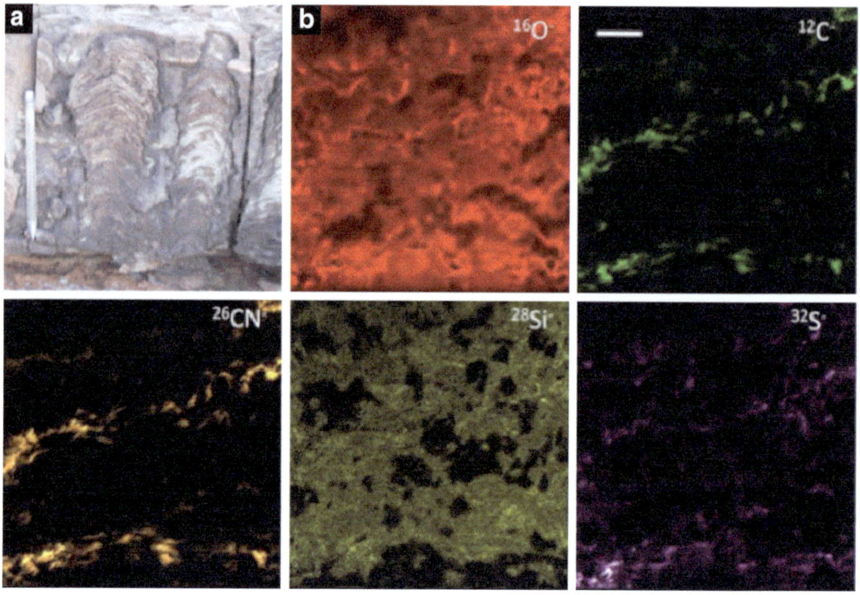

Figure 4. Ancient stromatolites from the ~2.72 Ga Tumbiana Formation, Western Australia. (**a**) Field photograph of columnar stromatolites from the Meentheena carbonate Member of the Tumbiana Formation (courtesy of the Geological Survey of Western Australia); (**b**), NanoSIMS ion images of oxygen, carbon, nitrogen (measured as CN^-), silicon and sulphur from a 60×60 μm^2 area of a Tumbiana stromatolite. Similar elemental distributions to the modern examples are observed, with laminations defined by spatial correlation and intensity of carbon, nitrogen and sulphur (partly). Increased colour intensity reflects increased relative abundance of each ion. Scale: pen in (**a**) is 12 cm long; scale bar in (**b**) is 10 μm.

3. An ancient stromatolite from the 2.72 Ga Tumbiana Formation of Western Australia (Fig. 4). Diverse stromatolites (domical, columnar, branching and stratiform) are found in the Meentheena Carbonate Member of the Tumbiana Formation, part of the Fortescue Group, Pilbara, Western Australia (Walter, 1972). The dominant microstructure consists of groups of laminations organised as bulbs, varying in size from several hundred microns to several centimetres in height (Lepot et al., 2008). Micritic calcium carbonate layers alternate with thinner siliceous layers. These stromatolites formed mainly by in situ precipitation (Grotzinger and Knoll, 1999) in either shallow seawater (e.g. Sakurai et al., 2005) or a large lake (Buick, 1992).

   Recent work has demonstrated clusters of "cell-like" carbonaceous globules associated with nano-aragonite crystals in bulbous Tumbiana stromatolites (Lepot et al., 2008); these carbonaceous globules contain many of the same functional groups as modern bacteria. This, taken together with the morphological complexity and close comparison to modern stromatolites of the

Tumbiana examples, provides strong evidence for their biogenicity. Hence, this sample was chosen for NanoSIMS study as an example of one of the oldest widely accepted biogenic stromatolites.

2.1.2. Stromatolite Results

2.1.2.1. Lake Thetis Stromatolite

Several ~ $50\times50\ \mu m^2$ laminated areas of the Lake Thetis stromatolite were mapped using both the Cs^+ and O^- primary ion beams. Similar chemical patterns were observed in each area and representative images are shown in Fig. 2. The most important result is the close correlation of the carbon, nitrogen and sulphur images. These images highlight chemically the laminations running from left to right across the field of view (Fig. 2). The near identical spatial distribution of these three elements, together with changes in intensity that parallel one another (i.e. a very bright area on the carbon map correlates with very bright areas on the nitrogen and sulphur maps), appears to be a characteristic of microbial material preserved as wavy laminations within the carbonate matrix. Under the optical microscope (e.g. Fig. 1d), it is difficult to distinguish the morphology of an individual lamina. However, the high spatial resolution of the NanoSIMS allows us to see in greater detail; we see that laminae commonly comprise a mixture of networks of anastomosing microbial material and some larger clumps or grains of microbial material. In addition, we can clearly observe small deviations and undulations in laminae and one area, in particular, where microbial material appears to join adjacent laminae (just north-east of the centre of Fig. 2). This could, for example, represent filamentous microbes that were growing up through trapped sediment towards the mat surface.

It is important to confirm that the elemental distributions of C, N and S are due to organics/microbial matter and not other mineral phases. This can be done by comparing the C, N and S maps with those for oxygen, which will highlight carbonate and sulphate minerals, and with images of the cations contained in known matrix minerals (in this case, calcium from calcium carbonate). The O and Ca images correlate well with one another but anti-correlate with C, N and S indicating that the C, N and S distributions are indeed due to preserved microbial material in this stromatolite. In ancient stromatolites, silicon will be a useful element to analyse as many ancient stromatolites are partially or entirely silicified. In this case, Si is merely an accessory phase, but it does show some correlation with C, N and S. This may be due to the trapping of windblown silica rich particles by the microbes or may be due to precipitation catalysed by the metabolic activity of the microbes. Further investigation is needed to distinguish these processes.

It is worth noting that NanoSIMS images give only the relative concentrations of each of the elements; it is not possible to obtain an absolute concentration. Additionally, it is very difficult to obtain meaningful ratios of elements present in a sample. For example, one may be interested in the N/C ratio of the microbial material to provide information about the parts of microbes that are preserved in stromatolites. The apparent N/C ratio of a sample in SIMS is controlled not only by its true N/C ratio but also by SIMS instrumental parameters that control CN^- and C^- ion yields, and by the matrix that the N- and C-rich material lies in. Such ratios were not attempted in this study.

2.1.2.2. Shark Bay Stromatolite

Several ~ 50×50 μm² areas of the Shark Bay stromatolite were mapped using both the Cs^+ and O^- primary ion beams. Since this stromatolite was not laminated when viewed under the optical microscope (Fig. 3b), we decided to target the peloidal to oolitic trapped grains to investigate the distribution of microbial material in these areas. Representative NanoSIMS ion images from the coatings of these grains are shown in Fig. 3c. A similar elemental distribution is observed to that seen in the Lake Thetis stromatolite. Carbon, nitrogen and sulphur distributions are closely correlated spatially, and oxygen shows almost exact anti-correlation. Again, this appears characteristic of microbial material interspersed with precipitated carbonate cement. In this case, some of the laminae are almost entirely organic, some are almost entirely carbonate cement and some (e.g. just left of centre in Fig. 3c) appear to be a mixture of microbial material and carbonate. The sulphur distribution does not correlate exactly with carbon and nitrogen. This may be due to the presence of some sulphate minerals or an as-yet-unidentified sulphur phase.

2.1.2.3. 2.72 Ga Stromatolite

The ~2.72 Ga Tumbiana Formation stromatolite was investigated in an identical manner to the modern examples from Shark Bay and Lake Thetis. Representative NanoSIMS ion images are shown in Fig. 4b. Despite the great age of this stromatolite, similar elemental distributions are observed to those found in the modern examples. Once again carbon and nitrogen correlate spatially with one another and highlight laminations running from left to right across the field of view (Fig. 4b). Sulphur correlates, in part, with carbon and nitrogen, but the correlation is not as well matched as in the modern examples. Again, we see that the laminations are not simple; they consist of intertwined strands and anastomosing networks of organic material. The silicon image shows that the original carbonate structure has been partly silicified and this appears to have masked much of the (anti)correlation of C, N and S with oxygen. Although the elemental patterns observed here are very similar to those in the modern examples, for stromatolites of this great age we must test that the signals we see are truly ancient and biological. Nitrogen is a key biological element that is concentrated by cellular products, along with carbon, via biotic fixation processes (e.g. Frausto Da Silva and Williams, 2001). Although the mere presence of nitrogen cannot be used to infer biogenicity, a close correlation on a sub-micron scale of nitrogen with carbon has been used as an indicator of biological processing in well-preserved carbonaceous microfossils from both the ~1,900 Ma Gunflint Chert of Canada (Robert et al., 2005) and the ~850 Ma Bitter Springs Chert of central Australia (Oehler et al., 2006), where the carbon and nitrogen chemical maps show direct correlation with optical images of the microfossils. In contrast, an abiogenic source of nitrogen in organic matter requires specific conditions and reactants. Nitrogen has been incorporated into organic matter abiogenically by high temperature Fischer Tropsch synthesis, using CO, H_2, NH_3 and a metal catalyst, in laboratory experiments attempting to replicate the formation of organic matter in meteorites (Kung et al., 1979; Hayatsu et al., 1972). However, field observations of the Meentheena Carbonate Member, indicating deposition in a lacustrine or shallow marine environment with little hydrothermal influence (e.g. Lepot et al., 2008), are inconsistent with conditions required for such abiogenic nitrogen synthesis. Sulphur is likewise concentrated by biotic fixation processes (Frausto Da Silva and Williams, 2001) and while its mere presence may be due to both biological and abiological concentration, a positive correlation with carbon and nitrogen lends further support to a biological formation mechanism.

The syngeneity of the Tumbiana chemical signals is unquestionable since the laminations with which they directly correlate have been shown to be primary (bio)sedimentary structures (e.g. Buick, 1992). NanoSIMS analysis of the Tumbiana stromatolites adds further support to a biological origin for these structures (cf. Lepot et al., 2008) and provides chemical signals that may act as a baseline for the study of other Precambrian stromatolites of unknown biogenicity.

2.2. SULPHUR ISOTOPE ANALYSIS

Here, we describe a method we have developed for measuring S isotope ratios in micron-sized grains using the high-resolution imaging capability of the NanoSIMS.

2.2.1. NanoSIMS Setup

All sulphur isotopic measurements were obtained using a Cs^+ primary ion beam focused to about 100 nm. The secondary ions $^{12}C^-$, $^{12}C^{14}N^-$, $^{32}S^-$ and $^{34}S^-$ were recorded simultaneously. The smallest pyrite grains (Type 1, see below) typically measure 2–3 μm in size, which are too small to be analysed by a static beam individually with sufficient stability and precision. It was therefore necessary to raster the primary ion beam over an area of 12 × 12 or 15 × 15 μm, producing an image from which the isotope data could be extracted. The images were obtained at a resolution of 256 × 256 pixels, each pixel measuring 47–59 nm, with a count time of 5 ms per pixel. Each pixel records the counts for each secondary ion, corrected for dead time (44 ns) and detector yield. Cameca NanoSIMS Image processing software was used to ratio the $^{34}S^-$ counts to the $^{32}S^-$ counts for the pixels defining each individual grain in an image. Variations in the S isotope ratio at the pixel-to-pixel scale are overshadowed by the large counting error on each pixel. It was therefore necessary to use the mean ratio value for the group of pixels defining a particular grain in the ion images. Charge compensation with the electron gun was not necessary as the pyrites maintained electrical conductivity throughout the analysis.

2.2.2. Factors Affecting Accuracy and Precision

2.2.2.1. Detector Ageing

The high electron affinity of S means that the negative secondary ion yield is very high under Cs^+ bombardment. On the one hand, this is good for generating enough counts to get good counting statistics, but has the disadvantage of generating so many counts that detector ageing and quasi-simultaneous arrivals (QSAs) at the electron multiplier (EM) become important issues (Slodzian et al., 2001). Detector ageing occurs when the secondary ion signal is so intense that the first dynode of the EM begins to degrade. The detector response changes over the course of its lifetime, but with intense signals (more than 400,000 counts per second) the effect is much more rapid. In analyses using two EMs where one EM has a very intense signal, the change in response of the detector with the high intensity signal can artificially influence the measured isotope ratio. To avoid this, the secondary ion flux can be kept low by cutting the beam with slits and apertures, which also increases the mass resolution to give better peak shapes. If necessary, a correction can be applied to the isotope ratio by monitoring the detector response during the analysis. Faraday cup (FC) detectors, which do not suffer from ageing effects or QSA, could not be used for this type of analysis because the response time of the FC is too slow and the background noise is too great to produce the good quality images needed for isotope ratio measurements.

2.2.2.2. QSA Effects

QSA presents a far more difficult artefact to overcome. This occurs when the ionisation efficiency is so high that a single primary ion may sputter two or more secondary ions, which travel through the mass spectrometer together and hit the first dynode of the EM more or less simultaneously. As EMs count the pulses generated by the incoming ions, simultaneous arrivals are measured as single pulses. This will, of course, have an effect on the isotope ratio, as part of the signal will not be detected. While QSA is not dependent on count rate, the magnitude of the effect is. High count rates at the detector can lead to contributions of up to 100‰ (10%) from QSA alone. This phenomenon is particularly problematic when measuring natural abundance isotopes with NanoSIMS. A theoretical correction model exists for a measured ratio (R): $R_{meas} = R_{true} \times (1 + \beta \times K)$, where K is the ratio of secondary to primary ion current and $\beta = 0.5$ (Slodzian et al., 2004; Hillion et al., 2008). However, experiments have shown that this value of β deviates significantly from 0.5, with observed values between 0.5 and 1 (e.g. $\beta = 0.69$ in Slodzian et al., 2004). In our preliminary analyses, neither the theoretical value of $\beta = 0.5$ nor the experimental value of $\beta = 0.69$ was found to adequately correct for QSA effects observed at our analytical conditions in the NanoSIMS 50.

Our modified experimental approach uses an empirical correction derived from measuring the change in isotope ratio at different primary to secondary ion current ratios. This is achieved by measuring the $^{34}S/^{32}S$ isotope ratio with a constant primary ion current, while varying the secondary ion current with slits and/or energy offsets. Plotting the $^{34}S/^{32}S$ ratio against K for a particular sulphide grain produces a linear relationship. The slope of the linear regression may then be used as the correction coefficient, β, in the data processing (cf. Slodzian et al., 2004; Hillion et al., 2008). This approach was used on each grain measured, producing an individual, empirically derived correction factor for each measurement. Another critical aspect of the QSA correction is that the relationship between the isotope ratio and the primary/secondary ion current only becomes linear after the secondary ion yield has reached a steady state.

For pyrite, this requires a primary ion implantation dose of $\sim 8 \times 10^{16}$ cm^{-2}. Under these analysis conditions, this requires approximately 150 min (1.4 pA primary beam current, 12×12 μm raster size). Figure 5 explains the correction of our raw data for QSA effects and presents the uncertainty associated with this correction.

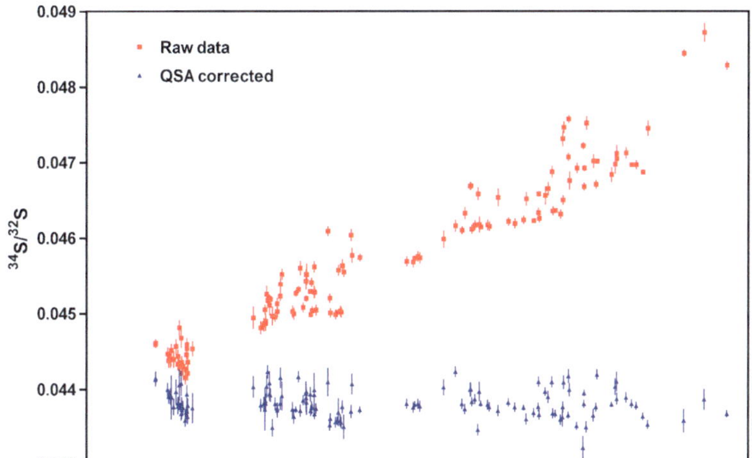

Figure 5. A comparison of raw NanoSIMS $^{34}S/^{32}S$ data and QSA corrected $^{34}S/^{32}S$ data. The raw data shows a strong positive correlation with ^{32}S count rate and a large spread of $^{34}S/^{32}S$ ratios; this correlation is merely an instrumental artefact due to QSA. The raw data is corrected by applying a series of linear regressions through data from each sulphide grain (i.e. one regression line through data for the CP1 standard, one through data for Type 2 pyrite, one through data for Type 3 pyrite, and one through data from each individual Type 1 pyrite grain). The errors associated with the QSA correction are dependent on the quality of fit of each regression line. These errors are propagated (added to the Poisson counting errors) to obtain the error bars shown on the QSA corrected data. The uncertainties associated with the QSA correction are generally only slightly larger than the Poisson counting errors. Hence, the spread in $^{34}S/^{32}S$ for QSA corrected data can be considered real. The sample data is then calibrated to V-CDT using the data obtained from the CP1 standard, which has a known $^{34}S/^{32}S$ ratio, a known V-CDT value and known analytical errors (-4.6 ± 0.2‰).

2.2.2.3. Poisson Counting Precision

As mentioned above, EMs are ion-counting detectors, which means their precision is dependent on Poisson counting statistics. The internal precision of a measurement derived from two EM detectors is given by:

$$\text{Precision}\,(\%) = \sqrt{(1/N_1 + 1/N_2)} \times 100$$

where N_1 and N_2 are the total counts on ion species 1 (^{32}S) and 2 (^{34}S), respectively. As ^{32}S represents 95.0% of the total S abundance (versus only 4.2% for ^{34}S), the counts from ^{32}S are over 20 times higher than ^{34}S. This makes the ^{32}S contribution to the counting precision largely negligible. Therefore, the precision can be approximated as $1/\sqrt{N_2}$, the counts on the lower abundance isotope. To achieve an isotope ratio measurement with an internal precision of 1‰ (0.1 %), one million counts from the lower abundance isotope are needed. For our analyses, multiple images were acquired sequentially, which were then added together to improve the counting statistics. For the standard (CP1) and Type 2 and 3 pyrites, where the grain filled the entire image frame, five images were acquired with a total acquisition time of 13 min. For the Type 1 pyrites, fifteen images were acquired, with a total time of 35 min per measurement, resulting in several million counts for ^{34}S over a typical 12 × 12 μm image. A typical 2–3 μm-sized grain, from the same raster size and under the same analytical conditions, would yield between 100,000 and 500,000 counts on ^{34}S, giving precisions of between ~3 and 1.5‰, respectively.

2.2.2.4. Reproducibility

Poisson counting statistics only represent the theoretical "best" precision possible. More critical for the NanoSIMS analyses is the reproducibility, or external precision, of the isotope measurements. This can be difficult to ascertain when there is inherent variability between grains in the sample. For the micron-sized pyrites (Type 1), it is necessary to confirm that any apparent natural grain-to-grain variation in the isotope ratio was not simply due to poor reproducibility in the measurements.

As we were unable to measure the reproducibility by measuring multiple Type 1 grains to determine the standard deviation, we demonstrated that the uncertainty due to the Poisson counting statistics is a close approximation of the external precision. To illustrate this, we acquired images of the CP1 standard and large grains (Type 2) in the sample where the grains were big enough to fill the entire imaged area. Using image processing software, we measured the isotope ratio for small sub-regions (discrete groups of pixels) to create areas similar in size to the smaller Type 1 pyrite grains, and compared the isotope ratios from the different areas (Fig. 6). The ratios calculated for each individual area (Fig. 6b–e) had essentially the same value as the ratio for the whole image (Fig. 6a), producing a small standard deviation between areas. The internal precision for the whole area, propagating Poisson counting error, uncertainty on the QSA correction, and conversion to per mil was 1.2‰, while the internal precision of the sub-regions was

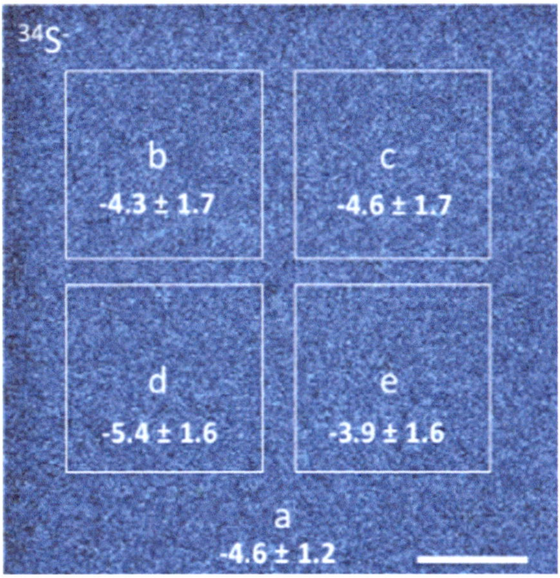

Figure 6. NanoSIMS ion image of $^{34}S^-$ from the CP1 standard, with $\delta^{34}S_{V\text{-}CDT}$ (‰) values superimposed to illustrate data reproducibility within a single analysis. 10–12 μm analysis areas (a) were split into sub-areas (b–e) of approximately equal size to Type 1 pyrites. The standard deviation (reproducibility) between these sub-areas is 0.6‰, which is smaller than the average Poisson error due to counting statistics (1.2–1.7‰). This gives $\chi^2 < 1$ and indicates that the reproducibility was good and the analyses are of sufficient quality. See text for further details. Scale bar is 2 μm.

1.6–1.7‰ due to the smaller number of pixels, and thus a lower number of counts, in the sub-regions. The standard deviation of the sub-regions (the reproducibility) was 0.6‰. As the Poisson counting error is the theoretical limit of counting precision, the 'true' uncertainty of the measurements here is given by the propagated counting statistics, i.e. the larger of the two errors. The quality of the analysis can then be expressed using the χ^2 relationship:

$$\chi^2 = (\text{error mean} / \text{Poisson})^2$$

A "good" measurement will have a reproducibility that is equal to or better than the counting statistics, giving a χ^2 of 1 or less. All measurements on the standard satisfied this test. Hence, if variations in $\delta^{34}S$ that are larger than the Poisson error are seen on a grain-to-grain scale amongst the Type 1 pyrites, they can be considered a real, natural variation in S-isotope fractionation.

2.2.2.5. Accuracy

The accuracy of an isotopic measurement is a function of the mass fractionation imparted by the instrument (instrumental mass fractionation or IMF), which incorporates effects due to the sputtering, ionisation, extraction, transmission and detection of the secondary ions (Winterholler et al., 2008). IMF effects associated with the sputtering and ionisation of the sample include matrix effects and sample preparation factors such as topography. IMF effects due to the extraction and transmission of secondary ions generally remain constant throughout an analytical session. The IMF effects associated with the detection and measurement of secondary ions relate to issues of QSA and dead time, which have already been addressed. For our flat, polished samples topography is not significant, and as the standard is mounted alongside the sample, effects due to movement on the sample holder are minimal.

To determine IMF, we calibrated our CP1 standard and Strelley Pool sample using LA-ICP-MS. After correcting for dead time and QSA effects, we used the standard calibration formula:

$$\delta^{34}S_{\text{IMF}} = \left[\left[({}^{34}S/{}^{32}S)_{\text{CP1 - NanoSIMS}} / ({}^{34}S/{}^{32}S)_{\text{CP1 - LA - ICP - MS}}\right] - 1\right] \times 1{,}000$$

to obtain a NanoSIMS IMF of −23‰ relative to the known absolute $^{34}S/^{32}S$ ratio of CP1. This is of a similar magnitude to the −20‰ IMF reported by Graham and Valley (1992) using a Cameca IMS 4f instrument.

With regard to matrix effects, Riciputi et al. (1998) investigated the matrix effects associated with different sulphide chemical compositions in SIMS. Their data showed a maximum of ~3‰ fractionation between seven different sulphide minerals when using the Cs^+ primary ion beam. Specifically, the fractionation between pyrite (FeS_2; our sample) and chalcopyrite ($CuFeS_2$; our standard) was <1‰ (Riciputi et al., 1998, Table 5 and Fig. 6). We confirmed that there is only a small matrix effect between pyrite and chalcopyrite by measuring several Type 3 pyrites using the same LA-ICP-MS instrumentation as for CP1 (Wacey et al., 2009). After the 23‰ IMF correction, there was an added discrepancy of ~2‰ between the average Type 3 $\delta^{34}S$ measured by LA-ICP-MS and that measured by NanoSIMS. This discrepancy may be due to some small matrix effect (1–2‰) but it is also of the same magnitude as the analytical error associated with the NanoSIMS measurements. Hence, we conclude that in our experiment the change in matrix between pyrite and chalcopyrite does not significantly affect IMF.

All $^{34}S/^{32}S$ ratios were converted to $\delta^{34}S$, in ‰, using the following formula:

$$\delta^{34}S(\%) = \left[({}^{34}S/{}^{32}S\text{sample}) - ({}^{34}S/{}^{32}S\text{standard})/({}^{34}S/{}^{32}S\text{standard})\right] \times 1{,}000$$

and all $\delta^{34}S$ values were normalised to the CP1 standard of $\delta^{34}S_{\text{V-CDT}} = -4.6‰$. Errors associated with the known LA-ICP-MS value for CP1 (±0.2‰) were propagated to obtain a final ‰ value for each sample grain. Isotope measurements of Strelley Pool sample pyrite grains were bracketed between CP1 standard measurements.

2.2.3. Samples

The samples used for developing a NanoSIMS sulphur isotope methodology come from a silicified sandstone at the base of the ~3.43 billion-year-old Strelley Pool Formation from the Pilbara of Western Australia (Wacey et al., 2006). The Strelley Pool Formation is a regionally extensive marker horizon that separates the thick, dominantly volcanic successions of the 3.525–3.427 Ga Warrawoona and 3.35–3.315 Ga Kelly Groups (Hickman, 2008). It is internationally acknowledged as one of the world's most important early Archean formations for the

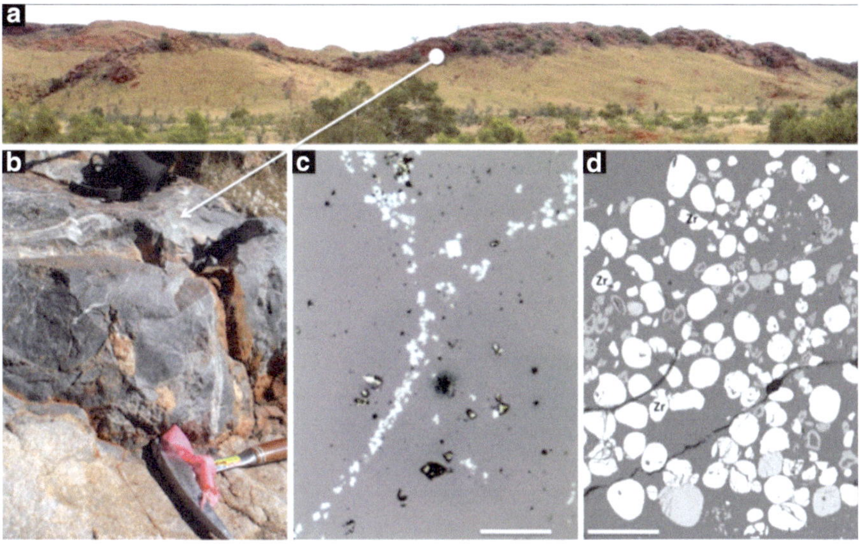

Figure 7. Pyrite samples for sulphur isotope analysis. (**a**) Field photograph looking south showing a ridge of the ~3.43 Ga Strelley Pool Formation, approximately 5 km east of Strelley Pool, Western Australia. The Formation dips almost vertically away from the viewer (to the south); (**b**) outcrop photograph of well-preserved grey-black sandstone, containing Type 1 and 2 pyrites, from the base of the Strelley Pool Formation; (**c**) reflected light photomicrograph of Type 1 pyrite (*white*) coating large, rounded detrital quartz grains (*grey*). A cube of Type 2 pyrite can be seen just above the centre of this image; (**d**) Reflected light photomicrograph of Type 3 detrital pyrite grains (*white*) occurring in a placer-type deposit with chromite (*light grey*), rutile (*light grey rings*) and zircon (Zr). Scale bar is 50 μm for (**c**) and 500 μm for (**d**).

study of early life on Earth. The sandstone unit exhibits low angle cross-bedding, channel bedforms and relatively high textural and compositional sediment maturity, consistent with deposition in a high energy regime (Wacey et al., 2006, 2010a). At the sample locality (Fig. 7a, b), near to Strelley Pool, the sandstone lies unconformably on eroded 3.515 Ga Coonterunah Subgroup volcanics, providing one of the earliest preserved land surfaces in the rock record (Lowe, 1983; Buick et al., 1995). The basal sandstone was lithified by the time that the overlying laminated carbonates (including putative stromatolites) of the Strelley Pool Chert were deposited. Early silicification is clearly indicated by the presence of rounded, intraformational clasts of the sandstone within higher beds of the sandstone itself as well as within the overlying chert. This is an important observation that helps to constrain the age of the pyrite within the sandstone. Four types of pyrite can be identified within the Strelley Pool sandstone. Type 1 pyrites are micron-sized grains coating many of the detrital framework sandstone grains (Fig. 7c). These are found in areas where the sandstone is black and evidently very well preserved, and also in black intraformational clasts higher up in the sandstone member.

They are thus interpreted as approximately syn-depositional (~3.43 Ga) in age. Type 2 pyrites are larger, cubic pyrite crystals (Fig. 7c, square just above centre). These likely formed by the dissolution and re-crystallisation of Type 1 pyrites during low-grade metamorphism sometime prior to the last significant metamorphic event in this area at ~2.9 Ga. Type 3 pyrites are well rounded and partially eroded detrital pyrite (older than 3.43 Ga) occurring in spatially restricted placer deposits with other heavy minerals such as zircon, chromite and rutile (Fig. 7d). Type 4 pyrite consists of small cubic or rhombic crystals sealed within detrital siliceous grains. These are rather rare and were not analysed for $\delta^{34}S$ in this study, although they occur in close proximity to areas analysed for $\delta^{13}C$ (see Sect. 2.3 below).

2.2.4. Sulphur Isotope Results

Type 1 pyrite: Sixty-nine Type 1 pyrite measurements, from four separate analysis areas of the thin section, have a range of $\delta^{34}S_{V\text{-}CDT}$ from −11.7‰ to 6.3‰. The mean value for these grains is −2.1‰.

Type 2 pyrite: Forty-four measurements from Type 2 pyrites have a range of $\delta^{34}S_{V\text{-}CDT}$ from −6‰ to 3.4‰. The mean value for these grains is −2.9‰.

Type 3 pyrite: Eight measurements from Type 3 pyrites have a range of $\delta^{34}S_{V\text{-}CDT}$ from −1.3‰ to 3.9‰. The mean value for these grains is 1.5‰.

The analytical errors for the three types of pyrite were of similar magnitude and ranged from <1‰ to ~4‰, most commonly around 2‰. While these errors are clearly larger than those obtained by conventional gas source mass spectrometry, they are sufficiently small to trace most natural fractionations of sulphur isotopes by biological and non-biological means in the rock record.

Two areas of Type 1 pyrites are illustrated in Fig. 8. It is not possible to demonstrate the $^{34}S/^{32}S$ ratio as an image because the differences are only apparent on a grain-to-grain scale, and not on a pixel-to-pixel scale. The grains appear largely homogenous, with little evidence of zonation across a grain. The grain edges are not well defined due to small amounts of instrumental drift during the acquisition. Hence, the pixels at the edges of the grains were not selected for the isotope ratio measurements.

The mean of Type 1 pyrites (−2.1‰) is very similar to that of Type 2 (−2.9‰). This similarity in means between Type 1 and Type 2 pyrites demonstrates the amount of information that would have been lost had the pyrites not been analysed by NanoSIMS. For example, using laser ablation it would likely only have been possible to obtain one measurement from this entire area (using conventional powdering of the rock sample the spatial resolution would be many, many times worse); this measurement would have returned a value approximately equal to the mean of Type 1 and Type 2 pyrites (i.e. around −2.5‰). This value is not far removed from that for magmatic mantle-derived H_2S [+1 ± 1‰ (Ohmoto and Rye, 1979)]. Since the isotopic composition of non-biological magma-derived sulphides may range between 4–5‰ of this mantle value (Schneider, 1970), the most logical conclusion from bulk data would be that both Type 1 and Type 2 pyrites had formed non-biologically from hydrothermal or volcanogenic processes.

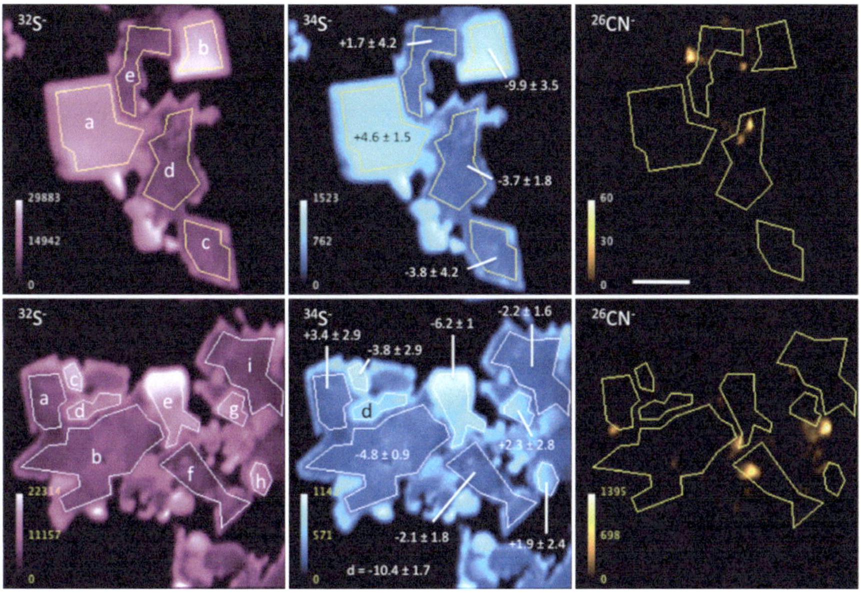

Figure 8. $\delta^{34}S$ measurement from Type 1 pyrite. NanoSIMS secondary ion images of $^{32}S^-$, $^{34}S^-$ and $^{26}CN^-$ associated with two areas of micron-sized (Type 1) pyrites within the Strelley Pool sandstone. *Black area* is background quartz. Cameca NanoSIMS Image processing software was used to isolate the pixels defining different grains (e.g. **a**–**e** on top row) in each image and then to ratio the $^{34}S^-$ counts to the $^{32}S^-$ counts for each grain. The $^{26}CN^-$ image shows organic matter associated with the pyrite. The mean $\delta^{34}S_{V\text{-}CDT}$ values (‰) for each grain with fully propagated errors are superimposed on the ^{34}S images. Scale bar is 2.5 μm. Colour scale is calibrated in counts per second.

In contrast, using NanoSIMS (Fig. 9), we see that the Type 1 pyrite exhibits a spread in $\delta^{34}S$ of 18‰ over less than 1 mm within the thin section, and a maximum ^{34}S depletion of about −15‰ from inferred early Archean seawater sulphate [$\delta^{34}S \sim +3$‰ (Strauss, 2003)]. This is of a similar magnitude to putative reports of biological sulphur processing in other early Archean sediments (Ohmoto et al., 1993; Shen et al., 2001; Shen and Buick, 2004; Philippot et al., 2007). This data does not give definitive evidence for biological participation in the formation of Type 1 pyrite, because non- biological processes such as reduction of sulphate by Fe^{2+}-bearing minerals (Ohmoto and Goldhaber, 1997) and thermochemical reduction by hydrocarbons (Machel, 2001) can impart fractionations of a similar magnitude. However, when combined with other evidence (e.g. close association of carbon and nitrogen with the Type 1 pyrites (Fig. 8), and field evidence for low temperature, syn-depositional to early diagenetic timing of formation of Type 1 pyrite), a biological formation mechanism appears the most plausible scenario. The modelling of this data to potentially determine the microbial metabolism and concentrations of sulphate in the Archean ocean is beyond the scope of this chapter; instead details can be found in Wacey et al. (2010b).

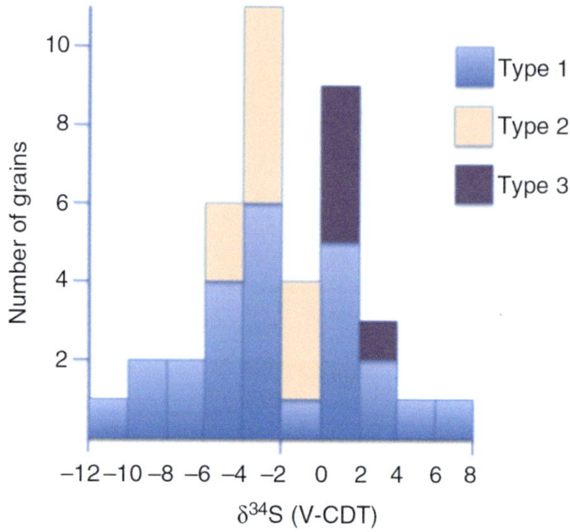

Figure 9. Stacked histogram of δ^{34}S data from three types of pyrite within the ~3.43 Ga Strelley Pool sandstone. Sixty-nine analyses on 25 individual grains were made for Type 1 pyrite, 44 analyses on ten grains for Type 2, and eight analyses on five grains for Type 3. This histogram reports the mean value for each individual grain.

The similarity in the mean values for Type 1 and Type 2 pyrites suggests that Type 2 may have formed from dissolution and recrystallisation of Type 1 during low-grade metamorphism. Type 3 pyrites exhibit δ^{34}S signatures (mean 1.5‰) within the range of values for igneous sulphides formed using H_2S derived from Archean magma (Ohmoto et al., 1993). These Type 3 detrital pyrites were likely eroded from a >3.43 Ga igneous body and then concentrated as placer deposits in the Strelley Pool sandstone.

2.3. CARBON ISOTOPE ANALYSIS

Carbon isotope measurements were made using an integration method, as the features were mostly large enough to completely fill the field of view. A slightly defocused Cs^+ primary beam was rastered over an area of $10 \times 10\ \mu m^2$, made up of 64×64 pixels, and the counts per pixel were integrated over the total number of pixels per frame. To achieve the one million counts from the less abundant isotope, necessary for a precision of 1‰, 500 frames were measured for each analysis. Only counts from the central 44×44 pixels were recorded to eliminate the contribution from crater edge effects.

The mass spectrometer was tuned to allow the ^{13}C peak to be resolved from ^{12}C^{1}H on mass 13, while maintaining good counting statistics ($M/\Delta M = 3{,}000$, with 200k cps on ^{12}C). Counts for ^{12}C and ^{13}C were recorded simultaneously using EMs, and the ratio was calculated from the total counts recorded. The counts were corrected for instrumental effects, such as dead time and EM yield. Under these analytical conditions, detector ageing was negligible. QSA was considered negligible because the count rates for the standard and sample were very similar, so any QSA correction should not affect the relative δ^{13}C values between standard and sample. This point still requires some clarification, however, as the extent of the QSA effect on carbon has yet to be accurately determined.

The carbon-rich material in our samples is largely kerogenous in nature, but its great age and metamorphism mean that it approaches a graphitic composition (although there is still some disorder seen in transmission electron microscopy and laser Raman spectra); thus, the NanoSIMS was calibrated using the graphite standard USGS-24, which has a $\delta^{13}C_{V\text{-}PDB}$ value of −16‰. $^{13}C/^{12}C$ raw ratios were converted to $\delta^{13}C$ values (in ‰) using the formula:

$$\delta^{13}C_{V\text{ - }PDB}\% = [(^{13}C/^{12}C_{sample}) - (^{13}C/^{12}C_{standard}) \times 1{,}000$$

To reduce instrumental effects caused by the movement of the ion beam between sample holders, the grains of USGS-24, typically around 10 μm in diameter, were mounted directly within the sample thin-section close to the analysis area. The NanoSIMS was then tuned to give good spot-to-spot reproducibility on the standard before moving to the sample kerogen. Multiple analyses of several grains of USGS-24 gave $\delta^{13}C$ values with a 1σ external precision for the standard of 3‰.

2.3.1. Samples

Our NanoSIMS carbon isotope methodology was developed using kerogen from ambient inclusion trails (AITs) from a silicified sandstone at the base of the ~3.43 Ga Strelley Pool Formation (see Sect. 2.2.1 for geological setting). AITs are poorly understood microtubular structures that appear to be created by the migration of mineral crystals through a lithified or partly lithified substrate (Tyler and Barghoorn, 1963). One potential driving force behind this phenomenon is the decomposition of organic material attached to the mineral crystals (Knoll and Barghoorn, 1974). If such organic material could be shown to be biological, then AITs may be used as a biosignature for investigating early life on Earth. Wacey et al. (2008) have investigated AITs from the Strelley Pool Formation in some detail, this is summarised below.

AITs in the Strelley Pool sandstone are strictly substratum specific, confined to ~5% of rounded clasts, comprising well-preserved microcrystalline silica that contain small cubes of pyrite (Type 4 pyrite above). The microtubes are 1–15 μm in diameter, up to 300 μm in length and can be either hollow or filled with silica, ferrous phosphate, aluminium phosphate or jarosite (likely a younger contaminant). They exhibit a range of morphologies from straight, to curved, twisted or even helical, and are occasionally branched. Their diameter appears to remain constant along their length, and they commonly cluster around clots of organic matter. These have been interpreted as AIT for the following reasons (Wacey et al., 2008): they only occur in clasts rich in pyrite (or pseudomorphs after pyrite) and commonly also rich in organic matter; some still contain propelled terminal pyrite crystals (Fig. 10a); many microtubes originate from the centre of the silica grains and can move both inwards and outwards through the grain. In contrast, endolithic borings would only penetrate inwards from the edge of a grain; many of the microtubes have polygonal cross sections, consistent with the geometry of

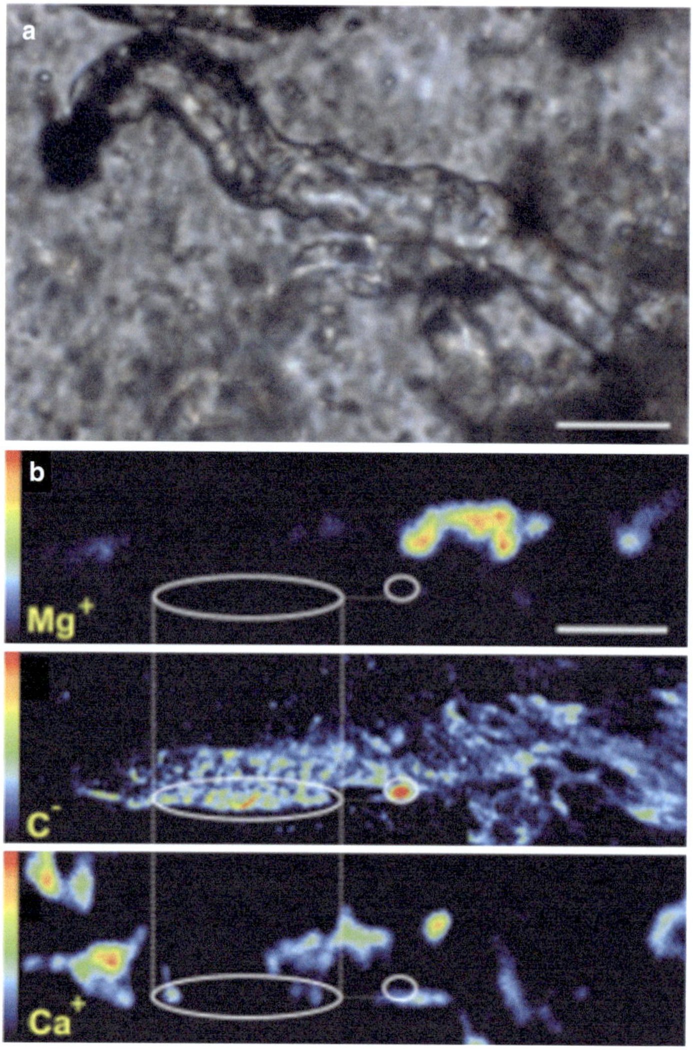

Figure 10. An ambient inclusion trail (AIT) from the ~3.43 Ga Strelley Pool sandstone. (**a**) Thin section photomicrograph showing an AIT with terminal pyrite crystal that has been propelled from the bottom right to the top left of the image. Note *dark areas* of organic material within the tubular structure; (**b**) NanoSIMS ion images of magnesium, carbon and calcium from an adjacent AIT, showing that carbon distribution is not correlated with either calcium or magnesium within the AIT. This indicates that the carbon is present in organic material and not a mineral phase such as carbonate. Oval areas indicate identical areas on each image. Scale bars are 10 μm. Part (**b**) is taken from Wacey et al. (2008).

a mineral grain, but inconsistent with microfossils or microbial borings. Our $\delta^{13}C$ measurements were made on carbon found within these AITs, mostly occurring as linings along tube edges.

A slightly modified version of our NanoSIMS carbon isotope methodology has also been used for solidified oil (pyrobitumen) and kerogen in ~2.63 Ga shales from the Jeerinah Formation, Pilbara Craton, Western Australia (see Rasmussen et al., 2008 for details).

2.3.2. Carbon Isotope Results

Using NanoSIMS, Wacey et al. (2008) found significant enrichments in carbon and nitrogen in AIT from the centre of microcrystalline silica grains in the Strelley Pool Formation. These enrichments corresponded to silica and oxygen depletion in the host grain and were often greatest in a narrow band at the outer edge of the AIT. Ca and Mg were present in some AITs but they did not correlate with carbon (Fig. 10b) indicating that the carbon was organic, not merely bound within a carbonate phase. Other potentially bio-limiting elements [e.g. P, S, K, Co, Fe, Ni, Zn (Frausto Da Silva and Williams, 2001)] were also correlated with carbon or concentrated in some way within the AIT infillings (Wacey et al., 2008).

NanoSIMS carbon isotope analysis was performed directly on the carbon rich interior of two AIT structures (AIT 1 and AIT 2). Five analysis points, each comprising 500 measurements, from AIT 1 resulted in a mean $\delta^{13}C_{V\text{-}PDB}$ value of −27.6‰; the errors for each analysis point were ~3.0‰. Four analysis points, again each comprising 500 measurements, from AIT 2 resulted in a mean $\delta^{13}C_{V\text{-}PDB}$ value of −24.0‰, and errors of 3.1–3.8‰. The errors on these analyses are around an order of magnitude higher than those obtainable by conventional gas-source mass spectrometry (e.g. Van Zuilen et al., 2002); clearly, NanoSIMS is not the technique to use in cases where very high precision is needed. However, there are many situations where spatial resolution is more important than high precision, and errors of ~3‰ do not compromise the interpretation of the data. This is often the case in the field of Precambrian palaeobiology where biological fractionations are relatively large [20–40‰ in favour of ^{12}C for enzymatic carboxylation reactions (Schidlowski, 1988)] but the microbes responsible are very small. In the case highlighted here, NanoSIMS is the only technique capable of measuring isotopes from the small amounts of carbon associated with these AIT structures, and gives data that is consistent with the biological C3 "Calvin cycle" carbon fixation pathway. Other carbon fixation pathways (e.g. C4 dicarboxylic acid or reductive acetyl coenzyme A pathways) do not fractionate carbon isotopes to the extent of the C3 Calvin Cycle (Schidlowski, 1988), but such biological signatures should still be resolvable using NanoSIMS.

3. Summary

NanoSIMS facilitates the in situ elemental and isotopic analysis of very small objects and the correlation of this data with morphological features observed in petrographic thin section, and with data from other techniques such as laser

Raman or electron microscopy. Applications of NanoSIMS are widespread through the geological and biological sciences; here, we have demonstrated its applicability for improving our understanding of chemical signals typical of life in both modern and ancient environments. The spatial resolution of the NanoSIMS allows detection and mapping of elements and isotopes within individual laminations of stromatolites, biominerals, trace fossils and microfossils. Quantitative isotope data can be obtained with 1–3‰ precision on objects as small as ~2 μm.

4. Acknowledgements

Our thanks go to Owen Green and Martin Brasier for provision of stromatolite samples, and to Adrian Boyce for laser ablation sulphur isotope analyses. DW is supported by a UPRF from The University of Western Australia. Both authors acknowledge the facilities, scientific and technical assistance of the Australian Microscopy and Microanalysis Research Facility at the Centre for Microscopy, Characterisation and Analysis, The University of Western Australia, a facility funded by The University, State and Commonwealth Governments. The authors thank P. Hoppe and an anonymous reviewer for comments that improved the quality of this chapter, and V. Tewari and J. Seckbach for the invitation to contribute to this book.

5. References

Allen, M.A., Goh, F., Burns, B.P. and Neilan, B.A. (2009) Bacterial, archaeal and eukaryotic diversity of smooth and pustular microbial mat communities in the hypersaline lagoon of Shark Bay. Geobiology **7**: 82–96.

Allwood, A.C., Walter, M.R., Kamber, B.S., Marshall, C.P. and Burch, I.W. (2006) Stromatolite reef from the Early Archaean era of Australia. Nature 441: 714–718.

Baumgartner, L.K., Reid, R.P., Dupraz, C., Decho, A.W., Buckley, D.H., Spear, J.R., Przekop, K.M. and Visscher, P.T. (2006) Sulfate reducing bacteria in microbial mats: changing paradigms, new discoveries. Sediment. Geol. **185**: 131–145.

Brasier, M., McLoughlin, N., Green, O. and Wacey, D. (2006) A fresh look at the fossil evidence for early Archaean cellular life. Philos. Trans. R. Soc. B **361**: 887–902.

Brown, R.G. (1983) Sea level history over the past 15,000 years along the Western Australian coastline, In: D. Hopley (ed.) *Australian Sea Levels in the Last 15,000 Years: A Review.* Department of Geography, James Cook University of North Queensland, Occasional Paper No. 3, pp. 28–36.

Buick, R. (1992) The antiquity of oxygenic photosynthesis – evidence from stromatolites in sulfate-deficient Archean lakes. Science **255**: 74–77.

Buick, R., Thornett, J.R., McNaughton, N.J., Barley, M.E. and Savage, M. (1995) Record of emergent continental crust ~3.5 billion years ago in the Pilbara craton of Australia. Nature **375**: 574–577.

Burns, B.P., Goh, F., Allen, M. and Neilan, B.A. (2004) Microbial diversity of extant stromatolites in the hypersaline marine environment of Shark Bay, Australia. Environ. Microbiol. **6**: 1096–1101.

Frausto Da Silva, J.J.R. and Williams, R.J.P. (2001) *The Biological Chemistry of the Elements: The Inorganic Chemistry of Life*, 2nd edn. Oxford University Press, Oxford.

Graham, C.M. and Valley, J.W. (1992) Sulfur analyses of pyrite. Chem. Geol. **101**: 169–172.

Grey, K., Moore, L.S., Burne, R.V., Pierson, B.K. and Bauld, J. (1990) Lake Thetis, Western Australia: an example of saline lake sedimentation dominated by benthic microbial processes. Aust. J. Mar. Freshwater Res. **41**: 275–300.

Grotzinger, J.P. and Knoll, A.H. (1999) Stromatolites in Precambrian carbonates: evolutionary mileposts or environmental dipsticks? Ann. Rev. Earth Planet. Sci. **27**: 313–358.

Grotzinger, J.P. and Rothman, D.H. (1996) An abiotic model for stromatolite morphogenesis. Nature **383**: 423–425.

Hayatsu, R., Studier, M.H., Matsuoka, S. and Anders, E. (1972) Origin of organic matter in early solar system – VI. Catalytic synthesis of nitriles, nitrogen bases and porphyrin-like pigments. Geochim. Cosmochim. Acta **36**: 555–571.

Hickman, A.H. (2008) Regional review of the 3426–3350 Ma Strelley Pool Formation, Pilbara Craton, Western Australia. Geol. Surv. West. Aust. Rec. **2008**(15): 27p.

Hillion, F., Daigne, F., Girard, F. and Slodzian, G. (1994) A new high performance instrument: the CAMECA NanoSIMS 50, In: A. Benninghoven, Y. Nihei, R. Shimizu and H.W. Werner (eds.) Secondary Ion Mass Spectrometry: SIMS IX. Wiley, New York, pp. 254–256.

Hillion, F., Kilburn, M.R., Hoppe, P., Messenger, S. and Weber, P.K. (2008) The effect of QSA on S, C, O and Si isotopic ratio measurements. Geochim. Cosmochim. Acta 72: A377.

Hofmann, H.J., Grey, K., Hickman, A.H. and Thorpe, R.I. (1999) Origin of 3.45 Ga Coniform Stromatolites in the Warrawoona Group, Western Australia. Bull. Geol. Soc. Am. **111**: 1256–1262.

Knoll, A.H. and Barghoorn, E.S. (1974) Ambient pyrite in Precambrian chert: new evidence and a theory. Proc. Nat. Acad. Sci. U.S.A. **71**: 2329–2331.

Kung, C., Hayatsu, R., Studier, M.H. and Clayton, R.N. (1979) Nitrogen isotope fractionation in the Fischer-Tropsch synthesis and in the Miller-Urey reaction. Earth Plan. Sci. Lett. **46**: 141–146.

Lepot, K., Benzerara, K., Brown Jr, G.E. and Philippot, P. (2008) Microbially influenced formation of 2,724-million-year-old stromatolites. Nat. Geosci. **1**: 118–121.

Logan, B.W., Hoffman, P. and Gebelein, C.D. (1974) Algal mats, cryptalgal fabrics, and structures, Hamelin Pool, Western Australia. Am. Assoc. Pet. Geol. Mem. **22**: 140–194.

Lowe, D.R. (1980) Stromatolites 3,400-Myr old from the Archean of Western Australia. Nature **284**: 441–443.

Lowe, D.R. (1983) Restricted shallow-water sedimentation of early Archaean stromatolitic and evaporitic strata of the Strelley Pool chert, Pilbara block, Western Australia. Precambrian Res. **19**: 239–283.

Lowe, D.R. (1994) Abiological origin of described stromatolites older than 3.2 Ga. Geology **22**: 387–390.

Machel, H.G. (2001) Bacterial and thermochemical sulfate reduction in diagenetic settings – old and new insights. Sediment. Geol. **140**: 143–175.

McCollum, T.M. and Seewald, J.S. (2006) Carbon isotope composition of organic compounds produced by abiotic synthesis under hydrothermal conditions. Earth Planet. Sci. Lett. **243**: 74–84.

McLoughlin, N., Wilson, L. and Brasier, M.D. (2008) Growth of synthetic stromatolites and wrinkle structures in the absence of microbes – implications for the early fossil record. Geobiology 6: 95–105.

Oehler, D.Z., Robert, F., Mostefaoui, S., Meibom, A., Selo, M. and McKay, D.S. (2006) Chemical mapping of Proterozoic organic matter at submicron spatial resolution. Astrobiology **6**: 838–850.

Ohmoto, H. and Goldhaber, M.B. (1997) Sulphur and carbon isotopes, In: H.L. Barnes (ed.) *Geochemistry of Hydrothermal Ore Deposits*, 3rd edn. Wiley, New York, pp. 517–611.

Ohmoto, H. and Rye, R.O. (1979) Isotopes of sulphur and carbon, In: H.L. Barnes (ed.) *Geochemistry of Hydrothermal Ore Deposits*, 2nd edn. Wiley, New York, pp. 509–567

Ohmoto, H., Kakegawa, T. and Lowe, D.R. (1993) 3.4-billion-year-old pyrites from Barberton, South Africa: sulfur isotope evidence. Science **262**: 555–557.

Papineau, D., Walker, J.J., Mojzsis, S.J. and Pace, N.R. (2005) Composition and structure of microbial communities from stromatolites of Hamelin Pool in Shark Bay, Western Australia. Appl. Environ. Microbiol. **71**: 4822–4832.

Philippot, P., Van Zuilen, M., Lepot, K., Thomazo, C., Farquhar, J. and Van Kranendonk, M.J. (2007) Early Archean microorganisms preferred elemental sulfur, not sulfate. Science **317**: 1534–1537.

Playford, P.E. (1990) Geology of the Shark Bay area, Western Australia, In: P.F. Berry, S.D. Bradshaw and B.R. Wlison (eds.) *Research in Shark Bay*. Report of the France-Australe Bicentenary Expedition Committee, Western Australian Museum, pp. 13–33.

Rasmussen, B., Fletcher, I.R., Brocks, J.J. and Kilburn, M.R. (2008) Reassessing the first appearance of eukaryotes and cyanobacteria. Nature **455**: 1101–1104.

Reid, R.P., James, N.P., Macintyre, I.G., Dupraz, C.P. and Burne, R.V. (2003) Shark Bay stromatolites: microfabrics and reinterpretation of origins. Facies **49**: 299–324.

Riciputi, L.R., Paterson, B.A. and Ripperdan, R.L. (1998) Measurement of light stable isotope ratios by SIMS: matrix effects for oxygen, carbon, and sulfur isotopes in minerals. Int. J. Mass Spectrom. 178: 81–112.

Robert, F., Selo, M., Hillion, F. and Skrzypczak, A. (2005) NanoSIMS images of Precambrian fossil cells. LPSC XXXVI, abstract 1314.

Runnegar, B., Dollase, W.A., Ketcham, R.A., Colbert, M. and Carlson, W.D. (2001) Early Archaean sulphates from Western Australia first formed as hydrothermal barites not gypsum evaporites. Geol. Soc. Am. Abstr. Prog. **33**: A-404.

Sakurai, R., Ito, M., Ueno, Y., Kitajima, K. and Maruyama, S. (2005) Facies architecture and sequence-stratigraphic features of the Tumbiana Formation in the Pilbara Craton, northwestern Australia: implications for depositional environments of oxygenic stromatolites during the Late Archean. Precambrian Res. **138**: 255–273.

Schidlowski, M. (1988) A 3,800-million-year isotopic record of life from carbon in sedimentary rocks. Nature **333**: 313–318.

Schneider, A. (1970) The sulfur isotope composition of basaltic rocks. Contrib. Mineral. Petrol. **25**: 95–124.

Schopf, J.W. (2006) The first billion years: when did life emerge? Elements **2**: 229–233.

Semikhatov, M.A., Gebelein, C.D., Cloud, P., Awramik, S.M. and Benmore, W.C. (1979) Stromatolite morphogenesis: progress and problems. Can. J. Earth Sci. **16**: 992–1015.

Shen, Y. and Buick, R. (2004) The antiquity of microbial sulfate reduction. Earth Sci. Rev. **64**: 243–272.

Shen, Y. Buick, R. and Canfield, D.E. (2001) Isotopic evidence for microbial sulphate reduction in the early Archean era. Nature **410**: 77–81.

Slodzian, G., Chaintreau, M., Dennebouy, R. and Rousse, A. (2001) Precise in situ measurements of isotopic abundances with pulse counting of sputtered ions. Eur. Phys. J. Appl. Phys. **14**(3): 199–231.

Slodzian, G., Hillion, F., Stadermann, F.J. and Zinner, E. (2004) QSA influences on isotopic ratio measurements. Appl. Surf. Sci. **231–232**: 874–877.

Strauss, H. (2003) Sulphur isotopes and the early Archaean sulphur cycle. Precambrian Res. **126**: 349–361.

Tice, M.M. and Lowe, D.R. (2004) Photosynthetic microbial mats in the 3,416-Myr-old ocean. Nature **431**: 549–552.

Tyler, S.A. and Barghoorn, E.S. (1963) Ambient pyrite grains in Precambrian cherts. Am. J. Sci. **261**: 424–432.

Ueno, Y., Isozaki, Y., Yurimoto, H. and Maruyama, S. (2001) Carbon isotopic signatures of individual Archean microfossils (?) from Western Australia. Int. Geol. Rev. **43**: 196–212.

Van Zuilen, M.A., Lepland, A. and Arrhenius, G. (2002) Reassessing the evidence for the earliest traces of life. Nature **418**: 627–630.

Wacey, D. (2009) *Early Life on Earth: A Practical Guide*. Springer, Amsterdam, 285 p. ISBN 978-1-4020-9388-3.

Wacey, D., McLoughlin, N., Green, O.R., Stoakes, C.A. and Brasier, M.D. (2006) The ~3.4 billion-year-old Strelley Pool Sandstone: a new window into early life on Earth. Int. J. Astrobiol. **5**: 333–342.

Wacey, D., Kilburn, M.R., McLoughlin, N., Parnell, J., Stoakes, C.A., Grovenor, C.R.M. and Brasier, M.D. (2008) Use of NanoSIMS in the search for early life on Earth: ambient inclusion trails in a c. 3400 Ma sandstone. J. Geol. Soc. London **165**: 43–53.

Wacey, D., McLoughlin, N., Stoakes, C.A., Kilburn, M.R., Green, O.R. and Brasier, M.D. (2010a) The 3426–3350 Ma Strelley Pool Formation in the East Strelley greenstone belt – a field and petrographic guide. Geol. Surv. West. Aust. Rec. **2010**(10), 64p.
Wacey, D., McLoughlin, N., Whitehouse, M.J. and Kilburn, M.R. (2010b) Two coexisting sulfur metabolisms in a ca. 3400 Ma sandstone. Geology **38**: 1115–1118.
Walsh, M.M. and Lowe, D.L. (1999) Modes of accumulation of carbonaceous matter in the early Archean: a petrographic and geochemical study of the carbonaceous cherts of the Swaziland Supergroup, In: D.R. Lowe and G.R. Byerly (eds.) *Geologic Evolution of the Barberton Greenstone Belt, South Africa.* Geol. Soc. Am. Spec. Pap. 329. Boulder, Colorado, pp. 167–188.
Walter, M.R. (1972) *Stromatolites and the Biostratigraphy of the Australian Precambrian and Cambrian.* The Palaeontological Association, London.
Walter, M.R. (1976) Introduction, In: M.R. Walter (ed.) *Stromatolites.* Elsevier, Amsterdam, pp. 1–3.
Walter, M.R., Buick, R. and Dunlop, J.S.R. (1980) Stromatolites, 3,400–3,500 Myr old from the North Pole area, Western Australia. Nature **284**: 443–445.
Winterholler, B., Hoppe, P., Foley, S. and Andreae, M.O. (2008) Sulfur isotope ratio measurements of individual sulfate particles by NanoSIMS. Int. J. Mass Spectrom. **272**: 63–77.

Biodata of **Professor Vinod Chandra Tewari**, author of ***"Stromatolites, Organic Walled Microorganisms, Laser Raman Spectroscopy and Confocal Laser Scanning Microscopy of the Meso-Neoproterozoic Buxa Formation, Ranjit Window, Sikkim Lesser Himalaya, NE India"***

Professor Vinod C. Tewari is currently the Head of the Sedimentology Group at Wadia Institute of Himalayan Geology, Dehradun, Uttarakhand State, India. He has been a Regular and Senior Associate and also TRIL Fellow of International Centre for Theoretical Physics, Trieste, Italy, between 1998 and 2009. He was born in Almora, India, in 1954 and obtained his higher education including Ph.D. in *Geology* from the University of Lucknow in 1986. He joined Wadia Institute in 1978 and continued his research in the Lesser Himalayan stromatolites, microbiota, and isotope chemostratigraphy. Dr. Tewari taught Geology at Kumaon University, Nainital, Uttarakhand (U.K.), India, as Professor of Geology. Dr. Tewari worked on stable isotopes in Biogeochemistry department of the Max-Planck Institute, Mainz, Germany, as Visiting Professor in 1995. Professor Tewari's scientific interests are in the areas of Precambrian and Phanerozoic stromatolites, sedimentation, carbon isotope chemostratigraphy, genesis, early evolution, and diversification of life and its astrobiological significance. He was a visiting scientist in the CSEOL, University of California, Los Angeles, USA, in 2007 and used Laser Raman Spectroscopy and CLSM techniques for the Proterozoic Buxa microbiota. He has been associated with several International Geological Correlation Programme (I.G.C.P.) Projects such as Stromatolites, Biosedimentology of the Microbial Buildups, Phosphorites, Precambrian – Cambrian Boundary, Global bioevents, and The Rise and Fall of Vendian Biota. He has eighty research papers published in reputed journals to his credit and has edited several volumes of *Himalayan Geology Journal, India, and Journal of Nepal Geological Society*, Kathmandu, Nepal. His current interests are Cretaceous – Paleogene boundary, global extreme and complex climatic events from Neoproterozoic Snowball Earth to Holocene period, India – Asia collision, and evolution of Himalaya. Professor Tewari has organized first *Indo-Soviet Symposium on Stromatolites and Stromatolitic Deposits*

V.C. Tewari and J. Seckbach (eds.), *STROMATOLITES: Interaction of Microbes with Sediments*, Cellular Origin, Life in Extreme Habitats and Astrobiology 18, 495–522
DOI 10.1007/978-94-007-0397-1_22,

and other IGCP meetings in India and abroad related to stromatolites and phosphorites etc.. He has been one of the organizers of the *World Summit on Ancient Microscopic Fossils* held in University of California, Los Angeles, USA, in 2008. Professor Tewari is the member of the Editorial Board of the International Journal of Astrobiology published from New York, USA.

E-mail: **vtewari@wihg.res.in**

STROMATOLITES, ORGANIC-WALLED MICROORGANISMS, LASER RAMAN SPECTROSCOPY, AND CONFOCAL LASER SCANNING MICROSCOPY OF THE MESO-NEOPROTEROZOIC BUXA FORMATION, RANJIT WINDOW, SIKKIM LESSER HIMALAYA, NE INDIA

VINOD CHANDRA TEWARI
Wadia Institute of Himalayan Geology, Dehradun 248001, Uttarakhand, India

Abstract The Buxa Dolomite, a Meso-Neoproterozoic formation composed largely of stromatolitic cherty carbonate, is well developed in the Ranjit River section, south and west Sikkim, NE Lesser Himalaya. The formation is particularly well exposed in the Ranjit Tectonic Window of Sikkim, where its ~800-m-thick stratigraphic sequence has been studied in great detail. The assemblage of stromatolites recorded from exposures in the Ranjit Window is diverse, including *Colonnella columnaris, Conophyton garganicus, C. cylindricus, Gymnosolen ramsayi, Kussiella kussiensis, Minjaria uralica*, and forms assigned to *Aldania, Baicalia, Boxonia, Colleniella, Collumnaefacta, Jurusania, Nucleella*, and *Stratifera*. Diverse Neoproterozoic microfossils are also known from bedded carbonaceous Buxa cherts in Sikkim and Arunachal Pradesh, where they include numerous coccoidal taxa, occurring both as isolated cells and in colonies (*Eogloeocapsa, Eosynechococcus, Glenobotrydion, Huronispora, Myxococcoides, Palaeoanacystis*, and *Paratetraphycus*); filamentous fossils, both cellular trichomes and microbe-encompassing cylindrical sheaths (*Archaeotrichion, Cephalophytarion, Obruchevella, Oscillatoriopsis, Palaeolyngbya, Polythrichoides, Siphonophycus, Veteronostocale*, and *Volyniella*); and diverse acritarchs (*Archaeohystrichosphaeridium, Granomarginata, Leiosphaeridia, Lophosphaeridium, Margominuscula, Melanocyrillium, Micrhystridium, Navifusa, Trachysphaeridium, Trachyhystrichosphaera*, and *Vandalosphaeridium*). Coupled with optical microscopic studies, Raman spectroscopy and confocal laser scanning microscopy (CLSM) of selected microfossils from the Buxa cherts of the Ranjit Window document their three-dimensional preservation and carbonaceous (kerogenous) composition. The earlier study by Schopf et al. (Astrobiology **8**(4): 735–746, 2008) and the present additional investigations, based on a minute amount of fossiliferous rock – two thin sections, having a total area of 5 cm^2 (less than that of a US postage stamp), a volume of 0.04 cm^3 (less than half a grain of rice) and a weight of only ~0.1 g shows that even a minuscule amount of rock can contain definitive evidence of ancient life. The astrobiological implication of the present study has been discussed.

Keywords Stromatolites • Buxa Dolomite • Organic-walled microfossils • Laser Raman spectroscopy • Confocal laser scanning microscopy • Meso-Neoproterozoic • Lesser Himalaya • India • Sikkim • Paleobiology • Depositional environment

• Tidal flat • Oolites • Kerogen • Permineralized • Raman Preservation Index • Isotopes • Astrobiology

1. Introduction

The Buxa Formation in the northeastern Himalaya is a Meso-Neoproterozoic succession of dolomites, limestones, cherty stromatolitic, oolitic–intraclastic dolomites, calcareous quartzite, and black slates. The age of the Buxa Formation is tentatively assigned Meso-Neoproterozoic based on only available evidence of stromatolites and organic-walled microfossils. The microbial buildups of stromatolites in the Buxa Dolomite have been discussed. The detailed occurrence of the microorganisms from the lower part of the Buxa Dolomite in Ranjit river valley, Sikkim, NE Lesser Himalaya (Figs. 1 and 2), has been described. Laser Raman Spectroscopy and Confocal Laser Scanning Microscopy of the organic-walled microfossils from the Buxa cherts document their three dimensional preservation and carbonaceous (kerogenous) composition.

1.1. GEOLOGICAL SETTING AND STRATIGRAPHY

The area of investigation lies in the Northeastern Lesser Himalaya of India in Sikkim state. Sikkim state has international political boundaries with China and Bhutan in the east and Nepal in the west and West Bengal state of India in the south. The location of the Ranjit window in the Ranjit river section, Sikkim, is shown in Fig. 1. The geological map (Fig. 1 after Raina, 1976) shows the Ranjit tectonic window in the Sikkim Lesser Himalaya, which is a folded sequence of Lower Gondwana and the Meso-Neoproterozoic Buxa Formation. The Buxa Dolomite is the youngest lithounit of the Daling Group in the Sikkim Lesser Himalaya. The base of the Daling Group is known as Gorubathan Formation and consists of metamorphosed volcanoclastic sediments. The middle Reyang Formation is characterized by the orthoquartzites, variegated slates, cherts, conglomerate, and impersistent limestone. The Daling Group has a tectonic contact (Main Central Thrust) in the south with Central Crystallines (Sikkim/Darjeeling Group) and an unconformable/tectonic contact with the Lower Gondwana Ranjit (Lower Permian) in the north. Table 1 shows the tectonostratigraphy of the Ranjit window and the adjoining area. The Buxa Group is subdivided into three units in Buxa-Duars and the adjoining Bhutan Himalaya, namely Sinchula Quartzite, Jainti Quartzite, and Buxa Dolomite (Gansser, 1964). Acharyya (1974) subdivided the Buxa Group in the type area into two units, the lower Sinchula Formation and the upper Jainti Formation. The Buxa Dolomite in the type area comprises dolomite – orthoquartzite-variegated slates. In Arunachal Pradesh, the Miri Quartzite of Subansiri and Siang district overlie the Buxa Dolomite. The age of the Miri Quartzite conformably overlying the Buxa Dolomite is assigned Lower Cambrian on the basis of the presence of well-preserved ichnofossils

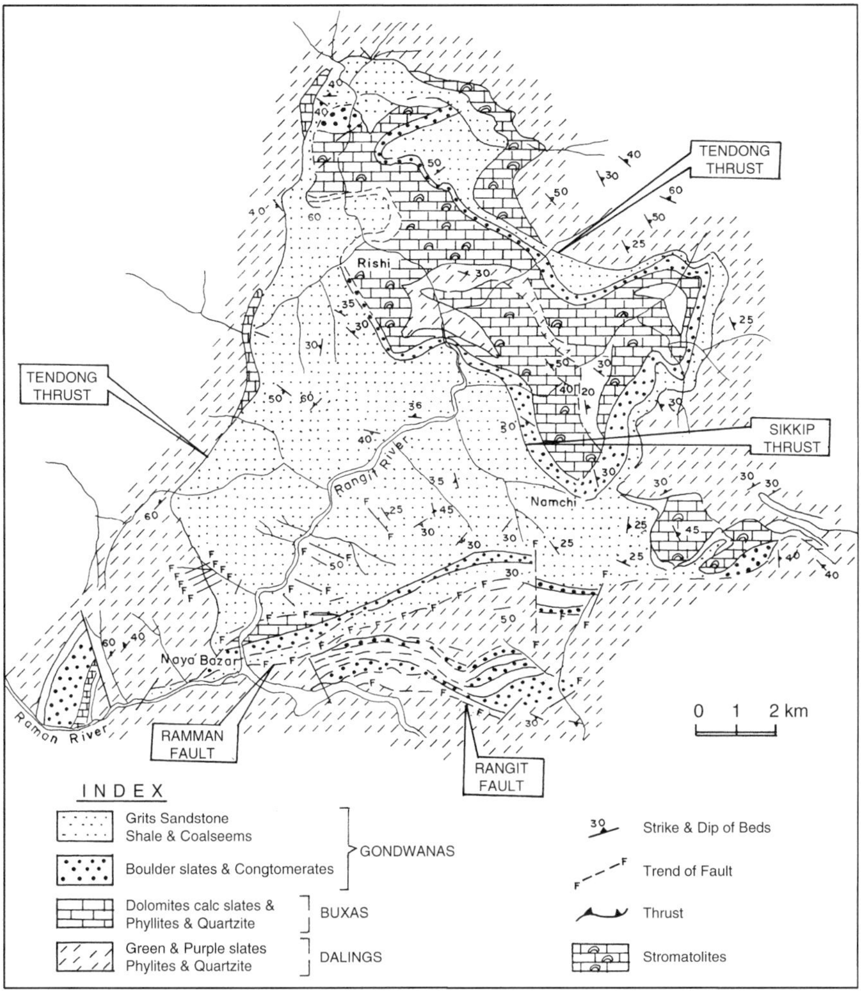

GEOLOGICAL MAP OF RANGIT WINDOW, SIKKIM N.E. LESSER HIMALAYA

Figure 1. Geological map of the Ranjit Window, Sikkim, NE Lesser Himalaya, India (after Raina, 1976).

(Tewari, 2002, 2003). Dolomites, limestones, cherty stromatolitic dolomite, oolitic–intraclastic dolomite, and calcareous quartzite represent the Buxa Group in the Arunachal Lesser Himalaya. The detailed geology and stratigraphic position of various sedimentary formations of NE Lesser Himalaya has been discussed in the recent literature (Kumar, 1997; Srinivasan, 2001; Tewari, 1998, 2001, 2002, 2003). The Buxa-Miri sediments (Terminal Neoproterozoic – Lower Cambrian) in the Arunachal area have a thrust contact with the underlying Lower Permian Gondwana Group.

Table 1. Tectonostratigraphy of the Ranjit window, Sikkim Lesser Himalaya.

Gondwana	Damuda Formation (=Namchi Formation)
(Ranjit) Group	Coarse to fine-grained gray sandstone, carbonaceous shale, powdery coal beds with *plant fossil*
Lower (permian)	Ranjit (Pebble Slate) Formation (=Tatapani (Pebble Slate) Formation)
Unconformity/tectonic	
Daling group (middle to late proterozonic)	Buxa subgroup (= Buxa formation, Acharyya, 1989)
	Blue-gray stromatolitic dolomite, pyritous black slate
	Reyang subgroup/formation
	Orthoquartzite, variegated slate, chert, conglomerate impersistant limestone
	Gorubathan Subgroup/Formation
	Metamorphosed volcanogenic and normal sediments
Tectonic contact (M.C.T)	
Darjeeling group	Kanchenjunga Migmatite Gneiss Darjeeling Gneiss interbanded with Calc, silicate gneiss, quartzite and amphibolites
Sikkim group	Chungthang group/formation

2. Fossil Locality

Situated in the northeastern Lesser Himalaya in Sikkim, a state of India that shares boundaries with the state of West Bengal to the south and with Bhutan and China to the east (Fig. 1), the Meso-Neoproterozoic Buxa Formation is predominantly dolomite. The litholog (Fig. 2) shows that the sampled locality occurs in the Ranjit tectonic window, a surviving remnant of Precambrian strata that is surrounded by much younger, Permian-age, tectonically folded Gondwana units. Within this window of Proterozoic deposit, the Buxa Formation, underlain by the Daling Formation (Fig. 1), is the youngest unit of the Daling Group.

3. Previous Work

3.1. SEDIMENTOLOGICAL AND PALEOBIOLOGICAL

As is summarized in Fig. 2, the Buxa Formation, which extends eastward from Sikkim past Bhutan into the Indian state of Arunachal Pradesh, is composed of cherty stromatolitic, oolitic, and intraclastic dolomite, limestones, calcareous

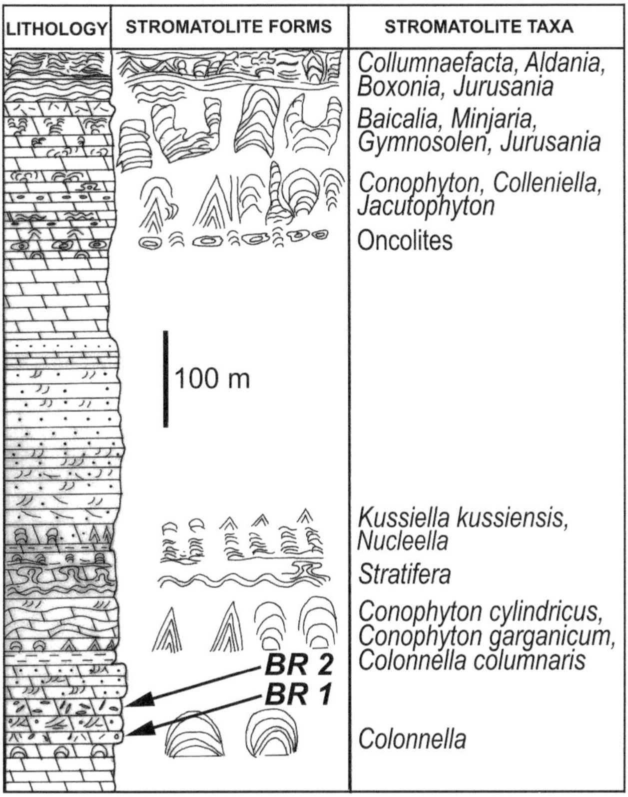

Figure 2. The lithocolumn of the Buxa Dolomite Formation showing distribution of stromatolitic buildups, *black chert bands*, oolites, and sedimentary structures. (Schopf et al., 2008, reproduced with permission from Astrobiology.)

quartizites, and black shales, slates, and phyllites (Acharyya, 1974; Raina, 1976; Sinha Roy, 1980; Tewari, 2001, 2003). Within the Ranjit tectonic window, the ~800-m-thick Buxa Formation has been measured stratigraphically (by VCT) along the Ranjit River from the locality sampled here to the village of Rishi, some 2 km to the southeast (Fig. 1c). The lower part of the formation, the source of the cherts studied here, is composed of a ~250-m- thick sequence of well-bedded gray cherty dolomite (e.g., Fig. 3a) that contains stromatolites of the groups *Colonnella, Conophyton, Kussiella, Nucleella, and Stratifera* (Fig. 2). Two lenses of well-bedded carbonaceous chert within this sequence (BR1 and BR2, Fig. 2) were sampled for the present study. The middle third of the formation, stratigraphically above the sampled horizons, consists of a 300-m-thick succession of finely laminated silicified gray dolomites, which contains thin bands of black chert, and olive gray

Figure 3. Field photographs of the *Conophyton garganicus* (*upper left and right*), *Colonnella columnaris* (*below left*), and *Jurusana* sp. (*below right*).

shales. The upper third of the Buxa Formation, its uppermost 250 m, is composed of interbedded dolomites and subsidiary dark gray shales, the dolomitic beds of which contain oncolites as well as stromatolites of the groups *Aldania, Baicalia, Boxonia, Colleniella, Collumnaefacta, Conophyton, Gymnosolen, Jurusania, and Minjaria* (Fig. 2). Shukla et al. (2006) recovered organic-walled microfossils comprising 36 taxa of cyanobacterial, acritarchs, and VSM, in which 17 taxa of cyanobacteria belong to Chroococcaceae, Nostocaceae, and Oscillatoriaceae, 18 taxa of Acritarchs belong to Sphaeromorphida, Scaphomorphida, and Sphaerohystrichomorphida subgroups, and one VSM (Vase shaped microfossils) from the Buxa cherts in the Arunachal Lesser Himalaya compares well with the known Late Neoproterozoic assemblages from other parts of the world.

3.1.1. Age of the Buxa Formation

The radiometric age of the Buxa Formation has yet to be determined. Earlier works, on the bases of lithostratigraphy and the presence of the stromatolite assemblages noted above, suggested ages that range from Late Precambrian to Late Paleozoic (Acharyya, 1974; Raina, 1976; Sinha Roy, 1980; Valdiya, 1980). Though neither of these lines of evidence provided firm evidence of the age of the unit, the more recent geologic studies and the findings of ministromatolites (reported also as "microstromatolites") and of characteristically Neoproterozoic organic-walled microfossils in the Buxa Formation in Arunachal Pradesh state (Tewari, 2001, 2003, 2009; Shukal et al., 2006). The lithology and stromatolite assemblages of the Buxa Formation in the Ranjit tectonic window (N.E. Lesser Himalaya) are notably similar to those of dolomites of the Meso-Neoproterozoic Shali–Deoban–Gangolihat carbonate belt of the N.W. Lesser Himalaya, a comparability regarded as suggesting that units of the two regions may be stratigraphically correlative (Tewari, 1989, 1993, 2002, 2004a, b, 2007, 2009). The assemblage of Neoproterozoic microfossils recently reported from the Buxa Formation in Arunachal Pradesh (Shukal et al., 2006) and from the N.W. Lesser Himalaya (Tewari, 2004a, b, and references therein) are consistent with this possibility, as is the composition of the biota recorded by Schopf et al (2008) including the occurrence in such assemblages of *Obruchevella*, a helically coiled fossil cyanobacterium similar to the modern oscillatoriacean *Spirulina* and a taxon that is particularly widespread in latest Proterozoic to earliest Cambrian strata.

4. Stromatolite

Microbial (stromatolitic) buildups are well developed in the Buxa Dolomite and show a variety of morphological diversity from bottom to top of the sequence (Fig. 2). The important buildups recognized are (1) Gray dolomite with *Colonnella columnaris, Kussiella kussiensis*, and *Conophyton garganicus* (Fig. 3). The cross-bedded intraformational pebbles, intraclastic–oolitic dolomite, and sandy–oolitic dolomite are intimately associated with this buildup (Fig. 4). This buildup was formed in the high energy subtidal and intertidal environment. (2) Dark gray cherty intraclastic–oolitic dolomite with microbial mats and stratified buildups (*Stratifera*) and *Nucliella* structures. This buildup was formed in subtidal environment. (3) A thick microbial buildup with diversified assemblage (*Baicalia* sp., *Kalpanaella* f. nov. Tewari, *Jurusania*, *Colloniella*, *Minjaria*, *Gymnosolen*,

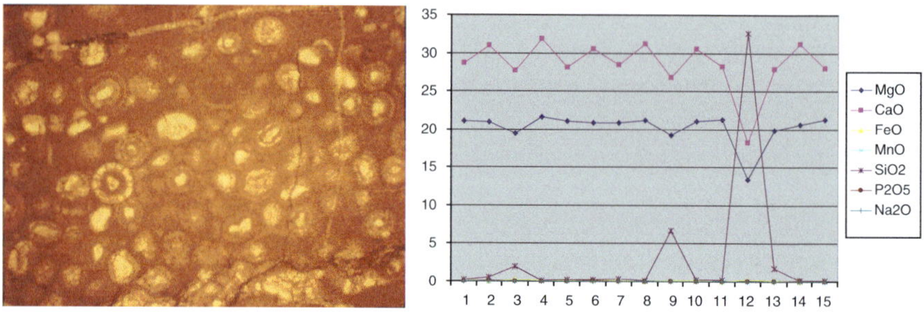

Figure 4. Microfacies of the oolitic Buxa Dolomite showing various stages of diagenesis (*left*) and Electron Probe Micro Analysis (EPMA) of the oolites (*right*).

Tungussia, *Aldania*, domal, *Nucliella* (Figs. 5 and 6), and linked conical-stratified and columnar-stratified forms). This buildup represents peritidal depositional environment. (4) A cyclic buildup of digitate microbialites, intraformational pebbles, crenulated microbialites, and columnar and domal forms is developed. There is a cyclicity in the intraformational pebbles and microbialites. This buildup is a product of high-energy intertidal depositional environment. (5) Shaly dolomite buildup with large domal columnar and *Nucleilla* forms (Fig. 7). The intraformational pebbles are found associated with shaly dolomite. Five cycles of columnar and domal buildups have been recorded. All these microbial buildups of the Buxa Dolomite recorded between Reshi and Tatapani in Ranjit river section suggest a shallow marine (high-energy tidal flat) depositional environment. The microbialite (stromatolitic) assemblage of the Buxa Dolomite suggests a Lower Riphean to Upper Riphean-Vendian (Meso-Neoproterozoic) age. The relationship between the stromatolite morphology and the depositional environment is shown in the sedimentological model (Fig. 8).

5. Materials and Methods

5.1. MATERIALS STUDIED

The permineralized microfossils were studied in petrographic thin sections BR-1 and BR-2, which were prepared from rocks collected (by V.C.T. and jointly studied with J.W Schopf and A.B. Kudrayvtsev in UCLA, in 2007 and 2008) from outcrops of 10 to 12 cm thick chert beds, BR1 and BR2 situated, respectively, ~70 and 100 m stratigraphically above the base of the predominantly dolomite Proterzoic Buxa Formation in the Ranjit tectonic window, Sikkim, northeastern India (Fig. 1). The locality sampled in ~10 km northwest of Namchi and ~2 km northwest of Rishi, Sikkim, at 88°18′E long, and 27°15′N lat., in the N.E. Lesser Himalaya (Fig. 1). The two thin sections (BR-1: ~1.15 × 2.0 cm, ~50 μm thick; and BR-2: ~1.35 × 2 cm, ~110 μm thick) have a total area of ~5 cm^2 and a combined volume

Figure 5. *Baicalia* sp. (*top and middle*) and *Kalpanaella* f. nova Tewari (*below*) buildup in the Buxa Dolomite, Ranjit window.

of ~0.04 cm^3 that, given the density of silica of ~2.6 g/cm^3, represents a total rock weight of ~0.1 g. The fossils were located by optical microscopy and documented by photomicrography; selected specimens were subsequently imaged by confocal

Figure 6. Field occurrence of bioherms of *Kussiella kussiensis* showing overhanging and bridging of the laminae. *Scale bar* is 10 cm.

laser scanning microscopy (Schopf et al., 2006) or were analyzed by Raman spectroscopy or imagery (Schopf et al., 2005, 2008; Schopf and Kudryavtsev, 2005) (Fig. 9).

6. Optical Microscopy

Studied in the unpolished petrographic thin section specified above (each finished by use of a slurry of 600 mesh carborundum and covered by a ~1-µm- thick veneer of Type FF, fluorescence-free, microscopy immersion oil; R.P. Cargille Laboratories, Inc., Cedar Grove, NJ), transmitted light optical photomicrographs of the specimens were obtained by use of UCLA #0026535 Leitz Orthoplan 2 microscope (Leitz, Wetzlar, Germany) equipped with an Olympus DP 12 Microscope Digital Camera (Olympus, Melville, NY). Microscope stage coordinates for specimens illustrated from thin sections BR-1 and BR-2.

7. Confocal Laser Scanning Microscopy

Three-dimensional confocal fluorescence images were obtained by use of an Olympus Fluo View 300 confocal laser scanning biological microscope system equipped with two Melles Griot lasers, a 488 nm 20 mW-output ion argon laser and 633 nm 10 mW-output ion helium-neon laser (Melles Griot, Carbsbad CA).

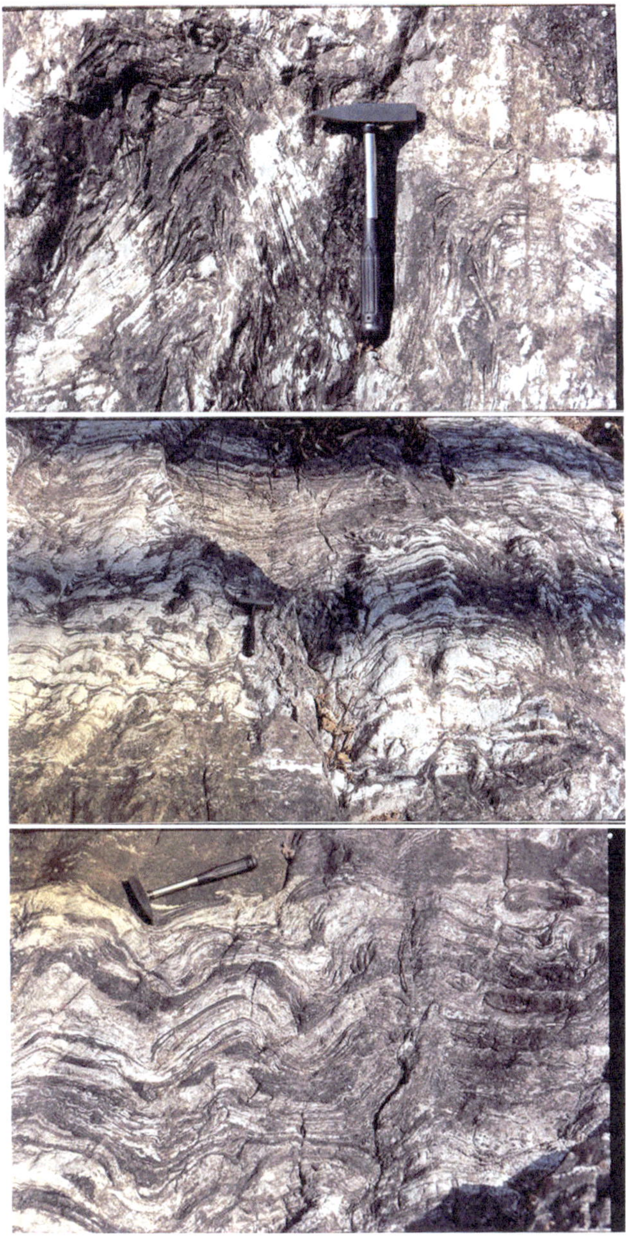

Figure 7. Field occurrence of the domal, stratified and conical stromatolites.

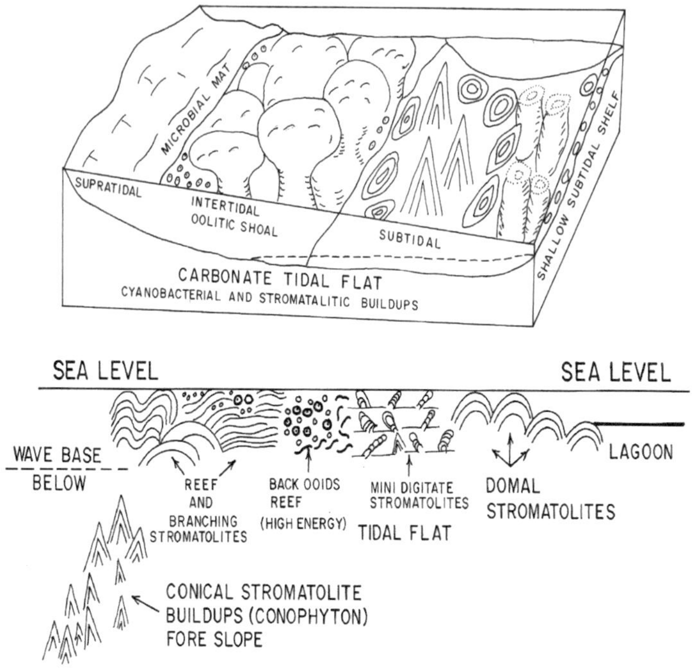

Figure 8. Sedimentological model for the deposition of the Buxa Dolomite, stromatolites, and oolites.

CLSM images were acquired via a 60× oil-immersion objective (numerical aperture = 1.4), with use of the fluorescence-free microscopy immersion oil noted above and filters in the light-path of the system that permitted transmission to its detector only of wavelengths>510 nm (for 488 laser excitation) and >660 nm (for 633 nm laser excitation) from the kerogen-derived fluorescence emitted by the specimens.

8. Raman Spectroscopy and Imagery

Raman spectra of the kerogen that comprises the permineralized fossils were obtained by the use of a T64000 (JY Horiba, Edison NJ) triple-stage laser-Raman system having macro-Raman and confocal micro-Raman capabilities. This system permitted acquisition both of individual point spectra and of Raman images that display the two dimensional spatial distribution of molecular–structural components of the specimens and their associated mineral matrices, with the varying intensities in such images corresponding to the relative concentrations of the molecular structures detected. Due to the confocal capability of the system, the use of a 50× objective (having an extended working distance of 10.6 mm and a numerical aperture of 0.5) provided a horizontal resolution of ~1.5 μm and a

Figure 9. View of the Ranjit River (*top*), black cherty lenses yielding microbiota (*middle*), and the thin bedded cherty black bands (*lower*, reproduced with permission from Astrobiology).

vertical resolution of 2–3 μm. The use of a Coherent Innova (Santa Clara, CA) argon ion laser in this system to provide excitation at 457.9 nm permitted measurements to be made over a range from <500 to <3,100 cm^{-1}, using a single spectral window centered at 1,800 cm^{-1}. Spectra were acquired simultaneously of the major bands (at ~1,365 and ~1,604 cm^{-1}) and the second-order band (at ~2,800 cm^{-1}) of the kerogen that comprises the fossils as well as of the major band of the permineralizing quartz (at ~465 cm^{-1}) and that of subsidiary calcite (at 1,087 cm^{-1}). For imaging, the region of a thin section containing a specimen to be analyzed was covered by a thin veneer fluorescence-free microscopy immersion oil (the presence of which has been shown to have no discernable effect on the Raman spectra acquired; Schopf et al., 2005), and the area of the fossils studied was centered in the path of the laser beam projected through the microscope of the system. The laser power used was ~1–8 mW over a ~μm spot, an instrumental configuration well below the threshold that results in radiation damage to specimens such as those studied here (Schopf et al., 2005, 2008).

9. Organic-Walled Microfossils

Organic walled microfossils have been reported from the Buxa black cherts from the Arunachal Lesser Himalaya (Tewari, 2001, 2009; Schopf et al., 2006). A diverse acritach assemblage has been found in the Buxa cherts from the Ranjit window section (Fig. 13). Schopf et al (2008, Figs. 10, 11, and 12) studied the microfossils in detail

9.1. PETROGRAPHY OF THE MICROFOSSILIFEROUS CHERTS

Studied in thin sections, both the chert samples were seen to be composed of interlocking mosaics of fine-grained quartz in which occur closely packed, millimeter-sized, subangular to rounded silicified fossil-bearing detriatal clasts and subsidiary, euhedral, carbonate rhombs ~20- to 40-μm in long dimension. Figure 3b, c shows examples of two such clasts, virtually all of which appear to be fragments of originally coherent, mucilage-embedded microbial mats that have been removed from their site of formation, redeposited, and then silicified to form the bedded cherts. The quartz grains in which the fossils are permineralized are generally 10–15 μm in size, with some grains as much as 40 μm in long dimension. The maximum size of these evidently somewhat recrystallized grains, appreciably larger than the 6–12 μm sized quartz grains typical of fossil-bearing unmetamorphosed Neoproterozoic cherts (e.g., those of the Bitter Springs Formation of central Australia; Schopf, 1968; Schopf and Blacic, 1971), is consistent with the tectonic setting in which the Buxa cherts occur. Similarly, the geochemically rather mature state of the kerogen that comprises the fossils (discussed below), which is

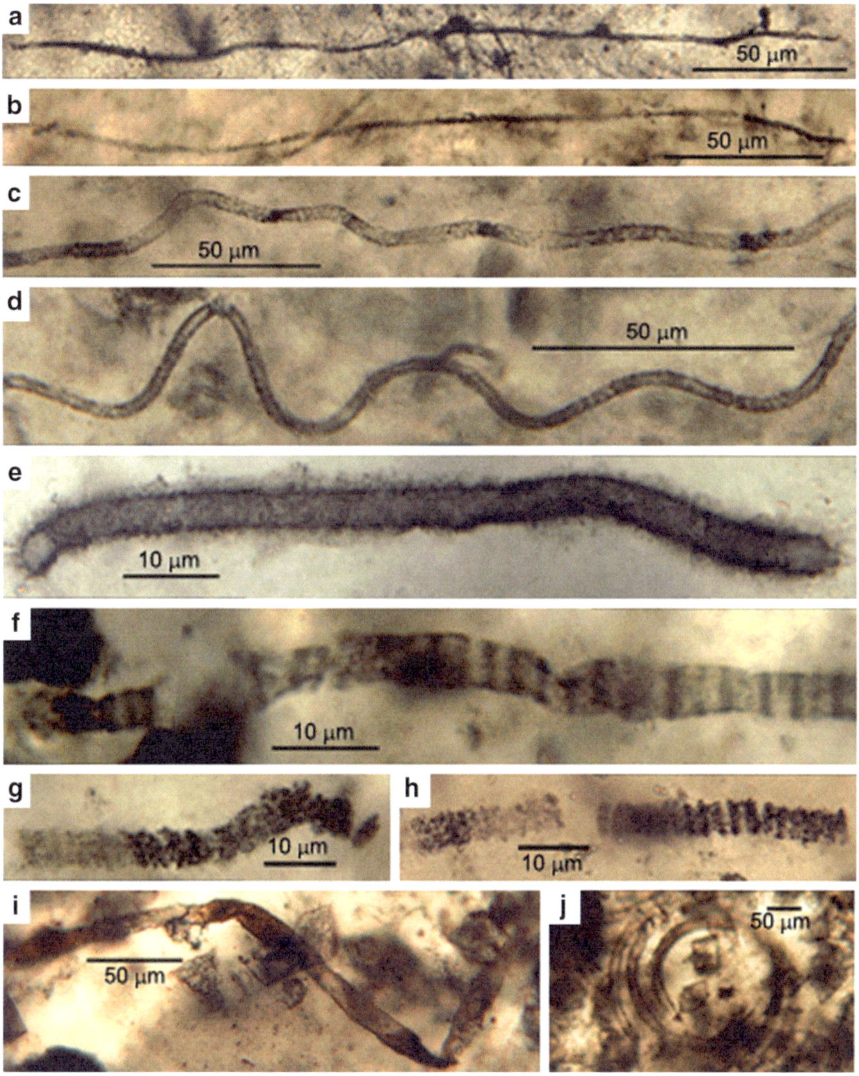

Figure 10. Optical photomicrographs showing filamentous chert-permineralized microfossils in petrographic thin section of the Proterozoic Buxa Formation of the Ranjit Window, Sikkim, northeastern India. Parts (**a**–**i**) show photomontages, necessitated by the three-dimensional sinuosity of the figured specimens. (**a**, **b**) Narrow (~1- to 3-μm diameter) bacterial and/or cyanobacterial filaments cf. *Archaeotrichion* (**a**) thin section BR-1, microscope stage coordinates 49.5 × 99.9; (**b**) section BR-1, 35.6 × 94.9). (**c**–**e**) Tubular ~3–5 μm diameter filaments (cf. *Eomycetopsis*), the cylindrical extracellular sheaths of oscillatoriacean cyanobacteria (**c**) section BR-1, 42.5 × 100.5; (**d**) section BR-1, 35.0 × 92.8; (**e**) section BR-2, 34.4 × 103.7. f: Septate ~3–5 μm diameter oscillatoriacean trichome (cf. *Cephalophytarion*) composed of *box-shaped* cells (section BR-2, 33.5 × 96.5). (**g**, **h**) Septate ~5 μm diameter oscillatoriacean trichomes (cf. *Oscillatoriopsis*) composed of disc-shaped cells (both specimens are from the same area of section BR-2, 31.6 × 97.8). (**i**) Broad (~10 μm diameter) ribbon like filament (cf. *Siphonophycus*), the flattened originally tubular extracellular sheath of an oscillatoriacean cyanobacterium (section BR-2, 32.9 × 102.6). (**j**) Helically coiled ~15–20 μm diameter filament cf. *Obruchevella*, a fossil oscillatoriacean similar to modern *Spirulina* (section BR-2, 41.0 × 99.4). (Schopf et al., 2008, reproduced with permission from Astrobiology).

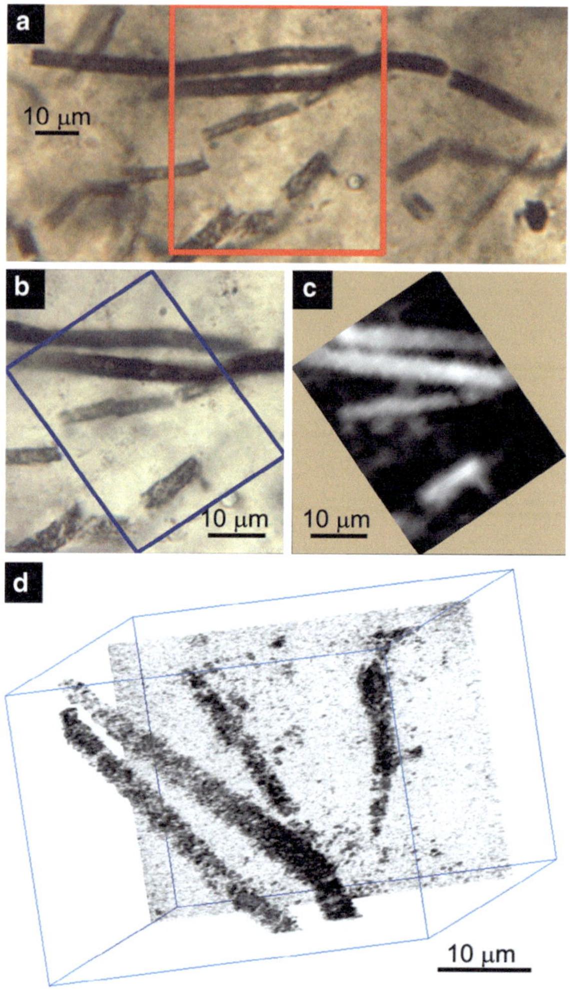

Figure 11. Optical photomicrographs, Raman image, and CLSM image of chert-permineralized tubular cyanobacterial (oscillatoriacean) sheaths in a petrographic thin section of the Proterozoic Buxa Formation of the Ranjit Window, Sikkim, northeastern India. (**a**, **b**) Photomicrographs of the specimens (*Eomycetopsis* sp.) with the *red rectangle* in (**a**) denoting the area shown in (**b**) and (**d**), and the *blue rectangle* in (**b**) denoting the area shown in (**c**). (**c**) Raman image acquired in a spectral window centered at the ~1,600 cm^{-1} band of kerogen that documents the carbonaceous composition of the fossils. (**d**) CLSM image of the fossils, rotated to show the specimens from the underside of the thin section (the gray focal plane at the back of the figure being situated toward the top of the section) and a perspective for viewing of the fossils uniquely provided by CLSM, demonstrating the three dimensionality of the filaments (Schopf et al., 2008, reproduced with permission from Astrobiology).

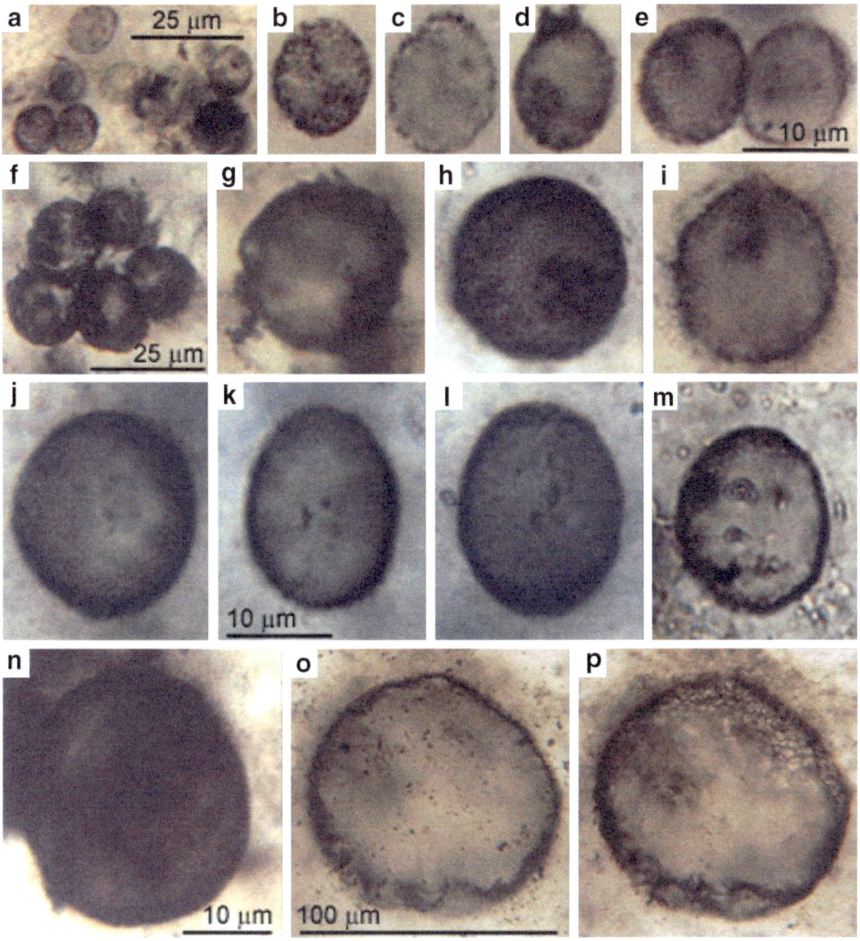

Figure 12. Optical photomicrographs showing chert permineralized colonial coccoidal microfossils in petrographic thin sections of the Proterozoic Buxa Formation of the Ranjit window, Sikkim, north-eastern India. (**a**–**i**) Colonial ~ 3–5 μm diameter chroococcacean cyanobacterial coccoidal cells (cf. *Palaeoanacystis*) irregularly distributed within an originally mucilaginous envelope; scale in (**f**) applies to (**b**) through (**i**); all specimens are from the same area of thin section BR-2, microscope stage coordinates 42.2 × 103.5. (**j**–**s**) Dyads and tetrads [in (**j**) shown in face view, in (**m**) shown at an oblique angle] of ~5–7 μm diameter "lima bean shaped" chroococcacean coccoids cf. *Eogloeocapsa*; scale in (**k**) applies to (**j**) through (**s**); all specimens from section BR-1 (**j**) 46.0 × 98.6; (**k**) 47.6 × 101.8; (**l**)46.1 × 98.7; (**m**) 46.2 × 99.4; (**n**) 48.4 × 101.8; (**o**) 46.3 × 99.5; (**p**) 46.3 × 99.6 (Schopf et al., 2008, reproduced with permission from Astrobiology).

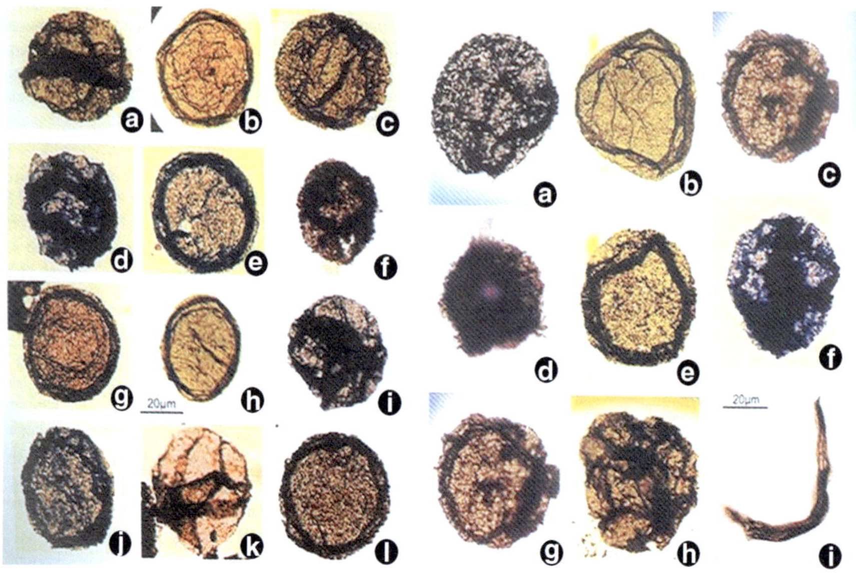

Figure 13. The diversified microbiota and acritarchs from the Buxa Dolomite, Ranjit Window, Sikkim Lesser Himalaya.

indistinguishable from that of particulate kerogen in their surrounding matrices, reflects preservation in a moderately metamorphosed terrain.

9.1.1. Diversity and Abundance of Microfossils

As viewed in the two petrographic thin sections, the fossil-bearing detrital clasts of the Buxa cherts range from small ovate bodies that average about 0.5 mm in diameter (in thin section BR1, ranging from ~0.4 to 1.8 mm, and in BR2, ~0.1–2 mm) to larger elongate shards, shreds, and pieces of originally mucilage-bound microbial mats (in BR1, ~0.3–2 mm broad and ~1–8 mm long, and in BR2, ~0.1–0.3 mm broad and ~1–4 mm long). Each of the thin sections contains ~200 such clasts, which comprise 70–80% of their total area, the remainder being composed of interstitial fine-grained quartz in which occur abundant carbonate rhombs. In each of the two thin sections, the abundance of identifiable fossils varies markedly both among and within their component clasts. In thin section BR1, ~60% of such clasts are fossil bearing, whereas less than 20% of the generally smaller clasts in section BR2 are fossiliferous, and both sections contain clasts that, though fossil bearing in one region, are devoid of fossils elsewhere. As is typical of Precambrian microbial communities, the concentration of structurally preserved microbes in the clasts is patchy. Quantified in terms of thin section area, some clasts are completely devoid of fossils or include only 1–2 fossils/mm^2 (equivalent to ~10^8 µm^3), whereas others contain localized packets of narrow, 1- to 3- µm- diameter filaments (e.g., Fig. 10a, b), which are density interlaced and present in concentrations of a few to several hundred fossils per mm^2. Numerous clasts contain entangled networks of subparallel ~3–5 μm diameter filaments (e.g., Fig. 10c–e) that have a density of tens to more than a hundred specimens/mm^2. In general, filamentous fossils tend to be particularly abundant in the elongate clasts, whereas the relatively rare, smaller-celled, coccoidal fossils (e.g., Fig. 12a–m) typically occur in localized regions, whether in ovate or elongate clasts, where they are present singly, in small clusters of a few to ~30 cells, or in larger colonies composed of a few hundred cells. It should be noted that the foregoing data regarding fossil abundance, despite being based on systematic measurements of clast sizes and fossils concentrations, are approximations. The patchy millimetric-scale distribution of the Buxa fossils is typical both of fossilized microbial communities and of their modern equivalents (where it is a result of founder effects – e.g., colonization by a particularly rapidly proliferating species – and of local environmental conditions such as the ability of light and nutrients, initial growth conditions, competition with near neighbors, and so forth). Moreover, for fossilized communities, assessment of "microbial density" can be based only on morphologically identifiable fossils remnants, which because of post-mortem degradation represent just a fraction of the assemblage originally present. Such assessment cannot take into account organic debris that does not exhibit identifiable evidence of its biologic origin (for the Buxa cherts, like that of all such deposits, comprising ≥95% of the organic matter preserved), nor can it include the tiny bits and pieces of cellular debris – true fossils, but optically nondescript – that occur in such deposits.

9.1.2. Microbial Assemblage

As is typical of permineralized Neoproterozoic microbial assemblages, the Buxa cherts studied contain both filamentous forms and include narrow specimens, 1–3 μm in diameter, which are similar to fossils of the genus *Archaeotrichion* of bacterial or cyanobacterial affinity (Fig. 10a, b); tubular 3–5 μm diameter filaments cf. *Eomycetopsis*, a fossil taxon composed of the cylindrical extracellular sheaths of oscillatoriacean cyanobacteria (Fig. 10c–e); and the similarly sized septate filaments that closely resemble species of *Cephalophytarion* (Fig. 10f) and Oscillatoriopsis (Fig. 10g, h), taxa of fossilized cellular oscillatoriacean trichomes. Also present are broad, 10 μm diameter nonseptate filaments, the encompassing sheaths of relatively large-diameter oscillatoriaceans (Fig. 10i), and 15–20 μm-diameter helically coiled specimens cf. *Obruchevella* (Fig. 10j), a fossil taxon composed of specimens similar to the modern oscillatoriacean *Spirulina*. The uniseriate nature of the cellular trichomes (Fig. 10f–h) evidences their formation by repeated cell divisions perpendicular to the long axes of the filaments, whereas CLSM studies of such filaments, results of a representative example of which are shown in Fig. 11d, demonstrate their three dimensional form. Of the various morphotypes detected – all of which are well known from numerous Neoproterozoic assemblages (e.g., Schopf, 1992a, b) – the *Eomycetopsis*-like extracellular sheaths are particularly abundant, whereas well-preserved cellular trichomes, such as that shown in Fig. 10f, are rare. Included among the coccoidal fossils of the assemblage are small, 3–5 μm diameter, originally mucilaginous envelope-enclosed colonial coccoids that closely resemble those of the fossil chroococcacean cyanobacterium *Palaeoanacystis* and dyads and tetrads composed of 5–7 μm diameter "lima bean-shaped" cells similar to those of the fossil chroococcacean taxon *Eogloeocapsa*. The assemblage also includes colonial, 5–10 μm diameter, 10–15 μm diameter (Fig. 12a, i), and 15–18 μm diameter (Fig. 12f–m) chroococcaceans that resemble species of the fossil genus *Myxococcoides*; larger ellipsoidal solitary unicells, 25–30 μm in size (Fig. 12n); and large shaeromorph acritarchs ~95 μm in breadth (Fig. 12o, p). The occurrence of such coccoids irregularly distributed in envelope-enclosed colonies or in closely packed colonial aggregates (Fig. 12a, f), as well as their presence in dyads and tetrads, demonstrates that they were produced by the standard processes of cell division. Like the filamentous forms noted above, all of these coccoidal morphotypes are well known from other Neoproterozoic cherts and, with the exception of the phytoplanktonic acritarchs, all appear to have been benthic members of mat-building microbial communities (Fig. 13).

10. Raman Index of Preservation (RIP) of the Buxa Fossils

The molecular–structural composition of the carbonaceous kerogen that comprises the Buxa fossils has been documented by Raman spectropscopy. As is shown by the representative example illustrated in Fig. 11c, two-dimensional Raman imagery demonstrates that such specimens (in this example, *Eomycetoposis*-like tubular cyanobacterial sheaths) are composed of kerogen. Moreover, the Raman spectra of the carbonaceous matter acquired in such studies can themselves be analyzed to yield the Raman Index of Preservation of such materials (Schopf et al., 2005), RIP values that provide an established basis for assessment of organic geochemical maturity (Schopf et al., 2005). Figure 14 shows seven Raman spectra acquired from organic-walled Neoproterozoic microfossils preserved at various stages of geochemical maturation (Schopf et al., 2005, 2008). As is there illustrated, the two major Raman bands of kerogen change markedly as a function of increasing geochemical alteration: the leftmost "D" band becomes increasingly more peaked (and, correspondingly, less broad and "bumpy") as the rightmost "G" band becomes increasingly narrow and ultimately bifurcated. Such data, obtainable from organic-walled fossils permineralized in rocks subjected even to greenschist facies metamorphism (Schopf et al., 2002, 2005), can provide definitive evidence of the fidelity of organic preservation that is unavailable by any other means. As is shown in Fig. 14, Buxa fossils from the Ranjit window, which have an RIP value of 5.1, exhibit an intermediate grade of organic maturation, being neither as well preserved as fossils of the Bitter Springs (the uppermost spectrum in Fig. 14) and other relatively little-altered Precambrian units but exhibiting an appreciably greater fidelity of geochemical preservation than fossils preserved in more metamorphosed Neoproterzoic deposits.

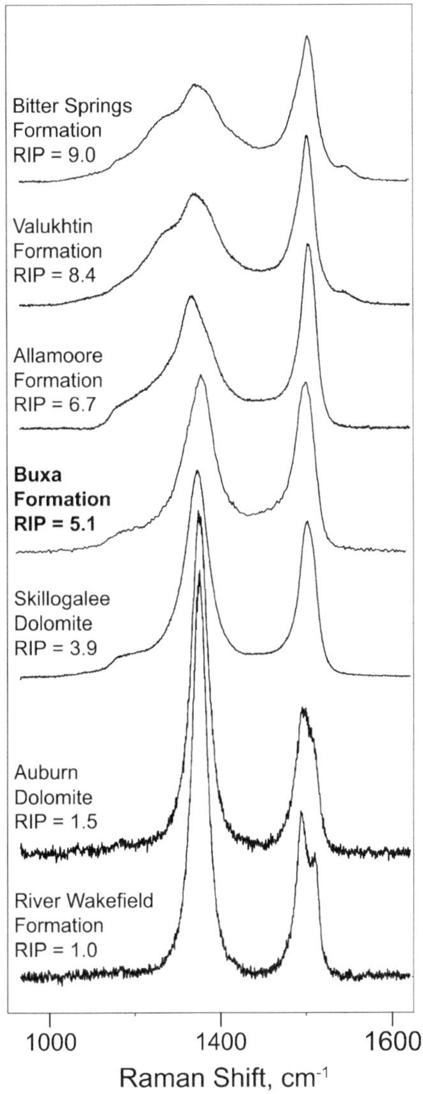

Figure 14. Raman spectra of chert-permineralized Neoproterozoic carbonaceous microfossils. Ordered by their RIP values (Schopf et al., 2005a) from less (*top*) to more (*bottom*) geochemically nature, the spectra shown are those of the kerogenous walls of fossils preserved in cherts of the ~750 Ma-old Bitter Springs Formation, the ~1,050-Ma-old Valukhtin and Allamoore Formations, the ~760-Ma-old Skillogalee and ~720-Ma-old Auburn Dolomites, and the ~775-Ma-old River Wakefield Formation (Schopf et al., 2005), compared with that of a representative fossil filament (*Eomycetopsis*, cf. Figs. 10c–e and 11) from the Buxa Formation. (Schopf et al., 2008, reproduced with permission from Astrobiology.)

11. Discussion and Conclusions

A very thick sequence of the Buxa Dolomite (800 m) is exposed between Reshi and Tatapani in the Ranjit Valley West Sikkim, NE Lesser Himalaya. Microbial (stromatolitic) buildups are well developed in the Buxa Dolomite and show a variety of morphological diversity from bottom to top of the sequence. The important buildups recognized are (1) Gray dolomite with *Colonnella columnaris*, *Kussiella kussiensis*, and *Conophyton garganicus*. The cross-bedded intraformational pebbles, intraclastic–oolitic dolomite, and the sandy–oolitic dolomite are intimately associated with this buildup. This buildup was formed in the high-energy subtidal and intertidal environment. (2) Dark gray cherty intraclastic–oolitic dolomite with microbial mats and stratified buildups (*Stratifera*) and *Nucleilla* structures. This buildup was formed in subtidal environment. (3) A thick microbial buildup with diversified assemblage (*Jurusania*, *Colleniella*, *Minjaria*, *Gymnosolen*, *Tungussia*, *Jacutophyton*, *Baicalia*, *Aldania*, domal, *Nucleilla*, and linked conical-stratified and columnar-stratified forms). This buildup represents peritidal depositional environment. (4) A cyclic buildup of digitate microbialites, intraformational pebbles, crenulated microbialites, and columnar and domal forms is developed. There is a cyclicity in the intraformational pebbles and microbialites. This buildup is a product of high-energy intertidal depositional environment. (5) Shaly dolomite buildup is characterized by large domal columnar and *Nucleilla* forms, and intraformational pebbles are found associated with shaly dolomite. Five cycles of columnar and domal buildups have been recorded. All these microbial buildups of the Buxa Dolomite recorded between Reshi and Tatapani in Ranjit river section suggest a shallow marine (high energy tidal flat) depositional environment. The microbialite (stromatolitic) assemblage of the Buxa Dolomite suggests a Lower Riphean to Upper Riphean–Vendian (Meso-Neoproterozoic) age.

As is documented above, both of the Buxa thin sections studied here contain diverse organic-walled microscopic fossils, the first such Precambrian fossils to be reported from the Ranjit tectonic window of Sikkim, northeastern India. Despite the rather poor state of cellular preservation of some of these chert permineralized microbes and the moderately advanced geochemical maturity of the kerogen of which they are composed (Fig. 14), they are, in fact, *bona fide* fossils. Primary among the numerous lines of evidence that establish biogenicity are (1) their morphology and carbonaceous composition, (2) the diversity of the fossils detected, (3) the presence of many examples of the various morphotypes (exhibiting varying stages of cellular degradation and, in some instances, evidence of cell division), and (4) their close similarity both to fossils known from other Proterozoic deposits and to microorganisms living today. Indeed, the evidence of life contained in these two tiny thin sections is so voluminous and compelling that, had only one of the sections been available for study – decreasing by half the source of the evidence presented here – the conclusion drawn would have been the same: microbial life was present and abundant when and where these rocks were formed. It is notable – and to some perhaps surprising – that in this study, it has proven possible to establish the existence of past life on the basis of such a minuscule amount (~0.1 g) of rock material. In fact, however, comparable results might have been obtained from such studies of small fossil-bearing samples of numerous other Precambrian deposits (including virtually all of the 25 Proterozoic cherty units from which fossils are illustrated in Schopf, 1992b). Because the microoraganisms of microbial assemblages are so minute, commonly only a few microns to less than a micron in size, and because they typically are preserved in clumped or intertwined masses composed of prodigious numbers of individuals, even small slivers of rock can contain huge numbers of specimens. Nevertheless, the preservation of intact fossil microbes in such deposits, even in rocks of the same mineralogy form, a single locality and stratigraphic horizon, can vary greatly. In this respect, the success of this study of the Buxa cherts was fortuitous. After this study was completed, two additional samples of chert collected from stratigraphically higher strata of the Buxa Formation in the Ranjit River section were examined; although each was highly carbonaceous, both were found to be barren of fossils. Even in this stromatolite-rich stratigraphic section, now demonstrated to be assuredly microfossiliferous, not all bedded cherts at all horizons contain cellularly identifiable remnants of life.

The Mesoproterozoic to Terminal Proterozoic (1,000–570 Ma) microbialite diversity has been reported from the Buxa Dolomite for the first time in a single section in Ranjit window, Sikkim Lesser Himalaya. The organic-walled microfossils such as *Siphonophycus*, *Eomycetopsis*, *Obruchevella*, and *Myxococcoides* have been recorded. The microfacies of the dolomite are oodolomicrite, oodolosparite, and micrite. The ooids are concentric, concentric cum radial, radial, and composite types showing various stages of diagenesis. The biosedimentological studies have suggested that the Buxa Dolomite was deposited in intertidal to subtidal environment. The stable carbon and oxygen isotope data from the Buxa Dolomite (Meso-Neoproterozoic) well developed in the Ranjit river valley, southern Sikkim, NE lesser Himalaya, India, is shown in Fig. 15. The Buxa Dolomite comprises stromatolitic dolomite, cherty dolomite, intraclastic–oolitic dolomite, and minor siliciclastic sediments. The stromatolite assemblage and the recently discovered microbiota from the Buxa Dolomite suggest a Meso-Neoproterozoic age for this dolomite (Tewari, 2003, 2004a, b; Tewari, 2007). Carbon isotopic data show fluctuations in D13 C (PDB) in Buxa Dolomite from −1.42 to +1.04. Oxygen isotope values grade from 19 to 23.9 (D18 0 SMOW). The lower part of the Buxa Dolomite shows mostly positive trend of excursion and may be result of increased rate of organic matter burial in a shallow carbonate platform. It is interpreted based on isotopic data combined with sedimentological and paleobiological studies that Buxa Dolomite was deposited in a carbonate platform well connected with the ocean. The environment was highly favorable for the luxuriant growth of the microbialites, and the cyanobacterial microbial communities were flourishing in the photic zone. The positive near-zero value of the Buxa Dolomite indicates that the environment of deposition was shallow marine (peritidal/subtidal to intertidal).

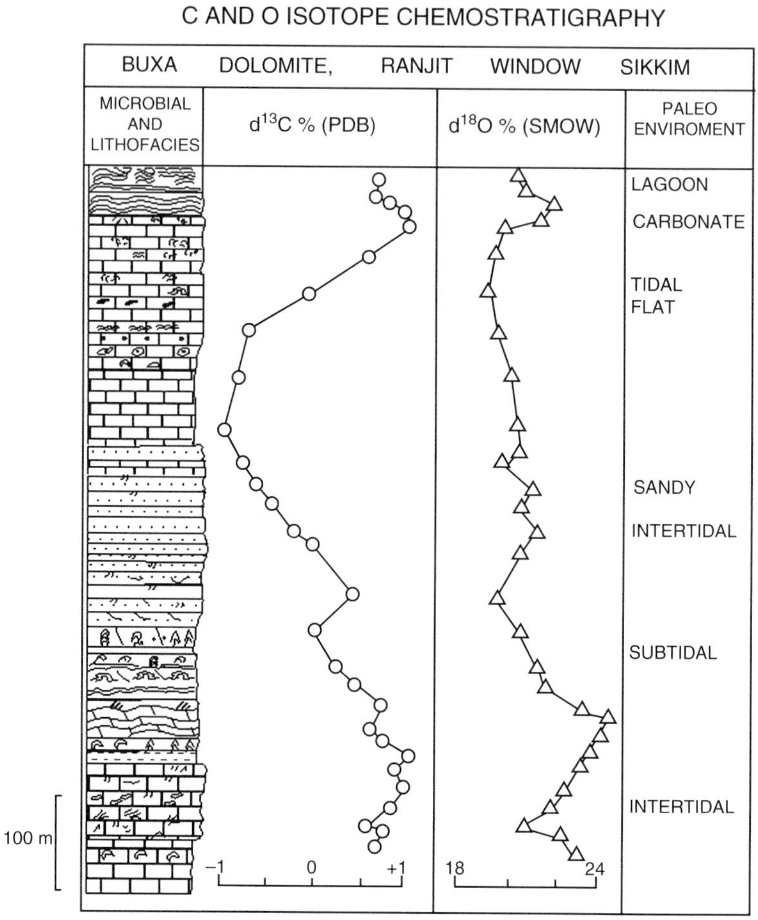

Figure 15. Carbon and oxygen isotope chemostratigraphy of the Buxa Dolomite.

12. Astrobiological Implications (Mars Sample Return) of the Buxa Microfossils

Interpretation of the evidence of life afforded by ancient rock samples on Earth is based on half a century of experience and the understanding of the early fossil record thus produced, as well as on deep knowledge of modern microbes, the communities they comprise, and the understanding of microbial evolution provided by molecular biology. Such data, of course, are unavailable for any other planet. Nevertheless, the evidence presented here has come from studies of a returned Martian sample. Most knowledgeable scientists would be convinced of the presence of past life on Mars. Other workers, however, would no doubt retain a degree of skepticism, perhaps calling for a search for nonbiological means by which to explain the seemingly compelling biological signals. And all could be expected to clamor for additional rock samples hosting additional examples of fossil microbes to be returned from Mars in order to build a body of paleobiologic data from that planet like that available of the Precambrian Earth. Whether ultimately relevant to Mars samples or not, what this study does show is that under appropriate circumstances, firm evidence of early life on Earth can be adduced from even a minuscule amount of fossil-bearing ancient rock – in this instance, the first Precambrian microfossiliferous samples to be discovered were in the Ranjit tectonic window of the northeastern Himalaya (Schopf et al., 2008).

13. Acknowledgments

I am grateful to Professor J.W. Schopf, A.B. Kudryavtsev, and J. Shen-Miller for discussions and helpful comments on the manuscript. The research on Laser Raman Spectroscopy and the Confocal Scanning Microscopy (CLSM) presented here was carried out at the IGPP Center for the Study of Evolution and the Origin of Life (CSEOL) at the University of California, Los Angeles, USA, in 2007. I thank CSEOL for providing support for my research visit to UCLA in 2007 as a Visiting Scientist, and in 2008, for participation in the World Summit on Ancient Microscopic Fossils as a member of the Organizing Committee. Dr. B.R. Arora, Director, Wadia Institute of Himalayan Geology, Dehradun, India, is thanked for facilities and permission to publish the chapter. Shri Girish Chauhan is thanked for ably typing the manuscript and fixing the figures. The Journal Astrobiology is thanked for granting permission to reproduce some figures and matter earlier published by us.

14. References

Acharyya, S.K. (1974) Stratigraphy and sedimentation of the Buxa Group, Eastern Himalaya. Himal. Geol. **4**: 102–116.

Kumar, G. (1997) Geology of Arunachal Pradesh. Special Publication Geological Society of India, **53**: 217p.

Raina, V.K. (1976) The Ranjit tectonic window: stratigraphy, structure and tectonic interpretation, and its bearing on the regional stratigraphy. Geol. Surv. India Misc. Publ. **41**: 36–42.

Schopf, J.W. (1968) Microflora of the bitter springs formation, late precambrian, central Australia. J. Paleontol. **42**: 651–688.

Schopf, J.W. (1992a) Proterozoic prokaryotes: affinities, geologic distribution, and evolutionary trends, In: J.W. Schopf and C. Klein (eds.) *The Proterozoic Biosphere, A Multidisciplinary Study*. Oxford University Press, Oxford, pp. 195–218.

Schopf, J.W. and Blacic, J.M. (1971) New microorganisms from the Bitter Springs Formation (Late Precambrian) of the Amadeus Basin, Australia. J. Paleontol. **45**: 925–961.

Schopf, J.W. and Kudryavtsev, A.B. (2005) Three dimensional Raman imagery of Precambrian microscopic organisms. Geobiology **3**: 1–12.

Schopf, J.W., Kudryavtsev, A.B., Agresti, D.G., Wdowiak, T.J. and Czaja, A.D. (2002) Laser-Raman imagery of Earth's earliest fossils. Nature **416**: 73–76.

Schopf, J.W., Kudryavtsev, A.B., Agresti, D.G., Czaja, A.D. and Wdowiak, T.J., (2005) Raman imagery: A new approach to assess the geochemical maturity and biogenicity of permineralized Precambrian fossils. Astrobiology **5**: 333–371.

Schopf, J.W., Tripathi, A. and Kudryavtsev, A.B. (2006) Three dimensional confocal optical microscopy of Precambrian microscopic organisms. Astrobiology **6**: 1–16.

Schopf, J.W., Tewari, V.C. and Kudryavtsev, A.B. (2008) Discovery of a new chert-permineralised microbiota in the Proterozoic Buxa formation of the Ranjit Window, Sikkim, Northeast India and its Astrobiological implications. Astrobiology **8(4)**: 735–746.

Shukla, M., Tewari, V.C. and Yadav, V.K. (1987) Late Precambrian microfossils from Deoban Limestone, Lesser Himalaya, India. Palaeobot. **35**: 347–356.

Shukla, M., Tewari, V.C., Babu, R. and Shrama, A. (2006) Microfossils from the Neoproterozoic Buxa Dolomite, West Siang district, Arunachal Lesser Himalaya, India and their significance. J. Pal. Soc. India **51**: 57–73.

Sinha Roy, S. (1980) Stratigraphic relations of the Lesser Himalayan formations of eastern Himalaya, In: K.S. Valdiya and S.B. Bhatia (eds.) *Stratigraphy and Correlation of Lesser Himalayan Formations*. Hindustan Publishing, Delhi, pp. 242–254.

Srinivasan, A. (2001) Stratigraphy and structure of low grade metasedimentaries in eastern Bhutan and Western Arunachal Pradesh. Himal. Geol. **22**: 83–98.

Tewari, V.C. (1989) Upper Proterozoic-Lower Cambrian stromatolites and Indian stratigraphy. Himal. Geol. **13**: 143–180.

Tewari, V.C. (1993) Precambrian and Lower Cambrian stromatolites of the Lesser Himalaya, India. Geophytology **23(1)**: 19–39.

Tewari, V.C. (1998) Prospects of delineating terminal proterozoic and precambrian-cambrian boundary in the northeastern Himalaya. Geosci. J. 19(2): 109–114.

Tewari, V.C. (2001) Discovery and sedimentology of microstromatolites from Menga Limestone (Neoproterozoic/Vendian), Upper Subansiri District, Arunachal Pradesh, northeastern Himalaya, India Curr. Sci. **80**(11): 1440–1444.

Tewari, V.C. (2002) Lesser Himalayan stratigraphy, sedimentation and correlation from Uttaranchal to Arunachal, In: C.C. Pant and A.K. Shrama (eds.) *Aspects of Geology and Environment of the Himalaya*. Gyanodaya Prakashan, Nainital, pp. 63–88.

Tewari, V.C. (2003) Sedimentology, palaeobiology and stable isotope chemostratigraphy of the Terminal Neoproterozoic Buxa Dolomite, Arunachal Pradesh, NE Lesser Himalaya. Himal. Geol. **24**: 1–18.

Tewari, V.C. (2004a) Palaeobiology and biosedimentology of the stromatolitic Buxa Dolomite, Ranjit Window, Sikkim, NE Lesser Himalaya, India. J. Seckbach, et al. (eds.) *Life in the Universe*. Kluwer Academic Publishers, Dordrecht, Netherlands

Tewari, V.C. (2004b) Microbial diversity in Meso-Neoproterozoic formations, with particular reference to the Himalaya, In: J. Seckbach (ed.) *Origins*. Kluwer Academic, Dordrecht, pp. 515–528.

Tewari, V.C. (2009) Proterozoic unicellular and multicellular fossils from India and their implications, In: J. Seckbach and M. Walsh (Eds) *From Fossils to Astrobiology: Records of Life on Earth and the Search for Extraterrestrial Biosignatures*. Springer, Berlin, pp. 119–139.

Tewari, V.C., Schopf, J.W. and Kudryavtsev, A.B. (2008) Neoproterozoic microfossils from the Buxa Dolomite, NE Himalaya, India: Analyses by Raman Spectroscopy and Optical and Confocal Laser Scanning Microscopy. World Summit on Ancient Microscopic Fossils, University of California, Los Angeles, USA (27Jul–02 Aug, 2008 9 abstract), p. 36.

Valdiya, K.S. (1980) Lesser Himalayan stromatolites: their biostratigraphic implications. Geol. Surv. India Misc. Publ. **44**: 117–127.

PART 5: GEOCHEMISTRY AND GEOMICROBIOLOGY OF STROMATOLITES AND MICROBIOTA

Brookfield
Coniglio
Glasauer
Rieu
S. Baskar
R. Baskar
Tewari
Thorseth
Øvreås
Lee
Routh
Dillon
Jenkins
Hikida
Lokho
Perri
Spadafora
Préat
de Jong
De Ridder
Gillan
Sumina
Sumin
Strauss

Biodata of **Dr. Michael E. Brookfield, Professor Mario Coniglio, Dr. Susan Glasauer**, and **Dr. Ruben Rieu**, authors of ***"Petrology, Elemental and Isotopic Geochemistry, and Geomicrobiology of Carbonate Infillings and Biofilms Lining Cracks Below the Neoproterozoic (Sturtian) Cap Carbonate in the Mirbat Inlier, Southernmost Oman"***

Dr. Michael E. Brookfield is a visiting scholar at the Institute of Earth Sciences, Academia Sinica, Taipei, Taiwan. His main interests are sedimentology, stratigraphy, tectonics and paleobiology. Over the last 40 years, he has done fieldwork in the Sahara and other deserts, the Canadian western Cordillera, the Himalaya and Central Asia, usually traveling alone and staying with the local inhabitants. He is an international fellow of the Explorers Club of New York, USA.

E-mail: **mbrookfi@hotmail.com**

Professor Mario Coniglio is a Professor in the Department of Earth and Environmental Sciences at the University of Waterloo and is currently the Associate Dean of Science for Undergraduate Studies. His research focuses on diagenesis of carbonate rocks. His field work has led him to the Caribbean, South America, Egypt, Oman, USA and all across Canada.

E-mail: **coniglio@uwaterloo.ca**

Dr. Susan Glasauer is an assistant Professor at the University of Guelph in the School of Environmental Sciences. Her research is in low temperature biogeochemistry, with particular emphasis on mineral–bacteria–aqueous interface geochemistry. She studies mineral signatures of bacteria in ancient environments, the geochemical cycling of minerals and metals mediated by microbes in subsurface environments, and the concentration of metals and biogenic minerals inside and on bacteria.

E-mail: **glasauer@uoguelph.ca**

Michael E. Brookfield

Mario Coniglio

Susan Glasauer

V.C. Tewari and J. Seckbach (eds.), *STROMATOLITES: Interaction of Microbes with Sediments*, Cellular Origin, Life in Extreme Habitats and Astrobiology 18, 525–540
DOI 10.1007/978-94-007-0397-1_23, © Springer Science+Business Media B.V. 2011

Dr. Ruben Rieu obtained an MSc in Geology from the University of Utrecht, The Netherlands, and a Ph.D. from the Swiss Federal Institute of Technology (ETH-Zurich) for which he was involved in the Neoproterozoic geology of Oman. Ruben currently works as a Reservoir Geologist in the petroleum industry where his main interest is in developing reservoir computer models.

E-mail: **Ruben.RIEU@ep.total.nl**

PETROLOGY, ELEMENTAL AND ISOTOPE GEOCHEMISTRY AND GEOMICROBIOLOGY OF CARBONATE INFILLINGS AND BIOFILMS LINING CRACKS BELOW THE NEOPROTEROZOIC (STURTIAN) CAP CARBONATE IN THE MIRBAT INLIER, SOUTHERNMOST OMAN

MICHAEL E. BROOKFIELD[1], MARIO CONIGLIO[2], SUSAN GLASAUER[3], AND REUBEN RIEU[4]

[1]Institute of Earth Sciences, Academia Sinica, P.O. Box 1-55 Nankang, Taipei 11529, Taiwan

[2]Department of Earth and Environmental Sciences, University of Waterloo, Waterloo, ON, Canada

[3]Department of Land Resource Science, University of Guelph, Guelph, ON, N1G 2W1, Canada

[4]Total E&P Nederland, Bordewijklaan 18, Den Haag, The Netherlands

Abstract Cryptic biofilms line cracks in granite basement below a Neoproterozoic cap carbonate in southern Oman. Their depth in the narrow cracks indicates that they grew in a dark non-photic environment, and signs of upward flow within the cracks suggest that they may have grown in waters expelled from depth, which may not have been in isotopic equilibrium with contemporary ocean waters. This study tested this hypothesis and disproved it. The biofilms and associated detrital carbonate were precipitated in isotopic equilibrium with contemporary seawater. The study also confirms the ^{13}C secular trends previously observed and attributed to seawater changes during post-glacial transgression and accumulation of the cap carbonate.

Keywords Petrology • Elemental, isotopic • Geochemistry • Geomicrobiology • Carbonate • Infillings • Biofilms • Basement • Cracks • Neoproterozoic • Sturtian • Cap • Carbonate • Mirbat, Oman

1. Introduction and Previous Work

Until recently, the prevailing view on subsurface life processes was that they are dominated by heterotrophic consumption of surface-derived carbon (Pedersen, 2001). It is now recognized, however, that reactive mineral surfaces and solute-rich ground waters provide potential energy sources sufficient for chemolithoautotrophic bacterial growth in the subsurface (Stevens, 1997), such as in caves and fissures (Engel et al., 2004). These energy sources are particularly important in deeper subsurface environments. Such possibilities have so far been ignored in discussion of the chemical and isotopic variations where the carbonate is produced by bacteria (Paerl et al., 2001). Here, we document a possible chemolithoautotrophically produced subsurface Neoproterozoic biofilm, its characteristics and its relationship to an overlying cap carbonate.

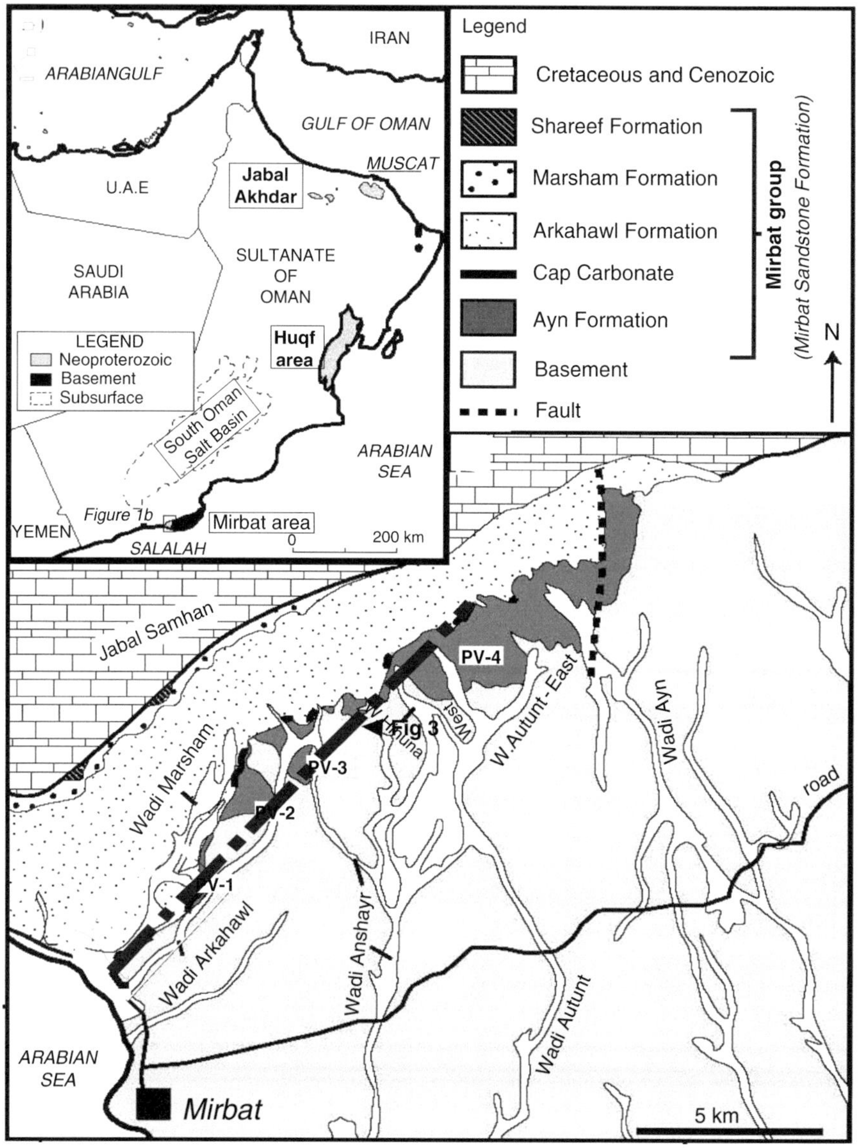

Figure 1. Location map of Neoproterozoic outcrops in Oman (inset) and map of Mirbat Group around Mirbat (from Allen, 2007).

The Huqf Supergroup is extensively exposed in Oman (Allen, 2007) (Fig. 1 inset).

In the Mirbat area (Fig. 1), carbonate laminae line carbonate-filled cracks, often as narrow as a few millimeters, in Precambrian basement up to 10 m below an overlying contemporary Neoproterozoic cap carbonate. The depth of the laminae below the basement rock surface, and the improbability of light reaching such depths along narrow cracks, suggests a non-photosynthetic origin. The fact that the laminae in some cases overlie mixed carbonate-clastic sediment in the crack indicates alternating carbonate-clastic deposition and carbonate encrustation during crack development.

The <723 Ma Mirbat Group consists of a largely undeformed, marine siliclastic succession framed by two Neoproterozoic glaciations (Rieu et al., 2006, 2007a, b; Rieu and Allen, 2008). The older glaciation is a discontinuous predominantly clastic glaciogenic succession (Ayn Formation) unconformably overlying 700–750 Ma basement (Worthington, 2005; Mercolli et al., 2006) (Fig. 2) and correlates with the Sturtian glaciation between about 710 and 667 Ma (Fanning and Link, 2004).

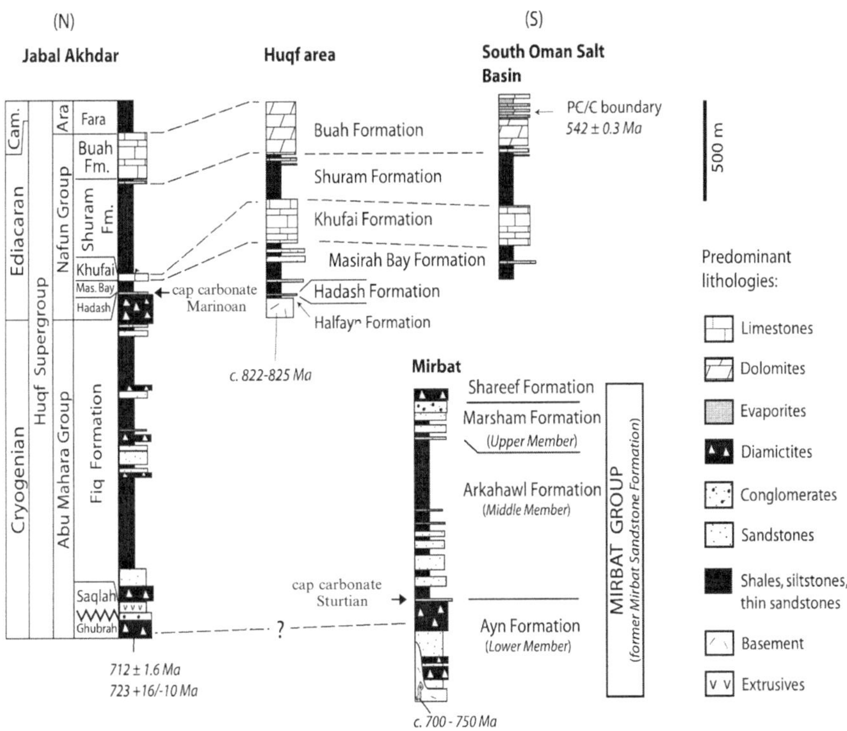

Figure 2. Comparative stratigraphy of Neoproterozoic units in Oman showing stratigraphic position of Mirbat Group (after Rieu et al., 2007b).

The Ayn Formation consists of glaciomarine rain-out diamictites and sediment gravity flows alternating with fluvial and deltaic sandstones and conglomerates, which fill valleys cut into the underlying granite (Kellerhals and Matter, 2003; Rieu et al., 2006) (Fig. 3). Above the glaciogenic clastics of the Ayn Formation is a transgressive, post-glacial cap carbonate, which at the basin margin onlaps on granodioritic paleohighs. A cap carbonate is the uppermost layer of a glacial sequence reflecting a major glaciation: it can be markedly diachronous, as in the Al Ayn example here, depending on the underlying topography and rate of transgression (Hoffman et al., 2007). The Al Ayn cap carbonate is locally overlain by sandstones but more commonly by about 40 m of transgressive shales (Fig. 2).

The Al Ayn cap carbonate is laterally discontinuous and varies considerably in thickness and facies; from in situ domal stromatolites through masses of structureless sparry calcite to re-sedimented and mass flow carbonate conglomerates to basement boulder beds with carbonate matrices (Rieu et al., 2006. The cap carbonate is locally well developed above paleohighs, where wave-rippled carbonate grainstones indicate deposition in shallow water and where stromatolites are prominently developed in the dolomitic tops of the carbonates (Fig. 4a) Overlying the Ayn Formation in the deeper parts of the basin, i.e., in the paleovalleys, the

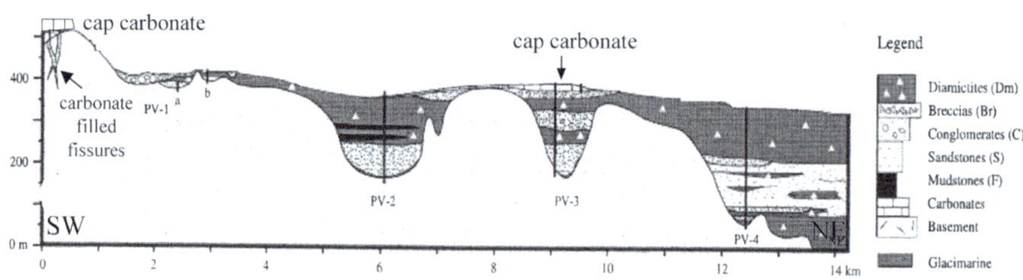

Figure 3. Facies variation in Al Ayn Formation (from Rieu et al., 2006) (location on Fig. 1).

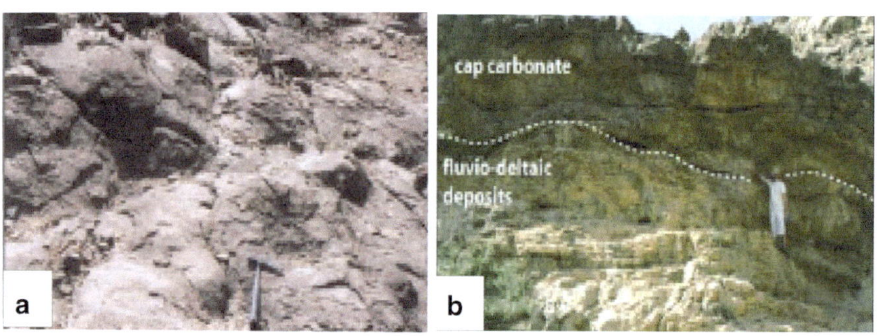

Figure 4. Variability of cap carbonate at Mirbat: (**a**) domal stromatolites in thick cap carbonate (paleovalley PV-3, Fig. 3); (**b**) thick cap carbonate overlying pebbly sandstones (paleovalley PV-2, Fig. 3).

cap carbonate is absent or represented by thin discontinuous mixed carbonate-siliclastic mass flow deposits (Fig. 4b).

The carbonates vary in $\delta^{13}C$ from −5.4‰ to +5.8‰ both vertically and laterally (Rieu et al., 2006). The detailed $\delta^{13}C$ composite isotope profile through the carbonate (Fig. 5) shows an upward trend from negative to positive (−3.5‰ to +5.8‰), though with a decrease in the topmost levels. The lack of correlation between diagenetic indicators (Mn/Sr ratios, $\delta^{18}O$) and variations in $\delta^{13}C$ suggests that the systematic variations observed in $\delta^{13}C$ have not been significantly altered by diagenesis (Rieu et al., 2006).

In the western part of the study area, where the cap carbonate directly overlies granodioritic basement, a network of predominantly very narrow carbonate veins, filled with detrital carbonate (mostly dolomite) mixed with minor amounts of siliclastics derived from the granodiorite, extends downwards into the granodiorite basement for up to10 m. (Fig. 6). One megafissure extends along the

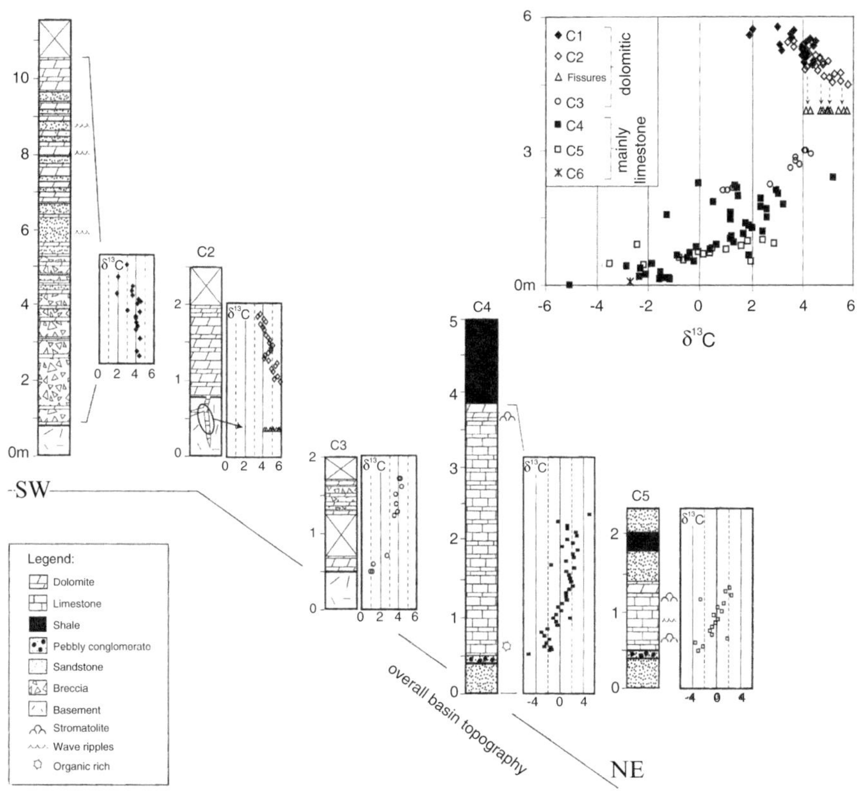

Figure 5. Correlation of carbonate sections from basement high to paleovalley with $\delta^{13}C$ values: inset shows composite $\delta^{13}C$ profile (from Rieu et al., 2006).

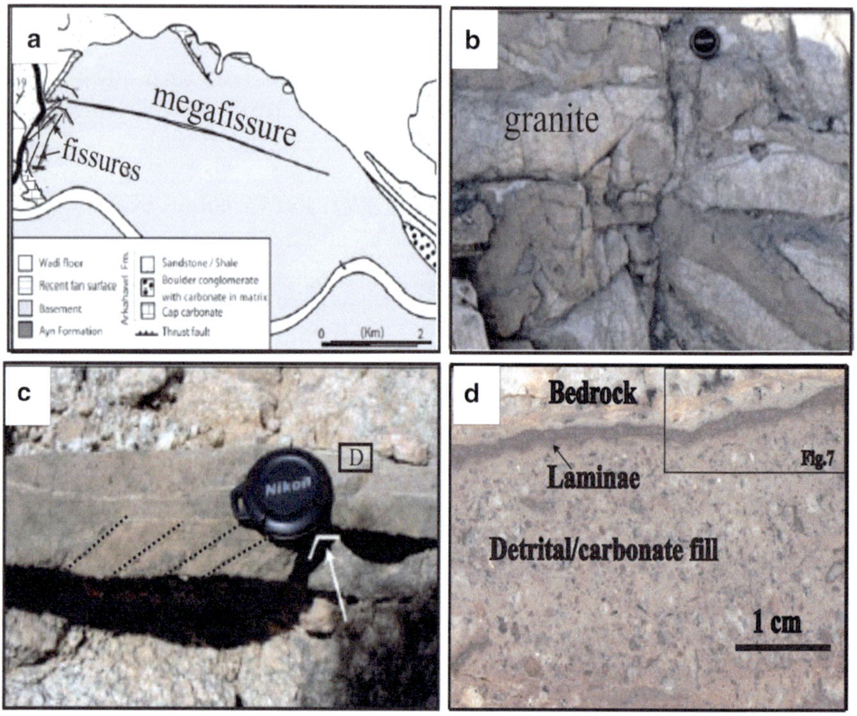

Figure 6. Fissures: (**a**) geological map of western part of area showing cap carbonates directly overlying fissure fills in granite basement, note megafissure stretching about 5 km; (**b**) carbonate fissure fills (*darker*) in basement; (**c**) mixed carbonate/clastic fissure fill showing miniforesets directed upwards above a negative step in fracture wall and location of analyzed slab D; (**d**) polished slab of upper edge of fill in (**c**) showing encrusting laminae and location of Fig. 7.

outcrop for more than 5 km (Fig. 6a) (Rieu et al., 2006). Occasionally, the veins have void-filling carbonate cement indicating further opening and/or solution widening after filling (Rieu et al., 2006). This article summarizes a multi-method study of one vein filling with a thin laminated lining, and relates it to the environment in which it developed (Fig. 6c, d).

2. Methods

2.1. PETROLOGICAL ANALYSIS

A slab of rock, incorporating the granite basement, the laminated layer and the detrital sandy dolomite vein filling, was divided up into eight petrographic domains for standard transmitted light petrography, cathodoluminescence and stable isotope

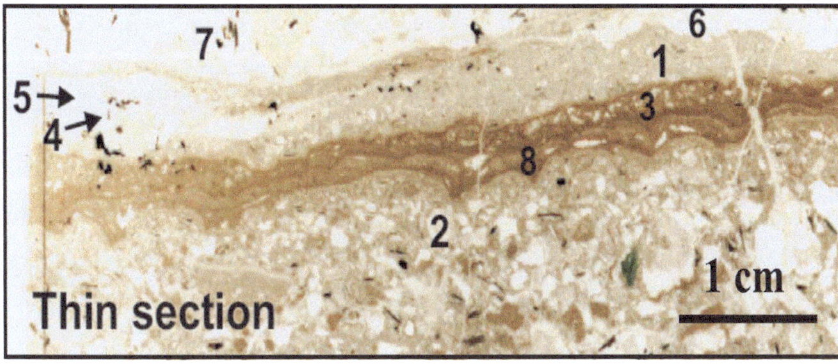

Figure 7. Thin section of edge of fissure fill showing domains analyzed.

analysis (Fig. 7). Domains 1 and 2 are detrital domains underlying and overlying the laminated layers (domains 3 and 8). A thin section cut-off was polished and examined with cathodoluminescence using a Reliotron cathodoluminescence instrument. Optimal results were obtained using a beam current of ~0.600 mA, a minimum voltage of 10 kV and a beam diameter of ~1 cm.

2.1.1. Scanning Electron Microscope Analysis

For basic scanning electron microscope (SEM) analysis, a polished thin section was coated with gold. A LEO FESEM 1530 instrument was operated at a voltage of 20 kV to yield resolution greater than 1.5 μm. Quantitative elemental compositions were determined using an integrated energy dispersive X-ray (EDX) spectrometer.

For geochemical mapping, two SEMs were used for imaging and EDX analyses. Acid etched-rock sections were prepared by exposing cut rock faces to 10% HCl for 30 s. Rock pieces were mounted on an aluminum stub and sputter-coated with a 60:40 gold/palladium alloy. Samples were examined using a Hitachi S-4500 field emission SEM, using a 5 kV accelerating voltage and a beam current of 10 A (located at Laboratory Services, University of Guelph). Element maps were generated using a LEO 440 SEM equipped with a Quartz Xone EDX analyzer (located at Surface Science Western, University of Western Ontario). A 25 kV electron beam was used to obtain back-scattered images, EDX spectra and EDX intensity maps using full spectral imaging. EDX analysis can detect all elements above atomic number 5 and has a minimum detection limit of 0.5 wt% for most elements.

2.1.2. Isotopic Analysis

For stable isotope analysis: four domains across the laminae (1, 2, 3, 8) were analyzed (Fig. 7). Areas 1 and 2 contain detrital siliclastics and surround the

Table 1. Carbon and oxygen isotope measurements of calcite (cc) and dolomite (dol) from domains 1, 2, 3, 8.

Sample identification	$^{13}C_{VPDB}$ (‰)	$^{18}Ov_{PDB}$ (‰)	$^{18}O_{VSMOW}$ (‰)
1 – cc	4.30	−11.52	19.03
1 – dol	5.07	−7.18	23.51
2 – cc	5.37	−8.82	21.81
2 – dol	6.91	−1.13	29.74
3 – cc	5.05	−7.68	22.99
3 – dol	5.30	−6.43	24.28
8 – cc	5.20	−6.77	23.93
8 – dol	4.90	−5.95	24.77

laminae of areas 3 and 8. Areas 4 and 5 are calcite veins within bedrock and areas 6 and 7 are bedrock. Samples of carbonate for stable carbon and oxygen isotope analysis were extracted using a modified dental drill and were analyzed at the Stable Isotope Laboratory at the University of Windsor. Approximately 3.0 mg of untreated powder was reacted with 100% phosphoric acid at 25°C and 50°C for calcite and dolomite, respectively, for at least 4 h and the evolved CO_2 gas was analyzed for $^{18}O/^{16}O$ and $^{13}C/^{12}C$ using a Finnigan-Mat Delta Plus mass spectrometer. Results are reported in conventional per mil notation relative to VPDB using standard correction procedures (Table 1).

Precision, based on replication of international standards and duplicate samples from this study, is better than 0.03‰ and 0.04‰ for C and O, respectively. All analyses were carried out in two stages, to separate the gas from calcite and from dolomite. Only the dolomite results are considered to be meaningful here, since the relatively small amount of calcite in the samples makes the analyses unreliable.

3. Descriptions

3.1. FIELD AND HAND SPECIMEN

Although the cracks in the basement generally follow a weak foliation, cracks also commonly cross-cut one another and one population defines a conjugate set perpendicular to the regional bedding (Rieu et al., 2006) (Fig. 6a, b). Cracks do not penetrate the overlying cap carbonate and thus predate and/or are contemporaneous with cap carbonate deposition. However, small-scale folds and small thrusts do displace the basement, cap carbonate and lowermost few meters of the overlying siliclastics, suggesting that deformation continued for some time after the formation and filling of the crack systems. Although the characteristics of the carbonate filling the cracks are identical to those of the overlying cap carbonate, in one place at least, cross bedding behind a "step" in a crack indicates that the mixed siliclastic and detrital carbonate filling was affected by groundwater flowing upward through the cracks (Fig. 6c). The laminae which line some cracks and overlap the detrital fillings indicate that the cracks were open during deposition of the overlying cap carbonate (Fig. 6d). The analyzed specimen shows thin laminae lining one side of a steeply dipping crack and encrusting an already deposited thin carbonate-siliclastic layer (Fig. 7).

3.2. PETROGRAPHY

The analyzed thin section was divided up into eight domains. Domain 1 is the first carbonate infilling of the crack. It has up to 50% poorly sorted siliclastics (40–120 μm) in a peloidal (micritic) carbonate matrix with crystals ranging in size from 80 to 200 μm. Domain 2 is a later infilling and consists of a basal dolomicrite-rich layer with sparse fine-grained siliclastic grains passing gradually upwards into a coarser siliclastic-rich dolomicrite. This coarser layer contains poorly sorted angular granodiorite clasts up to 3.2 mm (more typically 200–400 μm) and derived mineral clasts (consisting of altered feldspar, biotite, quartz and rare chlorite) in a peloidal (micritic) clotted/peloidal dolomicrospar matrix with crystals ranging from 80 to 200 μm, but mostly 40–120 μm in size). Some of the larger siliclastic grains have microfractures filled with carbonate. Domain 3 is the lower part of the laminated layer. It contains carbonate clasts up to 80 μm in size surrounded by a brown, inclusion-rich, carbonate (mostly dolomite) crystal mosaic with crystals up to 200 μm in size. Based on inclusion distribution, some of these crystals may be spherulitic. Domain 4 is an elongate angular basement rock fragment separated from the basement by a calcite vein of domain 5. Domain 5 is a vein with calcite cement mosaic consisting of zoned pore-filling crystals up to1,200 μm in size at its center. Domain 6 is an offset carbonate vein with abundant siliclastics and has clear anhedral calcite crystals between 50 and 100 μm in size. Domain 7 is the granodiorite bedrock, consisting of quartz, zoned plagioclase and chloritized biotite. Domain 8, the upper part of the laminated layer, consists of non-isopachous and discontinuous dolomitic carbonate laminae (400–1,200 μm wide), differentiated by unresolvable very thin, dark micritic laminae. The individual laminae consist of inclusion-rich equant and fibrous dolomite crystals up to 20–30 μm in size and contain small elongate fenestrae up to 750 μm long filled with clear fibrous dolomite cement up to 100 μm long and 20 μm wide. The very latest pore filling associated with the fibrous dolomite-filled fenestrae is unzoned calcite cement. Most calcite in the dolomite matrix is patchy and likely post-dolomite. Although this cannot be confirmed petrographically, the fact that calcite follows fibrous dolomite in the fenestrae supports the idea that the calcite postdates (and replaces?) dolomite. Domain 8 is sharply overlain by the lower part of domain 2. In thin section, the laminae (domains 3 and 8) show up as a thin (about 1 mm thick) lining with uneven base and pustular top (Fig. 7). It is this contrast between the basal and uppermost laminae that suggests a biological rather than an inorganic origin (Sabater, 2000; Andres and Reid, 2006), a topic which is further investigated in Sect. 3.4. In cathodoluminescence, the carbonate is moderately bright red/orange except for yellow zones in the carbonate-filled crack of domain 5, between bedrock domain 7 and siliclastic/dolomite domain 1, and the early part of the fringe of fibrous dolomite crystals filling fenestrae in the upper part of the laminae domain 8. Calcite has the same luminescence as the dolomite, so neither can be differentiated on the basis of cathodoluminescence alone, as is often possible in other situations. The differentiation is, however, straightforward in SEM backscatter with the dolomite/calcite ratio ranging from 70 to 90/30 to 10.

3.3. ELEMENT MAPPING

The element mapping variation mostly reflects the compositional differences between the siliclastic and carbonate constituents and does not add much to other analyses, so only one example is given (Fig. 8). The distribution of Si, Al, Fe and Na is obviously caused by the detrital siliclastic material that is readily visible petrographically, while the inverse relationship of Ca and Mg is caused by the mixed dolomite/calcite mineralogy. Any organic remnants have obviously been obliterated, as shown by the almost uniform P and C maps.

3.4. SEM SCANS OF LAMINAE

SEM analyses of the laminae (domain 8) reveal pitted areas, which resemble structures typically produced by globular microorganisms of 0.5–2.00 μm in size such as bacteria and archaea (Koch, 1996; Nealson, 1997) (Fig. 9). These strengthen the interpretation of the laminae as biologically secreted based on gross morphology.

3.5. ISOTOPE GEOCHEMISTRY

All the vein carbonates are re-crystallized microspars with variable degrees of dolomitization, and the isotopic systems are therefore undoubtedly altered to some degree (Table 1). Furthermore, the spread in value of the dolomite, for both $\delta^{13}C$ and $\delta^{18}O$, cannot be interpreted with any confidence, since the samples are heterogeneous mixtures of carbonate and siliclastics (see especially samples #1 and 2 with high siliclastic content). However, apart from these, the laminae have $\delta^{18}O$ values ranging from −6.0 to −7.2 per mil. Considering only dolomite, the $\delta^{13}C$ values from all the domains are comparable, having distinct positive values ranging from 4.9‰ to 6.9‰ (Table 1). The laminae have low variation in both $\delta^{13}C$ and $\delta^{18}O$, both here and by Rieu et al. (2006) (Table 1, Fig. 5). For $\delta^{13}C$, the

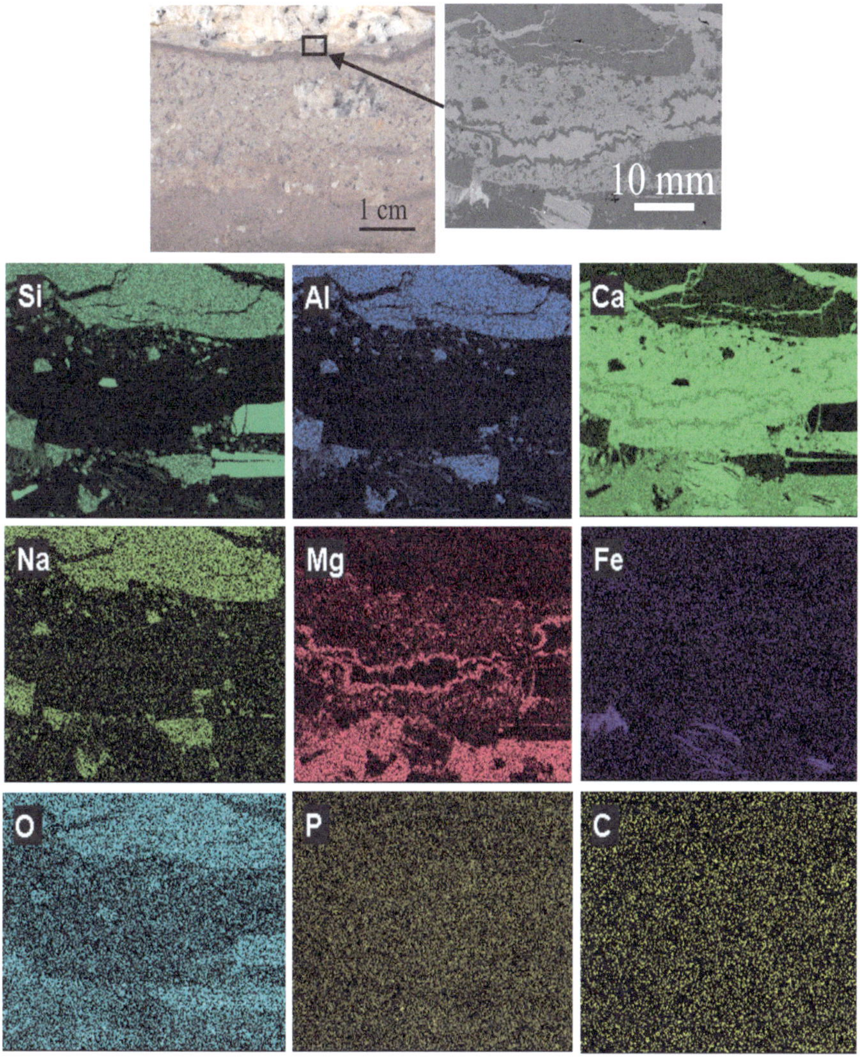

Figure 8. Semi-quantitative element mapping. Si, Al, Na, 0 correlate with bedrock clasts; Ca, Mg correlate with carbonate – note Mg rich dolomite lining to vug in center and domain 3 below; Fe correlates with biotite, P, C show no correlation.

values are the most positive of the entire cap carbonate. Such positive trends are unusual in cap carbonates (Hoffman et al., 2007).

If the $\delta^{18}O$ of the calcites are meaningful, the value of −11.5‰ for #1 is notable and suggests high temperature, either during precipitation or, more, likely, during later diagenesis: it is hard to imagine how else such a high negative value

Figure 9. SEM zoom detail of part of laminae, domain 8.

of $\delta^{18}O$ could be derived. This is, however, only one data point. The $\delta^{13}C$ data overwhelmingly conform that the analyzed crack fill has the same value as those from other crack fills analyzed by Rieu et al. (2006) (Fig. 5), which represents the most positive values recorded from all cap carbonate analyses in the area.

4. Interpretation

That the cracks in the basement granodiorite were open during and after cap carbonate deposition is shown by the mixed siliclastic/detrital carbonate fillings, carbonate laminae and the later void-filling cements. That the carbonate laminae are the result of biogenic secretion by aphotic bacteria or archaea is suggested by the improbability of there being sufficient light penetration to 5 m depth below the bedrock surface, given the narrow and irregular shape of the cracks, though calculations of the amount of light penetration (which we attempted) are hindered by the complexity of the fissures.

The fissure carbonates and overlying cap carbonate have comparable, though very variable C and O isotopic values with the highest values from the fissure fills and contemporary cap carbonates of section C2 (Fig. 5, Table 1), which suggests that carbonate precipitation was in isotopic equilibrium with contemporary seawater circulating through the cracks.

The laminae lining the crack was likely secreted by aphotic bacteria or archea, based on morphology, at the same time as the growth of stromatolites above (the same $\delta^{13}C$ values): there is no evidence that the laminae were formed under extreme conditions (see Satyanarayana et al., 2005). The dominant-mixed siliclastics/carbonate filling of the cracks were probably infiltrated from above, given their identical isotopic composition with the upper stromatolite layers, even if they were at times reworked by upwardly flowing water, especially since the stromatolites rest on basement in places.

5. Conclusions

Cryptic biofilms line cracks in basement below a Neoproterozoic (Sturtian) cap carbonate in southern Oman. The depth in the cracks indicates that they grew in dark non-photic environments. Signs of upward flow of water suggested a hypothesis that they may have grown in hydrothermal waters in isotopic disequilibrium with contemporary ocean waters. This detailed petrological and isotopic study negated this hypothesis and indicates that the crack carbonates were precipitated in isotopic equilibrium with contemporary seawater or with groundwater in equilibrium with contemporary seawater. It confirms the $\delta^{13}C$ trends previously observed and attributed to secular change in seawater composition $\delta^{13}C$ during post-glacial transgression and accumulation of the cap carbonate. The positive values of both stromatolites and laminae compared to the cap carbonate sections underlying the stromatolites suggests a change in isotopic composition of seawater during the climax of the post-glacial transgression, which is opposite to other younger cap carbonates (Hoffman et al., 2007).

6. Acknowledgments

M.B. thanks the Institute of Earth Sciences, Academia Sinica, Taiwan (via its director Bor-ming Jahn) for support. M.C. and S.G. are supported by NSERC Canada. R.R. was supported by ETH, Zurich. We thank Graham Shields for evaluation of an earlier draft of the manuscript and Vinod C. Tewari for inviting us to contribute this chapter and also for careful editing.

7. References

Allen, P.A. (2007) The Huqf Supergroup of Oman: basin development and context for Neoproterozoic glaciation. Earth Sci. Rev. **84**: 139–185.

Andres, M.S. and Reid, R.P. (2006) Growth morphologies of modern marine stromatolites: a case study from Highborne Cay, Bahamas. Sediment. Geol. **185**: 319–328.

Engel, A.S., Porter, M.L., Stern, L.A., Quinlan, S. and Bennett, P.C. (2004) Bacterial diversity and ecosystem function of filamentous microbial mats from aphotic (cave) sulfidic springs dominated by chemolithoautotrophic "Epsilonproteobacteria". FEMS Microbiol. Ecol. **51**: 31–53.

Fanning, C.M. and Link, P.K. (2004) U-Pb SHRIMP ages of Neoproterozoic glaciogenic Pocatello Formation, southeastern Idaho. Geology **32**: 881–884.

Hoffman, P.F., Halverson, G.P., Domack, E.W., Husson, J.M. and Schrag, D.P. (2007) Are basal Ediacaran (635 Ma) post-glacial "cap dolostones" diachronous? Earth Planet. Sci. Lett. **258**: 114–131.

Kellerhals, P. and Matter, A. (2003) Facies analysis of a glaciomarine sequence, the Neoproterozoic Mirbat Sandstone Formation, Sultanate of Oman. Eclogae Geol. Helv. **96**: 49–70.

Koch, A.L. (1996) What size should a bacterium be? A question of scale. Ann. Rev. Microbiol. **50**: 317–348.

Mercolli, I., Briner, A.P., Frei, R., Schonberg, R., Nagler, T.F., Kramers, J. and Peters, T. (2006) Lithostratigraphy and geochronology of the Neoproterozoic crystalline basement of Salalah, Dhofar, Sultanate of Oman. Precambrian Res. **145**: 182–206.

Nealson, K.H. (1997) Nannobacteria: size limits and evidence. Science **276**: 1776.

Paerl, H.W., Steppe, T.F. and Reid, R.P. (2001) Bacterially mediated precipitation in marine stromatolites. Environ. Microbiol. **3**: 123–130.

Pedersen, K. (2001) Exploration of deep intraterrestrial microbial life: current perspectives. FEMS Microbiol. Lett. **185**: 9–16.

Rieu, R. and Allen, P.A. (2008) Siliclastic sedimentation in the interlude between two Neoproterozoic glaciations, Mirbat area, southern Oman: a missing link in the Huqf Supergroup? GeoArabia **13**: 45–72.

Rieu, R., Allen, P.A., Etienne, J.L., Cozzi, A. and Weichert, U. (2006) A Neoproterozoic glacially influenced basin margin succession and 'atypical' cap carbonates associated with bedrock palaeovalleys, Mirbat area, southern Oman. Basin Res. **18**: 471–496.

Rieu, R., Allen, P.A., Plotze, M. and Pettke, T. (2007a) Compositional and mineralogical variations in a Neoproterozoic glacially influenced succession, Mirbat area, south Oman: implications for paleoweathering conditions. Precambrian Res. **154**: 248–265.

Rieu, R., Allen, P.A., Cozzi, A., Kosler, J. and Bussy, F. (2007b) A composite stratigraphy for the Neoproterozoic Huqf Supergoup of Oman: integrating new litho-, chemo- and chronostratigraphic data of the Mirbat area, southern Oman. J. Geol. Soc. Lond. **164**: 997–1009.

Sabater, S. (2000) Structure and architecture of a stromatolite from a Mediterranean stream. Aquat. Microb. Ecol. **21**: 161–168.

Satyanarayana, T., Chandralata, R. and Shivaji, S. (2005) Extremophilic microbes: diversity and perspectives. Curr. Sci. **89**: 78–90.

Stevens, T. (1997) Lithoautotrophy in the subsurface. FEMS Microbiol. Rev. **20**: 327–337.

Worthing, M.A. (2005) Petrology and geochronology of a Neoproterozoic dyke swarm from Mirbat, south Oman. J. African Earth Sci. **41**: 248–265.

Biodata of **Dr. Sushmitha Baskar, Dr. Ramanathan Baskar, Professor Vinod Chandra Tewari, Dr. Ingunn H. Thorseth, Dr. Lise Øvreås, Dr. Natuschka M. Lee**, and **Dr. Joyanto Routh**, authors *of "Cave Geomicrobiology in India: Status and Prospects"*

Dr. Sushmitha Baskar is presently associated with the Department of Environmental Science and Engineering, Guru Jambheshwar University of Science and Technology, Hisar. She was awarded Ph.D. in 2002 by the Guru Jambheshwar University of Science and Technology, Hisar. She has been awarded various prestigious national fellowships (CSIR–SRF, CSIR-RA, UGC Dr. DS Kothari postdoctoral fellowship) and international fellowships. She was awarded Swiss Government Fellowship (2002–2004) to work with Professor Judith Ann McKenzie in the Geomicrobiology laboratory at the Department of Earth Sciences, Swiss Federal Institute of Technology (ETH), Zurich, Switzerland where she started her research in cave geomicrobiology. She was recently awarded Norwegian Government fellowship to work with Dr. Ingunn H. Thorseth, Dr. Lise Øvreås and Professor Rolf Birger Pedersen, at the Centre for Geobiology, University of Bergen, Norway (2008–2009). Further, she is awarded Swedish Institute Grant to work with Dr. Joyanto Routh at the Department of Geology and Geochemistry, Stockholm University, Sweden (2009–2010). She has made significant contributions in reputed journals in the field of geomicrobiology, is a reviewer of scientific journals, presented papers in various national and international conferences held in India, Switzerland, Germany and Norway. She has authored three books on Environmental Sciences for Engineering undergraduates (2007), Environmental Sciences for undergraduates (2007), Natural Hazards (2009) along with her husband Dr. Ramanathan Baskar. Her research concentrates on cave geomicrobiology, in an attempt to understand the influence of microorganisms on mineral dissolution/precipitation and to understand the water–rock–microbe interactions.

E-mail: **rbaskargjuhisar@yahoo.com**

V.C. Tewari and J. Seckbach (eds.), *STROMATOLITES: Interaction of Microbes with Sediments*, Cellular Origin, Life in Extreme Habitats and Astrobiology 18, 541–569
DOI 10.1007/978-94-007-0397-1_24, © Springer Science+Business Media B.V. 2011

Dr. Ramanathan Baskar is an Associate Professor in the Department of Environmental Science and Engineering, Guru Jambheshwar University of Science and Technology, Hisar. He earned his Ph.D., from Punjab University, Chandigarh in the field of geology. He has been awarded various fellowships throughout his academic career such as prestigious Dr. K.S Krishnan Department of Atomic Energy research fellowship and fellowships funded by the various Government of India agencies such as the Council of Scientific and Industrial Research, University Grants Commission, Department of Ocean Development and Department of Science and Technology. Prior to joining the University in 1995, he worked as a research associate at Wadia Institute of Himalayan Geology, Dehradun and the School of Environmental Sciences, Jawaharlal Nehru University, New Delhi. His research interests are in the field of environmental geochemistry and geomicrobiology, where he has made significant contributions and participated in several national and international conferences, workshops, training programs held in India, Switzerland, Germany, Sri Lanka, France and Norway. He has been awarded many national major research projects, established several international collaborations and is a reviewer of scientific journals. He was awarded the ICSC World Laboratory scholarship to carry out research in the Department of Earth Sciences, University of Paris-Sud, France (1990–1991) during his doctoral program. He was invited as an Academic Guest in the Department of Earth Sciences, Swiss Federal Institute of Technology, Zurich, Switzerland (2003–2004) and Norwegian Centre for Excellence, Centre for Geobiology, Norway, Bergen (2008–2009). His research has recently taken a new direction into one of the fast growing frontier areas of science – geomicrobiology (scientific revolution at the boundary of geology and biology), in an attempt to understand to influence of microbes on mineral dissolution/precipitation and environmental geochemistry of speleothems to generate fundamental knowledge about the complex water–rock–microbe interactions, which has direct applications on issues of environmental management.

E-mail: **rbaskargjuhisar@yahoo.com**

Professor Vinod C. Tewari is currently the Head of the Sedimentology Group at Wadia Institute of Himalayan Geology, Dehradun and a Senior Associate of International Centre for Theoretical Physics, Trieste, Italy. He obtained his Ph.D. from the University of Lucknow in Geology in 1986 and continued his research at Wadia Institute. Dr. Tewari taught Geology at Kumaon University, Nainital, Uttarakhand, India as Professor of Geology. Professor Tewari's scientific interests are in the areas of Precambrian stromatolites, sedimentation, carbon isotope chemostratigraphy, genesis, early evolution and diversification of life and its astrobiological significance. His current interests are Speleothems, paleoclimate and monsoon study of Himalayan cave systems. He is associated with the International Geological Correlation Programme (I.G.C.P.) Project 493 on The Rise and Fall of Vendian Biota. He has eighty research papers published to his credit, and edited several volumes of Himalayan Geology, India and Journal of Nepal Geological Society, Kathmandu, Nepal. Professor Tewari has organized first Indo-Soviet Symposium on Stromatolites and Stromatolitic Deposits and other IGCP meetings in India. He has been one of the organizers of the World Summit on Ancient Microscopic Fossils held in University of California, Los Angeles, USA in 2008.

E-mail: **vtewari@wihg.res.in**

Dr. Ingunn H. Thorseth is Associate Professor in geochemistry/geomicrobiology at the Department of Earth Science, and one of the senior researchers at the Centre of Geobiology, at the University of Bergen, Norway. She received her Dr. Scient degree in geology from the University of Bergen in 1995, and was then postdoctoral fellow at the Oregon State University, USA. Since 1997, she has been at the University of Bergen. Her main research fields are low-temperature alteration of minerals and glasses, water–rock–microbe interactions, microbial biomineralisation and biosignatures. She was one of the first researchers to focus on the deep biosphere of the ocean crust, and her research include interdisciplinary studies of both recent seafloor basalts from mid-ocean spreading ridges and subseafloor basalts from older ocean crust, seafloor hydrothermal vent fields, and in situ and laboratory experiments. A major part of her research has also been concentrated on deterioration and preservation of rock art, and on weathering effects and biomineralisation of lichens. She was central in the development and establishment of the first interdisciplinary research group in geomicrobiology, and later in the Norwegian centre of excellence in Geobiology at UoB.

E-mail: **Ingunn.Thorseth@geo.uib.no**

Dr. Lise Øvreås received her MSc and DSc (1998) in microbial ecology from the University of Bergen, Norway. She has worked as a research scientist in Unifob (University research in Bergen) from 1998 to 2006 and has been an active member in a research group that has pioneered whole community analysis in soil, water and sediment ecosystems. She has a solid research experience in environmental microbiology, and in developing and applying molecular techniques for use in natural ecosystems. The basic molecular methods and diversity analyses are established routines in her laboratory. She was appointed as associated Professor in Department of Biology in 2006, and was promoted as full Professor in geomicrobiology in 2007. She is the group leader in the geomicrobiology research group and she is also leader in the thematic group "The deep biosphere" at the Centre of Excellence (COE), Centre of Geobiology (CGB) at the University of Bergen. At present, she is project leader and coordinator of an SIU funded NUFU project together with collaborators in Ethiopia and South Africa to study the microbial diversity in alkaline lakes in the Ethiopian rift valley. She has experience in project management and scientific leadership from research projects funded by the Norwegian research council (NRC), Nordic Academy of Science (NorFA), Statoil (Oil Company) as well as EU-funded projects. Her research interests are genetic diversity and population dynamics of microorganisms in their natural environments, population ecology and community ecology; natural variation in the microbial community composition and also the regulation of biodiversity and adaptability of a microbial community to external stress factors. Current work is focused on studies of microbial diversity in aquatic systems, Arctic soil systems, and the microbial diversity in the deep biosphere of the ocean crust.

E-mail: **lise.ovreas@bio.uib.no**

Dr. Natuschka M. Lee received her Master of Science Degree in Biology and Chemistry, and then Ph.D. (1996) in Environmental Biotechnology at Lund University in Sweden. She was then awarded grants for postdoc research at the Aalborg University in Denmark and at Technische Universität München in Munich, Germany, to learn and apply advanced aspects of radioactive and molecular tools in microbial ecology, taxonomy and phylogeny on different eco-systems. Since 2003, she has been the group leader of the lab for microbial systems ecology at the Department of microbiology at TUM. Her research areas deal with the microbial biodiversity and molecular ecology of extreme ecosystems such as the subsurface in pristine and contaminated eco-systems in different parts of the world, with links to astrobiology, symbiotic biology, bioremediation and the development of novel molecular tools for in situ detection of microorganisms in environmental systems. She teaches microbiology, molecular ecology and organises international courses on in situ detection of microbial structure and function in the environment.

E-mail: **natuschka.lee@microbial-systems-ecology.de**

Dr. Joyanto Routh is an Associate Professor in Biogeochemistry at the Department of Geology and Geochemistry at Stockholm University (Sweden). He did his Ph.D. in organic matter characterization and microbial geochemistry in Eocene sediments from Texas. After working in the petroleum industry, he joined Stockholm University in 2000. Routh is involved in various environmental issues related to groundwater pollution, acid mine drainage and eutrophication. He is also involved in research on paleoclimate and focuses on lake, mangrove and speleothem archives. He mainly uses biomarkers to study the climate or human-induced changes preserved in sedimentary records. He is leading projects funded by the European Union, Swedish Research Council and Sida and currently works in India, South Africa, Siberia, and Sweden.

E-mail: **joyanto.routh@geo.su.se**

CAVE GEOMICROBIOLOGY IN INDIA: STATUS AND PROSPECTS

SUSHMITHA BASKAR[1,7], RAMANATHAN BASKAR[1,7], VINOD CHANDRA TEWARI[2], INGUNN H. THORSETH[1,3], LISE ØVREÅS[1,4], NATUSCHKA M. LEE[5], AND JOYANTO ROUTH[6]

[1]*Centre for Geobiology, University of Bergen, P.O. Box 7803, 5020 Bergen, Norway*

[2]*Wadia Institute of Himalayan Geology, Dehradun 248001, Uttarakhand, India*

[3]*Department of Earth Science, University of Bergen, P.O. Box 7803, 5020 Bergen, Norway*

[4]*Department of Biology, University of Bergen, P.O. Box 7803, 5020 Bergen, Norway*

[5]*Department of Microbiology, Technische Universität München, 85354 Freising, Germany*

[6]*Department of Geology and Geochemistry, Stockholm University, S10691, Stockholm, Sweden*

[7]*Department of Environmental Science and Engineering, Guru Jambheshwar University of Science and Technology, Hisar 125001, Haryana, India*

Keywords Cave geomicrobiology • Microorganism • Stromatolites • Stable isotopes • Carbonate deposition • Sahastradhara • Mawsmai • Meghalaya • India

1. Introduction

The subsurface of the Earth is one of the major habitats and contains a significant proportion of microbial life (Whitman et al., 1998; Ghiorse, 2008; Roussel et al., 2008). However, our overall knowledge about the life forms and biogeochemical processes contained within it is rather scarce, mainly because of the difficulties in approaching this habitat. One relatively easy way to approach this habitat is to investigate karst terrains, which expand over ~20% of the Earth's subsurface (Ford and Williams, 2007). Since caves are one of the most prominent features of karst terrain, they may serve as noteworthy entries and virtual "windows" into subsurface habitats (e.g. Engel et al., 2008). It is widely recognized that caves can also host a wide spectrum of fascinating life forms, starting from biofilms harbouring different types of microorganisms to different types of cave-dwelling animals such as snails, worms, spiders, leeches, crickets, cockroaches, scorpions, fishes and bats. Cave geobiology is therefore a fascinating discipline for exploring different basic aspects of the subsurface eco-systems and their interactions with the eco-systems of the surface. Cave geomicrobiology deals specifically with the microorganisms, other life forms and their interactions with minerals and provides us with information about the past geomicrobiological interactions.

Cave geomicrobiology is rather a young discipline. Shoji and Folk (1964) were the first to report on microbial inclusions within rock and speculate the possible role of microorganisms in carbonate deposition. This was confirmed two decades later by Folk and Chafetz (1980) and the role of microorganisms in carbonate and phosphorite precipitation in stromatolites was confirmed by others (Logan et al., 1964; Chafetz and Folk, 1984; Tewari, 1995; Krajewski et al., 1994). Thereafter, cave geomicrobiological studies have been expanded tremendously in different parts of the world (e.g. Palmer, 1991; Banfield and Nealson, 1997; Woods et al., 1999; Saiz-Jimenez, 1999; Groth et al., 1999, 2001; Laiz et al., 1999, 2003; Davis, 2000; Hose et al., 2000; Neuweiler et al., 2000; Melim et al., 2001; Boston et al., 2001; Hammes and Verstraete, 2002; Frisia et al., 2002; Galy et al., 2002; Schabereiter-Gurtner et al., 2002; Bosak and Newman, 2003; Sanchez-Moral et al., 2003; Tooth and Fairchild, 2003; Engel et al., 2003, 2004; Cacchio et al., 2004; Barton et al., 2004; Chelius and Moore, 2004; Barton and Luiszer, 2005; Gonzalez et al., 2006; Canaveras et al., 2006; Meisinger et al., 2007; Engel et al., 2008, 2009). Excellent reviews have been published on different aspects of geomicrobiology (Northup and Lavoie, 2001; Northup et al., 1997; Barton and Northup, 2007; Barton, 2006; Ehrlich, 1998; Banfield et al., 1998; Jones, 2001; Newman and Banfield, 2002; Barton and Jurado, 2007; Whitman et al., 1998).

Almost all cave geomicrobiological studies have shown that cave environments should usually be regarded as extreme environments for life as the majority of the cave eco-systems are resource limited due to dark and aphotic conditions. Thus, most cave eco-systems must therefore depend on allochthonous organic matter and chemolithoautotrophy for energy and metabolism (e.g. Poulson and Lavoie, 2000; Simon et al., 2003; Engel et al., 2003, 2004). Several studies have shown that different types of dissolved substances in springs and ground water discharges in caves can serve as high energy yielding substrates for some groups of microorganisms (Sarbu et al., 1996; Egemeier, 1981; Angert et al., 1998; Hose et al., 2000). Furthermore, energy may also be acquired through aromatic compounds, fixing gases or oxidising/reducing metals within rocks (Ehrlich, 1998). Due to these microbial interactions with rocks and minerals for energy requirement, microorganisms may therefore play an important role in reshaping the mineral environment of caves and contribute significantly to various cave deposits such as speleothems. Since caves may function as natural windows into the subsurface, research on the microbial diversity, community compositions, metabolic activities and their interactions with geological materials through time may help us to explore the possible extent of microbial activities and detect novel aspects of biogeochemical cycles.

Several investigations on cave deposits have emphasized association of microorganisms with carbonates and speleothems, and contribute to the cave geology/geochemistry in different ways (Northup et al., 1997). One important parameter is their contribution to different types of chemical reactions such as precipitation and dissolution of different minerals (Castanier et al., 2000; Engel et al., 2003, 2004; Baskar et al., 2005, 2006, 2007; Tewari, 2008). Laboratory experiments have shown that bacterial species isolated from caves are able to precipitate carbonates under controlled conditions (Danielli and Edington, 1983; Rivadeneyra et al., 1993; Baskar et al., 2006). Furthermore, at least two other microbiological influences are (1) metabolic processes such as ammonification, denitrification, sulphate reduction and photosynthesis, which may increase the alkalinity and thus facilitate carbonate precipitation; and (2) production of extracellular polymeric substances, which may serve as important sediment trapping and binding agents and provide nucleation sites for the precipitation of minerals (Riding, 2000).

While these observations clearly point to the relevant roles of microbial activities in caves, many questions still remain to be answered. For example, the extent and influence of the total biodiversity of unculturable microorganisms, and how different types of processes (known versus possibly novel) interact with each other under different conditions, different time scales and on different cave sites. The cave geomicrobiological research on different cave sites all over the world should reveal interesting insights into different geomicrobiological interactions with various types of minerals. However, to achieve a holistic cave systems geobiology approach, different methodologies should be used, ranging from abiotic scientific disciplines such as geochemistry, stable isotope chemistry, paleolontology and geophysics to biological disciplines such as ecology, evolution, organism biology of different biological species and different molecular biological approaches to trace and link the activities and dynamics of culturable and unculturable biological species. With this approach, interesting insights into the diversity and resilience of different life forms in caves may be revealed and how to recognize biosignatures for subsurface life on other planetary bodies (Boston et al., 1992, 2001; McKay et al., 1994; Cunningham et al., 1995; Tewari, 1998, 2001; Boston, 2000). For example, it has been speculated that microorganisms may exist in hidden caves in Mars, or on the subsurface regions below the Martian polar ice cap (Tewari, 2004). However, since no single feature can be taken as conclusive evidence for past or present life forms, the development of a consortium of possible biosignatures based on extensive paleoastrobiological and geomicrobiological research on different sites on our planet Earth may help us identify circumstantial evidences of present or past life forms on other areas in outer space (Allen et al., 2000). The cave geobiology research may also lead to an increased understanding of how to facilitate the preservation of historical monuments and sculptures that may be endangered by microbially precipitated calcite coatings (Rodriguez-Navarro et al., 2003; Hoppert et al., 2004). In this review, a summary of the cave geomicrobiological research in the Indian sub-continent, which contains a large amount of interesting cave sites but where unfortunately most of them have not yet been investigated, has been attempted with a view to further explore them in the years to come.

2. Geomicrobiological Studies of Caves in the Indian Subcontinent

More than 1,545 cave sites (Deshmukh, 1994) have been discovered in the Indian sub-continent. However, only four of them have been hitherto examined:

- The Sahastradhara caves in Dehradun (Baskar et al., 2005, 2006; Tewari, 2008)
- The Brahmakhal cave in Uttarkashi, Garhwal (Tewari, 2008, 2009)
- The Borra caves in Vishakapatanam, Andhra Pradesh (Baskar et al., 2007, 2008)
- The Khasi and Jaintia Hill cave systems in Meghalaya (Baskar et al., 2009a)

These caves were investigated in different ways, ranging from traditional to a molecular microbiological point of view and, geochemical, mineralogical, stable isotope studies to explore the extent of microbial processes and their impact on precipitating carbonate deposits in these caves. The recognition of the importance of speleothems as archives for paleoenvironmental changes has increased over the last three decades (Hendy, 1971). The mechanisms of speleothem deposition have been proven to be sensitive to external, often climatic-driven processes that respond to annual–decadal short- as well as long-term changes (Wang et al., 2004). Speleothems can therefore be regarded as valuable archives of climatic conditions in the continents. Compared to other continental climate proxy recorders from lake sediments and peat cores, speleothems offer several advantages (Ghosh et al., 2006). These include worldwide distribution of caves containing speleothems, high precision uranium-series dating of speleothems and highly resolved time series data, and mostly easier accessibility. Based on this, the speleothem records from caves across Asia have contributed with new knowledge to our current understanding of many of the factors that control inter-annual to millennial scale variability in Asian monsoon precipitation (Wang et al., 2008; Sinha et al., 2005; Fleitmann et al., 2007) and provided important constraints for climate modelling scenarios (Overpeck and Cole, 2007).

3. Sahastradhara Caves, Dehradun

The cave systems in Sahastradhara are situated in the Dehradun Valley, a crescent-shaped intermontane valley formed within the Siwalik Formations in Garhwal Himalaya (Fig. 1). They are situated on the Krol carbonates and enclosed by the rivers Ganga in the east and Yamuna in the west. The caves are rather small in size

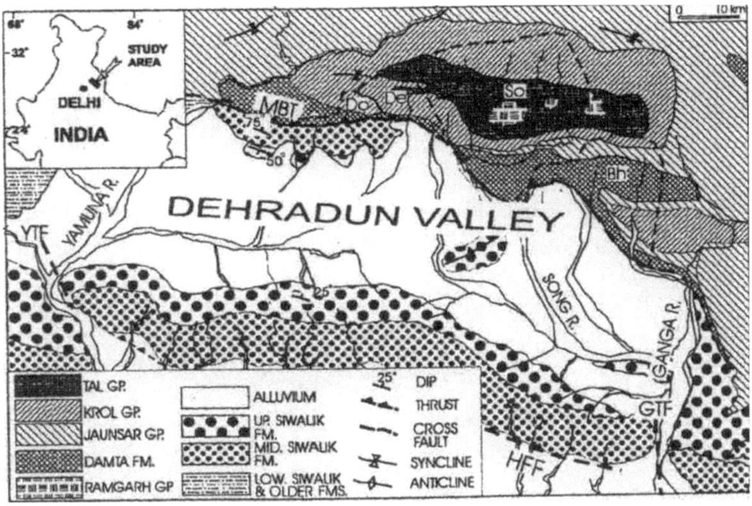

Figure 1. Geological map of Dehradun Valley. *YTF* Yamuna Tear Fault; *GTF* GangaTear Fault; *MBT* Main Boundary Thrust; *HFF* Himalayan Frontal Fault; Drainage areas of the Donga fan (Do), Dehradun fan (De), Song River (So) and Bhogpurfan (Bh) (Singh et al., 2001).

(10 m long, 2 m wide), and are well known for their pH neutral springs that are believed to cure various kinds of skin diseases. The word Sahastradhara means the place of the "1,000-fold spring". The areas around the caves are frequently visited by people throughout the year, mainly due to the therapeutic value of the springs. However, nobody enters into these caves due to their small sizes. The geology of the Dehradun Valley has been extensively studied (Singh et al., 2001; Tewari, 1989, 2007).

The speleothems found in three cave systems in Sahastradhara, Dehradun (Fig. 2a, b) (Tewari, 2008, 2009) were studied for mineralogy, geochemistry and microbial ecology to better understand the role of microorganisms in mineral formation (Baskar et al., 2005, 2006). Calcite was the dominant mineral in the stalactites. Thin-section petrography revealed that the stalactites consisted of microcrystalline calcite or micrite that occurred in chains, attributable to mineralized bacterial cells. *The microfacies of the Sahastradhara stalactites show radial fibrous calcite* (Fig. 3c, d) *indicative of diagenetic imprints* (Tewari, 2008). Quantification of total cell numbers by DNA-specific DAPI staining (Amann et al., 1995) of fixed cells revealed the presence of a large number of microbial cells (9×10^5 cells, g sed^{-1} dry wt) in the caves. The fluorescence in situ hybridization (FISH) techniques showed the presence of a large number of tentatively active microbial cells (~55% of the total cell number), and microbial community was dominated by species within *Bacteria*, such as the sulphate-reducing bacteria, as well as species within *Archaea* (Baskar et al., 2005, 2006). The high fraction of tentatively active cells indicated a high probability for their participation in bio-mineralization processes

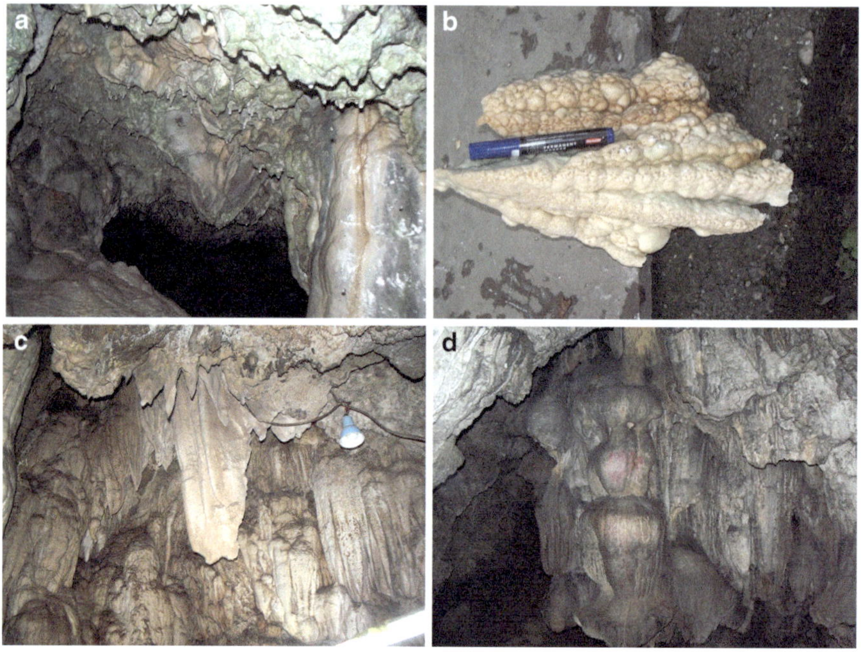

Figure 2. Speleothems in different cave systems in India. (**a**) Dripping water, stalactites, stalagmites (Sahastradhara caves) studied for oxygen isotopes during monsoon. (**b**)Column from the Sahastradhara cave studied for carbonate microfacies and C and O isotope ratios. (**c**) Speleothems in the Brahamakhal (Prakateshwar) cave studied for microfacies and C and O isotope ratios. (**d**) Speleothems in the Mawsmai cave showing pillar-like structures near Cherrapunji, Meghalaya.

involved in the cave formations (Baskar et al., 2005, 2006). In addition, laboratory-based culture experiments using bacterial strains (identified by 16S rRNA gene amplification and sequencing as *Bacillus thuringiensis* and *Bacillus pumilis*) isolated from stalactites showed that they are able to form $CaCO_3$ crystals (Fig. 4a, b) (Baskar et al., 2006).

3.1. BRAHMAKHAL (PRAKATESHWAR) CAVE, UTTARKASHI, UTTARAKHAND

The Brahmakhal (Prakateswar) caves were discovered in 1978 and are situated near the village Mehar Gaon in Uttarkashi district of the Garhwal Himalaya (30°23′145″ N; 78°07′743″ E). Stalactites and stalagmites are well developed in the cave (Fig. 2c) (Tewari, 2008). The thin sections of the stalactites have been studied for mineralogical and microfacies analysis. Radiating fibrous calcite and micro-lamination of calcite and organic-rich microbial laminae have been recorded in the Prakteshwar speleothems (Fig. 3a, b) (Tewari, 2008, 2009). Geomicrobiological investigations will be initiated in this cave soon.

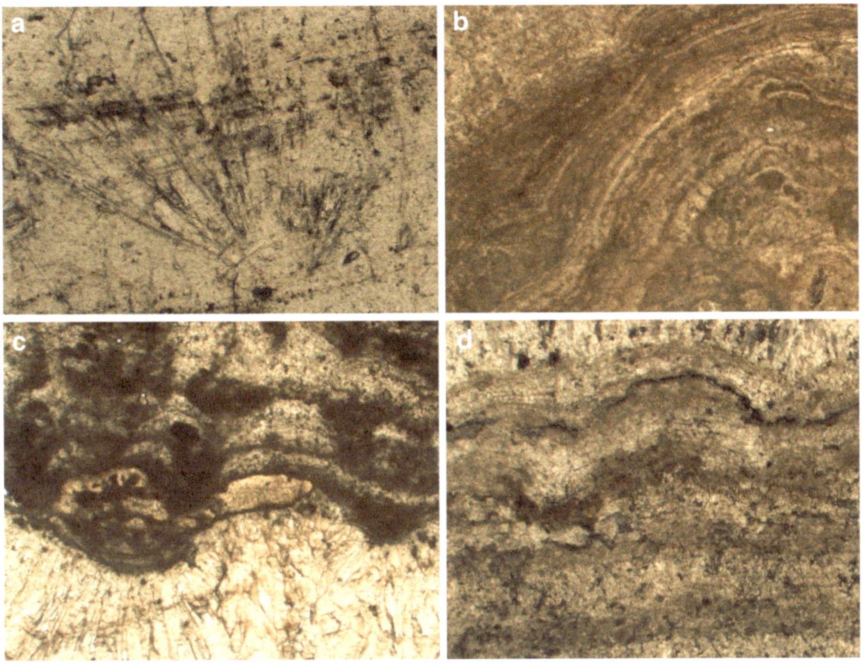

Figure 3. Microfacies of the speleothems from the Himalaya, India. (**a**) Microphotograph of radiating fibrous calcite in Brahmakhal (Prakateshwar) cave, Uttarkashi, Garhwal Himalaya. (**b**) Finely laminated stalactite with alternating calcite (*white*) and *dark* (organic) laminae from the Brahmakhal (Prakateshwar) cave, Uttarkashi. (**c**) Microphotograph of fibrous calcite in Sahastradhara stalactite showing radiaxial fabric (*lower part*) and alternating *dark* (organic/microbial) and *light* (calcite) microstromatolitic laminae (*upper part*). (**d**) Microphotograph of microstromatolitic *light* and *dark* laminae with filamentous and cell-like structures indicating microbial influenced precipitation of carbonate in Sahastradhara cave, Dehradun.

3.2. BORRA CAVES, VISHAKAPATANAM, ANDHRA PRADESH

The Borra Caves are the second largest caves in the Indian sub-continent, and were discovered by William King George of the Geological Survey of India in 1807. These caves have a rich cultural, religious, eco-tourism and are of significant historical importance. They are situated in the Araku Valley, ~90–95 km from Visakhapatnam (18°15′N; 83°3′E) on the east coast of India (Fig. 5) (Bas et al., 2002). It is also one of the most important hill regions of the Eastern Ghats, known not only for its biodiversity, but also for its rich mineral deposits. The geology of the region is represented by the Khondalite suite of rocks (garnetiferrous sillimanite gneisses and quartzo-feldsphatic garnet gneisses) of Archaen age. Quaternary deposits consist of red bed sediments, laterites, pediment fans, colluvium, alluvium and coastal sands (Bhowmik et al., 1995; Bhattacharya and Kar, 2004). All these rocks lie within calc-silicate granulites and garnet-sillimanite

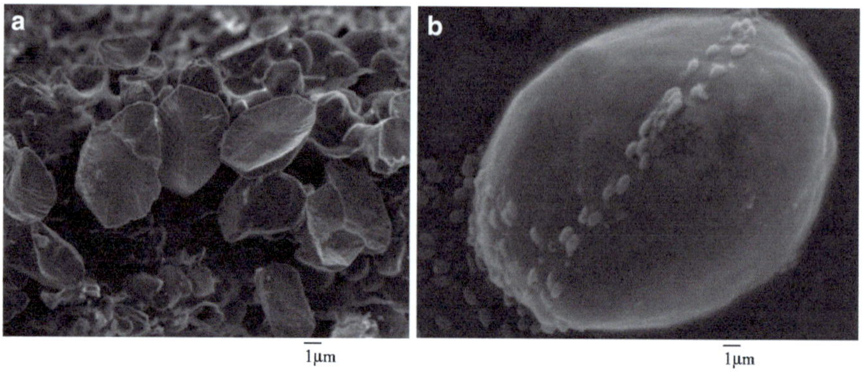

Figure 4. (**a**, **b**) Calcite crystals precipitated in vitro by isolates, which are related to *Bacillus pumilis* (**4a**), to *Bacillus thuringiensis* (**4b**) isolated from Sahastradhara caves, Dehradun, India.

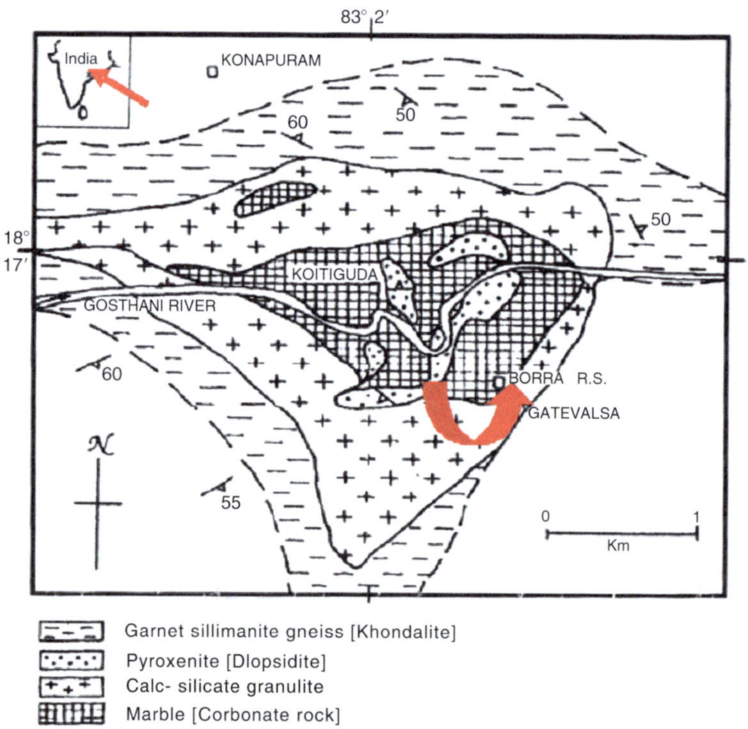

Figure 5. Geological map of Borra cave area. Map based on Bhowmik et al. (1995). *Blank areas* are quartzo-feldspathic gneisses (*R.S.* Railway station). *Inset map* shows location of Borra (*arrows*).

gneisses (khondalites), which have been repeatedly folded and metamorphosed at high grades and form a part of the Eastern Ghats granulite belt (Bhowmik et al., 1995; Bhattacharya and Kar, 2004). The annual temperature of Araku Hills is ~25°C and the annual rainfall is 950 mm. Coarsely pure white, crystalline carbonate rocks are found at Borra; deformed, banded marbles extend over a triangular area km^2 in the cave. The caves are surrounded by diopside–scapolite–feldspar calc-granulites. The pyroxenite outcrops at Borra are dark and massive and include discontinuous calc-silicate bands, with brown mica and calcite (Bas et al., 2002).

The caves host a variety of speleothems and deep inside are springs with thick orange to reddish brown microbial mats (Fig. 6a) (Baskar et al., 2008). The Gosthani River flows between the spectacular stalactites and stalagmites, and also out of the caves. Through the whole length of the cave, there is a twilight zone with limited light penetration and a deeper totally aphotic zone can be observed.

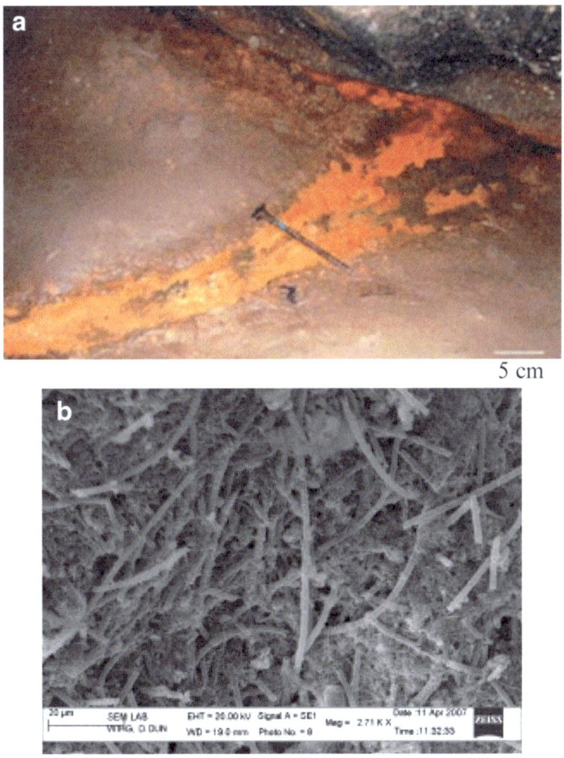

Figure 6. Organic mat on the spring water, Borra caves, Vishakapatanam, India. (**a**) Orange organic mat/biofilm on the spring, Borra caves. (**b**) SEM of organic mat, Borra caves, showing filamentous *Leptothrix*-like bacterial sheaths and Fe precipitates.

The cave is approximately 2 km long, 12 m high and is located 1,300 m above sea level. The deep aphotic inner cave wall has an average temperature of approximately 16°C. Inside the cave, there are discharges from five springs. One of these springs has developed a thick iron-rich, orange microbial mat in the cave passages distributed over a distance of about 150 m.

Mineralogy and thin sections of stalactites revealed micrite, occurring as laminated features of clotted with chocolate-brown blebs identical to microbial structures in modern and ancient stromatolitic carbonates (Baskar et al., 2007). Scanning electron microscopic (SEM) studies revealed the presence of calcified prokaryotic-like cells, microrods and needle calcite. The microfabric features/textures preserved in the speleothems indicate the presence of microbes and the role of microorganisms in the genesis of the Borra speleothem carbonates (Baskar et al., 2007). In addition, the deep aphotic regions of the cave exhibit a significant amount of unexplored microbial mats in spring waters. Baskar et al. (2008) reported the possible impact of microorganisms in these organic mats on the cave formation, focusing on their role on iron precipitation. The pH-neutral spring water contained iron and the organic mat floated on the spring. This organic mat (solid material) had a total organic carbon (TOC) content of ~5.4 wt% and 427 ppb iron. The different prokaryotes distinguished are (1) *Leptothrix*-like organisms, entombed in bacterial mineral sheaths, (2) a few twisted stalks resembling those of *Gallionella*-like organisms and (3) some additional (rod-shaped) morphotypes of un-identifiable type (Fig. 6b; Baskar et al., 2008). The study indicated that the formation of these iron-rich organic mats is associated with the presence of abundant active communities of iron oxidising and precipitating bacteria and that the elements concentrated in the mats could serve as an energy source to the microbial community inhabiting the mats. Consequently, microbial colonization of different areas of the springs may be controlled by elemental substrate concentration. The geochemical composition of the mats appears to reflect the type of the ions and in situ metabolism by bacteria that may be concentrating the iron in the mats (Baskar et al., 2008).

3.3. MEGHALAYA CAVES, MEGHALAYA

Meghalaya is one of the northeastern states of India and the state harbours more than 1,000 caves, a few of them form some of the longest caves in the Indian sub-continent. "Meghalaya" means "the abode of clouds", derived from Sanskrit language – "*megh*" meaning cloud and "*alaya*" meaning home. Geomicrobiological studies have so far been carried out only on two of the more than 1,000 cave systems in Meghalaya: the Mawsmai (25°07′N, 91°21′E) and the Krem Phyllut caves (25°41′N, 92°09′E) are located on the East Khasi Hills (Fig. 7). The Khasi Hills is an uplifted Precambrian crystalline complex and forms the northeastern extension of the Indian Peninsular Shield. It is an E–W trending oblong horst block elevated about 600–1,800 m above the Bangladesh plains in south, and

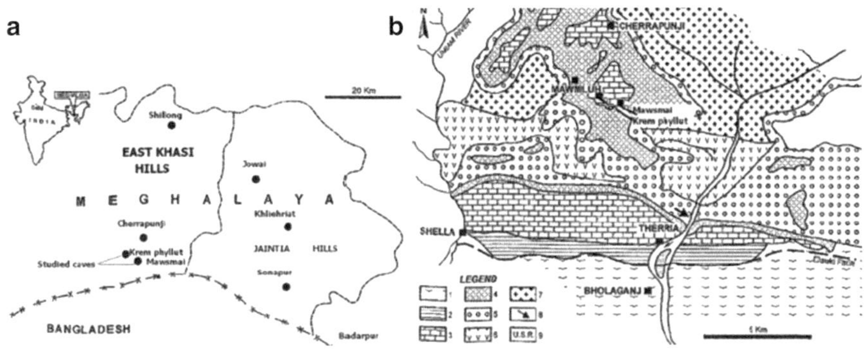

Figure 7. Geological map of East Khasi Hill caves and Jaintia Hill caves, Meghalaya, India. (**a**) Location map of the study area. *Inset* shows its position in India. (**b**) Geological map of the studied cave area (Ghosh, 1940; Garg and Jain, 1995) 1. Alluvium, 2. Kopili formation (shale/sandstone), 3. Sylhet, 4. Therria formation (calcareous bands, sandstone, shale), 5. Langpar formation and Um Sohryngkew formation (calcareous shale, marl and mudstone), 6. Sylhet trap, 7. Archaen, 8. K/T Sect., 9. Um Sohryngkew river.

separated from the Peninsular India by the Rajmahal-Garo gap (Ghosh et al., 2005). The Khasi group consists of sandstone and conglomerate of the Jadukata formation, and overlies the feldspathic sandstone of the Mahadek Formation. The Proterozoic metasedimentary Shillong Group and the basement Gneissic Complex make up most of the Meghalaya Plateau (Ghosh et al., 2005). The state has huge deposits of limestone and abundant rainfall, which is one of the main reasons for formation of these magnificient Karst caves. The important limestone deposits in the East Khasi Hills are in the Mawmluh–Mawsmai Hills of Paleocene Lakadong Limestone deposited after the Cretaceous–Plaeogene (K/Pg) boundary in Meghalaya Plateau (Tewari, 2009). The cave (Fig. 2d) (Tewari, 2008) lies south of the wettest place in the world – Cherrapunji town now renamed as Sohra in local language.

These caves are not so easily accessible to public since they are situated in a hilly, uninhabited area. Mawsmai caves are located in a thick forested area and are quite small (160 m long, 15 m high, 4–10 m wide), but the inner parts are large enough to facilitate movement in the passage ways. The main entry to these caves is located close to the Mawsmai village and the entry is a fairly narrow (1.8 m) vertical opening. The cave is totally aphotic and has many stalagmites and stalactites. The average annual temperature of the inner cave is ~15–19°C. The stalactites range in sizes from 7–10 cm length and 8–15 cm diameter (small) to 50–150 cm length and 50–100 cm diameter (large). The Krem Phyllut caves have a large fossil passageway. This cave is relatively large (total length 1,003 m, width 4.5 m, height 15 m) and has three entrances (approximately of 2 m height, 2.5 m width). The deep aphotic inner cave wall had an average annual temperature of ~15–17°C. Two springs run about 500 m through the cave. The length and diameter of the stalactites range from 6–7 cm in length and 25–30 cm in diameter (small) to 30–40 cm in length and 50 cm in diameter (large). The columns are 40–45 cm long and the diameter of the upper end is 40 cm and that of the lower end is 60 cm. The speleothems are larger at the entrance and smaller towards the interior. The main cavern extends to about 50–60 m and narrows to a much smaller tunnel wherein gypsum deposits are found on the cave walls. The characteristic cave fauna consisted of sporadic amounts of prawns in the water and rats on the ground.

The first observations and hypotheses by Baskar et al. (2009a) reported the possible influence of microorganisms on the speleothem formations in the two so far investigated caves in Meghalaya. The spring waters have a pH of 7.4–7.8 and the speleothems have a TOC content of ~0.37–1.98 wt%. The water (spring, pool and dripping) contained concentrations of 37 ppb Sr, 1.5–2.6 ppb Si, 3–12 ppb Cr, 4–5 ppb Ba, 1–4 ppb Cu, 3–28 ppb Zn, 2.3–5.8 ppb Mg, and in addition to calcium, the speleothems contained 0.12–1.69 Fe_2O_3, 0.31–0.91% MgO, 0.002–0.131% Na_2O, 0–3.24% Al_2O_3 and 46–501 ppm Sr. Thin sections showed alternating white and dark-laminated bands and voids and the presence of lithified structures and internal fabrics similar to microbialites along with some organic inclusions (Baskar et al., 2009a). Dark-coloured clotted peloidal fabrics were interpreted as calcified bacterial aggregates since such formations have been reported in stromatolite, thrombolite, travertine and reefs (Chafetz and Buczynski, 1992; Pedley, 1992; Tewari and Joshi, 1993; Riding, 2000; Baskar et al., 2007). Calcite in speleothems and gypsum in cave wall deposits are the dominant minerals. Microfabric features showed strong resemblance to fossilised bacteria, calcified filaments, needle calcite and numerous nano-scale calcite crystals, highly weathered and disintegrated crystals of calcite pointing a significant microbial influence on calcite genesis (Baskar et al., 2009a). The classification of microrod calcite has been reported earlier, and its origin has been attributed to rapid precipitation under high supersaturation states during evaporation in soils or calcification of bacilli-like organisms (Loisy et al., 1999). Similar microfabrics have been observed in marine environments and cave speleothems elsewhere, supporting the hypothesis on microbial involvement in speleothem precipitation (Chafetz, 1986; Melim et al., 2001; Baskar et al., 2007, 2009a; Tewari, 2008, 2009). Around 30 aerobic, heterotrophic isolates were obtained from both caves by conventional microbiological isolation procedures and further evaluated by different molecular techniques. The isolates were closely affiliated to *Bacillus cereus*, *Bacillus mycoides*, *Bacillus licheniformis*, *Micrococcus luteus* and *Actinomycetes*. In addition to this, a number of strains were obtained that have not yet been finally identified, such as cocci, red-pigmented strains similar to *Streptomyces* and some strains of gypsum precipitating bacteria (Figs. 8 and 9, Baskar et al., 2009a).

The bacterially precipitated calcium differs morphologically from the pure inorganic formed crystals. Similar types of bio-minerals could also be observed in the speleothem samples, and the resemblance provided additional support for biogenic influence on the formations of these cave systems in Meghalaya. The study confirmed that (1) carbonate precipitation occurred during microbial

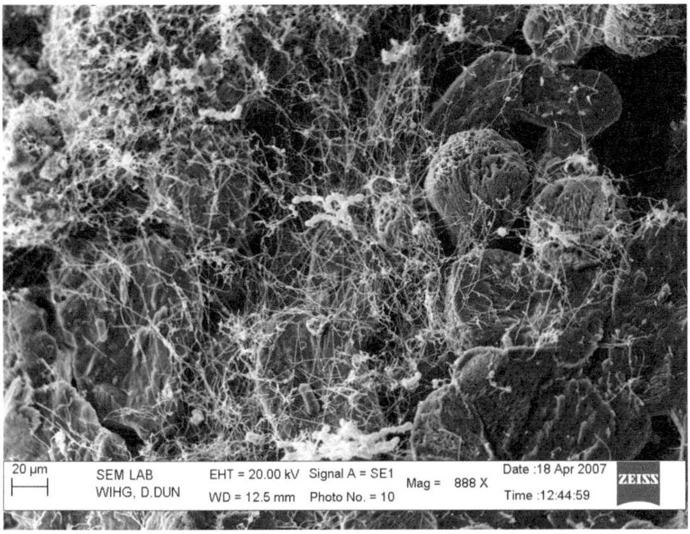

Figure 8. SEM image showing *Actinomycetes*-like bacteria, unidentified cocci, a variety of tentatively identified *Bacillus*-like bacteria isolated from caves in Meghalaya precipitating calcite in vitro.

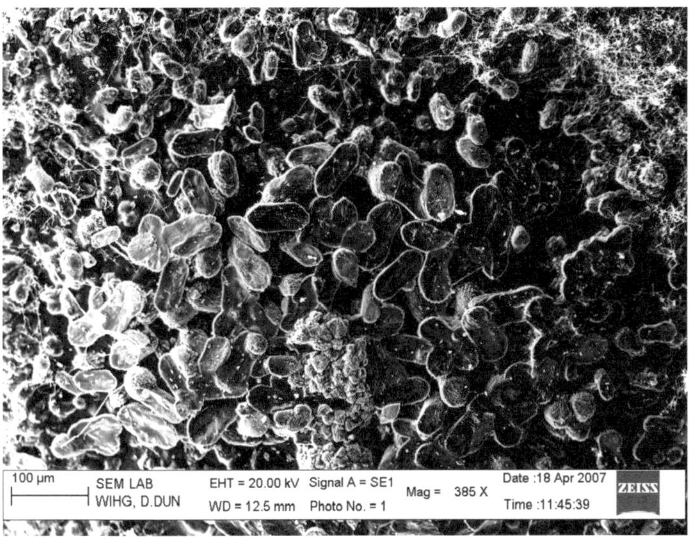

Figure 9. SEM image of unidentified bacteria from Mawsmai caves, Meghalaya precipitating gypsum in vitro.

$CaCO_3$ precipitation by a closely related group of bacteria following the ammonification/nitrate reduction pathway; and (2) gypsum precipitates were formed by microbial-mediated geochemical reactions. It therefore appears to be likely that numerous processes have acted together to induce the precipitation of carbonates and gypsum. These geomicrobiological and inorganic transformation processes may therefore explain some of the biogenic involvement in the formation of the speleothem deposits in cave systems.

4. Stable Isotope and Paleoclimate Studies in the Indian Caves

With an accurate chronology, the cave archives can be correlated with climatological events on land spanning <1 million year, potentially comparable to the chronological records found in the deep sea sediments and continental ice records (McDermott et al., 1999). For example, carbonate speleothems may serve as proxy indicators of paleoclimate, and provide information on annual temperature, rainfall, atmospheric circulation and vegetation changes (McDermott, 2004). The annual growth layers in speleothems have provided a rainfall proxy record in Madagascar (Brook et al., 1999), monsoon patterns in the Indian subcontinent (Denniston et al., 2000; Yadava et al., 2004) and change in vegetation associated with warmer or cooler conditions in Botswana (Holmgren and Lauritzen, 1995).

Studies on Indian speleothems to estimate the past monsoon and climate are taking momentum, and very interesting results on the paleomonsoon variability preserved in stalagmites have been published (Yadava et al., 2004; Yadava and Ramesh, 2005; Ramesh, 2001; Sinha et al., 2005; Tewari, 2008, 2009). Speleothems may also provide archives of past precipitation since the oxygen isotopic ratio of stalagmite calcite is controlled by $\delta^{18}O$ values of precipitation. Thus, Yadava and Ramesh (2005) and Ramesh (2001) reconstructed the monsoon precipitation from $\delta^{18}O$ of cave stalagmite from Gupteshwar cave, Orissa in east India, back to 3,400 years B.P. Similarly, Sinha et al. (2005) studied the Timta cave in the Kumaon Himalaya, and the oxygen isotope ratio ranged from −5.2‰ to −8.4‰ for the period. The difference of about −3.2‰ was interpreted to reflect the varying strength of the Indian Summer Monsoon system. More recently, Tewari (2008, 2009) studied the carbon and oxygen isotopic signatures of a number of caves from the Garhwal Lesser Himalaya, North India (Fig. 2a, c). The $\delta^{13}C_{(PDB)}$ of the Sahastradhara stalactites range from −2.47‰ to −6.06‰ and $\delta^{18}O$ from 22.99‰ to 26.21‰. In Brahmakhal cave, the $\delta^{13}C_{(PDB)}$ vary from −5.34‰ to −7.65‰ and $\delta^{18}O$ vary from 20.67‰ to 22.84‰. The $\delta^{13}C_{(PDB)}$ value of Pratapnagar modern stalactite calcite varies from −4.85 to −9.23‰ and the $\delta^{18}O$ range from 19.50‰ to 21.57‰. As discussed above, variation of $\delta^{18}O$ in stalagmites is related to the $\delta^{18}O$ values in the precipitation and the amount of precipitation during the monsoon season. The $\delta^{18}O$ values of regional precipitation and the changes in calcite $\delta^{18}O$ over time reflect changes in monsoonal precipitation. Study of modern cave drip waters and stalagmites demonstrates that the stalagmites were deposited in or very near isotopic equilibrium. The $\delta^{18}O$ isotope data of drip water from Sahastradhara cave in Dehradun (Fig. 2a) during the monsoon season (August and

September, 2007) varies from −4.58 to −5.14‰ (PDB) Fleitmann et al. (2003, and the references therein) have interpreted that the western part of Indian Summer Monsoon in Oman and Yemen and showed variations in the amount of rainfall with more negative $\delta^{18}O$ values indicating higher precipitation. Negative $\delta^{18}O$ values have also been recorded in the Sahastradhara and Brahmakhal caves. Cave calcite also contains information about the isotopic composition of meteoric water from which calcite is precipitated and dated calcite from speleothem with U/Th disequilibrium series method. So far only the Sahastradhara cave has been examined for geomicrobiological as well as stable isotope studies. Thus, detailed high-resolution O isotope speleothem data from other cave systems in the Himalaya may yield interesting insights into the record of palaeomonsoon history.

5. Future Prospects of Geomicrobiological Studies of Indian Caves

Among >1,545 cave sites that have been discovered in Indian sub-continent, only three of them have been subjected to geomicrobiological studies: the Sahastradhara, Borra and Meghalaya caves. This chapter summarizes the so far performed studies on cave geomicrobiology, paleoclimate and monsoon of these Indian cave sites. The authors mainly focused on geochemical, mineralogical, traditional and molecular microbiological approaches to explore the qualitative extent of microbial involvement in cave calcite formations. This includes the documentation and recognition of mineralized microbes preserved in rocks, either by tracing nucleic acid-based signatures or bio-geochemical processes. These approaches proved that microorganisms played a significant role in cave eco-systems. The exciting research goals for the future will then be to explore the geomicrobiology in Indian cave eco-systems further through more rigorous, advanced, quantitative and interdisciplinary approaches (geology, chemistry, biology), towards a more advanced, holistic understanding of the specific roles of microbes in cave eco-systems (Baskar et al., 2009b). Because of the large number of unexplored caves, such as in the NW Himalaya and in Arunachal, Meghalaya, Mizoram and Manipur in the northeastern Himalaya (Tewari, 2009), it can be anticipated that many unique and interesting research projects may become identified. However, to achieve these research goals, the teaching policy in India must be expanded to promote co-ordinated teaching programmes in geology, chemistry and biology at a national level. Recently, the first steps have been taken towards this, for example the declaration programme "Increased understanding of the Earth processes as well as emerging newer concepts and methodologies that requires interactive research programmes involving geoscientists, physicists, chemists, biologists and mathematicians" by the Department of Science and Technology (DST), New Delhi, India (DST Vision, 1996). Furthermore, the Earth System Science (ESS) division of the DST has identified a National Programme on "Science of Shallow Subsurface Studies" but unfortunately, cave geomicrobiology has so far not yet been specifically included as a thrust area of research. However, we propose that cave geomicrobiology research in Indian caves has immense potential to generate fundamental knowledge about the subtle interplay between mutualism/competition and heterotrophy/autotrophy in terrestrial subterranean systems, which may result in numerous practical applications within for example medicine, human health and biotechnology, and provide a deeper understanding of the development of life not only on our planet but also in outer space.

6. Acknowledgments

Dr. D.D.S. Sandhu (Vice-Chancellor, Guru Jambheshwar University of Science and Technology, Hisar) is thanked for support and encouragement. SB and RB thank Centre for Geobiology, Norwegian Centre of Excellence, University of Bergen, Norway for invitation as academic guests and SB thanks the Research Council of Norway for financial support. RB thanks UGC for major research project. V.C. Tewari is thankful to Dr. B.R. Arora, Director, Wadia Institute of Himalayan Geology, Dehradun for facilities and permission to publish the paper. The reviewers are thanked for their comments and suggestions that helped to improve this article.

7. References

Allen, C.C., Albert, F.G., Chafetz, H.S., Combie, C., Graham, C.R., Kieft, T.L., Kivett, S.J., McKay, D.S., Steele, A., Taunton, A.E., Taylor, M.R., Thomas-Keprta, K.L. and Westall, F. (2000) Microscopic physical biomarkers in carbonate hot springs: implications in the search for life on Mars. Icarus **147**: 49–67.

Amann, R.I., Ludwig, W. and Schleifer, K.H. (1995) Phylogenetic identification and in situ detection of individual microbial cells without cultivation. Microbiol. Rev. **59**: 143–169.

Angert, E.R., Northup, D.E., Reysenbach, A.L., Peek, A.S., Goebel, B.M. and Pace, N.R. (1998) Molecular phylogenetic analysis of a bacterial community in Sulphur River, Parker Cave, Kentucky. Am. Mineral. **83**: 1583–1592.

Banfield, J.F. and Nealson, K.H. (1997) Geomicrobiology: Interactions between microbes and minerals, In: Ribbe, P.H. (ed.) *Reviews in Minerology 35*. Mineralogical Society of America, Washington, D.C., 448 p.

Banfield, J.F., Welch, S.A. and Edwards, K.J. (1998) Microbes as geochemical agents. Geochem. News **96**: 11–17.

Barton, H.A. (2006) Introduction to cave microbiology: a review for the non-specialist. J. Cave Karst Stud. **68**: 43–54.

Barton, H.A. and Jurado, V. (2007) What's up down there? Microbial diversity in caves. Microbe **2**: 132–138.

Barton, H.A. and Luiszer, F. (2005) Microbial metabolic structure in a sulfidic cave hot spring: potential mechanisms of biospeleogenesis. J. Cave Karst Stud. **67**: 28–38.

Barton, H.A. and Northup, D.E. (2007) Geomicrobiology in cave environments: past, current and future perspectives. J. Cave Karst Stud. **69**(1): 163–178.

Barton, H.A., Taylor, M.R. and Pace, N.R. (2004) Molecular phylogenetic analysis of a bacterial community in an oligotrophic cave environment. Geomicrobiol. J. **21**: 11–20.

Bas, M.J.L., Subbarao, K.V. and Walsh, J.N. (2002) Metacarbonatite or marble? – the case of the carbonate, pyroxenite, calcite – apatite rock complex at Borra, Eastern Ghats, India. J. Asian Earth Sci. **20**: 127–140.

Baskar, S., Baskar, R., Mauclaire, L. and McKenzie, J.A. (2005) Role of microbial community in stalctite formation, Sahastradhara caves, Dehradun, India. Curr. Sci. **88**: 1305–1308.

Baskar, S., Baskar, R., Mauclaire, L. and McKenzie, J.A. (2006) Microbially induced calcite precipitation in culture experiments: possible origin for stalactites in Sahastradhara caves, Dehradun, India. Curr. Sci. **90**: 58–64.

Baskar, S., Baskar, R. and Kaushik, A. (2007) Evidences for microbial involvement in the genesis of speleothem carbonates, Borra Caves, Vishakapatanam, India. Curr. Sci. **92**(3): 350–355.

Baskar, S., Baskar, R., Lee, N., Kaushik, A. and Theophilus, P.K. (2008) Precipitation of iron in microbial mats of the spring waters of Borra Caves, Vishakapatnam, India: some geomicrobiological aspects. Environ. Geol. **56**(2): 237–243.

Baskar, S., Baskar, R., Lee, N. and Theophilus, P.K. (2009a) Speleothem formations of Mawsmai caves and Krem Phyllut caves, Meghalaya, India: some evidences for biogenic activities. Environ. Geol. **57**: 1169–1186.

Baskar, S., Baskar, R. and Barton, H.A. (2009b) Cave geomicrobiology as a thrust area of research in the Indian context. Curr. Sci. **97**(5): 621–622.

Bhattacharya, S. and Kar, K. (2004) Alkaline intrusion in a granulite ensemble in the Eastern Ghats belt, India: Shear zone pathway and a pull-apart structure. Proc. Indian Acad. Sci. Earth Planet. Sci. **113**: 37–48.

Bhowmik, S.K., Dasgupta, S., Hoernes, S. and Bhattacharya, P.K. (1995) Extremely high-temperature calcareous granulites from the Eastern Ghats, India: evidence for isobaric cooling, fluid buffering, and terminal channelized fluid flow. Eur. J. Mineral. **7**: 689–703.

Bosak, T. and Newman, D.K. (2003) Microbial nucleation of $CaCO_3$ in the Precambrian. Geology **31**: 577–580.

Boston, P.J. (2000) Life below and life "Out There." Geotimes **45**: 14–17.

Boston, P.J., Ivanov, M.V. and McKay, C.P. (1992) On the possibility of chemosynthetic ecosystems in subsurface habitats on Mars. Icarus **95**: 300–308.

Boston, P.J., Spilde, M.N., Northup, D.E., Melim, L.A., Soroka, D.S., Kleina, L.G., Lavoie, K.H., Hose, L.D., Mallory, L.M., Dahm, C.N., Crossey, L.J. and Schelble, R.T. (2001) Cave biosignature suites: Microbes, minerals and Mars. Astrobiol. J. **1**(1): 25–55.

Brook, G.A., Rafter, M.A., Railsback, L.B., Sheen, S.W. and Lundberg, L. (1999) A high-resolution proxy record of rainfall and ENSO since AD 1550 from layering in stalagmites from Anjohibe Cave, Madagascar. The Holocene **9**: 695–705.

Cacchio, P., Contento, R., Ercole, C., Cappuccio, G., Martinez, M.P. and Lepidi, A. (2004) Involvement of microorganisms in the formation of carbonate speleothems in the Cervo Cave (L'Aquila – Italy). Geomicrobiol. J. **21**: 497–509.

Canaveras, J.C., Cuezva, S., Sanchez-Moral, S., Lario, J., Laiz, L., Gonzalez, J.M. and Saiz-Jimenez, C. (2006) On the origin of fiber calcite crystals in moonmilk deposits: Naturwissenschaften **93**: 27–32.

Castanier, S., Le M'etayer-Levrel, G. and Perthuisot, J.P. (2000) Bacterial roles in the precipitation of carbonate minerals, In: R.E. Riding and S.M. Awramik (eds.) *Microbial Sediments*. Springer, Heidelberg, pp. 32–39.

Chafetz, H.S. (1986) Marine peloids: a product of bacterially induced precipitation of calcite. J. Sediment Petrol. **56**: 812–817.

Chafetz, H.S. and Buczynski, C. (1992) Bacterially induced lithification of microbial mats. Palaios **7**: 277–293.

Chafetz, H.S. and Folk, R.L. (1984) Travertines: depositional morphology and the bacterially constructed constituents. J. Sediment. Petrol. **54**: 289–316.

Chelius, M.K. and Moore, J.C. (2004) Molecular phylogenetic analysis of Archaea and bacteria in Wind Cave, South Dakota. Geomicrobiol. J. **21**: 123–134.

Cunningham, K.I., Northup, D.E., Pollastro, R.M., Wright, W.G. and LaRock, E.J. (1995) Bacteria, fungi and biokarst in Lechuguilla Cave, Carlsbad Caverns National Park, New Mexico. Environ. Geol. **25**: 2–8.

Danielli, H.M.C. and Edington, M.A. (1983) Bacterial calcification in limestone caves. Geomicrobiol. J. **3**: 1–16.

Davis, D.G. (2000) Extraordinary features of Lechuguilla Cave, Guadalupe Mountains, New Mexico. J. Cave Karst Stud. **62**: 147–157.

Denniston, R.F., González, L.A., Asmerom, Y. and Reagan, M.K. (2000) Speleothem records of early and late Holocene vegetation dynamics in the Ozark Highlands, U.S.A. Quatern. Int. **67**: 21–27.

Deshmukh, M. (1994) Influence of geology on the localization of ancient caves. J Geol. Soc. India **44**: 213–217.

DST Vision (1996) Earth sciences. Curr. Sci. **71**: 820–823.

Egemeier, S.J. (1981) Cavern development by thermal waters. Bull. Natl. Speleol. Soc. **43**: 31–51.

Ehrlich, H.L. (1998) Geomicrobiology: its significance for geology. Earth Sci. Rev. **45**: 45–60.

Engel, S.E., Lee, N., Porter, M.L., Stern, A.L., Bennett, P.C. and Wagner, M. (2003) Filamentous Epsilonproteobacteria dominate microbial mats from sulfidic cave springs. Appl. Environ. Microbiol. **69**: 5503.

Engel, A.S., Porter, M.L., Stern, L.A., Quinlan, S. and Bennett, P.C. (2004) Bacterial diversity and ecosystem function of filamentous microbial mats from aphotic (cave) sulfidic springs dominated by chemolithoautotrophic "Epsilonproteobacteria". FEMS Microbiol. Ecol. **51**: 31–53.

Engel, A.S., Meisinger, D.B., Porter, M.L., Baskar, S., Baskar, R. and Lee, N.M. (2008) Cave microbiology as a window into the subsurface. CAREX meeting, Sant Feliu de Guixols, Spain, 30 November – 2 December.

Engel, A.S., Meisinger, D.B., Porter, M.L., Payn, R.A., Schmid, M., Schleifer, K.H. and Lee, N.M. (2009) Linking phylogenetic and functional diversity to nutrient spiraling in microbial mats from Lower Kane Cave (USA). ISME J. **4**(1): 98–110.

Fleitmann, D., Burns, S.J., Mudelcee, M., Neff, U., Kramer, J., Mangini, A. and Matter, A. (2003) Holocene forcing of the Indian monsson recorded in a stalagmite from Southern Oman. Science **300**: 1737–1739.

Fleitmann, D., Burns, S.J., Mangini, A., Mudelsee, M., Kramers, J., Villa, I., Neff, U., Al-Subbary, A.A., Buettner, A., Hippler, D. and Matter, A. (2007) Holocene ITCZ and Indian monsoon dynamics recorded in stalagmites from Oman and Yemen (Socotra), Quat. Sci. Rev. **26**: 170–188.

Folk, R.L. and Chafetz, H.S. (1980) Quaternary travertine of Tivoli (Roma), Italy: bacterially constructed carbonate rock. Geol. Soc. Am. Abst. Prog. **12**: 428.

Ford, D. and Williams, P. (2007) *Karst Hydrogeology and Geomorphology*. Wiley, New York.

Frisia, S., Borsato, A., Fairchild, I.J., McDermott, F. and Selmo, E.M. (2002) Aragonite-calcite relationships in speleothems (Grotte De Clamouse, France): environment, fabrics and carbonate geochemistry. J. Sediment. Res. **72**: 687–699.

Galy, A., Bar-Matthews, M., Halicz, L. and O'Nions, R.K. (2002) Mg isotopic composition of carbonate: insight from speleothem formation. Earth Planet. Sci. Lett. **201**: 105–115.

Garg, R. and Jain, K.P. (1995) Significance of the terminal cretaceous calcareous nannofossil marker Micula prinsii at the Cretaceous-Tertiary boundary in the Um Sohryngkew River section, Meghalaya, India. Curr Sci **66**(12): 1012–1017.

Ghiorse, W.C. (2008) Thirty years of subsurface microbiology. *Conference Proceedings from 7th International Symposium for Subsurface Microbiology*, 16–21 November 2008, Shizuoka, Japan.

Ghosh, S., Fallick, A.E., Paul, D.K. and Potts, P.J. (2005) Geochemistry and origin of Neoproterozoic Granitoids of Meghalaya, Northeast India: implications for linkage with amalgamation of Gondwana Supercontinent. Gondwana Res. **8**(3): 421–432.

Ghosh, A.M.N. (1940) The stratigraphical position of the Cherra sandstone, Assam. Rec GSI **75**: 1–19.

Ghosh, P., Adkins, J., Affek, H., Balta, B., Guo, W.F., Schauble, E.A., Schrag, D. and Eiler, J.M. (2006) ^{13}C-^{18}O bonds in carbonate minerals: a new kind of paleothermometer. Geochim. Cosmochim. Acta. **70**(6): 1439–1456.

Gonzalez, J.M., Portillo, M.C. and Saiz-Jimenez, C. (2006) Metabolically active Crenarchaeota in Altamira Cave. Naturwissenschaften **93**: 42–45.
Groth, I., Vettermann, R., Schuetze, B., Schumann, P. and Saiz-Jimenez, C. (1999) Actinomycetes in karstic caves of northern Spain (Altamira and Tito Bustillo). J. Microbiol. Methods **36**: 115–122.
Groth, I., Schumann, P., Laiz, L., Sanchez-Moral, S., Canaveras, J.C. and Saiz-Jimenez, C. (2001) Geomicrobiological study of the Grotta dei Cervi, Porto Badisco, Italy. Geomicrobiol. J. **18**: 241–258.
Hammes, F. and Verstraete, W. (2002) Key roles of pH and calcium metabolism in microbial carbonate precipitation: Rev. Environ. Sci. Biotechnol. **1**: 3–7.
Hendy, C.H. (1971) The isotopic geochemistry of speleothems – 1. The calculation of the effects of different modes of formation on the isotopic compositions of speleothems and their applicability as paleoclimatic indicators. Geochim. Cosmochim Acta **35**: 801–824.
Holmgren, K. and Lauritzen, S.E. (1995) Th-230/U-234 and C-14 Dating of a Late Pleistocene Stalagmite in Lobatse-II Cave, Botswana. Q. Sci. Rev. **13**(2): 111–119.
Hoppert, M., Flies, C., Pohl, W., Gunzl, B. and Schneider, J. (2004) Colonization strategies of lithotrophic microorganisms on carbonate rocks. Environ. Geol. **46**: 421–428.
Hose, L.D., Palmer, A.N., Palmer, M.V., Northup, D.E., Boston, P.J. and DuChene, H.R. (2000) Microbiology and geochemistry in a hydrogen-sulfide-rich karst environment. Chem. Geol. **169**: 399–423.
Jones, B. (2001) Microbial activity in caves – a geological perspective: Geomicrobiol. J. **18**: 345–357.
Krajewski, K.P., Cappellen, P.V., Trichet, J., Kuhn, O., Lucas, J., Algarra, A.M., Prevot, L., Tewari, V.C., Knight, T. and Lamboy, M. (1994) Biological processes and apatite formation in sedimentary environment. Eclogae. Helv. **87**(3): 701–745.
Laiz, L., Groth, I., Gonzalez, I. and Saiz-Jimenez, C. (1999) Microbiological study of the dripping water in Altamira Cave (Santillana del Mar, Spain). J. Microbiol. Methods **36**: 129–138.
Laiz, L., Pinar, G., Lubitz, W. and Saiz-Jimenez, C. (2003) Monitoring the colonization of monuments by bacteria: Cultivation versus molecular methods. Environ. Microbiol. **5**: 72–74.
Logan, B.W., Rezak, R. and Ginsburg, R.N. (1964) Classification and environmental significance of algal stromatolites. J. Geol. **72**: 68–83.
Loisy, C., Verrechia, E.P. and Dufour, P. (1999) Microbial origin for pedogenic micrite associated with a carbonate paleosol (Champagne, France). Sediment. Geol. **126**: 193–204.
McDermott, F. (2004) Palaeo-climate reconstruction from stable isotope variations in speleothems: a review. Q. Sci. Rev. **23**(7–8): 901–918.
McDermott, F., Frisia, S., Huang, Y., Longinelli, A., Spiro, S., Heaton, T.H.E., Hawkesworth, C., Borsato, A., Keppens, E., Fairchild, I., van Borgh, C., Verheyden, S. and Selmo, E. (1999) Holocene climate variability in Europe: evidence from delta^{18}O, textural and extension-rate variations in speleothems. Q. Sci. Rev. **18**: 1021–1038.
McKay, C.P., Ivanov, M. and Boston, P.J. (1994) Considering the improbable: life underground on Mars. Planet Rep. **14**: 13–15.
Meisinger, D.B., Zimmermann, J., Ludwig, W., Schleifer, K.H., Wanner, G., Schmid, M., Bennett, P.C., Engel, A.S. and Lee, N.M. (2007) In situ detection of novel Acidobacteria in microbial mats from a chemolithoautotrophically-based cave system (Lower Kane Cave, WY, USA). Environ. Microbiol. **9**(6): 1523–1534.
Melim, L.A., Shinglman, K.M., Boston, P.J., Northup, D.E., Spilde, M.N. and Queen, J.M. (2001) Evidence for microbial involvement in pool finger precipitations, Hidden cave, New Mexico. Geomicrobiol. J. **18**: 311–329.
Neuweiler, F., Rutsch, M., Geipel, G., Reimer, A. and Heise, K.H. (2000) Soluble humic substances from in situ precipitated microcrystalline $CaCO_3$, internal sediment, and spar cement in a Cretaceous carbonate mud-mound. Geology **28**: 851–854.
Newman, D.K. and Banfield, J.F. (2002) Geomicrobiology: molecular-scale interactions underpin biogeochemical systems. Science **296**: 1071–1077.
Northup, D.E. and Lavoie, K.H. (2001) Geomicrobiology of caves: a review. Geomicrobiol. J. **18**: 199–222.

Northup, D.E., Reysenbach, A.L. and Pace, N.R. (1997) Microorganisms and speleothems, In: C.A. Hill, P. Forti (eds.) *Cave Minerals of the World*. National Speleological Society, Huntsville, pp. 261–266.

Overpeck, J.T. and Cole, J.E. (2007) Climate change: lessons from a distant monsoon. Nature **445**: 270–271.

Palmer, A.N. (1991) Origin and morphology of limestone caves. Geol. Soc. Am. Bull. **103**: 1–21.

Pedley, M. (1992) Freshwater (phytoherm) reefs: the role of biofilms and their bearing on marine reef cementation. Sediment. Geol. **79**: 255–274.

Poulson, T.L. and Lavoie, K.H. (2000) The trophic basis of subsurface ecosystems, In: D.C. Wilkens, D.C. Culver and W.F. Humphreys (eds.) *Ecosystems of the World 30*. Elsevier, Amsterdam, pp. 231–249.

Ramesh, R. (2001) High resolution Holocene monsoon records from different proxies: an assessment of their consistency. Special Section: Climate, Monsoon and India's Water. Curr. Sci. **81**(11): 1432–1436.

Riding, R. (2000) Microbial carbonates: the geological record of calcified bacterial–algal mats and biofilms. Sedimentology **47**: 179–214.

Rivadeneyra, M.A., Delgado, R., Delgado, G., Del Moral, A., Ferrer, M.R. and Ramos-Cormenza, A. (1993) Precipitation of carbonate by Bacillus sp. isolated from saline soils. Geomicrobiol. J. **11**: 175–184.

Rodriguez-Navarro, C., Rodriguez-Gallego, M., Chekroun, K.B. and Gonzalez-Munoz, M.T. (2003) Conservation of ornamental stone by Myxococcus xanthus-Induced carbonate biomineralization. Appl. Environ. Microbiol. **69**: 2182–2193.

Roussel, E.G., Cambon Bonavita, M.A., Querellou, J., Cragg, B.A., Webster, G., Prieur, D. and Parkes, R.J. (2008) Extending the sub-seafloor biosphere. Science **320**: 1046.

Saiz-Jimenez, C. (1999) Biogeochemistry of weathering processes in monuments: Geomicrobiol. J. **16**: 27–37.

Sanchez-Moral, S., Canaveras, J.C., Laiz, L., Saiz-Jimenez, C., Bedoya, J. and Luque, L. (2003) Biomediated precipitation of $CaCO_3$ metastable phases in hypogean environments: a short review. Geomicrobiol. J. **20**: 491–500.

Sarbu, S.M., Kane, T.C. and Kinkle, B.K. (1996) A chemoautotrophically based cave ecosystem. Science **272**: 1953–1955.

Schabereiter-Gurtner, C., Saiz-Jimenez, C., Pinar, G., Lubitz, W. and Rolleke, S. (2002) Cave paleolithic paintings harbour complex and partly unknown microbial communities. FEMS Microbiol. Lett. **211**: 7–11.

Shoji, R. and Folk, R.L. (1964) Surface morphology of some limestone types as revealed by electron microscope. J. Sediment. Res. **34**: 144–155.

Simon, K.S., Benfield, E.F. and Macko, S.A. (2003) Food web structure and the role of epilthic biofilms in cave streams. Ecology **84**: 2395–2406.

Singh, A.K., Parkash, B., Mohindra, R., Thomas, J.V. and Singhvi, A.K. (2001) Quaternary alluvial fan sedimentation in the Dehradun Valley piggyback basin, NW Himalaya: tectonic and paleoclimatic implications. Basin Res. **13**: 449–471.

Sinha, A., Cannariato, K.G., Stott, L.D., Cheng, H., Edwards, R.L., Yadava, M.G., Ramesh, R. and Singh, I.B. (2005) Variability of southwest Indian summer monsoon precipitation during Bolling – Allerod. Geol. Soc. Am. **33**(10): 813–816.

Tewari, V.C. (1989) Upper Proterozoic – Lower Cambrian stromatolites and Indian stratigraphy. Him. Geol. **13**: 143–180.

Tewari, V.C. (1995) Controls of phosphorite formation superimposed on biological activity in the Lesser Himalaya, India. Geosci. J. **14**(12): 135–153.

Tewari, V.C. (1998) Earliest microbes on earth and possible occurrence of stromatolites on Mars, In: *Proceedings of the Vth Trieste Conference on Chemical Evolution, I.C.T.P., Trieste, Italy*. Kluwer Academic, Dordrecht, pp. 381–289.

Tewari, V.C. (2001) Origins of life in the universe and earliest prokaryotic microorganisms on earth, In: *Proceedings of the VIth Trieste Conference on Chemical Evolution, I.C.T.P., Trieste, Italy.* Kluwer Academic, Dordrecht, pp. 251–254.

Tewari, V.C. (2004) Microbial diversity in Meso- Neoproterozoic Formations, with particular reference to Himalaya, India, In: J. Seckbach (ed.) *Origins*. Kluwer Academic, Dordrecht, pp. 515–528.

Tewari, V.C. (2007) The Rise and Decline of Ediacaran Biota: Paleobiological and Stable Isotopic Evidence from the NW and NE Lesser Himalaya, India, In: P. Vickers Rich and P. Komarower (eds.) *Special Publication 286*. Geological Society of London, London, pp. 77–101.

Tewari, V.C. (2008) Speleothems from the Himalaya and the Monsoon: a preliminary study. Earth Sci. India **1**(IV): 231–142.

Tewari, V.C. (2009) Speleothems from the Himalaya, paleoclimate and monsoon, In: S.C. Tripathi (ed.) *Advances in Earth Science*. Satish Serial, Delhi, India.

Tewari, V.C. and Joshi, M. (1993) Stromatolite microstructures: A new tool for biostratigraphic correlation of Lesser Himalayan carbonates. J. Him. Geol. **4**(2): 229–239.

Tooth, A.F. and Fairchild, I.J. (2003) Soil and karst aquifer hydrological controls on the geochemical evolution of speleothem-forming drip waters, Crag Cave, southwest Ireland. J. Hydrol. **273**: 51–68.

Wang, Z., Schauble, E.A. and Eiler, J.M. (2004) Equilibrium thermodynamics of multiply-substituted isotopologues of molecular gases. Geochim. Cosmochim. Acta **68**: 4779–4797.

Wang, Y.J., Cheng, H., Edwards, R.L., Kong, X.G., Shao, X., Chen, S., Wu, J.Y., Jiang, X.Y., Wang, X.F. and An, Z.S. (2008) Millennial- and orbital-scale changes in the East Asian monsoon over the past 224,000 years, Nature **451**: 1090–1093.

Whitman, W.B., Coleman, D.C. and Wiebe, W.J. (1998) Prokaryotes: the unseen majority. Proc. Natl. Acad. Sci. U.S.A. **95**: 6578–6583.

Woods, A.D., Bottjer, D.J., Mutti, M. and Morrison, J. (1999) Lower Triassic large sea-floor carbonate cements; their origin and a mechanism for the prolonged biotic recovery from the end-Permian mass extinction. Geology **27**: 645–648.

Yadava, M.G. and Ramesh, R. (2005) Monsoon reconstruction from radiocarbon dated tropical Indian speleothems. The Holocene **15**(1): 48–59.

Yadava, M.G., Ramesh, R. and Pant, G.B. (2004) Past monsoon rainfall variations in peninsular India recorded in a 331 year old speleothem. Holocene **14**: 517–524.

Biodata of **Professor Jesse G. Dillon**, author of ***"The Role of Sulfate Reduction in Stromatolites and Microbial Mats: Ancient and Modern Perspectives"***

Professor Jesse G. Dillon is currently an Associate Professor in the Department of Biological Sciences at California State University, Long Beach. He obtained his Ph.D. from the University of Oregon in 2000 and continued his studies and research as a Postdoctoral fellow at the University of Washington until 2003. Professor Dillon's scientific interests are in the areas of: microbial mats, thermophiles, halophiles, cyanobacteria and sulfate-reducing microorganisms.

E-mail: **jdillon@csulb.edu**

V.C. Tewari and J. Seckbach (eds.), *STROMATOLITES: Interaction of Microbes with Sediments*, Cellular Origin, Life in Extreme Habitats and Astrobiology 18, 571–590
DOI 10.1007/978-94-007-0397-1_25, © Springer Science+Business Media B.V. 2011

THE ROLE OF SULFATE REDUCTION IN STROMATOLITES AND MICROBIAL MATS: ANCIENT AND MODERN PERSPECTIVES

JESSE G. DILLON
Department of Biological Sciences, California State University, Long Beach, CA 90840, USA

Abstract Sulfate reduction is an evolutionarily ancient process and sulfate-reducing microorganisms were likely key members of Precambrian stromatolite communities, as they are in modern photosynthetic microbial mats. Some of the highest rates of sulfate reduction ever measured have been observed in hypersaline microbial mats, supporting the view that sulfate respiration is a dominant carbon mineralization process in these communities. Sulfate consumption and the alkalinity that results from carbon utilization have also been linked to carbonate precipitation in lithified mats. Diverse groups of sulfate-reducing bacteria (SRB), primarily members of the Deltaproteobacteria, have been found to live in stratified zones in microbial mats, some localized near the surface despite high levels of oxygenic photosynthesis by cyanobacteria. Culture studies have shown that some SRB can switch to aerobic metabolism under microaerophilic conditions; however, it is not known how SRB tolerate the very high levels found in situ. Possible strategies involve aggregation and diel migration. Recent application of technologies such as nanometer-scale secondary ion mass spectrometry (nanoSIMS) and metagenomics to mats have enabled ultra fine-scale mapping of sulfate reduction activity and have broadened our understanding of how sulfur metabolism fits into the broader picture of microbial diversity and functionality.

Keywords Microbial mat • Hypersaline • Sulfate-reducing bacteria (SRB) • Deltaproteobacteria • *Desulfonema–Desulfosarcina–Desulfococcus* (Dn–Ds–Dc) group • Microcoleus • Cyanobacteria • Oxycline • Lithification • Extracellular polymeric substances (EPSs) • Aragonite • Dolomite • Guerrero Negro • Solar Lake • Highborne Cay • 16S rRNA • Dissimilatory sulfite reductase (dsrAB) • CARD-FISH • nanoSIMS

1. Introduction

Modern photosynthetic microbial mats (both lithifying and soft) are found in a range of coastal and hypersaline habitats and are thought to be analogous to fossilized stromatolites from the Precambrian (Stal, 1994; Walter, 1994; Des Marais, 1995). These mats are typically complex communities including all

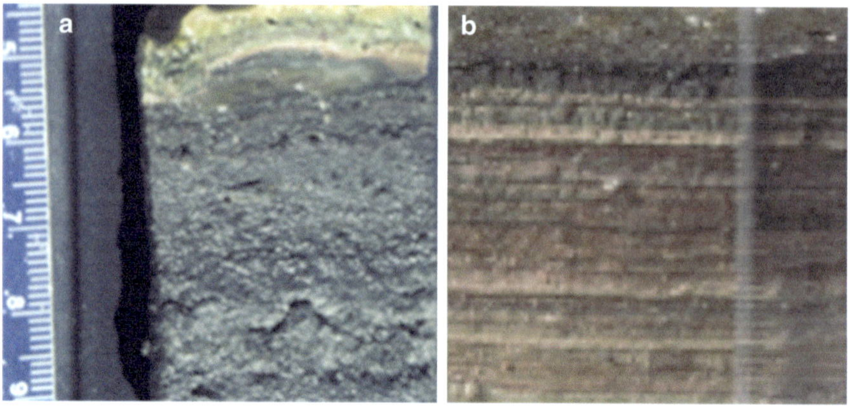

Figure 1. Cross-sections of mat from Pond 4 Near 5 of the ESSA facility in Guerrero Negro, MX. Panel (**a**) shows freshly excised mat, (**b**) shows the mat after it was left in air overnight to oxidize.

major biogeochemical processes including carbon, oxygen, and sulfur cycling (Canfield and Des Marais, 1993). Cyanobacterial (oxygenic) photosynthesis is often a dominant process near the surface of the mat; however, a range of anaerobic heterotrophic processes including fermentation and anaerobic sulfate respiration are also important. Indeed, even the common hypersaline mat-forming cyanobacterium, *Microcoleus chthonoplastes*, will switch to perform fermentation in the dark consuming photosynthate produced during the day (Moezelaar et al., 1996). It is thought that a significant fraction of the excreted products (e.g., organic acids) from cyanobacterial fermentation in mats is consumed by sulfate-reducing bacteria (SRB) (Stal, 2001). The effects of sulfate reduction are obvious in many mats where the mat or underlying sediments appear black below the oxycline due to the accumulation of metal sulfides (e.g., pyrite). This is especially true in non-lithifying mats such as those found in evaporation ponds in the Exportadora de Sal (ESSA) saltworks in Guerrero Negro, Baja CA, MX, where the black precipitates can obscure the mat's layering, which only becomes apparent after the mat is reoxidized (Fig. 1). Overall, the interactions of cyanobacteria and SRB play central roles in the metabolic and geochemical functioning of modern microbial mats; roles that these two groups have likely been playing for billions of years.

2. Early Evolution of Sulfate Reduction

The extent of metabolic diversity of microbial communities on the early Earth is a central question in microbial evolution. Since stromatolites are thought to represent some of the earliest communities that arose during the Archaean (Walter, 1994), metabolic activity present in those communities is of particular interest. Significant attention has focused on the importance of phototrophy, especially anoxygenic photosynthesis and later oxygenic cyanobacterial photosynthesis (Olson and Pierson, 1986; Nisbet and Fowler, 1999; Des Marais, 2000). Methanogenesis was likely an important process on the early Earth and may have produced greenhouse warming, which would have facilitated the rise of microbial life (Catling et al., 2001; Kasting and Siefert, 2002). However, the available carbon fixed by autotrophic community members would also have enabled the rise of other metabolic types. Because of the reducing atmospheric conditions, anaerobic heterotrophic pathways would likely have become important; their time of origin is a key question.

The relatively low sulfate levels in the Archaean ocean (<200 μM) are predicted to have limited the activity of SRB compared to modern seawater communities, likely favoring methanogenic archaea with whom SRB compete for substrates (Canfield et al., 2000; Habicht et al., 2002). However, stable isotopic ($\delta^{34}S$) analyses of barite deposits indicate that sulfate-reducing activity may have evolved as early as ~3.5 Ga (Shen et al., 2001; Shen and Buick, 2004; Ueno et al., 2008). At that time, sulfate reduction may have centered in areas of localized sulfate concentration such as evaporation ponds. This very early origin of sulfate respiration has been disputed (Philippot et al., 2007); however, additional evidence exists for the origins being at least 2.8–3.1 Ga (Schidlowski, 1979). Recent isotopic studies have found evidence of sulfate reduction within stromatolites from 2.7 Ga (Kakegawa and Nanri, 2006). This suggests that sulfate reducers were important metabolic contributors to early microbial mat communities.

The chemical and geological evidence corresponds with molecular phylogenetic studies that support the early evolution of sulfate-reducing microorganisms (Wagner et al., 1998; Klein et al., 2001). Indeed, a recent phylogenomic analysis found that Deltaproteobacteria, whose common ancestor may have been an SRB, likely preceded the emergence of cyanobacteria (Blank, 2004).

3. Sulfate Reduction Rates in a Range of Mats

Sulfate reduction activity in mats and their associated sediments have traditionally been measured via $^{35}SO_4^{2-}$ radiotracer measurements (Jørgensen and Cohen, 1977; Skyring, 1984). These methods continue to be used either via bulk measurements in cores or slurries (Teske et al., 1998; Visscher et al., 1998) or via application of silver foil/disks, which allows the detection of localized utilization within the mats (Visscher et al., 2000; Fike et al., 2007). An overview of sulfate reduction rates reported from different mats and stromatolites is shown in Table 1. Overall, a wide range of sulfate reduction from below detection to some of the highest rates ever measured has been found in mats, both lithified and not (Table 1).

Some of the earliest sulfate reduction assays were made in mats found in Hamelin Pool, Shark Bay, W. Australia (Bauld et al., 1979). Sulfate respiration was undetectable in the sediments below intertidal pustular mats (*Entophysalis*-dominated), but modest rates similar to those in sediments between the stromatolites

Table 1. Sulfate reduction rates reported in a range of lithifying and nonlithifying microbial mats.

Location	Mat type	References	Maximal SRR (nmol cm^{-3} d^{-1})[a]	Depth integrated rates (µmol cm^{-2} d^{-1})
Intertidal, Shark Bay, AUS	Soft, *Microcoleus*	Bauld et al. (1979)	100	0.5
Intertidal, Spencer Gulf, AUS	Soft, *Microcoleus*	Skyring (1984)	1,880	0.2–10.4
Subtidal, Highborne Cay, Exumas, BAH	Lithified, *Schizothrix*	Visscher et al. (1998)	552	0.18–0.29
Salt Pond, Solar Lake, Sinai, EGY	Partially lithified, *Microcoleus*/*Halothece*	Jørgensen and Cohen (1977) and Teske et al. (1998)	3,127–5,400	6.72
Salt Pond, Guerrero Negro, Baja CA, MEX	Soft, *Microcoleus*	Canfield and Des Marais (1991)	~14,000	n.d.
Transplanted Solar Lake Mat, Eilat, ISR	Partially lithified, *Microcoleus*/*Halothece*	Fründ and Cohen (1992)	~200	0.86–1.43
Salt Pond, Eilat, ISR	Endolithic gypsum crust, *Halothece*	Sørensen et al. (2004)	720	1.9
Salt Pond, Cabo Rojo, PUR	Soft, lightly lithified, *Microcoleus*	Casillas-Martinez et al. (2005)	6,336	n.d.
Mushroom Hot Spring, YNP, USA	Soft, *Synechococcus*	Dillon et al. (2007)	14,000	0.83–5.92

[a]Assumes 1 cm^{-3} = 1 g = 1 ml for conversions from reported values

were detected in sediments below smooth mats (*Microcoleus*-dominated). Generally, high rates have been detected in the well-studied soft or partially lithified mats in Solar Lake and ESSA (Table 1). Lower rates were detected in mats from the Solar Lake system that were transplanted to an artificial experimental setting in Eilat, Israel (Fründ and Cohen, 1992). These studies also found significant diel variation in sulfate reduction rates with higher overall per cm^3 or per cell rates during daytime and typically higher rates in surface layers than at depth (Canfield and Des Marais, 1991; Fründ and Cohen, 1992; Teske et al., 1998).

What is known about sulfate reduction rates in modern carbonate-lithified systems is primarily known from the work of Peter Visscher and colleagues in studying the subtidal stromatolites in Highborne Cay in the Bahamas. This group has detected lower rates of sulfate respiration than in soft mats (Table 1; Visscher et al., 1998). By using the silver foil method, they have been able to identify localized bands of sulfate reduction near the surface and at 3–4 mm depth in the stromatolites (Visscher et al., 2000). These bands were found to correspond to lithified micritic horizons within the mat suggesting a role in carbonate precipitation (see Sect. 7 below). Another study in an endolithic gypsum crust in a saltern in Israel found comparable rates to those in the carbonate mats (Table 1; Sørensen et al., 2004). Slurry experiments assaying the effect of variable salinity on rates showed sulfate respiration was maximal at ~10–12% salt and was significantly inhibited at the in situ saltern salinities (>20%). This suggests that SRB may not be optimally adapted to the gypsum crust environment; however, the rates measured in the slurries were 5–10 times lower than the crust incubations, so these results should be interpreted with caution.

The effect of other physicochemical milieu changes in mats has also been investigated. Seasonal changes in mats where climatic conditions vary across the year have been documented. A study of a lightly lithified mat in a salt pond in Puerto Rico found a tenfold increase in sulfate respiration during the wet season (Aug–Dec) compared with the dry (Jan–Apr), which correlated with increases in salinity, temperature and pH (Casillas-Martinez et al., 2005). In another study, sulfate levels were manipulated for sections of mat from the ESSA field site transported to NASA Ames in Mountain View, CA, USA where they were incubated in flumes in a rooftop greenhouse (Bebout et al., 2002, 2004). In flumes with reduced sulfate (<200 nM) meant to simulate early Earth levels, they found a threefold decrease in sulfate respiration rates compared to samples under in situ sulfate levels, resulting in an tenfold increase of methanogenesis. This supports the generally held view that methanogens are poorer competitors for carbon substrates than SRB, but their existence in the mat suggests that methanogens play a role in the mats in situ as well, a finding consistent with recent molecular and biochemical analyses of the ESSA mat (Orphan et al., 2008).

Some studies have also looked at the stimulative affect on sulfate reduction of adding carbon substrates. In slurries of stromatolite, sulfate reduction as measured by sulfide gas release was found to increase upon addition of glycolate, organic acids (e.g., acetate), ethanol as well as more exotic substrates such as cyanobacterial EPSs and sulfonates (e.g., taurine, cysteate) (Visscher et al., 1998, 1999). In the Solar Lake mat, glycolate addition (a typical photosynthate component) was shown to enhance rates for surface populations, whereas ethanol and lactate (common fermentation products) were found to stimulate activity in deeper layers (Fründ and Cohen, 1992). Overall, these findings suggest that, despite the large amount of living biomass in these mats, because of the very rapid mineralization rates, usable carbon is actually limiting in these systems.

The importance of sulfate reduction has not only been observed in high sulfate, hypersaline mats. Past studies have shown the importance of sulfate respiration in microbial mats found in hot springs such as the high sulfate Bath Lake in Yellowstone National Park (YNP), WY, USA (Ward and Olson, 1980). A recent study in the ~60°C outflow of Mushroom Spring, a low sulfate (<200 μM) hot spring in YNP detected surprisingly high rates of sulfate reduction in the nonlithifying cyanobacterial mat, comparable with the highest rates measured in high sulfate hypersaline mats (Dillon et al., 2007). This system may more closely mirror conditions on the early Earth when sulfate levels were lower and temperatures may have been higher.

4. Role of SRB in Carbonate Precipitation in Modern Stromatolites

Stromatolites are lithified accretion structures that grow by trapping and binding of sediments and by precipitation of calcium carbonate (Bosak and Newman, 2003). These are only found in a few places on the modern Earth including Shark Bay, Australia and the Exumas Islands in the Bahamas (Bauld et al., 1979; Reid et al., 2000). The precise mechanisms governing the precipitation of carbonates including the possible involvement of microorganisms have long been a matter of study, but have not been fully elucidated (Krumbein, 1979; Chafetz, 1986; Grotzinger and Knoll, 1999). The following are descriptions of two of the best-studied types of stromatolites, the subtidal aragonite stromatolites in Highborne Cay, Bahamas and the dolomite stromatolites in South Australia and Brazil.

4.1. ARAGONITE PRECIPITATING HABITATS

Sulfate reduction activity has been linked with precipitation of aragonite carbonates in modern stromatolites found in Highborne Cay, Bahamas (Visscher et al., 1998, 2000; Reid et al., 2000; Andres et al., 2006). In those studies, high rates of sulfate reduction and greater abundance of SRB correlated with fine layers of micrite lithification (Visscher et al., 1998, 2000). The process of sulfate reduction mediates $CaCO_3$ precipitation via the reaction:

$$2[CH_2O] + SO_4^{2-} + Ca^{2+} \rightarrow CaCO_3 + CO_2 + HS^- + H^+ + H_2O \qquad (1)$$

where $[CH_2O]$ indicates organic carbon, typically photosynthate (Visscher et al., 2000).

There is a close association between cyanobacteria and SRB in this system as the carbon in this system is largely derived from photosynthetic carbon fixed by the dominant cyanobacterium (*Schizothrix* sp.) found in these stromatolites. Overall, sulfate reduction is thought to be responsible for nearly half of the carbon remineralization in these stromatolites (Visscher et al., 1998).

The metabolic activity of stromatolite microorganisms, especially cyanobacteria and SRB, is thought to enhance carbonate precipitation in a number of ways. An increase in alkalinity can enhance carbonate deposition. Traditionally, it was thought that cyanobacterial photosynthesis was the primary activity driving this process (Riding, 1982). In this view, photosynthesis (CO_2 fixation) results in localized CO_2 depletion, which shifts the inorganic C equilibrium toward CO_3^{2-}, increased pH and carbonate precipitation (Paerl et al., 2001).

However, sulfate respiration also results in increased alkalinity and likely plays a role in this process (Lyons et al., 1984).

Another important process in carbonate deposition is the heterotrophic breakdown of organic compounds that increases the availability of free calcium ions (Dupraz and Visscher, 2005). Particularly important could be the breakdown of extracellular polymeric substances (EPSs), which are often found in copious amounts in microbial mats and stromatolites. EPS is produced by a number of key mat and stromatolite-forming groups including both cyanobacteria and SRB (Stal, 2003; Decho et al., 2005). The initial formation of EPS in microbial mats and stromatolites can limit calcification due to the binding of minerals, especially magnesium and calcium (Decho et al., 2005; Braissant et al., 2007). This initial inhibition is greatly reduced when EPS is modified either biotically or abiotically (Dupraz and Visscher, 2005; Braissant et al., 2009), leading to subsequent carbonate precipitation. The rates of EPS production and degradation were found to be key features in the heterogenous, fine-scale localization of mineral precipitation in the Highborne Cay stromatolites (Decho et al., 2005). A study of cultured SRB isolated from the Bahamian stromatolites found that SRB produced copious amounts of EPS and highlighted their potential influence on the morphology and mineralogy of the precipitated carbonate (Braissant et al., 2007).

A study in the Highborne Cay stromatolites used microautoradiography to explicitly investigate the relative importance of photosynthetic activity compared to that of heterotrophic bacteria (Paerl et al., 2001). That study found that the majority of carbonate precipitation was physically associated, not with the cyanobacterial filaments, but rather with heterotrophic bacteria (which would include SRB) and suggested that primary productivity is only indirectly responsible for deposition by serving as the primary source of organic photosynthate (Paerl et al., 2001). This has been further supported with $\delta^{13}C$ isotopic signatures of dissolved inorganic carbon (DIC) preserved in the aragonite (Andres et al., 2006).

4.2. DOLOMITE PRECIPITATING HABITATS

The formation of the magnesium-rich carbonate mineral, Dolomite, is much less common today compared with ancient rocks (Vasconcelos et al., 1995). It is thought that it is kinetically unfavorable under current conditions (e.g., temperature) in most locations on Earth. However, dolomite precipitation has been observed in association with stromatolites found in hypersaline coastal lagoons and ephemeral lakes in Brazil and Australia (Wright, 1999; van Lith et al., 2002) and has been detected in Holocene age (~6,400 years) deposits off the coast of Belize that may still be undergoing dolomitization (Teal et al., 2000). The removal of sulfate ions has been proposed to reduce the kinetic inhibition of dolomite formation; microbial sulfate reduction has been invoked as the primary process in these examples (Wright, 1999; Teal et al., 2000; van Lith et al., 2002). SRB-mediated precipitation is also partly caused by an increase in pH/alkalinity as described above. This model is supported by pure culture studies that showed that SRB-meditated dolomite precipitation is possible under anoxic conditions even at low temperatures (Vasconcelos et al., 1995; Warthmann et al., 2000).

In both the coastal lagoons in Lagoa Vermelha, Brazil and the ephemeral, alkaline lakes in Coorong, S. Australia, there is a dramatic seasonal increase in salinity and the Mg/Ca ratio. In both systems, an increase in salinity as temperature increases in late spring/summer correlates with dolomite precipitation. Different mechanisms driving the Mg/Ca ration have been proposed. In Brazil, it was proposed that SRB take up magnesium, which forms an ion pair with sulfate in the water (Vasconcelos and McKenzie, 1997). The sulfate is then converted to sulfide, while excess magnesium beyond that needed for the microorganisms' physiological requirements is released back into the local environment. In Australia, it was proposed that the magnesium is released from stromatolitic cyanobacteria as the ephemeral lakes evaporate (Wright, 1999). Elevated magnesium is a prerequisite of dolomite formation. However, these changes in salinity and Mg/Ca ratio were not found to be sufficient to overcome the chemical limitations of dolomite formation. Instead, localized sulfate depletion, combined with the increase in alkalinity and the decrease in hydration energy of magnesium caused by sulfate respiration, was proposed as the most likely explanation overcoming the kinetic inhibition of dolomite formation (Wright, 1999; van Lith et al., 2002).

One additional way that all microbes in these settings, not just SRB, may facilitate dolomite precipitation is by serving as nucleation sites (Bosak and Newman, 2003). This may also play an important role in the preservation of microfossils, as bacterial cells are encased as dolomite grains in these habitats (Wright, 1999; van Lith et al., 2003).

5. Diversity of SRB in Microbial Mats and Stromatolites

Two main culture-independent sequencing approaches have been taken to study the diversity of microorganisms in mats and stromatolites. General 16S rRNA-based clone libraries broadly targeting entire Domains such as Bacteria have revealed great and perhaps unexpected genetic diversity in these communities including a number of novel phyla (Burns et al., 2004; Papineau et al., 2005; Sørensen et al., 2005; Ley et al., 2006; Allen et al., 2009). In general, significantly higher diversity has been observed in nonlithifying mats compared with lithified stromatolites even within the same lagoon (e.g., Hamelin Pool) (Allen et al., 2009).

These 16S rRNA gene-based studies, while showing significant levels of photosynthetic groups such as cyanobacteria and Chloroflexibacteria, have typically found the greatest abundance to be Proteobacterial sequences (Ley et al., 2006; Allen et al., 2009). Within this broad bacterial group, the Deltaproteobacteria, a group dominated by SRB, typically represent ~5–10% of the overall clonal abundance, suggesting a prominent role in these communities. Recently, a preliminary study of Deltaproteobacterial communities in lithifying vs. nonlithifying mats in a salt pan in the Bahamas reported differences in SRB communities in the two types of mats and lower overall diversity in the lithified system (Baumgartner et al., 2006). Bacterial 16S rRNA gene clone libraries recovered from gypsum crusts in the ESSA and Eilat Salterns, where detectable sulfate reduction activity has been detected (see above), showed similar relative abundances of Deltaproteobacterial sequences (Sørensen et al., 2004; Sahl et al., 2008).

The second approach comprises targeted investigations of SRB in particular. These involve either using SRB group-specific 16S rRNA probes (Risatti et al., 1994; Teske et al., 1998) or sequencing specific functional markers, for example, the dissimilatory sulfite reductase (*dsrAB*) gene, which codes for a key enzyme in the sulfate reduction pathway (Minz et al., 1999a, b). These targeted approaches have been applied in the hypersaline mats in ESSA and Solar Lake, respectively. Overall, SRB diversity was found to be dominated by Deltaproteobacterial SRB found in all layers, even those with high oxygen during the daytime due to cyanobacterial photosynthesis. In general, these studies found that surface layers tend to be dominated by members of the Desulfovibrionaceae, while the region bounding the oxycline (2–3 mm depth in daytime) tends to be dominated by members of the *Desulfonema–Desulfosarcina–Desulfococcus* (Dn–Ds–Dc) and Desulfobacteriaceae clades (Risatti et al., 1994; Minz et al., 1999a, b). Overall, these studies have shown strong vertical stratification in SRB community structure with highly localized bands of abundance near the oxycline with some phylotypes only found below the oxycline in permanently anoxic layers. This vertical variability likely reflects a number of factors including gradients of carbon availability (e.g., complete vs. incomplete oxidizers) as well as diel fluctuations of oxygen.

More recently, the functional gene approach has been applied in the ESSA mat as well (Dillon et al., unpublished data). Figure 2 shows a phylogram comparing the diversity of *dsrA* gene sequences recovered from ESSA with the Solar Lake sequences. As predicted from prior work in the ESSA mat, SRB diversity was dominated by Deltaproteobacterial lineages. Indeed, the ESSA *dsr* gene study recovered members of all of the groups of sequences recovered by Minz et al. (1999b); however, the relative abundance was not always similar. For example, ESSA clade 1, corresponding to part of the Solar Lake clade C and represented by one sequence in that study, contained almost half of the over 400 sequences recovered in Baja. In addition to the groups in common with Solar Lake, the ESSA study found seven additional groups, including deeply branching lineages not closely related to cultured SRB.

These studies have expanded our understanding of SRB diversity in soft or partially lithified mats; however, similar studies using specific targeted 16S or functional gene approaches in lithifying systems are needed to compare the diversity and better elucidate the role of diverse groups of SRB in mat formation and mineralization.

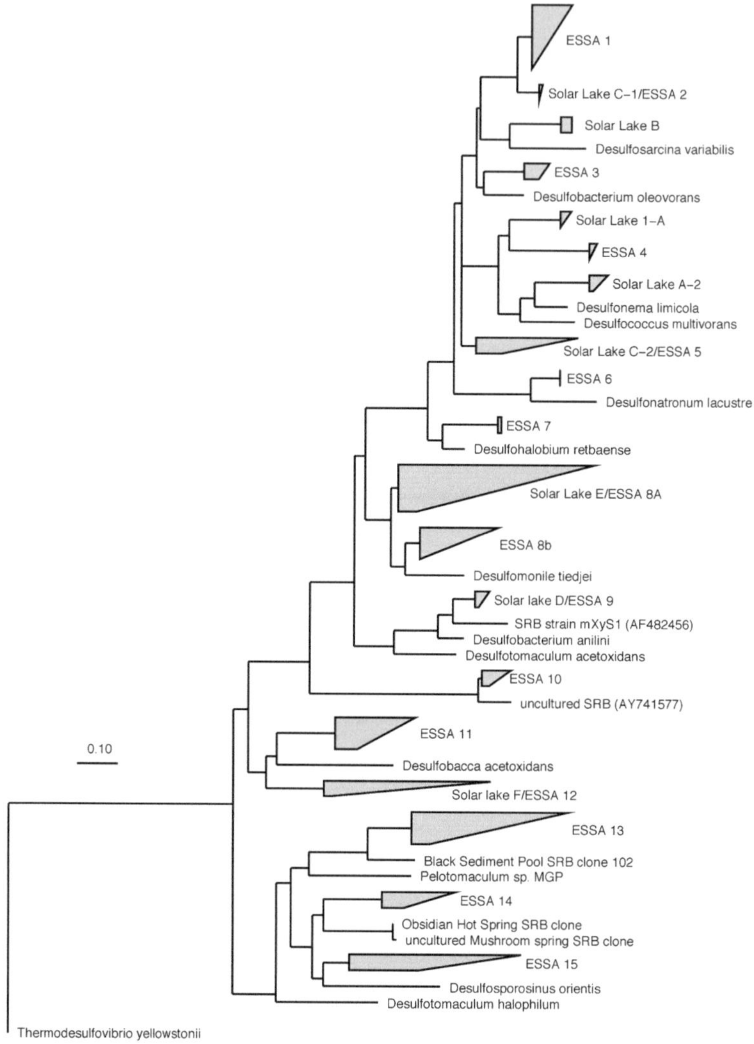

Figure 2. Unrooted neighbor-joining phylogram depicting relationships among DsrA amino acid sequences of clones obtained directly from the ESSA 4 near 5 microbial mat. Sequences from the Solar Lake mat (Minz et al., 1999) were included for comparison. Vertical wedge size corresponds to number of clones in each grouping.

6. Interaction of SRB with Oxygen

The traditional view, based primarily on culture studies, was that sulfate reduction is a strictly anaerobic process due to the inactivation of key enzymes in the sulfate reduction pathway by O_2 (Postgate, 1959; Widdel, 1988). This is logical considering the early evolution of SRB at a time when the atmosphere was reducing and oxygen tolerance mechanisms would have been unnecessary (Shen and Buick, 2004). It is also supported by visual evidence with the black, sulfidic layers found below the pigmented photosynthetic layers at the surface (Fig. 1a) (Baumgartner et al., 2006). However, a number of culture studies have revealed a more complex relationship with oxygen ranging from mere tolerance to switching metabolic pathways to use oxygen in some groups of SRB (Cypionka et al., 1985; Cypionka, 2000; Sigalevich et al., 2000; Jonkers et al., 2005). This is particularly true of *Desulfovibrio* spp., which have been found to obtain ATP from using oxygen as a terminal electron acceptor under microaerophilic conditions (Dilling and Cypionka, 1990) and in the case of *D. oxyclinae*, an isolate obtained from the oxic layer of the Solar Lake microbial mat, cell growth occurred (Krekeler et al., 1997; Sigalevich et al., 2000). Oxygen tolerance of SRB in microbial mat communities in the field was actually identified long before those studies (Jørgensen and Cohen, 1977). These early results were expanded by Canfield and Des Marais (1991), who showed that some of the highest rates of sulfate reduction ever measured occurred in the oxygenated portion of the *Microcoleus*-based mat in Guerrero Negro (Table 1). In fact, the highest rates of sulfate reduction measured in that study occurred above the oxycline, where total rates are likely to be underestimates since some of the sulfide produced may have been reoxidized (Teske et al., 1998). This suggests that SRB are capable of utilizing photosynthate during the day, potentially obviating the need to wait until darkness to consume primary production carbon and allowing them to compete more effectively with aerobic heterotrophs. However, it should be noted that none of the culture studies described above has been found to tolerate the very high levels of oxygen found in these mats nor has aerobic sulfate respiration been observed (Baumgartner et al., 2006). This indicates that there are emergent properties found in the complex mat environment that cannot be replicated in pure culture (Des Marais, 2003).

Other studies in subtidal stromatolites and hot springs have found that rates were higher at night, but that significant sulfate reduction occurred at or above the oxycline during the day (Visscher et al., 1998; Dillon et al., 2007). Thus, different types of mat communities may display different relationships with oxygen. Indeed, even within a mat, different groups of SRB are likely to be differentially oxygen tolerant and a variety of oxygen tolerance mechanisms have been proposed including enzymatic defenses (e.g., superoxide dismutase), clumping to create anaerobic microniches, and localized oxygen removal by active aerobic respiration and migration (Krekeler et al., 1998; Baumgartner et al., 2006). The co-occurrence of SRB with sulfur-oxidizing bacteria (SOB) that reoxidize the sulfide in the presence of oxygen may also play a role.

7. Role of SRB Migration in Microbial Mats

The role of migration of phototrophs and SOB in microbial mats, especially nonlithified, hypersaline mats, has been well documented (Garcia-Pichel et al., 1994; Bebout and Garcia-Pichel, 1995; Kruschel and Castenholz, 1998; Jonkers et al., 2003; Villanueva et al., 2007). Migration is often a response to changes in irradiance including harmful UV wavelengths. SRB have been shown to migrate in mats as well. Teske et al. (1998) probed DGGE bands and found evidence for migration of the *Desulfonema* group down into the mat during the daytime, presumably to avoid rising oxygen levels. Interestingly, these are members of the same Dn–Ds–Dc clade, which was found to localize near the oxycline in the Minz et al. (1999a) study above. Two recent studies have used terminal restriction fragment length polymorphism (T-RFLP) to track migrating populations in mats over a diel period (Fourçans et al., 2006; Dillon et al., 2009). The former study of a *Microcoleus*-type soft mat in a saltern (Salins-de Giraud, Camargue, France) found evidence of both upward and downward migration among members of the Dn–Ds–Dc clade and *Desulfobacter*, another group found to colocalize with the oxic–anoxic interface. The latter found evidence for extensive migration among numerous bacterial groups, including Deltaproteobacterial SRB, while others appeared to remain in the same location in the ESSA mat. These findings demonstrate that the complex laminated structure of mats is highly dynamic over the diel cycle.

8. Recent Developments

The application of cutting-edge technology to microbial mats has produced exciting results. A recent study has combined both new and old approaches by using catalyzed reporter deposition fluorescent in situ hybridization (CARD-FISH) analyses to the mat in combination with applying nanometer-scale Secondary Ion Mass Spectrometry (nanoSIMS) to the silver foil/disk approach for measuring sulfide production in the mat (Fike et al., 2007). With the nanoSIMS approach, no radiolabeling is necessary because the Ag_2S deposited on metal surfaces can be mapped and analyzed isotopically. This enabled simultaneous micrometer-scale mapping of both ^{32}S and ^{34}S, which enables calculation of $\delta^{34}S$. That study found bands of sulfide deposition that were consistent in size and location with CARD-FISH probing for Dn–Ds–Dc SRB populations. Evidence of sulfate reduction activity and the presence of SRB were found dispersed throughout the mat, but were concentrated in bands and aggregations within the upper ~2 mm in association with autofluorescent cyanobacteria (Fike et al., 2007) reminiscent of the findings in the Highborne Cay lithified system (Visscher et al., 2000). These results confirm past findings regarding the localization and potential aggregation strategies to tolerate oxygen, but provide a much more fine-scale view of mat activity. Interestingly, some of the sulfur fractionations (up to 55 per mil) were higher than expected for sulfate reduction (Detmers et al., 2001), suggesting that sulfur disproportionation may be important in this system as well.

Another use of new technology goes beyond the single gene sequencing approaches described above. The recent development of metagenomic analyses of mixed genomes in environmental samples has enabled a more holistic view of microbial communities (Steele and Streit, 2005). Metagenomic sequencing was initially applied in fairly simple microbial biofilms composed of only a few species such as those found in acid mine drainages (Tyson et al., 2004), but has more recently been applied to the highly complex, hypersaline Pond 4 mat in ESSA (Kunin et al., 2008). Not surprisingly, the latter study found pronounced gradients of genetic diversity with depth in the mat. Deltaproteobacterial sequences showed a vertical gradient comprising <5% of the sequence assignments at the surface, but increased to ~10% for the remaining depths (Kunin et al., 2008). Future analyses of this large genomic dataset specifically targeting genes in the sulfate respiratory pathway are warranted.

More direct analysis of the sulfur cycle was performed in another metagenomic-based study (Breitbart et al., 2009). That study combined metagenomic analyses with isotopic analyses of freshwater stromatolites in the Cuatro Cienegas site in the Chihuahuan desert of north-central Mexico. These extensive networks of geothermal springs, pools and streams are high in nitrate and sulfate, yet very low in phosphorus, which minimizes grazing and allows the formation of a range of stromatolitic communities that have been the site of intensive study in recent years (Garcia-Pichel et al., 2004; Souza et al., 2006; Falcón et al., 2007; Desnues et al., 2008). In spherical oncolites in the Rio Mesquites and in domal thrombolites in the Pozas Azules II site, evidence of genes for dissimilatory sulfate reduction activity coincides with negative sulfur isotopic depletion (−20–25 per mil $\delta^{34}S$) (Breitbart et al., 2009). The sulfur-isotopic and gene sequencing evidence supports an active sulfur cycle in this system, and $\delta^{13}C$ measurements suggest that, similar to the Highborne Cay system, respiratory remineralization of photosynthate is the main process responsible for carbonate precipitation. The low sulfur-isotopic depletion of biogenic sulfides is consistent with sulfate reduction without sulfur reoxidation (Canfield and Thamdrup, 1994). Interestingly, no evidence of sulfur oxidation genes were detected, unlike in most other mat systems where SRB and SOB co-occur.

9. Acknowledgments

The author acknowledges the support and mentorship of Dr. David Stahl who introduced me to the amazing world of sulfate-reducing bacteria. The author's work in Dr Stahl's lab, upon which some of this is based, was supported by NSF-IGERT grant (DGE-9870713), NSF grant (DEB-0213186) and NASA NAI grant (NCC2-1273). He also acknowledges the editors Dr. Vinod C. Tewari and Dr. Joseph Seckbach for their kind invitation to contribute this chapter as well as the insightful suggestions for improvement by two peer reviewers Dr. Andreas Teske and Dr. Harald Strauss.

10. References

Allen, M.A., Goh, F., Burns, B.P. and Neilan, B.A. (2009) Bacterial, archaeal and eukaryotic diversity of smooth and pustular microbial mat communities in the hypersaline lagoon of Shark Bay. Geobiology **7**: 82–96.

Andres, M.S., Sumner, D.Y., Reid, R.P. and Swart, P.K. (2006) Isotopic fingerprints of microbial respiration in aragonite from Bahamian stromatolites. Geology **34**: 973–976.

Bauld, J., Chambers, L.A. and Skyring, G.W. (1979) Primary productivity, sulfate reduction and sulfur isotope fractionation in algal mats and sediments of Hamelin Pool, Shark Bay, Western Australia. Aust. J. Mar. Freshw. Res. **30**: 753–764.

Baumgartner, L.K., Reid, R.P., Dupraz, C., Decho, A.W., Buckley, D.H., Spear, J.R., et al. (2006) Sulfate reducing bacteria in microbial mats: changing paradigms, new discoveries. Sediment. Geol. **185**: 131–145.

Bebout, B.M. and Garcia-Pichel, F. (1995) UV-B-induced vertical migrations of cyanobacteria in a microbial mat. Appl. Environ. Microbiol. **61**: 4215–4222.

Bebout, B.M., Carpenter, S.P., Des Marais, D.J., Discipulo, M., Embaye, T., Garcia-Pichel, F., et al. (2002) Long-term manipulations of intact microbial mat communities in a greenhouse collaboratory: simulating Earth's present and past field environments. Astrobiology **2**: 383–402.

Bebout, B.M., Hoehler, T.M., Thamdrup, B., Albert, D., Carpenter, S.P., Hogan, M., et al. (2004) Methane production by microbial mats under low sulfate concentrations. Geobiology **2**: 87–96.

Blank, C.E. (2004) Evolutionary timing of the origins of mesophilic sulphate reduction and oxygenic photosynthesis: a phylogenomic dating approach. Geobiology **2**: 1–20.

Bosak, T. and Newman, D.K. (2003) Microbial nucleation of calcium carbonate in the Precambrian. Geology **31**: 577–580.

Braissant, O., Decho, A.W., Dupraz, C., Glunk, C., Przekop, K.M. and Visscher, P.T. (2007) Exopolymeric substances of sulfate-reducing bacteria: interactions with calcium at alkaline pH and implication for formation of carbonate minerals. Geobiology **5**: 401–411.

Braissant, O., Decho, A.W., Przekop, K.M., Gallagher, K.L., Glunk, C., Dupraz, C. and Visscher, P.T. (2009) Characteristics and turnover of exopolymeric substances in a hypersaline microbial mat. FEMS Microbiol. Ecol. **67**: 293–307.

Breitbart, M., Hoare, A., Nitti, A., Siefert, J., Haynes, M., Dinsdale, E., et al. (2009) Metagenomic and stable isotopic analyses of modern freshwater microbialites in Cuatro Cienegas, Mexico. Environ. Microbiol. **11**: 16–34.

Burns, B.P., Goh, F., Allen, M. and Neilan, B.A. (2004) Microbial diversity of extant stromatolites in the hypersaline marine environment of Shark Bay, Australia. Environ. Microbiol. **6**: 1096–1101.

Canfield, D.E. and Des Marais, D.J. (1991) Aerobic sulfate reduction in microbial mats. Science **251**: 1471–1473.

Canfield, D.E. and Des Marais, D.J. (1993) Biogeochemical cycles of carbon, sulfur, and free oxygen in a microbial mat. Geochim. Cosmochim. Acta **57**: 3971–3984.

Canfield, D.E. and Thamdrup, B. (1994) The production of ^{34}S-depleted sulfide during bacterial disproportionation of elemental sulfur. Science **266**: 1973–1975.

Canfield, D.E., Habicht, K.S. and Thamdrup, B. (2000) The Archean sulfur cycle and the early history of atmospheric oxygen. Science **288**: 658–661.

Casillas-Martinez, L., Gonzalez, M.L., Fuentes-Figueroa, Z., Castro, C.M., Nieves-Mendez, D., Hernandez, C., et al. (2005) Community structure, geochemical characteristics and mineralogy of a hypersaline microbial mat, Cabo Rojo, PR. Geomicrobiol. J. **22**: 269–281.

Catling, D.C., Zahnle, K.J. and McKay, C. (2001) Biogenic methane, hydrogen escape, and the irreversible oxidation of early earth. Science **293**: 839–843.

Chafetz, H.S. (1986) Marine peloids; a product of bacterially induced precipitation of calcite. J. Sediment. Res. **56**: 812–817.

Cypionka, H. (2000) Oxygen respiration by *Desulfovibrio* species. Annu. Rev. Microbiol. **54**: 827–848.

Cypionka, H., Widdel, F. and Pfennig, N. (1985) Survival of sulfate-reducing bacteria after oxygen stress, and growth in sulfate-free oxygen-sulfide gradients. FEMS Microbiol. Ecol. **31**: 39–45.

Decho, A.W., Visscher, P.T. and Reid, R.P. (2005) Production and cycling of natural microbial exopolymers (EPS) within a marine stromatolite. Palaeogeogr. Palaeoclimatol. Palaeoecol. **219**: 71–86.

Des Marais, D.J. (1995) The biogeochemistry of hypersaline microbial mats, In: J. Jones (ed.) *Advances in Microbial Ecology*. Plenum Press, New York, pp. 251–274.

Des Marais, D.J. (2000) When did photosynthesis emerge on Earth? Science **289**: 1703–1705.

Des Marais, D.J. (2003) Biogeochemistry of hypersaline microbial mats illustrates the dynamics of modern microbial ecosystems and the early evolution of the biosphere. Biol. Bull. **204**: 160–167.

Desnues, C., Rodriguez-Brito, B., Rayhawk, S., Kelley, S., Tran, T., Haynes, M., et al. (2008) Biodiversity and biogeography of phages in modern stromatolites and thrombolites. Nature **452**: 340–343.

Detmers, J., Bruchert, V., Habicht, K.S. and Kuever, J. (2001) Diversity of sulfur isotope fractionations by sulfate-reducing prokaryotes. Appl. Environ. Microbiol. **67**: 888–894.

Dilling, W. and Cypionka, H. (1990) Aerobic respiration in sulfate-reducing bacteria. FEMS Microbiol. Lett. **71**: 123–127.

Dillon, J.G., Fishbain, S., Miller, S.R., Bebout, B.M., Habicht, K.S., Webb, S.M. and Stahl, D.A. (2007) High rates of sulfate reduction in a low-sulfate hot spring microbial mat are driven by a low level of diversity of sulfate-respiring microorganisms. Appl. Environ. Microbiol. **73**: 5218–5226.

Dillon, J.G., Miller, S., Bebout, B., Hullar, M., Pinel, N. and Stahl, D.A. (2009) Spatial and temporal variability in a stratified hypersaline microbial mat community. FEMS Microbiol. Ecol. **68**: 46–58.

Dupraz, C. and Visscher, P.T. (2005) Microbial lithification in marine stromatolites and hypersaline mats. Trends Microbiol. **13**: 429–438.

Falcón, L., Cerritos, R., Eguiarte, L. and Souza, V. (2007) Nitrogen fixation in microbial mat and stromatolite communities from Cuatro Cienegas, Mexico. Microb. Ecol. **54**: 363–373.

Fike, D., Ussler, W., Eiler, J., Guan, Y.B. and Orphan, V. (2007) Micron-scale resolution of sulfur cycling in a microbial mat. Geochim. Cosmochim. Acta **71**: A278–A278.

Fourçans, A., Solé, A., Diestra, E., Ranchou-Peyruse, A., Esteve, I., Caumette, P. and Duran, R. (2006) Vertical migration of phototrophic bacterial populations in a hypersaline microbial mat from Salins-de-Giraud (Camargue, France). FEMS Microbiol. Ecol. **57**: 367–377.

Fründ, C. and Cohen, Y. (1992) Diurnal cycles of sulfate reduction under oxic conditions in cyanobacterial mats. Appl. Environ. Microbiol. **58**: 70–77.

Garcia-Pichel, F., Mechling, M. and Castenholz, R.W. (1994) Diel migrations of microorganisms within a benthic, hypersaline mat community. Appl. Environ. Microbiol. **60**: 1500–1511.

Garcia-Pichel, F., Al-Horani, F.A., Farmer, J.D., Ludwig, R. and Wade, B.D. (2004) Balance between microbial calcification and metazoan bioerosion in modern stromatolitic oncolites. Geobiology **2**: 49–57.

Grotzinger, J.P. and Knoll, A.H. (1999) Stromatolites in Precambrian carbonates: evolutionary mileposts or environmental dipsticks? Annu. Rev. Earth Planet. Sci. **27**: 313–358.

Habicht, K.S., Gade, M., Thamdrup, B., Berg, P. and Canfield, D.E. (2002) Calibration of sulfate levels in the Archean ocean. Science **298**: 2372–2374.

Jonkers, H.M., Ludwig, R., De Wit, R., Pringault, O., Muyzer, G., Niemann, H., et al. (2003) Structural and functional analysis of a microbial mat ecosystem from a unique permanent hypersaline inland lake: 'La Salada de Chiprana' (NE Spain). FEMS Microbiol. Ecol. **44**: 175–189.

Jonkers, H.M., Koh, I.O., Behrend, P., Muyzer, G. and de Beer, D. (2005) Aerobic organic carbon mineralization by sulfate-reducing bacteria in the oxygen-saturated photic zone of a hypersaline microbial mat. Microb. Ecol. **49**: 291–300.

Jørgensen, B.B. and Cohen, Y. (1977) Solar Lake (Sinai). 5. The sulfur cycle of the benthic cyanobacterial mats. Limnol. Oceanogr. **22**: 657–666.

Kakegawa, T. and Nanri, H. (2006) Sulfur and carbon isotope analyses of 2.7 Ga stromatolites, cherts and sandstones in the Jeerinah Formation, Western Australia. Precambrian Res. **148**: 115–124.

Kasting, J.F. and Siefert, J.L. (2002) Life and the evolution of Earth's atmosphere. Science **296**: 1066–1068.

Klein, M., Friedrich, M., Roger, A.J., Hugenholtz, P., Fishbain, S., Abicht, H., et al. (2001) Multiple lateral transfers of dissimilatory sulfite reductase genes between major lineages of sulfate-reducing prokaryotes. J. Bacteriol. **183**: 6028–6035.

Krekeler, D., Sigalevich, P., Teske, A., Cypionka, H. and Cohen, Y. (1997) A sulfate-reducing bacterium from the oxic layer of a microbial mat from Solar Lake (Sinai), *Desulfovibrio oxyclinae* sp. nov. Arch. Microbiol. **167**: 369–375.

Krekeler, D., Teske, A. and Cypionka, H. (1998) Strategies of sulfate-reducing bacteria to escape oxygen stress in a cyanobacterial mat. FEMS Microbiol. Ecol. **25**: 89–96.

Krumbein, W.E. (1979) Photolithotropic and chemoorganotrophic activity of bacteria and algae as related to beachrock formation and degradation (Gulf of Aqaba, Sinai). Geomicrobiol. J. **1**: 139–203.

Kruschel, C. and Castenholz, R.W. (1998) The effect of solar UV and visible irradiance on the vertical movements of cyanobacteria in microbial mats of hypersaline waters. FEMS Microbiol. Ecol. **27**: 53–72.

Kunin, V., Raes, J., Harris, J.K., Spear, J.R., Walker, J.J., Ivanova, N., et al. (2008) Millimeter-scale genetic gradients and community-level molecular convergence in a hypersaline microbial mat. Mol. Syst. Biol. **4**: 198.

Ley, R.E., Harris, J.K., Wilcox, J., Spear, J.R., Miller, S.R., Bebout, B.M., et al. (2006) Unexpected diversity and complexity of the Guerrero Negro hypersaline microbial mat. Appl. Environ. Microbiol. **72**: 3685–3695.

Lyons, W.B., Long, D.T., Hines, M.E., Gaudette, H.E. and Armstrong, P.B. (1984) Calcification of cyanobacterial mats in Solar Lake, Sinai. Geology **12**: 623–626.

Minz, D., Fishbain, S., Green, S.J., Muyzer, G., Cohen, Y., Rittmann, B.E. and Stahl, D.A. (1999a) Unexpected population distribution in a microbial mat community: sulfate-reducing bacteria localized to the highly oxic chemocline in contrast to a eukaryotic preference for anoxia. Appl. Environ. Microbiol. **65**: 4659–4665.

Minz, D., Flax, J.L., Green, S.J., Muyzer, G., Cohen, Y., Wagner, M., et al. (1999b) Diversity of sulfate-reducing bacteria in oxic and anoxic regions of a microbial mat characterized by comparative analysis of dissimilatory sulfite reductase genes. Appl. Environ. Microbiol. **65**: 4666–4671.

Moezelaar, R., Buvank, S.M. and Stal, L. (1996) Fermentation and sulfur reduction in the mat-building cyanobacterium *Microcoleus chthonoplates*. Appl. Environ. Microbiol. **62**: 1752–1758.

Nisbet, E.G. and Fowler, C.M.R. (1999) Archaean metabolic evolution of microbial mats. Proc. R. Soc. Lond. B **266**: 2375–2382.

Olson, J.M. and Pierson, B.K. (1986) Photosynthesis 3.5 thousand million years ago. Photosynth Res. **9**: 251–259.

Orphan, V.J., Jahnke, L.L., Embaye, T., Turk, K.A., Pernthaler, A., Summons, R.E. and Des Marais, D.J. (2008) Characterization and spatial distribution of methanogens and methanogenic biosignatures in hypersaline microbial mats of Baja California. Geobiology **6**: 376–393.

Paerl, H.W., Steppe, T.F. and Reid, R.P. (2001) Bacterially mediated precipitation in marine stromatolites. Environ. Microbiol. **3**: 123–130.

Papineau, D., Walker, J.J., Mojzsis, S.J. and Pace, N.R. (2005) Composition and structure of microbial communities from stromatolites of Hamelin Pool in Shark Bay, Western Australia. Appl. Environ. Microbiol. **71**: 4822–4832.

Philippot, P., Van Zuilen, M., Lepot, K., Thomazo, C., Farquhar, J. and Van Kranendonk, M.J. (2007) Early Archaean microorganisms preferred elemental sulfur, not sulfate. Science **317**: 1534–1537.

Postgate, J. (1959) Sulphate reduction by bacteria. Annu. Rev. Microbiol. **13**: 505–520.

Reid, R.P., Visscher, P.T., Decho, A.W., Stolz, J.F., Bebout, B.M., Dupraz, C., et al. (2000) The role of microbes in accretion, lamination and early lithification of modern marine stromatolites. Nature **406**: 989–992.

Riding, R. (1982) Cyanophyte calcification and changes in ocean chemistry. Nature **299**: 814–815.

Risatti, J.B., Chapman, W.C. and Stahl, D.A. (1994) Community structure of a microbial mat: the phylogenetic dimension. Proc. Natl. Acad. Sci. U.S.A. **91**: 10173–10177.

Sahl, J.W., Pace, N.R. and Spear, J.R. (2008) Comparative molecular analysis of endoevaporitic microbial communities. Appl. Environ. Microbiol. **74**: 6444–6446.

Schidlowski, M. (1979) Antiquity and evolutionary status of bacterial sulfate reduction: sulfur isotope evidence. Orig. Life Evol. Biosph. **9**: 299–311.

Shen, Y. and Buick, R. (2004) The antiquity of microbial sulfate reduction. Earth Sci. Rev. **64**: 243–272.

Shen, Y., Buick, R. and Canfield, D.E. (2001) Isotopic evidence for microbial sulphate reduction in the early Archaean era. Nature **410**: 77–81.

Sigalevich, P., Baev, M.V., Teske, A. and Cohen, Y. (2000) Sulfate reduction and possible aerobic metabolism of the sulfate-reducing bacterium *Desulfovibrio oxyclinae* in a chemostat coculture with *Marinobacter* sp. Strain MB under exposure to increasing oxygen concentrations. Appl. Environ. Microbiol. **66**: 5013–5018.

Skyring, G.W. (1984) Sulfate reduction in marine sediments associated with cyanobacterial mats in Australia, In: Y. Cohen, R. Castenholz and H. Halvorson (eds.) *Microbial Mats: Stromatolites*. Alan R. Liss, New York, pp. 265–275.

Sørensen, K.B., Canfield, D.E. and Oren, A. (2004) Salinity responses of benthic microbial communities in a solar saltern (Eilat, Israel). Appl. Environ. Microbiol. **70**: 1608–1616.

Sørensen, K.B., Canfield, D.E., Teske, A.P. and Oren, A. (2005) Community composition of a hypersaline endoevaporitic microbial mat. Appl. Environ. Microbiol. **71**: 7352–7365.

Souza, V., Espinosa-Asuar, L., Escalante, A.E., Eguiarte, L.E., Farmer, J., Forney, L., et al. (2006) An endangered oasis of aquatic microbial biodiversity in the Chihuahuan desert. Proc. Natl. Acad. Sci. U.S.A. **103**: 6565–6570.

Stal, L.J. (1994) Microbial Mats in coastal environments, In: L. Stal and P. Caumette (eds.) *Microbial Mats*. Springer-Verlag, Berlin, pp. 21–32.

Stal, L.J. (2001) Coastal microbial mats: the physiology of a small-scale ecosystem. S. Afr. J. Bot. **67**: 399–410.

Stal, L.J. (2003) Microphytobenthos, their extracellular polymeric substances, and the morphogenesis of intertidal sediments. Geomicrobiol. J. **20**: 463–478.

Steele, H.L. and Streit, W.R. (2005) Metagenomics: advances in ecology and biotechnology. FEMS Microbiol. Lett. **247**: 105–111.

Teal, C.S., Mazzullo, S.J. and Bischoff, W.D. (2000) Dolomitization of Holocene shallow-marine deposits mediated by sulfate reduction and methanogenesis in normal-salinity seawater, northern Belize. J. Sediment. Res. **70**: 649–663.

Teske, A., Ramsing, N.B., Habicht, K., Fukui, M., Kåver, J., Jørgensen, B.B. and Cohen, Y. (1998) Sulfate-reducing bacteria and their activities in cyanobacterial mats of Solar Lake (Sinai, Egypt). Appl. Environ. Microbiol. **64**: 2943–2951.

Tyson, G.W., Chapman, J., Hugenholtz, P., Allen, E.E., Ram, R.J., Richardson, P.M., et al. (2004) Community structure and metabolism through reconstruction of microbial genomes from the environment. Nature **428**: 37–43.

Ueno, Y., Ono, S., Rumble, D. and Maruyama, S. (2008) Quadruple sulfur isotope analysis of ca. 3.5 Ga Dresser Formation: new evidence for microbial sulfate reduction in the early Archean. Geochim. Cosmochim. Acta **72**: 5675–5691.

van Lith, Y., Vasconcelos, C., Warthmann, R., Martins, J.C.F. and McKenzie, J.A. (2002) Bacterial sulfate reduction and salinity: two controls on dolomite precipitation in Lagoa Vermelha and Brejo do Espinho (Brazil). Hydrobiologia **485**: 35–49.

van Lith, Y., Warthmann, R., Vasconcelos, C. and McKenzie, J.A. (2003) Microbial fossilization in carbonate sediments: a result of the bacterial surface involvement in dolomite precipitation. Sedimentology **50**: 237–245.

Vasconcelos, C. and McKenzie, J.A. (1997) Microbial mediation of modern dolomite precipitation and diagenesis under anoxic conditions (Lagoa Vermelha, Rio de Janeiro, Brazil). J. Sediment. Res. **67**: 378–390.

Vasconcelos, C., McKenzie, J.A., Bernasconi, S., Grujic, D. and Tiens, A.J. (1995) Microbial mediation as a possible mechanism for natural dolomite formation at low temperatures. Nature **377**: 220–222.

Villanueva, L., Navarrete, A., Urmeneta, J., White, D.C. and Guerrero, R. (2007) Analysis of diurnal and vertical microbial diversity of a hypersaline microbial mat. Arch. Microbiol. **188**: 137–146.

Visscher, P.T., Reid, R.P., Bebout, B.M., Hoeft, S.E., Macintyre, I.G. and Thompson, J.A. (1998) Formation of lithified micritic laminae in modern marine stromatolites (Bahamas): the role of sulfur cycling. Am. Mineral. **83**: 1482–1493.

Visscher, P.T., Gritzer, R.F. and Leadbetter, E.R. (1999) Low-molecular-weight sulfonates, a major substrate for sulfate reducers in marine microbial mats. Appl. Environ. Microbiol. **65**: 3272–3278.

Visscher, P.T., Reid, R.P. and Bebout, B.M. (2000) Microscale observations of sulfate reduction: correlation of microbial activity with lithified micritic laminae in modern marine stromatolites. Geology **28**: 919–922.

Wagner, M., Roger, A.J., Flax, J.L., Brusseau, G.A. and Stahl, D.A. (1998) Phylogeny of dissimilatory sulfite reductases supports an early origin of sulfate respiration. J. Bacteriol. **180**: 2975–2982.

Walter, M.R. (1994) Stromatolites: the main geological source of information on the evolution of the early benthos, In: S. Bengston (ed.) *Early Life on Earth*. Columbia University Press, New York, pp. 270–286.

Ward, D.M. and Olson, G.J. (1980) Terminal processes in the anaerobic degradation of an algal-bacterial mat in a high-sulfate hot spring. Appl. Environ. Microbiol. **40**: 67–74.

Warthmann, R., van Lith, Y., Vasconcelos, C., McKenzie, J.A. and Karpoff, A.M. (2000) Bacterially induced dolomite precipitation in anoxic culture experiments. Geology **28**: 1091–1094.

Widdel, F. (1988) Microbiology and ecology of sulfate- and sulfur-reducing bacteria, In: A. Zehnder (ed.) *Biology of Anaerobic Microorganisms*. Wiley, New York, pp. 469–575.

Wright, D.T. (1999) The role of sulphate-reducing bacteria and cyanobacteria in dolomite formation in distal ephemeral lakes of the Coorong region, South Australia. Sediment. Geol. **126**: 147–157.

Biodata of **Robert Gwyn Jenkins** and **Yoshinori Hikida**, authors of ***"Carbonate Sediments Microbially Induced by Anaerobic Oxidation of Methane in Hydrocarbon-Seeps"***

Dr. Robert Gwyn Jenkins is currently Research Fellow of the Japan Society for the Promotion of Science at the Yokohama National University in Yokohama, Japan and Bavarian State Collection for Palaeontology and Geology in Munich, Germany. He obtained his Ph.D. from the University of Tokyo in 2006 and continued his studies and research in that university and in Yokohama National University. Dr. Jenkins' scientific interests are in the area of the evolution of chemosynthesis-based ecosystem, the taxonomy of chemosynthetic bivalves, and microbial sediments in hydrocarbon seep environments.

E-mail: **robertgj@ynu.ac.jp**

Dr. Yoshinori Hikida is currently employed by the Nakagawa Museum of Natural History, Japan. He obtained his Ph.D. from Hokkaido University in 1997 and continued his studies and research at the Museum. Dr. Hikida's scientific interests are in the areas of: Cretaceous to Tertiary molluscan fauna and chemosynthetic communities in Hokkaido Island, and in the shell micostructural evolution of molluscs.

E-mail: **nmhikida@coral.ocn.ne.jp**

Robert Gwyn Jenkins

Yoshinori Hikida

V.C. Tewari and J. Seckbach (eds.), *STROMATOLITES: Interaction of Microbes with Sediments*, Cellular Origin, Life in Extreme Habitats and Astrobiology 18, 591–605
DOI 10.1007/978-94-007-0397-1_26, © Springer Science+Business Media B.V. 2011

CARBONATE SEDIMENTS MICROBIALLY INDUCED BY ANAEROBIC OXIDATION OF METHANE IN HYDROCARBON-SEEPS

Examples from Late Cretaceous Hydrocarbon-Seep Deposits in Hokkaido, Japan

ROBERT GWYN JENKINS[1] AND YOSHINORI HIKIDA[2]

[1]Faculty of Education and Human Sciences Yokohama National University, 79-2 Tokiwadai, Hodogaya-ku, Yokohama, 240-8501, Japan
and
Bavarian State Collection for Palaeontology and Geology, Richard-Wagner Str. 10, 80333 Munich, Germany
[2]Nakagawa Museum of Natural History, 28-9 Yasukawa, Nakagawa Town, Hokkaido 098-2626, Japan

Abstract Stromatolites, laminated benthic microbial deposits, and other microbialites have been known for more than three billion years since the very early periods of the history of life on Earth. Most of them are thought to have been formed as a result of photosynthetic processes. In contrast to such microbialites, chemosynthetically produced microbialites, mainly carbonate minerals, have been found in deep sea settings where active hydrocarbon seeps (synonymous with cold seeps and methane seeps) existed. Seep carbonates are formed under the influence of the anaerobic oxidation of methane (AOM) mediated by anaerobic methane-oxidizing archaea and sulfate-reducing bacteria. Some internal textures of these chemosynthetically mediated carbonates are similar to those of photosynthetically mediated microbialites. But the chemical signatures of the two types are different from each other. The most distinguishable signature is that the seep carbonate has ^{13}C-depleted carbonate cements and lipid biomarkers. Here, we show lithological and chemical features of Late Cretaceous hydrocarbon seep carbonates, which were found in fore-arc basin deposits in Hokkaido, Japan, as examples of chemosynthetically mediated microbialites.

Keywords Methane seep • Hydrocarbon seep • Cold seep • Gas hydrate • Chemosynthesis • Anaerobic oxidation of methane (AOM) • Anaerobic methane-oxidizing archaea • Sulfate reduction • Seep carbonate • Carbon isotope • Biomarker • Fractured sediment • Coated grain • Networked cement • Tube worm • Cretaceous

1. Introduction

Benthic microbial sediments mediated by photosynthetic processes have been known since the very early period of Earth's history (e.g., Awramik and Sprinkle, 1999; Riding, 2000). In contrast to these photosynthetically induced microbialites, chemosynthetically induced microbialites have been found in deep-sea hydrocarbon seep environments (e.g., Greinert et al., 2002). The term "hydrocarbon seep" designates pore water, which contains reduced compounds such as methane, other light hydrocarbons or hydrogen sulfide, all generated at a subsurface level, and which gushes out from the sea floor. Reduced compounds contained in cold seep fluid result in substantial biogeochemical reactions. A key reaction in seep environments is anaerobic oxidation of methane (AOM) as expressed below.

$$CH_4 + SO_4^{2-} \rightarrow CO_3^{2-} + H_2S + H_2O \quad (1)$$

This designates methane anaerobically oxidized by sulfate and generates carbonate ion and sulfide. It has recently been revealed that consortia of anaerobic methane-oxidizing archaea (ANME) and sulfate-reducing bacteria (SRB) are responsible for the AOM reaction (e.g., Orphan et al., 2001). The reaction leads in turn to increasing alkalinity and the subsequent precipitation of carbonate minerals (e.g., Aharon, 2000). Due to the fact that the AOM reaction occurs in anoxic conditions, the carbonates generally precipitate within sediments, although some situations, e.g., when the sea bottom water is anoxic or when the sea floor is covered by microbial mat, cause precipitation on the sea floor (Michaelis et al., 2002; Teichert et al., 2005).

Hydrocarbon seeps are widely distributed along the active and passive continental margins of the world's oceans (Fig. 1; Campbell, 2006). Many carbonates

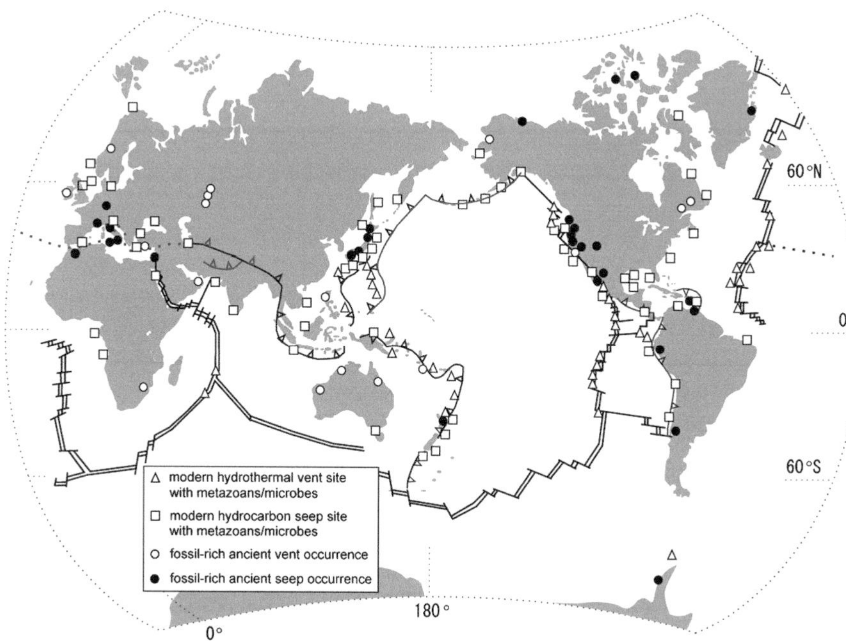

Figure 1. Distribution of modern and ancient cold seeps and hydrothermal vents. After Campbell (2006).

are associated with the seeps, and investigations of the carbonates have revealed that some lithological textures similar to the photosynthetically mediated textures, for example, clotted micrite, stromatolite, and oncoid, are found (e.g., Greinert et al., 2002). Such carbonates have significant geochemical signatures, such as strongly negative carbon isotopic compositions of carbonate minerals (usually less than −30‰ vs. PDB) and extremely negative carbon isotopic compositions of microbial lipids (ca. nearly or lower than −100‰ vs. PDB; Peckmann and Thiel, 2004).

The lithological texture and chemical signatures were potentially imprinted in the carbonate rocks. Thus, the hydrocarbon seep signatures can be matched to the geological record to enable ancient hydrocarbon seep deposits to be recognized (Campbell et al., 2002; Peckmann and Thiel, 2004). Here, we show as examples lithological and geochemical signatures of ancient hydrocarbon seep deposits, which were found in a Cretaceous (145 to 65 million years ago) deposit in Hokkaido, Japan.

2. Ancient Hydrocarbon Seeps in Japan

Many dozens of post-Cretaceous hydrocarbon seeps have been found in different parts of the Japanese archipelagoes (Majima et al., 2005). Cretaceous seep deposits are accumulated particularly in Hokkaido island, northern Japan (Fig. 2) apart from two localities in southwestern Japan, Sada in Shikoku Island (Nobuhara et al., 2008), and Goshoura in Kyushu Island (RGJ, unpublished data). The Cretaceous seep deposits in Hokkaido crop out within the Yezo Group and are distributed along the meridional belt (Fig. 2). The Yezo Group is composed of marine siliciclastic deposits, which were sedimented in the Yezo forearc basin along the western margin of a subduction zone in the circum Pacific region during the Cretaceous Period (Takashima et al., 2004). It is noteworthy that the oldest seep, dated as upper Albian (ca. 100 million years ago), is located in the middle of the belt and the seeps become younger as one moves northward (Fig. 2). Between those Cretaceous seeps in Hokkaido, the youngest seeps (Campanian) exposed in the Nakagawa area, northern Hokkaido, have been investigated in detail by means of lithological, geochemical, and paleontological aspects. Thus, this chapter emphasizes a review of the Campanian (ca. 80 million years ago) seeps in the Nakagawa area.

3. Campanian (Late Creteceous) Hydrocarbon Seeps in Nakagawa

3.1. GEOLOGICAL SETTINGS

At least five Late Cretaceous hydrocarbon seeps have been exposed in the Nakagawa area (Kaim et al., 2009). This is one of the best localities for investigating ancient hydrocarbon seeps, because of its well-exposed condition, the existence of several types of seep deposits, minor occurrences of diagenetic stresses, and well-preserved fossils. Two are well investigated, i.e., Yasukawa and

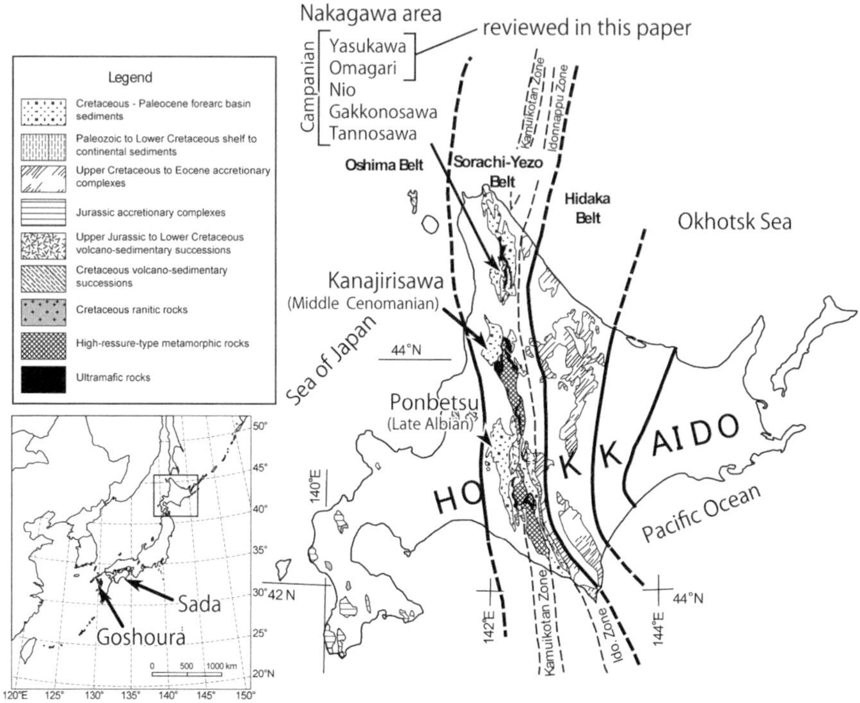

Figure 2. Distribution of Cretaceous seep deposits in Japan with special emphasis on Hokkaido. The geological map is modified from Takashima et al. (2004).

Omagari seeps (Fig. 2; Hikida et al., 2003; Jenkins et al., 2007b). The Yasukawa and Omagari seep deposits are exposed along the Abeshinai River. The Yasukawa seep is located about 3 km downstream from the Omagari seep. The two seeps are embedded in sediments of the upper part of the Omagari Formation, Yezo Group, and have been dated as Lower Campanian (Takahashi et al., 2007; ca. 80 million years ago). The Omagari Formation, which is composed of alternating greenish sandstone and sandy siltstone with some intercalations of gravel beds, was deposited on the continental-shelf margin to the continental slope (Jenkins et al., 2007b; Takahashi et al., 2007).

3.2. LITHOLOGICAL AND CHEMICAL SIGNATURES

3.2.1. Yasukawa Seep Carbonate

The Yasukawa locality crops out within tens of meters of the bank of the Abeshinai River. The locality consists of four carbonate bodies exceeding 50 cm in maximum diameter, and among these, we targeted the largest carbonate body, measuring 2 m in width (Fig. 3a, b). The lateral extension of the body is composed of siliciclastic

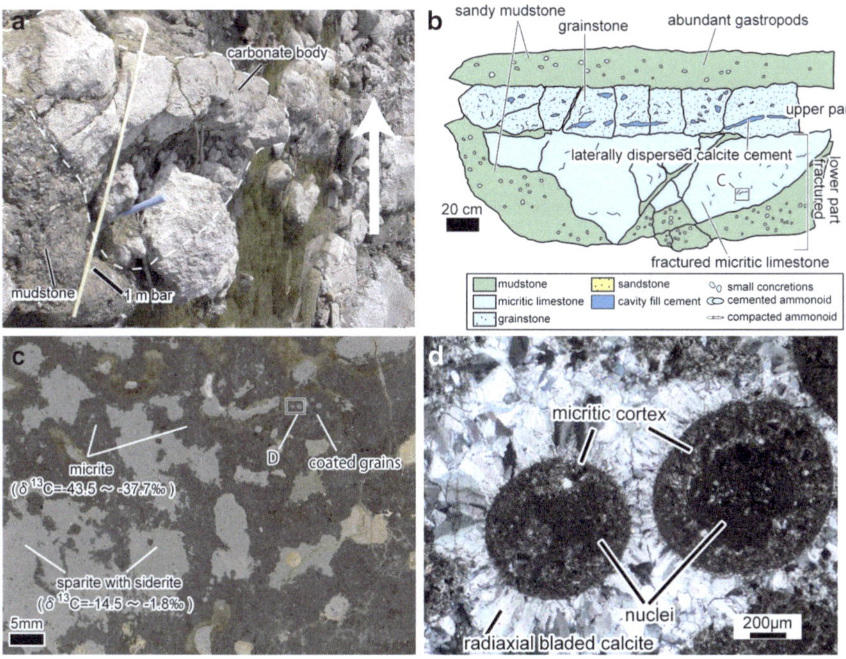

Figure 3. Photographs and sketch of the largest Yasukawa carbonate body. (**a**) Outcrop photograph of the carbonate body. The *white arrow* indicates an upward direction. (**b**) Sketch of the carbonate body. (**c**) Photographs of polished slab sample obtained from the lower part of the carbonate body. Sampling point is indicated in (**b**). The photograph exhibits characteristic fractured fabric, which can be seen elsewhere within the lower part of the carbonate body. (**d**) Thin-section microphotograph of "ooid-like" coated grains. Counterpart area of the sample for the thin-section is indicated in (**c**). The coated grains have siliciclastic nuclei and a micritic cortex. The grains are surrounded by radiaxial bladed calcite cement. Cross-polarized light.

sediments with small carbonate concretions (less than 10 cm in diameter). The body, together with its lateral flank, is representative of the Yasukawa seep as a whole. Using lithological and geochemical signatures as our criteria, we can say that the carbonates were formed under the influence of AOM in a hydrocarbon seep environment (Jenkins et al., 2007b, 2008).

The above carbonate body at the Yasukawa site can be divided into lower and upper parts (Fig. 3b). The most conspicuous texture in the lower part is composed of fractured dark gray to grayish black micritic layers and fractured sandstone and tuffaceous siltstone layers (Fig. 3c). The fractures are often covered by pinkish to brownish networked calcite cement (Figs. 3c and 4b), which are composed of radiaxial bladed calcite, and connected to each other. Micritic parts show clotted fabric (Fig. 4c, d). Sandstone and tuffaceous siltstone are cemented by microsparry calcite, sometimes with siderite crystals. It is noteworthy that the lower part contains three layers, which are composed of accumulations of ooid-like

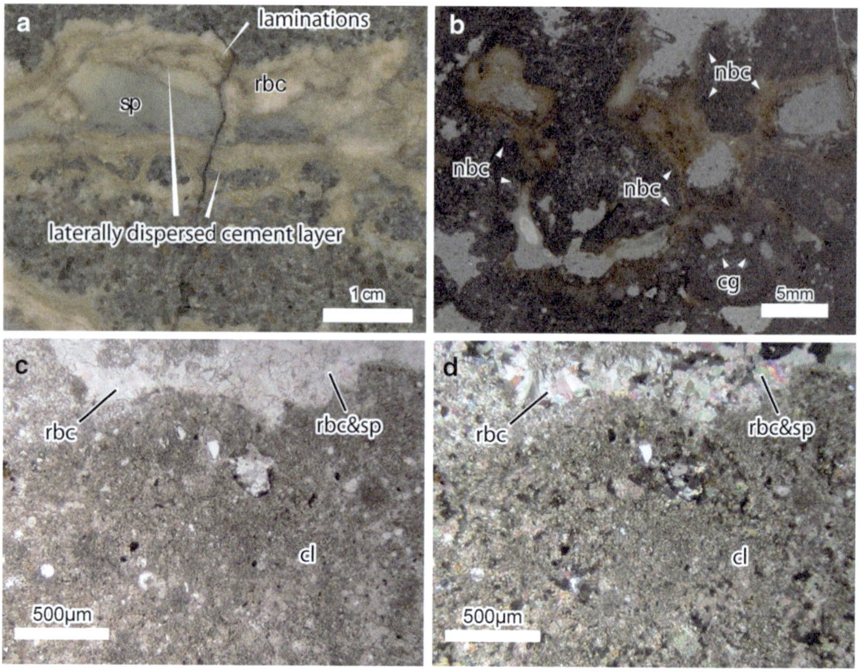

Figure 4. Photographs of the Yasukawa carbonate body. (**a**) Polished slab sample obtained from the upper part. Laterally dispersed cement layer composed of radiaxial bladed calcite is well visible. (**b**) Polished slab sample obtained from the lower part. (**a**, **b**) Upward direction of photographs is the same as the stratigraphic upward direction. (**c**, **d**) Clotted fabric is characteristically well visible in the micritic part of the lower part of Yasukawa carbonate. (**c**) Plane- and (**d**) cross-polarized light. *cl* clotted fabric, *cg* coated grain, *nbc* networked radiaxial bladed calcite, *rbc* radiaxial bladed calcite, *sp* sparry calcite cement.

coated grains (Fig. 3c, d; Jenkins et al., 2008). The original composition of the upper part, i.e., before cementation, was medium- to coarse-grained sandstone. Characteristic cements in the upper part consist of laterally dispersed vein-like calcitic cement layers, ca. 1–2 cm in thickness, which have laminations (Figs. 3b and 4a). The marginal part of the layer is whitish to pinkish lamina, composed of radiaxial bladed calcite, while the central part of the cements is yellowish sparry calcite. Void spaces can sometimes be seen in the centermost part of the vein-like structures. Sandy siltstone overlying the carbonate body contains many small concretions (ca. 1 cm in diameter). Small concretions surrounding the carbonate body are also found. Looking at the Yasukawa carbonate overall, we can say, on the basis of thin-section observations, that the clotted micrite and radiaxial bladed calcite were precipitated earlier and the sparry calcite was precipitated later.

The earlier precipitated minerals have strongly negative carbon isotopic compositions, ca. −40‰, whereas the later precipitated minerals have almost 0‰

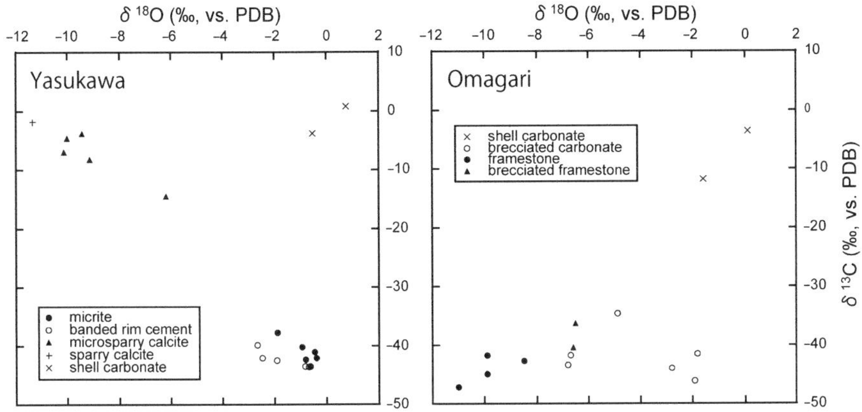

Figure 5. Cross-plots of carbon and oxygen isotopic compositions of the Yasukawa and Omagari carbonate bodies.

(vs. PDB; Jenkins et al., 2007b; Fig. 5). Archaeal lipid biomarkers, e.g., pentamethylicosane (PMI), with extremely depleted ^{13}C, less than −100% (vs. PDB), have also been detected on the basis of an analysis of the carbonate body (Jenkins et al., 2008). These negative carbon isotopic compositions of such minerals and lipids are comparable with those of thermogenic and biogenic methane, which have generally less than −30‰ (Whiticar, 1999). The only archaea, which has such negative carbon isotopic composition, is anaerobic methane-oxidizing archaea (ANME; e.g., Michaelis et al., 2002). This archaea uses methane as the carbon source for its cell synthesis and performs AOM with SRB.

3.2.2. Omagari Seep Carbonate

The Omagari carbonate body, which measures about 10 m wide and 5 m high, is exposed along the Abeshinai River as an islet (Fig. 6a). The surrounding sediments were eroded during the last half century, making it difficult to know the exact dip angle. Geopetal structures which can be observed in the vestimentiferan tubes indicate that the carbonate body is roughly dipping eastward, at an angle between 20° and 60°. The body can be classified mainly into two lithofacies, i.e., boundstone facies and brecciated facies (Fig. 6b).

The boundstone facies is located mainly, relatively speaking, in the upper part, while the brecciated facies is located in the lower part. However, the boundary between the facies intersects the estimated bedding plane, which was estimated by reference to the geopetal structures, i.e., the facies undergoes a lateral change. Even so, it can be said that in general terms, the brecciated facies is underlying and the boundstone facies is overlying.

The brecciated facies is composed of brecciated gray micritic limestone showing a clotted fabric and a brown to pinkish white networked cement layer. The facies includes a considerable amount of siderite. This facies is distributed in

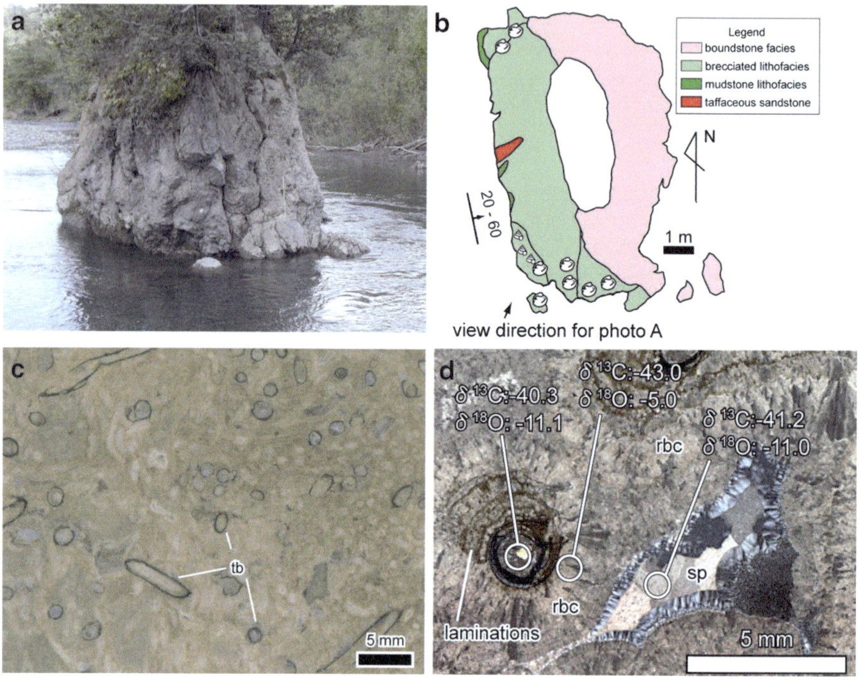

Figure 6. Photographs of the Omagari carbonate. (**a**) Outcrop. (**b**) Sketch of plane view of the Omagari carbonate. (**c**) Polished slab samples obtained from boundstone facies. (**d**) Thin-section photograph of the boundstone facies with carbon and oxygen isotopic compositions of each sampling point. Worm tubes are covered by radiaxial bladed calcite. Silica and sparry calcite fill the void space. Cross-polarized light.

the central to western parts of the body. Many peloids and oncoids can be seen in the facies. The boundstone facies is composed of framestone and boundstone. The framestone consists of abundant worm tubes as a rigid frame and radiaxial bladed calcite cement directly covering the worm tubes (Figs. 6c, d and 7a). Isopachous cement composed of radiaxial bladed cements without worm tubes, which can be classified within the boundstone category, can also be found. Several weak laminae can be observed within the radiaxial bladed cement (Figs. 6d and 7a). The layer becomes conspicuously visible when the carbonate is etched by weak hydrochloric acid. On the basis of its brownish color and resistance to acids, we concluded that the laminae were originally microbial mats. It is noteworthy that the facies contains almost no siliciclastic sediments. Framboidal pyrite is frequently found around the worm tubes within the calcite cement (Fig. 7e, f). The paragenetic sequence of calcite precipitation is similar to that of Yasukawa carbonate, i.e., micrite and radiaxial bladed calcite, are precipitated earlier, and sparry calcite is precipitated later. The carbon isotopic composition of the carbonate minerals is around −40 to −45‰ (vs. PDB; Hikida et al., 2003; RGJ unpublished data; Figs. 5 and 6d). Biomarker analyses detected archaeal lipid

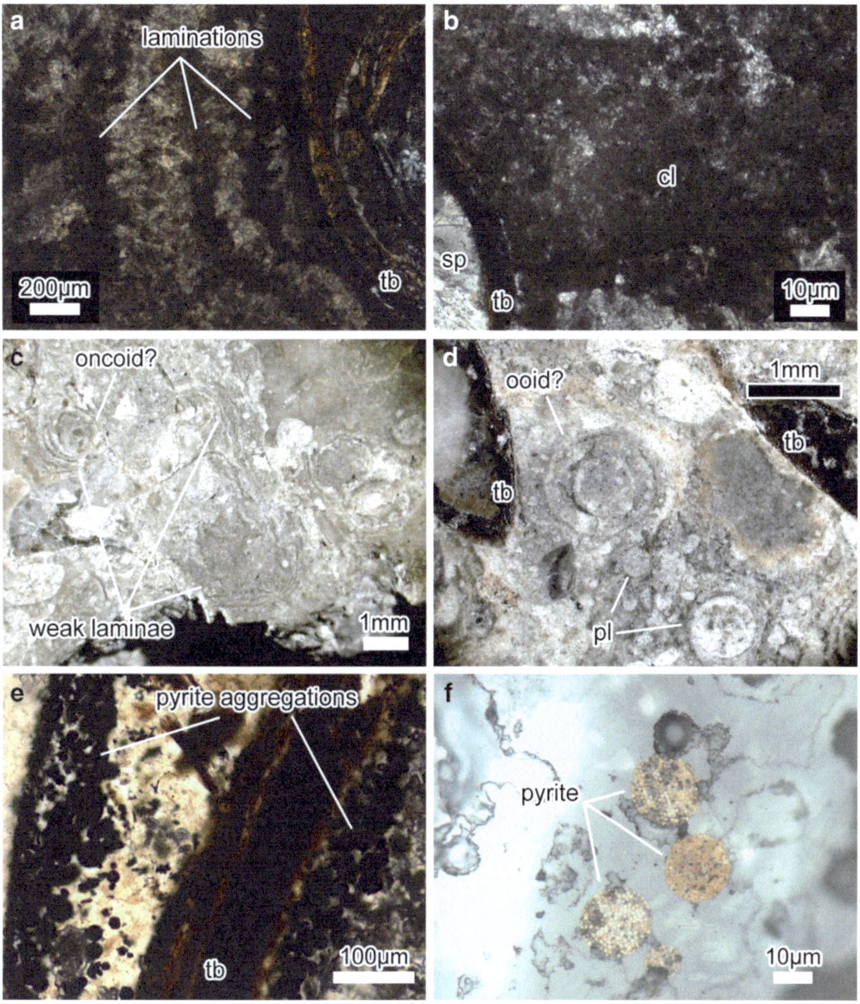

Figure 7. Photographs of internal textures of the Omagari carbonate body. (**a**) Thin-section microphotograph of concentric calcitic laminae covering a delaminated worm tube. Cross-polarized light. (**b**) Thin-section microphotograph of typical clotted fabric. Cross-polarized light. (**c**) Photograph of a polished slab sample. Stromatolite-like laminated structures can be observed. (**d**) Photograph of a polished sample. Peloids and ooid-like grains are observed. (**e**) Aggregations of framboidal pyrite. Plane-polarized light. (**f**) Close-up microphotograph of framboidal pyrites. Reflected light. *cl* clotted fabric, *oo* ooid-like grain, *pl* peloid, *rbc* radiaxial bladed calcite, *sp* sparry calcite, *tb* worm tube.

biomarkers, e.g., PMI, with extremely depleted in ^{13}C as well as Yasukawa carbonate (Ogihara, 2005; RGJ unpublished data). Thus, these lines of evidence again indicate that the Omagari carbonate body was formed under the influence of AOM activities mediated by anaerobic methane-oxidizing archaea. Furthermore, the existence of framboidal pyrites in earlier precipitated carbonate phases suggests that sulfate reduction occurred at the same time as AOM. This is a reasonable assumption because hydrogen sulfide produced as a result of AOM may combine with reduced iron to form monosulfide, i.e., a precursor of pyrite. Subsequent reaction of the precursor with elemental sulfur or sulfides promotes the formation of pyrite. The sulfate reduction during AOM is responsible for SRB. The SRB is commonly associated with anaerobic methane-oxidizing archaea in hydrocarbon seeps. Thus, it is possible to say, on the basis of associations of SRB and anaerobic methane-oxidizing archaea, that the Omagari seeps may also have experienced such microbial activities.

3.2.3. Comparison Between the Yasukawa and Omagari Seeps

The major feature distinguishing the Omagari and Yasukawa carbonates is that the Omagari has carbonate cements growing directly on the worm tubes with a very minor amount of sediment. This texture is not found in the Yasukawa. The presence of the texture evokes the presumption that the facies was formed above the sea floor at the Omagari. However, it is difficult to think that the carbonate precipitation could be mediated by AOM reaction in an open water column. This is because the AOM occurs only in anoxic conditions. Thus, such AOM-derived carbonate cements are usually formed under the sea floor apart from cases, where the bottom water is anoxic as in the Black Sea (Michaelis et al., 2002). Recently, Teichert et al. (2005) proposed that a layer of microbial biomats covering the sea floor made it possible for carbonate precipitation mediated by AOM to take place on the sea floor. In the case of Omagari, the alleged worm tubes, which probably inhibited the insertion of oxic water from open water column, and the fact that worm tubes were covered by the microbial biomats, which also inhibited the insertion of oxic water, created conditions that were preferably suited to AOM. Furthermore, on the basis of a larger carbonate body size, framestone with very minor sediments, and much larger scaled fracturing in the Omagari carbonate, it can be said that the Omagari seep was much more effusive than the Yasukawa seep. The effusive fluid would cause the oxic–anoxic boundary to rise up. These factors would be responsible for creating thin layers of anoxic bottom water within the level of the alleged worm tubes and created conditions that were favorable for the precipitation of carbonate minerals on the worm tubes above the sea floor.

The differences in the strength of fluid flows of the two seeps might be the cause of different faunal associations (Jenkins et al., 2007a, b; Kaim et al., 2009). The vestimentiferan tube worms and associated limpet *Serradonta*, which lived on the tubes are abundant in the Omagari but are extremely rare in Yasukawa. The gastropod *Provanna* sp., a common genus in modern hydrothermal vents and hydrocarbon seeps (Warén and Bouchet, 2001), can be found in both sites but is more common in Omagari. The gastropod *Hokkaidoconcha hikidai*, related to provannid lineage, is absent in Omagari but abundant in Yasukawa. *Acharax,* a solemyid bivalve which is known as low-sulfide specialist in recent hydrocarbon seep (Sahling et al., 2002), is rare in Omagari and the central part of Yasukawa whereas it is commonly found in the periphery of the Yasukawa seep. Faunal changes of this kind are observed in recent hydrocarbon seeps (e.g., Knittel et al., 2005).

4. Summary

The Omagari and Yasukawa carbonate bodies show various textures, which are similar to the texture found in photosynthetically mediated microbialites, for example, stromatolite-like laminated carbonates, ooid-like grains, and oncoids. However, the geochemical evidences, for example, carbon isotopic compositions and lipid biomarkers, clearly characterize the Omagari and Yasukawa carbonates as having been formed under the influence of AOM mediated by anaerobic methane-oxidizing archaea and SRB. Thus, such textures can be formed not only as a result of photosynthetic microbial activities but also as a result of microbial activities based on chemosynthetic energy as found in hydrocarbon seeps. Before the appearance of macro-benthic animals in hydrocarbon seep environments (i.e., before Silurian), the hydrocarbon seep might have supported chemosynthetic life forms excluding macro-benthic animals and might have formed authigenic carbonate sedimentary structures similar to those formed by photosynthetic microbes. Thus, some of the ancient stromatolites that existed before Slurian might be misinterpreted as being stromatolites formed by photosynthetic microbes.

In summary, there is a very strong need for comprehensive techniques, for example, observations of the mode of occurrence of carbonates and their internal textures, and analyses of geochemical signatures such as carbon isotopic compositions and lipid biomarkers, to be carried out to understand what kind of microbial activities produced each stromatolite and related microbial sediments.

5. Acknowledgments

We deeply appreciate the suggestion by Professor Vinod C. Tewari that we should be invited to write this chapter. The paper benefited from peer reviews by Professor V.C. Tewari and Atsushi Yamamoto (University of Tsukuba). Heartfelt gratitude is due to Kazushige Tanabe (University of Tokyo) for his support at all stages of this research, and to Andrzej Kaim (Institute of Paleobiology, PAN, Poland) for helpful discussions and information on fossils and their recent counterparts. Thanks are extended to M.E. Jenkins for his linguistic improvements of this article. The research leading by RGJ was financially supported by the Nippon Foundation−Hadal Environmental Science Education Program (HADEEP), the Fukada Geological Institute, scientific research grant support from the Japan Society for the Promotion of Science (JSPS; grant no. 18403013, leader Kazushige Tanabe), and Grant-in-Aid for JSPS Fellows.

6. References

Aharon, P. (2000) Microbial processes and products fueled by hydrocarbons at submarine seeps, In: R.E. Riding and S.M. Awramik (eds.) *Microbial Sediments*. Springer, Heidelberg, pp. 270–281.

Awramik, S.M. and Sprinkle, J. (1999) Proterozoic stromatolites: the first marine evolutionary biota. Hist. Biol. **13**: 241–253.

Campbell, K.A. (2006) Hydrocarbon seep and hydrothermal vent paleoenvironments and paleontology: past developments and future research directions. Palaeogeogr. Palaeoclimatol. Palaeoecol. **232**: 362–407.

Campbell, K.A., Farmer, J.D. and Des Marais, D. (2002) Ancient hydrocarbon seeps from the Mesozoic convergent margin of California: carbonate geochemistry, fluids and palaeoenvironments. Geofluids **2**: 63–94.

Greinert, J., Bohrmann, G. and Elvert, M. (2002) Stromatolitic fabric of authigenic carbonate crusts: result of anaerobic methane oxidation at cold seeps in 4,850 m water depth. Int. J. Earth Sci. **91**: 698–711.

Hikida, Y., Suzuki, S., Togo, Y. and Ijiri, A. (2003) An exceptionally well-preserved fossil seep community from the Cretaceous Yezo Group in the Nakagawa area, Hokkaido. Paleontol. Res. **7**: 329–342.

Jenkins, R.G., Kaim, A. and Hikida, Y. (2007a) Antiquity of the substrate choice among acmaeid limpets from the Late Cretaceous chemosynthesis-based communities. Acta Palaeontol. Pol. **52**: 369–373.

Jenkins, R.G., Kaim, A., Hikida, Y. and Tanabe, K. (2007b) Methane-flux-dependent lateral faunal changes in a Late Cretaceous chemosymbiotic assemblage from the Nakagawa area of Hokkaido, Japan. Geobiology **5**: 127–139.

Jenkins, R.G., Hikida, Y., Chikaraishi, Y., Ohkouchi, N. and Tanabe, K. (2008) Microbially induced formation of ooid-like coated grains in the Late Cretaceous methane-seep deposits of the Nakagawa area, Hokkaido, northern Japan. Island Arc. **17**: 261–269.

Kaim, A., Jenkins, R.G. and Hikida, Y. (2009) Gastropods from Late Cretaceous Omagari and Yasukawa hydrocarbon seep deposits in, the Nakagawa area, Hokkaido, Japan. Acta Palaeontol. Pol. **54**: 463–490.

Knittel, K., Losekann, T., Boetius, A., Kort, R. and Amann, R. (2005) Diversity and distribution of methanotrophic archaea at cold seeps. Appl. Environ. Microbiol. **71**: 467–479.

Majima, R., Nobuhara, T. and Kitazaki, T. (2005) Review of fossil chemosynthetic assemblages in Japan. Palaeogeogr. Palaeoclimatol. Palaeoecol. **227**: 86–123.

Michaelis, W., Seifert, R., Nauhaus, K., Treude, T., Thiel, V., Blumenberg, M., Knittel, K., Gieseke, A., Peterknecht, K., Pape, T., Boetius, A., Amann, R., Jorgensen, B.B., Widdel, F., Peckmann, J., Pimenov, N.V. and Gulin, M.B. (2002) Microbial reefs in the Black Sea fueled by anaerobic oxidation of methane. Science **297**: 1013–1015.

Nobuhara, T., Onda, D., Kikuchi, N., Kondo, Y., Matsubara, K., Amano, K., Jenkins, R.G., Hikida, Y. and Majima, R. (2008) Lithofacies and fossil assemblages of the Upper Cretaceous Sada Limestone, Shimanto City, Kochi Prefcture, Shikoku, Japan. Fossils **84**: 47–60. (Japanese with English abstract).

Ogihara, S. (2005) The evolution of chemosynthetic biological community at the site of cold-seep carbonate precipitation. Fossils **78**: 40–46. (Japanese with English abstract).

Orphan, V.J., House, C.H., Hinrichs, K.U., McKeegan, K.D., and DeLong, E.F. (2001) Methane-consuming archaea revealed by directly coupled isotopic and phylogenetic analysis. Science **293**: 484–487.

Peckmann, J. and Thiel, V. (2004) Carbon cycling at ancient methane-seeps. Chem. Geol. **205**: 443–467.

Riding, R. (2000) Microbial carbonates: the geological record of calcified bacterial-algal mats and biofilms. Sedimentology **47**: 179–214.

Sahling, H., Rickert, D., Lee, R.W., Linke, P. and Suess, E. (2002) Macrofaunal community structure and sulfide flux at gas hydrate deposits from the Cascadia convergent margin, NE Pacific. Mar. Ecol. Prog. Ser. **231**: 121–138.

Takahashi, A., Hikida, Y., Jenkins, R.G. and Tanabe, K. (2007) Stratigraphy and megafauna of the Upper Cretaceous Yezo Supergroup in the Teshionakagawa area, northern Hokkaido, Japan. Bull. Mikasa City Museum Nat. Sci. **11**: 25–59.

Takashima, R., Kawabe, F., Nishi, H., Moriya, K., Wani, R. and Ando, H. (2004) Geology and stratigraphy of forearc basin sediments in Hokkaido, Japan: Cretaceous environmental events on the north-west Pacific margin. Cretaceous Res. **25**: 365–390.

Teichert, B.M.A., Bohrmann, G. and Suess, E. (2005) Chemoherms on Hydrate ridge – unique microbially-mediated carbonate build-ups growing into the water column. Palaeogeogr. Palaeoclimatol. Palaeoecol. **227**: 67–85.

Warén, A. and Bouchet, P. (2001) Gastropoda and Monoplacophora from hydrothermal vents and seeps; new taxa and records. Veliger **44**: 116–231.

Whiticar, M. (1999) Carbon and hydrogen isotope systematics of bacterial formation and oxidation of methane. Chem. Geol. **161**: 291–314.

Biodata of **Dr. (Mrs.) Kapesa Lokho** and **Professor Vinod Chandra Tewari**, authors of *"Biostratigraphy, Sedimentation and Chemostratigraphy of the Tertiary Neotethys Sediments from the NE Himalaya, India"*

Dr. (Mrs.) Kapesa Lokho is currently working as Scientist-C in Wadia Institute of Himalayan Geology, Dehradun India. She obtained her doctoral degree in Micropaleontology from Nagaland University, Kohima, India, in 2005 and is continuing her research work in WIHG, Dehradun. Presently, she is working on Paleogene and Neogene sediments on foraminiferal biostratigraphy and paleo-environments of Indo-Myanmar ranges. She likes gardening and listening music in her leisure time.

E-mail: **kapesa@wihg.res.in**

V.C. Tewari and J. Seckbach (eds.), *STROMATOLITES: Interaction of Microbes with Sediments*, Cellular Origin, Life in Extreme Habitats and Astrobiology 18, 607–630
DOI 10.1007/978-94-007-0397-1_27, © Springer Science+Business Media B.V. 2011

Professor Vinod C. Tewari is currently the Head of the Sedimentology Group at Wadia Institute of Himalayan Geology, Dehradun and a Senior Associate of International Centre for Theoretical Physics, Trieste, Italy. He obtained his Ph.D. from the University of Lucknow in Geology in 1986 and continued his research in Wadia Institute. Dr. Tewari taught Geology at Kumaun Univerisity, Nainital, Uttarakhand (U.K.), India as Professor of Geology. Professor Tewari's scientific interests are in the areas of Precambrian stromatolites, sedimentation, carbon isotope chemostratigraphy, genesis, early evolution and diversification of life and its astrobiological significance. He has been associated with the International Geological Correlation Programme (I.G.C.P.) Projects on Stromatolites, Biosedimentology of Microbial Build-ups and currently on *The Rise and Fall of Vendian Biota*. He has eighty research papers published to his credit, and edited several volumes of *Himalayan Geology*, India and *Journal of Nepal Geological Society*, Kathmandu, Nepal. Professor Tewari has visited UCLA, CSEOL, USA in 2007 and 2008 to work on Laser Raman Spectroscopy with Professor J.W. Schopf. He organised first *Indo-Soviet Symposium on Stromatolites and Stromatolitic Deposits* and other IGCP meetings in India. He has been one of the members of the organising committee of the World Summit on Ancient Microscopic Fossils organised by Professor J.W. Schopf, held in University of California, Los Angeles, USA in 2008. He is a member of the Editorial Board of the international journal *Astrobiology*.

E-mail: **vtewari@wihg.res.in**

BIOSTRATIGRAPHY, SEDIMENTATION AND CHEMOSTRATIGRAPHY OF THE TERTIARY NEOTETHYS SEDIMENTS FROM THE NE HIMALAYA, INDIA

KAPESA LOKHO AND VINOD CHANDRA TEWARI
Wadia Institute of Himalayan Geology, Dehradun 248001, Uttarakhand, India

Keywords Biostratigraphy • Sedimentation • Chemostratigraphy • Tertiary • Neotethys • Assam • Arunachal Pradesh • Nagaland • Manipur • Meghalaya • NE Himalaya • India

1. Introduction

The Assam–Arakan Basin in the northeastern region of India (22°N–28°N: 90°E–96°E) is one of the largest sedimentary basins encompassing the states of Assam, part of Arunachal Pradesh, Meghalaya, Tripura, Nagaland, Manipur and Mizoram (Fig. 1). The exposed sedimentary sequence in the geosyncline ranges in age from Late Mesozoic to Cenozoic and the bulk of the succession comprises the Tertiary strata. The sediments are represented in two distinct shelf and basin lithofacies. The distinction is more pronounced in the Paleogene sequence than in the Neogene. The rocks of the shelf facies are exposed in the Garo Hills, Khasi–Jaintia Hills, part of North Cachar and Mikir Hills and also in the subsurface below the alluvium of Upper Assam. The basinal facies are developed in the Naga Hills, Manipur, Surma Valley of south Cachar, Tripura and Mizoram. Due to rarity of fossils, particularly microfossil markers in the region, lithostratigraphic classification formed the basis for correlation. The basin has drawn the attention of explorationists for search of oil ever since the discovery of petroleum in Assam at Digboi in the year 1889.

Occurrences of Neotethys Tertiary Basins and sediments in the Northeast India are described separately for each region.

(a) Arunachal Pradesh
(b) Assam Shelf
(c) Nagaland–Manipur
(d) Meghalaya
(e) Tripura–Mizoram Fold Belt

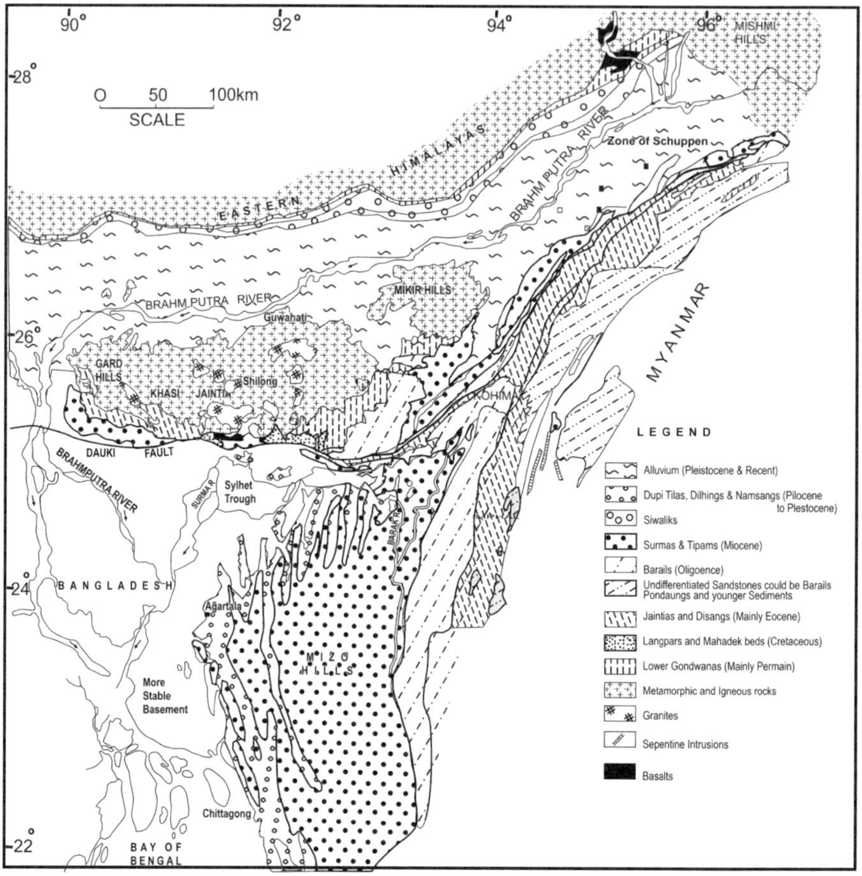

Figure 1. Geological map of Assam and surroundings, NE Himalaya, India showing occurrences of the fossiliferous formations (simplified after Das Gupta, 1977).

2. Arunachal Pradesh

Arunachal Pradesh comprises four geotectonic blocks, viz., (1) the Himalaya, (2) The Mishmi Hills, (3) Naga–Patkai ranges of the Arakan–Yoma mountain, (4) The Brahmaputra plain, each characterised by distinct stratigraphy and structure, and separated from each other by major tectonic lineaments, and therefore, have different geological history beginning from Proterozoic to Cenozoic (Kumar, 1997; Tewari, 1998). The oldest succession is of Paleoproterozoic consisting of Sela Group, Bomdila Group and granites. Mesoproterozoic to Neoproterozoic and Early Paleozoic sediments are mainly shallow marine quartzite–carbonate and conglomerate deposits (Tewari, 1998, 2001, 2002, 2003).

Tewari (2001) discovered microstromatolites and associated filamentous microbiota from the Buxa Dolomite well exposed in the Menga window, Subansiri valley. The new revised stratigraphic succession for the Arunachal Lesser Himalaya from Kameng in the west to Siang in the east has been proposed by Tewari (2002, 2003). Carbon isotope chemostratigraphy of the Buxa Dolomite has been established for the first time in Arunachal Lesser Himalaya, which is suggestive of Neoproterozoic marine sea water chemistry (Tewari, 2002, 2003, 2007). Late Neoproterozoic microfossils were described for the first time from the Buxa cherts by Shukla et al. (2006) from West Siang district and a global correlation of the microfossil assemblages has been discussed. Tewari (2009a) further described additional paleomicrobial community of the Neoproterozoic age. The significant assemblages described are 31 benthic and planktic forms. The cyanobacterial remains are *Huronisporapsilata, Eosynechoccusmoorei, Glenobotrydionaenigmatis, Myxococcoides minor, Oscillatoriopsis breviconvexa, O. robusta, Siphonophycus typicum, Obruchevella parva* and a variety of acritarchs. Stromatolites and microbiota have been studied from the Buxa Dolomite in Ranjit valley, Sikkim Lesser Himalaya by Tewari (2004a,b). Schopf et al. (2008) have done Laser Raman Spectroscopy (LRS) and Confocal Laser Scanning Microscopy (CLSM) of the micro-organisms for the first time and detected organic matter (kerogen) in the cellular structures. They have also discussed the astrobiological implications of the LRS and CLSM studies on Buxa micro-organisms. Early Permian marine Gondwana sediments were deposited under marginal marine and glacial conditions in the Arunachal and Sikkim Himalaya. Maithy et al. (2006) reported *Gangamopteris cyclopteroides, G. intermedia, Glossopteris angustifolia, G. communis, G. damudica, G. intermittens,* from the Bhareli Formation, West Kameng district, Arunachal Pradesh. Based on these Lower Gondwana plant fossils, Early Permian age to the Bhareli Formation has been suggested. After a major hiatus with a duration of 193 My, the transgressive Late Paleocene to Middle Eocene sediments of Yinkiong Group consisting of shallow marine sediments and volcanics were deposited. A hiatus of 22 My occurs between Bartonian and Chattian. Sediments of Siwalik Group deposited under fluvial conditions overlying Late Oligocene–Early Miocene igneous rocks (Raju, 2008).

Singh (1983) reported foraminifera *Nummulites atacicus, N. dandotica, N. daviesi, N. laminose, N. globulus, N. spinosa,* etc. and Prasad and Dey (1986) reported palynomorphs assemblage *Dandotiaspora, Cyathidites, Malayaeaspora, Foveotriletes, Spinainaperturites,* etc. from the Dalbuing Formation. Based on these findings, Early Eocene age is assigned to the Dalbuing Formation. The foraminiferal assemblage is indicative of shallow marine environment. Sinha et al. (1986) suggested the Eocene basin was initially marked by the freshwater regime, which gradually turned into a marine basin. Tripathi and Mamgain (1986) recorded *Nummulites cf. N. globules, Orbitolites cf. O. complanata* and *Discocyclina* sp. from Dalbuing Formation, assigned Early Eocene age and also compared it with the Subathu Group of the northwestern Himalaya.

3. Assam

The state of Assam is geologically divided into Mikir Hills, Dhansiri Valley and Upper Assam. The review of fossiliferous horizons, age and depositional environment is discussed separately.

3.1. MIKIR HILLS

The Shillong Plateau, Mikir Hills and the Assam Valley are an autochthons consisting of crystalline rocks partly covered by nearly horizontal or low dipping Tertiary and younger sediments (Mathur and Evans, 1964). The oldest rocks of the Mikir Hills (Fig. 1) are the Mikir Traps of Early Cretaceous age and are overlain by Tura Formation of Late Paleocene–Early Eocene with a hiatus of 51 My. Tura Formation is overlain by a succession of Sylhet Limestone deposited under shallow marine platform conditions. Whiso et al. (2006) reported *Planorotalites palmerae* from the Mikir Hills and suggest latest part of Early Eocene to Late Eocene age for the Sylhet Group of rocks. Kopilli Formation deposited under marginal marine to outer shelf, Barail Group deposited under marginal marine to non-marine followed by a major hiatus from Late Oligocene to Early Miocene, Bokabil Formation followed by Khoraghat Sandstone deposited under shallow inner shelf to marginal marine environment. They are overlain by a non-marine succession of Tipam Group, Namsang Formation, Dihing Formation and alluvium (Raju, 2008).

3.2. DHANSIRI VALLEY

Dhansiri Valley is a part of Upper Assam platform in the Assam–Arakan basin. The sedimentary fill in this shelf part of the basin is about 7,000 m (Murty, 1983). The oldest rocks are of Early Cretaceous and non-marine sediments. The succession from Late Paleocene to Holocene includes Tura, Sylhet, Kopili, Barail, Bokabil, Tipam and Girujan formations (Venkatachala and Berry, 1985). Sylhet Limestone was deposited under inner shelf. Lower Kopilli Formation contains uvigerinids suggestive of middle shelf condition. Upper part of Kopilli Formation is richly fossiliferous. The Barails in Tynyphe Valley yielded arenaceous foraminifera (Raju, 2008).

3.3. UPPER ASSAM

The Upper Assam shelf represents the platform part of Assam–Arakan geosyncline, encompassing an area of 57,000 km^2 consisting of Assam Valley including a part of the Himalayan frontal deep (Sahai, 1985). The oldest sediments reported are of Early Permian age. They are overlain by Tura Formation with a major hiatus of 193 Ma duration. Bardalaya et al. (2002) recognised three 3rd order sequences within Tura Formation and it is both source and reservoir rock for the hydrocarbons. Deb et al. (2007) recognised four unconformities suggestive of five sequences within the succession of Late Eocene to Pleistocene (Raju, 2008). The Disang shale, Kopili shale and shales of Barail coal-shale sequence are the main source rocks in the area south of Brahmaputra River.

4. Naga Hills (Nagaland and Manipur)

The Naga Hills comprise the Manipur and Nagaland states (Figs. 1 and 2), which form an integral part of the Indo-Myanmar Ranges (IMR) of Northeast India, have evolved as an accretionary prism due to subduction of the Indian plate below the Myanmar plate during the Alpine Himalayan tectogenesis (Desikachar, 1974; Acharyya et al., 1986). The oldest group of rocks, the metamorphic complex lies on the extreme eastern part of the states while the Ophiolite Melange Zone comprising the ophiolite suite and its associated sediments such

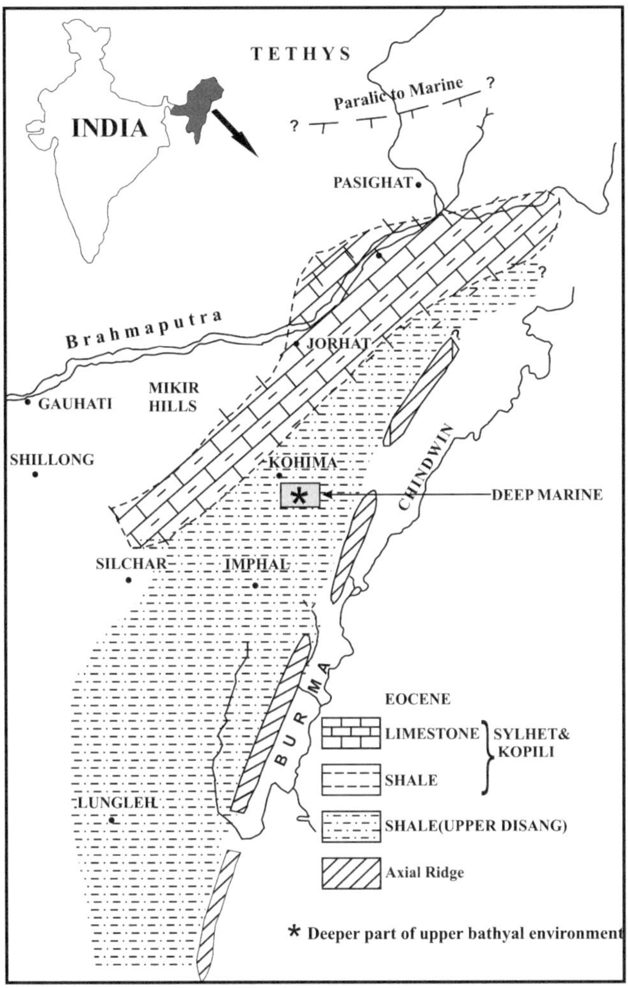

Figure 2. Reconstruction of paleogeographic map of Eocene, NE Himalaya (modified after Ranga Rao, 1983).

as chert, limestone, shale and sandstone of the Late Cretaceous age occupies the eastern part of the state. Chungkham and Jafar (1998) reported Santonian to Maastrichtian foraminifera and nannofossil assemblage from pelagic limestone of Melange Zone of Manipur Ophiolite Belt (MOB). The recovered foraminifers are *Abathomphalus mayaroensis, Contusotruncana contusa, Dicrinella asymetrica, Gansserina gansseri, Globotruncana aegyptica, G. ventricosa, Globotruncanita angulata, G. conica, G. elevata, Margiotruncana sinuosa, Pseudoguembelina excolata, Globotruncana arca, G. ganssiri, G. staurti* and *Heterohelix globulosa.* Lower Disang consists of low-grade metamorphics and Upper Disangs are significant from the view point of hydrocarbon exploration in the western part of Nagaland. Lokho et al. (2004b) reported *Hantkenina alabamensis*', *Hantkenina liebusi, Cribrohantkenina inflata, Pseudohastigerina naguewichiensis, P. micra, P. barbadoensis, Turborotalia cerroazulensis pomeroli, T. cerroazulensis cocoaensis, T.c. cerroazulensis, Chiloguembelina martini, C. cubensis, C.* cf. *tenuis, Chiloguembelina* sp., *Globigerinatheka semiinvoluta, Globigerina* sp., *Nummulites chavannesis, N. pengaronensis, Baggina cojimarensis, B. dominicana, B. dentate, B. subinaequalis, Cibicidoides* sp., *Cancris mauryae, Nonionella* sp., *Cibicidoides* sp., *Cyclammina* sp., *Pseudonodosaria* sp., *Cibicides* sp., *Lagena acutiscosta var proboscidialis, Lagena striata, Lagena* sp., *L. sulcata,* var. *spicata, Lenticulina* sp., *Pyrgo* sp., *Praebulimina* sp., *Osangularia* sp., *O. plummerae, Bolivina* sp., *Rectobolivina* sp., *Dentalinoides* sp., *Elphidiella* sp., *Miliola* sp., *Triloculina* sp., *Turrilina robertsi, Turrilina* sp., *Parafissurina* sp., *Uvigerina continuosa, U. cf. eocaena, U. longa, U.* cf. *steyeri, U. cocoaensis, U. Moravia, U. vicksburgensis, U.* cf. *steyeri, U. jacksonensis, U. glabrans* from the Upper Disang Formation exposed in and around Pfutsero of southcentral Nagaland. A collection of fossil pteropods, including some unidentified species, provisionally referable to the families Limacinidae, Creseidae and Clioidae is reported from the late Middle Eocene–Late Eocene beds of the Upper Disang Formation exposed near the town of Pfutsero, Phek District, South Central Nagaland, Assam– Arakan Basin, northeastern India (Lokho and Kumar, 2008; Fig. 3).

In the south central part of Nagaland, the Upper Disangs were deposited under upper bathyal indicating anoxic condition with the findings of uvigerinids and pteropods with its pyritised tests (Lokho et al., 2004a; Lokho and Kumar, 2008). Baruah et al. (1987) reported *Nummulites pengaronensis, N. discorbinus, Discocyclina dispansa, D. eamesi, Pellatispira madaraszi, P. inflata, Hantkenina alabamensis, Globorotalia centralis, Globorotalia cerroazoalensis, Globigeriniata ampliapertura, Pseudohastigerina barbadoensis,* etc. from near Heningkunglwa and Lotsu village, SW of Dimapur, Nagaland. Based on these findings, Barail Group is suggested as Late Oligocene age and deposited under brackish water environment of deposition for the Barails. The Barails are unconformably overlain by the Surma Group of Miocene age with transitional characters from flysch to molasses sediments (Ibotombi, 1998). The Surma Group is followed by the Tipam, Girujan, Namsang, Dihing Formations and alluvium all deposited under non-marine condition during Late Miocene to Holocene (Fig. 10).

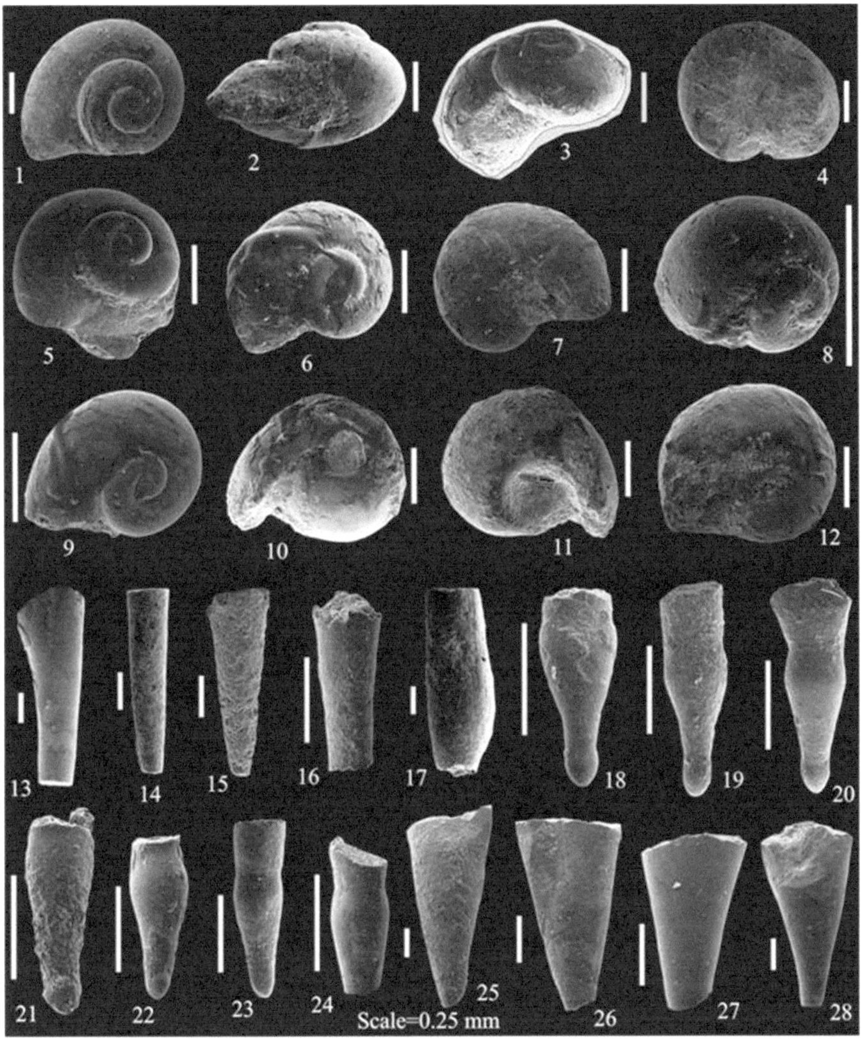

Figure 3. Fossil Eocene pteropods from the NE Himalaya, India (Lokho and Kumar, 2008).

5. Meghalaya (Shillong Plateau)

The northeastern extension of Indian Peninsular shield is represented by Shillong Plateau of Meghalaya (Fig. 5). The Plateau shows an extensive development of Cretaceous-Eocene sedimentary rock sequence in its southern part (Evans, 1932). As a result of tectonics in the Himalayan and in the adjoining Arakan-Yoma region, the Shillong Plateau occurs as an uplifted part of the basement bounded by nearly east–west aligned normal faults towards the northern and southern boundaries

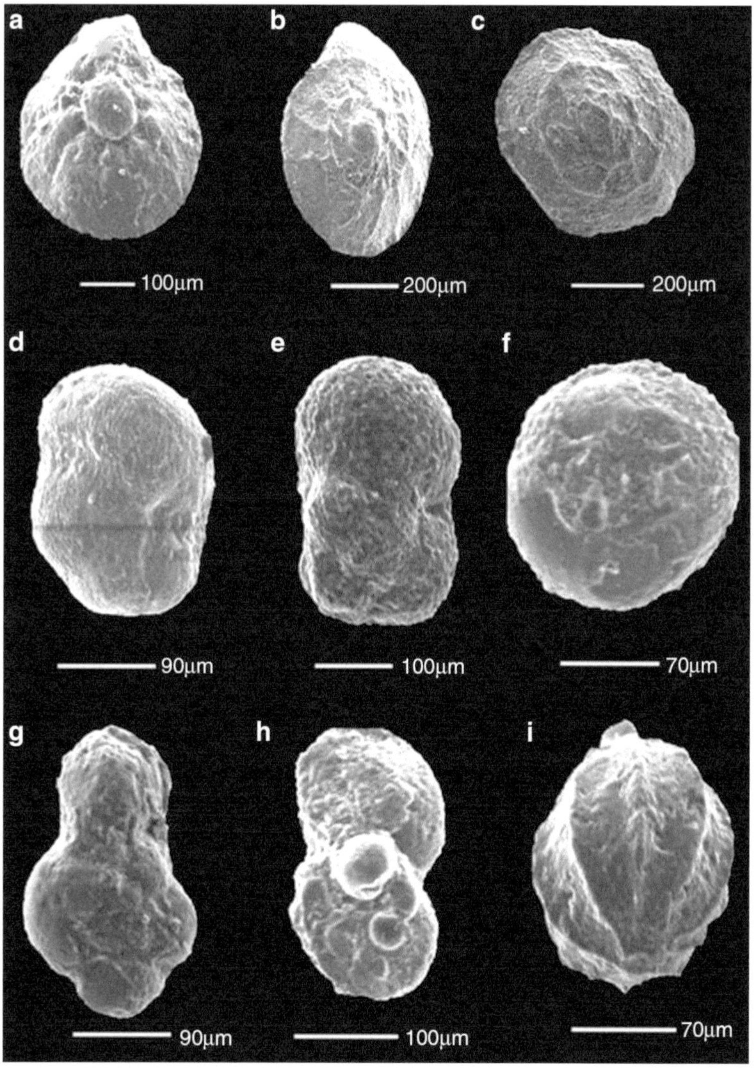

Figure 4. Miocene foraminifera from Bhuban Formation, Mizoram (Lokho and Raju, 2008).

of the Plateau. Cenozoic transgression in the Shillong Plateau began in the Danian (Nagappa, 1959; Garg and Jain, 1995), which was followed by a very large transgression during the Late Paleocene when an arm of sea from the western Tethys covered extensively the Shillong Plateau via the Middle East, Pakistan and Tibet (Jauhri et al., 2006; Tewari et al., 2008, 2010). In the Northeast India, K/T boundary is delineated at Um Sohryngkew river section (Pandey, 1980, 1990; Bhandari et al., 1987;

Garg and Jain, 1995; Tewari, 2004c, 2009b; Tewari et al., 2007). Iridium enrichment at Cretaceous/Tertiary boundary in Meghalaya (Pandey, 1980; Bhandari et al., 1987) suggest Maastrichtian age to the Mahadeo Formation and Paleocene age to the Langpar Formation based on the foraminiferal assemblages in Um Sohryngkew river section. Kak and Subramanyam (2002) on the basis of field, petrography and textural studies show that Lower Mahadeo (Mahadek) sediments are of fluvial and Upper Mahadek sediments are of marine origin in the Mahadeo (Mahadek) Formation of Wahblei river section, West Khasi Hills, Meghalaya. Jauhri et al. (2006) identified two facies associations in the carbonate sequence. (1) Facies A characterising the lower 50 m part and (2) Facies B characterising the upper 50 m part. The facies A corresponds to the Thanetian sedimentation cycle in which deposition occurred in a relatively low-energy environment on a shallow subtidal ramp (protected lagoon). The Facies B correlates with the Ilerdian sedimentation cycle during which deposition occurred under low-to-moderate energy conditions on a relatively deeper mid-uppermost outer ramp, which allowed development of sediment-binding coralline algae and foraminifera such as discocyclinids and *Ranikothalia* sp.

Tewari et al. (2008) carried out carbon and oxygen isotope analysis of Paleocene algal–foraminiferal Lakadong Limestone samples from Mawsmai and Mawmluh section profiles for the first time (Fig. 9). The C and O isotope ratios and algal–foraminiferal assemblage suggest shallow marine carbonate build-up depositional environment for the Lakadong Limestone. Tewari et al. (2010) have also carried out a detailed microfacies analysis of the Lakadong Limestone and have interpreted that they are tidal flat deposits (Figs. 6, 8).

6. Cretaceous–Paleogene Boundary in the Shillong Plateau Meghalaya

The Cretaceous–Paleogene and Paleocene–Eocene boundaries represent global catastrophic and palaeoclimatic events well documented by the bolide impact evidence and faunal extinction. Tewari (2009b) and Tewari et al. (2007) have discussed this global event in detail. The Um Sohryngkew River and other sections in Shillong Plateau, Meghalaya, northeast India are very promising areas to study these events in India for regional and global correlations. Several potential Cretaceous–Paleogene and Paleocene–Eocene boundary sections such as Cherrapunji (Sohra)–Shillong Road section, Mawmluh Limestone Quarry section, Komorrah Limestone Quarry section, Shella River section, Mawsmai-Shella Road section, Mawlong–Mashmak–Sohra Road section, Ranikaur–Dirang road section and Mawsynram section have been studied in the south Shillong Plateau in parts of East and West Khasi Hill districts, Meghalaya. Coevals of some of these sections were also studied in West Khasi Hills for vertebrate and other biotas. Particular emphasis was laid on Late Cretaceous–Paleocene sediments, which are represented by the Mahadeo, Langpar and Lakadong Formations in ascending order. The sequences are rich in well preserved foraminifers, echinoids, molluscs, plant fossils, trace fossils, etc. Among the vertebrate fossils, some dinosaur remains in the Mahadeo Formation and fragmentary fish scales in the Lakadong Formation have been collected. Marine transgressive and continental environment was recognised in the Mahadeo Formation during Late Cretaceous period. Cretaceous anoxic event (reducing environment) is represented by the black shale–pyrite and fossil wood found in the Mahadeo Formation. The lithostratigraphy, fossil content and age of the sedimentary formations of the south Shillong Plateau, Meghalaya is shown in Table 1. Petrographic thin sections of the Lakadong Limestone from Mawsynram and Mawmluh sections were studied for the microfacies analysis of the algal–foraminiferal limestone. The Lakadong Limestone is mainly composed of calcite, dolomite and smaller and larger microfossils. The X-ray diffraction has confirmed the presence of dominantly carbonate minerals calcite and dolomite; however, minor peaks of quartz, mica and clay minerals have been recognised. The microfacies identified include micrites, intramicrites, sparite, oosparite and intrasparite. The calcareous algal–foraminiferal assemblage recognised (Fig. 7) include coralline algae *Lithophyllum, Dasycladaceae* and *Miscellania* sp., *Rotalia* sp., *Ranikothalia* sp., *Miliolids, Dentaloides, Textularia* sp., *Alveolina* sp., *Discocyclina* sp., *Nummulites* foraminifera. The Palaeocene–Lower Eocene age is indicated by these microfossil assemblages. The limestone was deposited in shallow marine conditions and tidal influence is recorded since reworked clasts, oolites, intraclasts, broken shell fragments are intimately associated with algae and foraminifera. Algal facies indicate deposition in carbonate reef environment.

The carbon and oxygen isotopic analysis of some representative Paleocene algal–foraminiferal Lakadong Limestone samples from Mawsmai and Mawmluh section profiles was carried out for the first time. Figure 2 shows that the $\delta^{13}C$ ratio vary from −0.05‰ (PDB) to +1.83 ‰ (PDB) and $\delta^{18}O$ ratio vary from 16.62‰

Table 1. Lithostratigraphy, fossil content and age of the sedimentary formations in the south Shillong Plateau, Meghalaya (Tewari et al., 2008).

Lithounits/formations		**Dominant lithology**	**Fossil content**	**Age**
Kopili formation		Shale, sandstone, marl	Foraminifera	Late Eocene
	Prang limestone	Limestone	Foraminifera	Middle–Late Eocene
Sylhet	Nurpuh sandstone	Hard sandstone	Foraminifera?	Middle Eocene
Limestone	Umlatodoh limestone	Limestone	Foraminifera	Middle Eocene
	Lakadong sandstone	Soft friable sandstone	Plant impressions	Early Eocene
	Lakadong limestone	Dolomitic and pure limestone	Foraminifera, Algae, trace fossils	Late Paleocene–Early Eocene
Therria formation		Coarse sandstone	Plant impressions	Late Paleocene
Langpar formation		Shale and limestone	Foraminifera	Early Paleocene
Mahadeo formation		Sandstone	Foraminifera, plant remains, dinosaur bones, trace fossils	Late Cretaceous
Bottom formation				Late Cretaceous
Jadukuta formation				Late Cretaceous
Sylhet traps		Volcanics		Cretaceous–Jurassic

(SMOW) to 23.49‰ (SMOW). The C and O isotope ratios and algal–foraminiferal assemblage suggest shallow marine carbonate build-up depositional environment for the Lakadong Limestone. The sedimentary facies, trace fossils, microfossils and carbon isotope profiles from the Southern Alps in Italy (Western Adriatic Tethys Platform) is tentatively correlatable with the South Shillong Plateau (Tewari, 2009b; Tewari et al., 2009).

7. Paleocene Eocene Thermal Maxima

Prasad et al. (2007) carried out palynofacies studies for Paleocene Eocene Thermal Maxima (PETM) from two shallow marine successions encompassing the Paleocene–Eocene transition from Jathang and Cherrapunji areas of Khasi Hills, Meghalaya. The studies conclude that high sediment supply as a result of excessive warm and humid climate of PETM in low equatorial zone during slowly rising sea level resulted in the conversion of carbonate ramp into clastic dominated coastal marine setup with extensive development of freshwater marshes brackish lagoons and estuarine and bay fill deposits within a transgressive system tract. Carbon isotopic studies are in progress for the PETM excursion in the NE region. A global paleogeographic map (Fig. 12) of the world shows the distribution of the Late Cretaceous–Eocene marine deposits.

8. Mizoram

Mizoram is a part of the Neogene Surma Basin comprising a belt of elongated folds having submeridonal trend and arcuate shape with westward convexity (Figs. 1 and 2). The fold belt is elongated in the N–S direction almost parallel to the suture zone of the Arakan-Yoma subduction (Karunakaran, 1974; Ganguly, 1975, 1983; Nandy, 1982; Nandy et al., 1983). The Hills comprises an uninterrupted succession of the order of 5,000 m of Tertiary clastic sediments, consisting of a hybrid association of arenaceous and argillaceous fractions. The oldest outcrops belong to the Surma Group (Miocene) that comprises mainly alternating shale/clay stone with sandstones and siltstones. The Surma Group is divided into the Bhuban and Bokabil formations. On the basis of landsat imagery, five lithostratigraphic units, i.e. Lower Bhuban, Middle Bhuban, Upper Bhuban, Bokabil and Tipam are identifiable in the Neogene. The Upper Bhuban beds occur in the anticlinal part along the western Mizoram but in the central and eastern Mizoram they are confined to synclinal lows. Jauhri et al. (2003) reported two species of foraminifera viz. *Pseudotaberina malabarica* and *Borelis pygmaeus* from Upper Bhuban Formation of Mizoram. Lokho and Raju (2007) reported the foraminiferal assemblage of *Ammonia umbonata*, *Baggina* sp., *Clavatorella sturanii*, *Lagena* sp., *Orbulina bilobata, Praeorbulina glomerosa circularis* and *Praeorbulina transitoria* and placed them in the planktonic foraminiferal Zone N9 (Fig. 4). These data

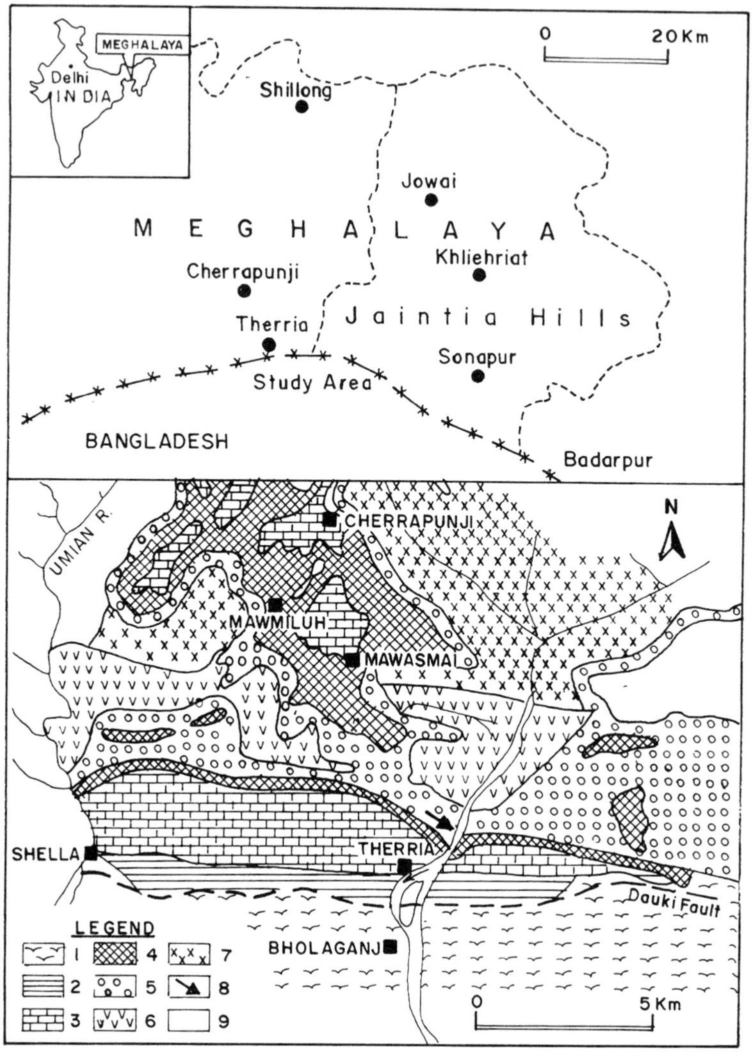

Figure 5. Geological map of Shillong Plateau, Meghalaya, NE India.

suggest that two maximum flooding surfaces (MFSs) can be recognised within Upper Bhuban. Tiwari et al. (2007) delineated a total of seven Normals and seven reverse magneto zones in Middle Bhuban and suggested an age of 21.77 Ma to 15.16 Ma. These ages correspond to Middle Aquitanian to Earliest Langhian. Available data from planktonic foraminifera and magnetostratigraphy relate well with standard stages. Singh (2007) reported *P. escornebovenesis, A. umbonata,*

Figure 6. Field photograph of the Lakadong Sand stone, Mawmluh Quarry, Meghalaya, NE India.

Cibicides sp., *Anomalinoides* sp., *Elphidium* sp., costate *uvigerina* sp., *Austrotrillina* fragments, *Globigerina praebulloides,* etc. The findings suggest a Burdigalian age for the Upper Bhuban member depositing under an inner to middle shelf paleoenvironment condition.

9. Tripura–Cachar Fold Belt

The Tripura–Cachar Fold belt represents the frontal fold belt of the Assam–Arakan Basin along the convergent Indo-Burmese plate margin. This area constitutes the Surma sub-basin of the Assam–Arakan Basin. The belt comprises long linear tight N–S trending anticlines separated by broad synclinal troughs (Ram and Venkataraman, 1984). Subsequent subduction of Indian plate margins underneath the Burmese plate to the southeast formed a trench for deposition of alternating fine-to-medium grain size siliciclastic sediments, which range in age from Eocene to Recent (Singh et al., 2007).

In Tripura, the outcrop section consists of Barail of Oligocene age, Bhuban, Bokabil of Surma Group, Jaipur and Gobindpur Formation of Tipam, Dupitila and Dihing formations.

Bhatia (1996) studied a detailed foraminiferal biostratigraphy scheme for Tripura–Cachar region and led to the recognition of a marine biozones and six

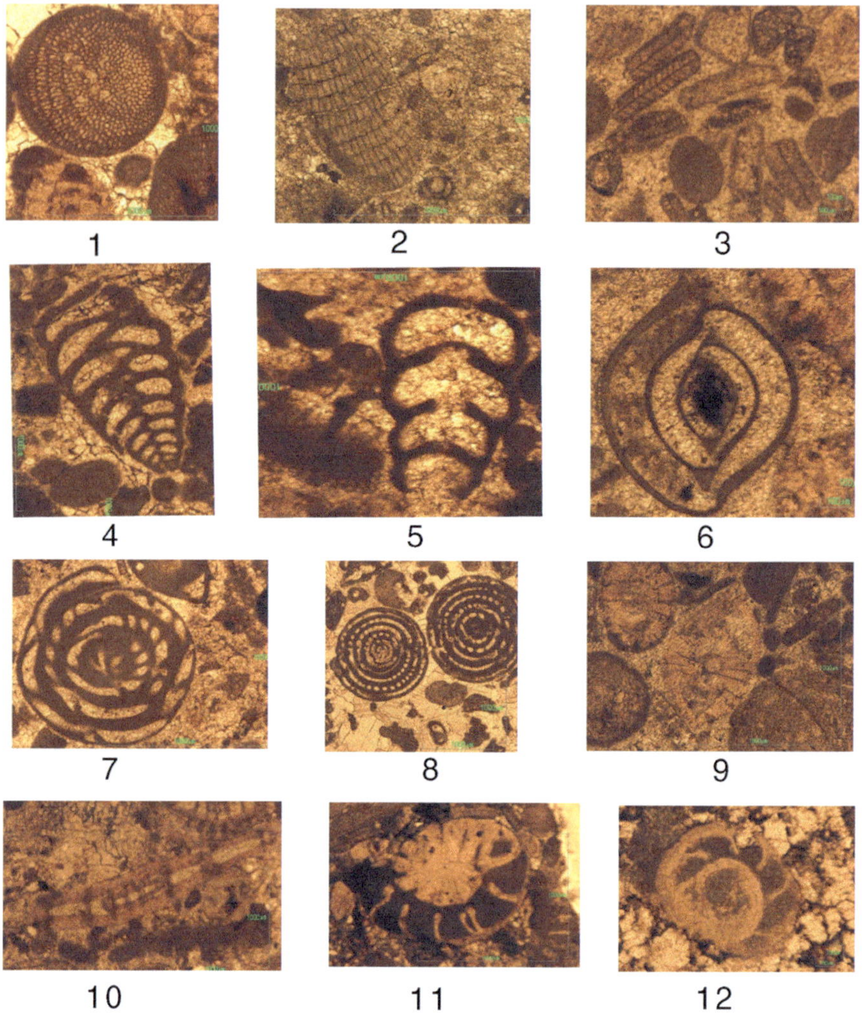

Figure 7. (1, 2) *Corallina* algae, MQ-2, MQ-3; (3) *Distichoplax* sp., MQ-12; (4, 5) *Textularia* sp.?, MQ-9,4; (6) *Spiroloculina*? sp., MQ-0; (7, 8) *Nummulites*, MQ-9,4; (9) *Miscellanea* sp., MQ-0; (10) *Assilina* sp., MQ-5; (11) *Rotalia* sp., MQ-14; (12) *Operculina* sp., MQ-12 (Tewari et al., 2010).

subzones ranging in age from Late Oligocene to Early Pliocene. Kale et al. (2007) built a Mio-Pliocene stratigraphy of Tripura. They recognised four 3rd/4th order sequences based on primary sequence boundary and surfaces of major onlap identified on seismic data constraint with seismic, wireline logs, paleontological, palynological and sedimentological inputs. Das (2003) recognised six genetic sequences in Middle, Upper Bhuban and Bokabil formations. These sequences are bounded by MFSs. Baruah et al. (1996) reported *Globorotalia praebulloides,*

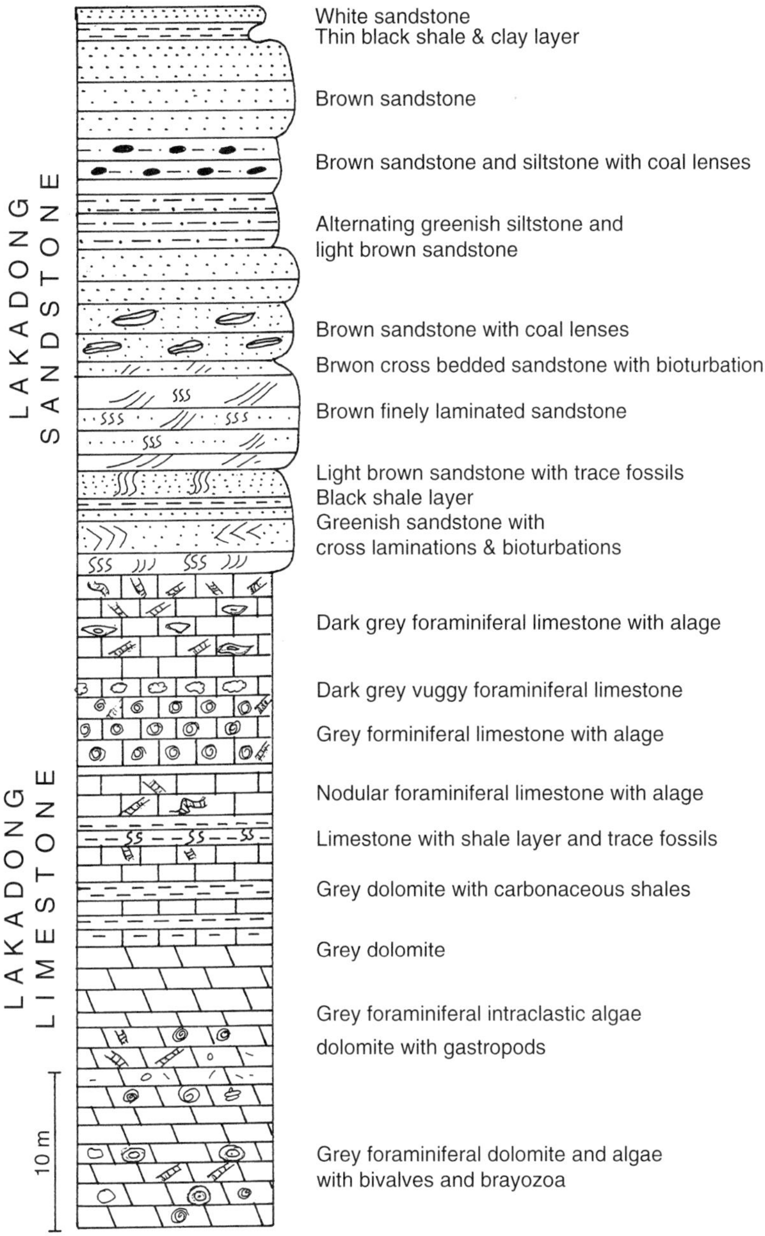

Figure 8. Litho column of the Lakadong Limestone and Sandstone, Meghalaya (Tewari et al., 2010).

G. obesa, G. peripheronda, G. siakensis, Neogloboquadrina continuosa, Globogerina falconensis, Globigerinoides immaturus, G. trilobus from the Middle Bhuban Formation and a species of *Globigerinoides* from Upper Bhuban unit at Well

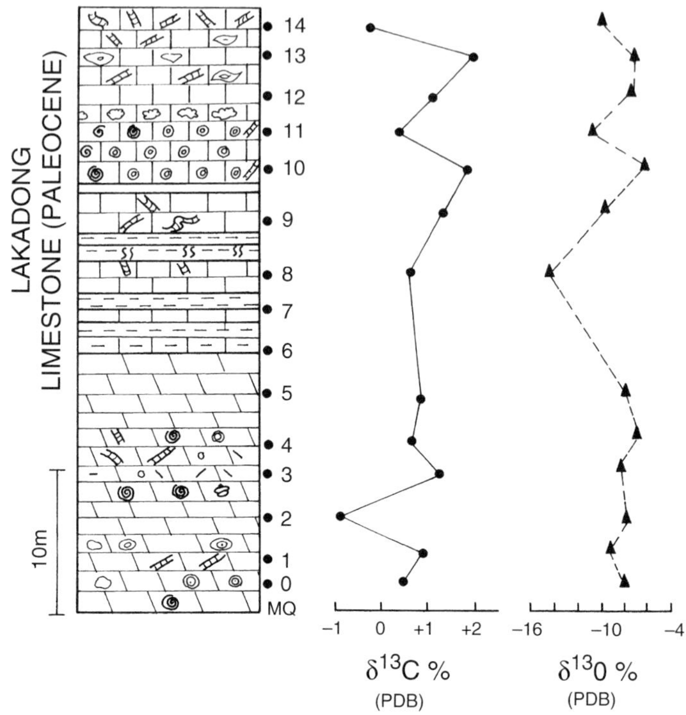

Figure 9. Carbon and oxygen isotope chemostratigraphy of Lakadong Limestone (Tewari et al., 2010).

Bhubandar-A from Cachar area, Assam. Ramesh (2004) reported abraded benthic foraminifera of *Planulina* sp., *Uvigerina flinti, Lenticulina* sp., *Baggina* sp., *Ammonia umbonata, Roephax, Haplophragmoides suborbicularis, H. stomatus, Karrerulina conica and Trochamminoides folius* from lower Bhubans unit at well Badarpur-L (BP-L). He also reported foraminifera of *Globorotalia mayeri, Gl. continusa, Globogerinoides bisphericus, Globoqudrina altispira, Globigerina praebulloides, Lenticulina,* rare *Uvigerina javana, Baggina, Recurvoides, Cyclammina, Bathysiphon, Gerochammina, Reophax* and radiolarian at well Badarpur-L (BP-L) from Middle Bhuban unit from subsurface of Cachar area, Assam. Baruah et al. (1996) reported *Globorotalia obesa, G.* cf. *acostaensis, Globigerinoides trilobus* and *Streptochilus globigerum* from upper part of Bokabil Formation at Patharia-A in Cachar area in Assam. They described fluvio-deltaic for most of the Bokabil time. The sea receded gradually in the top part of the Bokabil Formation from the entire Cachar area. Judging from the lateral facies variations in the marine sequences in different wells, they suggest that all the marine incursions encroached from west (Figs. 10–12).

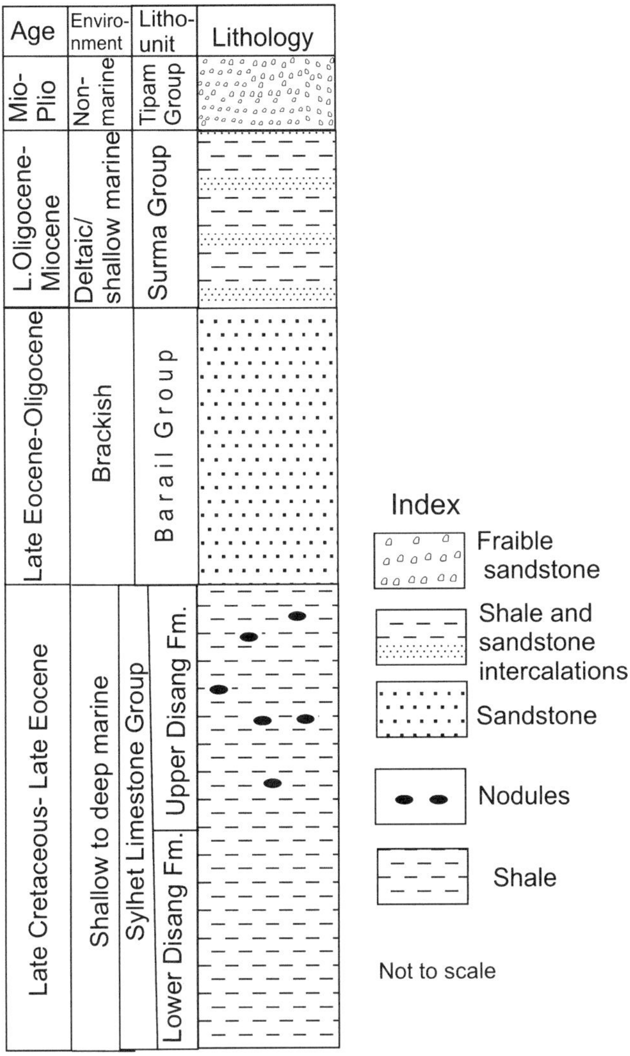

Figure 10. Generalised lithostratigraphy (not to scale) of Paleogene and Neogene sediments of Northeast India.

10. Discussion and Conclusions

The Northeast comprises two major segments, a shelf region bordering the Shillong Plateau and Mikir Hills (Fig. 1). This shelf may be divided into three segments – South Shillong, Mikir Hills and Upper Assam (Figs. 1 and 2). The Shelf is bordered in the south by the Dauki Fault (Fig. 5) and in the east by the Naga Thrust (Fig. 1) is often referred as Schuppen Belt made up of imbricate thrusts.

Formations	Series	Age	Bio-zones	South central Nagaland N.E. India Present work	Western Nagaland N.E. India Baruah et al., 1987	Assam, N.E. India Samanta 1973	Cauvery Basin, S. India Raju, 1971	Rakhi Nala, Pakistan Samanta, 1973.	Tropical/subtropical areas: Blow, 1969 Berggren & Couvering (1974)	Tropical/subtropical areas: Blow, 1979
Disang Formation (upper)	Late Eocene	Priabonian	*Turborotalia cerroazulensis* zone P17	*Turborotalia cerroazulensis*	*N. pengaroenensis* -*P. madaraszi* -*D. dispansa*	*Globigerina gortanii*	*Globorotalia cerroazulensis*		*Globigerina gortanii/ Globorotalia centralis*	*Globigerina gortanii gortanii/ Goborotalia T. centralis*
			Cribrohantkenina inflata Zone P16	*Cribrohantkenina inflata*	*C. inflata* -*H. alabamensis*	*Cribrohantkenina inflata*		*Globigerina officinalis*	*Cribrohantke-nina inflata*	*Cribrohantk-enina inflata*
			Globigerinatheka semiinvoluta Zone P15	*G.semiinvoluta*		*Globigerapsis semiinvoluta*	*Globigerapsis mexicana* = (*G. semiinvoluta*)		*Globigerapsis mexicana* = (*G. semiinvoluta*)	*P. Semiin-volutus*

Figure 11. Correlation of foraminiferal zones (Venkatachalapathy et al., 2007).

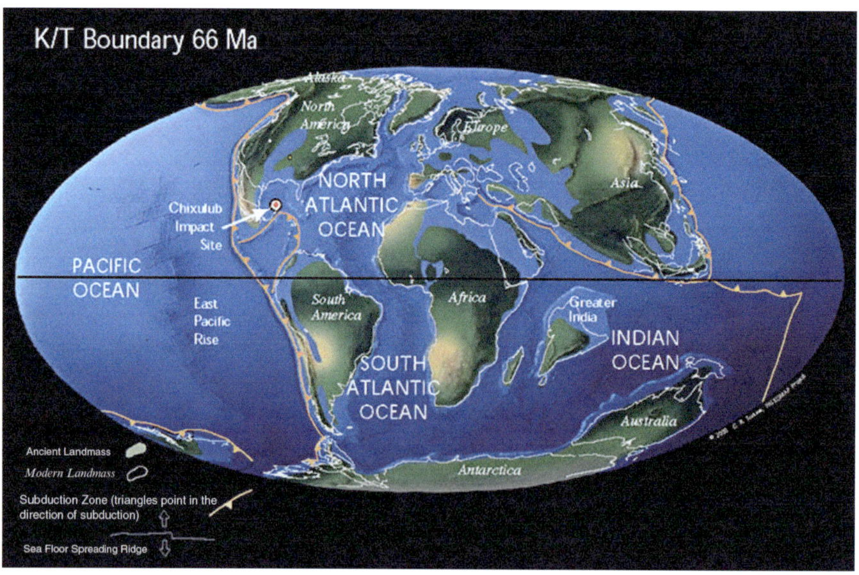

Figure 12. Global palaeogeography of the Cretaceous-Tertiary boundary (http://www.scotese.com).

The entire Schuppen Belt is 25–30 km wide. To the east of Schuppen Belt lies the Nagaland, which is continuous with the similarly folded and faulted sedimentaries of Manipur (Figs. 1 and 2). To the southeast and west of this fold belt, it continues in Tripura where the folded and faulted sediments of mostly of Neogene age are recorded. In the Cachar–Tripura area, sediments ranging in age from Oligocene to Pleistocene are found (Pandey and Dave, 1998). The stratigraphic successions are described with special emphasis to age, environment of deposition and fossils in this chapter to give a general idea about the shelf in the NE region. Assam–Arakan region falls in the Category I basins of the Oil and Natural Gas Corporation (ONGC, Dehradun, India, Pandey and Dave, 1998). The biostratigraphy of the Tethyan sediments has been based on rich fossil assemblages. Recently, Lokho et al. (2004a) have prepared detailed biostratigraphic correlation charts for the northeastern fossiliferous basins. Tewari (2009b) and Tewari et al. (2008, 2009) have interpreted that the Paleocene–Eocene sediments are shallow marine (tidal flat) based on carbon isotope ratios in Shillong Plateau, Meghalaya. However, the hydrocarbon exploration in the Tethyan Himalayan zone is difficult since there is absence of long anticlinal features expected to be petroliferous. The steep heights and inaccessibility of the regions prevents the oil exploration in the Himalayan Tethyan basins.

11. Acknowledgements

We thank the Director, Wadia Institute of Himalayan Geology, Dehradun, Uttarakhand for encouragement and permission to publish this chapter for this book. Dr. D.S.N. Raju (O.N.G.C. Consultant, Dehradun) has kindly reviewed the chapter and we are indebted to him for his valuable suggestions and comments for the improvement of the chapter. The anonymous referee is thanked for the critical comments. V.C. Tewari is grateful to Professor Katica Drobne, Ljubljana, Slovenia for discussions.

12. References

Acharyya, S.K., Roy, D.K. and Mitra, N.D. (1986) Stratigraphy and palaeontology of the Naga Hills Ophiolite Belt. Mem. Geol. Surv. India **119**: 64–74.

Bardalaya, J., Sur, P., Prasadhan, U.C. and Goel, S.M. (2002) Sequence stratigraphy and hydrocarbon potential of Tura Formation in Upper Assam: a promising play for the future. Proceedings of 1st Conference and Exhibition on Strategic Challenges Basins 1, pp. 417–422.

Baruah, R.M., Singh, N.P. and Rao, D.C. (1987) Foraminiferal biostratigraphy of Disang and Barail Groups of a part of Nagaland. Proc. Natl. Sem. Tert. Orogeny 305–327.

Baruah, R.M., Datta, K. and Murthy, M.S. (1996) Biofacies, lithofacies and depositional environment of subsurface Surma sediments in Cachar area, Assam. Proceedings XV Indian Colloquium of Micropaleontology and Stratigraphy, Dehradun, pp. 305–316.

Bhandari, N., Shukla, P.N. and Pandey, J. (1987) Iridium enrichment at Cretaceous/Tertiary Boundary in Meghalaya. Curr. Sci. **56**(19): 1003–1005.

Bhatia, M.L. (1996) A palaeontological approach to the stratigraphic investigation of Neogene sediments of Tripura- Cachar region. Proceedings of XV Indian Colloquium of Micropaleontology and Stratigraphy, Dehradun, pp. 283–294.

Chungkham, P. and Jafar, S.A. (1998) Late Cretaceous (Santonian Maastrichtian) integrated Coccolith–Globotruncanid biostratigraphy of pelagic limestone from the accretionary prism of Manipur, Northeastern India. Micropalaeontology **44**(1): 69–83.

Das Gupta, A.B. (1977) Geology of Assam-Arakan Region. Quart. Jour. Geol. Met. Soc. Ind., **49**: 1–50.

Das, S.K. (2003) Genetic sequences, stratigraphic analysis of Surma Group of sediments in Western Tripura. Proceedings Petrotech, pp. 8–12.

Deb, S.S., Baruah, L., Bharti, B., Boroghohain, R. and Talukdar, T.N. (2007) Tectonics, depositional setting and reservoir characteristics of sedimentary sequence in North bank of river Brahmaputra of Upper Assam Basin. APG.Bull. **1**: 27–33.

Desikachar, S.V. (1974) A review of the tectonics and geologic history of Eastern India in terms of Plate Tectonics theory. J. Geol. Soc. India **15**: 137–149.

Evans, P. (1932) Tertiary succession in Assam. Trans. Min. Geol. Inst. India **27**: 155.

Ganguly, S. (1975) Tectonic evolution of Mizo Hills. Bull. Geol. Min. Met. Soc. India **48**: 28–40.

Ganguly, S. (1983) Geology and hydrocarbon prospects of Tripura – Cachar – Mizoram region. J. Petrol. Asia **6**(4): 105–109.

Garg, R. and Jain, K.P. (1995) Significance of the terminal Cretaceous calcareous nannofossil marker *Micula princii* at the Cretaceous-Tertiary boundary in the Um Sohryngkew River section, Meghalaya, India. Curr. Sci. **66**(12): 1012–1017.

Ibotombi, S. (1998) On the geology of Manipur. Ninth Manipur Science Congress, 12–17.

Jauhri, A.K., Mandaokar, B.D., Mehrotra, R.C., Tiwari, R.P. and Singh, A.P. (2003) Corals and foraminifera from the Miocene (Upper Bhuban Formation) of Mizoram, India. J. Pal. Soc. India **48**: 135–138.

Jauhri, A.K., Misra, P.K., Kishore, S. and Singh, S.K. (2006) Larger foraminifera and calcareous Algal facies in the Lakadong Formation of the South Shillong Plateau. J. Pal. Soc. India **51**(2): 51–61.

Kak, S.N. and Subramanyam, A.V. (2002) Depositional environment and age of Mahadek Formation of Wahblei River section, West Khasi Hills, Meghalaya. J. Geol. Soc. India **60**: 151–162.

Kale, A.S., Sinha, P.K., Agarwalla, R.C., Ray, G.K. and Baruach, R.M. (2007) Miocene–Pliocene Sequence stratigraphy of Tripura, India and its implication on hydrocarbon exploration. APG Bull. **1**: 45–81.

Karunakaran, K. (1974) Geology and Mineral resources of the Northeastern states of India. Misc. Publ. Geol. Surv. India **30**(4): 93–101.

Kumar, G. (1997) Geology of Arunachal Pradesh. Geological Society of India, Bangalore, pp. 1–217.

Lokho, K. and Kumar, K. (2008) Fossil Pteropods (Thecosomata holoplanktonic Mollusca) from The Eocene of Assam – Arakan Basin, northeastern India. Curr. Sci. **94**(5): 647–652.

Lokho, K. and Raju, D.S.N. (2007) Langhian (early Middle Miocene) foraminiferal assemblage from Bhuban Formation, Mizoram, NE India. J. Geol. Soc. India **70**: 933–938.

Lokho, K., Raju, D.S.N., Kumar, G. and Venkatachalapathy, R. (2004a) Stratigraphic tables for North-east Basins of India: with brief notes compiled by D.S.N. Raju. Indian J. Petrol. Geol. **13**(1): 1–7.

Lokho, K., Venkatachalapathy, R. and Raju, D.S.N. (2004b) Uvigerinids and associated foraminifera and their value as direct evidence for shelf and deep marine paleoenvironments during Upper Disang of Nagaland, Eastern Himalaya and its implication in hydrocarbon exploration. Indian J. Petrol. Geol. **13**(1): 1–7.

Maithy, P.K., Bindal, C.M., Bhusan, S.K. and Sharma, S. (2006) Lower Gondwana plant fossils from Arunachal Lesser Himalaya and their age. J. Geol. Soc. India **68**: 316–326.

Mathur, I.P. and Evans, P. (1964) Oil in India. International Geological Congress, 22nd Session, New Delhi, 58 p.

Murty, K.N. (1983) Geology and hydrocarbon prospects of Assam Shelf: recent advances and present status. Petrol. Asia J. **1**: 1–14.

Nagappa, Y. (1959) Foraminiferal biostratigraphy of the Cretaceous–Eocene succession in the India–Pakistan –Burma region. Micropalaeontology **5**(2): 145–192.

Nandy, D.R. (1982) Geological set up of the Eastern Himalaya and the Patkai – Naga – Arakan Yoma (India–Burma) Hill Ranges in relation to the Indian Plate movement. Misc. Publ. Geol. Surv. India **41**: 205–213.

Nandy, D.R., Gupta, S.D., Sarkar, K. and Ganguly, S. (1983) Tectonic evolition of Tripura–Mizoram Fold Belt, Surma Basin Nort East India. Q. J. Geol. Min. Met. Soc. India **35**(4): 186–194.

Pandey, J. (1980) Cretaceous foraminifera of Um Sohryngkew River section, Meghalaya. J. Pal. Soc. India **25**: 53–74.

Pandey, J. (1990) Cretaceous-Tertiary boundary, iridium anomaly and foramimifer breaks in the Um Sohryngkew River section, Meghalaya. Curr. Sci. **69**(11): 570–575.

Pandey, J. and Dave, A. (1998) Stratigraphy of Indian petroliferous basins. Presidential Address, XVI Indian Colloquium on Micropalaeontology and Stratigraphy, N.I.O., Goa, India, pp. 1–44.

Prasad, B. and Dey, A.K. (1986) The occurrence of Eocene sediments in Arunachal Pradesh: a palynological evidence. Bull. ONGC **23**(2): 67–74.

Prasad, V., Garg, R. and Ateequzzaman, K. (2007) Climate and relative sea level changes as reflected in the Palynofloral and palynofacies distribution during PETM event in Meghalaya. XXI Indian Colloquium on Micropaleontology and Stratigraphy, BSIP, Lucknow, pp. 136–137.

Raju, D.S.N. (2008) The preserved rock record, stratigraphic gaps and cycles of sea level fluctuations/palaeoenvironments in the Proterozoic and Phanerozic sequences of India: an overview with a mega chart. Mem. Geol. Soc. India **71**: 1–44.

Ram, J. and Venkataraman, B. (1984) Tectonic framework and hydrocarbon prospects of Mizoram. Petrol. Asia J. **2**: 60–65.

Ramesh, P. (2004) Reassessment depositional environments during sedimentation of Lower and Middle Bhuban Formation subsurface of Cachar area, Assam – Arakan Basin, India. Second Assoc. Petrol. Geol. Conference and Exhibition, Khajuraho, pp. 1–12.

Ranga Rao (1983) Geology and hydrocarbon potential of a part of Assam – Arakan basin and its adjacent region. Petroliferous Basins of India. Petrol. Asia J. **6**(4): 127–169.

Sahai, S. (1985) Upper Assam shelf–hydrocarbon of area under ONGC exploration activities. Petrol. Asia J. **8**: 17–55.

Schopf, J.W., Tewari, V.C. and Kudrayavtsev, A.B. (2008) Discovery of a new chert–permineralised Microbiota in the Proterozoic Buxa Formation of the Ranjit window, Sikkim, NE Lesser Himalaya, India and its astrobiological implications. Astrobiology **8**(4): 735–746, Mary Ann Liebert, Inc., 140 Huguenot Street, New Rochelle, NY 10801, USA.

Shukla, M., Tewari, V.C., Babu, R. and Sharma, A. (2006) Microfossils from the Neoproterozoic Buxa Dolomite, West Siang district, Arunachal Lesser Himalaya, India and their significance. J. Palaeo. Soc. India **51**(1): 57–73.

Singh, T. (1983) Late early Eocene larger foraminiferids from Siang district, Arunachal Pradesh, India and their significance. Geosci. J. **4**(2): 141–156.

Singh, S.D. (2007) Foraminifera from the Bhuban Formation of Hari River section, South Shillong, Meghalaya, India. Petrotech **1985**: 1–6.

Singh, R.K., Agarwalla, R., Bora, M.A.S. and Ray, G.K. (2007) Hydrocarbon prospectivity of Miocene sands in Cachar area, Assam, Assam–Arakan fold belt, India. Assoc. Petrol. Geol. Bull. **1**: 35–41.

Sinha, N.K., Satsangi, P.P. and Mishra, U.K. (1986) Paleontology of Permian and Eocene rocks of Siang district, Arunachal Pradesh. Geol. Surv. India Records **114**(4): 53–60.

Tewari, V.C. (1998) Prospects of delineating terminal Proterozoic and Precambrian–Cambrian Boundary in the Northeastern Himalaya. Geosci. J. **xix**(2): 109–113.

Tewari, V.C. (2001) Discovery and sedimentology of microstromatolites from Menga Limestone (Neoproterozoic/Vendian), Upper Subansiri district, Arunachal Pradesh, Northeastern Himalaya, India. Curr. Sci. **80**(1): 1440–1444.

Tewari, V.C. (2002) Lesser Himalayan Stratigraphy, Sedimentation and Correlation from Uttaranchal to Arunachal. Gyanodaya Prakashan, Nainital, pp. 63–88.

Tewari, V.C. (2003) Sedimentology, palaeobiology and stable isotope chemostratigraphy of the Terminal Proterozoic Buxa Dolomite, Arunachal Pradesh, NE Lesser Himalaya. Him. Geol. 24(2): 1–18.

Tewari, V.C. (2004a) Microbial diversity in Meso-Neoproterozoic formations, with particular reference to the Himalaya, In: J. Seckbach (ed.) *Origins*. Kluwer Academic, Dordrecht, pp. 515–528.

Tewari, V.C. (2004b) Palaeobiology and biosedimentology of the stromatolitic Buxa Dolomite, Ranjit Window, Sikkim, NE Lesser Himalaya, India, In: J. Seckbach, et al. (eds.) *Life in the Universe*. Kluwer Academic, Dordrecht, pp. 249–250.

Tewari, V.C. (2004c) Extraterrestrial impact on Earth and extinction of life in the Himalaya, In: J. Seckbach, et al. (eds.) *Life in the Universe*. Kluwer Academic, Dordrecht, pp. 245–248.

Tewari, V.C. (2007) The rise and decline of Ediacaran biota: palaeobiological and stable isotopic evidence from the NW and NE Lessser Himalaya, India, In: P. Vickers Rich and P. Komarower (eds.) *The Rise and Fall of the Ediacaran Biota. Special Publication, 286.* Geological Society, London, pp. 77–102.

Tewari, V.C. (2009a) Proterozoic unicellular and multicellular fossils from India and their implications, In: J. Seckbach and M. Walsh (eds.) *From Fossils to Astrobiology*. Springer, Heidelberg, pp. 119–139.

Tewari, V.C. (2009b) Cretaceous-Tertiary boundary (K/TB) global event of paleoclimate and tectonics in the Shillong Plateau, Meghalaya and Indo-Myanmar region. *Invited lecture in the Earth System Science.* 96th Indian Science Congress, 3–7 Jan 2009, Northeastern Hill University, Shillong, Meghalaya, India, pp. 52–54.

Tewari, V.C., Barbara, S., Katica, D., Nevio, P., Rodolfo, R. and Dolence, T. (2007) Peritidal sedimentary depositional facies and carbon isotope variation across K/ T Boundary carbonates from NW Adriatic platform. Paleogeogr. Paleoclimatol. Paleoecol. **255**: 77–86.

Tewari, V.C., Sial, A.N., Kumar, K., Lokho, K. and Siddaiah, N.S. (2008) Late Cretaceous–Paleocene sedimentation, carbon isotope chemostratigraphy and basin evolution in the South Shillong Plateau. Indo-Myanmar Ranges in the Tectonic Framework of the Himalaya and Southeast Asia. Manipur University, Imphal, pp. 52–54, 27–29 Nov 2008 (abstract).

Tewari, V.C., Kumar, K., Lokho, K. and Siddaiah, S. (2010) Lakadong Limestone: Paleocene–Eocene boundary carbonate sedimentation in Meghalaya, northeastern India. Curr. Sci. **98**(1): 88–95.

Tiwari, R.P., Malsawma, J., Sangode, S.J. and Arora, B.R. (2007) Magnetostratigraphy of a part of Middle Bhuban sequence (Surma Group), Aizawl, Mizoram. J. Geol. Soc. India **70**: 667–674.

Tripathi, C. and Mamgain, V.D. (1986) The larger foraminifera from the Yinkiong Formation (Early Eocene) of East Siang District, Arunachal Pradesh. J. Paleontol. Soc. India **31**: 76–84.

Venkatachalapathy, R., Lokho, K. and Whiso, K. (2007) A study of microfossils in identifying the source rocks for oil in Nagaland, NE India. Gond. Geol. Mag. Spl. Vol. **9**: 33–45.

Venkatachala, B.S. and Berry, C.M. (1985) Source rock palynology of Upper Assam and Dhansiri valley-priliminary assessment. Petroliferous Basins of India **8**(2): 44–55.

Whiso, K., Venkatachalapathy, R., Ramesh, P. and Kachhara, R.P. (2006) Occurrence and age significance of *Planorotalites palmerae* (Cushman and Bermudez) in the Dillai Parbat area of Assam, NE India, In: D.K. Sinha (ed.) *Micropaleontology: Application in Stratigraphy and Paleooecanography*. Narosa, New Delhi, pp. 141–144.

Biodata of **Dr. Edoardo Perri** and **Dr. Alessandra Spadafora**, authors of *"Evidence of Microbial Biomineralization in Modern and Ancient Stromatolites"*

Dr. Edoardo Perri (Ph.D.) is an Assistant Professor in the Department of Earth Science at the Calabria University in Rende, Italy. His research and teaching interests include sedimentology and biofacies of carbonate platform sedimentary systems, and petrology and geochemistry of carbonate minerals. In the last 5 years, he has particularly focused his research on the processes of formation and diagenesis of fossil and modern microbial carbonates.

E-mail: **eperri@unical.it**

Dr. Alessandra Spadafora (Ph.D.) is an Executive Geologist in the Environmental Protection Agency of Calabria (ARPACal), Italy. She earned her Ph.D. at the Calabria University in 2007 and spent a post-doctoral period at the Geomicrobiology Laboratory of ETH, Zurich. In the last 5 years, she has been involved in research and projects focused on the microbe–mineral interaction and stratigraphy and biofacies of carbonate platforms. Her research centres on the biomineralization process in modern and ancient microbialites, role of microbes on dolomite precipitation, sedimentology and diagenesis of carbonate rocks.

E-mail: **a.spadafora@arpacal.it**

Edoardo Perri

Alessandra Spadafora

V.C. Tewari and J. Seckbach (eds.), *STROMATOLITES: Interaction of Microbes with Sediments*, Cellular Origin, Life in Extreme Habitats and Astrobiology 18, 631–649
DOI 10.1007/978-94-007-0397-1_28, © Springer Science+Business Media B.V. 2011

EVIDENCE OF MICROBIAL BIOMINERALIZATION IN MODERN AND ANCIENT STROMATOLITES

EDOARDO PERRI[1] AND ALESSANDRA SPADAFORA[2]
[1]*Dipartimento di Scienze della Terra, Università della Calabria, 87036 Rende, Italy*
[2]*Dipartimento Provinciale di Cosenza, ARPACal. Agenzia Regionale Protezione Ambiente Calabria, 87100 Cosenza, Italy*

Abstract Carbonate microbialites representing microbial mat and/or stromatolite development are remarkable fossils from the early Earth's environment. Even though their biogenicity can be questionable, since they often lack any proof of the original microorganisms in the form of fossils and rarely retain their primary biogeochemical signatures, recent studies of subfossil and fossil stromatolites demonstrate that microbialites can preserve evidence of microbial biomineralization processes.

Modern lithifying microbial mats produce a range of carbonate precipitates resulting from the interplay of the biological activities of microorganisms and the environmental conditions. The mechanisms of crystal nucleation and growth, such as the final texture of the precipitates and the fossilization of organic components, are recognizable in fine-grained Holocene stromatolites which have been subject to little diagenetic alteration. These stromatolites show diversified fabrics characterized by thin or crude lamination and/or thrombolitic clotting, but they exhibit a pervasive peloidal microfabric consisting of dark micritic aggregates surrounded by limpid crystals of aragonite arranged in spherulitic aggregates. The co-existence of submicron flat and filamentous mucus-like remains of degraded EPS and fossilized bacteria, strictly associated with the carbonate phases, implies that the organic matter and the microbial metabolism played a fundamental role in the precipitation of the minerals. The initial products of calcification are nanoglobules (60–200 nm) that subsequently merged into larger spheres (200–500 nm), playing a primary role as centres of nucleation for succeeding crystal growth. Submicron crystals, resulting from the coalescence of carbonate nanoglobules around nuclei of degraded organic matter (EPS and/or bacterial bodies), grew to form larger aggregates that constitute dark peloids. Precipitation of aragonitic spherulites around peloids then occurs filling inter-peloidal space. This process results in the formation of a structure that has an intimate mixture of micrite and microspar; this is comparable with the microstructures of Triassic stromatolites, which also preserve fossilized bacteria and EPS. These features confirm a biological origin for these ancient microbialites through similar microbial-mediated biomineralization processes that occur in the modern lithifying microbial mats.

Keywords Carbonate microbialite • Stromatolite • Biomineralization • Microbial mat • Thrombolites • Extracellular polymeric substance • Mineral nanoglobule

1. Introduction

Biological, chemical and physical processes interacting together produce a large variety of organo-sedimentary deposits within sedimentary systems. These include stromatolites, which provide evidence of life in the early Earth. Important processes operating to generate these typically laminated deposits include trapping of sediment grains by organic substrates, mineralization of autochthonous microorganisms and precipitation of carbonate minerals, either inorganically or through biological microbial mediation. In modern microbial mats, a complex biological and biochemical organization leads to several zones of photoautotrophic organisms with layers of aerobic and anaerobic heterotrophs metabolizing within a variable amount of extracellular polymeric substance (EPS). Modern lithifying microbial mats produce a range of carbonate precipitates resulting from the interplay of the biological activities of microorganisms and the environmental control on organic matter consumption, and the saturation state of the solution; these are the crucial factors that drive precipitation (Dupraz and Visscher, 2005). Even when stromatolite accretion is strongly influenced by a large amount of trapped grains, the processes of biologically induced mineral precipitation follow similar biochemical paths and are indisputably fundamental for the formation of the microbialitic deposit (Reid et al., 2000).

Macroscopic lamination and growth-form are the best-known features of ancient and modern stromatolites, which are indubitably comparable in terms of morphology. Identification of components and mineral fabrics helps to understand the organisms and processes involved in stromatolite formation, even if they form at the microscale or nanoscale. Trapped grains, fossilized prokaryotes and micro-eukaryotes, micrite and microsparite are fundamental components of stromatolites. The relative abundance of grains and fossils distinguishes agglutinated from skeletal stromatolite end-members, whereas only a micrite/microsparite mineral fabric characterizes the laminae of "fine-grained" stromatolites (Riding, 2000). Many ancient and modern stromatolites lack any concrete proof, in the form of actual fossils, for the original microbes and rarely retain their primary geochemical signatures. For these reasons, proving the biogenicity of stromatolites can be very difficult and contentious. Moreover, although the biochemical processes leading to carbonate precipitation in the microbial mat communities have been measured in the field and reproduced in lab cultures, the mechanisms of crystal nucleation and growth structures, such as the final textures and processes of fossilization, are less investigated. Here are reported results of studies on modern (from Brazil) and subfossil Holocene (from Australia) fine-grained stromatolites, compared with fossil Triassic counterpart (from Italy) (Perri and Tucker, 2007).

These stromatolites show original fossil structures, mineral textures and geochemical signatures that are very similar to some modern carbonate-producing microbial mats.

2. Stromatolite Mineral Texture

Stromatolites are forming in Lagoa Vermelha, a small shallow (<2 m) hypersaline lagoon located 100 km east of Rio de Janeiro and situated in a semi-arid climatic setting (Vasconcelos et al., 2006). The bottom surface of the lagoon is dominated by a widespread microbial mat that covers, in the shallower areas, low-relief lithified subfossil stromatolites, which form individual heads ~10 cm high and ~30 cm in diameter (Fig. 1).

The internal fabric of the stromatolites is composed of alternating continuous and regular millimetre-thick, convex-upward laminae intercalated with sub-centimetric, massive, broadly laminated layers and millimetre-size elongate cavities. Allochthonous sand-size particles and larger bioclasts (small bivalves, gastropods, foraminifers and ostracods) are locally incorporated into the laminae, but these are generally rare. In thin section, the stromatolite reveals a peloidal and aphanitic fabric (Fig. 1b, c). The peloidal texture originates from the 20–50 μm size aggregates of dark micrite surrounded by microspar (layer 1; Fig. 1c). Inside the thinner laminae, this texture passes gradually from peloidal to more solid and homogeneous (aphanitic) micrite (layer 2; Fig. 1c). Within the poorly laminated

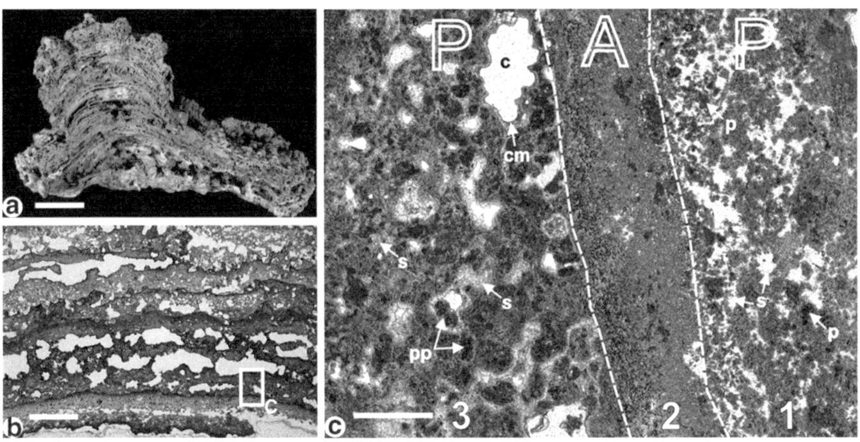

Figure 1. (**a**) Cross-section of a domal stromatolite from Lagoa Vermelha, Brazil. (**b**) Subregular laminae of the stromatolite in thin section. *White areas* are cavities. (**c**) Close-up view of laminae, *dashed lines* indicate the trend of lamination. Thin laminae (1 and 2) are followed by a massive thrombolitic layer (3) with large cavities forming crudely laminated layer. *Bright white areas* are primary porosity. *P* peloidal microfabric, *A* aphanitic microfabric, *p* peloids composed of dark micrite, *s* microspar, *pp* clots of peloidal aggregate, *c* cavities, *cm* primary marine cements. Scale bars: (**a**) 2 cm; (**b**) 2 mm; (**c**) 250 μm.

layers, a clotted thrombolitic fabric, containing many microcavities, is recognizable (layer 3; Fig. 1c). Clots, 50–100 μm in size, are composed of aggregated or single peloids and typically show a relatively evenly spaced distribution in a matrix composed of microsparite (Fig. 1c). Microcavities are lined by thin (~30–50 μm) isopachous fringes of primary marine cements, clearly formed after the peloidal texture. They locally contain intra-bioclastic sediment or are empty, lacking later diagenetic cements. Peloidal texture is formed by oval-shaped crystal aggregates from 5 to 10 μm, together with microsparite that appears as spherulites and fringes of acicular limpid crystals, with a typical thickness of 1–3 μm and a length of 10–20 μm (Figs. 2 and 3). The oval-shaped masses originate from the iso-oriented assemblage of small trigonal polyhedrons, which are ~1 μm wide and ~0.3 μm thick (Fig. 3b). Mineralogical and compositional analyses indicate that the polyhedrons of ovoid crystals are high Mg-calcite with a 25–35 mol.% of Mg and Na around 1–2 mol.%. The acicular crystals are aragonite with Ca 94–96 mol.% and only traces of Mg, Na and Sr.

Higher magnification observations of the crystal surfaces show the presence of much smaller spheroidal structures or nanoglobules, 50–100 nm in size, aggregated to form a granular texture (Fig. 3b). These finer-scale structures are patchily distributed, but they are particularly abundant in the dark micritic calcite. They are, however, absent in the aragonitic microsparite.

Figure 2. SEM pictures of the stromatolite crystalline microfabric, composed of *oval-shaped* aggregates of crystals of high Mg-calcite, constituting micrite peloids, surrounded by spherulitic fans of acicular crystals composed of aragonite, corresponding with microsparite. Lagoa Vermelha stromatolite, Brazil. Scale bar 25 μm.

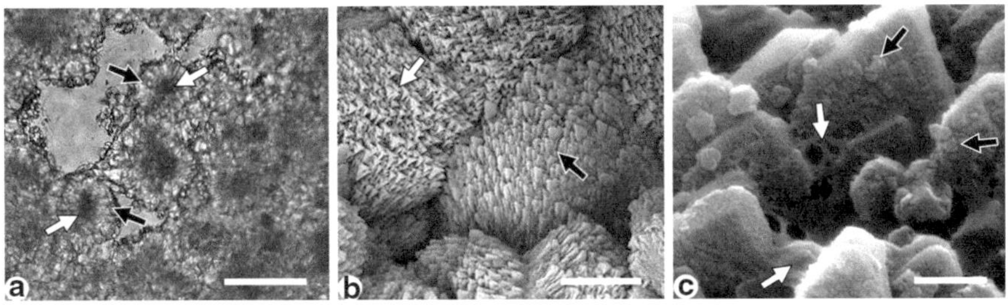

Figure 3. (**a**) Dark micrite (*white arrows*) forming peloids, formed by cloudy irresolvable crystals, and microsparite (*black arrows*) composed of acicular fans of limpid crystals. Thin section. (**b**) SEM view of *oval-shaped* aggregate of high Mg-calcite showing iso-oriented assemblage of trigonal polyhedrons (*white arrow*: *c*-axis view; *black arrow*: *a*/*b*-axis view). (**c**) Granular nanometre-scale texture (*black arrows*) on the surface of high Mg-calcite crystals. *White arrows* indicate EPS remains, see below. Lagoa Vermelha stromatolite, Brazil. Scale bars: (**a**) 50 μm; (**b**) 15 μm; (**c**) 0.7 μm.

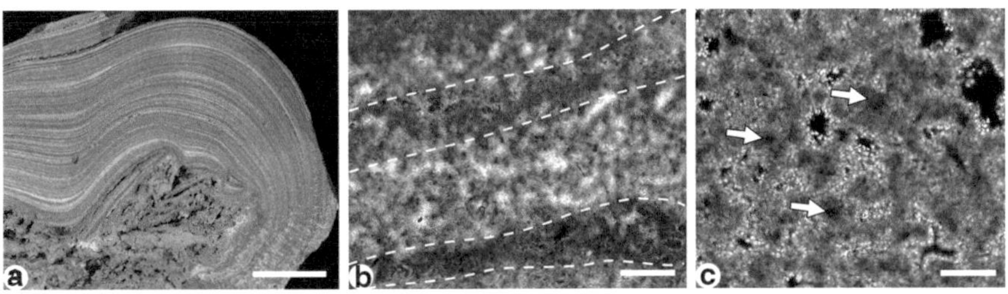

Figure 4. (**a**) Cut surface of a hemispheroidal stromatolite from Marion Lake, the internal lamination, is characterized by an extremely fine alternation of *light* and *dark-grey* laminae which grew on porous carbonate. (**b**) Microfabric of laminae (*dashed lines*) composed of a prevalent peloidal texture. Thin section. (**c**) Dark micrite forming peloids (*white arrows*) surrounded by fine fringes of acicular microspar (*white crystals*). Thin section, crossed polars. Scale bars: (**a**) 1 cm; (**b**) 200 μm; (**c**) 50 μm.

Subfossil stromatolites also occur on a regressive marginal flat surrounding Marion Lake on the Yorke Peninsula, South Australia. The depression last became inundated by the sea during the Holocene transgression (~6,500 years ago), when a protected lagoonal marine environment was initiated and after aragonite/gypsum deposition (between 5,000 and 3,000 years ago) the stromatolite-rich layer was formed (Von Der Borch et al., 1977). Stromatolites formed hemispheroids that are stacked or laterally linked (Fig. 4a); oncolites are also common, notably irregular forms, which encrust a variety of substrate irregularities.

Internal lamination is characterized by an extremely fine sub-millimetric alternation of porous and dense laminae, light-grey and dark-grey,respectively (Fig. 4b). Cavities in the porous laminae are very small (tens of microns) and empty. Trapped grains are absent. The microfabric of the laminae is composed of a widespread peloidal texture, although in many finer and darker laminae peloids coalesce to form uniform aphanitic layers (Fig. 4b, c). Aggregates of very fine anhedral to subhedral micrite (crystal size <1 μm) constitute the peloids, which

Figure 5. (**a**) Spherulites formed by radial oriented acicular crystals. (**b**) Inner view of a spherulite showing the micritic nanoglobular crystals in the nucleus (forming peloids) and the gradual transition to the acicular ones. (**c**) Close-up view of the micrite in the nuclei, which composed of nanocrystals with globular (*arrows*) to polyhedral shape (*dashed lines*). Scale bars: (**a**) 25 μm; (**b**) 5 μm; (**c**) 500 nm. Marion Lake stromatolite, Australia.

are 10–20 μm in size and surrounded by fine fringes of acicular microspar; the latter creates the general spherulitic organization of these mineral phases (Fig. 5a, b). SEM observations reveal that micrite crystals in the spherulite nuclei and the aphanitic layers are made up of coalescing nanoglobules, which can pass to small polyhedrons (Fig. 5c). Acicular crystals show a prismatic elongate shape, broadly hexagonal in transverse section. Close to the spherulite nuclei acicular crystals present a granular texture on their external surface (Fig. 5b). XRD bulk analyses of the rock give a general aragonitic composition but, through EDS spot analyses, Ca in the two mineral phases varies from 94–95 mol.% in the acicular crystals to 89–90 mol.% in the nanocrystal aggregates. The remaining is represented by 2–3 mol.% of Sr and Na, plus ~1 mol.% of S in the acicular crystals, and 2–4 mol.% of Mg, Sr and Na, plus 1–2 mol.% of S and Si in the nanocrystals. The quite high content of minor elements, mainly in the nanocrystals, distinguishes this aragonite from normal primary marine cements.

Upper Triassic (~210 My) stromatolites, formed within restricted hypersaline peritidal marine environments, crop out in NW Calabrian Arc (Italy) (Perri et al., 2003; Mastandrea et al., 2006; Perri and Tucker, 2007). These deposits have been affected by high-pressure and low- or medium-temperature metamorphism during the Alpine–Apennine orogeny. Two types of growth structure are recognizable in the stromatolites: low-relief planar-domal and high-relief columnar forms (Fig. 6a). Lamination consists of an alternation of layers of light-grey microspar (planar anhedral to subhedral crystals 5–60 μm) and homogeneous and/or peloidal dark-grey micrite (planar euhedral crystals <5 μm) (Figs. 6b and 7a, b). Laminae vary from isopachous, even and smooth, to an alternation of thin dark irregular, even strongly crinkled, micritic layers intercalated with thicker and lighter-coloured microsparitic layers (Fig. 6b). The laminated macrofabric commonly passes to thrombolitic layers, which can form laterally continuous beds several centimetres thick, and also low-relief small-scale mound structures (Fig. 6a).

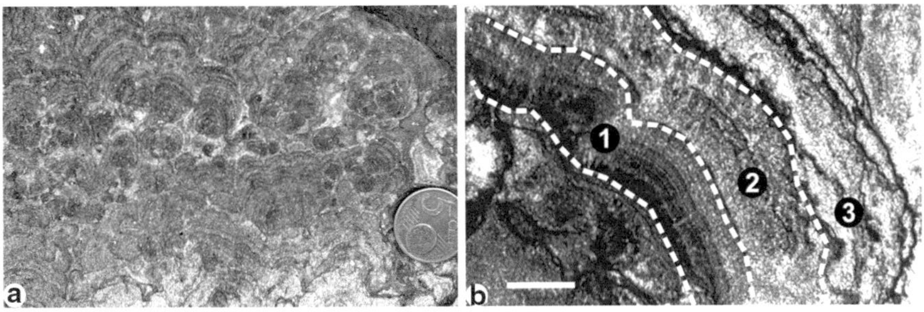

Figure 6. (**a**) Field view of Triassic stromatolite, Italy. (**b**) Various types of microfabric in the stromatolitic laminae (1) isopachous and smooth thin micro-laminae, (2) peloidal/aphanitic, (3) sparse crinkled and irregular laminae. Scale bar: 2 mm, thin section.

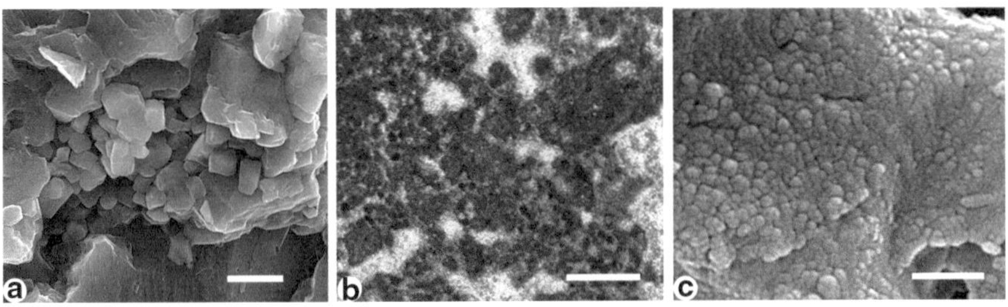

Figure 7. (**a**) SEM view of a micritic area surrounded by microsparite. (**b**) Well-developed peloidal/aphanitic microfabric. Thin section. (**c**) Granular texture, consisting of nanoscale spheroidal bodies on the crystal surface. SEM view, gentle acid etching, after Mastandrea et al. (2006). Scale bars: (**a**) 2 μm; (**b**) 1.2 mm; (**c**) 600 nm. Triassic stromatolite, Italy.

Small gastropods, ostracods, green algae and foraminifera, as well as other sand-size particles, were rarely trapped. Few cavities are present in the stromatolitic framework, and these are generally rimmed by isopachous fringes of fibrous or micritic cement, and later filled by sparry dolomite cement. A granular texture, consisting of spheroidal bodies ranging in size from 70 to 150 nm, is commonly present on the micrite crystal surface (Fig. 7c). These features are absent in the coarser microsparitic and sparitic crystals. They were observed on the surfaces of both fresh, untreated samples and acid-etched samples. Stromatolites have an almost wholly dolomite mineralogy. Spot analyses show that micrite crystals have an average of 53 mol.% Ca, whereas larger crystals have an excess of Mg up to 52 mol.%. Stable isotope measurements give an average $\delta^{18}O = +0.3‰$ (PDB) and $\delta^{13}C = +2.7‰$ (PDB). Under cathodoluminescence, the micritic laminae are non-luminescent, whereas microsparry and sparry dolomite have a dull luminescence. This suggests little or no later alteration of the dolomite crystals comprising the stromatolitic laminae.

3. Biogenic Organo-Mineral Structures

3.1. BACTERIAL FOSSIL REMAINS

Mineralized bacteria-like fossil remains were found closely associated with mineral phases in all studied stromatolites. They mainly consist of filamentous and subspherical coccoid forms preserved as either empty moulds or mineralized bodies.

In the Lagoa Vermelha samples, filamentous and rod-like fossils are slightly curved, apparently solid, tubes (Fig. 8). They have a uniform diameter of ~2 μm and have been observed as isolated fragments or closely associated, appearing to form colony-like clusters (Fig. 8a). The observed mineralized filaments apparently have retained their original size, since the tubular moulds, locally preserved with the mineralized bodies, show an internal diameter a little larger than the filaments (Fig. 8b). The coccoid fossils mainly consist of spherical and ovoidal, solid or empty mould, structures (Fig. 9). According to their dimensions, it is possible to recognize two distinct classes, which are around 2 and 4 μm in diameter and are often seen as isolated forms (Fig. 9a, b). Another type of submicron-sized coccoid mould with a diameter of 200–300 nm is recognizable, forming colonial-like clusters, closely stacked or thinly grouped (Fig. 9c). Both filamentous and coccoid fossils show a fine granular texture formed by nanocrystal globular aggregates of calcite or aragonite. It is commonly observed that the coccoids are often the nucleation centres for growth of aragonite crystal spherulites and fans.

A less variegated mineralized bacteria-like fossil biota was found in the stromatolites of Marion Lake. It consists of empty moulds and mineralized bodies of subspherical coccoid forms, composed of aragonite, varying in dimension from 0.5 to 11 μm. In terms of dimensions, it is possible to distinguish three

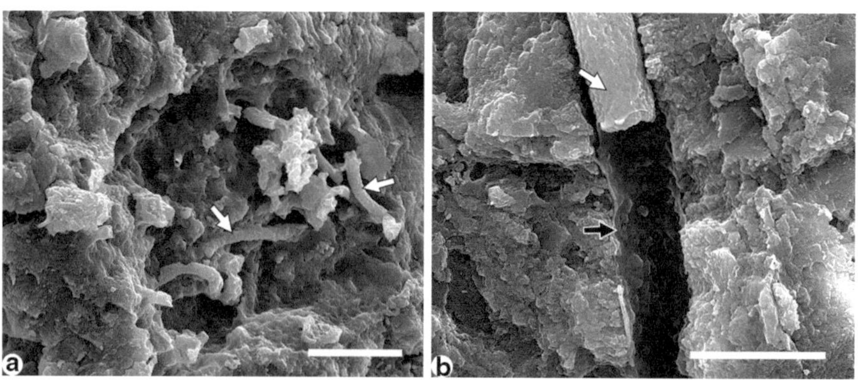

Figure 8. SEM photomicrographs of filamentous bacterial fossils. (**a**) Closely associated slightly curved mineralized (high Mg-calcite) tubes (*white arrows*) forming a colony-like cluster. (**b**) Close-up view of a filament (*white arrow*) embedded in a mucus-like organic membrane (*black arrow*). Scale bars: (**a**) 10 μm; (**b**) 5 μm. Lagoa Vermelha stromatolite, Brazil.

Figure 9. Coccoid bacterial fossils. (**a**) Single mineralized ovoidal bacteria (*arrow*) in the centre of aragonite crystals. (**b**) Mineralized spherical form (*arrow*) encased in high Mg-calcite. In both (**a**) and (**b**) fossils present a nanoglobular texture. (**c**) Dwarf empty moulds (*arrows*) in high Mg-calcite. All SEM views. Scale bars: (**a**) 5 μm; (**b**) 1.8 μm; (**c**) 1 μm. Lagoa Vermelha stromatolite, Brazil.

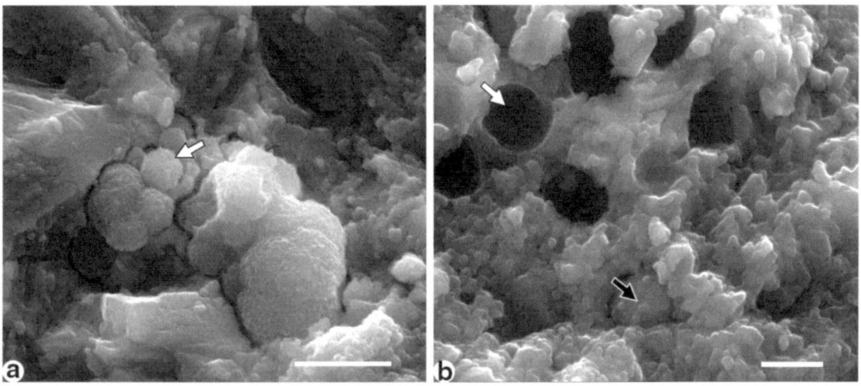

Figure 10. Spherical bacteria-like forms. (**a**) Cluster of mineralized solid internal moulds (*arrow*). (**b**) Empty (*white arrow*) and solid (*dark arrow*) moulds. Note in both pictures the nanoglobular texture of the fossil bodies and of the encasing aragonite. SEM views. Scale bars: (**a**) 1 μm; (**b**) 2 μm. Marion Lake stromatolite.

classes of abundance: 0.5–1 μm (50%), 5.5–6 μm (45%) and 2.5 μm (5%). Where visible, a granular-nanoglobular texture characterized these mineralized bacterial cells (Fig. 10a).

In the Triassic stromatolites, fossil remains of autochthonous filamentous microorganisms, including *Girvanella* and *Ortonella* sp. (probably cyanobacteria) and encrusting forms (red algae) are recognizable in thin section (Fig. 11a). SEM observations of untreated surfaces reveal the presence of spheroidal structures, particularly within the dark micritic laminae. The majority of these remains consist of submicron-sized fragments of hollow spheroids with 100–150-nm-thick walls, in some cases enclosed within dolomite crystals and protruding from their surfaces. Spheroids occurring between crystals within the intercrystalline pore space are commonly collapsed or crushed between flanking crystals. Rarely, whole spheroidal structures are preserved, with a uniform diameter of around

Figure 11. (**a**) Filamentous fossil remains (*arrows*) of autochthonous microorganisms, *Girvanella*. Thin section. (**b**) Well-preserved spheroids within the intercrystalline space interpreted as mineralized bacteria in the Triassic stromatolite. (**c**) Submicron-sized spheroidal structures interpreted as mineralized dwarf bacteria (*arrows*). (**b**) and (**c**) SEM views after Perri and tucker (2007). Scale bars: (**a**) 50 μm; (**b**) 2 μm; (**c**) 500 nm.

1.0 μm (Fig. 11b). Spheroid walls have a fine granular texture formed by nanocrystal aggregates of Ca-rich dolomite (Fig. 11b) with an average Ca content of 55.4 mol.%, reaching in some cases nearly 70 mol.%, and C-enriched organic matter. Higher magnification observations of the same micritic laminae show a second type of spheroidal–ovoidal structure of submicron size. These are 150–400 nm in diameter, isolated or in small clusters, and usually enclosed within the dolomite crystals (Fig. 11c).

3.2. EXTRACELLULAR SUBSTANCES FOSSIL REMAINS

In the Lagoa Vermelha stromatolites, the dark micrite that constitutes the peloids, isolated or aggregated, is characterized by a strong autofluorescence indicating the diffuse presence of organic matter (Fig. 12). In contrast, aragonite crystals that surround the peloids show very low autofluorescence. Moreover, autofluorescence reveals delicate folded layers lining the cavities and tubular forms, otherwise almost transparent in transmitted light (Fig. 12a). SEM analyses confirm the diffuse presence of organic matter, the majority of which consists of planar or sheet-like features (Fig. 12b). Gently folded sheets, 0.5–1 μm thick, line the cavities and/or envelop the crystal aggregates extending for several tens of microns. These sheets locally form a subpolygonal honeycomb-like network and/or gradually pass from flat sheets to filaments of a mucus-like substance (Fig. 12b). Moreover, filamentous tubular organic moulds of bacteria-like bodies are associated with the mucus-like planar sheets. As they lack any mineral affinity, in terms of shape and composition, the nature of these substances is clearly organic, see later discussion. Submicron flat and filamentous organic structures commonly occur between the polyhedrons of high Mg-calcite, suggesting a close association with this mineral phase (Fig. 12c). Aragonite crystals, which are locally partially coated with large sheets, lack these small-scale organic structures. Thin organic membranes and filaments cover mineralized fossils (Fig. 8b).

Figure 12. (**a**) Peloidal microfabric of Lagoa Vermelha stromatolites under UV light: calcite is strongly autofluorescent (*dark arrows*) (green fluorescence) compared with the less intense response of aragonite (*white arrows*) (compare with Fig. 3a). Note also the thin organic fluorescent remains in the cavities (*dashed arrow*). (**b**) Thick sheets of organic-rich mucus-like remains (EPS) forming subpolygonal honeycomb-like network (*white arrows*, SEM view). (**c**) Flat and filamentous mucus-like organic structures (*arrows*) between the polyhedrons of calcite. Note the granular texture particularly abundant on the surface of the polyhedrons of calcite (see also Fig. 3c). Scale bars: (**a**) 50 μm; (**b**) 5 μm; (**c**) 500 nm.

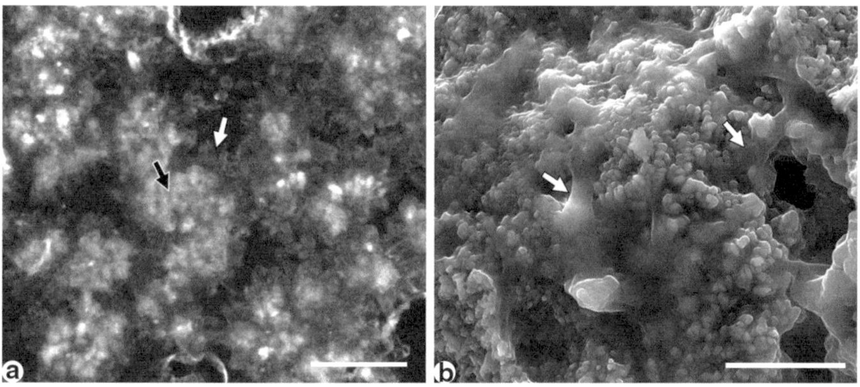

Figure 13. (**a**) Marion Lake stromatolite microfabric with UV confocal microscope reveals a strong green autofluorescence of micritic nuclei (*dark arrow*) and very low fluorescence of the acicular crystals (*white arrow*). *Dark areas* are empty cavities. Compare with Fig. 4c. Scale bar: 30 μm (**b**) Mucus-like organic matter remains (*arrows*) enveloping microcrystals in a spherulite nucleus. Scale bar: 5 μm. SEM view.

Similar to Lagoa Vermelha, in the Marion Lake stromatolites the dark micrite that constitutes the peloids shows a strong autofluorescence, due to organic matter remains, in comparison with the acicular crystals that show very low autofluorescence (Fig. 13a). Under SEM, planar or sheet-like structures with features indicating an original viscous mucus-like behaviour commonly coat cavities and envelop crystals (Fig. 13b). A honeycomb-like network has not been observed, but organic membranes and filaments may envelop mineralized bacterial fossils.

In the Triassic stromatolites, associated with the bacterial fossil structures, are planar or sheet-like features. They consist of 100–200-nm-thick gently folded sheets that occur within crystals and cross between crystals; they also envelop the

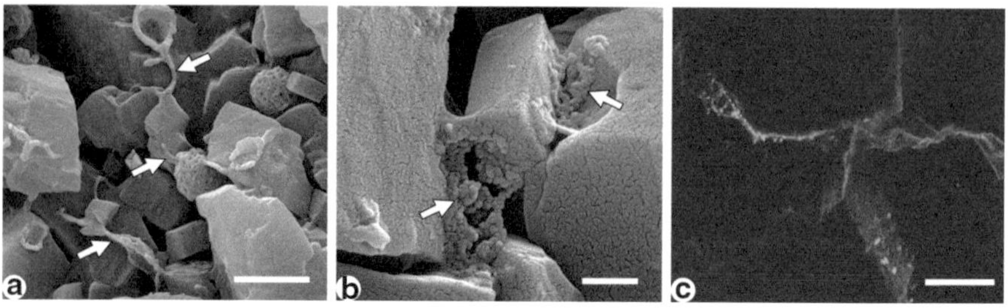

Figure 14. (**a**) Gently folded sheets occur within crystals and extend across them; they envelop the larger spheroids (*arrows*). (**b**) Planar structures (*arrows*) connecting dolomite microcrystals and showing a nanoglobular textured surface. (**a**) and (**b**) SEM views. (**c**) Stromatolitic fabric under UV light using a confocal microscope. Organic matter relics show a draped arrangement of delicately folded surfaces and submicron-sized spheroidal granular bodies around the external sides of the crystals (*dark areas*). Scale bars: (**a**) 2 μm; (**b**) 500 nm; (**c**) 15 μm. Triassic stromatolite, Italy, after Perri and tucker (2007).

larger spheroids (Fig. 14). Generally, they are seen as even, isopachous ribbons extending for several microns, but they are also fragmented, showing an irregular granular-textured surface, enveloping nanoglobules <100 nm (Fig. 14b). The planar, sheet-like structures have the same (quite Ca-rich) dolomite composition as the adjacent and enclosing crystals, even if they appear strongly enriched in organic carbon. In fact, dark micritic dolomite laminae show strong autofluorescence (Fig. 14c). A draped arrangement of a delicate folded layer between and upon the dolomite crystals that surrounds the submicron-sized spheroidal bodies is often seen. The location, form and dimensions of these organic matter relicts, inferred from their autofluorescence, coincide exactly with the spheroidal and planar-sheet structures seen with the SEM, confirming their biological origin (Fig. 14).

4. Discussion and Conclusion

Holocene subfossil and Triassic fine-grained stromatolites show distinctive fabrics, characterized by thin or crude lamination and/or thrombolitic clotting and exhibit a pervasive peloidal or subordinately aphanitic microfabric of the micrite. In the former case, the micrite texture consists of dark aggregates of carbonate (peloids) (high Sr–Na–S–aragonite and high Mg–Na–calcite) that originated from the random to iso-oriented assemblage of submicron nanocrystals and organic matter remains. Limpid acicular crystals of aragonite (depleted in minor elements) are arranged in spherulites surrounding these aggregates. The crystal assemblage characterized by the aggregation of nanocrystals, the widespread presence of benthic fossil microorganisms clearly embedded in the micrite structure and the antigravity assemblage of peloids indicate the autochthonous origin of the micritic aggregates that constitute the laminae. Thus, their origin is related to some in situ process of precipitation that also probably represents the main mechanism for the stromatolite formation. Similarly, Upper Triassic stromatolites consist of peloidal and aphanitic aggregates of dark-grey micrite (<5 μm) alternating with light-grey microspar (5–60 μm). Later diagenetic processes generated coarse, near-stoichiometric to quite Mg-rich, fabric-destructive dolomite crystals (>60 μm) that contrast with the fine crystalline texture of the stromatolitic laminae containing the fossil remains composed of Ca-rich dolomite. It is thus likely that the mineralized organic remains, together with the smallest dolomite crystals, have retained their original composition. Moreover, the $\delta^{18}O$ values of micrite crystals indicate that there has been little resetting of the isotope signatures through later re-crystallization; the non-luminescent nature of the micritic dolomite supports this interpretation (Mastandrea et al., 2006). The limpid acicular crystals of aragonite that surround the micritic aggregates in the subfossil stromatolites possibly correspond, after burial diagenesis, to the limpid anhedral microspar in the fossil forms.

The purely descriptive term "peloid" refers to micritic aggregates of uncertain origin including allochthonous grains with various origins (Flügel, 2004). They have often been regarded as being autochthonous, benthic and microbial in origin, since they constitute the most common microfabric of modern and ancient microbialites (Monty, 1976; Riding, 2000; Riding and Tomás, 2006). However, it has been demonstrated that very similar micro-morphologies can be created by purely abiotic mechanisms (Bosak et al., 2004). Chafetz (1986) and Riding (2002) suggested that peloids could be calcified aggregates resembling bacterial microcolonies in Phanerozoic fossil biofilms. Following studies of modern microbial mat lithification, the formation of peloidal carbonate precipitates has been associated with the metabolic activities of bacteria; see below.

The (optically) dark micritic carbonates, forming peloid nuclei and more uniform aphanitic microlayers in all stromatolites, show a consistent presence of organic matter remains, as indicated by their strong autofluorescence in comparison with microsparite crystals. These organic matter remains are observed as submicron, flat and filamentous mucus-like structures inside the interspaces of the crystals and are mainly interpretable as remains of degraded (mineralized or not) EPS. Such mucus-like structures occasionally closely resemble the subpolygonal honeycomb arrangement of the natural EPS (Défarge et al., 1996). Holocene stromatolites lacking burial diagenesis preserve EPS, but possibly only after drying in the absence of degrading organisms. EPS from hot springs is stable to temperatures of at least 72°C and can partially retain the three-dimensional geometry as the water cools (Allen et al., 2000). On the other hand, during ageing the EPS can be preserved by a variety of elements and compounds, and the textures and structures, as well as some of its organic breakdown products, can be even mineralized (Camoin et al., 1999; Westall et al., 2000). The presence of fossil EPS strictly associated with the smallest crystal units (polyhedrons and/or nanoglobules) of carbonate implies that organic matter played a primary role in the precipitation of the mineral that forms the micrite. At the same time, EPS is virtually absent in the later, microsparitic (i.e. aragonite) phase, indicating different mechanism of precipitation.

In all the discussed stromatolites, there are many micron-sized spheroidal bodies, interpreted as the fossilized remains of original bacterial cells, which are strictly associated with carbonate mineral phases. They mainly consist of subspherical, rod-like and filamentous forms, isolated and in colony-like clusters, and these are easily referable to the bacterial fauna which is well documented in the living carbonate-producing mats. The observed dimensions of the mineralized bodies should correspond with the original width of the bacterial cell since the moulds, locally preserved with the mineralized bodies, show a slightly larger internal diameter. This also could indicate an early process of mineralization affecting bacterial bodies before a reduction of their dimensions due to loss of fluid. Fossilization of the bacterial bodies is possible since, during entombment, the bacterial cells adsorb Mg^{2+} and Ca^{2+} ions that then combine with bicarbonate to precipitate crystals on the surface of the cellular membrane (van Lith et al., 2003).

The nanospheres in the range of 150–400 nm could be interpreted as fossilized dwarf bacteria-like life forms, since living dwarf bacteria are recognized as significant components of stromatolite-producing microbial mat communities (Stolz et al., 2001). However, the identification and even existence of living and fossil sub-bacteria-sized (<300 nm) microorganisms (known as "nannobacteria", "nanobacteria" or "nanobes") remain controversial. Nevertheless, spheroidal mineralized objects <300 nm in size have often been interpreted as fossilized microbes in sulphides, oxides, clays and particularly in microbial carbonates (Vasconcelos et al., 1995; Camoin et al., 1999; Folk, 1999; Sprachta et al., 2001; Dupraz et al., 2004). Moreover, living "nanobacteria", in the range of 100–300 nm, have been found in several living organic tissues from which they have been isolated and cultured (Sommer et al., 2004). An additional biological explanation for the nanospheres is that they could be interpreted as mineralized ultramicrobacteria (Southam and Donald, 1999). An alternative explanation is that the nanospheres are the result of decomposition, dehydration and shrinkage of micron-sized bacterial forms.

Spheroids or nanoglobules less than 200 nm commonly characterize the early stage crystalline units of Holocene and Triassic stromatolites. They have been found extensively in sedimentary environments and have controversially been interpreted as nanobacteria (Folk, 1993) or as mineralized macromolecules (Southam and Donald, 1999). Very similar nanoscale globules have been observed in microbial carbonate precipitation experiments (Aloisi et al., 2006; Bontognali et al., 2008). They also form as a result of enzyme-driven bacterial decay of organic tissues (Schieber and Arnott, 2003). In any case, nanoscale calcified spheres commonly occur both in nature and in nucleation experiments, suggesting that nanoglobules represent an important early stage in microbial-induced mineralization, including fossilization of bacterial bodies.

The relationship between microbial communities, organic matter and carbonate precipitation has been described in several papers (see Dupraz et al., 2008 for a review). The processes of EPS degradation are particularly significant in modern microbial mats because they commonly lead to carbonate precipitation, whereas the bacterial bodies themselves seem less involved (Trichet et al., 2001; Dupraz et al., 2004; Gautret et al., 2004). It has been shown that nanospheres (200–500 nm) with a high organic matter content replace the EPS structure (Dupraz et al., 2004) and even bacterial bodies (Sprachta et al., 2001; van Lith et al., 2003). Nevertheless, the initial products of calcification are nanoglobules (60–200 nm) released from the microbial cell wall in the surrounding aquatic environment (Aloisi et al., 2006; Bontognali et al., 2008). These subsequently are merged into larger spheres, which apparently play a primary role as centres of nucleation for succeeding crystal growth. Moreover, nanospheres, in a more advanced stage of mineralization, coalesce to form larger aggregates followed by precipitation of microsparite and sparite in the remaining spaces.

In conclusion, the co-existence of submicron flat and filamentous mucus-like remains of degraded or mineralized EPS and fossilized rod-like and filamentous bacteria, strictly associated with the nanoglobular micrite crystals, implies that the organic matter and the microbial metabolism played a fundamental role in the precipitation of the minerals. At the same time, the complex processes that induce mineral precipitation in living microbial mats seem well recorded in the studied lithified stromatolites; the bacterial metabolism and the organic matter degradation promote the precipitation of micron-scale carbonate aggregates resulting from the agglutination of nanometer-scale nanoglobules formed during the degradation processes. Carbonate precipitation processes progressively substitute, eventually replacing, the degraded organic structure (EPS and microorganisms). The continuation of this aggregation process forms larger subhedral crystals (small polyhedrons), which subsequently merge to form isolated or connected peloids. All these steps may represent the first stage of stromatolite formation. The subsequent gradual formation of acicular crystals around primary micrite aggregates may occur as the organic matter availability for heterotrophs progressively decreases, and this creates ideal conditions for purely inorganic precipitation (Dupraz et al., 2004; Riding and Tomás, 2006) or for the establishment of different metabolic activities in a distinct bacterial community, able to induce mineral precipitation (Arp et al., 2003). Ultimately, early primary marine cements eventually fill remaining cavities. The final product is the formation of the fabric documented in the subfossil stromatolites. Burial diagenesis, inducing re-crystallization and geochemical equilibrium, produces crystal growth and inevitably partial (or complete) destruction of original mineral and organo-mineral structures; organic matter remains evolve into kerogenic compounds and new precipitation of late cements may take place.

5. Acknowledgements

We thank J. Seckbach and V.C. Tewari for the kind invitation to contribute to this volume. We are grateful to M.E. Tucker for the preliminary review of the manuscript and to G. Aloisi and V.C. Tewari for their constructive final revision.

6. References

Allen, C.C., Albert, F.G., Chafetz, H.S., Combie, J., Graham, C.R., Kieft, T., Kivett, S.J., McKay, D., Steele, A., Taunton, A.E., Taylor, M.R., Thomas-Keprta, K.L. and Westall, F. (2000) Microscopic physical biomarkers in carbonate hot springs: implications in the search for life on Mars. Icarus **147**: 49–67.

Aloisi, G., Gloter, A., Wallmann, K., Guyot, F. and Zuddas, P. (2006) Nucleation of calcium carbonate on bacterial nanoglobules. Geology **34**: 1017–1020.

Arp, G., Reimer, A. and Reitner, J. (2003) Microbialite formation in seawater of increased alkalinity, Satonda Crater Lake, Indonesia. J. Sediment. Res. **73**: 105–127.

Bontognali, T.R.R., Vasconcelos, C., Warthmann, R.J., Dupraz, C., Bernasconi, S.M. and McKenzie, J.A. (2008) Microbes produce nanobacteria-like structures, avoiding cell entombment. Geology **36**: 663–666

Bosak, T., Souza-Egipsy, V. and Newman, D.K. (2004) A laboratory model of abiotic peloid formation. Geobiology **2**: 189–198.

Camoin, G.F., Gautret, P., Montaggioni, L.F. and Cabioch, G. (1999) Nature and environmental significance of microbialites in Quaternary reefs: the Tahiti paradox. Sediment. Geol. **126**: 271–304.

Chafetz, H.S. (1986) Marine peloids: a product of bacterially induced precipitation of calcite. J. Sediment. Res. **56**: 812–817.

Défarge, C., Trichet, J., Jaunet, A.M., Michel, R., Tribble, J. and Sansone, F.J. (1996) Texture of microbial sediments revealed by cryo-scanning electron microscopy. J. Sediment. Res. **66**: 935–947.

Dupraz, C. and Visscher, P.T. (2005) Microbial lithification in marine stromatolites and hypersaline mats. Trends Microbiol. **13**: 429–438.

Dupraz, C., Visscher, P.T., Baumgartner, L.K. and Reid, R.P. (2004) Microbe-mineral interactions: early carbonate precipitation in a hypersaline lake (Eleuthera Island, Bahamas). Sedimentology **51**: 745–765.

Dupraz, C., Reid, R.P., Braissant, O., Decho, A.W., Norman, R.S. and Visscher, P.T. (2008) Processes of carbonate precipitation in modern microbial mats. Earth Sci. Rev. **96**(3): 141–162.

Flügel, E. (2004) *Microfacies of Carbonate Rocks: Analysis, Interpretation and Application.* Springer, Berlin, pp. 976.

Folk, R.L. (1993) SEM imaging of bacteria and nannobacteria in carbonate systems and rocks. J. Sediment. Res. **63**: 990–999.

Folk, R.L. (1999) Nanobacteria and the precipitation of carbonate in unusual environments. Sediment. Geol. **126**: 47–55.

Gautret, P., Camoin, G., Golubic, S. and Sprachta, S. (2004) Tracing automicrite formation and its biochemical setting in modern lagoonal microbialites. J. Sediment. Res. **74**: 462–478.

Mastandrea, A., Perri, E., Russo, F., Spadafora, A. and Tucker, M.E. (2006) Microbial primary dolomite from a Norian carbonate platform, northern Calabria, southern Italy. Sedimentology **53**: 465–480.

Monty, C.L.V. (1976) The origin and development of cryptalgal fabrics, In: M.R. Walter (ed.) *Stromatolites*. Elsevier, Amsterdam, pp. 193–249.

Perri, E. and Tucker, M.E. (2007) Bacterial fossils and microbial dolomite in Triassic stromatolites. Geology **35**: 207–210.

Perri, E., Mastandrea, A., Neri, C. and Russo, F. (2003) A micrite-dominated Norian carbonate platform from Northern Calabria (Southern Italy). Facies **49**: 101–118.

Reid, R.P., Visscher, P.T., Decho, A., Stolz, J.K., Bebout, B.M., Dupraz, C., MacIntyre, I.G. and Paerl, H.W. (2000) The role of microbes in accretion, lamination and early lithification of modern marine stromatolites. Nature **406**: 989–992.

Riding, R. (2000) Microbial carbonates: the geological record of calcified bacterial-algal mats and biofilms. Sedimentology **47**: 179–214.

Riding, R. (2002) Biofilm architecture of Phanerozoic cryptic carbonate marine veneers. Geology **30**: 31–34.

Riding, R. and Tomás, S. (2006) Stromatolite reef crusts, Early Cretaceous, Spain: bacterial origin of in situ-precipitated peloid microspar? Sedimentology **53**: 23–34.
Schieber, J. and Arnott, H.J. (2003) Nanobacteria as a byproduct of enzyme-driven tissue decay. Geology **31**: 717–720.
Sommer, A.P., Pretorius, A.M., Kajander, E.O. and Oron, U. (2004) Biomineralization induced by stressed nanobacteria. Cryst. Growth Des. **4**: 45–46.
Southam, G. and Donald, R. (1999) A structural comparison of bacterial microfossils vs. "nanobacteria" and nanofossils. Earth Sci. Rev. **48**: 251–264.
Sprachta, S., Camoin, G., Golubic, S. and Le Campion, Th. (2001) Microbialites in a modern lagoonal environment: nature and distribution (Tikehau atoll, French Polynesia). Palaeogeogr. Palaeoclimatol. Palaeoecol. **175**: 103–124.
Stolz, J.F., Feinstein, T.N., Salsi, J., Visscher, P.T. and Reid, R.P. (2001) TEM analysis of microbial mediated sedimentation and lithification in modern marine stromatolites. Am. Mineral. **86**: 826–833.
Trichet, J., Défarge, C., Tribble, J., Tribble, G. and Sansone, F. (2001) Christmas Islands lagoonal lakes: models for the deposition of carbonate evaporite-organic laminated sediments. Sediment. Geol. **140**: 177–189.
van Lith, Y., Warthmann, R., Vasconcelos, C. and McKenzie, J.A. (2003) Microbial fossilization in carbonate sediments: a result of the bacterial surface involvement in dolomite precipitation. Sedimentology **50**: 237–245.
Vasconcelos, C., McKenzie, J.A., Bernasconi, S., Grujic, D. and Tien, A.J. (1995) Microbial mediation as a possible mechanism for dolomite formation. Nature **377**: 220–222.
Vasconcelos, C., Warthmann, R., McKenzie, J.A., Visscher, P.T., Bittermann, A.G. and van Lith, Y. (2006) Lithifying microbial mats in Lagoa Vermelha, Brazil: modern Precambrian relics? Sediment. Geol. **185**: 175–183.
Von der Borch, C.C., Bolton, B. and Warren, J.K. (1977) Environmental setting and microstructure of subfossil lithified stromatolites associated with evaporates, Marion Lake, South Australia. Sedimentology **24**: 693–708.
Westall, F., Steele, A., Toporski, J., Walsh, M., Allen, C., Guidry, S., McKay, D., Gibson, E. and Chafetz, H. (2000) Polymeric substances and biofilms in terrestrial and extraterrestrial materials. J. Geophys. Res. **105**: 511–528.

Biodata of **Professor Alain R. Préat, Dr. Ing. Jeroen de Jong, Dr. Chantal De Ridder**, and **Dr. David C. Gillan**, authors of "***Possible Fe Isotope Fractionation During Microbiological Processing in Ancient and Modern Marine Environments***"

Professor Alain R. Préat is currently codirector of the Biogeochemistry-Earth System Laboratory in the Université Libre de Bruxelles (ULB), Belgium. He obtained his Ph.D. from the ULB in Geology on sedimentology of Paleozoic carbonates in 1985 and became Professor in 2000. His scientific interests are in the areas of carbonate sedimentology and geochemistry, and particularly in the study of the origin of the red pigmentation of matrices of Phanerozoic limestones. He carries out this research with marine and microbiologists of the ULB. Recently, he started new research on African Precambrian carbonates.

E-mail: **apreat@ulb.ac.be**

Ing. Jeroen de Jong is currently research associate at the Université Libre de Bruxelles (ULB), Belgium. He became engineer of analytical environmental chemistry in the Netherlands in 1988. Between 1990 and 2000, he carried out research at the Royal Netherlands Institute for Sea Research in the field of marine biogeochemistry of iron and other trace metals with a focus on the Southern Ocean. Since 2001, he has been a lab manager of the MC-ICP-MS facility at ULB, combining his research interests in analytical chemistry, marine trace metal chemistry of the polar regions, and isotope geochemistry.

E-mail: **jdejong@ulb.ac.be**

Alain R. Préat

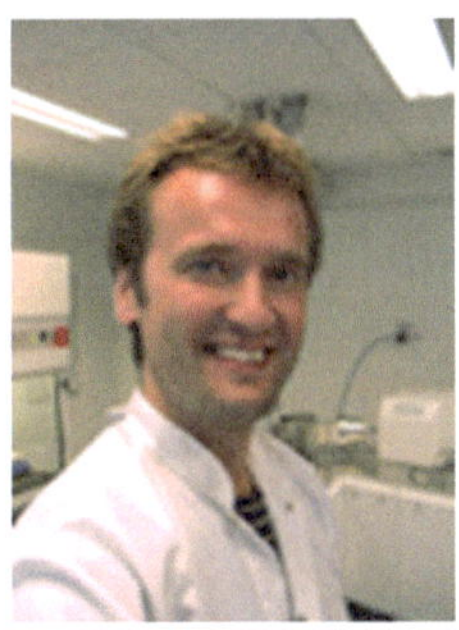

Ing. Jeroen de Jong

V.C. Tewari and J. Seckbach (eds.), *STROMATOLITES: Interaction of Microbes with Sediments*, Cellular Origin, Life in Extreme Habitats and Astrobiology 18, 651–673
DOI 10.1007/978-94-007-0397-1_29, © Springer Science+Business Media B.V. 2011

Dr. Chantal De Ridder is currently Professor in biology at the Université Libre de Bruxelles (ULB) in Belgium. Since 1980, her research focuses on the feeding behaviors, structures, and processes of echinoids. In 1990, she started to carry works on symbioses involving echinoid hosts. She is presently interested in the adaptive features of symbioses and conduces researches on the role the gut microflora of regular echinoids from two extreme environments (sunken woods, Antarctic seas) and on the lifecycle ectoparasite crabs (Pinnotheridae) living on Caribbean echinoids.

E-mail: **cridder@ulb.ac.be**

Dr. David C. Gillan is currently Professor in microbiology at the Mons University (UMons) in in Belgium. Since 1994, his research focuses on microbial ecology, including metazoan–microbial symbioses, with a particular interest for biogeochemistry and microbial communities living in sediments (past and present-day environments). He is presently studying the relationship between benthic microbial communities and fluxes of metallic contaminants.

E-mail: **david.gillan@umons.ac.be**

Chantal De Ridder

David C. Gillan

POSSIBLE FE ISOTOPE FRACTIONATION DURING MICROBIOLOGICAL PROCESSING IN ANCIENT AND MODERN MARINE ENVIRONMENTS

ALAIN R. PRÉAT[1], JEROEN T. M. DE JONG[1], CHANTAL DE RIDDER[2], AND DAVID C. GILLAN[3]

[1]*Department of Earth and Environmental Sciences, Université Libre de Bruxelles, 50 av. F.D. Roosevelt, 1050 Brussels, Belgium*

[2]*Department of Marine Biology, Université Libre de Bruxelles, 50 av. F.D Roosevelt, 1050 Brussels, Belgium*

[3]*Department of Biology, Mons University, 6 av. du Champ de Mars, 7000 Mons, Belgium*

Abstract Eight iron (Fe) isotopic compositions of iron deposits in biofilms and granules found in two recent burrowing marine invertebrates (the sea urchin *Echinocardium cordatum* and the bivalve *Montacuta ferruginosa*) were obtained by Multiple-Collector Inductively Coupled Plasma Mass Spectrometry (MC-ICP-MS). δ^{56}Fe values ranged between −1.78‰ and −0.74‰. The lightest δ^{56}Fe is associated with the iron granules in the intestinal wall of *E. cordatum* and may be due to the abiotic oxidation of source Fe(II) with an isotopic composition reflecting that of light reduced Fe in sediment porewater. This lightest value could represent the best value for the pristine value. Fe in the biofilms was typically heavier by up to +1‰, mean ~ +0.7‰. These results are compared with Fe isotopic composition of 17 Jurassic limestones from the Rosso Ammonitico Veronese (Italy) containing red and gray hemipelagic facies. The red facies show clear evidence of iron bacteria and fungi, which are interpreted as a possible equivalent of the iron microbial communities associated with the recent organisms. Pronounced Fe isotope fractionation was observed in the Jurassic red hardground levels and in the more condensed red facies where bacteria and fungi lived and have accumulated, with values typically lighter by −1‰ than the gray facies where microorganisms were absent. This fractionation probably involved the passive accumulation of originally light porewater Fe in the exopolymeric substances (EPS) produced by filamentous bacteria, thereby favoring heavier Fe isotopes. Alternating stages of oxidation Fe(II)/Fe(III) occurred near the sediment/water interfaces as a consequence of microenvironmental changes in the marine porewaters and caused the red/gray facies interlayering. The comparison of the Fe isotopic compositions of the "biominerals" in the recent organisms and in the iron minerals of the red and gray Jurassic facies suggests an isotopic biofractionation of at least ~+0.7‰. Both studied organisms (the sea urchin and the bivalve) thrive in similar microenvironmental conditions as the microorganisms of the condensed red facies. Their Fe isotope compositions are the same, as is the range of the probable biofractionation.

Keywords Iron isotopes • Microbial mediation • Italian Jurassic Rosso Ammonitico • Recent sea urchins and bivalves

1. Introduction

We have recently investigated the iron (Fe) isotopic composition of 17 Jurassic limestones from the Italian Rosso Ammonitico Veronese (Fig. 1) on a millimeter–centimeter scale between thinly interstratified red and gray–yellow facies (Préat et al., 2008a).

Four localities were investigated for Fe isotope analyses and concerned Voltascura, Castelletto, San Sisto, and Forte di Campo Luserna. In the first one (Voltascura), the interbedding of gray–yellow and red facies is due to the occurrence of neptunian sills filled with different sediments, known as Lumachella a *Posidonia alpina* (white-to-pink thin-shelled bivalve wackestone to grainstone) of Plienbachian–

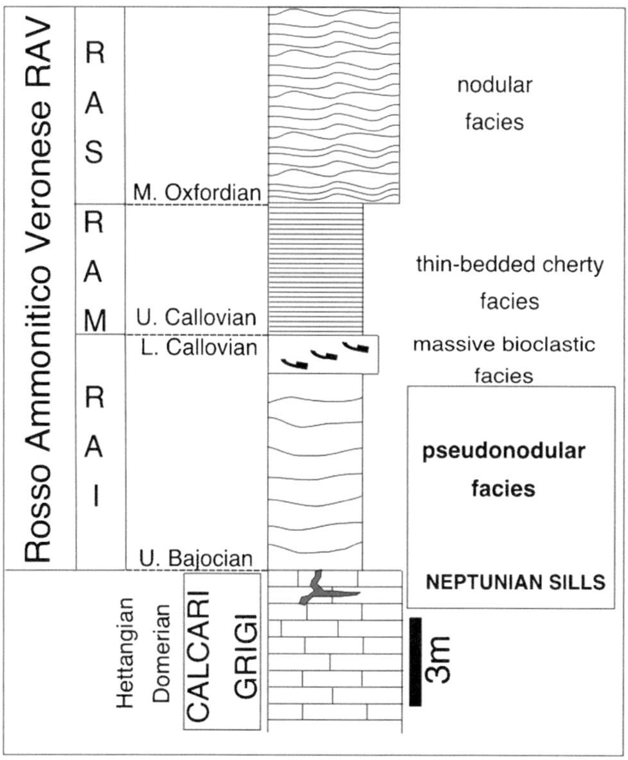

Figure 1. Stratigraphic framework of the Rosso Ammonitico Veronese (modified from Martire, 1996). *RAI* Rosso Ammonitico Inferiore, *RAM* Rosso Ammonitico Medio, *RAS* Rosso Ammonitico Superiore. Fe isotopic composition was determined from *gray* and *red* matrices of the RAI and *grayish yellow* and *red* matrices at the contact of the neptunian sills with the RAI (Préat et al., 2008a, b).

Bajocian age. This long interval is related to subaerial and conspicuous submarine erosion before renewal of sedimentation in a pelagic environment with red facies (Rosso Ammonitico Inferiore). In the three other localities, the samples came from the Rosso Ammonitico Inferiore (lower to upper part in the pseudonodular facies, Fig. 1) and the gray–red interlayering occurs on a millimeter–centimeter scale. However, one gray–red interlayered sample (sample RAV94 in Préat et al., 2008a, b) has been taken from the Rosso Ammonitico Inferiore (1 m from its base). The paleoenvironment was hemipelagic to pelagic and characterized by scarcity or absence of light, as indicated by the lack of phototrophic algae and corals, scarcity of oxygen (variable bioturbation in the red matrices), and a low sedimentation rate as indicated by numerous ferruginous hardgrounds. Scanning electron microscopy (SEM) observations showed that the red matrices contain numerous fragile microfilaments attributed to iron-oxidizing bacteria and fungi (Mamet and Préat, 2003). These observations, along with the iron sheaths observed in the outer wall of the micrometric filaments suggested a simple mechanism for the accumulation of the Fe (oxyhydr)oxides (now hematite) in the samples. The proposed mechanism includes the following steps (Préat et al., 1999):

1. Ferric iron mineral precipitation in the exopolymeric substances (EPS) of the microorganisms that formed a benthic community at the water–micrite interface or within the sediment.
2. Strong bacterial upward growth in the form of small microtufts of unorientated biofilms around various carbonate particles with further degradation supplying the micritic sediment with submicronic iron hydroxides.
3. Progressive burial of the mineral-encrusted microorganisms.
4. Lysis of these microorganisms and EPS degradation liberating further the submicronic iron hydroxides (now hematite) in the matrix.

These red matrices display four different homogeneous fabrics (1) dense micrite, (2) "peloidal" clotted, (3) finely laminar or stromatolitic, and (4) fine microspar. Finally, the micrite matrix of the red facies has been influenced by bacterial filaments, which have been considered microaerophilic (Préat et al., 2006) i.e., they are gradient bacteria with regard to oxygen concentration in neutral pH environments (Fenchel and Finlay, 1995; Emerson and Moyer, 1997; Hallberg and Ferris, 2004; Fortin and Langley, 2005). In contrast, grayish samples are linked with increasingly coalescent structure of anhedral micrite and microsparitic crystals. They consist mainly of peloidal packstone recording higher energy levels. Microfossils of iron bacteria and fungi have not been observed in these gray samples. The presence of different colored facies in the Jurassic Rosso Ammonitico of Verona is linked with variation in oxygenation gradients. Bottom oxygenated conditions favored the spreading of burrowers and prevented the development of iron-oxidizing bacterial biofilms, which require a poorly oxygenated environment (Fenchel and Finlay, 1995).

Our previous iron isotope study of this Jurassic unit showed that there is a clear iron isotopic difference or "fractionation" on a millimeter–centimeter scale in the red facies relative to the gray ones. The red residues, mainly composed of hematite

(X-ray diffraction or XRD), displayed $\delta^{56}Fe$ down to −1.46‰, while $\Delta\delta^{56}Fe$ between red and gray residues ranged from −1.52‰ to −0.06‰ (median −0.97‰). This difference is especially intense in the hardground levels where bacteria and fungi lived and have accumulated (Préat et al., 2008a). Two hypotheses have been considered to explain this light Fe isotopic compositions in the red facies (1) a possible "vital effect" related to passively adsorbed iron or precipitated iron, or both, within the bacterial EPS (sheaths or extracellular matrix) in the bacterial sheaths or inside ferruginous microstromatolites or (2) an effect of assimilation of a $Fe(II)_{aq}$ source with a negative Fe isotopic signature followed by a microbially mediated iron precipitation by oxidation characterized by a positive Fe isotopic signature. In the gray facies, the iron isotopic compositions of the residues are quite constant ($\delta^{56}Fe$ values vary from −0.34‰ to +0.23‰) and constitute a "baseline" relative to the Fe isotopic values of the red facies. In contrast to the red samples, the gray Fe isotopic data match the Fe isotopic signature of the terrestrial baseline established for igneous rocks and low-C_{org} clastic sedimentary rocks (Beard et al., 2003a, b). The Fe isotopic compositions of the gray laminations could reflect the detrital input of iron minerals in the paleoenvironment, where no redox processes and no iron bacterial occurrences have been recorded at the water interface, as the oxygen level was too high.

The preceding study did not conclude on the mechanism responsible for the measured iron isotopic compositions. The fact that this fractionation is observed on a millimeter–centimeter scale suggests that it occurred in the early marine diagenetic iron cycle. The aim of this paper is to shed new light on the possible fractionations in abiotic and biotic systems occurring between oxidized and reduced phases of Fe in marine sediments. For this purpose, two recent marine organisms (the sea urchin *Echinocardium cordatum* and the bivalve *Montacuta ferruginosa*), which carry iron microbial communities (De Ridder, 1994; Gillan and De Ridder, 1997; Gillan, 2003), were investigated to determine the iron isotopic composition of the minerals they have produced. Understanding the Fe isotope composition of these recent iron-encrusted microbial communities may help to decipher fossil counterparts, here the microbial-mediated Fe redox cycling in the Jurassic Rosso Ammonitico of Verona.

2. Background and Methodology

2.1. STRUCTURE OF THE BIOFILM MATS AND THEIR MICROENVIRONMENTS

The sea urchin *Echinocardium cordatum* and its commensal bivalve, *Montacuta ferruginosa* were sampled in December 2008 at the lower intertidal level on the sandy beach of Wimereux (France). The animals were buried in fine-to-medium grained sediments at ca. 15 cm depth and above the level of redox potential discontinuity (RPD) layer. Nodules and intestinal wall isolated from living *E. cordatum* and entire individuals of *M. ferruginosa* were kept frozen at −80°C until the mineral separation.

2.1.1. The Oncoid-Like Nodules in Echinocardium cordatum

Nodules built by bacteria and structured as oncoids occur in the intestinal cecum of several species of burrowing sea urchins including *E. cordatum* (De Ridder, 1994). They consist of a centrally located particle (the nucleus), usually hard debris of animal or plant origin (shell, fish skeletal piece, algal, or wood fragments), wrapped by a bacterial mat where filamentous bacteria predominate (De Ridder et al., 1985) (Fig. 2a, b). Anaerobic conditions prevail in the nodules allowing the development of fermentative and sulfate-reducing bacteria (Thorsen, 1998). Three main bacterial groups occur in the nodule mat, the δ-Proteobacteria, the Bacteroidetes, and the Firmicutes (Gomes et al., 2006), but large filamentous δ-Proteobacteria, belonging to the genus *Desulfonema,* are recurrent members of the nodule microflora (Thorsen et al., 2003; Gomes et al., 2006). These gliding bacteria live in the peripheral layer of the nodule and play a conspicuous role by structuring the mat. They produce an extracellular matrix (EPS) that persists after the bacterial lysis. This results in downward accumulation of EPS and in the formation of concentric EPS layers around the nucleus (Fig. 2b, d). Mature (old) nodules usually reach 0.5 cm in diameter, the mat thickness and the nucleus width determining the nodule size. Soluble reduced iron (Fe^{2+}aq, originating from porewater in the ingested sediment) appears to be immobilized by the negatively charged EPS, similar to the process described for iron–manganese nodules (Libes, 1992). In the sea urchin nodules, the adsorbed cations could react with sulfide produced by sulfate-reducing bacteria to form iron sulfide (FeS) as noticed by Thorsen (1998). Iron sulfide supposedly accumulates in the nodule inner layers being “trapped” in the extracellular polymers (De Ridder and Brigmon, 2003). Occasionally, the inner layer of the nodules present a rusty color suggesting that iron can be oxidized owing to the sporadic presence of oxygen in the close vicinity of the nodule (De Ridder and Brigmon, 2003).

2.1.2. Iron Granules in the Intestinal Wall of Echinocardium cordatum

The connective tissue layer of the intestine of *E. cordatum* and of *Brissopsis lyrifera* (and of several other related spatangoid species) contains ferric iron granules (Fig. 2e) (Buchanan et al., 1980; De Ridder and Jangoux, 1993). These spherical granules consist of insoluble ferric phosphate giving the intestinal wall a typical yellowish-to-orange color (Buchanan et al., 1980). Their number and size increase across the digestive wall (from the digestive epithelium toward the mesothelium) (Fig. 2e). Their diameter ranges from 0.1 to 1.0 μm in *B. lyrifera* (Buchanan et al., 1980) and from 30 to 350 nm in *E. cordatum* (De Ridder and Jangoux, 1993). They also accumulate with the age of the individuals, accounting up to 30% of the dry weight of the intestinal wall for *B. lyrifera* (Buchanan et al., 1980). Because bacteria are totally lacking in the digestive wall of the echinoid, they cannot be involved in oxidation processes that occur in the connective tissue layer (Buchanan et al., 1980; De Ridder and Jangoux, 1993). Soluble iron (Fe^{2+}) originating from the anoxic intestinal lumen presumably circulates (via diffusion or water movements) across the digestive wall. On the way, it precipitates into insoluble oxides in the connective tissue layer where it meets oxygen, i.e., where it is brought closer to the well-oxygenated coelomic cavity (Buchanan et al., 1980; De Ridder and Jangoux, 1993).

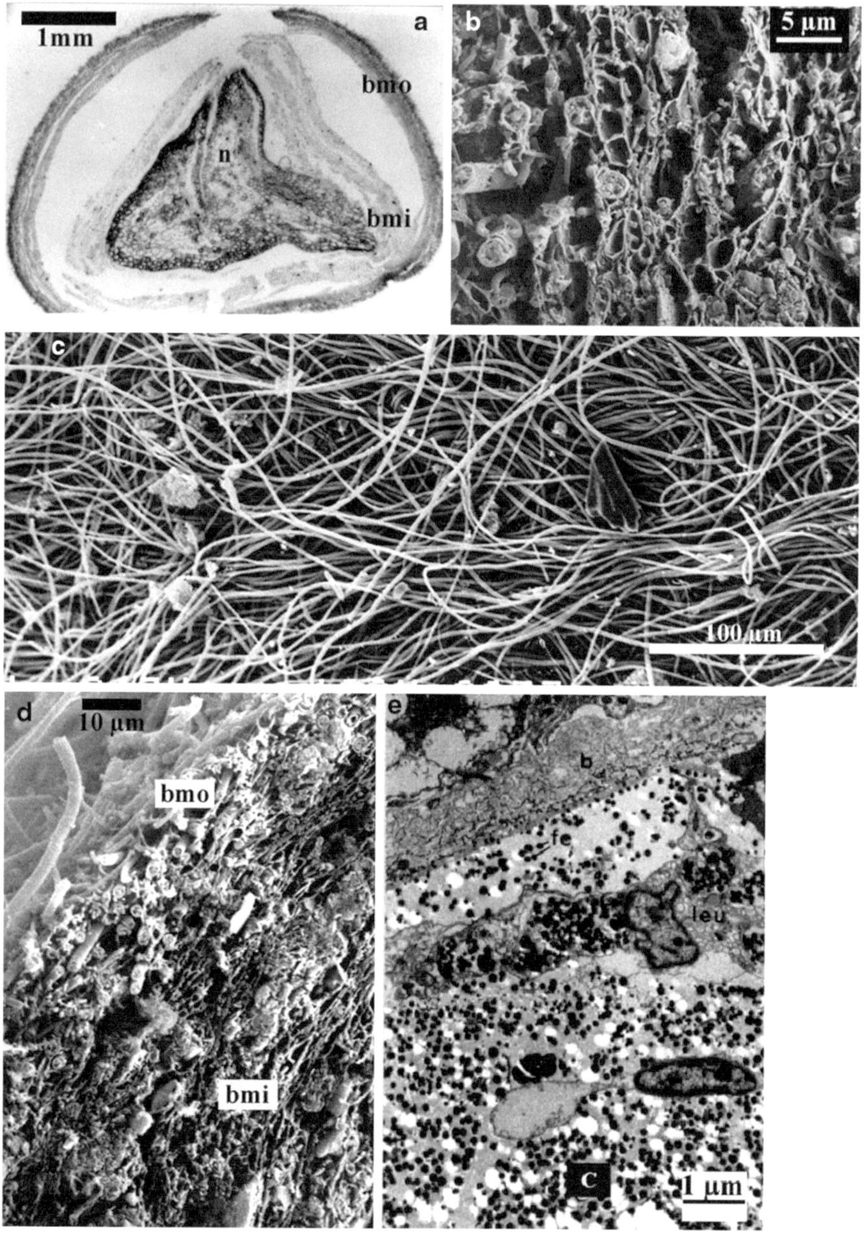

Figure 2. (**a**) Section through a bacterial nodule (optical microscopy). (**b**–**d**) Bacterial mat (SEM): (**b**) detail of the inner layer showing released EPS, (**c**) detail of filamentous bacteria in the nodule outer layer, (**d**) section in the outer and inner layer of a nodule, showing the progressive accumulation of EPS in the inner layer. (**e**) Section in the intestinal wall showing iron granules in the connective tissue (TEM). *b* basal lamina of the digestive epithelium, *bmi* inner layer of the bacterial mat (EPS), *bmo* outer layer of the bacterial mat (living filamentous bacteria), *c* connective tissue layer, *fe* ferric iron granules, *leu* leukocyte in the connective tissue layer, *n* nucleus of the nodule (vegetal fragment).

This is supported by the occurrence of an oxygen gradient across the digestive wall demonstrated by Thorsen (1998).

2.1.3. The Iron-Encrusted Microbial Community of Montacuta ferruginosa

M. ferruginosa is a marine bivalve that lives in the burrow of the echinoid *E. cordatum* (Gage, 1966). The shell of the bivalve is covered by an iron-encrusted biofilm (Fig. 3). Under the SEM, the biofilm of *M. ferruginosa* appears as a structured microbial mat with three separate layers (Gillan and De Ridder, 1995, 1997; Gillan et al., 1998). The inner layer is relatively thick (about 50 μm) and made of densely packed mineral granules that adhere to the shell. Remains of filamentous bacteria (ferric iron-encrusted empty sheaths and extracellular material) are occasionally found in this layer. The intermediary layer is relatively thin (up to 10 μm) and consists of various types of ferric iron-encrusted microorganisms. The outer layer consists of living microorganisms that are similar to those of the intermediary layer. The filamentous bacteria are particularly abundant, they extend up to 3 mm away from the bivalve. Filamentous bacteria related to Beggiatoaceae are predominant in all parts of the biofilm (Gillan and De Ridder, 1997). The genesis of this mineral-microbial mat could be partly explained by a three-step process (1) ferric iron deposition within microbial EPS in the outer layer; (2) release and accumulation of heavily ferric iron-encrusted EPS after lysis of the bacteria in the intermediary layer; (3) degradation of EPS and accumulation of ferric iron minerals in the inner layer.

XRD and energy-dispersive X-ray (EDAX) analyses have indicated that the mineral forming in the biofilm is amorphous and rich in iron, phosphorus, and calcium with traces of silicon and magnesium (Gillan and De Ridder, 1995, 1997). The mineral forms granules whose diameter ranges from 0.05 to 1 μm in the colloidal range. Infrared absorption spectra of the mineral were obtained by applying the KBr disk method. Phosphate absorption bands due to P–O stretching vibrations were observed at 1,100 and 1,020 cm^{-1} (Gillan and De Ridder, 2001). This result, as well as the similarity of spectra with other well-known (bio)minerals, suggests that the mineral is an amorphous iron oxyhydroxide gel with phosphate

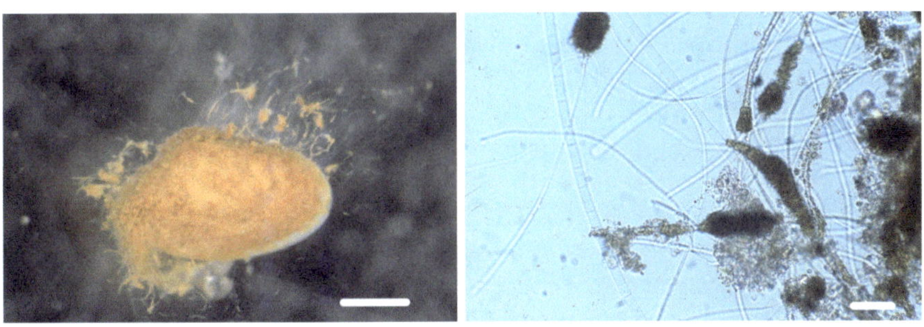

Figure 3. *Left*: General view of *Montacuta ferruginosa*; binocular, scale bar 1 mm. *Right*: Filamentous bacteria encrusted with ferric iron minerals in the biofilm of *Montacuta ferruginosa*; differential interference contrast, scale bar 20 μm.

sorbed on its surface rather than a pure ferric phosphate. Iron oxyhydroxide gels are minerals that may form abiotically in seawater (Byrne and Kester, 1976). However, transmission electron microscopy observations have suggested that the mineral is not the result of a simple microbial bioaccumulation but the result of an *in situ* microbial biomineralization process taking place in the EPS.

Two microbial iron precipitation pathways are possible in the biofilm, namely microbial iron oxidation and microbial degradation of organic Fe(III) complexes (Gillan et al., 2000). The iron-oxidizing activity was assayed spectrophotometrically following the formation of the dye Wurster blue in biofilm extracts. An iron-oxidizing activity was effectively detected in extracts obtained by oxalic acid treatment of biofilm fragments. On the contrary, extracts obtained without oxalic acid treatment, heated extracts, or extracts supplemented with $HgCl_2$ did not show any activity. This suggests that an iron-oxidizing factor (IOF), possibly an enzyme, was coprecipitated with the iron mineral (oxalic acid dissolves iron precipitates). Additional information gathered using sodium dodecyl sulfate polyacrylamide gel electrophoresis (SDS-PAGE), gel filtration chromatography, and ultraviolet (UV) spectrophotometry indicate that the IOF would be a small peptide or glycopeptide (MW 1,350 Da). Microbial degradation of organic Fe(III) complexes was assayed with living biofilm fragments incubated in a media containing ferric citrate (1 mM). Analysis of the supernatants at various intervals of time using atomic absorption spectrometry revealed that the iron complex was degraded by living microorganisms much faster than in the heat-killed negative controls. We conclude that ferric iron precipitation in the biofilm, in addition to other processes, may proceed via microbial Fe(II) oxidation as well as microbial degradation of organic Fe(III) complexes.

2.2. GEOLOGICAL SETTING OF LIMESTONES

The Rosso Ammonitico Veronese (Middle-Upper Jurassic) is a thin lithostratigraphic unit, consisting of red, mostly nodular, pelagic limestones deposited on a current-swept structural high, in the Southern Alps (Clari and Martire, 1996). Sedimentation rates were very low, and omission surfaces and mineralized hardgrounds were common. Facies are varied and show evidence of early lithification near the sediment/water interface (Jenkyns, 1974; Massari, 1981; Clari et al., 1984; Clari and Martire, 1996; Martire, 1996; Préat et al., 2006). Many microfacies are similar to those observed in other Paleozoic and Mesozoic red carbonates, with an abundance of hematite bioconstructions (Mamet and Préat, 2006). We have shown that the origin of the pigmentation is similar in all the studied cases and due to the activity of iron-oxidizing bacteria. The dispersion of the omnipresent hematite in the matrix causes the pigmentation grading from pink to deep red. Among the eight major recognized hematite concentration types in nearly all the studied cases (Mamet and Préat, 2006; Préat et al., 2008b), various types of stromatolites are observed and consist of microstromatolites, crenulated stromatolites, endolithic stromatolites, and oncolites (Martire, 1996; Clari and Martire, 1996).

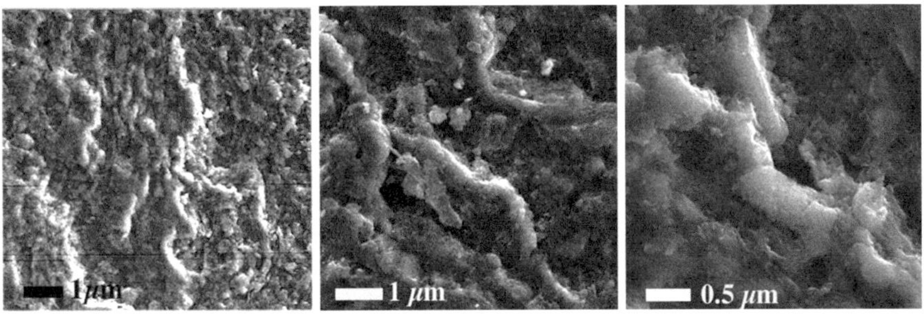

Figure 4. SEM photomicrographs from the Castelletto section (Lower Rosso Ammonitico, sample RAV135 in Préat et al., 2006) showing presumed iron oxidizing bacteria filaments on submicronic to micronic micrite. They appear not to be fungi, because they are much smaller than typical fungi, or cyanobacteria, because they are coated by iron. They can constitute up to 20% of the micrite matrix.

Nevertheless, two notable differences are observed in the Ammonitico Rosso Veronese: the presence of manganese and the existence *of in situ* bacterial–fungal mats in the matrix. These microbial mats can represent up to 20% of the sediment (Mamet and Préat, 2003) (Fig. 4).

Their excellent preservation (absence of packing or crushing) is due to the slow sedimentation rate of the pelagic sediments or of the hardgrounds. The paragenesis of the carbonates of the Jurassic Ammonitico Rosso Veronese shows the following sequence (Préat et al., 2006) (1) early stage of submicronic deposition with picoeucaryotes, nanodebris, and iron bacteria; (2) first matrix recrystallization with formation of neomicrite *sensu* Loreau (1972); (3) microsparitization; (4) residual pores progressively filled by three types of calcite cements, nonluminescent, luminescent, and dull; and (5) fractures filled by luminescent cement. At a larger scale, diagenetic conditions changed from oxidizing (sea floor) to weakly reducing (very shallow burial) and to more intensively reducing (deeper burial) conditions (Martire, personal communication).

In the studied cases, including the Ammonitico Rosso Veronese, red pigmentation was not related to alteration or contamination (Mamet and Préat, 2006). Iron bacteria were present between the initial micritic components (coccoliths and others) and the latter recrystallized neomicrite. Bacteria thrived in open marine water near the sediment/water interface. There is no evidence for a relation with warmer mineralized fluids. Under cathodoluminescence, all the cements are dull marine suggesting that during sedimentation slightly reducing conditions prevailed and that no meteoric cementation occurred. Iron bacteria colonized dysoxic–anoxic interfaces that produced various microaerophilic environments. The limiting factor was the oxygen content, which is always low in these very quiet and relatively deep environments. This oxygen probably presented steep gradients over short distances (Fenchel and Finlay, 1995; Little et al., 1997; Munn, 2004). The studies by Gillan and De Ridder (1995, 1997) on recent sediments highlight the importance of steep dysoxic–anoxic gradients on a millimeter–centimeter scale, indicating that when these interfaces are available, they are quickly colonized by iron bacteria. This is particularly the case in the Castelletto, San Sisto, and Forte di Campo Luserna sections where no stratigraphic gap is present. In Voltascura, the color interlayering is not due to true interstratification (except for sample RAV94) since a large stratigraphic gap is present (Martire, 1996), but the red facies were also dysoxic as they contain iron bacteria.

2.3. IRON ISOTOPE ANALYSIS

Analytical procedures have been described elsewhere in more detail (Préat et al., 2008a; De Jong et al., 2007, 2008), but will be briefly repeated here.

Reagents. We used 14M HNO_3 24M HF and 6M HCl acids purified by subboiling distillation in Teflon stills, as well as Merck Suprapur 30% H_2O_2.and ultrahigh purity water (UHP) (18.2 MΩ grade). Thoroughly acid cleaned Savillex® beakers (Savillex, Minnetonka, MN, USA) were used throughout.

Biological sample treatment samples. Five deep frozen *M. ferruginosa* bivalves were thawed in filtered seawater. The animals were removed and the shells with biofilm (weighing approximately 15 mg each) were put in Savillex beakers, dried and weighed. Deep frozen intestinal nodules (weighing 4.4 and 8.7 mg) as well as one intestinal wall (40 mg) of the sea urchin *E. cordatum* were thawed for careful rinsing with filtered seawater to remove sediment particles. The tissue was put in Savillex beakers, dried, and weighed.

Limestones sample treatment. Powdered subsamples were extracted by means of a small hand drill (Proxxon Micromot, Germany) with a titanium drill head. Approximately 25 mg of powder was collected into a Savillex beaker.

Leaching procedure limestones. A weak HCl leaching treatment was applied to dissolve the carbonate phase and to isolate the Fe-bearing mineral phases.

Total dissolution: Biological samples, as well as the limestone leachates and residues were digested in a mixture of 14M HNO_3, 30% H_2O_2, and 24M HF in closed Savillex beakers on a hot plate at ~125°C over a period of 24 h, and dried down at 100°C. The efficiency of the acid-digestion method was verified by Isotope Dilution MC-ICP-MS, using the reference material IAEA-392 (Algae, International Atomic Energy Agency). We found 505 ± 2 $\mu g\ g^{-1}$ ($n=2$), in excellent agreement with the certified value (497 ± 13 $\mu g\ g^{-1}$).

Iron separation and purification. Fe fractions were isolated from the sample matrix by a one-step anion-exchange chromatographic separation following the procedure by Maréchal et al. (1999) using BioRad AG-MP-1 resin. Coeluted organics from sample or resin may cause matrix effects (Shiel et al., 2009). Therefore, the samples were digested by refluxing the residue with concentrated HNO_3. To verify quantitative yield of the chromatographic separation, hence absence of isotopic artifacts, concentration data were obtained on sample splits before and after separation. An alkalic ocean island basalt Kerguelen LVLK-132 was recovered at 95.8 ± 2.6% ($n=3$). Three different Rosso Ammonitico limestones

(RAV100, 102 and 160: both red and gray facies) yielded 97.7 ± 4.1% ($n=6$). RAV102G and the reference material IAEA-392 (Algae) were measured with Isotope Dilution MC-ICP-MS and yielded 99.8% and 99.5% recoveries, respectively. Total procedural blanks were typically <10 ng Fe.

Mass spectrometry. Isotopic measurements were performed on a Nu Plasma MC-ICP-MS (Nu Instruments). For the limestones, a combination of standard sample bracketing and elemental doping (Cu) was applied, with the instrument operated at low resolution(~300 M/ΔM) (De Jong et al., 2007). To minimize interfering argide formation, a Cetac Aridus desolvating sample introduction system was used. All measurement solutions (400 ppb Fe) were made up in 0.05 M HNO_3. Internal precisions (1SE) are typically on the order of ±0.03‰ on δ^{56}Fe and ±0.04‰ on δ^{57}Fe.

The biological samples were analyzed in static mode without elemental doping. A Nu Instruments DSN-100 desolvating sample introduction system was used. The mass spectrometer was upgraded with an E2M80 high capacity primary pump and high sensitivity "B" cones, and operated in pseudo-high resolution (~1,500 M/ΔM), with high voltage lens "5" set at 0 V to improve the edge resolution power (resolution of leading edge at 5% and 95% peak height) (Abraham et al., 2008). All analytical solutions contained 2 ppm Fe. Internal precisions (1SE) are typically on the order of ±0.02‰ on δ^{56}Fe and ±0.03‰ on δ^{57}Fe.

Isotopic data are reported here in delta notation (‰) for ratios 56/54 and 57/54, relative to the international iron standard IRMM-014 (Taylor et al., 1992). Regularly measured as quality control samples are: IRMM-014, basalt LVLK-132, an in-house atomic absorption Fe standard ("Merck"), and hematite and iron chloride ("Fe salt") in-house standards obtained from the Swiss Federal Institute of Technology (ETH, Zürich). Results on accuracy and reproducibility of the Nu Plasma are listed in Table 1. IRMM-014 yielded a δ^{56}Fe of 0 within error for both low- and high-resolution measurements. The basalt LVLK-132 values agree well

Table 1. Accuracy and reproducibility.

	Mode	δ^{56}Fe (‰)	SD	δ^{57}Fe (‰)	SD	*n*
IRMM-014	LR	0.00	0.08	0.01	0.10	189
IRMM-014	HR	−0.02	0.05	−0.01	0.08	22
Kerguelen basalt LVLK-132	LR	0.14	0.07	0.21	0.11	31
Merck std	LR	0.40	0.11	0.60	0.14	71
ETH hematite	LR	0.55	0.05	0.82	0.08	13
ETH hematite	HR	0.53	0.05	0.80	0.09	4
ETH "Fe salt"	LR	−0.74	0.09	−1.08	0.13	7
ETH "Fe salt"	HR	−0.71	0.06	−1.01	0.08	11
Avg. terrestrial igneous rock[a]		0.09	0.05	0.11	0.07	54
ETH hematite, "certified"[b]		0.55	0.04	0.82	0.05	>500
ETH "Fe salt," "certified"[c]		−0.73	0.05	−1.07	0.08	>400

[a] Data from Beard et al. (2003a)

[b] Average value compiled from literature (see text for references)

[c] J. Wiederhold personal communication

with the terrestrial igneous rock average (Beard et al., 2003a). The Fe isotopic compositions of the ETH hematite and "Fe salt" in-house standards are for both the low- and high-resolution analyses in good agreement with reported values from expert laboratories equipped with various types of MC-ICP-MS instruments (Fehr et al., 2008; Dideriksen et al., 2006; Teutsch et al., 2005; Poitrasson and Freydier, 2005; Williams et al., 2004).

3. Results

Fe isotopic compositions (δ^{56}Fe) from Fe_{tot} solid phase (amorphous minerals) trapped in the microbial mats *of E. cordatum* and *M. ferruginosa* are negative and range between −1.78‰ and −0.74‰. Solid phase Fe_{tot} concentrations range from 10.10% to 0.75% as determined by isotope dilution (Table 2). The δ^{56}Fe isotopic compositions of the amorphous minerals related to the bacterial mats associated with the sea urchin and the bivalves are similar to the compositions of the Fe-oxide (hematite) of the red facies in the Rosso Ammonitico of Verona (median −0.84‰, range: −1.46 to + 0.26‰) (Préat et al., 2008a). As for the results of the Rosso Ammonitico, the observed Fe fractionations are not correlated with the total iron concentrations in the analyzed fractions.

4. Discussion

4.1. ORIGIN OF THE IRON

Different pathways have been proposed to explain the origin of the iron responsible for the red pigmentation of various red limestones including (1) a telogenetic alteration (Miller and Folk, 1955; Walker et al., 1967); (2) a detrital input (Krynine,

Table 2. Iron isotopic composition of iron-encrusted biofilms of *M. ferruginosa* (samples M1–M5) and iron-encrusted nodules and intestinal wall of *E. cordatum* (samples E1, E2, E3). First column gives the iron content (%) of the samples. Sample description in text. See Préat et al. (2008a, b) for comparison with iron isotopic composition of Rosso Ammonitico Veronese samples.

		δ^{56}Fe (‰)		δ^{57}Fe (‰)		
	% Fe	Avg	SD	Avg	SD	*n*
Bivalve M1	1.61	−0.84	0.04	−1.25	0.07	2
Bivalve M2	0.80	−0.74	0.01	−1.08	0.00	2
Bivalve M3	0.78	−1.66	0.04	−2.44	0.03	3
Bivalve M4	1.72	−1.14	0.04	−1.71	0.04	3
Bivalve M5	0.75	−1.15	0.01	−1.64	0.00	2
E1 nodule	2.73	−1.06	0.01	−1.54	0.01	2
E2 nodule	2.72	−0.76	0.06	−1.08	0.09	2
E3 wall	10.10	−1.78	0.02	−2.64	0.01	2

1949, 1950; van Houten, 1961, 1973); and (3) the activities of iron bacteria (Boulvain et al., 2001; Préat et al., 2006; Préat et al., 2008b). The reader will find a detailed discussion of this problem in Préat et al. (2008a), who concluded that regardless of the Fe content, one can reasonably attribute the primary origin of the hematite iron of the Rosso Ammonitico to incorporation of exchangeable iron from the clay fraction as suggested by Hofstetter et al. (2003) and Bensing et al. (2005). Hofstetter et al. (2003) documented the ability of clay minerals to transport iron as Fe^{2+}, bound by ion exchange, which is not readily released and made available for oxidation. Bensing et al. (2005) investigated the Permian Abo Formation in New Mexico and concluded that the detrital iron-bearing clay-size material is the primary source of iron in the studied Formation, and not the iron-bearing sand and silt grains, which are typically considered to be the principal source of iron for hematite precipitation in the red beds. These authors considered also the possibility of colonization of the microporous network in the clay fraction by microbes (because of the greater content in microbes and greater reactive surface of the clay) that caused the release of iron trapped in the clays.

Detailed understanding of the pathways by which biological processing of Fe occurs has been reviewed by Johnson et al. (2004) and include (1) lithotrophic or phototrophic metabolism; (2) dissimilatory Fe(III) reduction and (3) assimilatory Fe metabolism. Bacteria may cycle Fe through valence changes in various situations. For example, anoxygenic photosynthetic bacteria or chemolithotrophic microbes could derive energy from Fe(II) oxidation (Staubwasser et al., 2006). High pore fluids $Fe(II)_{aq}$ contents have been reported by Severmann et al. (2006) in eastern boundary of the North Pacific from California near Monterey Bay at a water depth of 450 m or more where bottom water oxygen is very low (±10 μM O_2, Zengh et al., 2000).

4.2. IRON ISOTOPIC COMPOSITION

4.2.1. Marine Porewaters and Recent Organisms

Marine pore fluids and sedimentary solid phases displayed negative δ^{56}Fe along two sediment profiles at the California continental margin (Severmann et al., 2006). Light Fe isotope compositions of $Fe(II)_{aq}$ in porewaters (minima of −3.0‰ and −1.8‰ for each of the profiles) were observed by these authors near the sediment surface, which increases to heavier values with burial depth. These negative values are the result of various diagenetic pathways that include significant dissimilatory iron reduction, bacterial sulfate reduction and in some cases production of sulfide (FeS and pyrite). Analysis of porewaters allowed these authors to assess the Fe isotope composition of the benthic Fe flux in a classical shelf setting. Our Fe isotope compositions of the modern organisms (*E. cordatum* and *M. ferruginosa*) provide therefore important constraints on the mechanisms involved in producing Fe isotope variation in the formation of induced biominerals through the activity of microbial communities (iron bacteria). Moreover, Fantle and DePaolo (2004) and Staubwasser et al. (2006) demonstrated that marine porewater Fe_{aq} has a low $\delta^{56}Fe$ value complementary to the heavier residual reactive Fe(III) that remains in the sediment column. This is particularly true for anoxic and low-oxygen sediments on some shelf and continental slopes at present day.

Our sample "E3 Wall" is of particular importance in determining the pristine iron isotope composition in the marine porewaters of the studied site. This sample that has no relation with the bacteria corresponds to the chemical oxidation of $Fe(II)_{aq}$ into ferric iron granules. The conditions inside the digestive tube are anoxic allowing intratissue migration of dissolved $Fe(II)_{aq}$ and its subsequent oxidation when reaching oxygenated area within the intestinal wall (Buchanan et al., 1980). The $\delta^{56}Fe$ value of −1.78‰ (sample E3 Wall) can thus be considered as reflecting the $\delta^{56}Fe(II)_{aq}$ from porewaters in the ingested sediment. The Fe isotope composition of the granule is probably very close to the original $Fe(II)_{aq}$ since the oxidation process occurred very fast and may not introduce an additional fractionation. Our observed $\delta^{56}Fe$ value is in good agreement with those expected at circumneutral pH of most low-temperature aqueous systems for Fe^{2+} (Johnson et al., 2008). Since *E. cordatum* did not experience burial (at a geological scale) no negative isotope composition imprint has been added to the iron.

Ferrous iron (FeS?), in the nodules of *E. cordatum* (Thorsen, 1998), and ferric iron, in the epibiotic mat on *M. ferruginosa* (Gillan, 2003), present similar isotope compositions ranging from −0.74‰ to −1.66‰. In both cases, iron seemingly accumulates more specifically in EPS produced by filamentous bacteria. In both cases too, iron seemingly presents alternating stages of oxidation (Fe^{2+}/Fe^{3+}) owing to punctual variations in the surrounding conditions. The nodules usually constitute an anoxic environment where iron occurs in a reduced form (Thorsen, 1998); sometimes however, iron is oxidized presumably in the sporadic presence of oxygen (De Ridder and Brigmon, 2003). Oxygen is usually present in the surrounding of *M. ferruginosa*, and ferric iron occurs in its epibiotic bacterial mat; anoxic conditions may, however, occasionally have induced reduction (Gage, 1966; Gillan and De Ridder, 1997). The minerals observed in the studied recent organisms have of course been formed during the early marine diagenesis and their iron isotope compositions are related to the redox changes of mobilized $Fe(II)_{aq}$. As demonstrated by our results, the produced isotope variations are considerable, up to ~ +1‰ (sample M2) relative to the pristine E3 value and around 0.6‰ (samples M4 and M5) or less (samples M3). Fractionation of this size and direction can be attributed to several factors such as microbial assimilation (Wasylenki et al., 2007; Schoemann et al., 2008), adsorption to microorganisms (Brantley et al., 2001, 2004), adsorption to Fe(hydr)oxides (Teutsch et al., 2005; Crosby et al., 2005; Icopini et al., 2004). These data suggest that our reported Fe-isotope values could be regarded as a direct or indirect "biosignature," which produced the isotopic fractionation in the minerals formed through metabolic processes related to the microbial activity observed in *E. cordatum* and *M. ferruginosa*. However, we have to remain cautious since similar Fe isotopic fractionations may be produced by abiological processes (review in Beard et al., 2003a), but Fe isotope composition of sample E3, which has not been altered by bacterial activities, supports that the other measured Fe isotopes have been modified by the iron bacteria, since at this local scale the origin of the iron is the same.

4.2.2. Implications for the Jurassic Rosso Ammonitico

The Rosso Ammonitico Veronese "red" samples show Fe isotopic values ranging between −1.52‰ and –0.06‰ (Préat et al., 2008a). These values (median –0.84‰) are in the same range as reported here for the recent organisms. There are also other important similarities between the Jurassic and the recent materials, the most important being the activity of iron bacteria at the dysoxic–anoxic sediment/water interfaces. In both cases, iron bacteria thrived in the superficial part of unconsolidated sediment in microaerophilic environments with steep oxygen gradient over short distances. These microenvironments have no paleogeographic or depth connotation, the main point being the presence of dysoxic–anoxic interfaces on millimeter to centimeter scales (Mamet and Préat, 2006; Gillan et al., 2000). The Fe isotopic values of the Jurassic unit is also very similar to those reported on Fe oxyhydroxides in shallow sediment profiles from reducing and oxic environments in the Arabian Sea margin ranging from −0.77‰ to −0.19‰ (Staubwasser et al., 2006).

We have shown that an iron isotope fractionation of the "red" samples of the Rosso Ammonitico Veronese could be deduced when the values are compared on a millimeter–centimeter scale to the iron isotopic values of the associated "gray" samples in Castelletto, San Sisto, and Forte di Campo Luserna sections (Préat et al., 2008a). A similar isotope fractionation has been reported in the Voltascura section between the gray–yellow limestones of the shallow platform and the overlying red Rosso Ammonitico. Despite that about 20 My separate both facies, an iron isotope fractionation has been established, which is of the same type as those of the preceding sections and as sample RAV94 belonging to the lower part of the Rosso Ammonitico Inferiore. As mentioned earlier, clear evidence of iron bacteria fossils has been observed in the red matrix (Fig. 4), whereas the gray matrix is characterized by the absence of iron bacteria fossils. In gray facies, the Fe isotopic compositions of the residues are quite constant (δ^{56}Fe values vary from −0.34‰ to 0.23‰) and constitute a "baseline" relative to the Fe isotopic values of the red facies. The gray data match the Fe isotopic signature of the terrestrial baseline established for igneous rocks and low-C_{org} clastic sedimentary rocks (Beard et al., 2003a, b). The Fe isotopic compositions of the gray laminations could reflect the detrital input of iron minerals in the paleoenvironments where no redox processes and no iron bacterial occurrences have been recorded at the water interface, as the oxygen level was too high (Préat et al., 2008a), as was the case, for example, in the gray facies of the shallow platform. Upon introduction into seawater, iron reacts with oxygen and is converted into insoluble oxides (Libes, 1992).

One way to explain the observed differences of the Fe isotopic compositions of the red and gray samples is to consider the early marine diagenetic iron cycle as recently studied by Staubwasser et al. (2006). Within the sequence of early marine diagenetic redox reactions, Fe oxyhydroxides may be reduced in the sediment column along two competing pathways (1) microbial reduction or dissimilatory Fe reduction and (2) abiotic reduction by dissolved sulfide. Following these authors, when Fe is diagenetically recycled between reducing sediments at depth and an oxic top layer, the process may result in the accumulation of light Fe in top layer while the complementary heavier residual is buried. Fe diffusing from the seafloor back into the ocean should reflect the low δ^{56}Fe diagenetic source of dissolved Fe. Another consequence of this recycling is that δ^{56}Fe of the leached reactive Fe is considerably lighter than that of the bulk sediment (Staubwasser et al., 2006).

The gray matrix in the Ammonitico Rosso Veronese (δ^{56}Fe near 0‰) could be regarded as an analog of bulk sediment and the EPS of the Jurassic iron bacteria could be the analog of the "oxic top layer" on the Recent Arabian Sea margin. Chemolithotrophic Fe oxidation by microbes may have affected Fe isotope fractionation during oxidation of porewater $Fe(II)_{aq}$. This process is therefore suggested for the red samples of the Ammonitico Rosso Veronese and the biofilm of *Montacuta ferruginosa*, where ferric oxyhydroxides and ferric phosphates were formed. The precise processes involved in these Fe isotope fractionations have to be investigated by new analyses. It is well established that the largest fractionations in biotic and abiotic systems occur between oxidized and reduced phases of Fe, where ferric Fe species tend to have the highest δ^{56}Fe values (see Severmann et al., 2006 and references therein). Canfield et al. (1993) estimated that individual Fe atoms may undergo hundreds of oxidation–reduction cycles before ultimate burial. Equilibrium fractionation by kinetic effects during mineral precipitation also plays an important role in the process, slow microbial oxidation of Fe(II) that might constitute a biosignature could effectively be overprinted by relatively rapid isotopic equilibration between $Fe(II)_{aq}$ and $Fe(III)_{aq}$ pools (Balci et al., 2006).

5. Conclusions

Despite our somewhat limited number of observations on modern living organisms, this paper offers the preliminary results of a fascinating geomicrobiological investigation of modern biological and ancient geological samples. The following conclusions are proposed on the basis of consistent ecological (biology) and sedimentological observations (geology), and on Fe isotopic analyses of amorphous minerals present in *E. cordatum* and *M. ferruginosa*, and iron minerals associated with the red and gray laminar limestones from the Mid-Jurassic of the Trento Plateau (Lower Rosso Ammonitico Veronese).

The Fe isotopic compositions of the gray facies display values in a narrow range in δ^{56}Fe (around 0‰) and reflect the conservative behavior of Fe during weathering under oxygenated surface conditions (Fig. 5a) (Beard and Johnson, 2004). All other Fe isotope compositions are negative and display significant Fe isotope variations. This is particularly the case for the red facies of the Rosso Ammonitico and it seems also the case for the studied recent marine organisms.

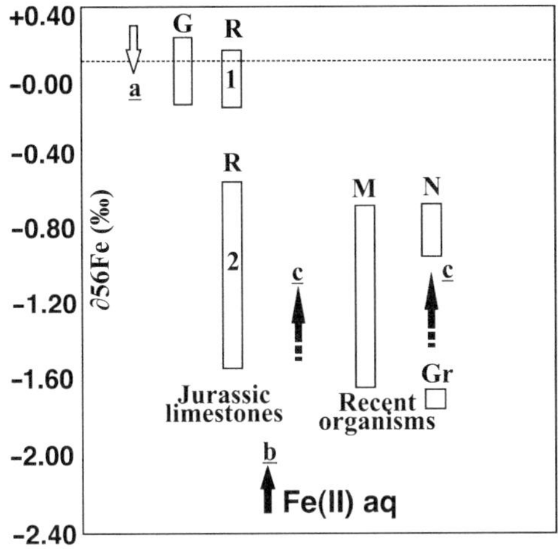

Figure 5. *Abbreviations: G* gray facies, Rosso Ammonitico; *R* red facies, Rosso Ammonitico with 1 for burrowed sediments and 2 for hardgrounds and condensed levels; *M M. ferruginosa*, *N* and *Gr* nodules and granules of *E. cordatum*; *Fe(II)aq* reduced iron in the porewaters. The Fe isotopic composition values are from the iron in the associated minerals; (**a**), (**b**) and (**c**) see text for explanation.

These results are preliminary and need support from future additional analyses. The pristine and the lightest δ^{56}Fe (Fig. 5 Gr) is associated with the iron granules in the intestinal wall of *E. cordatum*, which were probably related to the chemical oxidation of Fe(II) into Fe(III) without the activity of iron microbial communities. This lightest Fe isotopic composition could have resulted of former fractionation in the interstitial fluids present in the porewaters of the silty quartzoze (Fig. 5b) sediment through dissimilatory Fe(III) reduction or other processes as those invoked by Severmann et al. (2006) in the marine porewaters along the California Coast.

Ferrous iron in the nodules of *E. cordatum*, ferric iron in the amorphous minerals of *M. ferruginosa* and in the hematite of the red facies of the Rosso Ammonitico Veronese show an iron isotopic fractionation (at least up to +1‰, Fig. 5c) on a local scale (in the same burrow for the recent organisms and on a millimeter–centimeter scale in the red facies relative to the gray ones) through the activity of iron bacteria. In these cases, iron accumulated passively more specifically in the EPS produced by the filamentous bacteria. Alternating stages of oxidation Fe(II)/Fe(III) occurred near the sediment/water interfaces as a consequence of microenvironmental changes in the marine porewaters. The resulting Fe isotope fractionation is more pronounced in the Jurassic red facies with values lighter than −0.7‰ observed in the hardground levels and in the more condensed facies where bacteria and fungi lived and have accumulated (Préat et al., 2008a) and less pronounced (between 0% and −0.20%) in the red burrowed facies. The oxygen content is probably one of the main controlling factors (Fenchel and Finlay, 1995). Both modern organisms (the sea urchin and the bivalve) studied here thrive in similar microenvironmental conditions as those of the condensed red facies. Their Fe isotope compositions are the same, as is the range of the probable fractionation.

This contribution highlights the role of the EPS in the microbial biofilms. They potentially may provoke a biosignature through the Fe isotope composition of the minerals they induce. This is of particular importance in studying microbial communities and stromatolites in the geologic record. Our study also shows that the iron communities have no paleogeographic and depth significances: the intertidal sandy beach (*E. cordatum* and *M. ferruginosa)* and the hemipelagic carbonate (Rosso Ammonitico) environments display a similar Fe isotopic cycling during early diagenesis despite completely opposite contexts (different lithologies, different depths). We believe the key is EPS.

6. Acknowledgments

This study was financially supported by the Belgian Fonds National de la Recherche Scientifique (FNRS) (FRFC grant no. 2.4.578.08F) to A. Préat and FRFC grant no. 2.4594.07 to C. De Ridder. The FNRS also supported the development of the MC-ICP-MS facility at the ULB, and we are thankful to N. Mattielli and C. Maerschalk for maintaining the facility and its associated labs in excellent shape. F. Poitrasson and J. Wiederhold are thanked for providing the ETH in-house standards. This is a contribution of the Centre Interuniversitaire de Biologie Marine (CIBIM). Comments by L. Martire (U. of Torino, Italy) are gratefully acknowledged.

7. References

Abraham, K., Opfergelt, S., Fripiat, F., Cavagna, A.-J., de Jong, J.T.M., Foley, S., André, L. and Cardinal, D. (2008) δ^{30}Si and δ^{29}Si determinations on USGS BHVO-1 and BHVO-2 reference materials with a new configuration on a Nu Plasma multi-collector ICP-MS. Geostand. Geoanal. Res. **32**: 193–202.

Balci, N., Bullen, T.D., Witte-Lien, K., Shanks, W.C., Motleica, M. and Mandernack, K.W. (2006) Iron isotope fractionation during microbially stimulated Fe(II) oxidation and Fe(III) precipitation. Geochim. Cosmochim. Acta **70**: 622–639.

Beard, B.L., Johnson, C.M., Skulan, J.L., Nealson, K.H., Cox, L. and Sun, H. (2003a) Application of Fe isotopes to tracing the geochemical and biological cycling of Fe. Chem. Geol. **195**: 87–117.

Beard, B.L., Johnson, C.M., Von Damm, K.L. and Poulson, R.L. (2003b) Iron isotope constraints on Fe cycling and mass balance in oxygenated Earth oceans. Geology **31**: 629–632.

Beard, B.L. and Johnson, C.M. (2004). Fe isotopes variations in the Modern and Ancient earth and other planetory bodies. In Geochemistry of non-traditional stable isotopes, Reviews in Mineralogy and Geochemistry, Mineralogical Society of America **55**: 319–357.

Bensing, J.P., Mozley, P.S. and Dunbar, N.W. (2005) Importance of clay in iron transport and sediment reddening: evidence from reduction features of the Abo Formation, New Mexico, U.S.A. J. Sediment Res. **75**: 562–571.

Boulvain, F., De Ridder Ch., Mamet, B., Préat, A. and Gillan, D. (2001) Iron microbial communities in Belgian Frasnian carbonate mounds. Facies **44**: 47–60.

Brantley, S.L., Liermann, L.J. and Bullen, T. (2001) Fractionation of Fe isotopes by soil microbes and organic acids. Geology **29**: 535–538.

Brantley, S.L., Liermann, L.J., Guynn, R.L., Anban, A., Icopini, G.A. and Barling, J. (2004) Fe isotopic fracionation during mineral dissolution with and without bacteria. Geochim. Cosmochim. Acta **68**: 3189–3204.

Buchanan, J.B., Brown, B.E., Coombs, T.E., Pirie, B.J.S. and Allen, J.A. (1980) The accumulation of ferric iron in the guts of some spatangoid echinoderms. J. Mar. Biol. Assoc. UK **60**: 631–640.

Byrne, R.H. and Kester, D.R. (1976) Solubility of hydrous ferric oxide and iron speciation in seawater. Mar. Chem. **4**: 255–274.

Canfield, D.E., Thamdrup, B. and Hansen, J.W. (1993) The anaerobic degradation of organic matter in Danish coastal sediments: Iotn reduction, manganese reduction, and sulfate reduction. Geochim. Cosmochim. Acta **57**: 3867–3883.

Clari, P.A. and Martire, L. (1996) Interplay of cementation, mechanical compaction, and chemical compaction in nodular limestones of the Rosso Ammonitico Veronese (Middle-Upper Jurassic, Northeastern Italy). J. Sediment Res. **66**: 447–458.

Clari, P.A., Marini, P., Pastorini, M. and Pavia, G. (1984) Il Rosso Ammonitico Inferiore (Bajociano-Calloviano) nei Monti Lessini settentrionali. Riv. It. Paleont Strat. **90**: 15–86.

Crosby, H.A., Johnson, C.M., Roden, E.E. and Beard, B.L. (2005) Coupled Fe(II)-Fe(III) electron and atom exchange as a mechanism for Fe isotope fractionation during dissimilatory iron oxide reduction. Environ. Sci. Technol. **39**: 6698–6704.

Gomes da Silva, S., Gillan, D., Dubilier, N. and De Ridder, C. (2006) Characterization by 16S rRNA gene analysis and *in situ* hybridization of bacteria living in the hindgut of a deposit-feeding echinoid (Echinodermata). J. Mar. Biol. Assoc. U.K. **86**: 1209–1213.

de Jong, J.T.M., Schoemann, V., Tison, J.-L., Becquevort, S., Masson, F., Lannuzel, D., Petit, J., Chou, L., Weis, D. and Mattielli, N. (2007) Precise measurement of iron isotopes in marine samples by multi-collector inductively coupled plasma mass spectrometry. Anal. Chim. Acta **589**: 105–119.

de Jong, J., Schoemann, V., Lannuzel, D., Tison, J.-L. and Mattielli, N. (2008) High-accuracy determination of iron in seawater by isotope dilution multiple collector inductively coupled plasma mass spectrometry (ID-MC-ICP-MS) using nitrilotriacetic acid chelating resin for preconcentration and matrix separation. Anal. Chim. Acta **623**: 126–139.

De Ridder, C. (1994) Symbioses between spatangoids (Echinoidea) and *Thiothrix*-like bacteria (Beggiatoales), In: B. David, A. Guille, J.P. Feral and M. Roux (eds.) *Echinoderms Through Time*; *Proceedings of 8th International Echinoderm Conference, Dijon 1993.* Balkema, Rotterdam, pp. 619–625.

De Ridder, C. and Brigmon, R.L. (2003) "Farming" of microbial mats in the hindgut of echinoids, In: W.E. Krumbein, D.M. Paterson and G.A. Zavarzin (eds.) *Fossil and Recent Biofilms: A Natural History of Life on Earth*. Kluwer Academic, Boston, pp. 217–225.

De Ridder, C. and Jangoux, M. (1993) The digestive tract of the spatangoid echinoid *Echinocardium cordatum* (Echinodermata): morphofunctional study. Acta Zool. **74**: 337–351.

De Ridder, C., Jangoux, M. and De Vos, L. (1985) Description and significance of a peculiar intradigestive symbiosis between bacteria and a deposit-feeding echinoid. J. Exp. Mar. Biol. Ecol. **91**: 65–76.

Dideriksen, K., Baker, J.A. and Stipp, S.L.S. (2006) Iron isotopes in natural carbonate minerals determined by MC-ICP-MS with a 58Fe–54Fe double spike. Geochim. Cosmochim. Acta **70**: 118–132.

Emerson, D. and Moyer, C. (1997) Isolation and characterization of novel iron-oxidizing bacteria that grow at circumneutral pH. Appl. Environ. Microbiol. **63**: 4784–4792.

Fantle, M.S. and DePaolo, D.J. (2004) Iron isotopic fractionation during continental weathering. Earth Planet. Sci. Lett. **228**: 547–562.

Fehr, M.A., Andersson, P.S., Hålenius, U. and Mörth, C.-M. (2008) Iron isotope variations in Holocene sediments of the Gotland Deep, Baltic Sea. Geochim. Cosmochim. Acta **72**: 807–826.

Fenchel, T. and Finlay, J. (1995) *Ecology and Evolution in Anoxic Worlds*. Oxford Series in Ecology and Evolution. Oxford University Press, Oxford, pp. 276.

Fortin, D. and Langley, S. (2005) Formation and occurrence of biogenic iron-rich minerals. Earth Sci. Rev. **72**: 1–19.

Gage, J. (1966) Observations on the bivalves *Montacuta substriata* and *M. ferruginosa*, "commensals" with spatangoids. J. Mar. Biol. Assoc. UK **46**: 49–70.

Gillan, D. (2003) The study of a Recent iron-encrusted biofilm in the marine environment, In: W.E. Krumbein, D.M. Paterson and G.A. Zavarzin (eds.) *Fossil and Recent Biofilms: A Natural History of Life on Earth*. Kluwer Academic, Boston, pp. 241–248.

Gillan, D. and De Ridder, C. (1995) The microbial community associated with *Montacuta ferruginosa*, a commensal bivalve of the echnoid *Echinocardium cordatum*, In: R.H. Emson, A.B. Smith and A.C. Campbell (eds.) *Proceedings of 4th European Echinoderm Conference, London*. Balkema, Rotterdam, pp. 71–76.

Gillan, D. and De Ridder, C. (1997) Morphology of ferric iron-encrsuted biofilm forming on the shell of a burrowing bivalve (Mollusca). Aquat. Microb. Ecol. **12**: 1–10.

Gillan, D.C. and De Ridder, C. (2001) Accumulation of a ferric mineral in the biofilm of *Montacuta ferruginosa* (Mollusca, Bivalvia). Biomineralization, bioaccumulation, and inference of paleoenvironments, Chem. Geol. **177**: 371–379.

Gillan, D.C., Speksnijder, A.G.C.L., Zwart, G. and De Ridder, C. (1998) Genetic diversity of the biofilm covering *Montacuta ferruginosa* (Mollusca, Bivalvia) as evaluated by denaturing gradient gel electrophoresis analysis and cloning of PCR-amplified gene fragments coding for 16S rRNA. Appl. Environ. Microbiol. **64**: 3464–3472.

Gillan, D.C., Warnau, M., De Vrind-de-Jong, E.W., Boulvain, F., Préat, A. and De Ridder, C. (2000) Iron oxidation and deposition in the biofilm covering *Montacuta ferruginosa* (Mollusca, Bivalvia). Geomicrobiol. J. **17**: 147–151.

Hallberg, R. and Ferris, F.G. (2004) Biomineralization by *Gallionella*. Geomicrobiol. J. **21**: 325–330.

Hofstetter, T.B., Schwarzenbach, R.P and Haderlein, S.B. (2003) Reactivity of Fe(II) species associated with clay minerals. Environ. Sci. Technol. **37**: 519–528.

Icopini, G.A., Anbar, A.D., Ruebush, S.S., Tien, M. and Brantley, S.L. (2004) Iron isotope fractionation during microbial reduction of iron: the importance of adsorption. Geology **32**: 205–208.

Jenkyns, H.C. (1974) Origin of red nodular limestones (Ammonitico Rosso, Knollenkalke) in the Mediterranean Jurassic: a diagenetic model, In: K.J. Hsü and H.C. Jenkyns (eds.) *Pelagic Sediments: On Land and Under Sea 1*. International Association Sedimentologists Special Publication, pp. 249–271.

Johnson, C.M., Beard, B.L., Roden, E.E., Newman, D.K. and Nealson, K.H. (2004) Isotopic constraints on biogeochemical cycling of Fe. Rev. Min. Geochem. **55**: 359–408.

Johnson, C.M., Beard, B.L. and Roden, E.E. (2008) The iron isotope fingerprints of redox and biogeochemical cycling in Modern and Ancient Earth. Annu. Rev. Earth Planet. Sci. **36**: 457–493.

Krynine, P.D. (1949) The origin of red beds. N.Y. Acad. Sci. Trans. Ser. II **2**: 60–68.

Krynine, P.D. (1950) Petrology, stratigraphy, and origin of the Triassic sedimentary rocks of Connecticut. CT. Geol. Surv. Bull. **73**: 273.

Libes, S.M. (1992) An introduction to marine biogeochemistry. Wiley, New York, 734 pp.

Little, B.J., Wagner, P.A. and Lewandowski, Z. (1997) Spatial relationships between bacteria and mineral surfaces, In: J.F. Banfield and K.H. Nealson (eds.) *Interaction Between Microbes and Minerals.* Rev. Min. **35**: 123–159.

Loreau, J.P. (1972) Pétrographie des calcaires fins au microscope électronique à balayage: introduction à une classification des micrites. C.R. Acad. Sci. Paris **274**: 810–813.

Mamet, B. and Préat, A. (2003) Sur l'origine de la pigmentation de l'Ammonitico Rosso (Jurassique, région de Vérone, Italie du Nord). Rev. Micropal. **46**: 35–46.

Mamet, B. and Préat, A. (2006) Iron-bacterial mediation in Phanerozoic red limestones: state of the art. Sediment. Geol. **185**: 147–157.

Maréchal, C., Télouk, P. and Albarède, F. (1999) Precise analysis of copper and zinc isotopic compositions by plasma source mass spectrometry. Chem. Geol. **156**: 251–273.

Martire, L. (1996) Stratigraphy, facies and synsedimentary tectonics in the Jurassic Rosso Ammonitico Veronese (Altopiano di Asiago, NE Italy). Facies **35**: 209–236.

Massari, F. (1981) Cryptalgal fabrics in the Rosso Ammonitico sequences of Venetian Alps, In: A. Farinaci and S. Elmi (eds.) *Rosso Ammonitico Symposium Proceedings*. Tecnoscienza, Roma, pp. 435–469.

Miller, D.N. and Folk, R.L. (1955) Occurrence of detrital magnetite and ilmenite in red sediments: new approach to significance of redbeds. Am. Assoc. Petrol. Geol. Bull. **39**: 338–395.
Munn, C.B. (2004) *Marine Microbiology: Ecology and Application*. Garland, London, 282p.
Poitrasson, F. and Freydier, R. (2005) Heavy iron isotope composition of granites determined by high resolution MC-ICP-MS. Chem. Geol. **222**: 132–147.
Préat, A., Mamet, B., Bernard, A. and Gillan, D. (1999) Bacterial mediation, red matrices diagenesis, Devonian, Montagne Noire (southern France). Sediment. Geol. **126**: 223–242.
Préat, A., Morano, S., Loreau, J.P., Durlet, C. and Mamet, B. (2006) Petrography and biosedimentology of the Rosso Ammonitico Veronese (Middle-Upper Jurassic, Northeastern Italy). Facies **52**: 265–278.
Préat, A., de Jong, J., Mamet, B. and Mattielli, N. (2008a) Stable isotope and microbial mediation in red pigmentation of the Rosso Ammonitico (Mid-Late Jurassic, Verona Area, Italy). Astrobiology **8**: 841–857.
Préat, A., El Hassani, A. and Mamet, B. (2008b) Iron bacteria in Devonian carbonates (Tafilalt, Anti-Atlas, Morocco). Facies **54**: 107–120.
Schoemann, V., de Jong, J.T.M., Lannuzel, D., Tison, J.-L., Dellile, B., Chou, L., Lancelot, C. and Becquevort, S. (2008) Microbiological control on the cycling of Fe and its isotopes in Antarctic sea ice. Geochim. Cosmochim. Acta **72**(12): A209.
Severmann, S., Johnson, C.M., Beard, B.L. and McManus, J. (2006) The effect of early diagenesis on the Fe isotope compostions of porewaters and authigenic minerals in continental margin sediments. Geochim. Cosmochim. Acta **70**: 2006–2022.
Shiel, A.E., Barling, J., Orian, K.J. and Weis, D. (2009) Matrix effects on the multi-collector inductively coupled plasma mass spectrometric analysis of high-precision cadmium and zinc isotopes. Anal. Chim. Acta **633**: 29–37.
Staubwasser, M., von Blanckenburg, F. and Schoenberg, R. (2006) Iron isotopes in the early marine diagenetic iron cycle. Geology **34**: 629–632.
Taylor, P.D.P., Maeck, R. and De Bièvre, P. (1992) Determination of the absolute isotopic composition and atomic weight of a reference sample of natural iron. Int. J. Mass Spectrom. **121**: 111–125.
Teutsch, N., von Gunten, U., Porcelli, D., Cirpka, Q.A. and Halliday, A.N. (2005) Adsorption as a cause for iron isotope fractionation in reduced groundwater. Geochim. Cosmochim. Acta **69**: 4175–4185.
Thorsen, M.S. (1998) Microbial activity, oxygen status and fermentation in the gut of the irregular sea urchin *Echinocardium cordatum* (Spatangoida: Echinodermata). Mar. Biol. **132**: 423–433.
Thorsen, M.S., Wieland, A., Ploug, H., Kragelund, C. and Nielsen, P.H. (2003) Distribution, identity and activity of symbiotic bacteria in anoxic aggregates from the hindgut of the sea urchin *Echinocardium cordatum*. Ophelia **57**: 1–12.
van Houten, F.B. (1961) Climate significance of red beds, In: A.E.M. Nairn (ed.) *Descriptive Paleoclimatology*. Interscience, New York, pp. 89–139.
van Houten, F.B. (1973) Origin of red beds. A review. Annu. Rev. Earth Planet. Sci. **1**: 39–61.
Walker, T.R., Ribbe, P.H. and Hoena, R.M. (1967) Geochemistry of hornblende alteration in Pliocene red beds, Baja California, Mexico. Geol. Soc. Am. Bull. **78**: 1055–1060.
Wasylenki, L.E., Anbar, A.D., Liermann, L.J., Mathur, R., Gordon, G.W. and Brantley, S.L. (2007) Isotope fractionation during microbial metal uptake measured by MC-ICP-MS. J. Anal. Atom. Spectrom. **22**: 905–910.
Williams, H.M., McCammon, C.A., Peslier, A.H., Halliday, A.N., Teutsch, N., Levasseur, S. and Burg, J.-P. (2004) Iron isotope fractionation and the oxygen fugacity of the mantle. Science **304**: 1656–1659.
Zengh, Y., Anderson, R.F., Van Geen, A. and Kuwabara, J. (2000) Authigenic molybdenum formation in marine sediments: a link to porewater sulfide in the Santa Barbara Basin. Geochim. Cosmochim. Acta **64**: 4165–4178.

Biodata of **Dr. Evgenia Leonidovna Sumina** and **Dmitryi Leonidovich Sumin**, authors of ***"New Representations on the Nature of Stromatolites"***

Dr. Evgenia Leonidovna Sumina received her degree from the biological faculty of the Moscow State Pedagogical Institute in 1985. Currently, she is working as a teacher in the Department of Paleontology of the Geological Faculty in Moscow State University. She was researcher in the laboratory of the most ancient organisms of Borissyak Paleontological Institute, RAS, Moscow. In 2008, she defended the dissertation for scientific degree of Candidate Biological Sciences. Her scientific interests are stromatolites, experimental study of cyanobacterial communities, and the evolutionary stages of ancient life. She has 20 publications to her credit.

E-mail: **stromatolit@list.ru**

Dmitryi Leonidovich Sumin received his degree from the Department of Paleontology of the Geological Faculty, Moscow State University, in 1988. His scientific interests are vertebrate paleontology and the theory of evolution. He is an independent researcher and has eight publications so far.

E-mail: **stromatolit@list.ru**

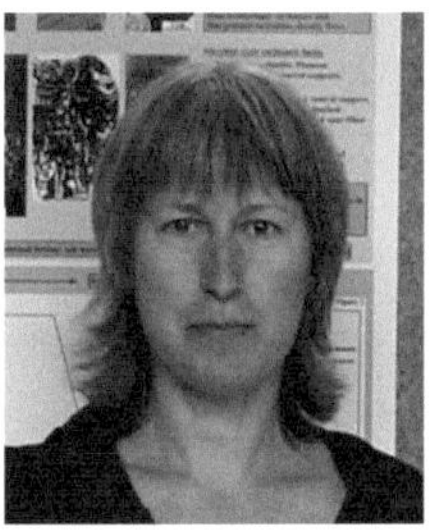

Evgenia Leonidovna Sumina

Dmitryi Leonidovich Sumin

V.C. Tewari and J. Seckbach (eds.), *STROMATOLITES: Interaction of Microbes with Sediments*, Cellular Origin, Life in Extreme Habitats and Astrobiology 18, 675–686
DOI 10.1007/978-94-007-0397-1_30, © Springer Science+Business Media B.V. 2011

NEW REPRESENTATIONS ON THE NATURE OF STROMATOLITES

EVGENIA LEONIDOVNA SUMINA[1] AND DMITRYI LEONIDOVICH SUMIN[2]
[1]*Borissyak Palaeontological Institute RAS, Profsouznaya Str., 123, Moscow, Russia*
[2]*Independent Researcher, Zhebrunova Str., 5.44, Moscow, Russia*

Keywords Stromatolites • Cyanobacteria • Filaments • Morphogenesis • Organism • Skeleton

According to fossil records stromatolites are known to have existed since 3.5 billion years ago. They are considered a product of the interaction of life (microorganisms) with the sediments. The use of stromatolites for stratigraphic purposes indicates their complexity in column forms and time-controlled variation in morphology. Such results are in contrast to the theory of the low-level integrity of the cyanobacterial community as an individual microorganic settlement. In some cases microorganic settlements cannot be distinguished by their stability and succession to total morphological characters over whole geological periods.

The main reason for the nature of stromatolite morphology may be explained by their ability to form a community, both in the presence and absence of mineral deposits. So the main objectives of our investigation are (1) experimental study of the formative reactions in recent microbial communities of filamentous cyanobacteria; (2) the detection of the integrity level of the community and; (3) the comparison of recent microbial community features with those of fossil stromatolites.

The pioneer workers on stromatolites have considered mainly two factors: stromatolite formation as a result of colony formation and as a microbial buildup. It means that stromatolite formation is the interaction of mineral matter with microbial fibers assembled in the colony. On the other hand, morphological features of stromatolites are described as genus and species on the basis of the coordinated activity of fibers and secondary presence of heredity coordination. The previously known concepts of stromatolite investigations do not answer the main question – What are the factors responsible for the nature of existence of definite forms of stromatolites and their morphological stability in time?

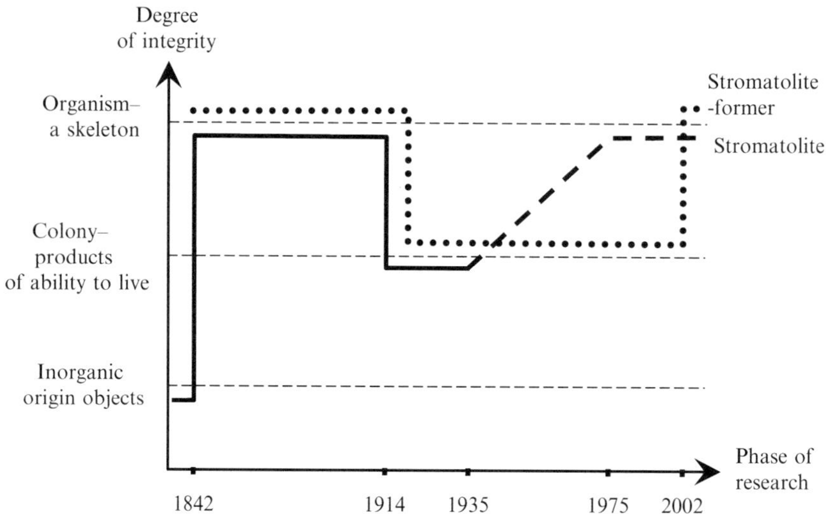

Figure 1. Scheme of changing opinions about former stromatolite and stromatolite.

The authors have attempted to explain this problem by comparing the data for stratigraphic distribution of stromatolites in geological sections and laboratory observations on the behavior of filamentous cyanobacteria and their communities.

Stromatolite investigation dates back to more than 150 years (Fig. 1). Earlier, stromatolites were considered a nonorganic species. Since the organic nature of these fossils was recognized, the attribute of sponges, corals, and stromatoporoids as skeletons of the multicellular eukaryotes or integrated organisms has been recognized. Walcott (1914) described the chains of spherical cells in layers of fossil stromatolites and compared them with filamentous blue-green algae. The main conclusions of Walcott's research established the blue-green algal nature of stromatolites. This point of view is confirmed by modern investigations of the presence of blue-green algae in modern stromatolites on Bahamas bank (Black, 1933).

In the early twentieth century, there was some attempt to use stromatolites for stratigraphy (Dingelshtedt, 1935). The complexes of stromatolites were established for identification of the biostratigraphical zonation. Stromatolites were also found useful for interregional and intercontinental correlations (Korolyuk, 1956; Krylov, 1963; Keller, 1966; Semikhatov, 1962; Komar, 1966; Semikhatov and Raaben, 2000). Therefore, the distribution of stromatolites in the Precambrian played the same role as skeletal organisms (fossils) in the Phanerozoic period.

Stromatolites are considered a tool for biostratigraphic reconstructions; therefore, hierarchy in structure, morphological changes within a single section, mutual correlation, adaptive nature, and irreversible time-change should occur. However, there is a controversy regarding the potential of stromatolites to form a colony – i.e., a combination of organisms, living densely together temporarily or permanently. If the colony status of stromatolites is true, they cannot be useful in stratigraphy. Heredity is an obligatory quality for biostratigraphy, but not necessarily for the formation of colonies. Only a combination of environmental influence and heredity fixing can be used for biostratigraphy (Menner, 1962). This is a point of contention between theoretical suggestions about the formation of stromatolites and their practical application in geology.

The time-controlled changes in the columnar stromatolites from many sites have shown some evolutionary variations. The Riphean stromatolite morphology changed regularly, slowly, and irreversibly. The change occurred simultaneously at three hierarchical levels of structure: form (whole buildup), structure (details of buildup), and microstructure (at the elementary level). The common tendencies are (1) complication in types of branching and outlines of columns (from passively branching to actively branching with a complex form of columns); (2) changing of sides of columns (from nonflat with falling edges to columns fully multilaminated); and (3) variation of microstructure (from single and flat to complex, dense and elongated). The same variation was also observed in the Riphean stromatolites in the Ural, East Siberia, and northern part of the Siberian platform (Krylov, 1963; Semikhatov, 1962; Komar, 1966).

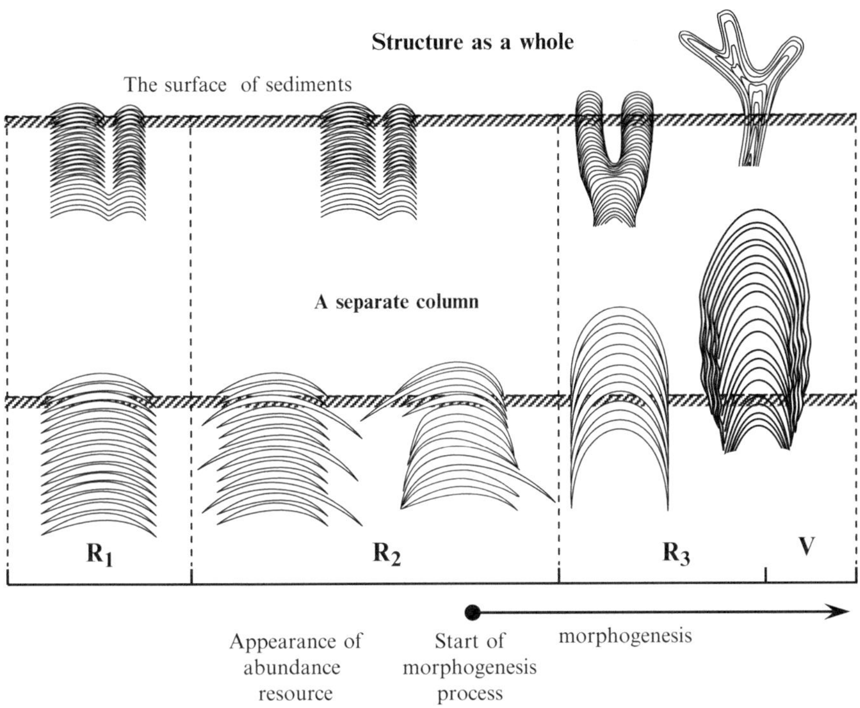

Figure 2. Scheme of morphogenetic transformations of Riphean stromatolites.

Morphogenesis in Riphean stromatolites is illustrated in Fig. 2. At Lower Riphean, the stromatolites were more abundant with nearly flat layers. In Middle Riphean, stromatolites with peaks and ridges were common. These morphological changes may be attributed to the evolution of specific photosynthesis and more intensive carbonate deposition. The appearance of additional features that characterized Middle Riphean helped the stromatolite to evolve side mantling instead of peak forming and layer steepness. This tendency increased in Upper Riphean when layer prominence evolved, and mantling also increased and became multilayer. Stromatolite taxa acquired species with the so-called "wall." Wall firmness allows stromatolite to construct columns with actively complex branching (Sumina and Sumin, 2006).

In summary, all such morphological variations provided photosynthetic assistance to the stromatolite to sufficiently increase its extent; due to the increased height of the columns, its elevation increased and permitted the film to cover lateral parts of the column in addition to the upper surface. The regular evolutionary changes have been observed by various authors for different groups of stromatolites and interpreted as a reflection of natural evolution. However, in recent studies on stromatolites, a controversy about the colony status of stromatolite formation exists.

The controversy is the result of assumptions about the nature of objects forming stromatolites, whose life-activity is to coordinate fossil regularities. They indicate symbiosis, master-species, various algae complexes, a stable community of algae reaching high cenosis, and even the presence of control mechanisms for the community. The definition leads to macroscopical discrepancy, with several parts and different mineral constituents appearing as a result of the assembly of macroscopic organisms. The concept of "colony," especially for prokaryotic organisms, is insufficient for describing a creation as complex as the stromatolite.

The results of the analysis of published reports on stromatolites were summarized in the preceding section. The next is devoted to the formation of stromatolites. The most probable early stromatolite appears as a filament cyanobacterial community, existing mainly as a film. So the terms "film" and "community" are used synonymously in this chapter. One needs to note that a "community" means the density of the settlement of independent elements or filaments. The film is a dense skin-type formation, consisting of randomly interwoven filament cyanobacteria.

It is typical for them to thicken up to 1 mm as a result of the sliding motion. In our experiments, we used the culture of thermophilic cyanobacterial genus *Oscillatoria terebriformis* as the main component that regulates the film structure and *Phormidium angustissimum* as the minor component. In some experiments, the culture of halophilic genus *Microcoleus chthonoplastes* was used.

The filaments inside of the film move slowly but constantly. Since the 1980s, such a lime-type community has been used in model experiments for stromatolite constructions, with Geological institute Russian Academy of Sciences (RAS) and Institute microbiology RAS. The model experiments were followed to obtain living and growing biomodels, morphologically identical to fossil stromatolites.

The main morphotypes of building, stratiform, actively branching, and passively branching (Orleanskij and Raaben, 1996, 1997) were obtained, but the mechanism of building formation could not be determined.

So the building of biological communities was observed, in the presence or absence of mineral deposits. The experiments established the ability of the community to morphogenesis, expressed by the formation of several structures such as the film, photosynthetic networks, envelopes of gas bubbles, multibeam units, fibers, rings, and capillary structures (Fig. 3). All of them differ in form, and the functions carried out arise originally in the homogeneous film; hence, the ability of the community to form them can be interpreted as a differentiation (Sumina, 2006). However, unlike the majority of eukaryotes, the differentiation in community cyanobacteria is reversible, that is, arising structures are not constant and by a regrouping of filaments can be transformed to a skin-like film. In experiments also, it has been shown that filaments react to various influences not as separate individuals but as parts of the complete structure. So filaments that are not damaged take part in the restoration of a damaged site. Besides, the phenomenon of redistribution of the substance in a film is noted. It is due to oxygen produced by cyanobacteria during photosynthesis.

Regarding stromatolites, it is possible that differences in time and type of microstructure of stromatolite layers, namely carbonate, cause interposition of grains as a result of the same processes of redistribution. Sedimentation of a deposit on the surface of a film delays the process of photosynthesis that, possibly, has led to the ability of filaments to rearrange mineral particles. To delay photosynthesis and reduce the time of crawling of filaments through a deposit, dropping out particles were formed by filaments in the form of compact accumulations, with free space between them. It enabled not only the individual filaments but also their groups to crawl through a layer of the dropped carbonate. It is necessary to note that this is only an assumption based on a comparison of the types of microstructures and their changes over time. Tewari and Joshi (1993) have studied the microstructures of the Riphean, Vendian, and Lower Cambrian stromatolites from the Lesser Himalayas of India and tried to establish that the major changes in the stromatolite microstructures are time-controlled.

To sum up the results of the experiments (1) it appears that filaments act not as independent elements but as parts of the whole, (2) there is the phenomenon of differentiation, and (3) there is the phenomenon of redistribution of substance. On this basis, it is possible to assume that the community of cyanobacteria does not have a colonial but an organism-like level of integrity.

Their organism-like behavior may be explained as follows. In the literature, replace "stromatolite former" with colony, biocenose, or symbiosis. It is clear that the same object cannot belong to all the listed categories because each of them reflects different levels of integrity. For a definition of the stromatolite status, it is necessary to compare the interaction character of the filaments in a film to those of the elements in each category. Proceeding from definitions of these terms, interaction and attitudes can be characterized as (1) an origin (a heredity) of elements, (2) their reaction against each other and on the system as a whole,

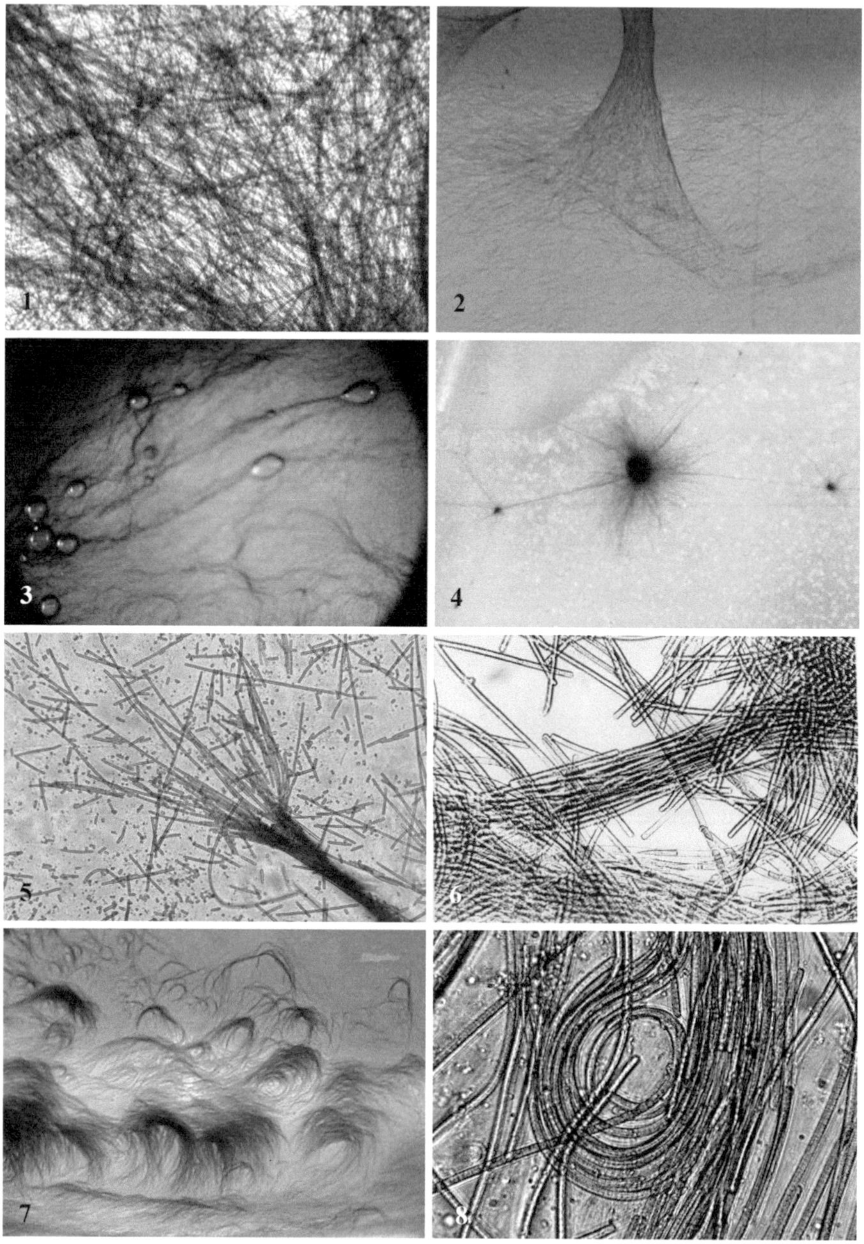

Figure 3. Structures formed from filaments by the (**1**) film, (**2**) photosynthetic networks, (**3**) envelopes of gas bubbles, (**4**) multibeam units, (**5,6**) fibers, (**7**) rings, and (**8**) capillary structures (thickness of filament 4 mcm).

(3) trophic interactions and (4) the ability for morphogenesis. In a film, (1) filaments, most likely, possess the general heredity as they do not have a sexual process; (2) filaments actively support integrity of a film, for example, restoration of its structural continuity after destruction, and allocation of the differentiated sites; (3) there is no competition between filaments, particularly for lighting and thus, the film can be carried to objects or organisms at a level of integrity, instead of to a colony, biocenose, or symbiosis; (4) the ability for morphogenesis is a property of organisms, and as the film possesses this ability, it can belong only to organism category.

It is necessary to note that a cyanobacterial film, seen in an organismic view, differs from "organisms" in a number of features. The film does not have irreversible differentiation, a constant size, or a developed internal content. In most cases, the mutual formation of structures in a film is not caused both in space and time. Dependence on the formation of mutually connected structures is observed only in such pairs as fibers – envelopes of gas bubbles and fibers as multibeam units. Correlation is noted also in consecutive stages of the formation of the film after deposit (Sumina, 2005). The existence of the internal content appears in all morphogenesis processes with the participation of a mineral deposit, resulting in stromatolite formation. It is expressed in interdependence of morphological signs of constructions and stability of morphotypes. The reason in favor of the presence of the rudimentary internal content might be the establishment of a fermentative mechanism for the regulation of a calcium carbonate by means of the presence of carboanhydrase, recently found in the matrix ("glycocalix") of the filament of cyanobacteria (Kupriyanova et al., 2004).

This allows us to assume that in the cyanobacterial community, an organism-level structure exists at the stage of morphogenesis, which precedes the occurrence of irreversible differentiations, steady correlations, and developed internal content.

After considering the stromatolite formation, it is necessary to also define the status of the mineral formation as a result of the interaction with a deposit, namely stromatolite. From the thesis about the organismic level of stromatolite-former integrity, the assumption follows that stromatolites can be considered a product skeleton performing the function of a cyanobacterial community.

To resolve whether such an assumption is acceptable, it is important to note two points, (1) the main part of the mineral in stromatolites is biogenic in nature; (2) the form and structure of the stromatolite is defined by a cyanobacterial film as a matrix, and the mechanism of action is enzyme dependent. Stromatolites have their own spatial organization and hierarchical connection of elements that is the final result of skeleton forming (Barskov, 1984). The biomineralization-type (Lowenstam, 1984) process occurs for stromatolites as a transition from an induced mineralization, which results in complication and microstructure ordering.

It is considered impossible to identify stromatolites with a skeleton as they are not produced by soft fabrics and "fabrics" are not generated yet. The cyanobacterial film also cannot be named "a soft body." But it is obvious that the film represents itself as a matrix in relation to the mineral layers of stromatolites. Layers are formed as a result of its live activity. The mineral layers perform the basic, supporting function for a mucous and soft film. Due to the presence of the form-building function, stromatolites have a certain, constant form, following adaptive laws. Stromatolite formation, therefore, can be considered an evidence of the morphogenetic opportunities of the community in conditions of deposit sedimentation. Morphology details, therefore, are equivalent to the structures of a film in the absence of a deposit.

The integrity of stromatolites as "skeletal" formations is shown in the features of their evolution: direction, irreversibility, and changes. This is most clearly expressed by columnar stromatolites, which are forms with evident and distinctly observable attributes. It is important to remember that it has been established that in the Riphean interval of geological history, that stromatolites were in abundance. Thus, the conclusions have been made for an interval when their morphological opportunities were expressed and fixed in terms of mineral records.

The form of construction during the Riphean indicates the structural and functional continuity that provides biological control, which is the characteristic of the film as a whole.

The indirect evidence of the mineral body as a skeleton may point to the relation between the number of groups and the number of taxons, for example, a genus rank (Fig. 4). This relation shows that there are no combinations of attributes and correspondence to the number of real taxons. The lack of combinations is the result of a soft body. It permits the comparison of stromatolites with fossil skeletons such as eukaryotes, for example, *Anthozoa*, *Bivalvia*, and *Ammonitida* (accordingly, 10, 10, 14 and 12) on this parameter. It indicates that there is a similar level of soft body control above the mineral formation.

Thus, stromatolites are formed entirely by the biological object. Their structure is hierarchical, that is, they are coordinated with each other and with the whole formation. The evolution of stromatolites seems stable and there is

Groups of organisms / Comparison parameters	Stromatolit-former 	Anthozoa 	Bivalvia 	Ammonitida 
Quantity of genera*	70	560	430	430
Group of characters	7	55	30	35
Relation	10	10	14	12

*Known from territory of the former USSR

Figure 4. Comparison of internal correlation degree of stromatolit-former and skeletons of some invertebrates.

orientation and stability over time, as well as an adaptive ability. Hence, stromatolites can be considered prokaryote analogs of the skeleton multicellular eukaryote.

A study of the geological history of the Archean period shows that even at that time, all the basic morphological signs of stromatolites were developed. They were characterized by a variety of laminations and constructions, including the branching forms (Hofmann, 2000). A comparison of the morphological varieties of the Archean and Proterozoic forms reveals that their structure is practically similar to the Riphean, despite fewer taxons in the Archean (Semikhatov and Raaben, 1996).

The presence of these processes implies that such a community possessed a certain integrity, an ability to adapt in relation to the environment, control over moving of elements inside the system, and the hierarchy of the elements composing it. These processes developed within the limits of formed prokaryote "macroorganic" cyanobacterial communities. It can be seen as an indication of the level of development of life in a given period of geological time. It is necessary to note that the development of such basic biological processes occurred under conditions of the regenerative hydrosphere as a whole, but the events occurred under microaerophilic conditions which developed due to the photosynthetic activity of cyanobacterial communities.

The conclusions can be summarized as follows:

1. It is experimentally shown that the cyanobacterial community that forms stromatolites possesses the complete macroscopic properties of an organism.
2. Stromatolites are formed due to an interaction of carbonate mineral formation (sediments) and cyanobacterial communities.
3. The biostratigraphic conclusions based on the study of stromatolites are debatable because the stromatolite-forming mechanisms maintain integrity and appear as co-subordinated morphological features of constructions.

1. Acknowledgments

We are grateful to Professor V.C. Tewari and Professor J. Seckbach, editors of the book, for inviting us to contribute this chapter. Review comments and suggestions from Professor Vinod C. Tewari were very useful to improve the chapter. The work was supported by the grant RFBR No. 080400484.

2. References

Barskov, I.S. (1984) Paleontological aspects of biomineralization. Reports 27 Intern. Geol. Congress. Section C.02. 2. Moscow: 61–66 (in Russian).

Black, M. (1933) The algal sediments of Andros island, Bahamas. Philos. Trans. Roy. Soc. Lond. Ser. B. **222**: 165–192.

Dingelshtedt, N.N. (1935) To a question on an origin dolomite and about age thicknesses of Southern Urals Mountains. Ann. Russ. Miner. Soc. **64**(2): 43–51 (in Russian).

Hofmann, H.J. (2000) Archean stromatolites as microbial archives, In: R.E. Riding and S.M. Awramik (eds.) *Microbial Sediments*. Springer, Heidelberg, pp. 315–326.

Keller, B.M. (1966) Divisions uniform stratigraphic scales Precambrian. Rep. Acad. Sci. USSR. **171**(6): 1405–1408 (in Russian).

Komar, V.A. (1966) *Stromatolites of Upper Precambrian Adjournment of the North of the Siberian Platform and their Stratigraphic Value*. M. Publishing, Ann Arbor, p. 122 (in Russian).

Korolyuk, I.K. (1956) Some stromatolites from Cambrian of the Irkutsk amphitheater. Proc. Inst. Oil Acad. Sci. USSR. **7**: 51–59 (in Russian).

Krylov, I.N. (1963) *Columnar Branching Stromatolites from Riphean Adjournment of Southern Urals Mountains and Their Importance for Stratigraphic Upper Precambrian*. M. Publishing, Ann Arbor, p. 133 (in Russian).

Kupriyanova, E.V., Markelova, A.G., Lebedeva, N.V., Gerasimenko, L.M., Zavarzin, G.A. and Pronina, N.A. (2004) Carboanhydrasa alkalophilic cyanobacteria Microcoleus chthonoplastes. Microbiology **73**(3): 307–311 (in Russian).

Lowenstam, H. (1984) Processes and products of a biomineralization and evolution of a biomineralization. Reports 27 Intern. Geol. Congress. Section C.02., 2. Moscow: 51–56 (in Russian).

Menner, V.V. (1962) Biostratigraphic bases of comparison of sea, lagoon and continental retinues. Proc. Geol. Inst. **65**: 375 (in Russian).

Orleanskij, V.K. and Raaben, M.E. (1996) Laboratory model analogue of oncoid and stromatolite. Algology **6**(1): 57–61.

Orleanskij, V.K. and Raaben, M.E. (1997) Laboratory model analogue of non-actively branching columnar stromatolites. Algology **7**(2): 185–188 (in Russian).

Semikhatov, M.A. (1962) *Riphean and Lower Cambrian of the Yenisei Range*. M. Publishing House, Ann Arbor, pp. 192–228 (in Russian).

Semikhatov, M.A. and Raaben, M.E. (1996) Dynamics of a global variety of stromatolites. Proterozoic Clause 2. Africa, Australia, Northern America and the general synthesis. Stratigr. Geol. Correl. **4**(1): 26–55 (in Russian).

Semikhatov, M. and Raaben, M. (2000) Proterozoic stromatolite taxonomy and biostratigraphy, In: R.E. Riding and S.M. Awramik (eds.) *Microbial Sediments*. Springer, Heidelberg, pp. 295–306.

Sumina, E.L. (2005) To a question on the nature of stromatolites. Paleobiology and detailed stratigraphy of Phanerozoic. To the 100 anniversary from the date of birth of Acad. V.V.Mennera. M.: the Russian Academy of Natural Sciences, 20–26 (in Russian).

Sumina, E.L. (2006) Behavior of filamentous cyanobacteria in laboratory culture. Microbiology **75**(4): 459–464.

Sumina, E.L. and Sumin, D.L. (2006) A superfluous resource as the factor morphogenesis. Problems of evolutionary morphology of animals. Theses Intern. Conf. Memories Acad. A.V.Ivanov. St.-Petersburg, pp. 114–115 (in Russian).

Tewari, V.C. and Joshi, M. (1993) Stromatolite microstructures: a new tool for biostratigraphic correlation of the Lesser Himalayan carbonates. Him. Geol. **4**(2): 19–29.

Walcott, C.D. (1914) Pre-cambrian algonkian algal flora. Smithsonian misc. Collections. **64:**(2) 77–156.

Biodata of **Professor Dr. Harald Strauss**, author of ***"Sulfur Isotopes in Stromatolites"***

Prof. Dr. Harald Strauss is currently a full Professor at the Institut für Geologie und Paläontologie, Westfälische Wilhelms-Universität Münster, Germany. He obtained his Ph.D. from the University in Göttingen in 1983. He was a Postdoctoral Fellow at Indiana University in Bloomington, USA (1986–1987), and at the University of California at Los Angeles, USA (1987–1988). He returned to Germany in 1988 to the Ruhr-Universität Bochum, where he obtained his Habilitation in 1994. Prof. Strauss' scientific interest is in the field of stable isotope research in sedimentary systems and the reconstruction of ancient biochemistries.

Email: **hstrauss@uni-muenster.de**

V.C. Tewari and J. Seckbach (eds.), *STROMATOLITES: Interaction of Microbes with Sediments*, Cellular Origin, Life in Extreme Habitats and Astrobiology 18, 687–701
DOI 10.1007/978-94-007-0397-1_31, © Springer Science+Business Media B.V. 2011

SULFUR ISOTOPES IN STROMATOLITES

HARALD STRAUSS
Institut für Geologie und Paläontologie, Westfälische Wilhelms-Universität Münster, Corrensstraße 24, 48149, Münster, Germany

Abstract Intense biological sulfur cycling characterizes modern marine microbial mats as well as their ancient counterparts – stromatolites. Respective microbially driven processes, such as bacterial sulfate reduction, sulfide oxidation, or the disproportionation of sulfur-bearing compounds of intermediate oxidation states, are associated with distinctive and frequently substantial fractionations of their four stable sulfur isotopes. These biosignatures are reflecting microbial sulfur cycling in modern and ancient sedimentary environments alike, whether they are stromatolitic or nonstromatolitic. Sulfur isotopes provide a unique opportunity to identify and quantify biological sulfur metabolism on Earth and possibly beyond.

Keywords Sulfur isotopes in stromatolites • Biological sulfur • Biosignatures • Chemofossils • Metabolism • Microbial mats • Precambrian • Microbes • Photosynthetic organism • Anaerobic heterotrophs • Chemolithoautotrophs • Geochemical environment

1. Introduction

Stromatolites provide a unique record of microbial life and respective activity throughout Earth history with the oldest example dating back to ~3.5 Ga (Allwood et al., 2006). Stromatolites, microbial mats, and biofilms – the complex terminologies in part already reflect the observation that microbial life has conquered all available environments, both in terms of marine and nonmarine habitats and in the modern and ancient world alike. Included are microbially mediated sedimentary deposits that originate through trapping and binding of particles or via active precipitation. In particular, throughout the Precambrian, stromatolites represent the most impressive evidence for life on our planet.

Although each occurrence is unique in its composition, the vertical architecture of most microbial mats in surface environments displays some overarching features, reflecting the principal functional groups of microbes: (a) oxygenic photosynthesizers being the primary producers, (b) anoxygenic photosynthetic organisms using sulfide as electron donor, (c) aerobic heterotrophs consuming organic carbon via aerobic respiration, (d) fermenters using organic carbon or sulfur compounds, (e) predominantly sulfate reducers as anaerobic heterotrophs, and (f) sulfide oxidizing organisms, many of which are chemolithoautotrophs (Dupraz and Visscher, 2005). A somewhat different picture emerges when considering microbial mats in environments devoid of light energy such as in cave deposits or deep sea communities.

The specific community structure at a given site will be dependent upon the geochemical environment in terms of the available element budget. As such, microbial mats reflect and archive pertinent element cycling of the respective depositional environment and offer the possibility for reconstructing environmental characteristics and microbially driven processes in the geological past. In marine settings, mat communities exhibit a complex suite of microbial sulfur metabolisms due to the abundance of dissolved oceanic sulfate. More so, microbial sulfur turnover plays the most important role in stromatolite formation due to the coupling of bacterial sulfate reduction and the precipitation of calcium carbonate.

Sulfur isotopes provide a unique analytical tool to unravel microbial sulfur metabolism and, hence, a possibility to track respective microbial activity through Earth's history. Spatial resolution ranges from the ecosystem level to high-resolution micron-scale mapping of variations in sulfur abundance and isotopic composition (Fike et al., 2008). The fact that different metabolic processes impart different sulfur isotope signatures represents the foundation for utilizing sulfur isotopes as a prominent biosignature. Its unequivocal identification requires an understanding of the respective isotope systematics, and a brief summary will be provided in the next section.

2. Sulfur Isotope Systematics

Different approaches in sulfur isotope biogeochemistry have been taken in order to qualitatively and quantitatively understand microbial sulfur cycling in modern and ancient settings. In part, this tracks the development of analytical instrumentation. Radioactive ^{35}S isotope work was introduced as an analytical into the field of biogeochemistry more than 30 years ago (e.g., Jorgensen, 1978). The use of radiolabeled sulfate for quantifying rates of microbial sulfur metabolism represents a classical example for this. Rates of microbial sulfur turnover, on the other hand, are dependent on environmental conditions and are expressed in different magnitudes of stable sulfur isotopic fractionation.

The application of stable isotope techniques to unravel microbial sulfur metabolism has a much longer tradition with early works dating back 50 years ago (e.g., Harrison and Thode, 1958; Thode et al., 1961; Kaplan and Rittenberg, 1964). Sulfur has four stable isotopes, ^{32}S, ^{33}S, ^{34}S, and ^{36}S, and sulfur isotope results are conventionally displayed in the standard delta notation:

$$\mathbf{d}^{3X}\mathrm{S}=\left({}^{3X}R_{\text{sample}}/{}^{3X}R_{\text{standard}}-1\right)\times 1000 \tag{1}$$

with $\delta^{3x}S$ representing 33, 34, or 36. Traditionally, sulfur isotope geochemistry has focused on $\delta^{34}S$ as its principal parameter, with *R* representing the ratio of $^{34}S/^{32}S$. The observation of high magnitude mass-dependent fractionation of sulfur isotopes ($^{34}S/^{32}S$) associated with bacterial sulfate reduction in modern marine settings has prompted its use as a biosignature ($\delta^{34}S$) for successfully tracing this process throughout Earth's history (e.g., Schidlowski, 1979; Strauss, 1997). More recently, the development of modern mass spectrometry and the discovery of substantial mass-independent sulfur isotope fractionation in sedimentary rocks older than 2.4 Ga (e.g., Farquhar et al., 2000) have extended sulfur isotope research toward the rare sulfur isotopes ^{33}S and ^{36}S. For ancient and modern environments, a specific focus is placed on the application of multiple sulfur isotope analyses (i.e., analyzing all four stable sulfur isotopes) to questions in microbial sulfur metabolism and geobiology (for some prominent reviews, see, e.g., Farquhar et al., 2003; Ono et al., 2006; Johnston et al., 2008; Ono, 2008). Results of multiple sulfur isotope measurements are expressed in the capital delta notation:

$$\Delta^{33}S = \mathbf{d}^{33}S - 1000 \times [(\mathbf{d}^{34}S/1000 + 1)^{0.515} - 1] \quad (2)$$

and

$$\Delta^{36}S = \mathbf{d}^{36}S - 1000 \times [(\mathbf{d}^{34}S/1000 + 1)^{1.90} - 1] \quad (3)$$

The exponent λ [values of 0.515 and 1.90 in (2) and (3)] in calculating the capital delta notation reflects the sulfur isotopic equilibrium between different forms of sulfur during low-temperature processes. It defines a mass-dependent relationship in a three-isotope-plot (Hulston and Thode, 1965).

Unraveling microbial sulfur metabolism via stable sulfur isotope analyses has been very successful for microbial mats and beyond, both in modern and ancient environments. This will be documented in the following sections with examples of quite different environmental settings selected from the literature.

3. The Isotope Record of Microbial Sulfur Cycling

3.1. THE MODERN WORLD

Our understanding of biological sulfur cycling and associated characteristic sulfur isotope fractionation is based on bacterial culture experiments as much as on results from diverse modern marine and nonmarine settings. Already, early workers (e.g., Kaplan and Rittenberg, 1964) acknowledged that bacterial sulfate reduction is associated with a distinct fractionation against the heavy ^{34}S isotope, resulting in variable yet strongly negative $\delta^{34}S$ values. A maximum magnitude in fractionation for this process of 45‰ was confirmed also in more recent culture experiments (e.g., Detmers et al., 2001), although natural populations do not always show this maximum shift in sulfur isotopes (e.g., Habicht and Canfield, 2001). In the early/mid 1990s, it became clear that in addition to bacterial sulfate reduction, modern (and ancient) marine settings were affected by the bacterial disproportionation of intermediate sulfur-bearing compounds such as elemental sulfur, thiosulfate, or sulfite. Associated isotope effects were quantified (e.g., Habicht et al., 1998) and it became clear that the sulfur isotopic characteristics of natural marine environments were reflecting multiple processes. Hence, the apparent isotope fractionation between seawater sulfate as the ultimate sulfur source and sedimentary sulfide as the final reaction product results from bacterial sulfate reduction, partial oxidation to sulfur intermediates, and subsequent disproportionation of these. This ultimately explained a discrepancy in the magnitude of fractionation observed early on between the culture experiments for bacterial sulfate reduction (maximum fractionation of 45‰) and the modern and ancient marine settings (e.g., Canfield, 2001). Consequently, the isotopic composition of sedimentary sulfide ($\delta^{34}S$) is mostly interpreted as a combination of bacterial sulfate reduction and additional disproportionation reactions, although different views are being discussed (e.g., Brunner and Bernasconi, 2005; Donahue et al., 2008). The formation of partially oxidized sulfur compounds, however, requires an oxidative step, and its antiquity could be instrumental for understanding environmental evolution. Based on multiple sulfur isotope data, disproportionation reactions are clearly discernible back in time until at least 1.3 Ga ago, possibly having started even earlier (Johnston et al., 2005a). In contrast to sulfate reduction and disproportionation of sulfur intermediates, it was observed that inorganic and microbial sulfide oxidation is generally only associated with a minor change in $\delta^{34}S$ (e.g., Fry et al., 1984, 1988).

Recent multiple sulfur isotope work related to microbial sulfur metabolism (e.g., Johnston et al., 2005a, b, 2008) has greatly enhanced our understanding of these processes but more so of our ability to distinguish them in modern and ancient environments. It is clear from culture experiments and work in modern environments that distinct differences in the exponent λ [see (2) and (3)] allow for distinguishing different metabolic pathways (Fig. 1).

In combination with the classical $\delta^{34}S$ value (Johnston et al., 2005b) or in a $^{33}\lambda$–$^{36}\lambda$-crossplot (Johnston et al., 2008), bacterial sulfate reducers can be separated from sulfur compound disproportionators. Moreover, recent work by Zerkle et al. (2008) demonstrated that sulfur oxidation also results in a distinct multiple sulfur isotope signature. Future studies will provide additional supportive evidence, but it is already clear, that the complex reaction network of microbial sulfur metabolism at the ecosystem level can best be deciphered by multiple sulfur isotope studies.

Sulfur isotopic fractionation associated with microbial sulfur metabolism in microbial mats of Solar Lake, Sinai, Egypt, was studied, e.g., by Habicht and Canfield (1996, 1997). At this hypersaline lagoon, the rate of bacterial sulfate reduction varied between 0.1 and 37 μmol cm^{-3} per day, representing some of the highest sulfate reduction rates ever reported from natural populations of sulfate

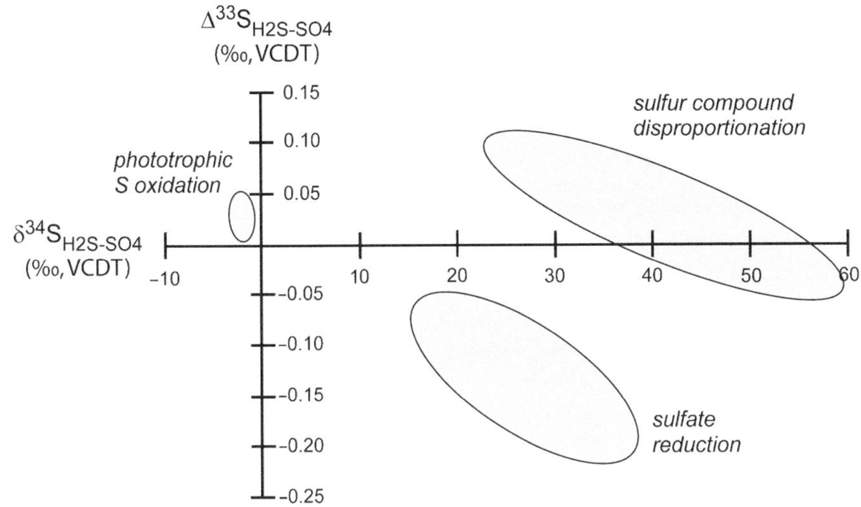

Figure 1. Multiple sulfur isotope signatures for different pathways of microbial sulfur cycling (adapted from Zerkle et al., 2008).

reducers. Fractionation was quantified not only along vertical sediment profiles but also during incubations at different sulfate concentrations and different incubation temperatures. Interestingly, the magnitude in isotope fractionation ($\Delta^{34}S$) between seawater sulfate and resulting hydrogen sulfide ranges between 16‰ and 42‰, with the largest ^{34}S depletions occurring when sulfate reduction rates (SRR) were lowest. However, although ^{34}S depletions were decreasing with increasing reduction rates, sulfur isotope fractionation remained constant at SRR above a certain threshold. No clear dependencies in the magnitude of sulfur isotope fractionation were apparent in respect to different sulfate concentrations and/or temperature. Overall, Habicht and Canfield (1997) concluded that the isotopic fractionation in ^{34}S associated with bacterial sulfate reduction in the microbial mats at Solar Lake was quite comparable to bacterial culture experiments. In addition to bacterial sulfate reducers, the authors were able to enrich cultures of elemental sulfur disproportionators from the microbial mats at Solar Lake and determine the associated sulfur isotopic fractionation. With some of the sedimentary sulfide being more depleted in ^{34}S than attributable to bacterial sulfate reduction alone, Habicht and Canfield (1997) suggested that the disproportionation of elemental sulfur contributes to the overall apparent isotope fractionation between seawater sulfate and resulting sedimentary sulfide, a conclusion that not only pertains to microbial mat settings but also represents an interpretation made very frequently for modern and ancient marine sedimentary environments (e.g., Canfield, 2001).

The microbial sulfur metabolism and associated sulfur isotopic fractionation at Solar Lake was revisited recently by Johnston et al. (2008), applying multiple sulfur isotope analyses. High magnitude fractionations between seawater

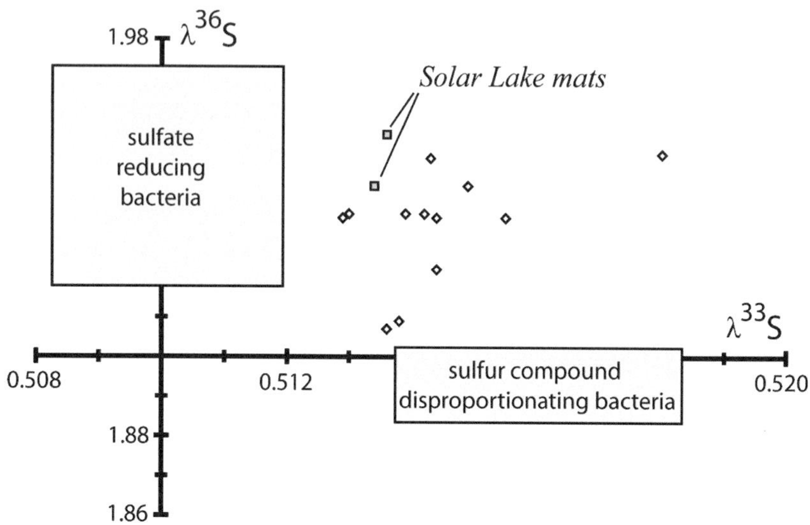

Figure 2. Different metabolic pathways of microbial sulfur cycling in modern marine environments based on the mass-dependent relationship expressed as $\lambda^{33}S$ and $\lambda^{36}S$. Average values for sulfate-reducing and sulfur-disportionating bacteria are shown for comparison (redrawn from Johnston et al., 2008).

sulfate and hydrogen sulfide with an average $\Delta^{34}S$ value at 43.2‰ were reproduced. Modern marine sediments, in general, including the Solar Lake microbial mats studied by Johnston et al. (2008), display variable $\Delta^{33}S$ and $\Delta^{36}S$ values. But more so, calculated $^{33}\lambda$ and $^{36}\lambda$ values exhibit a relationship that fall between the distinct fields defined by sulfate-reducing bacteria on the one hand and sulfur-disproportionating bacteria on the other (Fig. 2). Also, the multiple sulfur isotope results obtained for the microbial mat material is consistent with the data for pure cultures. This leads to the final conclusion that microbial sulfur metabolism in a modern marine sedimentary setting in general is controlled by bacterial sulfate reduction and sulfur-disproportionating reactions, including microbial mats. Hence, microbial sulfur metabolism and associated multiple sulfur isotope effects in modern mat environments are no different from settings devoid of microbial mats. By inference, this should also be true for ancient stromatolites.

3.2. THE PHANEROZOIC

Although studies of modern microbial mat systems reveal a substantial similarity in sulfur isotope fractionation between natural populations and culture experiments as well as between stromatolitic and nonstromatolitic environments, we will continue by examining an ancient stromatolite occurrence that was recently studied for its sulfur isotope geochemistry by Arp et al. (2008). The stromatolites of the Münder Formation at Thüste, Lower Saxony, Germany, are of Upper Jurassic age

(Tithonian, 146–151 Ma). They form one or two distinct horizons of dark colored and finely laminated stromatolite bioherms, each about 10 cm in thickness, which are located within a package of gray laminated marlstone overlying oolitic limestone. Traditionally, the stromatolites at Thüste were regarded to represent an intertidal cyanobacterial community (e.g., Jahnke and Ritzkowski, 1980). However, detailed petrography and particularly the elemental, organic, and isotope geochemical work by Arp et al. (2008) clearly reveals a different and rather complex microbial mat community in which anaerobic metabolic pathways dominate and in which sulfur metabolism plays a crucial role in mineral precipitation. This latter conclusion is based on the presence of pyrite, sometimes enriched in individual laminae, as well as of calcitic pseudomorphs thought to resemble original gypsum crystals.

Detailed carbonate carbon and oxygen isotope geochemistry identifies a substantial range in $\delta^{13}C$ and $\delta^{18}O$. In particular, the petrographically different carbonate phases, ranging from primary carbonate precipitates to diagenetic cements and including nonstromatolitic units as well as the stromatolite bioherms, display a variation in carbon isotope values from 0‰ to −30‰. Thereby, the stromatolitic bioherms, but more so the diagenetic calcites contained in pockets within the stromatolite build-ups, exhibit the most negative $\delta^{13}C$ values and suggest a high proportion of dissolved inorganic carbon that was derived from anaerobic oxidation of organic compounds and utilized subsequently for calcite precipitation.

The sulfur and oxygen isotopic compositions of carbonate-associated sulfate vary from 8.5‰ to 20.9‰ and from 17.8‰ to 22.3‰, respectively (Fig. 3). Moreover, both isotope systems exhibit a concomitant rise toward heavier δ-values

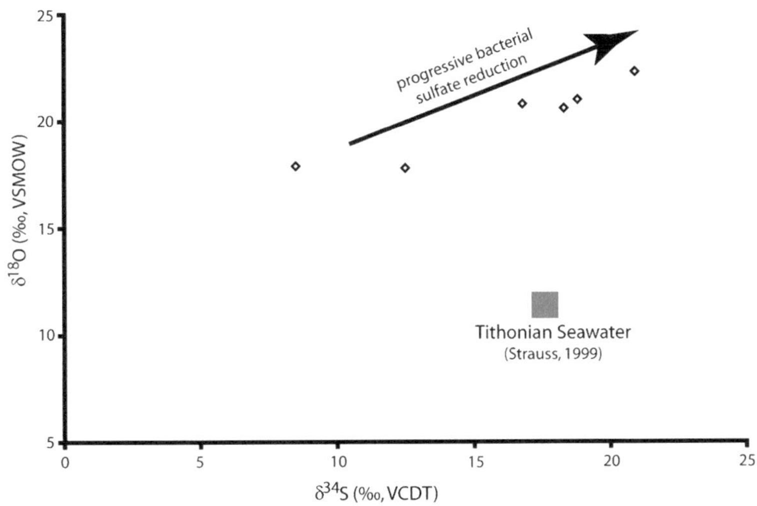

Figure 3. Sulfur and oxygen isotope data for carbonate-associated sulfate from Tithonian stromatolitic carbonates at Thüste, Germany (data from Arp et al., 2008).

starting from respective isotope values for Upper Jurassic seawater (e.g., Strauss, 1999; Kampschulte and Strauss, 2004). This clearly indicates bacterial sulfate reduction as the key process, which is also supported by the presence of pyrite. Variable $\delta^{34}S$ values between −3.0‰ and +3.8‰ for this pyrite point to its biological origin; yet, the rather positive sulfur isotope values suggest sulfate limitation. Bacterial sulfate reduction likely happened, at least in part, via anaerobic methane oxidation. Distinct biomarker evidence proves the presence of methanotrophic Archaea in addition to sulfate-reducing bacteria. Further supporting evidence for the presence and utilization of methane is provided by the highly ^{13}C depleted carbonate cements.

Based on their detailed petrographic and geochemical work, Arp et al. (2008) developed a clear understanding for this stromatolite occurrence that exhibits a complex biogeochemical interplay within a microbial mat environment, but that differs substantially from previous interpretations of the Thüste stromatolite locality. It is now clear that the microbial mat community is dominated by anaerobic and nonphotosynthetic metabolism, with sulfate and methane cycling being very prominent. Albeit, however, phototrophic organisms certainly provided the top layer in the mat community. Mats were impregnated by gypsum due to high evaporation in a marine saline lagoonal setting. Intense bacterial sulfate reduction resulted in carbonate precipitation lower down in the suboxic/anoxic part of the microbial mat build-up. Sulfate reduction was coupled to methane oxidation, and its intensity increased downwards as shown by the highly ^{13}C-depleted carbonates. Throughout this process, sulfate became limited as shown by progressively more ^{34}S- and ^{18}O-enriched isotope values for the carbonate-associated sulfate.

From this Phanerozoic example studied by Arp et al. (2008), conclusions drawn with respect to sulfur are comparable to those based on modern microbial mat communities. Microbial sulfur cycling, particularly bacterial sulfate reduction as well as the disproportionation of sulfur-bearing compounds with intermediate oxidation state (which could not clearly be identified at Thüste), plays an important role in element cycling and mineral formation in stromatolites. Detailed sulfur isotopic work provides crucial evidence for deciphering and understanding respective processes and environments. With stromatolites being much more common in the Precambrian, we will continue by exploring the really ancient sedimentary record for evidence of microbial sulfur cycling.

3.3. THE PRECAMBRIAN

The Precambrian sulfur isotope record ($\delta^{34}S$) has been reviewed extensively more recently, e.g., by Canfield (2004), Strauss (2002, 2004) or Lyons and Gill (2010), but no distinction was made in respect to microbial sulfur cycling and/or stromatolitic environments. Bacterial sulfate reduction was identified as the key process of microbial sulfur cycling, with its antiquity dating back to at least 2.7 Ga (e.g., Grassineau et al., 2006) if not 3.5 Ga (e.g., Shen et al., 2001, 2009, but see a different view proposed by Philippot et al., 2007).

The multiple sulfur isotope record as a whole, addressing particularly the aspect of temporal variations in mass-independent sulfur isotope fractionation [hereafter named S-MIF; the term refers to nonzero $\Delta^{33}S$ values – for calculation of $\Delta^{33}S$, see (2)] as a recorder of an evolving atmospheric chemistry (most notably in respect to the rise in atmospheric oxygen), was discussed, e.g., by Farquhar et al. (2000, 2007). In the context of stromatolitic versus nonstromatolitic environments, a facies-dependent triple sulfur isotope investigation (^{32}S, ^{33}S, ^{34}S) of the Neoarchean sedimentary succession (Transvaal Supergroup) on the Kaapvaal Craton, South Africa, by Ono et al. (2009) might serve as an excellent example. These authors were able to distinguish a facies-dependent biological signal that was superimposed on the atmospheric signature reflecting photochemical reactions during Neoarchean time. Of particular interest in the current discussion is clear evidence for extensive sulfur cycling in a microbial mat environment.

The discovery of S-MIF in both sedimentary sulfide and sulfate, in rock successions older than 2.35 Ga (Farquhar et al., 2000; Guo et al., 2009), guides our understanding of an Archean and early Paleoproterozoic global sulfur cycle that was apparently determined (or at least strongly affected) by UV-induced photochemistry of volcanogenic sulfur compounds in the higher atmosphere. As a principal process, the photochemical reaction of volcanogenic SO_2 to S_8 and H_2SO_4 was identified (e.g., Pavlov and Kasting, 2002). These compounds would be delivered as aerosols to Earth surface environments, and it was suggested that the former (S_8) carries large positive $\Delta^{33}S$ values, whereas the latter (H_2SO_4) would have delivered negative $\Delta^{33}S$ values. Photochemistry as well as the transfer of reaction products to the Earth surface environments is strongly dependent on the atmospheric composition, most notably in respect to a maximum concentration of oxygen proposed to be 10^{-5} PAL (present atmospheric level; Pavlov and Kasting, 2002). Of course, these atmospheric compounds would have been available to inorganic as well as microbially driven reactions, superimposing a mass-dependent fractionation (hereafter named S-MDF) on the atmospheric sulfur isotope signature. A treatment of respective issues has been provided, e.g., by Johnston et al. (2008) and references therein.

Triple sulfur isotope results obtained by Ono et al. (2009) for the Transvaal succession yield some clear relationships between $\delta^{33}S$ and $\delta^{34}S$ values and between $\delta^{34}S$ and $\Delta^{33}S$ values, an observation made by others before (e.g., Ono et al., 2003; Kaufman et al., 2007). Most notable is a linear array with $\delta^{33}S \approx 1.4 \times \delta^{34}S$ (and $\Delta^{33}S \approx 0.89 \times \delta^{34}S$), which appears to be characteristic for the Neoarchean triple sulfur isotope record. The total dataset measured for the Transvaal sediments, when plotted in a respective $\Delta^{33}S/\delta^{34}S$ diagram (Fig. 4), exhibits some scatter around the noted relationship. This discrepancy bears a signal superimposed on the purely atmospheric signature that could be derived from the mixing of different sulfur (isotope) reservoirs and/or caused by biological sulfur cycling.

Based on lithofacies, the entire sample set can be divided into two populations representing different water depths within a given slope environment (for

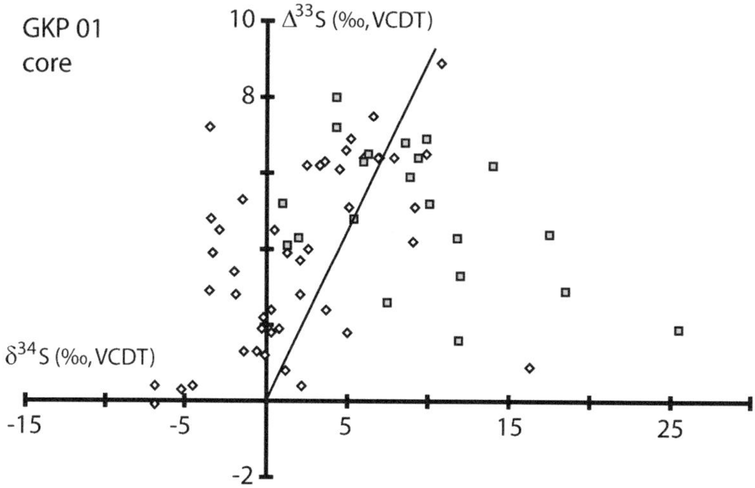

Figure 4. Sulfur isotopic composition of sedimentary pyrite from the Transvaal succession (*filled squares* are from stromatolic unit, *open diamonds* are from nonstromatolitic units, linear array: $\Delta^{33}S = 0.89 \times \delta^{34}S$); data from Ono et al. (2009).

details, see Ono et al., 2009, and references therein). In addition, largely siliciclastic units can be distinguished from carbonate-bearing units and from iron- and chert-rich lithologies. Finally, sediments of the Reivilo Formation and the lower part of the overlying Nauga Formation contain abundant microbialites. This latter part of the succession, i.e., the microbialites, can be clearly distinguished in their triple sulfur isotope signature, at least in the more shallow part of the slope (Fig. 4). Both sample populations from the deeper as well as the more shallow part of the basin display positive $\Delta^{33}S$ values. Given our current understanding, these positive $\Delta^{33}S$ values suggest their origin as an atmospheric S_8 compound (cf. Ono et al., 2003). A difference between both sample populations is discernible, however, when considering the $\delta^{34}S$ values. Here, samples from the deeper part of the basin display negative $\delta^{34}S$ values (data not shown). According to Ono et al. (2009), this indicates a mixture of S_8-derived sulfur with sulfur that originates from bacterial reduction of atmospheric sulfate. Bacterial sulfate reduction produces only mass-dependent sulfur isotope fractionation (for a review see, e.g., Canfield, 2001). Consequently, the $\delta^{34}S$ signature would be shifted to negative values (to the left in a $\Delta^{33}S$/ $\delta^{34}S$ diagram). In contrast, samples from the shallower environment are characterized by positive to strongly positive $\delta^{34}S$ values at a comparable range in $\Delta^{33}S$. The positive signature in $\delta^{34}S$ could be a sign of enhanced bacterial sulfate reduction, driving the pore water chemistry to sulfate limitation. Ono et al. (2009) acknowledge the high abundance of microbialites and a proposed rapid carbonate cementation of these sediments, all suggestive of high microbial activities including high rates of bacterial sulfate reduction. Still somewhat unclear is the origin of the positive $\Delta^{33}S$ values for pyrite, which suggest a positive $\Delta^{33}S$ value also for the seawater sulfate sulfur. This could point to the absence of carbonyl sulfide from the Neoarchean atmosphere (Ueno et al., 2009).

The study by Ono et al. (2009) clearly identifies the high potential of sulfur isotope measurements for understanding microbial sulfur metabolism in recent settings as well as in the geologic past, whether in a stromatolitic or a nonstromatolitic environment.

4. Conclusions and Future Research Directions

Stromatolites provide a unique record of microbial life and respective microbial activity throughout Earth's history, with the oldest example dating back to ~3.5 Ga (Allwood et al., 2006). Examples from modern and ancient microbial mat/stromatolite occurrences discussed in the previous sections yield some overarching conclusions. Most notable is the observation that in marine environments, intense microbial sulfur cycling dominates the mat communities. Respective evidence is archived in biogenic mineral matter and – for the modern world – in pore water. At least in modern and Phanerozoic environments, the isotopic signatures measured for sedimentary sulfur are quite comparable for culture experiments and natural populations. An additional, non mass-dependent (S-MIF) atmospheric signal dominates the picture for Archean and early Paleoproterozoic settings, providing a challenge for identifying the mass-dependent (S-MDF) biological signal. Our possibilities for deciphering the details of microbial sulfur cycling in the geological past will depend strongly on the progress achieved in modern settings. The availability of isotopic proxies has guided the search for biosignatures through Earth's history. This will continue. However, following the chronology in sulfur isotope research, it becomes clear that multiple sulfur isotope studies will be the ultimate approach in our quest to unravel and understand microbial sulfur metabolism in sedimentary environments.

5. Acknowledgments

The editors are thanked for inviting me to contribute. Continuing support of my research through the Deutsche Forschungsgemeinschaft (DFG) is gratefully acknowledged.

6. References

Allwood, A.C., Walter, M.R., Kamber, B.S., Marshall, C.P. and Burch, I.W. (2006) Stromatolite reef from the early Archaean era of Australia. Nature **441**: 714–718.

Arp, G., Ostertag-Henning, C., Yücekent, S., Reitner, J. and Thiel, V. (2008) Methane-related microbial gypsum calcitization in stromatolites of a marine evaporative setting (Münder Formation, Upper Jurassic, Hilc Syncline, north Germany). Sedimentology **55**: 1227–1251.

Brunner, B. and Bernasconi, S.M. (2005) A revised isotope fractionation model for dissimilatory sulfate reduction in sulfate reducing bacteria. Geochim. Cosmochim. Acta **69**: 4759–4771.

Canfield, D.E. (2001) Biogeochemistry of sulfur isotopes, In: J. Valley and D.R. Cole (eds.) *Stable Isotope Geochemistry*. Rev. Mineral. Geochem. **43**: 607–636.

Canfield, D.E. (2004) The evolution of the Earth surface sulfur reservoir. Am. J. Sci. **304**: 839–861.

Detmers, J., Brüchert, V., Habicht, K.S. and Küver, J. (2001) Diversity of sulfur isotope fractionations by sulfate-reducing prokaryotes. Appl. Environ. Microbiol. **67**: 888–894.

Donahue, M.A., Werne, J.P., Meile, C. and Lyons, T.W. (2008) Modeling sulfur isotope fractionation and differential diffusion during sulfate reduction in sediments of the Cariaco Basin. Geochim. Cosmochim. Acta **72**: 2287–2297.

Dupraz, C. and Visscher, P.T. (2005) Micobial lithification in marine stromatolites and hypersaline mats. Trends Microbiol. **13**: 429–438.

Farquhar, J., Bao, H.M. and Thiemens, M.H. (2000) Atmospheric influence of Earth's earliest sulphur cycle. Science **289**: 756–758.

Farquhar, J., Johnston, D.T., Wing, B.A., Habicht, K.S., Canfield, D.E., Airieau, S.A. and Thiemens, M.H. (2003) Multiple sulphur-isotopic interpretations of biosynthetic pathways: implications for biological signatures in the sulphur isotope record. Geobiology **1**: 15–27.

Farquhar, J., Peters, M., Johnston, D.T., Strauss, H., Masterson, A., Wiechert, U. and Kaufman, A.J. (2007) Isotopic evidence for Mesoarchaean anoxia and changing sulphur chemistry. Nature **449**: 706–709.

Fike, D.A., Gammon, C.L., Finke, N., Hoehler, T., Turk, K. and Orphan, V.J. (2008) Micron-scale mapping of sulfur cycling across the oxycline of a cyanobacterial mat. Geochim. Cosmochim. Acta **72**: A268.

Fry, B., Gest, H. and Hayes, J.M. (1984) Isotope effects associated with the anaerobic oxidation of sulfide by the purple photosynthetic bacterium *Chromatium vinosum*. FEMS Microbiol. Lett. **22**: 283–287.

Fry, B., Ruf, W., Gest, H. and Hayes, J.M. (1988) Sulfur isotope effects associated with oxidation of sulfide by O_2 in aqueous solution. Chem. Geol. **73**: 205–210.

Grassineau, N.V., Abell, P., Appel, P.W.U., Lowry, D. and Nisbet, E.G. (2006) Early life signatures in sulfur and carbon isotopes from Isua, Barberton, Wabigoon (Steep Rock), and Belingwe Greenstone Belts (3.8 to 2.7 Ga), In: S.E. Kesler and H. Ohmoto (eds.) *Evolution of Early Earth's Atmosphere, Hydrosphere, and Biosphere – Constraints from Ore Deposits*. Geol. Soc. Amer. Mem. **198**: 33–52.

Guo, Q., Strauss, H., Kaufman, A.J., Schröder, S., Gutzmer, J., Wing, B., Baker, M.A., Bekker, A., Kim, S.-T., Jin, Q. and Farquhar, J. (2009) Reconstructing Earth's surface oxidation across the Archean-Proterozoic transition. Geology **37**: 399–402.

Habicht, K.S. and Canfield, D.E. (1996) Sulphur isotope fractionation in modern microbial mats and the evolution of the sulphur cycle. Nature **382**: 342–343.

Habicht, K.S. and Canfield, D.E. (1997) Sulfur isotope fractionation during bacterial sulfate reduction in organic-rich sediments. Geochim. Cosmochim. Acta **61**: 5351–5361.

Habicht, K.S. and Canfield, D.E. (2001) Isotope fractionation by sulfate-reducing natural populations and the isotopic composition of sulfide in marine sediments. Geology **29**: 555–558.

Habicht, K.S., Canfield, D.E. and Rethmeier, J. (1998) Sulfur isotope fractionation during bacterial sulfate reduction and disproportionation of thiosulfate and sulfite. Geochim. Cosmochim. Acta **62**: 2585–2595.

Harrison, A.G. and Thode, H.G. (1958) Mechanism of the bacterial reduction of sulphate from isotope fractionation studies. Trans. Faraday Soc. **54**: 84–92.

Hulston, J.R. and Thode, H.G. (1965) Variations in S33 and S36 contents of meteorites and their relation to chemical and nuclear effects. J. Geophys. Res. **70**: 3475–3484.

Jahnke, H. and Ritzkowski, S. (1980) Die Fazies-Abfolge im Münder Mergel der Steinbrüche bei Thüste (Ober-Jura, Hilsmulde). Ber. Naturhist. Ges. Hannover **123**: 45–62.

Johnston, D.T., Wing, B.A., Farquhar, J., Kaufman, A.J., Strauss, H., Lyons, T.W., Kah, L.C. and Canfield, D.E. (2005a) Active microbial sulfur disproportionation in the Mesoproterozoic. Science **310**: 1477–1479.

Johnston, D.T., Farquhar, J., Wing, B.A., Kaufman, A.J., Canfield, D.E. and Habicht, K.S. (2005b) Multiple sulfur isotope fractionations in biological systems: a case study with sulfate reducers and sulfur disproportionators. Am. J. Sci. **305**: 645–660.

Johnston, D.T., Farquhar, J., Habicht, K.S. and Canfield, D.E. (2008) Sulphur isotopes and the search for life: strategies for identifying sulphur metabolisms in the rock record and beyond. Geobiology **6**: 425–435.

Jorgensen, B.B. (1978) A comparison of methods for the quantification of bacterial sulfate reduction in coastal marine sediments. III estimation from chemical and bacteriological field data. Geomicrobiol. J. **1**: 49–64.

Kampschulte, A. and Strauss, H. (2004) The sulfur isotopic evolution of Phanerozoic seawater based on the analysis of structurally substituted sulfate in carbonates. Chem. Geol. **204**: 255–286.

Kaplan, I.R. and Rittenberg, S.C. (1964) Microbiological fractionation of sulphur isotopes. J. Gen. Microbiol. **34**: 195–212.

Kaufman, A.J., Johnston, D.T., Farquhar, J., Masterson, A.L., Lyons, T.W., Bates, S., Anbar, A.D., Arnold, G.L., Garvin, J. and Buick, R. (2007) Late Archean biospheric oxygenation and atmospheric evolution. Science **317**: 1900–1903.

Lyons, T.W. and Gill, B.C. (2010) Ancient Sulfur Cycling and Oxygenation of the Early Biosphere. Elements **6**: 93–99.

Ono, S. (2008) Multiple-sulphur isotope biosignatures. Space Sci. Rev. **135**: 203–220.

Ono, S., Eigenbrode, J.L., Pavlov, A.A., Kharecha, P., Rumble, D. III, Kasting, J.F. and Freeman, K.H. (2003) New insights into Archean sulphur cycle from mass-independent sulphur isotope records from the Hamersley Basin, Australia. Earth Planet. Sci. Lett. **213**: 15–30.

Ono, S., Wing, B., Johnston, D., Farquhar, J. and Rumble, D. (2006) Mass-dependent fractionation of quadruple stable sulphur isotope system as a new tracer of sulphur biogeochemical cycles. Geochim. Cosmochim. Acta **70**: 2238–2252.

Ono, S., Kaufman, A.J., Fraquhar, J., Sumner, D.Y. and Beukes, N.J. (2009) Lithofacies control on multiple-sulfur isotope records and Neoarchean sulfur cycles. Precambr. Res. **169**: 58–67.

Pavlov, A.A. and Kasting, J.F. (2002) Mass-independent fractionation of sulphur isotopes in Archean sediments: strong evidence for an anoxic Archean atmosphere. Astrobiology **2**: 27–41.

Philippot, P., van Zuilen, M., Lepot, K., Thomazo, C., Farquhar, J. and van Kranendonk, M.J. (2007) Early Archaean microorganisms preferred elemental sulfur, not sulfate. Science **317**: 1534–1537.

Schidlowski, M. (1979) Antiquity and evolutionary status of bacterial sulphur reduction – sulphur isotope evidence. Orig. Life Evol. Biosph. **9**: 299–311.

Shen, Y.A., Buick, R. and Canfield, D.E. (2001) Isotopic evidence for microbial sulphate reduction in the early Archaean era. Nature **410**: 77–81.

Shen, Y., Farquhar, J., Masterson, A., Kaufman, A.J., Buick, R. (2009) Evaluating the role of microbial sulfate reduction in the early Archean using quadruple isotope systematics. Earth Planet. Sci. Lett. **279**: 383–391.

Strauss, H. (1997) The isotopic composition of sedimentary sulfur through time. Palaeogeogr. Palaeoclimat. Palaeoecol. **132**: 97–118.

Strauss, H. (1999) Geological evolution from isotope proxy signals: sulfur. Chem. Geol. **161**: 89–101.

Strauss, H. (2002) The sulfur isotopic composition of Precambrian sedimentary sulfides – seawater chemistry and biological evolution, In: W. Altermann and P. Corcoran (eds.) *Precambrian Sedimentary Environments: A Modern Approach to Ancient Depositional Systems. Special Publications 33 of IAS*. Wiley-Blackwell, New York, pp. 67–105.

Strauss, H. (2004) 4 Ga of seawater evolution: evidence from the sulfur isotopic composition of sulfate, In: J.P. Amend, K.J. Edwards and T.W. Lyons (eds.) *Sulfur Biogeochemistry: Past and Present*. Geol. Soc. Am. Spec. Pap. **379**: 195–205.

Thode, H.G., Monster, J. and Dunford, H.B. (1961) Sulphur isotope geochemistry. Geochim. Cosmochim. Acta **25**: 159–174.

Ueno, Y., Danielache, S.O., Johnson, M.S. and Yoshida, N. (2009) Carbonyl sulphide (OCS) in the Archean atmosphere. Geochim. Cosmochim. Acta **73**: A1358.

Zerkle, A.L., Farquhar, J., Johnston, D.T., Cox, R.P. and Canfield, D. (2008) Fractionation of multiple sulphur isotopes during phototrophic oxidation of sulphide and elemental sulphur by a green sulphur bacterium. Geochim. Cosmochim. Acta **73**: 291–306.

PART 6: ASTROBIOLOGY

Bonaccorsi
Chela-Flores
Tewari

Biodata of **Dr. Rosalba Bonaccorsi**, author of ***"Preservation Potential and Habitability of Clay Minerals- and Iron-Rich Environments: Novel Analogs for the 2011 Mars Science Laboratory Mission"***

Dr. Bonaccorsi is an environmental scientist currently working at NASA Ames Research Center as interdisciplinary scientist. In 2001, she obtained her Ph.D. in Geological, Marine and Environmental Sciences from the University of Trieste (Italy). As a NASA Postdoctoral Fellow (2005–2008), she was involved in a Robotic drilling simulation of near-surface and deep-subsurface biospheres (the MARTE Project). Dr. Bonaccorsi's scientific interests are in the areas of astrobiology, including Planetary Protection (bioburden detection) and remediation; geobiology of surfaces and subsurface biospheres; habitability of mineralogical Mars analogs, and hyper-arid desert regions worldwide (e.g., Antarctica, Atacama, Mojave).

E-mail: **rosalba.bonaccorsi-1@nasa.gov**

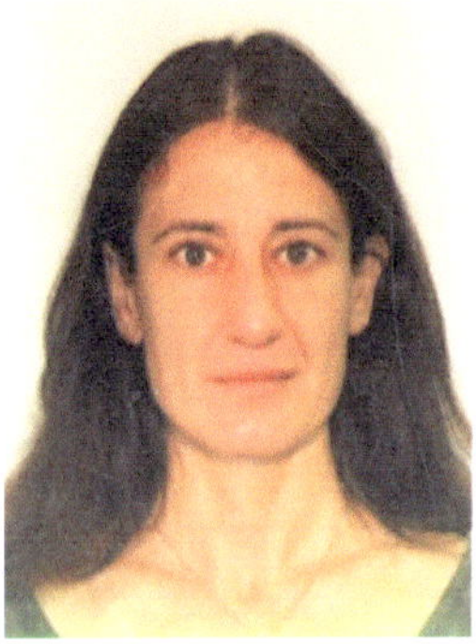

V.C. Tewari and J. Seckbach (eds.), *STROMATOLITES: Interaction of Microbes with Sediments*, Cellular Origin, Life in Extreme Habitats and Astrobiology 18, 705–722
DOI 10.1007/978-94-007-0397-1_32,

PRESERVATION POTENTIAL AND HABITABILITY OF CLAY MINERALS- AND IRON-RICH ENVIRONMENTS: NOVEL ANALOGS FOR THE 2011 MARS SCIENCE LABORATORY MISSION

ROSALBA BONACCORSI
NASA Ames Research Center/SETI Institute, MS. 245-3 Space Science and Astrobiology Division, Moffett Field, CA 94035, USA

Abstract Near-future planetary missions, including the US 2011 Mars Science Laboratory (MSL11) and the ESA 2016 Pasteur ExoMars, will primarily seek key information on the geological and biological history of Mars. In this context, studies addressing the potential for (1) preservation of organics and (2) habitability and microbial biomass in phyllosilicate-rich and ferric oxide-rich environments can provide key interpretive information for near-future Mars Science/life detection missions.

In the first part of this chapter, a case for enhanced and preferential preservation of organics (~0.23 wt.%) in phyllosilicate-rich (47–74%) zones within an organic-poor (C-org: 0.05 wt.%) ferric oxide-rich (42–94%) subsurface rock system is presented. In the second part, first-order observations on microbial biomass inhabiting clay-rich versus nonclays materials are given as an example to frame habitability potential in clay mineral-rich environments from arid (Death Valley, CA) and hyperarid (Atacama, Chile) deserts.

Keywords Astrobiology • Atacama Desert • Death Valley National Park • Habitability • Mars Science Laboratory • Microbial biomass • Organics • Phyllosilicate analogs • Preservation potential • Rio Tinto

1. Introduction

Hydrated clay minerals, or phyllosilicates, are unambiguously present on Mars and have been recently suggested to support preservation of organic matter (Ehlmann et al., 2008; Bishop et al., 2008; Chevrier, 2008; Bonaccorsi and Stoker, 2008). This, in analogy with studies of earth analogs, is relevant to the next decade planetary/ astrobiology missions including the US 2011 Mars Science Laboratory (MSL11) and the ESA 2016 Pasteur ExoMars that will primarily seek key information of the geological and biological history of Mars.

All of the four MSL11 landing site candidates include clay deposits and have been hierarchically ranked by relevance with respect to environmental context, mineralogical diversity, habitability, and preservation potential of organics in mineralogical/ geological environments, suggesting water activity (e.g., Beegle et al., 2007).

In the above context, and in this chapter, I outline an example of (a) the enhanced preservation of organics against oxidation in clay minerals from a hyperacidic subsurface environment; (b) total viable biomass content in clay mineral-rich versus clay mineral-poor and/or nonclay environments as analog models for habitability potential of mineral environments on Mars. Phyllosilicate and hematite deposits under study contain variable amounts of total organic carbon (TOC: <0.01 to ~3 wt.%) and are from a broad range of climate settings (mean annual precipitation, MAP <0.2 to ~700 mm/year) and regions, i.e., the Atacama Desert, Death Valley National Park (DVNP), and the California Coast.

2. Implications of Organics on Mars

Two of NASA's key goals over the next several decades of planetary exploration are to determine whether life developed on any planetary body in the solar system and to learn more about the origin of life on Earth by studying the evolution of complex organic molecules on other planetary bodies. Although Mars appears to have acquired the three key building blocks for life (water, an energy source, and a supply of organic molecules) in its geological history, organic carbon has not yet been detected.

Within the Mars Exploration Program Analysis Group (MEPAG), the high-level goal of the search for evidence of extinct or extant life (as we know it) and a complementary search for habitable zones for future missions involve a cascade of fundamental questions:

(1) Are exogenous/endogenous organics present on Mars? This requires the ability to detect and distinguish between organic and inorganic carbon in planetary samples. (2) Were/are these organics of biological origin (biogenicity)? (3) If organics are not observed, is it because they were never present, or because they were not preserved over time in geological material, or rather because they were not detected?

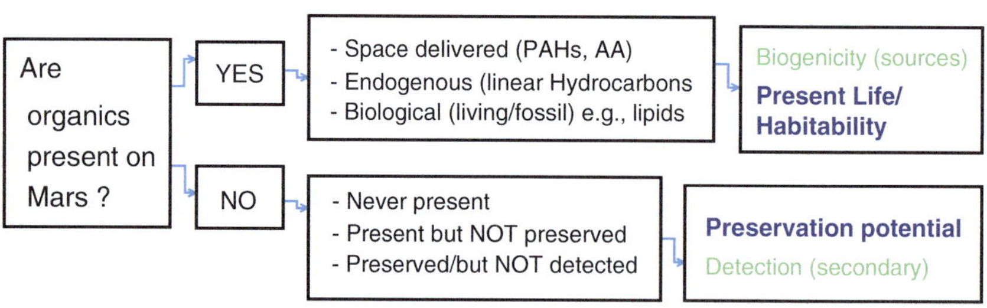

PAHs : Polycyclic aromatic Hydrocarbons
AA : Amino acids

The diagram below shows connections between these fundamental questions and key science objectives for the search for past and present Martian life.

Achieving these objectives rests on the ability to determine the potential for preservation of organics and habitability of key geological and mineral environments such as phyllosilicates and iron-rich materials, which are widespread on Mars.

Prebiotic and/or biologically produced organics, if life was present on an earlier Mars, may have been entrained but not permanently preserved in the Martian geological record (Sumner, 2004). Complex relationships between organics and mineral phases, e.g., carbonates, silicates, and oxides, may also affect or enhance the preservation potential of organics as well as their instrumental detection. However, the true amount of organic compounds in samples must be correctly determined before addressing the preservation potential of organics in Mars-relevant materials.

The NRC Task Group on organic environments in the Solar System recommends multiple strategies for the improvement of detection of biogenic organics on Mars today. For instance, further investigations should be performed to understand oxidation effects on organics in regolith analog materials as well as determine time scales involved in the oxidative alteration of organic materials (The NRC Task Group on organic environments in the Solar System, 2007, and references therein).

3. Background

3.1. PHYLLOSILICATE DEPOSITS AND MSL LANDING SITE CANDIDATES

Hydrated clay minerals, or phyllosilicates, which record multiple episodes of aqueous activity on early Mars (Ehlmann et al., 2008), have been identified on the surface of Mars by OMEGA, the visible near-infrared hyperspectral imager onboard Mars Express (Poulet et al., 2005; Bibring et al., 2006), and by the Spirit rover observation at West Spur, Columbia Hills, in Gusev Crater (Wang et al., 2006).

Phyllosilicate-rich fluvial–lacustrine deposits, e.g., Eberswalde, Holden, and Jezero craters, as well as in the Nili Fossae region, have been identified by the Mars Reconnaissance Orbiter (MRO) instruments, i.e., the high-resolution camera HiRISE and the visible-infrared imaging spectrometer CRISM (e.g., Bishop et al., 2008). Furthermore, phyllosilicates deposits buried near surface could exist on Mars (Catling, 2007; Chevrier, 2008). More recently, stratigraphic units comprising thick deposits of Fe/Mg–smectites, ferrous phases, hydrated silica, and Al–phyllosilicates were identified in Mawrth Vallis (Bishop et al., 2008; Ehlmann et al., 2008).

Furthermore, if our models for preservation potential in terrestrial samples can be translated to Mars, buried phyllosilicates on Mars would have a higher potential to preserve organics than ancient wetland and hematite sites, e.g., Meridiani Planum, where organics are unlikely to be preserved (Sumner, 2004; Davila et al., 2008).

Understanding the limit for organic preservation and habitability potential of these mineralogical analogs will provide critical information in support of landing site selection for the MSL and ExoMars missions. All of the four MSL landing site candidates include clay deposits and have been hierarchically ranked by relevance with respect to context, diversity, habitability, and preservation potential of organics (http://marsoweb.nas.nasa.gov/landingsites/index.html). Although habitability has been an ambiguous criteria to be defined, it will be the most discriminating criteria for the final selection. Three groups of sites are distinguished as follows (1) Mawrth Vellis with 100% of the landing ellipse surface covered by highly diversified phyllosilicates with the oldest, in-situ formed claystones; (2) Younger, Late Noachian/Early Hesperian crater fills, e.g., transported and collected into aqueous systems (Eberswalde and Holden), or aeolian layered deposits (Gale crater). These would have higher habitability but lower detectability because few clay mineral deposits occur and clays could be sparse/mixed to matrix; and (3) hematite-rich plains with older Noachian terrain containing few/sparse outcrops of clays final and sulfate–clay contacts reachable by a rover.

At the time this chapter is in press, the final landing site selection is under consideration by the landing site Science Steering Committee.

3.2. PRESERVATION OF ORGANICS ON MARS

The only information we have about organics on Mars derives from results of the Viking GC-MS experiments (e.g., Klein, 1992; Navarro-González et al., 2006) and, more recently, by the Phoenix Mission. Therefore, the search for organics on Mars will be one of the primary objectives of the MSL Mission (Mahaffy, 2008) and beyond. For instance, a wide range of organic compounds in Martian materials (Glavin et al., 2006) will be the target of the instrument payload Sample Analysis at Mars (SAM), including a pyrolysis–gas chromatograph–mass spectrometer (pyr–GC–MS). Its counterpart onboard ExoMars, the GC–MS Mars Organic Molecule Analyzer (MOMA), will detect molecular species at ppb to ppt level.

Meteoritic-delivered carbonaceous organic components (Van der Velden and Schwartz, 1977; Stoks and Schwartz, 1979) should be detectable in the Martian regolith (Benner et al., 2000) in the absence of oxidative decomposition. It is predicted that surface organics would not be preserved within the Martian regolith, owing to the production of oxidants by physicochemical interactions between the solar UV radiation and the soil itself down to ~5 m depth (Bullock et al., 1994; Zent, 1998; Lammer et al., 2003). Instead, buried organics may be well preserved in iron oxide-rich regolith against cosmic radiation (Aubrey et al., 2006). However, possible organic biosignatures within the Martian iron oxide-rich soils may be destroyed by metal-catalyzed oxidation even at depth (Sumner, 2004, and references therein; Apak, 2008).

3.3. PRESERVATION OF ORGANICS IN PHYLLOSILICATES

On Earth, organic matter is rarely preserved in highly oxidized minerals/oxidizing environments (e.g., Sumner, 2004 and references therein; Bonaccorsi and Stoker, 2008). This is particularly true for hematite sites associated with ancient wetland on Mars, e.g., Meridiani Planum (Sumner, 2004; Davila et al., 2008). Therefore, a better model for searching ancient organic compounds on Mars includes clay-rich deposits (e.g., Bonaccorsi and Stoker, 2008), owing to the ability of many clay minerals to sequester organics, inhibiting their decay. Shales preserve organics effectively due to their low permeability, while carbonates and cherts are less reactive to organic carbon (Sumner, 2004).

Phyllosilicate minerals, such as smectites and kaolinite, can interact with organic molecules in many ways. For instance, montmorillonite catalyzes polymerization of RNA-like oligomers (e.g., Ertem et al., 2007 and references therein; Ferris et al., 1989). In marine sediments, where the ambient environment is oxidizing, and organic carbon is present as <10% particulate and >90% dissolved forms (Keil et al., 1994), abundances are linearly correlated with sediment-specific surface area. This indicates that absorption and long-term stabilization of dissolved organics on mineral surfaces are the most effective protection against environmental oxidation (Hedges and Kiel, 1995; Cowie et al., 1995).

In clay-supported soils, stabilization of dissolved organics by sorption on mineral surfaces (D'Acqui et al., 1998; Kaiser and Zech, 2000; Kleber et al., 2005) as well as increased mean residence times of organics (Torn et al., 1997; Wattel-Koekkoek and Burman, 2004) are also key factors for enhancing preservation.

In the following section, I present a near surface-shallow chemostratigraphic model including the following elements (I) clay-rich materials, at contact with (II) Fe-oxy hydroxide-rich materials, and (III) common sources of organics to both materials. A geological environment where these elements coexist would represent an ideal MSL landing site candidate analog where instrumental detection of fossil and labile organic compounds, as well as microbes, is easily verifiable.

3.4. THE RIO TINTO ANALOG FOR NEAR-SUBSURFACE CLAY DEPOSITS

An interesting case for organics preferentially preserved in phyllosilicate-rich materials is described in Bonaccorsi and Stoker (2008). Shallow core samples were acquired within an oxidized remnant of a hyperacidic (pH 2.3) rock drainage system in a volcanically hosted massive sulfide deposit at Rio Tinto, Spain (Figs. 1 and 2), an important analog of the Sinus Meridiani hematite site on Mars (e.g., Stoker et al., 2008).

Organic geochemistry results indicated that phyllosilicate-rich zones preserve up to ten times more organic carbon (C-org: ~0.23 wt.% blue arrow) than the embedding hematite-rich rocks, e.g., C-org: ~0.05 wt.%, red arrow (Fig. 3).

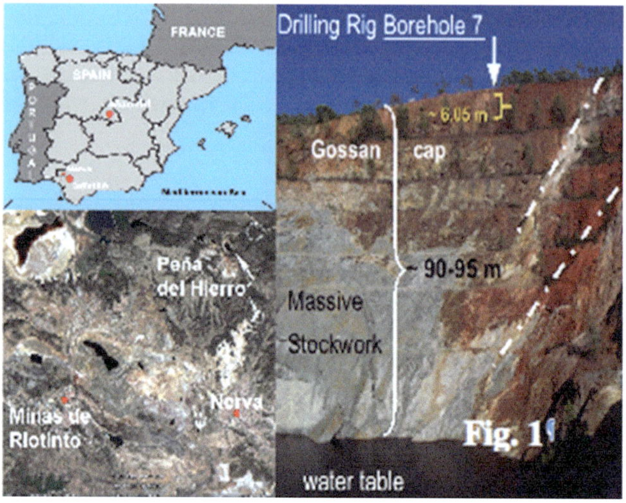

Figure 1. Borehole 7 drilling site at the Rio Tinto Region. Oxidized gossan cap (*red color*) overlying the massive ore deposit (*gray color*). After Bonaccorsi and Stoker, 2008.

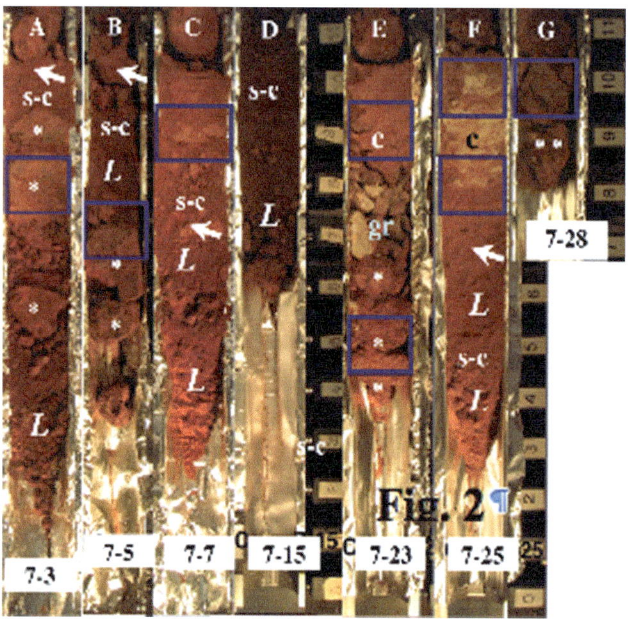

Figure 2. Selected cores including hematite/goethite-rich materials (*bright red* and *brown colors*) embedding pockets of phyllosilicates (*light-toned zones*). After Bonaccorsi and Stoker, 2008.

Interestingly, even the shallow porous oxidized rocks immediately beneath the biologically active, organic-rich surface soil horizon (3–11 wt.%, green arrow) appears to be a highly oxidizing environment (Fig. 3). Possibly, dissolved organic carbon

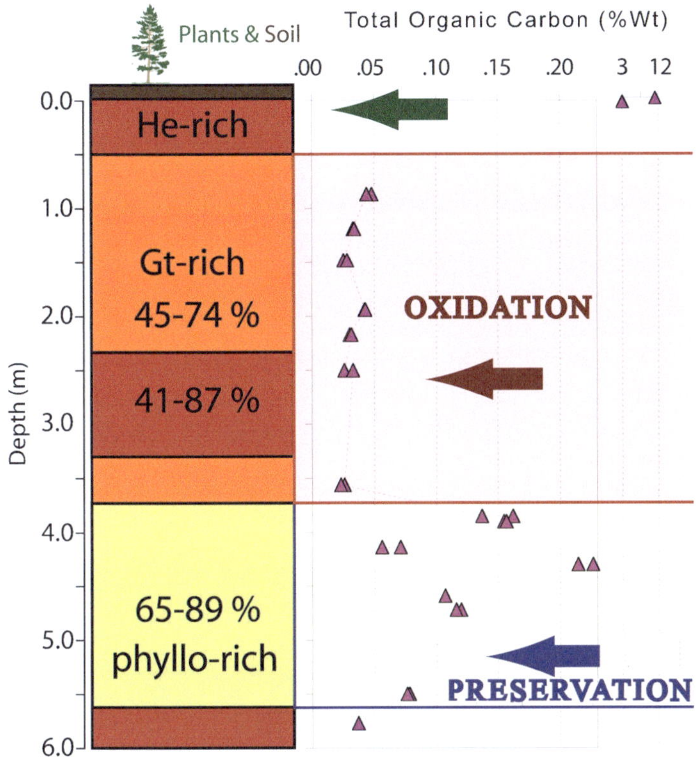

Figure 3. Chemostratigraphic model proposed for preservation (After Bonaccorsi and Stoker, 2008). Total organic carbon is surprisingly low at the surface-oxidized soil horizon beneath the biologically active organic-rich soil horizon (C-org: 3–11 wt.%). Note the organic carbon dropping from ~3 to 12 wt.% to less than 0.05 wt.% (<3 cm depth) where modern soil organic matter (SOM) and dissolved organic carbon (DOC) should be still abundant. Through the entire hole, down to ~6 m depth, the organic carbon is poorly preserved in hematite- (He: 41–87%) and goethite-rich (Gt: 45–74%) rocks.

(DOC) entrained by rain (500–700 mm/year) is absorbed and it can be "rapidly" oxidized within the shallow ironstone environment (Fig. 3), which is not conducive to preservation.

3.5. HABITABILITY OF PHYLLOSILICATES

In addition to the preservation potential, which is a key requirement for the detection of past life on Mars, habitability, i.e., the capability of supporting past and present microbial life, will be one of the most important criteria for the search for present (and past) life.

When addressing the question on whether clay-rich materials would represent a preferentially habitable environment versus nonclay, or clay-poor matrices, a simplified comparative approach may help. One approach could be that of looking at biological markers that can be easily measured in situ and that can be unequivocally used as proxies for living microbial biomass. In Sect. 3.6, I briefly describe some of the methods used for this application.

Performing habitability studies on a set of natural environmental samples is certainly more complex and less reproducible than using synthetic mineral samples under controlled laboratory (biological) conditions. However, such complexity can be mitigated by an in depth knowledge of the environmental context (e.g., water availability, organic carbon/ energy sources, potential biological sources) and by applying a comparative approach. In addition, direct study of habitability potential on complex environmental matrices would generate data to constrain a better model for the interpretation of an even more complex and still generally unknown Martian environment. Understanding such complexity will be, indeed, one of the most challenging tasks of the upcoming MSL11 and ExoMars missions.

An example of comparative investigation of clays and nonclay mineral analog environments is given in Sect. 4.

3.6. ANALYTICAL APPROACH

3.6.1. Total Viable Biomass

The portable hygiene monitoring system LIGHTING MVP (BioControl Systems, Inc., WA, USA) is used to assay levels of Adenosine 5′-triphosphate (ATP). As the ATP is the energy system carrier used by all living organisms, this technique is routinely applied to estimate the relative metabolic activity of cells and the viable biomass in soil samples (Cowan et al., 2002). The ATP assay-based biomass data are expressed as Relative Luminosity Units (RLUs) and calibrated vs. Phospholipid Fatty Acid (PLFA)-based total biomass (cells/g soil) on subaliquots of primary soil samples tested for ATP. Details on methods are described in full elsewhere (Miller et al., 2008; Bonaccorsi and Stoker, 2008, and references therein).

3.6.2. Gram-Negative Biomass

The endotoxin-producing microbial biomass in the rock and mineral samples can be determined with a portable system (Charles River PTS System), based on the Limulus Amebocyte Lysate (LAL) assay (Bruckner and Venkateswaran, 2007). Basically, this assay detects the level of endotoxin (EU/mL), also known as lipopolysaccarides (LPS), in a sample. LPS are present in the external cellular membrane of Gram-negative-like microorganisms, such as bacteria, cyanobacteria (Raetz and Whitfield, 2002), unicellular algae (Bedick et al., 2001), as well as vascular plants (Armstrong et al., 2006). LPS amounts can be then translated into bacterial biomass (cells/g), i.e., 1 EU/mL is equivalent to about 10^5 cells/mL, *Escherichia coli*-like cells.

The LAL assay can be successfully applied to determine the Gram-negative biomass in rocks and fine-grained materials with apparently no interference from clay-rich and ferric iron-rich matrix (Bonaccorsi et al., 2010). This assay has been routinely used to detect endotoxin contamination in environmental liquid samples and biopharmaceutical products, in Planetary Protection protocols (Procedural Requirements NASA, 2007) and, more recently, for astrobiology related field investigations (e.g., Allen et al., 2003; Eigenbrode et al., 2006; Bonaccorsi et al., 2010).

4. Approach and Study Areas

Clay-supported versus Fe-oxide-rich samples from the hyperarid Atacama Desert (Fig. 4) are compared with samples from the arid Death Valley (Fig. 5) and the moister California Coast (Pescadero State Beach) as the end-member with the most rich and diverse biomass (Fig. 6).

Bonaccorsi et al. (2010) have found that when comparing phyllosilicate-rich samples against an aridity/moisture gradient, the positive linear relationship observed in coarse-grained (nonclay) soil between microbial biomass, plant cover, and rainfall may not be applied to clay mineral-rich environment.

For clarity, study sites, location, and environmental features of these sites are outlined below and summarized in Table 1.

Figure 4. Atacama Desert, Brine-rich oxidized and clay deposits (**a**). Banded structure (**b**) and desiccation polygons (**c**).

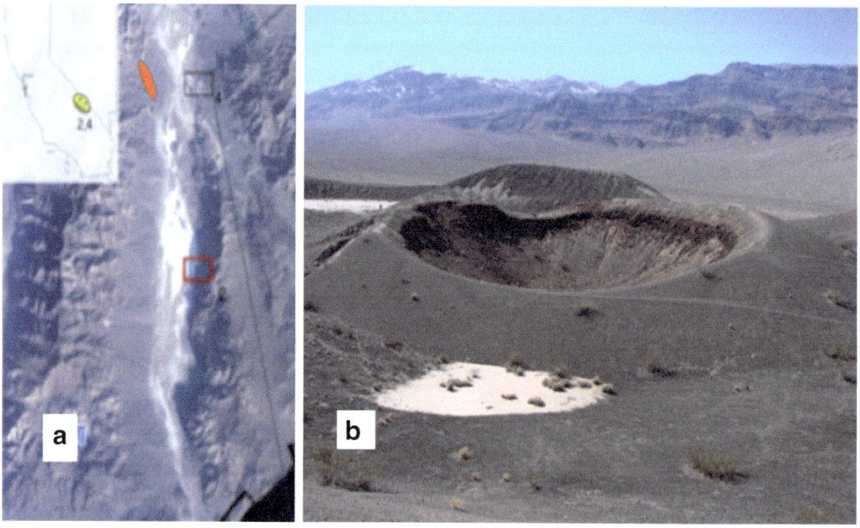

Figure 5. Locations of Death Valley National Park (**a**), and Little Hebe Crater (**b**).

Figure 6. Location of Pascadero State Beach (**a**) and overview of the studied deposits (**b**, **c**).

Table 1. Habitability tests performed with the LAL- and ATP-based assays. Modified after Bonaccorsi et al. (2010). See main text for more explanation.

Location	MAP (mm/year)	Sample name	Depth	Type	LAL EU/mL ($N = 2$)	LPS-based biomass cells/g	ATP assay RLUs (avg ± SD)	Description
Atacama, other arid (nonclay) sites	~12	AT06-03, 0-1	0–1	Soil	3.47	1.82×10^6	21,573 ± 812	Coarse grained, plant-colonized
	>2	AT06-05, 0-3	0–3	Soil	0.55	3.27×10^5	388 ± 35	Coarse grained, plant-barren
	<2	AT08-ATN-24	0–2	Soil	<0.05	$\sim 5.0 \times 10^4$	259 ± 39	Coarse grained, plant-barren
	~100	AT08-11	0–2	Fe-oxides	334	1.7×10^8	–	Hematite-rich, plant-colonized, top soil
Atacama Hyperarid desert core (Yungay)	<2	AT08-YU-42C	0–3	Clays (1)	<0.05	$\sim 6.0 \times 10^4$	–	Dessication polygons. Plant-barren
		AT08-YU-42B	0–2	Fe-oxides	<0.20	$\sim 1.2 \times 10^5$	–	Plant-barren
(DVNP) Little Hebe crater	~120	HUBE08-3C	0–1	Clays (2)	106	6.4×10^7	9,989 ± 246	Clay pond deposit. Plant-colonized
		HUBE08-3D	0–2	Fe-Oxides	<15	$\sim 9.8 \times 10^6$	870 ± 109	Hematite-rich and gypsum-bearing. Plant-barren
Pescadero State Beach	~750	PES08-2C	*	Fe-Oxides	31.4	1.9×10^7	2,858 ± 1,461	Goethite-rich oxidized sandstone
		PES08-2D	*	Clays	24.1	1.5×10^7	912 ± 200	Gray clays
		PES08-1B	*	Clays (3)	48.8	$\sim 3.0 \times 10^7$	1,475 ± 514	Green and blue clays

* Indicate surface samples from wall deposits (cfr. Figure 6)

4.1. HYPERARID ATACAMA DESERT

Phyllosilicate deposits in the hyperarid Atacama represent an improved ideal analog to constrain habitability potential in the most life-depleted setting.

The Atacama Desert has been extensively investigated as it represents the driest place on Earth, i.e., many times drier than other arid and hyperarid regions on Earth (McKay et al., 2003; Navarro-González et al., 2003; Warren-Rhodes et al., 2006). Coarse-grained Atacama soils are typical alluvial containing gypsum, quartz, feldspars, low amounts of clay, and very low levels of refractory organics (at the 0.01% wt.). In these soils, both biomass and preserved organics increase by a latitudinal/ altitudinal rainfall gradient (Navarro-González et al., 2006; Bonaccorsi and McKay, 2008, 2009).

Hematite rich (dark toned in Fig. 4a, b) and clay mineral deposits (Fig. 4, light toned) were collected in the Yungay Desert region for habitability studies (Bonaccorsi et al., 2010).

4.2. ARID DEATH VALLEY

Little Hebe is a young (300 years) and small spatter cone (formed by molten lava) located in the Ubehebe Volcanic Field (37°0′35 N, 117°27′1 W) in the Death Valley National Park, CA (Fig. 5a). The whole feld comprises over 13 maar volcanoes formed by subsequent phreatomagmatic explosions exposing thick sediments of Miocene age (Crowe and Fisher, 1973) and represents a geological analog for caldera structures identified on Mars, i.e., highland paterae (Crown and Greeley, 1993).

Oxidized (hematite-rich) and gypsum-bearing cemented sandstones were sampled from the Little Hebe crater's rim (Fig. 5b), which is plant-barren and arider because of exposure to strong wind blasting. Silty-clay, phyllosilicates-bearing light-toned sediments of alluvial/aeolian origin accumulate in the area and at the crater's bottom, which is colonized by sparse vegetation, e.g., Desert Holly (*Atriplex hymenelytra*) and Creosote bush (*L. tridentata*) as main source of microbial biomass.

4.3. MOIST COASTAL FOG REGION

Pescadero State Beach is a coastal site of California, characterized by Mediterranean climate (rain of 700 mm/year), fog, and low overcast year round.

Clays were sampled from an exposed outcrop (Fig. 6b, c) consisting of water-saturated estuarine mud (Holocene) predominantly gray, green, and blue clays (indicated by arrows in Fig. 6c) capped by an oxidized, goethite-rich, poorly consolidated sandstone unit (Brabb et al., 1998). Soils atop these units are extensively colonized by cordgrass (*Spartina sp.*) and pickleweed (*Salicornia* sp.).

Ocean spray and near-surface soil horizons represent an obvious diversified source of living biomass (including gram negative-like organisms) and dissolved organic matter to the exposed formations.

5. Viable Biomass in Phyllosilicate Mineral Environments

Microbial biomass contents have been determined in phyllosilicate-rich materials including mineral phases similar to those identified or inferred to exist on the Martian surface. This was done to explore the possible habitability of analog mineral deposits formed on a wetter/warmer Mars. Viable microbial biomass content in clays versus nonclay materials along a climate precipitation gradient is summarized in Table 1.

By examination of Table 1, three observations can be made: (1) Atacama desiccation polygons (Sample AT08-YU-42C, Fig. 4c) and contiguous hematite-rich deposits (sample AT08-YU-42B, Fig. 4a) contain similar amounts of biomass, which is even lower than that of coarse-grained soil from nearby sites (Samples AT08 and AT-06). The Atacama clays (muscovite and kaolinite) are three-orders of magnitude lower than (2) surface clays (montmorillonite, illite, and chlorite) from the Death Valley (Sample HUBE08-3C, Fig. 5b). Finally (3) the Gram-negative biomass content of the water-saturated clays (PES08-1B: kaolinite, illite, and vermiculite, Fig. 6c) from humid coastal sites (>700 mm/year) is about the same order of magnitude (~(or about) 10^7 cells/g), which is counter-intuitive, than that found in clays from the arid site (>10 to <100 mm/year rainfall).

6. Concluding Remarks

Comparisons across the rainfall gradient suggests that, at least for the clay analog environment studied here, increasing vegetation cover and moisture availability would not necessarily translate into higher content of living biomass/habitability.

Levels of living biomass inhabiting clays could be related not only to the potential sources of organics and biomass in the surrounding environment, but also to the interplay of several factors, including, but not limited to, clay type assemblages and abundances, grain size, precipitation, and/or environmental moisture.

Furthermore, it is unclear weather or not clay mineral-rich environments have a higher habitability potential with respect to that of background, clay mineral-barren environments.

Geochemical analyses are underway to determine the effective role of minerals in hosting viable biomass and/or preserving related organic biosignatures. This will complement information on the habitability of these relevant Mars analogs and in support of tactical decision-making by the integrated MSL science team.

A deeper understanding of preservation potential and habitability of phyllosilicates and hematite-rich materials, achieved by studying new analog sites where these minerals are simultaneously present, will therefore provide critical information to help identify the best locations to search for ancient and possibly modern organic biosignatures on Mars.

7. Acknowledgments

I thank Professor Vinod C. Tewari, W.I.H.G, Dehradun (India), for kindly inviting me to contribute this chapter for the book. I deeply thank Professor Julian Chela Flores, I.C.T.P., Trieste, Italy, and Professor Nevio Pugliese, University of Trieste (Italy), who reviewed the chapter critically and provided suggestions to improve the chapter. Special thanks also go to Chris McKay, Carol Stoker, and Aaron Zent (NASA Ames Research Center) for insightful discussion.

Research in Death Valley Note Park conducted under DEVA-SCI-0026.

8. References

Allen, C.C., Wainwright, N.R., Grasby, S.E. and Harvey, R.P. (2003) Life in the ice. Sixth International Conference on Mars, 2003 contribution no. 3138.

Apak, R. (2008) Life detection experiments of the Viking Missionon Mars can best be interpreted with a Fenton oxidation reaction composed of H_2O_2 and Fe^{2+} and iron-catalyzed decomposition of H_2O_2. Int. J. Astrobiol. **7**: 187–194.

Armstrong, M.T., Theg, S.M., Braun, N., Wainwright, N., Pardy, R.L. and Armstrong, P.B. (2006) Histochemical evidence for lipid A (endotoxin) in eukaryote chloroplasts. FASEB J. **20**: 2145–2146.

Aubrey, A., Cleaves, H.J., Chalmers, J.H., Skelley, A.M., Mathies, R.A., Grunthaner, F.J., Ehrenfreund, P. and Bada, J.L. (2006) Sulfate minerals and organic compounds on Mars. Geology **34**(5): 357–360.

Bedick, J.C., Shnyra, A., Stanley, D.W. and Pardy, R.L. (2001) Innate immune reactions stimulated by a lipopolysaccharide like component of the alga Prototheca (strain 289). Naturwissenschaften **88**: 482–485.

Beegle, L., Wilson, M., Abilleira, F., Jordan, J. and Wilson, G. (2007) Mars Astrobiology Field Laboratory and the search for signs of life. Astrobiology **7**(4): 545–577.

Benner, S.A., Devine, K.G., Matveeva, L.N. and Powell, D.H. (2000) The missing organic molecules on Mars. Proc. Natl. Acad. Sci. U.S.A. **97**: 2425–2430.

Bibring, J-P., Langevin, V., Mustard, J.F., Poulet, F., Arvidson, R., Gendrin, A., Gondet, B., Mangold, N., Pinet, P., Forget, F. and the Omega Team. (2006) Global mineralogical and aqueous Mars history derived from Omega/Mars Express. Science **312**: 400–404.

Bishop, J.L., Noe Dobrea, E.Z., McKeown, N.K., Parente, M., Ehlmann, B.L., Michalski, J.R., Milliken, R.E., Poulet, F., Swayze, G.A., Mustard, J.F., Murchie, S.L. and Bibring, J-P. (2008) Phyllosilicate diversity and past aqueous activity revealed at Mawrth Vallis, Mars. Science **321**: 830–833.

Bonaccorsi, R. and McKay, C.P. (2008) Total biomass and organic carbon along a N-S moisture gradient of the Atacama region, Chile [abstract 1489]. Thirty-Ninth Lunar and Planetary Science. Conference Abstracts, Lunar and Planetary Institute, Houston.

Bonaccorsi, R. and McKay, C.P. (2009) Microbial biomass in clays from arid/ hyper-arid regions: testing for Habitability in desert Mars Analogs, In: R. Reynolds and D.R. Jessey (eds.) *The 2009 Desert Symposium Field Guide and Proceedings*. California State University, Desert Studies Consortium and LSA Associates, April 2009.

Bonaccorsi, R. and Stoker, C.R. (2008) Science results from a Mars drilling simulation (Rio Tinto, Spain) and ground truth for remote observations. Astrobiology **8**(5): 967–985.

Bonaccorsi, R., McKay, C.P. and Chen, B. (2010) Biomass and habitability potential of clay minerals and iron-rich environments: Testing novel analogs for Mars Science Laboratory landing site candidates. Special Issue Layered Silicate Materials. Philosophical Magazine, **90**(17–18): 2309–2327.

Brabb, E.E., Graymer, R.W. and Jones, D.L. (1998) Geology of the onshore part of San Mateo County, California: a digital database; USGS. Open-File Report 98-137. http://pubs.usgs.gov/of/1998/of98-137/smgeo.txt.

Bruckner, J. and Venkateswaran, K. (2007) Overview of methodologies to sample and assess microbial burden in low biomass environments. Jpn. J. Food Microbiol. **24**(2): 61–70.

Bullock, M.A., Stoker, C.R., McKay, C.P. and Zent, A.P. (1994) A coupled soil-atmosphere model of H_2O_2 on Mars. Icarus **107**: 142–154.

Catling, D.C. (2007) Ancient fingerprint in the clay. Nature **448**: 31–32.

Chevrier, V. (2008) Organic burial site on Mars? Nat. Geosci. **1**: 348–350.

Crown, D.A. and Greeley, R. (1993) Volcanic geology of Hadriaca Patera and the eastern Hellas region of Mars. J. Geophys. Res. **98**: 3431–3451.

Cowan, D.A., Russell, N.J., Mamais, A. and Sheppard, D.M. (2002) Antarctic Dry Valley mineral soils contain unexpectedly high levels of microbial biomass. Extremophiles **6**: 431.

Cowie, G.L., Hedges, J.I., Prahl, F.G. and De Lange, G.J. (1995) Elemental and major biochemical changes across an oxidation front in a relict turdibite: a clear-cut oxygen effect. Geochim. Cosmochim. Acta. **59**: 33–46.

Crowe, B.M. and Fisher, R.V. (1973) Sedimentary structures in base-surge deposits with special reference to cross bedding, Ubehebe Craters, Death Valley. Geol. Soc. Amer. Bull. **84**: 663–682.

D'Acqui, L.P., Daniele, E., Fornasier, F., Radaelli, L. and Ristorri, G.G. (1998) Interaction between clay microstructure, decomposition of plant residues and humification. Eur. J. Soil Sci. **49**: 579–587.

Davila, A.F., Fairén, A.G., Duport, L.G., Stoker, C.R., Amils, R., Bonaccorsi, R., Zavaleta, J., Lim, D., Schulze-Makuch, D. and McKay, C.P. (2008) Subsurface formation of oxidant on Mars and implications for the preservation of organic biosignatures. EPSL **272**(1–2): 456–463.

Ehlmann, B.L., Mustard, J.F., Fassett, C.I., Schon, S.C., Head III, J., Des Marais, D.J., Grant, J.A. and Murchie, S.L. (2008) Clay minerals in delta deposits and organic preservation potential on Mars. Nat. Geosci. **1**: 355–358.

Eigenbrode, J., Benning, L.G., Maule, J., Wainwright, N., Steele, A., Amundsen, H.E.F. and the AMASE 2006 Team (2009) A field-based cleaning protocol for sampling devices used in life-detection studies. Astrobiology **9**(5): 455–465.

Ertem, G., Hazen, R.M. and Dworkin, J.P. (2007) Sequence analysis of trimer isomers formed by montmorillonite catalysis in the reaction of binary monomer mixtures. Astrobiology **7**: 715–722.

Ferris, J.P., Ertem, G. and Agarwal, V. (1989) The adsorption of nucleotides and polynucleotides on montmorillonite clay. Orig. Life Evol. Biosph. **19**: 153–164.

Glavin, D.P., Cleaves, H.J., Buch, A., Schubert, M., Aubrey, A., Bada, J.L. and Mahaffy, P.R. (2006) Sublimation extraction coupled with gas chromatography-mass spectrometry: a new technique for future in situ analyses of purines and pyrimidines on Mars. Planet. Space Sci. **54**: 1584–1591.

Hedges, J.I. and Kiel, R.G. (1995) Sedimentary organic matter preservation: an assessment and speculative synthesis. Mar. Chem. **49**: 81–115.

Kaiser, K. and Zech, W. (2000) Dissolved organic matter sorption by mineral constituents of subsoil clay fractions. J. Plant Nutr. Soil Sci. **163**: 531–535.

Keil, R.G., Tsamakis, E., Fuh, C.B., Giddings, C. and Hedges, J.I. (1994) Mineralogical and textural controls on organic composition of coastal marine sediments: hydrodynamic separation using SPLITT fractionation. Geochim. Cosmochim. Acta. **57**: 879–893.

Kleber, M., Mikutta, R., Torn, M.S. and Jahn, R. (2005) Poorly crystalline mineral phases protect organic matter in acid subsoil horizons. Eur. J. Soil Sci. **56**: 717–725.

Klein, H.P. (1992) The Viking biology experiments: epilogue and prologue. J. Orig. Life Evol. Biosph. **21**: 255–261.

Lammer, H., Lichtenegger, H.I.M, Kolb, C., Ribas, I., Guinan, E.F., Abart, R. and Bauer, S.J. (2003) Loss of water from Mars: implications for the oxidation of the soil. Icarus **165**: 9–25.

Mahaffy, P. (2008) Exploration of the habitability on Mars: development of analytical protocols for measurement of organic carbon on the 2009 Mars science laboratory. Space Sci. Rev. **135**: 255–268.

McKay, C.P., Friedmann, E.I., Gomez-Silva, B., Cáceres-Villanueva, L., Andersen, D.T. and Landheim, R. (2003) Temperature and moisture conditions for life in the extreme arid region of the Atacama Desert: four years of observations including the El Niño of 1997–1998. Astrobiology **3**: 393–406.

Miller, D.P., Bonaccorsi, R. and Davis, K. (2008) Design and practices for use of automated drilling and sample handling in MARTE while minimizing terrestrial and cross contamination. Astrobiology **8**(5): 947–965.

National Aeronautics and Space Administration. Procedural requirements NPR 5340 (2007) NASA standard procedures for the microbial examination of Space Hardware. http://nodis3.gsfc.nasa.gov/main_lib.cfm.

Navarro-González, R.F., et al. (2003) Mars-like soils in the Atacama Desert, Chile, and the dry limit of microbial life. Science **302**: 1018–1021.

Navarro-González, R.F., et al. (2006) The limitations on organic detection in Mars-like soils by thermal volatilization–gas chromatography–MS and their implications for the Viking results. Proc. Natl. Acad. Sci. U.S.A. **103**(44): 16089–16094. doi: 10.1073/pnas.0604210103.

Poulet, F., Bibring, J-P., Mustard, J.F., Gendrin, A., Mangold, N., Langevin, Y., Arvidson, R., Gondet, B., Gomez, C. and the Omega Team. (2005) Phyllosilicates on Mars and implications for early Martian climate. Nature **438**: 623–627.

Raetz, C.R.H. and Whitfield, C. (2002) Lipopolysaccharide endotoxins. Annu. Rev. Biochem. **71**: 635–700.

Stoker, C.R., Cannon, H.N., Dunagan, S.E., Lemke, L.G., Glass, B.J. and 24 others. (2008) The 2005 MARTE Robotic Drilling Experiment in Rio Tinto Spain: objectives, approach, and results of a simulated mission to search for life in the Martian subsurface. Astrobiology **8**(5): 921–945.

Stoks, P.G. and Schwartz, A.W. (1979) Uracil in carbonaceous meteorites. Nature **282**: 709–710.

Sumner, D.Y. (2004) Poor preservation potential of organics in Meridiani Planum hematite-bearing sedimentary rocks. J. Geophys. Res. **109**: E12007. doi:10.1029/2004JE002321.

Task Group on organic environments in the Solar System, National Research Council (2007) Exploring organic environments in the Solar System 124 pages http://www.nap.edu/catalog.php?record_id=11860.

Torn, M.S., Trumbore, S.E., Chadwick, O.A., Vitousek, P.M. and Hendricks, D.M. (1997) Mineral control of soil organic carbon storage and turnover. Nature **389**: 170–173.

Van der Velden, W. and Schwartz, A.W. (1977) Search for purines and pyrimidines in the Murchison meteorite. Geochim. cosmochim. Acta **41**: 961–968.

Wang, A., Korotev, R.L., Jolliff, B.L., Haskin, L.A., Crumpler, L., Farrand, W.H., Herkenhoff, K.E., de Souza, P.Jr, Kusak, G.A., Hurowitz, J.A. and Tosca, N.J. (2006) Evidence of phyllosilicates in Wooly Patch, an altered rock encountered at West Spur, Columbia Hills, by the Spirit rover in Gusev crater, Mars, J. Geophys. Res. **111**: E02S16.

Warren-Rhodes, K.A., Rhodes, K.L., Pointing, S.B., Ewing, S.E., Lacap D.C., Gómez-Silva, B., Amundson, R., Friedmann, E.I., McKay, C.P. (2006) Hypolithic cyanobacteria, dry limit of photosynthesis, and microbial ecology in the Hyperarid Atacama Desert. Microbial Ecol. **52**: 389–398.

Wattel-Koekkoek, E.J.W. and Burman, P. (2004) Mean residence time of kaolinite and smectite-bound organic matter in Mozambiquan Soils. Soil Sci. Soc. Am. J. **68**(1): 154.

Zent, A.P. (1998) On the thickness of the oxidized layer of the martian regolith. J. Geophys. Res. **103**: 31491–31498.

Biodata of **Julian Chela-Flores** and **Vinod Chandra Tewari**, authors of ***"The Sulfur Cycle on the Early Earth: Implications for the Search of Life on Europa and Elsewhere"***

Professor Julian Chela-Flores was born in Caracas, Venezuela, and studied in the University of London, England, where he obtained his Ph.D. in quantum mechanics (1969). He was a researcher at the Venezuelan Institute for Scientific Research (IVIC) and Professor at Simon Bolivar University (USB), Caracas, until his retirement in 1990. During his USB tenure, he was Dean of Research for 6 years. He is a Fellow of The Latin American Academy of Sciences, The Academy of Sciences of the Developing World, the Academy of Creative Endeavors (Moscow), and a Corresponding Member of the Venezuelan "Academia de Fisica, Matematicas y Ciencias Naturales." His current positions are Staff Associate of the Abdus Salam International Center for Theoretical Physics (ICTP), Trieste, Research Associate, Dublin Institute for Advanced Studies (DIAS), and Professor-Titular, Institute of Advanced Studies (IDEA), Caracas. His particular area of expertise is astrobiology, in which he is the author of numerous papers. He organized a series of Conferences on Chemical Evolution and the Origin of Life from 1992 till 2003. He is the author of "The New Science of Astrobiology From Genesis of the Living Cell to Evolution of Intelligent Behavior in the Universe" (Springer) and "A Second Genesis: Stepping stones Towards the Intelligibility of Nature" (WSP).

E-mail: **chelaf@ictp.it**

V.C. Tewari and J. Seckbach (eds.), *STROMATOLITES: Interaction of Microbes with Sediments*, Cellular Origin, Life in Extreme Habitats and Astrobiology 18, 723–736
DOI 10.1007/978-94-007-0397-1_33,

Professor Vinod C. Tewari was born in Almora, Uttarakhand, in India in 1954 and is currently a Senior Scientist and the Head of the Sedimentology Group at Wadia Institute of Himalayan Geology, Dehradun, India, and a Senior Associate of International Center for Theoretical Physics, Trieste, Italy. He obtained his Ph.D. from the University of Lucknow in Geology in 1986 and continued his research in Wadia Institute. Dr. Tewari taught Geology at Kumaon University, Nainital, Uttarakhand (U.K.), India, as Professor of Geology. Professor Tewari's scientific interests are in the areas of Precambrian stromatolites, sedimentation, carbon isotope chemostratigraphy, genesis, early evolution and diversification of life, and its astrobiological significance. He is associated with the International Geological Correlation Program (I.G.C.P.) Project 493 on The Rise and Fall of Vendian Biota and Ediacaran System. He has eighty five research papers published to his credit, and edited several volumes of Himalayan Geology, India, and *Journal of Nepal Geological Society*, Kathmandu, Nepal. Professor Tewari has organized first Indo-Soviet Symposium on Stromatolites and Stromatolitic Deposits and other IGCP meetings in India. He is one of the organizers of the World Summit on Ancient Microscopic Fossils held in the University of California, Los Angeles, USA, in 2008. He is a member of the Editorial Board of the international journal *Astrobiology*, New York, USA.

E-mail: **vinodt1954@yahoo.co.in**

THE SULFUR CYCLE ON THE EARLY EARTH: IMPLICATIONS FOR THE SEARCH OF LIFE ON EUROPA AND ELSEWHERE

JULIAN CHELA-FLORES [1,2] AND VINOD CHANDRA TEWARI [3]

[1] *The Abdus Salam International Centre for Theoretical Physics, Strada Costiera 11, 34014 Trieste, Italy*

[2] *Instituto de Estudios Avanzados (IDEA), Caracas 1015A, República Bolivariana de Venezuela*

[3] *Wadia Institute of Himalayan Geology, Dehradun 24800, Uttarakhand, India*

Abstract The search for life in the universe, especially on the Jovian satellite Europa, could benefit from our knowledge of the bacterial processing of sulfur on the early Earth. We know that sulfate respiring bacteria reduce sulfur and produce large fractionation between its isotopes, especially ^{32}S and ^{34}S. The presence of sulfur patches on the Europan surface, as revealed by the Galileo mission and confirmed by the New Horizons, may have some astrobiological implications. In principle, they could be related to sulfate-reducing bacteria and sulfur disproportionation on the ocean seafloor and its subsurface. The presence of pyrite in the oncolitic and stromatolitic laminae recorded from several Precambrian formations of the world reveal pyrite biomineralization in highly reducing conditions in the Archean and Proterozoic. A review of geological and biogeochemical data from the Precambrian demonstrates that both pyrite and evaporite formed biologically by dissimilatory sulfate reduction. In the present review, we maintain that S-isotope analysis is a most valuable tool for the exploration of the Solar System. In situ analysis of the Europan surficial icy patches should be targets for the future exploration of the Jovian System by the future worldwide effort to explore the Jovian System.

Keywords Sulfur cycle • Early Earth • Stromatolites • Europa • Mars • Astrobiology • Galileo probe • Pyrite biomineralization • Sulfur isotopes

1. Introduction

The possibility of life on Jovian satellite Europa and on Mars could be tested, based on the occurrence of biogenic chemical elements (C, H, O, N, S) on the early Earth (Chela-Flores, 2006). We know that sulfate-respiring bacteria reduce sulfur and produce large fractionations between the ^{32}S and ^{34}S isotopes. The presence of sulfur patches on the surfaces of Europa and Mars (including rich concentration of sulfur in the Martian meteorites) may have implications in our search in our solar system for biomarkers, both for the Galilean satellite and for the Red Planet. We discuss the role of microbial sulfur on the early Earth and its potential astrobiological significance. In the search for biomarkers, new techniques and instrumentation are being developed. They will facilitate the interpretation of biomarkers. Firstly, our capability to infer the presence of past life on terrestrial planets such as Mars, by questioning what amount of rock is needed to distinguish past life from nonlife (Schopf et al., 2008). Secondly, new instrumentation are available that aims at distinguishing past life (or extant life) on the icy bodies of the outer solar system such as the Galilean satellites Europa, Ganymede, and Enceladus in the Saturn System. The recent S isotope data from the 3.4-billion-year old (Ga) North Pole barite deposit in Australia provides the oldest evidence of microbial sulfate reduction.

It also demonstrates the presence of sulfate-reducing microbes in the early Archean (Shen and Buick, 2004). S isotopic ratios provide valuable clues regarding the presence of sulfur-dependent metabolic activity on the early Earth. Archaean oceans at 3.5 Ga were sulfate rich and sulfides were formed by sulfate-reducing bacteria. S isotopic compositions of sulfides are enriched in ^{32}S and may be biogenic in origin. In geological history, the major deposits of stromatolitic phosphorites (with pyrite and oncolites) occur in the Precambrian–Cambrian boundary succession of the Asian-Pacific region of the world (Tewari, 1991). The Precambrian–Cambrian phosphorite–stromatolite association from the Tal Formation, Lesser Himalaya, India (and elsewhere), shows the presence of pyrite in the oncolitic and stromatolitic microlaminae. The pyrite follows the original biolamination pattern indicating conditions of a reducing paleoenvironment (cf. Fig. 1 and Tewari, 1991).

A review of the geological and biochemical data from the Precambrian demonstrates that pyrites and evaporites were formed biologically by dissimilatory sulfate reduction. We conclude that S isotope analysis is most valuable for solar system exploration. In situ analysis of the Europan surficial icy patches of sulfur, together with carbon isotopic signatures, will inevitably be targets for future space missions to the Jovian system, now in their planning stages (the worldwide collaboration for a Europa-Jupiter System Mission, EJSM).

The search for life in the universe and the possibility of life on Jovian satellite Europa and the planet Mars could be related to the occurrence of sulfur on early earth (Seckbach and Chela-Flores, 2007; Chela-Flores et al., 2008). We know that sulfate-respiring bacteria on earth reduce sulfur and produce large fractionations between the ^{32}S and ^{34}S isotopes. The presence of sulfur patches on the surface of Europa as well as on the Martian surface and a rich concentration of sulfur in the Martian meteorites have astrobiological implications regarding the existence of microbial life on these two most promising candidates for the search for extraterrestrial life in our solar system. We are discussing the possible role of microbial sulfur on earth and their significance in the origin of life and astrobiology. The Precambrian–Cambrian phosphorite–stromatolite

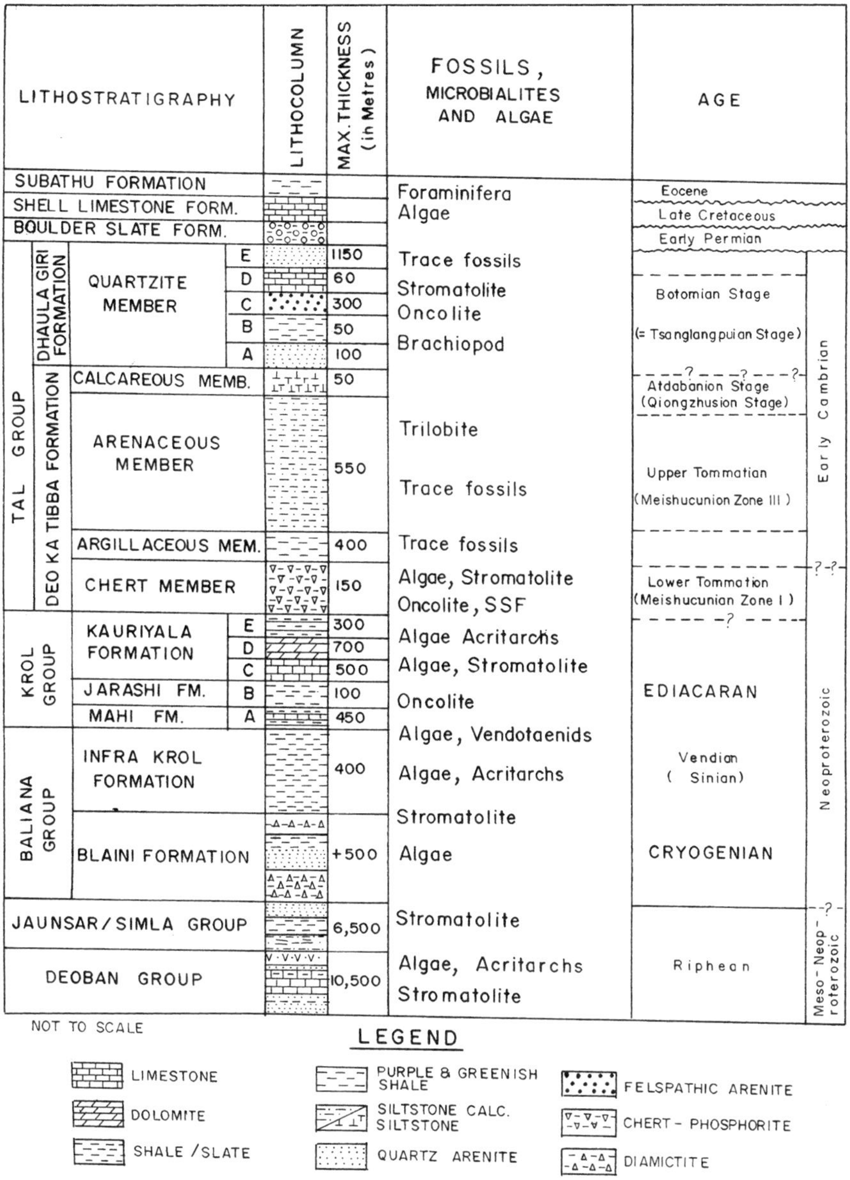

Figure 1. Precambrian–Cambrian boundary stromatolitic, oncolitic, pyritic phosphatic beds at the contact of the Krol-Tal Formations, Lesser Himalaya, India, represents global phosphogenic event (Tewari, 1991, 2007).

association from the Tal Formation, Lesser Himalaya, India, and elsewhere show the presence of pyrite in the oncolitic and stromatolitic microlaminae. The pyrite follows the original biolamination pattern, indicating conditions of reducing paleoenvironment (Tewari, 1991). A review of the geological and biochemical data from Precambrian demonstrates that pyrites and evaporates were formed biologically by dissimilatory sulfate reduction.

We conclude that S isotope analysis is the most valuable for planetary exploration. In situ analysis of the Europan surficial icy patches of sulfur together with carbon isotopic signatures will inevitably be targets for future space missions.

2. Earliest Evidence of Sulfate-Reducing Bacteria and Their Modern Analogs

In the previous section, we have mentioned the existence of evidence of sulfur isotope data from 3.47 Ga old North Pole Dome, Western Australia. Microscopic pyrite associated with barite deposits from North Western Australia. This pyrite is the oldest evidence of microbial sulfate reduction on primitive earth. Organic carbon isotope data from the oldest metasediments 3.8 Ga old Isua complex is consistent with the existence of autotrophic CO_2 fixation in to biomass (Schidlowski et al., 1983). During microbial sulfate reduction, the stable isotopes ^{32}S and ^{34}S are discriminated so that the daughter sulfides are isotopically fractionated with respect to the parent sulfate, with the sulfides being depleted in ^{34}S. The oldest terrestrial S-isotopic records come from highly metamorphosed and deformed ferruginous rocks resembling banded iron formations from the Isua Supracrustal Belt, Greenland (3.8 Ga), which shows narrow range with a mean value of + 0.5 ± 0.9‰ (Schidlowski et al., 1983). However, sedimentary sulfides in the 2.7 Ga old iron formations of Canada are highly depleted in $\delta^{34}S$ values as low as –17.5‰. This indicates that microbial sulfate reduction must have evolved by 2.7 Ga. The modern marine environments, sulfate reduction and pyrite formation occurs near the sediment surface where sulfur reduction rates are highest. The occurrence of pyrite and siderite in Archaean sedimentary rocks from Pilbara, Australia, indicates that oxygen was less in the Archaean atmosphere. The evolution of atmospheric oxygenation is linked to the Precambrian sulfur isotopic records. Shen and Buick (2004) have also interpreted that the stromatolites associated with the North Pole barite of Australia must have been formed by green and purple photoautotrophic sulfur-oxidizing bacteria of the *Cholorobiaceae* and *Chromatiaceae*. Sulfate reduction is a complex process requiring advanced membrane bound transport enzymes, proton motive force generation through the activities of ATPase and other proteins involved in charge separation, and the genetic synthesis through DNA and RNA. The giant sulfur bacterium *Thiomargarita namibiensis* occurs in high biomass in surface sediments off the coast of Namibia (Schultz and Schulz, 2005). This bacterium gains energy by oxidizing sulfide, which accumulates in anoxic marine sediments as a result of the degradation of organic matter by sulfate-reducing bacteria. Modern phosphorite formation has been reported from these sulfate-reducing bacteria.

3. Factors that Are Needed for Understanding the Precambrian Sulfur Cycle

A series of topics in Precambrian geology will be needed for gathering the insights that will orient us in the eventual search for other microbially driven sulfur cycles in the Solar System: Proterozoic–Cambrian sulfur isotopic ratios, pyrite formation in stromatolitic–Oncolitic–Phosphatic sedimentary environment, and sulfide microbial mineralization. On the other hand, we recall that Ediacaran (Neoproterozoic) rifting resulted by the breakup of the supercontinent Rodinia around 750–690 Ma (Tewari, 2007, 2008). The development of the Krol basin in the Lesser Himalaya (Fig. 1) is associated with this event and the glacial diamictites (Blaini Formation) succession was deposited at the base of the basin, which is correlated with the global Neoproterozoic glaciation (Marinoan/Blainian, Tewari, 2001a, b, 2004, 2007, 2008). The overlying pink limestone is cap microbial carbonate with highly depleted carbon isotope ratios (Tewari and Sial, 2007). The Krol–Tal Ediacaran–Lower Cambrian carbonate–stromatolitic– phosphatic–oncolitic–pyritic beds are located at the Precambrian–Cambrian boundary in the Lesser Himalaya as well as in the China, Mongolia, Oman, and Iran (Brasier et al., 2002; Tewari, 1989, 1993, 1994, 1999, 2007; Goldberg et al., 2005).

There was global oceanic anoxia near the Precambrian/Cambrian boundary. This was also the period when soft-bodied Ediacaran metazoans declined (extinction?) on earth and another biological diversification of Cambrian life took place. It is interesting that small shelly fauna (protoconodonts and conodonts, phosphatized oncolites, and stromatolites having pyritic and phosphatic microlaminae and algae) were restricted to the reducing (nonphotosynthetic) environment (Figs. 2–4 and Tewari, 1984, 1989, 1994, 1996, 2004, 2007).

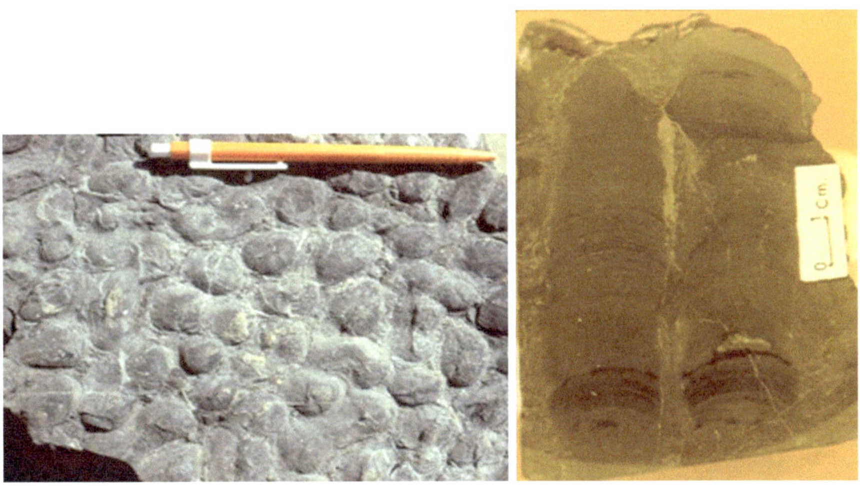

Figure 2. *Collumnaefacta vulgaris* stromatolite, Lower Cambrian, Lower Tal Formation, Lesser Himalaya, India, showing the top view heads of the phosphatic pyritic stromatolite (*left*) and the columnar structure with *dark black* phosphatic laminae and *golden yellow* pyritic laminae (Tewari, 1984, 1989, 1991).

Figure 3. Boxonia stromatolite from the Lower Cambrian Tal Formation, Lesser Himalaya, India, showing development of pyrite (*golden yellow color*) in the stromatolitic laminae as well as in the intercolumnar area and the thick wall between the columns (Tewari, 1989).

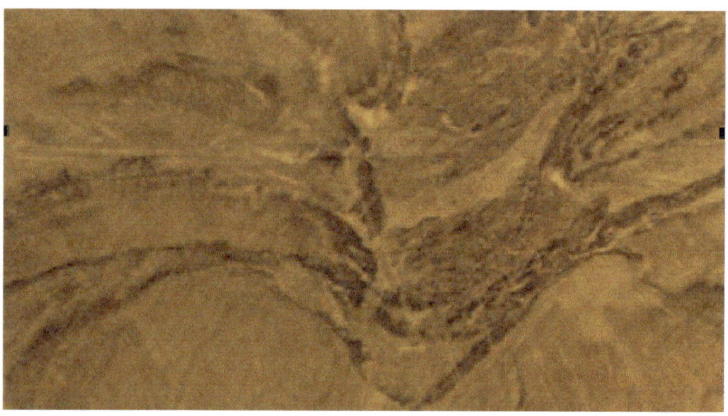

Figure 4. Phosphatic *dark black* laminae in the stromatolites of Meso- to Neoproterozoic Gangolihat Dolomite, Kumaon Lesser Himalaya, the intraclasts, pyritic pellets, and oolites are found in the intercolumnar area (Tewari, 1989).

The global sulfur isotopic trends suggest a major increase in the importance of sulfate-reducing bacteria with rising sulfate levels. The decreasing $\delta^{34}S$ values for sedimentary pyrite and increasing $\delta^{34}S$ values for sea water sulfate may be the result of widespread reduction under conditions similar to those of modern oceans (Lambert and Donnelly, 1992). The definitions of $\delta^{34}S$ are given in Sect. 4.

4. Bacterial Sulfate Reduction

Sulfur is an important element for all microorganisms, animals, and plants on Earth. The main sources of sulfur are bacterial sulfate reduction, plants, and soils. Sulfur isotopic ratios provide valuable clues regarding the presence of sulfur-based metabolic activity on the early earth. The redistribution of the primordial isotopic mixtures can be followed up in terms of the appropriate parameter, namely

$$\delta^{34}S = \left[(^{34}S/^{32}S)_{sa} / (^{34}S/^{32}S)_{st} - 1 \right] \times 10^3 [‰, CDT].$$

For simplicity, this function will be referred to as the $\delta^{34}S$ parameter, or simply as the delta parameter. Its value is close to zero when the sample coincides with the corresponding value of the Canyon Diablo meteorite that is a triolite (FeS), abbreviated as CDT. This parameter allows a comparison of a sample (sa) with the standard (st) CDT. The relevant terms are the dominant sulfur isotope (^{32}S) and the next in abundance (^{34}S). In fact, $(^{34}S/^{32}S)_{st}$ coincides with the average terrestrial fraction of the two most abundant isotopes of sulfur. We obtain positive values of the delta-parameter when, by comparison, we have a larger quantity of the less abundant isotope ^{34}S.

Sulfur isotopic values of $\delta^{34}S$ for sulfide (pyrite) and sulfate (barite) minerals in the early Archaean display a relatively narrow spread around $\delta^{34}S = 0 \pm 3‰$ for sulfides and delta $^{34}S = 4 \pm 1‰$ for sulfates (Strauss, 2003). Archaean oceans at 3.5 billion years ago (Ga) were sulfate rich and sulfides were formed by sulfate-reducing bacteria (Ohmoto, 1992). The average sulfate content of the mantle is about 300–400 ppm (Gehlen, 1992). Sulfur isotope ratios in mantle sulfur are close to meteorites (delta $^{34}S = 0.5‰$). Sulfur isotopic compositions of sulfides are enriched in ^{32}S and may be of biogenic origin. The presence of pyrite in black shales, chert, and phosphorite association in Proterozoic and Early Cambrian formations with $\delta^{34}S > +4‰$ indicates that sulfate- and sulfide-reducing bacteria were present in these depositional environments (Tewari, 1984, 1996; Krajewski et al., 1994). Precambrian–Cambrian boundary black shale-pyrite, stromatolite, and small shelly fauna indicate highly reducing paleoenvironment in lagoonal facies, where sulfate-reducing bacteria must have flourished (Tewari, 1984, 1994, 1996; Tewari and Qureshy, 1985). Neoproterozoic carbonates (1,000–540 million years before the present) of the world are characterized by positive $\delta^{13}C$ values (Tewari and Sial, 2007; Tewari, 2007), where it is defined as follows:

$$\delta^{13}C = \left[(^{13}C/^{12}C)_{sa} / (^{13}C/^{12}C)_{st} - 1 \right] \times 10^3 \left[‰, PDB \right].$$

The value of $\delta^{13}C$ is close to zero when the sample coincides with the PeeDee belemnite standard (PDB) in which $(^{13}C/^{12}C) = 88.99$ and $\delta^{13}C$ is defined as equaling 0.00‰. This parameter can be used as a good biosignature. On the Earth biota, for instance, there is ample evidence that photosynthetic bacteria, algae, and plants have typical significant deviations that yield values of up to –30 and beyond, due to biological processes (Schidlowski et al., 1983). But the main point that we have emphasized in the past is that negative values of the $\delta^{13}C$ parameter do not arise exclusively from biogenic sources. For this reason, we have mentioned in the present paper that sulfur is a better biomarker for the study of possible biosignatures.

5. Sulfur Patches on Europa: Is There Evidence for Biogenicity?

There are significant strategies for identifying those places where future landers could search for the biosignatures, such as the penetrators that are now being tested for the 2014 MoonLITE Mission (Smith et al., 2008) and subsequently for Europa (Gowen et al., 2009). The Jovian satellite Europa is the most appealing site for the discovery of extraterrestrial life in our cosmic neighborhood. A key factor in this enterprise has already been provided by the discovery of sulfur patches on the icy surface of this satellite by the Galileo mission. The discovery is significant due to several additional measurements that strongly suggest the presence of an internal deep ocean, a potential habitat for extremophilic (cryophilic) microorganisms. The Galileo Near-Infrared Mapping Spectrometer (NIMS) evidence for the presence of sulfur compounds has been discussed in detail in our previous paper (Chela-Flores, 2006). The most likely sites would be where the salt deposits, or organics, are concentrated, as suggested by the NIMS data. For instance, the search for biosignatures could focus on the area north of the equatorial region, between 0°N and 30°N and between the longitudes 240° and 270° (cf. McCord et al., 1998, Fig. 2a). But a more intriguing and smaller patch would be the narrow band with high-concentration of non-ice elements that lies east of the Conamara Chaos, between the Belus and Asterius lineae, namely between 18°N and 20°N, and longitudes 198°–202° (cf. McCord et al., 1998, Fig. 2d).

Definite answers can be searched in situ on the icy surface with GC-MS instrumentation for the corresponding measurements with the help of biogeochemistry, especially with the $\delta^{34}S$ parameter. Measurements by mass spectrometry are needed. In a feasible mission to Europa, they are possible as discussed earlier (Chela-Flores, 2006), due to miniaturized equipment that is already in existence. A specific example is provided by mass spectrometry on a possible future lander on Europa. At this stage, it is possible to suggest the best possible landing site. We have suggested that at the "patch" found in the Europan surface coordinates 200°W, 20°N (longitude and latitude), respectively, there is a scientific valid way of testing biogenicity through isotopic fractionation that may have occurred on sulfur patches on the Europan icy surface (Singer, 2003; Bhattacherjee and Chela-Flores, 2004).

6. The Antarctic Dry Valley Lakes: Possible Relevance in the Search for Biogenicity

The early stages of future missions may be initially tested on Earth, in environments that are similar to Europa, namely the dry valley lakes of southern Victoria Land of Antarctica (Doran et al., 1994; Parker et al., 1982; Priscu et al., 1998). One large lake lies underneath the Vostok Station, the Russian Antarctic base about 1,000 km from the South Pole. A lake, the size of Lake Michigan, was discovered beneath this Station in 1996 (Ellis-Evans and Wynn-Williams, 1996), after having drilled in that area since 1974. The lake lies under some 4 km of ice. Lake Vostok, as it is known, may harbor a unique microflora. The retrieval of biota from Lake Vostok will serve as a test for handling a larger aquatic medium, such as the proposed Europan submerged ocean that may be teeming with life.

At the time of writing, the lake itself has not been sampled, prevented by the bioethical principles of planetary protection. On the other hand, in the dry valley lakes, there is already a well-studied biota that consists of abundant microorganisms living underneath their iced surface. The estimated annual sulfur removal is over 100 kg in the case of Lake Chad in the dry valleys (Parker et al., 1982). Thus, endogenic sulfur and other chemical elements will be, at any time, found on the icy surface of the dry valley lakes. These environments will help us to decide on the experiments that should be performed with the help of the forthcoming Europa missions, such as the above-mentioned EJSM mission that is being considered by the major space agencies (Grasset et al., 2009).

7. Discussion and Conclusion

Recently, the significance of chemoautotrophs (microorganisms using inorganic or organic substances as energy source rather than light) has been discussed with special reference to the Jupiter's satellite Europa (Chela-Flores, 2006). The search for presence of sulfur in the core of Europan ocean, its chemistry and the possible presence of sulfur-reducing bacteria could be similar to the early stages of geological evolution on the Earth. They have emphasized that the sulfur patches on the icy surface of the Europa might contain biomarkers and should be aimed to study in future Europa mission for extraterrestrial life in the universe. We would like to restate the main conclusions of our previous work (Seckbach and Chela-Flores, 2007; Tewari and Chela-Flores, 2009). S isotope analysis is the most valuable for planetary exploration. In situ analysis of the Europan surficial patches of sulfur, together with carbon signatures, could yield a clearer interpretation of biosignatures.

In the present paper, we have indicated that geological and biogeochemical data from many sources of the Archaean and Proterozoic demonstrate that pyrites and evaporites were formed biologically by dissimilatory sulfate reduction (Shen and Buick, 2004; Schidlowski et al., 1983; Konhauser, 2007). Rocks of Archean age (older than 2.5 Gyr BP) provide the best evidence of early metabolic processes. Their study allows the reconstruction of the biogeochemical cycle for sulfur since the origin of life on Earth. The remarkable will inevitably be targets for future space missions that are expected to return to Europa in the next decade.

With landers, or low-cost penetrators that could first of all be tried out on the Lunar surface (Smith et al., 2008), we would be in a position to test the redox state of the Europan ocean. Alternatively, the imprint of the possibly biogenic signature of the surficial sulfur would be retained in the dust cloud that surrounds this singular Jovian satellite (Kruger et al., 2003). The arguments in the present paper continue to point towards mass spectrometry as the principal instrumentation for future probing of the Europan patches, either in orbit or with penetrators.

8. Acknowledgments

We are grateful to the Abdus Salam International Center for Theoretical Physics ICTP, Trieste, Italy, and Wadia Institute of Himalayan Geology, Dehradun, Uttarakhand, India, for this collaborative research. Vinod C. Tewari carried out research at ICTP between 2005 and 2008 as a senior associate.

9. References

Bhattacherjee, A.B and Chela-Flores, J. (2004) Search for bacterial waste as a possible signature of life on Europa, In: J. Seckbach, J. Chela-Flores, T. Owen and F. Raulin (eds.) *Life in the Universe, Cellular Origin and Life in Extreme Habitats and Astrobiology 7*. Springer, Dordrecht, pp. 257–260. http://www.ictp.it/~chelaf/ss27.html.

Brasier, M.D., Green, O.W., Jephcoat, A.P., Kleppe, A.K., Van Kranendonk, M.J., Lindsay, J.F., Steele, A. and Grassineau, N.V. (2002) Questioning the evidence for Earth's oldest fossils. Nature **416**: 76–81.

Chela-Flores, J. (2006) The sulphur dilemma: are there biosignatures on Europa's icy and patchy surface? Int. J. Astrobiol. **5**: 17–22.

Chela-Flores, J., Kumar, N., Seckbach, J. and Tewari, V.C. (2008) Distinguishing between signatures of past life and nonlife. *Geophysical Research Abstracts 10*. E.G.U. General Assembly, Vienna.

Doran, P.T., Wharton Jr., R.A. and Berry, L.W. (1994) Paleolimnology of the McMurdo Dry Valleys, Antarctica. J. Paleolimnol. **10**: 85–114.

Ellis-Evans, J.C. and Wynn-Williams, D. (1996) A great lake under the ice. Nature **381**: 644–646.

Gehlen, K. (1992) Sulfur in the Earth's mantle: a review, In: M. Schidlowski (ed.) *Early Organic Evolution: Implications for Mineral and Energy Resources*. Springer, Berlin, pp. 359–366.

Goldberg, T., Paulton, W.S. and Strauss, H. (2005) Sulphur and Oxygen isotope signatures of late Neoproterozoic to early Cambrian sulphate, Yangtze Platform, China: diagenetic constraints and sea water evolution. Precam. Res. **137**: 223–241.

Gowen, R., Smith, A., Ambrosi, R., Prieto Ballesteros, O., Barber, S., Barnes, D., Braithwaite, C., Bridges, J., Brown, P., Church, P., Collinson, G., Coates, A., Collins, G., Crawford, I., Dehant, V., Dougherty, M., Chela-Flores, J., Fortes, D., Fraser, G., Gao, Y., Grande, M., Griffiths, A., Grindrod, P., Gurvits, L., Hagermann, A., van Hoolst, T., Hussmann, H., Jaumann, R., Jones, A., Jones, G., Joy, K., Karatekin, O., Kargl, G., Macagnano, A., Mukherjee, A., Muller, P., Palomba, E., Pike, T., Proud, B., Pullen, D., Raulin, F., Richter, L., Ryden, K., Sheridan, S., Sims, M., Sohl, F.,

Snape, J., Stevens, P., Sykes, J., Tong, V., Stevenson, T., Karl, W., Wilson, L., Wright, I. and Zarnecki, J. (2009) Looking for astrobiological signatures with penetrators on Europa. Physical and Engineering Sciences Exploratory Workshops, W08-115: Biosignatures on Exoplanets; The Identity of Life, 22–26 June 2009, Mulhouse, France. http://www.ictp.it/~chelaf/ESFsummary.pdf.

Grasset, O., Lebreton, J.-P., Blanc, M., Dougherty, M., Erd, C., Greeley, R., Pappalardo, B. and the Joint Science Definition Team (2009) The Jupiter Ganymede Orbiter as part of the ESA/NASA Europa Jupiter System Mission (EJSM). *EPSC Abstracts 4, EPSC2009-784.* European Planetary Science Congress.

Konhauser, K. (2007) *Introduction to Geomicrobiology*. Blackwell, Malden, pp. 320, 342–343.

Krajewski, K.P., Cappallen, P.V., Trichet, J., Kuhn, O., Lucus, J., Algarra, A.M., Prevot, L., Tewari, V.C., Knight, T., Lamboy, M. (1994) Biological processes and apatite formation in sedimentary environment. Ecol. Geol. Helv. **87**(3): 701–745.

Kruger, H., Krivov, A.V., Sremcevi, M. and Grün, E. (2003) Impact-generated dust clouds surrounding the Galilean moons. Icarus **164**: 170–187.

Lambert, I.B. and Donnelly, T.H. (1992) Global oxidation and a supercontinent in the Proterozoic: evidence from stable isotopic trends, In: M. Schidlowski (ed.) *Early Organic Evolution: Implications for Mineral and Energy Resources*. Springer, Berlin, pp. 408–414.

McCord, T.B., Hansen, G.B., Clark, R.N., Martin, P.D., Hibbitts, C.A., Fanale, F.P., Granahan, J.C., Segura, N.M., Matson, D.L., Johnson, T.V., Carlson, R.W., Smythe, W.D., Danielson, G.E. and the NIMS Team (1998) Non-water-ice constituents in the surface material of the icy Galilean satellites from the Galileo near-infrared mapping spectrometer investigation. J. Geophys. Res. **103**: 8603–8626.

Ohmoto, H. (1992) Biogeochemistry of sulfur and the mechanisms of sulfide – sulfate mineralization in Archean oceans, In: M. Schidlowski (ed.) *Early Organic Evolution: Implications for Mineral and Energy Resources*. Springer, Berlin, pp. 378–397.

Parker, B.C., Simmons Jr., G.M., Wharton Jr., R.A., Seaburg, K.G. and Love, G.F. (1982) Removal of organic and inorganic matter from Antarctic lakes by aerial escape of blue-green algal mats. J. Phycol. **18**: 72–78.

Priscu, J.C., Fritsen, C.H., Adams, E.A., Giovannoni, S.J., Paerl, H.W., McKay, C.P., Doran, P.T., Gordon, D.A., Lanoil, B.D. and Pickney, J.L. (1998) Perennial Antarctic lake ice: an oasis for life in a polar desert. Science **280**: 2095–2098.

Schidlowski, M., Hayes, J.M. and Kaplan, I.R. (1983) Isotopic inferences of ancient biochemistries: carbon, sulfur, hydrogen, and nitrogen, In: J.W. Schopf (ed.) *Earth's Earliest Biosphere Its Origin and Evolution*. Princeton University Press, Princeton, pp. 149–186.

Schopf, J.W., Tewari, V.C. and Kudrayvtsev, A.B. (2008) Discovery of a new chert – permineralised microbiota in the Proterozoic Buxa Formation of the Ranjit window, Sikkim, N.E. Lesser Himalaya, India and its Astrobiological Implications. Astrobiol. J. **8**(4): 735–746.

Schultz, H.N. and Schulz, H.D. (2005) Large sulfur bacteria and the formation of phosphorite. Science **307**: 416–418.

Seckbach, J. and Chela-Flores, J. (2007) Extremophiles and chemotrophs as contributors to astrobiological signatures on Europa: a review of biomarkers of sulfate-reducers and other microorganisms, invited talk (6694-32) at instruments methods and missions for astrobiology X, 2007. SPIE Optics and Photonics Symposium, San Diego, California, USA, August 26–30, 2007.

Shen, Y. and Buick, R. (2004) The antiquity of microbial sulfate reduction. Earth-Sci. Rev. **64**: 243–272.

Singer, E. (2003) Vital clues from Europa. New Scientist Magazine **2414**(27 September): 22–23, http://www.ictp.it/~chelaf/VitalClues.pdf

Smith, A., Crawford, I.A., Gowen, R.A., Ball, A.J., Barber, S.J., Church, P., Coates, A.J., Gao, Y., Griffiths, A.D., Hagermann, A., Phipps, A., Pike, W.T., Scott, R., Sheridan, S., Sweeting, M., Talboys, D., Tong, V., Wells, N., Biele, J., Chela-Flores, J., Dabrowski, B., Flannagan, J., Grande, M., Grygorczuk, J., Kargl, G., Khavroshkin, O.B., Klingelhoefer, G., Knapmeyer, M., Marczewski, W., McKenna-Lawlor, S., Richter, L., Rothery, D.A., Seweryn, K., Ulamec, S., Wawrzaszek, R., Wieczorek, M. and Wright, I.P. (2008) LunarEX – a proposal to cosmic vision. Exp. Astron. **23**(3): 711–740. 10.1007/s10686-008-9109-6 (August 21). http://www.ictp.it/~chelaf/Penetrator.pdf.

Strauss, H. (2003) Sulfur isotopes and the early Archaean sulfur cycle. Precambr. Res. **126**: 349–361.

Tewari, V.C. (1984) Discovery of lower Cambrian stromatolites from the Mussoorie Tal Phosphorites, India. Curr. Sci. **53**(6): 319–321.

Tewari, V.C. (1989) Upper proterozoic –Lower Cambrian stromatolites and Indian stratigraphy. Him. Geol. **13**: 143–180.

Tewari, V.C. (1991) Palaeomicrobiology, palaeoenvironment and isotope geochemistry of the stromatolitic- carbonate- chert –phosphate association from Lesser Himalaya, India. Nat. Sem. Appl. Geomicrobiol. India. pp. 93–107.

Tewari, V.C. (1993) Ediacaran metaphytes from the Lower Krol Formation, Lesser Himalaya, India. Geosci. J. **14**(1, 2): 143–148.

Tewari, V.C. (1994) Sedimentology of the rocks of Deoban basin, Dhuraphat area, Saryu valley, Eastern Kumaon Lesser Himalaya. Geosci. J. **15**(2): 117–162.

Tewari, V.C. (1996) Controls of phosphorite formation superimposed on biological activity in the Lesser Himalaya, India. Geosci. J. **16**(2): 135–153.

Tewari, V.C. (1999) Vendotaenids: earliest megascopic multicellular algae on Earth. Geosci. J. **20**: 77–85.

Tewari, V.C. (2001a) Origins of life in the universe and earliest prokaryotic microorganisms on Earth, In: J. Chela-Flores, et al. (eds.) *First Steps in the Origin of Life in the Universe*. Kluwer Academic, Dordrecht, pp. 251–254.

Tewari, V.C. (2001b) Neoproterozoic glaciation in the Uttaranchal Lesser Himalaya and the global palaeoclimate change. Geol. Surv. India Spl. Publ. **65**(3): 49–56.

Tewari, V.C. (2004) Microbial diversity in Meso- Neoproterozoic Formations, with particular reference to the Himalaya, In: J. Seckbach (ed.) *Origins*. Kluwer Academic, Dordrecht, pp. 515–528.

Tewari, V.C. (2007) The rise and decline of the Ediacaran biota: palaeobiological and stable isotopic evidence from the NW and NE Lesser Himalaya, India, In: R.P. Vickers and P. Komarower (eds.) *Rise and Fall of the Ediacaran Biota. Special Publication 286*. Geological Society of London, London, pp. 77–101.

Tewari, V.C. (2008) Proterozoic unicellular and multicellular fossils from India and their implications, In: J. Seckbach (ed.) *From Fossils to Astrobiology. Cellular Origin, Life in Extreme Habitats and Astrobiology Series*. Springer, Dordrecht, pp. 119–139.

Tewari, V.C. and Chela-Flores, J. (2009) Possible role of sulfur on the early diversification of life on earth Astrobiological implications, In: K.L. Srivastava (ed.) *Economic Mineralization*. Scientific, Jodhpur, India, pp. 53–56.

Tewari, V.C. and Sial, A.N. (2007) Neoproterozoic–Early Cambrian isotopic variation and chemostratigraphy of the Lesser Himalaya, India, Eastern Gondwana. Chem. Geol. **237**: 64–88.

PART 7:
SUMMARY, CONCLUSIONS AND FUTURE PROSPECTS

Seckbach
Tewari

Biodata of **Joseph Seckbach** and **Vinod Chandra Tewari**, editors of this volume

Professor Joseph Seckbach is the founder and chief editor of book series **Cellular Origins, Life in Extreme Habitats, and Astrobiology (COLE** see www.springer.com/series/5775). He is the author of several chapters in this series. Dr. Seckbach earned his PhD from the University of Chicago, IL (1965), and spent his post-doctoral years in the Division of Biology at Caltech (Pasadena, CA). Then he headed a team for searching for extraterrestrial life at the University of California at Los Angeles (UCLA). Later, he was appointed to the faculty of the Hebrew University (Jerusalem, Israel) , where he performed algal research and taught biological courses until his retirement. Among his publications are books, scientific articles in the lines of phytoferritin, cellular evolution, acidothermophilic algae, and life in extreme environments. Dr. Seckbach is the co-editor of ***Proceeding of Endocytobiology VII Conference*** (Frieburg, Germany, 1998) and the ***Proceedings of Algae and Extreme Environments Meeting*** (Trebon, Czech Republic, 2000). His edited volume (with Richard Gordon) entitled ***Divine Action and Natural Selection: Science, Faith, and Evolution*** has been published by World Scientific Publishing Company (2008). His recent interest is in the field of enigmatic micro-organisms and life in extreme environments.

E-mail: **seckbach@huji.ac.il**

V.C. Tewari and J. Seckbach (eds.), *STROMATOLITES: Interaction of Microbes with Sediments*, Cellular Origin, Life in Extreme Habitats and Astrobiology 18, 739–742
DOI 10.1007/978-94-007-0397-1_34,

Professor Vinod. C. Tewari is currently the Head of the Sedimentology Group at Wadia Institute of Himalayan Geology, Dehradun, Uttarakhand, India. He was Regular and Senior Associate and also TRIL Fellow of International Center for Theoretical Physics, Trieste, Italy, between 1998 and 2009. He was born in Almora, India, in 1954 and obtained his higher education including PhD from the University of Lucknow in Geology in 1986. He joined Wadia Institute in 1978 and continued his research in the Lesser Himalayan stromatolites, microbiota, and isotope chemostratigraphy. Dr. Tewari taught Geology at Kumaon University, Nainital, Uttarakhand (U.K.), India, as Professor of Geology. Dr. Tewari worked on stable isotopes in Biogeochemistry Department of the Max-Planck Institute, Mainz, Germany, as Visiting Professor in 1995. Professor Tewari's scientific interests are in the areas of Precambrian and Phanerozoic stromatolites, sedimentation, carbon isotope chemostratigraphy, genesis, early evolution and diversification of life and its astrobiological significance. He has been associated with several International Geological Correlation Program (I.G.C.P.) Projects like Stromatolites, Biosedimentology of the Microbial Buildups, Phosphorites, Precambrian–Cambrian Boundary, Global Bioevents and The Rise and Fall of Vendian Biota. He has 80 research papers published in reputed journals to his credit and edited several volumes of *Himalayan Geology Journal*, India, and *Journal of Nepal Geological Society*, Kathmandu, Nepal. His current interests are Cretaceous–Paleogene boundary, global extreme and complex climatic events from Neoproterozoic Snowball Earth to Holocene period, India–Asia collision and evolution of Himalaya. Professor Tewari has organized the first ***Indo-Soviet Symposium on Stromatolites and Stromatolitic Deposits*** and other IGCP meetings in India and abroad related to stromatolites and phosphorites, etc. He has been one of the organizers of the ***World Summit on Ancient Microscopic Fossils*** held in University of California, Los Angeles, USA, in 2008. Professor Tewari is the member of the Editorial Board of the *International Journal of Astrobiology* published from New York, USA.

E-mail: **vtewari@wihg.res.in**

SUMMARY, CONCLUSIONS, AND FUTURE PROSPECTS

JOSEPH SECKBACH[1] AND VINOD CHANDRA TEWARI[2]
[1]P.O. Box 1132, Efrat 90435, Israel
[2]Wadia Institute of Himalayan Geology, Dehradun 248001, Uttarakhand, India

This COLE volume describes and discusses the interaction of microbes with sediments since the earliest stromatolites were formed on Earth, around 3,500 Ma ago. One of the most significant areas of sedimentary research today in the Earth Sciences is the interaction of microbes with sediments, a topic that is fascinating, complex, rewarding, and enlightening. In the primordial ecosystems of the Archean, the changing environment gradually led to adaptations of early microbes that radically altered the terrestrial phylogenetic tree. Such complexity is entangled with the intrinsic difficulty of identifying truly reliable fossils from which we perhaps might be able to reconstruct the early steps in the origin of life on Earth and the pathway of its evolution. By fortuitous coincidence, this volume was initiated in 2008, the 100th anniversary of the term "stromatolith" (Kalkowsky, 1908).

The terminology of stromatolites underwent considerable transformations since Kalkowsky's 1908 introduction of "stromatolith." In this book, stromatolites are treated as organo-sedimentary structures with a direct relation to life and thus also to Darwinian evolution. This, however, is not exclusively the case for all stromatolite definitions and not all stromatolite researchers would agree on such exclusivity of terminology (cf. Hofmann, 2000; Altermann, 2007).

Geomicrobiology of cave deposits is an area of research to interpret paleoenvironment and interaction of microbes with sediments giving some examples from the Sahastradhara and Mawsmai caves from India.

The study of stromatolites may contribute significantly to placing a few stepping stones firmly along the pathway of Darwinian evolution from the Archean up to the present. A great deal of the entire span of geologic time since the first appearance of stromatolites on Planet Earth is covered in this volume, and several authors devoted their papers to the evidence for biogenicity of Archean stromatolites. Such information was obtained owing to the fact that we have two reliable windows on the nature of the early habitats on Earth: the Pilbara Craton, especially with the Dresser Formation in Western Australia, and the Kaapvaal Craton in South Africa.

The Archean environment was prolific with life other than the stromatolite builders, namely – as today – the cyanobacteria. From the locations in Australia and South Africa, details of ancient hydrothermal vents have been retrieved. A complex set of both sulfur-reducing bacteria (Shen and Buick, 2004) and methanogens (Ueno et al., 2006) were already at play a mere billion years after the formation of the Earth itself, possibly before the best evidence of Archean stromatolites. Also focused upon in detail are the Meso-Neoproterozoic stromatolite diversity and associated organic-walled microfossils in the black cherts that represent the Proterozoic biosphere of the Earth before the Cambrian explosion of life. In more recent geologic strata, the evidence for microbial life garners less controversy in the interpretation from the rocks that have been studied. The Mesozoic Tethyan carbonate and phosphate stromatolites and biomineralization in the Alpine and Arctic regions to Japan and Korea explain the role of microbes and sedimentation in different environments. Since the present is the key to the past, the modern stromatolites from Shark Bay in Australia and from the Bahamas can reveal information on the depositional system and the sediment–microbe relationship in which the Archean–Proterozoic stromatolites were formed. Various authors concentrated on these aspects. In all of the foregoing, the acquisition of information benefitted from modern instrumental techniques such as laser Raman spectroscopy, confocal laser scanning microscopy (Schopf et al., 2008) and Nano-SIMS, cathodoluminescence and sulfur and carbon isotopes.

The overall picture that emerges from the interaction of microbes with sediments provides a significant basis for extrapolating to what could happen elsewhere in the Solar System. This is particularly timely as the science of astrobiology, namely the study of the origin, evolution, distribution, and destiny of life in the universe is going through its golden age.

1. References

Altermann, W. (2007) Accretion, trapping and binding of sediment in Archean stromatolites – morphological expression of the antiquity of life. Space Sci. Rev. **135**: 55–79.

Hofmann, H.J. (2000) Archean stromatolites as microbial archives, In: R.E. Riding and S.M. Awramik (eds.) *Microbial Sediments*. Springer, Berlin, pp. 315–327.

Kalkowsky, E. (1908) Oolith und stromatolith in Norddeutschen Buntsandstein. Z. Dtsch. Geol. Ges. **60**: 68–125.

Schopf, J.W., Tewari, V.C. and Kudryavtsev, A.B. (2008) Discovery of new Chert-Permineralised microbiota in the Proterozoic Buxa Formation of the Ranjit Window, Sikkim, Northeast India and it's astrobiological implications. Astrobiology **8**(4): 735–746.

Shen, Y. and Buick, R. (2004) The antiquity of microbial sulfate reduction. Earth Sci. Rev. **64**: 243–272.

Ueno, Y., Yamada, K., Yoshida, N., Maruyamaand, S. and Isozaki, Y. (2006) Evidence from fluid inclusions for microbial methanogenesis in the early Archaean era. Nature **440**: 516–519.

AUTHOR INDEX

V.C. Tewari and J. Seckbach (eds.), *STROMATOLITES: Interaction of Microbes with Sediments*, Cellular Origin, Life in Extreme Habitats and Astrobiology 18,
DOI 10.1007/978-94-007-0397-1,

SUBJECT INDEX

MIX
Papier aus verantwortungsvollen Quellen
Paper from responsible sources
FSC® C105338

If you have any concerns about our products,
you can contact us on
ProductSafety@springernature.com

In case Publisher is established outside the EU,
the EU authorized representative is:
Springer Nature Customer Service Center GmbH
Europaplatz 3, 69115 Heidelberg, Germany

Printed by Libri Plureos GmbH
in Hamburg, Germany